MECHANICAL BEHAVIOR
of MATERIALS

SECOND EDITION

MECHANICAL BEHAVIOR of MATERIALS

SECOND EDITION

Thomas H. Courtney

WAVELAND
PRESS, INC.
Long Grove, Illinois

For information about this book, contact:
Waveland Press, Inc.
4180 IL Route 83, Suite 101
Long Grove, IL 60047-9580
(847) 634-0081
info@waveland.com
www.waveland.com

10-digit ISBN 1-57766-425-6
13-digit ISBN 978-1-57766-425-3

Printed in the United States of America

8   7   6   5   4   3

To the youngest Courtneys I know: Kenan Patrick Courtney and Peyton Quinn Courtney. May they be as fortunate as I in discovering the ultimate teacher.

When Tom Casson contacted me late in 1997 relative to a second edition of this book, I leaped at the opportunity because immediately following publication of the first edition I realized how much better that edition could have been. I was not, in late 1997, even discouraged by the daunting schedule Tom specified. He was hoping that all of the rewriting could be accomplished in a one-year period. I more or less met his schedule. Now, exhausted by the enterprise, I realize the wisdom of that trite phrase, "Look before you leap."

This book has been substantially revised. A superficial comparison of the two editions will spot quickly some changes; example problems and certain chapters now sporting frontispieces being most prominent. The problems were added at the suggestion of reviewers of the proposed second edition. And the frontispieces were inserted to provide some appeal in this age of the required "visual impact." But there have also been more substantive changes made.

Even though the order of the book remains rather much as it was in the first edition, almost every chapter has been altered. The most prominent changes are to Chaps. 9 and 10, which deal with low-temperature fracture and these chapters have been completely rearranged. Chapter 9 (Fracture Mechanics) now focuses on the engineering science, and application of, fracture mechanics and auxiliary measures of fracture resistance such as the impact test. Chapter 10 (Toughening Mechanisms and the Physics of Fracture) now mainly deals with toughening mechanisms and fracture mechanisms in flaw-free materials. The sections on toughening mechanisms pay considerably more attention to ceramics and composites than was provided in the first edition.

Since material has been added to many of the chapters, in an (almost successful) attempt to keep the length of the book within bounds some material has been deleted. A litany-like listing of the changes made in the remaining chapters follows. In Chap. 1 (Overview of Mechanical Behavior), I have deleted much of the "mechanics" contained in the first edition, although discussion of yield criteria and a brief description of Mohr's circle are retained. A brief section on fracture toughness measurement has been added. In Chap. 2 (Elastic Behavior), moduli variations among the material classes are compared and the physical bases for these variations rationalized. The treatment of polymer elasticity has also been expanded. The changes made to Chap. 3 (Dislocations) are minor; discussion of twinning has been expanded and the evolution of dislocation arrangements with plastic strain is now treated in more detail. Chapter 4 (Plastic Deformation in Single and Polycrystalline Materials) now considers in more detail the plastic-flow behavior in the different material classes and compares plastic flow (e.g., wavy vs. planar slip) and work-hardening characteristics of the fcc and the bcc transition metals. The treatment of particle hardening in Chap. 5 (Strengthening of Crystalline Materials) has been expanded and is now more up-to-date. Here I have attempted to emphasize the similarity in the expressions for strength due to different particle hardening mechanisms and to also indicate their similarity to expressions for the strength arising from solid-solution hardening. A case study on the development of microalloyed steels has been added. Off-axis behavior of aligned fiber composites is now treated in more detail (Chap. 6, Composite Materials), and there is more discussion on "Modern" composites in this chapter. Chapter 7 (High-Temperature Deformation of Crystalline Materials) remains much the same. However, a major omission in the first edition—solute drag creep and the Portevin-LeChatelier effect—are included in the current edition. And a case study on tungsten light bulb filaments has been added. Details on polymer molecular architecture (which can be found in elementary materials science and engineering texts) have been deleted from Chap. 8 (Deformation of Noncrystalline Materials) and the rheological details of polymer deformation are not discussed to the same extent they were in the first edition. However, discussion of silicate and metallic glasses has been slightly expanded. Chapter 11 (High-Temperature Fracture) has been reduced in length. This was achieved by eliminating the detailed mathematical description of void growth. At the time of the first edition, I had thought that this promising approach to prediction of creep fracture times and strains would soon see engineering implementation. However, the lack of ancillary physical property data and our inability to realistically mimic void spacing and the like during creep suggested a more condensed treatment was in order. Chapter 12 (Fatigue of Engineering Materials) has, I believe, been improved by including a section treating design against fatigue fracture in "flawed" and in "flaw-free" materials. This chapter also now considers the relationships among the endurance limit, the fatigue threshold stress intensity factor, and material fracture toughness. How these relationships differ in the different material classes is also discussed. And a brief description of substructure evolution (e.g., persistent slip bands) during cyclical straining of metals has been added. Chapter 13 (Embrittlement) has not been changed much. However, corrosion fatigue in "flawed" and "flaw-free" materials is now treated. Chapter 14 (Cellular Solids) is new. This addition is justified because of the technological importance of these materials. And the mechanical behavior of cellular solids well exemplifies the interaction between mechanics and microstructure that is a characteristic feature of how a "materials person" approaches mechanical behavior.

This book, I hope, comprehensively treats the mechanical behavior of materials. The extended treatment is intentional. Individual faculty members deem certain topics more important than others do. However, (as in many academic matters) seldom do faculty members agree on what these more important topics are. Thus, the breadth and depth of this book is an attempt to permit individual instructors to select those topics they wish to emphasize and to do so at a level they consider appropriate. Because the book is comprehensive, to adequately cover all of the material in it at the level at which it is addressed in the text would likely require two full semesters of a typical three-credit course. Most curricula do not have the luxury of allocating this amount of time to mechanical behavior of materials. Some experience indicates the following types of scenarios are possible with the book. A two-quarter three-credit course or a four-credit semester course could address most of the topics covered in the book. This would require some selectivity on the part of the instructor, both with respect to chapters covered and within individual chapters as well. A logical "division" is that approximately half of the course would consider deformation and half of it fracture. A four-credit quarter course or a three-credit semester course would call for further culling in individual chapters and perhaps deletion of some material (e.g., Chaps. 13 and 14).

I believe the book can be used in either an undergraduate upper division course (or courses) or in an introductory graduate course. Of course, the flavor and emphases would differ between these situations. This, I think, is possible with this text. In the undergraduate courses I have taught using this book, I have de-emphasized mathematical developments in an attempt to inculcate in students an appreciation for the "material" aspects of mechanical behavior. And there are some topics (e.g., tensile fracture in flaw-free materials is but one of them) that would be covered in much less depth in an undergraduate course.

As in the first edition, a relatively large number of problems are provided with each chapter. They range in difficulty; some are straightforward whereas others are, should we say, "challenging." And some are not only challenging, but lengthy as well. These are easily spotted in the sections at chapter ends, and one might ask, What is their purpose? During the last several years I have departed from the practice of conventional examinations. Instead, students work in teams (typically two-student teams for a graduate course and three-student teams for an undergraduate course) on (usually) four assignments per quarter. Each assignment ordinarily consists of three problems with one or two of these problems being of the "lengthy and challenging" variety. I can't say that I have never pulled my hair out when grading some of the problem "solutions" handed in by the students. On the other hand, far more often the solutions presented (even by undergraduate juniors) have been so well done that I wanted to stand up and cheer.

As always in an undertaking such as this, numerous people have helped. Jeff Spencer, an undergraduate in our department at Michigan Tech, helped immeasurably with the photography. Emily Gray and Jean Lou Hess were persistent, albeit gently demanding at times, "cheerleaders." Professor A. K. Mukherjee of the University of California–Davis provided me with some timely reprints on the topics of creep and superplasticity. He also pointed out a critical flaw in one of the figures of the first edition. Likewise, Professor Lloyd A. Heldt, of Michigan Technological University, supplied me with several topical articles on stress corrosion cracking. I have also used some problems developed by Professor John A. Wert, a former colleague at the University of Virginia. At this point, I don't recall the specific problems that originated with him, but thanks anyway, John.

Thomas H. Courtney

BRIEF CONTENTS

TABLE OF CONTENTS

CHAPTER 1

Overview of Mechanical Behavior

1.1
INTRODUCTION

This book deals with the mechanical behavior of solids, particularly as this behavior is affected by processes taking place at the microscopic and/or atomic level. The response of a solid to external or internal forces can vary considerably, depending on the magnitude of these forces and the material characteristics. For example, if the forces are great the material may fracture. Lesser values of force may result in material permanent deformation without fracture and, if the forces are low enough, the material may deform only in an elastic way. The treatment of mechanical behavior in this book closely parallels these three possibilities.

While our aim is to relate the mechanical behavior of a solid to material structure at the microscopic and atomic level, this response is manifested macroscopically. Thus, to fulfill adequately the objective of this text, a reasonable background in the concepts of mechanical behavior as measured and assessed at a macroscopic level is required. Indeed, it is this coupling between material microstructure and bulk properties that constitutes one of the most fruitful areas of materials science and engineering.

Accordingly, for this book to be of maximum benefit to a student using it, prior exposure to the concepts of materials science and mechanics is required. As the text is expected to be used primarily by students of materials science and engineering, many of whom will need some refreshing in the concepts of mechanics and strength of materials, this chapter presents a brief overview of these topics. Readers with an adequate background and recollection of these topics as found, for example, in undergraduate statics and strength of materials courses may choose to skip this chapter. Alternatively, the chapter can be used as a source for reviewing macroscopic concepts as they are introduced throughout the book.

1.2
ELASTIC DEFORMATION

When a solid is subjected to external forces, it undergoes a change in shape. When the load is released, the shape may not return to what it was prior to the application of the force; under these circumstances we say that the material has deformed permanently. Forces less than those that cause permanent deformation deform the solid elastically; that is, when the force is subsequently removed the body assumes the dimensions it had prior to its application.

The elastic behavior of many materials can be represented by a form of Hooke's law. This is illustrated in Fig. 1.1, which indicates that the extension of a sample ($= l - l_0$ where l_0 is the sample length prior to, and l this length following application of the force) is linearly related to the force, F, i.e.,

$$(l - l_0) = \delta l \sim F \tag{1.1}$$

The extension also depends on sample dimensions. For example, as indicated in Fig. 1.1b, a doubling of the initial sample length leads to a doubling of the exten-

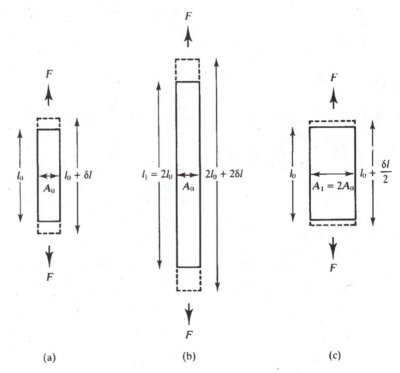

(a)   (b)   (c)

Figure 1.1
(a) Following application of a force (F), sample length increases from l_0 (solid lines) to $l_0 + \delta l$ (dotted lines). For linear elastic deformation, δl increases linearly with F. (b) If the applied force is kept the same, but l_0 is doubled, the extension is doubled. (c) If the force is kept constant and l_0 is the same as in (a), but the cross-sectional area is doubled, the extension is halved. These examples show that for linear elastic deformation, δl increases linearly with sample length and F, but varies inversely with initial sample cross-sectional area. (Not illustrated in the figure is the relatively lesser lateral contraction that accompanies sample extension.)

sion, whereas if the sample cross-sectional area (A) normal to the applied force (the transverse cross-sectional area) is doubled, the extension is halved (Fig. 1.1c). Other alterations in sample length or cross-sectional area lead to equivalent results in that extension is found to vary linearly with sample length and inversely with cross-sectional area. Thus,

$$(l - l_0) = \delta l \sim Fl_0/A \qquad (1.2)$$

Equation (1.2) is often written in normalized form. That is the "normalized" force is defined by F/A (with units of newtons per square meter (1 N/m^2 is also equivalent to 1 Pa) or pounds per square inch); this ratio is called the stress and given the symbol σ. The "normalized" extension, which is dimensionless, is defined as $\delta l/l_0$ and is called the strain and given the symbol ε. Thus, Eq. (1.2) can be written as

$$\varepsilon = \frac{\sigma}{E} \qquad (1.3)$$

where the proportionality constant E is a material property, called the Young's or tensile modulus. A material having a high value of the tensile modulus is stiff; i.e., it is resistant to *tensile deformation* of the kind just described.

Linear elasticity of this kind is observed in all classes of solids. It is the dominant mode of elastic deformation in all solids at low temperatures, in crystalline solids and inorganic glasses up to moderately high temperatures, and in noncrystalline polymers at low temperatures. The extent of linear elasticity is usually quite limited; that is, most materials are capable of being linearly elastically extended only to strains on the order of several tenths of a percent. As discussed in Chap. 2, linear elasticity represents the stretching (or compression/distortion) of atomic bonds, and for this reason E is a measure of a material's bond strength.

As shown in Fig. 1.2, a change in material shape can also be caused by *shear* stresses. These cause relative displacement of the upper and lower surfaces of the solid illustrated. The shear stress is defined as the magnitude of the shearing force divided by the cross-sectional area over which it acts; the shear stress is designated τ. The shear strain, γ (Fig. 1.2), is given by the relation $\tan \gamma = \delta l/l_0$ where δl is the relative displacement of the surfaces and l_0 their vertical separation. When the shear strain is a linear elastic one, it is usually small so that $\tan \gamma \cong \gamma$, and γ and τ are related through

$$\gamma = \frac{\tau}{G} \qquad (1.4)$$

The constant G in Eq. (1.4), the *shear modulus,* is also a material property. In a physical sense G can be viewed as a measure of the resistance to bond distortion

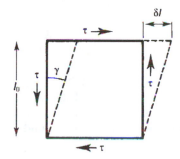

Figure 1.2
A shear stress (τ) distorts a body. The stress is defined as the applied force divided by the surface area on which it acts. Shear strain, γ, is defined by $\tan \gamma$ ($\cong \gamma$) = $\delta l/l_0$. For linear elasticity, τ and γ are related by $\tau = G\gamma$, where G is the material's shear modulus.

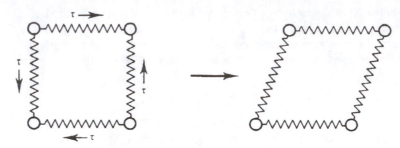

Figure 1.3
As shown here, a shear displacement results in bond distortion at the atomic level. Atoms are represented by the open circles and atomic bonds by the "springs." The shear strain arises from the "bending" of the atomic bonds.

within a solid. This can be visualized by considering the simple-cubic single crystal of Fig. 1.3. Here the shear force acts on a (001) plane in the [010] direction, with the plane of the drawing representing a (100) plane. The change in atomic positions due to the shear stress results from "bending" of atomic bonds.

A material's tensile modulus can be ascertained by careful measurement of the linear stress-strain relationship in a tension test, which is a test similar to that illustrated in Fig. 1.1. The shear modulus can be obtained indirectly from a tension test (Chap. 2), or directly by other testing procedures about which more is written later.

Almost all classes of solids also exhibit, at least over a certain temperature range, nonlinear and time-dependent elasticity. This *viscoelasticity*, as it is called, is most common to noncrystalline polymers, but also occurs to a much more limited extent in crystalline solids and inorganic glasses. One essential difference between this kind of elasticity and linear elasticity is illustrated in Fig. 1.4. This illustration shows that the strain in a linear elastic solid is a single-valued function of the stress; that is, the loading and unloading segments of the σ-ε relationship coincide (Fig. 1.4a). In contrast, the stress-strain relationship in a viscoelastic material (Fig. 1.4b) depends on the sense of loading. Moreover, the level of stress attained depends, too, on the rate at which a viscoelastic material is stretched (the strain rate). With increasing strain rate a viscoelastic material becomes stiffer; for example, the "average" modulus (e.g., σ_1/ε_1 in Fig. 1.4b) increases with strain rate. Viscoelastic behavior is also manifested by a strain that varies with time under conditions of a constant applied stress. That is, upon initial application of the stress some instantaneous (linear elastic) strain is first experienced, following which the material continues to extend, with the strain approaching some asymptotic value. On removal of the load, the linear elastic strain is instantaneously, and the viscoelastic strain sluggishly, recovered.

Nonlinear elasticity, of which viscoelasticity is one example, need not be time-dependent. For example, nonlinear time-independent elasticity is observed in certain fine, strong crystalline solids called whiskers. Whiskers typically have diameters on the order of micrometers, and when stretched in tension they deform in a linear elastic way up to strains on the order of half a percent. For elastic strains in excess of this (and whiskers are capable of such strains) the σ-ε relationship is nonlinear. An extreme example of nonlinear time-independent elasticity is found in

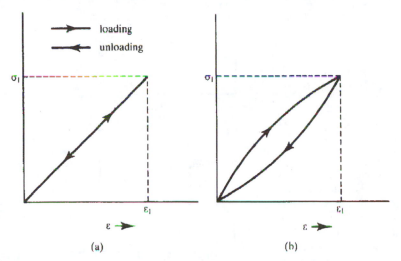

Figure 1.4
Illustrating one difference between linear elasticity and viscoelasticity. (a) Stress and strain in a linear elastic material are uniquely related when the material is subjected to repeated loading and unloading. That is, stress and strain follow the same path on unloading ($<$) as on loading ($>$). (b) This is not so for a viscoelastic solid. The unloading path differs from the loading one; the divergence between the two is generally greater the higher the strain rate. Higher rates of loading and unloading also typically lead to a higher overall stress level in a viscoelastic material.

elastomers. These are a special class of polymers that over a limited temperature range are capable of demonstrating extensive elastic strains (up to a thousand percent or so). This rubber elasticity is quite different from linear elasticity, which is, as mentioned, ordinarily limited and, as might be expected, the causes of rubber elasticity differ fundamentally from those of linear elasticity. Further discussion of elastic properties and behavior is given in Chap. 2.

1.3
PERMANENT DEFORMATION

A. The Tension Test

A material's response to uniaxial loading is assessed most often by means of a tension test (Fig. 1.5). In this test a material is usually stretched at a specific rate, and the force required to cause an extension δl is measured. Force is measured with a load cell (often a calibrated, stiff spring), and the extension is measured often by means of a device called an extensometer. Knowledge of F, δl, and sample geometry allows calculation of the stress and strain.

A material's elastic modulus can be measured by such a procedure, although to do so requires an accurate measurement of the extension, since linear elastic strains are limited. Some materials—brittle ones—manifest only macroscopic elastic deformation up to the stress at which they fracture. Examples include inorganic

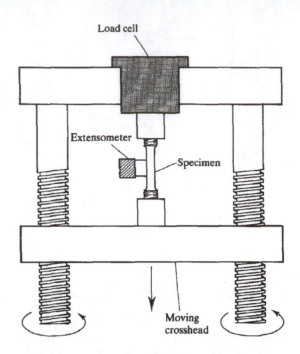

Figure 1.5
A schematic of a tension test.
The sample is elongated at a
specified rate, and the force
required to produce a given
elongation is measured via
the load cell. Elongation is
measured by an extensometer
or similar device. Knowing
sample dimensions, the stress
and strain can be calculated
by measurement of F and δl.

glasses, polycrystalline ceramics at room temperature, and some metals and their alloys at low temperatures. On the other hand, most metals at ordinary temperatures, and many ceramics at high temperatures, deform permanently before fracture. A schematic tensile stress-strain curve for such inorganic solids is given in Fig. 1.6.[1] If the material is loaded to a stress σ_1, and then the load is removed, some permanent or plastic strain (ε_{pl}) remains (Fig. 1.6b). Before unloading, the total strain at the stress σ_1 is $\varepsilon_{pl} + \varepsilon_{el}$ (with $\varepsilon_{el} = \sigma_1/E$). The transition from linear elastic to plastic behavior is gradual, and it is difficult to assess accurately the lower limiting stress below which no plastic deformation at all is found. Consequently, a different stress—one needed to cause a small but readily measurable plastic strain—is used to characterize a material's resistance to plastic deformation.

This stress is called the yield strength (σ_y) and can be measured in the manner shown in Fig. 1.6b. A line parallel to the initial (linear) modulus line is drawn from an offset on the strain axis (typically $\varepsilon_{pl} = 0.002 = 0.2\%$). The intersection of this line with the stress-strain curve defines the material's offset yield strength σ_y. Clearly σ_y depends on the specified offset strain, and so this must be stated when a yield strength is quoted. As mentioned, quite often the offset strain is 0.002, and this can be assumed in the absence of specification of an offset strain.

That σ_y represents the stress required to produce the offset strain can be deduced by recalling that on unloading the material recovers elastic strain according to Eq. (1.3). Thus, if the material were loaded to the stress σ_y and then unloaded, the resulting permanent strain would equal the offset strain. Hence, while σ_y could be determined by a repetitive loading and unloading procedure, involving measur-

---

[1]The tensile deformation behavior of polymerics, which differs considerably from that of inorganics, is discussed in Chap. 8.

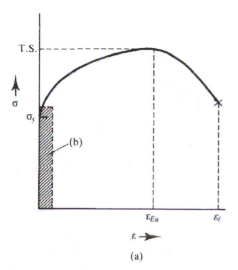

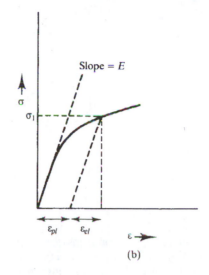

(a)

(b)

Figure 1.6
A schematic tensile stress-strain curve. (a) Following linear elastic deformation, plastic flow commences at a stress approximately equal to the yield strength, σ_y. Following yielding, the material work hardens; the stress required to continue deformation increases with increasing strain. The maximum engineering stress a material can withstand in a tensile test is the tensile strength, T.S. The strain at T.S. ($= \varepsilon_{Eu}$) represents the maximum strain for which plastic deformation is uniform along the sample length. For strains greater than this, stress decreases; the phenomenon is associated with nonuniform material deformation (necking). Fracture, denoted by X, takes place at the engineering strain, ε_f. (b) An expanded view of the low-strain (shaded) region of (a). Plastic flow initiates at a stress less than σ_1. At σ_1, the total strain is the sum of the elastic (ε_{el}) and plastic (ε_{pl}) components; ε_{el} is given by σ_1/E. It can be found graphically by subtracting from the total strain the strain not recovered; the latter is obtained by drawing a line of slope E downward from σ_1. The plastic, or permanent strain is represented by the intersection of the unloading line with the strain axis. The 0.2% yield strength can be obtained by offsetting a strain of 0.002 on the strain axis, and drawing a line parallel to the initial loading line. The intersection of this line with the stress-strain curve defines the stress required to cause a permanent strain of 0.002.

ing the progressively increasing permanent strain at each unloading step, the offset procedure accomplishes the same end with far less effort.

Following yielding, the stress required for further plastic flow (the flow stress) increases with increasing strain (Fig. 1.6). The positive slope of the stress-strain curve indicates that the material is made more resistant to plastic deformation by virtue of the deformation itself; that is, the material work hardens. During plastic deformation, material volume remains constant,[2] and if the deformation is uniform along the sample length, then the sample shape changes as shown in Fig. 1.7b. The

---

[2]This is not so for elastic deformation. Lateral contractions, along the two axes orthogonal to the tensile axis, accompany sample extension along the stress axis. For linear elastic deformation the ratio of (each) lateral strain to the tensile strain is $-\nu$, where ν is Poisson's ratio typically having a value of about ⅓ for metals. During permanent deformation, the ratio of lateral to tensile strain is $-½$, and this leads to constancy of volume. More is written about Poisson's ratio in Chap. 2.

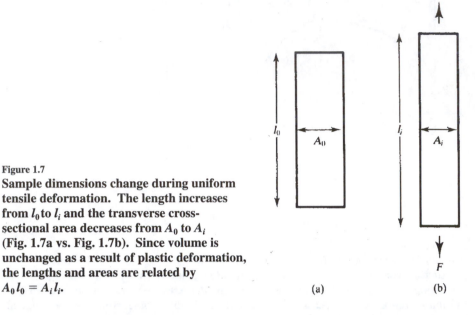

Figure 1.7
Sample dimensions change during uniform
tensile deformation. The length increases
from l_0 to l_i and the transverse cross-
sectional area decreases from A_0 to A_i
(Fig. 1.7a vs. Fig. 1.7b). Since volume is
unchanged as a result of plastic deformation,
the lengths and areas are related by
$A_0 l_0 = A_i l_i$.

original and strained dimensions are related through $A_0 l_0 = A_i l_i$ where A_0 and l_0 are the original transverse cross-sectional area and sample length, respectively, and A_i and l_i represent these quantities in the strained condition. As the cross-sectional area decreases with increasing strain, the sample experiences an effective stress greater than that suggested on the basis of the initial cross-sectional area. We can define a *true stress* (σ_T) as the ratio of force to the instantaneous area; i.e.,

$$\sigma_T = \frac{F}{A_i} \tag{1.5}$$

Henceforth the subscript T is used to designate true stress (or strain) and the subscript E to denote engineering stress (or strain), the engineering definition being the basis of previous definitions.

It is also worthwhile to reconsider our definition of strain when plastic deformation is appreciable. On the basis of the gage length remaining the same and equal to the initial one, each increment of plastic extension δl produces an equivalent strain. However, as sample length increases, one should consider the instantaneous gage length as a basis for calculating strain. If this is done, it is apparent that strain is overestimated by the initial definition. *True strain* is based on instantaneous sample length. It can be approximated by considering the total strain to result from a series of small, incremental extensions (δl) with the gage length at each increment being the instantaneous sample length. Thus,

$$\varepsilon_T = \frac{\delta l}{l_0} + \frac{\delta l}{l_1} + \frac{\delta l}{l_2} + \cdots = \sum_i \left(\frac{\delta l}{l_i}\right) \tag{1.6}$$

where $l_1 = l_0 + \delta l$, $l_2 = l_1 + \delta l$, etc. When expressed in differential form, Eq. (1.6) becomes exact; i.e.

$$d\varepsilon_T = \frac{dl}{l} \qquad (1.7)$$

On integrating Eq. (1.7) from $l = l_0$ to $l = l_i$,

$$\varepsilon_T = \ln \frac{l_i}{l_0} \qquad (1.8)$$

The constant-volume condition of plastic deformation allows relationships to be developed among the various stresses and strains, provided deformation along the gage length is uniform. Equation (1.5) can be rearranged as

$$\sigma_T = \frac{F}{A_i} = \frac{F}{A_0}\frac{A_0}{A_i} = \sigma_E\left(\frac{A_0}{A_i}\right) \qquad (1.9)$$

Since $(A_0/A_i) = (l_i/l_0)$ and $l_i = l_0 + \delta l$, we have $(A_0/A_i) = [1 + (\delta l/l_0)]$, so that

$$\sigma_T = \sigma_E (1 + \varepsilon_E) \qquad (1.10)$$

Clearly, $\sigma_T > \sigma_E$ for a tensile test, a result intuitively deduced previously. In contrast, if the material were compressed so that the cross-sectional area increased during deformation (with $\varepsilon_E < 0$), we would find $\sigma_E > \sigma_T$.

We also find (Prob. 1.3) that for uniform gage length deformation, ε_T and ε_E are related through

$$\varepsilon_T = \ln(1 + \varepsilon_E) \qquad (1.11)$$

Equation (1.11) shows that $\varepsilon_T < \varepsilon_E$ in a tension test (i.e., $\ln(1 + x) < x$). Thus, a point defining $\sigma_E - \varepsilon_E$ in a stress-strain diagram is displaced upwards and to the left to define the equivalent $\sigma_T - \varepsilon_T$ point. The difference between the true and engineering stresses and strains increases with plastic deformation. Thus, at low strains, $\sigma_T \approx \sigma_E$ and $\varepsilon_T \approx \varepsilon_E$, so that, for example, in discussion of elastic deformation there is no need to differentiate between engineering and true stress and strain.

As shown in Fig. 1.6a, engineering stress attains a maximum value at the strain ε_{Eu}. This maximum is called the material's tensile strength (T.S., also sometimes referred to as the ultimate tensile strength, U.T.S.). Continued straining beyond ε_{Eu} requires an ever-decreasing applied force. On the basis of this observation, it might be inferred that the material has exhausted its work-hardening capacity at ε_{Eu}, and has begun to "work soften" for strains greater than this. However, materials do not behave so capriciously. Instead, the tensile point is associated with a geometrical instability, and not with a fundamental alteration in material behavior.

The geometrical instability associated with the tensile strength can be discussed with reference to Fig. 1.8. Each and every tensile bar has inhomogeneities along its length; either within it (e.g., small inclusions or porosity) or on its surface (e.g., machining marks or a taper along the bar surface). Strain is localized in these regions, and this leads to a locally greater reduction in area (Fig. 1.8a). For strains less than ε_{Eu}, the increase in flow stress accompanying the greater strains is large enough to lead to removal of the incipient instability (Fig. 1.8b). This process occurs regularly and repeatedly during tensile loading, and could be monitored if sufficiently accurate instrumentation were available. The rate of work hardening decreases as deformation continues; that is, the increase in flow stress per unit strain becomes less with increasing strain. Thus, it becomes progressively more difficult to work

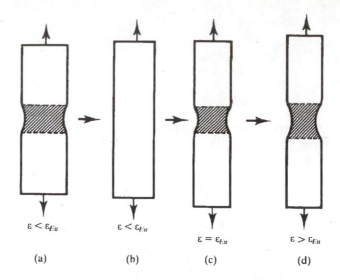

Figure 1.8
(a) During tensile deformation, strain is periodically localized at "weak links" (the shaded area) along the sample length. (b) For strains less than ε_{Eu}, the work hardening in these areas strengthens the material enough relative to the material outside of them so that the instability is removed. (c) The work-hardening rate decreases with strain. At ε_{Eu} the fractional decrease in cross-sectional area at a "weak link" exactly compensates the increase in flow strength due to work hardening there. Thus, a permanent instability is formed at a strain infinitesimally greater than ε_{Eu}, and this leads to neck formation. (d) Further deformation is localized in the instability region, and the neck becomes more and more pronounced.

harden an incipient instability sufficiently to remove it. At the tensile point, the work-hardening capacity has been diminished enough that an instability once formed continues to develop. As shown in Fig. 1.8c, in a cylindrical bar the instability takes the form of a neck.

Clearly, the criterion for necking is related to the material's work-hardening tendencies vis-à-vis those that initiate instability. The criterion can be expressed quantitatively by realizing that at T.S. the engineering stress or, equivalently, the force reaches a maximum, i.e., $dF = 0$. Using the definition of true stress (Eq. (1.5))

$$dF = 0 = \sigma_T dA_i + A_i d\sigma_T \qquad (1.12a)$$

or

$$\frac{d\sigma_T}{\sigma_T} = -\frac{dA_i}{A_i} \qquad (1.12b)$$

as the necking criterion. In other words the onset of necking is characterized by the fractional increase in flow stress (as defined by $d\sigma_T/\sigma_T$) being exactly balanced by the fractional decrease in load-bearing area (as measured by dA_i/A_i). Prior to necking, $|d\sigma_T/\sigma_T| > |dA_i/A_i|$ and incipient instabilities are removed. Following necking, the instability continues to develop (Fig. 1.8d).

After the neck has formed, further plastic deformation is constrained to its vicinity. Indeed, the remainder of the sample actually undergoes some elastic contraction as a result of the decrease in tensile force concomitant with neck formation.

Because of these considerations the engineering stress-strain curve is of little fundamental value when $\varepsilon > \varepsilon_{Eu}$. Nevertheless, other properties are often quoted from the results of a tensile test. One is the fracture strain (ε_f, Fig. 1.6a), often called percent elongation when ε_f is expressed in terms of a percentage (percent elongation = $[(l_f - l_0)/l_0] \times 100$, where l_f is the sample length at fracture). Percent elongation is often taken as a rough measure of a material's ductility. However, since deformation following necking is restricted to the necked region, percent elongation depends on initial sample length (Prob. 1.5), and therefore this must be specified for this parameter to have meaning. Although not quoted as often as percent elongation, ε_{Eu} is more of an inherent material property than ε_f. Indeed, in a real sense ε_{Eu} represents a material's resistance to neck development; that is, it is a measure of work-hardening capability. In this capacity uniform strain can be used to assess a material's suitability for many metal-forming operations that require a material to manifest a strong resistance to neck development. Finally, the uniform strain has one other significance; it is the strain at fracture that would be found in a sample of "infinite" gage length (Prob. 1.5); in this further respect, it is also more of a material property than ε_f.

Another measure of material ductility is reduction in area at fracture, usually expressed as percent R.A. (%R.A. = $[(A_0 - A_f)/A_0] \times 100$, where A_0 and A_f represent the initial and final sample cross-sectional areas, respectively). The final cross-sectional area is measured as the area of the neck following fracture. Since %R.A. is independent of sample gage length, it is more of a material property than percent elongation.

As a result of the nonuniform deformation following the onset of necking, true stress and strain cannot be calculated from engineering stress and strain via Eqs. (1.10) and (1.11). However, true stress can still be defined as the force divided by the instantaneous area, provided the latter is taken as the minimum (i.e., the neck) cross-sectional area. Some care must be taken when doing this, particularly at the later stages of neck development and at strains close to the fracture strain. A well-developed neck alters the stress state in the neck region from that of simple tension, and the effect is that $\sigma_T = (F/A_{neck})$ becomes only an approximation. Additionally, internal voids, which are precursors to fracture, form in the last stages of a tensile test, and this leads to an underestimate of σ_T when it is calculated in the above way.

By considering the neck as the deforming volume, true strain can also be redefined following necking. Before necking, $\varepsilon_T = \ln(l_i/l_0)$ (or, equivalently, $\varepsilon_T = \ln(A_0/A_i)$). Following necking, it is defined only on an area basis, that is, by the latter expression with A_i taken as the neck area. Because confusion often arises as to the conditions under which Eqs. (1.10) and (1.11) are appropriate, as well as when they are not, Table 1.1 synopsizes engineering and true definitions of stress and strain, and expressions for them appropriate to tensile flow before and after necking are also listed there.

A graph of true stress-true strain does not demonstrate anything unusual at the tensile strength (Fig. 1.9). This is additional evidence that necking is geometric in origin and does not reflect changes in material properties. One final point is in order. We have mentioned that, prior to necking, $\varepsilon_T < \varepsilon_E$. At some strain greater than ε_{Eu}, this is no longer so (Fig. 1.9). In effect, localized deformation leads to ε_E values that are no indication of the much greater strain found in the neck region; true strain, as calculated by $\ln(A_0/A_{neck})$, is not subject to such a shortcoming.

Parameter	Fundamental definition	Prior to necking	Following necking
Engineering stress (σ_E)	$\sigma_E = \dfrac{F}{A_0}$	$\sigma_E = \dfrac{F}{A_0}$	$\sigma_E = \dfrac{F}{A_0}$
True stress (σ_T)	$\sigma_T = \dfrac{F}{A_i}$	$\sigma_T = \dfrac{F}{A_i}$ $= \sigma_E(1 + \varepsilon_E)$	$\sigma_T = \dfrac{F}{A_{neck}}$
Engineering strain (ε_E)	$\varepsilon_E = \dfrac{\delta l}{l_0}$	$\varepsilon_E = \dfrac{\delta l}{l_0}$	$\varepsilon_E = \dfrac{\delta l}{l_0}$
True strain (ε_T)	$\varepsilon_T = \ln \dfrac{A_0}{A_{min}}$	$\varepsilon_T = \ln \dfrac{l_i}{l_0}$ $= \ln \dfrac{A_0}{A_i}$ $= \ln (1 + \varepsilon_E)$	$\varepsilon_T = \ln \dfrac{A_0}{A_{neck}}$

Table 1.1
Definitions of and relationships between true and engineering stress and strain

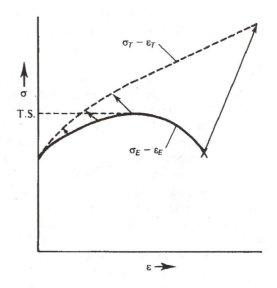

Figure 1.9
A schematic showing the relationship between tensile true stress-true strain (broken line) and engineering stress–engineering strain (solid line). For engineering strains less than ε_{Eu}, $\sigma_T > \sigma_E$ and $\varepsilon_T < \varepsilon_E$. The necking point ($\sigma_E$ = T.S.) has no particular significance in the true stress-true strain curve. At some strain greater than ε_{Eu}, ε_T (when calculated on the basis of neck area) becomes greater than ε_E, while σ_T remains greater than σ_E.

Equation (1.3) (i.e., $\varepsilon = \sigma/E$) is a constitutive equation relating strain and stress during linear elastic tensile loading. As discussed in Chap. 2, there is a fundamental basis—relating to chemical bond strength—that defines the form of this equation. Similar efforts have been made to develop fundamental constitutive equations relating stress and plastic strain. However, the diversity of phenomena taking place during plastic deformation, and the degree to which these vary among materials and material classes, has limited the usefulness of such descriptions. As a result, for the most part, empirical equations are used to describe the plastic-flow behavior of solids. A number of these have been put forth and have been found to describe plastic-flow behavior satisfactorily. One of the most common relates true stress and true strain by

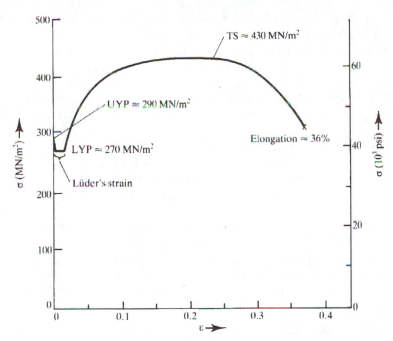

Figure 1.10
Engineering stress-strain diagram for a mild (1018 hot-rolled) steel, which exhibits a
yield point. Plastic flow initiates at the upper yield point (UYP). It continues at the
lower one (LYP). At LYP, permanent deformation is heterogeneously distributed
along the sample length. A deformation band, formed at the UYP, propagates along
the gage length at the LYP. The band occupies the whole of the gage length at the
Lüders strain. Beyond this point, work hardening commences. (*From K. M. Ralls,*
T. H. Courtney, and J. Wulff, An Introduction to Materials Science and Engineering,
***Wiley, New York, 1976.*)**

$$\sigma_T = K(\varepsilon_T)^n \qquad (1.13)$$

where n, the strain-hardening coefficient, is a measure of the material's work-
hardening behavior and K is called the strength coefficient. There is no physical
significance, per se, to K; it can be thought of simply as the true stress required to
cause a true strain of unity. On the other hand, and as expected, n correlates with a
material's resistance to necking (Prob. 1.7). For metals at ordinary temperatures, n
is in the range from ca. 0.02 to about 0.50.

   The stress-strain curve of Fig. 1.6a accurately schematizes the behavior of many
engineering solids, particularly metals at temperatures at which they exhibit time-
independent plastic flow. However, the initiation of plasticity in certain solids
(including some metals, polymers, and ceramics and depending on temperature,
strain rate and structural considerations) does not follow the scenario of Fig. 1.6. Instead these materials exhibit a yield point. The room temperature engineering stress-
strain curve of a mild steel is characterized by a yield point, as shown in Fig. 1.10.
Plastic flow commences at a stress equal to the upper yield point (UYP), and then
continues at a lower stress level (LYP, the lower yield point). We see that for this
steel, as well as for other materials manifesting a yield point, the stress required to

initiate plastic flow is greater than that required to continue it. This situation holds up to a certain strain (for steel this strain is the Lüders strain noted in Fig. 1.10). Plastic deformation is heterogeneously distributed along the gage length of the steel during this initial stage of plastic deformation. A small permanently deformed volume first forms at the UYP and spreads along the gage length at the LYP stress until the sample is characterized by a uniform permanent strain (the Lüders strain). Beyond this stage the stress-strain behavior is similar to that of materials that do not exhibit yield-point behavior.

For mild steel, the Lüders strain is limited. This is not so for the many polymers which often display yield-point behavior at ordinary temperatures. For these materials, the nonuniform strain may be on the order of a hundred percent or more. Ceramics, particularly in single-crystal form, can also manifest yield points under certain strain rate and temperature combinations.

The present discussion has described time-independent plastic flow. This is generally the situation for metals and ceramics at ordinary, but not elevated, temperatures. It is almost invariably not so for polymeric solids at temperatures at which they exhibit extensive permanent deformation. The time dependence of plastic flow is described briefly in the following section. But first we apply the concepts of this section to a sample problem.

EXAMPLE PROBLEM 1.1. The following true stress-true strain data were obtained for copper. (Elastic strains can be neglected in this problem.)

Stress (MPa)	0	141	202	252	290	319	343	360	373	390
Strain	0	.087	.172	.259	.339	.413	.482	.547	.608	.770

Forging, as indicated in the sketch below, resembles a compression test. A copper disk is forged to a reduction in height of 50%. The forging pressure, p, is defined as the compressive force divided by the *initial* cross-sectional area of the material. Plot p as a function of reduction in height. Assume, as would be the situation in a compression test, that there is no friction between the copper disk and the forging die.

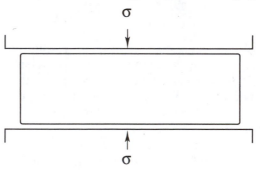

Solution. The forging pressure, defined as $p = F/A_0$, can be written as

$$p = \left(\frac{F}{A_i}\right)\left(\frac{A_i}{A_0}\right)$$

where A_i is the instantaneous transverse cross-sectional area of the disk and A_0 the corresponding initial area. The ratio, F/A_i, is the true stress. Further, the area ratio is also given by (h_0/h_i) with h_0 being the initial disk height and h_i the instantaneous one. Since $h_0/h_i = 1/(1 - \delta h/h_0)$, we have

$$p = \frac{\sigma_T}{(1 - \delta h / h_0)}$$

The true stress–true strain relationship is the same in tension and compression. For a given height reduction, we calculate the true strain from $\varepsilon_T = \ln(h_i/h_0)$ The resulting strain values are negative, but using their magnitude permits us to determine the corresponding true stress values from the data given. The procedure to solve the problem is as follows. First, specify a value of $\delta h/h_0$. Second, determine the true strain and, from a plot of true stress vs. true strain (provided below), determine the appropriate value of true stress. Finally, use the equation above to calculate the forging pressure.

$\delta h/h_0$	h_i/h_0	ε_T	σ_T (MPa)	p (MPa)
0.05	0.95	−0.051	105	111
0.10	0.90	−0.105	140	156
0.15	0.85	−0.163	190	224
0.20	0.80	−0.223	220	275
0.25	0.75	−0.288	250	333
0.30	0.70	−0.357	291	416
0.35	0.65	−0.431	315	485
0.40	0.60	−0.511	340	567
0.45	0.55	−0.598	370	673
0.50	0.50	−0.693	385	770

A graph of p vs. $\delta h/h_0$ is provided below. Note that the forging pressure increases rapidly with sample reduction in height. This is due to two factors. First, the material work-hardens during compression. Second, p is based on the initial cross-sectional area. In compression this area increases with plastic strain, in distinction to what happens in tension. Thus, the curve below which, in effect, is an engineering stress-engineering strain diagram in compression rises rapidly with engineering strain (i.e., reduction in height). This is in contrast to an engineering stress-engineering strain curve in tension where the level of the stress is reduced by the sample geometrical changes taking place as it is plastically deformed.

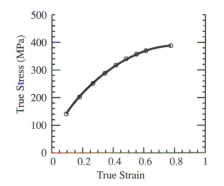
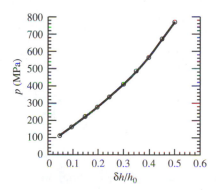

Example Problem 1.1
(a) True stress-true strain curve for copper. (b) Forging pressure as a function of reduction in height.

B. Strain-Rate Sensitivity

An increase in strain rate generally increases the flow stress of a material, although the degree to which it does so is a strong function of the temperature and is specific to the material. There are a number of reasons for this strain-rate sensitivity, and they all relate to the atomistic and/or microscopic mechanisms of permanent deformation. These are discussed in chapters dealing with dislocations (Chap. 3), high-temperature deformation (Chap. 7), and the flow of noncrystalline solids (Chap. 8). For the present we simply note that the strain-rate sensitivity of the flow stress is often adequately represented by the empirical equation

$$\sigma_T = K'(\dot{\varepsilon}_T)^m \tag{1.14}$$

where $\dot{\varepsilon}_T$ is the true *strain rate*, m the *strain-rate sensitivity*, and K' a constant that signifies it is the material flow stress at a true strain rate of unity. The strain-rate sensitivity varies between zero (in which case the material is not strain-rate sensitive) and unity (in which instance the stress increases linearly with strain rate and the material is called a viscous solid).[3] High values of m indicate resistance to neck development in tension, just as do high values of the strain-hardening coefficient in Eq. (1.13). Thus, some materials are capable of "strain rate" hardening. This can be illustrated by considering the behavior of a glass rod when it is stretched in tension at high temperature, or by the slow extension of "silly putty" at room temperature. In both cases, m is unity or close to it; the glass at the high temperature, and the silly putty at room temperature, draw down to a point (%R.A. = 100) before they mechanically fail, and any "neck" that develops is a diffuse one. This behavior is a direct result of these materials' high strain-rate sensitivities.

Metals and alloys at ordinary temperatures are not particularly strain-rate sensitive; e.g., for most of these materials $m \cong 0.00$–0.10 at room temperature. Their strain-rate sensitivities increase with temperature, and for certain metallic alloys m can be as high as ca. 0.8 over a limited range of temperature and stress. The associated resistance to neck development renders them "superplastic" at these stress-temperature combinations, and allows them to be extensively deformation-processed. Superplasticity is described further in Chap. 7.

The flow stresses of noncrystalline polymers at temperatures at which they can be permanently deformed are generally more strain-rate sensitive than those of metals. As with metals, m increases with temperature for polymers, and for polymers the strain-rate sensitivity approaches unity as the equilibrium melting temperature is reached.

Equation (1.14) describes strain-rate hardening in the absence of strain-hardening, whereas Eq. (1.13) describes strain-hardening in the absence of strain-rate sensitivity. When both effects are important, the equations can be combined; that is,

$$\sigma_T = K''(\varepsilon_T)^n (\dot{\varepsilon}_T)^m \tag{1.15}$$

We see that when the material does not strain harden ($n = 0$), Eq. (1.15) reduces to Eq. (1.14) and $K'' = K'$. Conversely, when $m = 0$, it becomes Eq. (1.13) with $K'' = K$. The form of Eq. (1.15) indicates clearly that hardening is caused by increases in both strain and strain rate.

---

[3]When $m = 1$, Eq. (1.14) is often expressed in the form $\sigma_T = 3\eta\dot{\varepsilon}_T$, where η is the material's viscosity. The viscosity is also linearly related to shear stress and shear strain rate by $\tau = \eta\dot{\gamma}$.

C. Yielding Under Multiaxial Loading Conditions

Descriptions of material behavior under uniaxial or tensile loading is convenient for elucidating the fundamentals of behavior in the absence of complexities associated with multiaxial loading. In fact, this is the approach most often taken in this book. Nonetheless, in actual service most materials are subjected to a variety of loading conditions (e.g., biaxial or triaxial stresses), and material response may be profoundly altered by such variations. As but one example, the relative tendency of a material to deform plastically rather than to fracture depends strongly on the state of stress. In this section we describe how the condition for plastic yielding depends on stress state.

In general, a material may be expected to be subject to a combination of external and internal normal (i.e., tensile or compressive) and shear stresses. However, a suitable set of orthogonal coordinates can be chosen so that the stress state in this reference frame consists only of normal stresses. We designate these axes as 1, 2, and 3 and the corresponding normal stresses as σ_1, σ_2, and σ_3.

One, algebraically simple, criterion for yielding under multiaxial loading is due to Tresca. It states that plastic yielding commences when the algebraic difference between the maximum and minimum normal stresses is equal to the material's tensile yield strength, σ_y, i.e.,

$$\sigma_{max} - \sigma_{min} = \sigma_y \qquad (1.16)$$

The Tresca yield criterion for conditions of biaxial loading (σ_1, $\sigma_2 \neq 0$, $\sigma_3 = 0$) is illustrated schematically in Fig. 1.11a. The yield condition is such that stress combinations lying within the yield locus shown in the figure do not lead to plastic flow, whereas those lying without it do. Yielding in the first quadrant (σ_1, $\sigma_2 \geq 0$) is defined when the greater of σ_1, σ_2 equals σ_y. This follows from Eq. (1.16), since $\sigma_3 = 0$; therefore σ_{max} (= greater of σ_1 or σ_2) = σ_y. Yielding in the second quadrant ($\sigma_1 < 0$, $\sigma_2 \geq 0$) is defined by $\sigma_2 - \sigma_1 = \sigma_y$, for in this case $\sigma_{max} = \sigma_2$ and $\sigma_{min} = \sigma_1$. This leads to the 45° line defining the yield criterion in the second quadrant. Except for the interchange of σ_1 and σ_2, the yield criterion in the fourth quadrant ($\sigma_1 > 0$, $\sigma_2 < 0$) is the same as in the second. In the third quadrant (σ_1, $\sigma_2 < 0$, $\sigma_3 = 0$), σ_3 is the algebraically maximum stress, and thus yielding is initiated when the magnitude of the algebraically minimum stress equals σ_y (i.e., $\sigma_y = \sigma_1$ if $|\sigma_1| > |\sigma_2|$, and vice versa).

There are several interesting aspects of the Tresca yield condition. One is that the yield criterion is the same in tension as in compression; e.g., for $\sigma_1 = \sigma_3 = 0$, $\sigma_2 = \pm\sigma_y$ defines yielding. We also note that for tensile biaxial loading (σ_1, $\sigma_2 > 0$) the yield condition is unaffected by the minor tensile stress. For example, if we initially have a stress state $\sigma_1 > 0$, $\sigma_2 = \sigma_3 = 0$, and then increase σ_2, the yield condition remains $\sigma_1 = \sigma_y$ so long as $\sigma_1 > \sigma_2$. This somewhat unexpected result is at variance with experimental studies, as is shown later. Indeed, a different yield criterion, the von Mises one, which predicts that yielding in the first quadrant is a function of both σ_1 and σ_2 is generally more accurate than the Tresca one for predicting yielding under multiaxial stress states.

The von Mises yield criterion is expressed as

$$\frac{1}{\sqrt{2}}[(\sigma_1 - \sigma_2)^2 + (\sigma_1 - \sigma_3)^2 + (\sigma_2 - \sigma_3)^2]^{1/2} = \sigma_y \qquad (1.17)$$

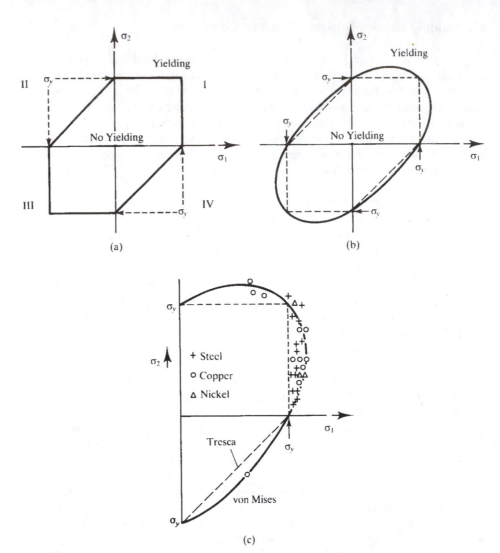

Figure 1.11
(a) The Tresca yield condition for biaxial loading. Stress combinations lying within the solid lines do not result in plastic flow; those without it do. In quadrants I and III, yielding occurs when the magnitude of the algebraically largest (in I) or smallest (in III) exceeds σ_y, the tensile yield strength. In quadrant II, ($\sigma_2 > 0$, $\sigma_1 < 0$, $\sigma_3 = 0$), yielding is defined by $\sigma_{max}(= \sigma_2) - \sigma_{min}(= \sigma_1) = \sigma_y$, and this results in the 45° line defining yielding. The yield criterion is similar in quadrant IV ($\sigma_1 > 0$, $\sigma_2 < 0$, $\sigma_3 = 0$), except that σ_1 and σ_2 are interchanged. (b) The von Mises yield condition (Eq. (1.17)) for biaxial loading is shown by the solid line. Stress combinations lying within this ellipse do not lead to plastic flow; those without it do. The Tresca condition (dotted line) is compared to the von Mises one in this figure. The Tresca is the more conservative criterion, and the two criteria are equivalent only for uniaxial ($\sigma_{1,2} > 0$ with $\sigma_{2,1} = \sigma_3 = 0$), and balanced biaxial ($\sigma_1 = \sigma_2$, $\sigma_3 = 0$) loading. (c) Comparison of experimental data for selected metals with the Tresca and von Mises criteria. The latter clearly fits the data better, although the difference between the two criteria is not great. (Data for biaxial loading from W. Lode, Zeit. für Physik, 36, 913, 1926. Torsion ($\sigma_1 = -\sigma_2$, $\sigma_3 = 0$) datum for copper from G. I. Taylor and H. Quinney, Phil. Trans. Roy. Soc., 230A, 323, 1931.)

The condition states that yielding will not take place for principal stress combinations such that the left-hand side of Eq. (1.17) is less than σ_y, but that flow will occur if it is greater than σ_y. The yield locus under biaxial loading for the von Mises condition is illustrated in Fig. 1.11b. We see that it is an ellipse in the σ_1, σ_2 plane, and the von Mises and the Tresca conditions are equivalent only for uniaxial loading ($\sigma_1 \neq 0$, $\sigma_2 = \sigma_3 = 0$ or $\sigma_2 \neq 0$, $\sigma_1 = \sigma_3 = 0$) and balanced biaxial loading ($\sigma_1 = \sigma_2$, $\sigma_3 = 0$).

As noted, the yield loci of Figs. 1.11(a–c) are appropriate to a situation where one of the principal stresses is zero. Both the Tresca and von Mises yield criteria also apply to the situation where this is not the case. When this is so, the criteria can be graphically displayed in three-dimensional principal stress space. The von Mises yield locus then is cylindrical in shape with the axis of the cylinder lying along the [111] direction in principal stress space. Thus, when this cylinder is "sliced" along the $\sigma_3 = 0$ plane, an ellipse is found (cf. Fig. 1.11b). Similarly, the Tresca yield surface is a regular hexagon translated along the [111] direction in principal stress space. When this shape is sectioned along the $\sigma_3 = 0$ plane, the distorted hexagon displayed in Fig. 1.11a is observed. Finally, we note that the three-dimensional Tresca yield surface is inscribed within the von Mises one, touching the latter at the hexagon vertices.

A number of ideas have been set forth attempting to rationalize the von Mises condition on fundamental grounds. However, it is essentially an empirical criterion that nonetheless more accurately describes yielding under multiaxial stress states than does the Tresca condition. This is shown in Fig. 1.11c, in which results obtained from biaxial yielding studies are shown.

The Tresca and von Mises criterion can also be compared in other ways. The Tresca criterion, even though not as accurate as the von Mises one, is more conservative. That is, excepting for the cases noted above, the Tresca yield locus lies within the von Mises one. It is often used for this reason, and it is used, too, in cases where its algebraic simplicity makes it convenient to apply. In a number of instances in this book we have occasion to describe flow under multiaxial stress states. Depending on the circumstance, one or the other of the above criterion will be used in doing so.

EXAMPLE PROBLEM 1.2
a. The tensile yield strength of a material is 400 MPa. The material is subjected to balanced biaxial compression ($\sigma = -150$ MPa) along the two axes orthogonal to the tensile axis. Determine the value of the tensile stress necessary to cause yielding according to the Tresca condition. How does this stress value compare to that predicted by the von Mises criterion?

b. Suppose a compressive stress of 150 MPa were applied in only one of the two directions orthogonal to the tensile axis. What would be the tensile yield stress according to the Tresca condition? According to the von Mises condition?

Solution
a. For the Tresca condition, the maximum stress will be the tensile stress required to cause yielding; we designate this stress, σ_{Tr}. The minimum stress (σ_{min}) is -150 MPa. Thus, from Eq. (1.16)

$$\sigma_{Tr} = 400 \text{ MPa} + \sigma_{min} = 400 \text{ MPa} - 150 \text{ MPa} = 250 \text{ MPa}$$

For the von Mises criterion, we use Eq. (1.17). We take $\sigma_2 = \sigma_3 = -150$ MPa, and designate σ_1 as σ_{vM}. Thus,

$$\frac{1}{\sqrt{2}}[(\sigma_{vM} + 150)^2 + (\sigma_{vM} + 150)^2]^{1/2} = 400 \text{ MPa}$$

or

$$\sigma_{vM} = 250 \text{ MPa}$$

Thus, for this loading arrangement, the Tresca and von Mises criteria give equivalent results.

b. For this loading condition, the Tresca criterion again predicts a tensile yield stress of 250 MPa since the value of σ_{min} remains -150 MPa. However, the von Mises criterion provides a different value of the required tensile stress. Using Eq. (1.17) with $\sigma_2 = 0$ and $\sigma_3 = -150$ MPa,

$$\frac{1}{\sqrt{2}}[(\sigma_{vM})^2 + (\sigma_{vM} + 150)^2 + (150)^2]^{1/2} = 400 \text{ MPa}$$

Squaring both sides of this equation and rearranging

$$\sigma_{vM}^2 + 150\sigma_{vM} + [(150)^2 - (400)^2] = 0$$

This quadratic equation has the solution (taking the positive root, because $\sigma_{vM} > 0$)

$$\sigma_{vM} = \frac{1}{2}\{-150 + [4(400)^2 - 3(150)^2]^{1/2}\} = 303 \text{ MPa}$$

Thus, for this loading arrangement the Tresca and von Mises conditions predict different values for the tensile yield stress with, as expected, the Tresca condition being the more conservative prediction.

D. Mohr's Circle

As is discussed further in Chaps. 3–5, plastic deformation in crystalline solids occurs by the shear (sliding) of atomic planes. Thus, while an applied tensile force can result in plastic deformation it is the resolution of this force into a shear stress acting on the planes on which the sliding displacement takes place that is the true cause of this plastic deformation. In this section, we review briefly a technique—the Mohr's circle representation—that permits graphical description of the relationships among the principal stress components and the shear stresses that result from them.[4]

Consider Fig. 1.12a, which depicts a solid subjected to a tensile force, F. The transverse cross-sectional area normal to the force is A_1, so that the tensile stress (σ_1) in the specified "1" direction is $\sigma_1 = F/A_1$. We wish to determine how F is manifested as a shear stress. The magnitude of such a shear stress depends on the orientation with respect to the tensile axis of the plane on which the shear stress acts. In Fig. 1.12a, we consider an arbitrary orientation; the plane considered is rotated counterclockwise by the angle θ with respect to the "1" axis.

As indicated in Fig. 1.12a, F can be resolved into a shear force (F_s) and a tensile force (F_t) on the plane considered. The shear force is $F_s = F \cos(\pi/2 - \theta) = F \sin \theta$. The area ($A_s$) of the plane considered is greater than A_1 by the factor $1/\cos \theta$. Thus, the shear stress acting on this plane is given by

---

[4]The Mohr's circle representation can also be used to transform a system of stresses in one coordinate system to that in another. Mechanics texts elaborate on this procedure which we do not need in our abbreviated discussion here.

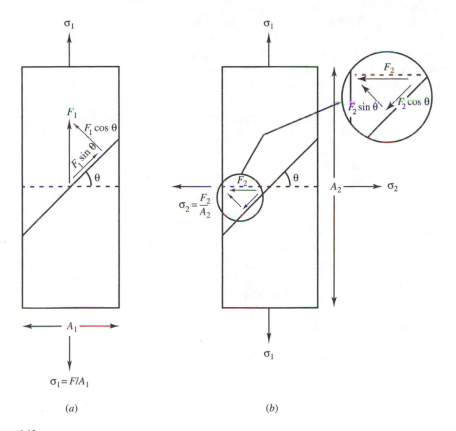

Figure 1.12
**(a) Resolution of a tensile stress into a shear and a tensile stress on a plane rotated
by an angle θ to the tensile axis. The shear force on this plane is equal to $F \sin \theta$, and
the tensile force is equal to $F \cos \theta$. The area of the rotated plane is equal to $A_1/\cos \theta$
where A_1 is the transverse cross-sectional area normal to the applied force (such that
the nominal tensile stress, $\sigma_1 = F/A_1$). Thus, the resolved shear stress on the rotated
plane is $\tau = (F/A_1) \sin \theta \cos \theta$; the corresponding tensile stress is $\sigma = \sigma_1 \cos^2 \theta$. (b) A
second principal component of stress can be considered. Here the nominal second
stress is $\sigma_2 = F/A_2$. This stress reduces the shear stress on the rotated plane and also
alters the tensile stress there. As discussed in the text, for this situation we have $\tau =
(\sigma_1 - \sigma_2) \sin \theta \cos \theta$ and $\sigma = (\sigma_1 + \sigma_2)/2 + [(\sigma_1 - \sigma_2)/2](2 \cos^2 \theta - 1)$. These rela-
tionships can be reformulated in terms of the angle 2θ (Eqs. (1.18)–(1.21)).**

$$\tau = \frac{F_s}{A_s} = \frac{F}{A_1} \sin \theta \cos \theta = \sigma_1 \sin \theta \cos \theta = \frac{1}{2} \sigma_1 \sin 2\theta \qquad (1.18)$$

where the formulation involving 2θ comes from the trigonometric identity, $\sin 2\theta =
2\sin\theta\cos\theta$. Similar reasoning is applied to determine the tensile stress on this plane.
The tensile force is $F \cos \theta$. The area on which it acts is A_s, so

$$\sigma = \frac{F}{A_1} \cos^2 \theta = \sigma_1 \cos^2 \theta = \frac{1}{2} \sigma_1 (1 + \cos 2\theta) \qquad (1.19)$$

The 2θ formulation comes from $\cos^2 \theta = (1 + \cos 2\theta)/2$.

Equations (1.18) and (1.19) can be graphically displayed in the form of a Mohr's circle appropriate to uniaxial tension, as shown in Fig. 1.13a. The abscissa of this diagram represents a tensile (or compressive) stress and the ordinate represents a shear stress. The diagram is laid out so that the stress σ_1 (the only principal stress in the case of a tensile test) is marked on the tensile stress axis. A circle of diameter having magnitude σ_1 is drawn. To determine the stress state in a plane rotated by the angle θ from the reference axis (the tensile axis), a diameter is drawn through the circle at an angle 2θ from the reference axis as shown in Fig. 1.13a. The intersection of this diameter defines the shear and tensile stress in the rotated system. We see that the maximum shear stress is found when $2\theta = 90°$; i.e., it occurs on a plane rotated at an angle of 45° from the tensile axis. Further, the maximum value of this shear stress, τ_{max}, is simply $\sigma_1/2$.

The Mohr's circle representation can also be used to characterize stress states other than uniaxial tension, as can be illustrated for biaxial tension (Fig. 1.12b). We note that the secondary tensile force, F_2, produces a shear stress acting in the opposite sense to that caused by F_1. Thus, a biaxial tensile stress state is less propitious for inducing shear displacements than is a uniaxial one. The principle of superposition of stress tells us that the stress state corresponding to biaxial tension is obtained by combining the results of analyses applied to the individual stress states. The details of the analysis can be found in mechanics texts; here only the results are presented. The stresses are expressed in terms of the principal stresses, σ_1 and σ_2. (Recall that σ_2 is calculated on the basis of the area denoted as A_2 in Fig. 1.12b.)

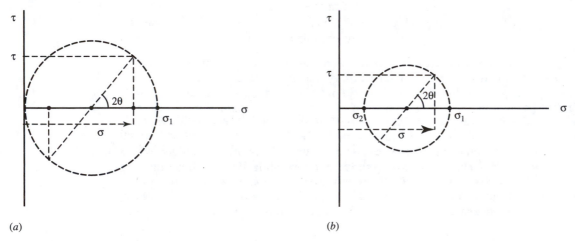

(a)   (b)

Figure 1.13
**Mohr's circle representation of the stress states corresponding to the situations of Figs. 1.12a and b.
(a) Uniaxial tension is represented by a circle with diameter equal to σ_1. The intercepts of the circle on the tensile stress axis are 0 and σ_1, respectively. A plane rotated by the angle θ with respect to the plane of principal stress has a shear stress acting on it as shown. The maximum shear stress has magnitude $\sigma_1/2$ and is found on a plane rotated 45° from the tensile axis. (b) The Mohr's circle representation of biaxial tension (Fig. 1.12b). The circle has a diameter equal to the difference between the magnitudes of the principal stresses. The intercepts of the circle along the tensile stress axis are σ_2 and σ_1, respectively. A plane rotated by an angle θ with respect to the original axis (the "1" direction) has the shear stress component illustrated. The maximum shear stress is again found on a plane rotated by 45°. However, in this case the maximum shear stress is $(\sigma_1 - \sigma_2)/2$.**

$$\tau = \frac{1}{2}(\sigma_1 - \sigma_2)\sin 2\theta \qquad (1.20)$$

$$\sigma = \frac{1}{2}\sigma_1(1 + \cos 2\theta) + \frac{1}{2}\sigma_2(1 - \cos 2\theta)$$

$$= \frac{1}{2}(\sigma_1 + \sigma_2) + \frac{1}{2}(\sigma_1 - \sigma_2)\cos 2\theta \qquad (1.21)$$

The Mohr's circle representation of biaxial tension (i.e., Eqs. (1.20) and (1.21)) is shown in Fig. 1.13b. Here the principal stresses, σ_1 and σ_2, are laid out on the tensile stress axis. Their difference defines the diameter of the Mohr's circle. The shear stress acting on a plane oriented at the angle θ with respect to the σ_1 axis is equal to the radius of this new Mohr's circle multiplied by $\sin 2\theta$ (Fig. 1.13b.) The tensile stress in this plane is given by the mean of the principal stresses plus the radius of the Mohr's circle times $\cos 2\theta$. We see that the plane on which the maximum shear stress operates is still rotated by the angle $45°$ from the initial, reference plane. However, the magnitude of this shear stress is reduced—by an amount directly proportional to the magnitude of the second tensile stress component—relative to that pertaining to uniaxial tension. (However, the maximum shear stress is equal to the radius of the Mohr's circle in both cases.) Thus, a second tensile component of stress reduces the magnitude of the maximum shear stress and, therefore, ought to delay the initiation of plastic flow. In contrast, if the secondary principal stress were a compressive stress, this would facilitate plastic deformation.

Three-dimensional stress states can also be represented by a Mohr's circle description, as shown in Fig. 1.14. Three separate Mohr's circles are constructed around the principal stress combinations $\sigma_1 - \sigma_2, \sigma_1 - \sigma_3, \sigma_2 - \sigma_3$ (by convention $\sigma_1 \geq \sigma_2 \geq \sigma_3$, algebraically). It can be shown that all possible stress states lie within the shaded area shown in the diagram, or on its bounding surfaces. The maximum shear stress developed in the 1-2 plane is indicated by τ_{max1-2} in the diagram and is equal to $(\sigma_1 - \sigma_2)/2$; similar relationships hold for the other planes. The greatest shear stress $[= (\sigma_1 - \sigma_3)/2 = (\sigma_{max} - \sigma_{min})/2]$ is developed in the 1-3 plane, and we see that the Tresca yield condition is equivalent to equating this stress to the *shear* yield strength in tension, $\tau_y (= \sigma_y/2)$. Additionally, the von Mises yield criterion is related to the three principal values of τ_{max} $[(\sigma_1 - \sigma_2)/2, (\sigma_1 - \sigma_3)/2, (\sigma_2 - \sigma_3)/2]$. This gives credence to the idea that plastic flow is caused by shear stresses, and is consistent, too, with experimental findings that such low-temperature flow in crystalline solids results from shear displacements.

Mohr's circle representations for several common loading arrangements are shown in Fig. 1.15. The convenience of the representation is evident in these examples. We also note that increases (or decreases) in σ_1, σ_2, and σ_3 by equal amounts do not change the maximum shear stresses developed, and thus do not alter either the Tresca or von Mises yield criterion. Such a variation of σ_1, σ_2, and σ_3 is equivalent to introducing a hydrostatic component of stress, and we therefore conclude that such a stress component has no influence on the plastic-flow criteria for solids. In contrast, the hydrostatic stress (as measured by the sum of the principal stresses or one-third of this sum, the mean pressure) can be related to fracture criteria in many solids. Or to put this in another way, *plastic flow is promoted by*

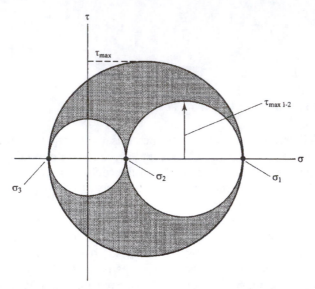

Figure 1.14
Mohr's circle description of stress state in three dimensions for principal stresses
$\sigma_1 > \sigma_2 > \sigma_3$. **Three separate Mohr's circles are constructed for the different princi-**
pal stress combinations. Allowable stress states are restricted to the shaded area
(or its boundaries) in the diagram. The value of the maximum shear stress in the 1-2
plane (τ_{max1-2}) is given by the radius of the circle determined by the $\sigma_1 - \sigma_2$ stress
combination and so forth. Thus, the greatest shear stress ($\tau_{max} = (\sigma_1 - \sigma_3)/2$) is
found in the 1-3 plane.

large algebraic differences in the principal stress components, whereas fracture is
promoted by a large algebraic sum of these stresses.

   This concludes our brief (and incomplete) description of Mohr's circle. We
reemphasize that plastic flow is induced by shear stresses, which in turn depend
upon differences in principal stresses. Before ending our discussion of plastic flow
we briefly discuss other test methods useful for characterizing plastic flow. We focus on the hardness and the torsion tests. The former is widely and conveniently
used to assess the general resistance of a material to plastic flow. In contrast, the
torsion test is not often used, because it is a test relatively difficult to conduct.
However, it is useful for describing flow behavior at large plastic strains, and in this
context has been of special importance in describing plastic flow at high temperatures at which the malleability of many metals is extensive.

E. The Hardness Test

Hardness is a somewhat imprecise term. However, it is basically a measure of a
material's resistance to surface penetration by an indenter having a force applied to
it. Since the indentation process takes place by plastic deformation in metals and alloys, hardness is inherently related to these materials' plastic flow resistance. However, brittle solids, such as glass and polycrystalline ceramics at room temperature,
can also be indentation-hardness tested. That they can be indicates that these mate-

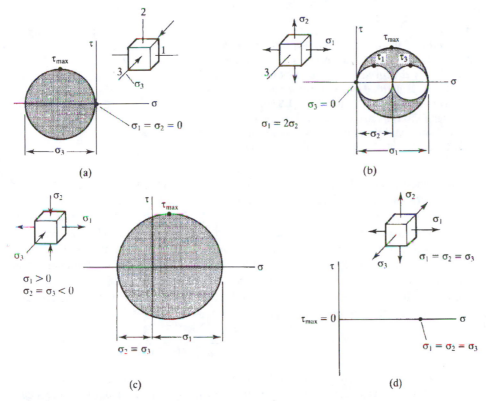

(a)

(b)

(c)

(d)

Figure 1.15
**Three-dimensional Mohr's circle description of stress for some common loading
schemes: (a) uniaxial compression ($\sigma_1 = \sigma_2 = 0$, $\sigma_3 < 0$); (b) biaxial tension ($\sigma_1 >
\sigma_2 > 0$, $\sigma_3 = 0$); (c) uniaxial tension plus balanced biaxial compression ($\sigma_1 > 0$, $\sigma_2 =
\sigma_3 < 0$); (d) hydrostatic tension ($\sigma_1 = \sigma_2 = \sigma_3 > 0$). For a given normal stress, the
maximum shear stress is the same in uniaxial tension and compression. From (b) we
note that superposition of a biaxial tensile stress ($\sigma_2 > 0$) onto a state of uniaxial loading does not alter the maximum shear stress. That is why the Tresca yield condition
has the shape it does in the first and third stress quadrants (cf. Fig. 1.11a). From
(c) we see that addition of compressive stresses to uniaxial loading increases the
value of the maximum shear stress. Thus, such a situation (often developed in
metal-working processes) facilitates plastic deformation. The Mohr's "circle" for
hydrostatic tension (d) is a point. Therefore, plastic flow cannot take place in such
circumstances as shear stresses are not developed. Even inherently ductile materials
fracture in hydrostatic tension before plastic flow takes place.**

rials are capable of some plastic flow, at least at the microscopic level. Indentation
hardness testing of brittle solids is often frequently accompanied by crack formation. This artifact permits the fracture propagation resistance of brittle solids to be
assessed, as is discussed in Sect. 1.4A.

As mentioned, indentation is accomplished by applying a load to an indenter
positioned on the material's surface. To achieve the purpose of the test, the indenter
material must be considerably harder than the material indented. Typical indenters

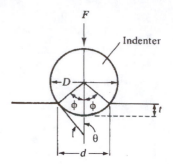

Figure 1.16
Schematic of the Brinell hardness test. A hardened
steel ball of diameter D is impressed into the test
material by the force F. The Brinell hardness num-
ber (BHN) is defined as the force divided by the sur-
face area of the indentation (Eq. (1.22)). For a given
F, softer materials are penetrated more deeply than
harder ones, and the indentation area is greater for
a softer material. Its BHN is therefore lower.

are hardened steel and diamond and, depending on the material tested and the type of test, various combinations of load and indenter are used in hardness testing.

A common hardness test is the Brinell test. In it, the indenter is a 10-mm diameter hardened steel ball, and the applied mass is usually 3000 kg; the test procedure is illustrated in Fig. 1.16. Since a "hard" material resists indentation more than a "soft" one, the average indentation depth, surface area, and projected area (Fig. 1.16) are small if the material is "hard" and relatively large if it is "soft." The Brinell hardness number (BHN) is defined as the load, F, divided by the *surface* area of the indentation. From Fig. 1.16 we have

$$\text{BHN} = \frac{2F}{\pi D[D - (D^2 - d^2)^{1/2}]} = \frac{F}{\pi Dt} \tag{1.22}$$

where D is the ball diameter, t the indentation depth, and d the indentation diameter. In performing the test, F and D are known *a priori,* so measurement of d allows calculation of the BHN. The BHN had dimensions of stress (e.g., N/m²), but hardness numbers are still often quoted in terms of mass, rather than force, per unit area (e.g., kg/mm²).

As mentioned, plastic flow takes place beneath the indenter for malleable materials. Thus, the hardness number is related to the average flow stress of the material in the volume subject to plastic deformation. The flow pattern in the region beneath the indenter is illustrated schematically in Fig. 1.17. Within this volume there exists a gradient in plastic strain; this strain is greatest near the indenter surface and least at the boundary between the elastic and plastic regions. The elastic region surrounding the plastic one produces a constraint on plastic deformation in the latter. As a consequence the average pressure (p) exerted by the indenter exceeds considerably the material's flow strength. Analysis shows that the hardness (which scales with p) is related to the yield strength, σ_y, through

$$\text{Hardness} \cong (2.5-3.0)\,\sigma_y \tag{1.23}$$

Equation (1.23) applies strictly only to materials that do not work harden. For work-hardening materials, σ_y in Eq. (1.23) is replaced by the flow stress corresponding to the average strain within the plastic zone; this strain is typically on the order of several percent. These complications notwithstanding, Eq. (1.23)—in which the constant is often taken as 3.0—is widely used to estimate material yield strengths from the results of hardness measurements.

It is clear that hardness is a measure of a material's plastic flow resistance, and is especially useful for this purpose when comparative assessments are made.

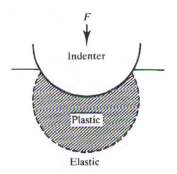

Figure 1.17

The pattern of deformation beneath a hardness indenter. The plastically deforming material is constrained from yielding by the elastic material surrounding it. As a result, the mean pressure needed to initiate plastic flow considerably exceeds the material's tensile yield strength. (*After M. C. Shaw and G. J. DeSalvo, Trans. A.S.M.E., Ser. B. J. Eng. Ind., 92, 469, 1970.*)

Moreover, since hardness tests are more convenient to carry out than tensile tests, the hardness test has found widespread use in research and industrial applications.

Other hardness tests are common. The Vickers hardness test uses a square-based diamond pyramid as the indenter, and applied loads vary from ca. 1 to 120 kg. The diamond pyramid hardness number (DPH, or alternatively, the Vickers hardness number, VHN) is determined by dividing the load by the indented surface area; thus, DPH has dimensions of stress. Since the load employed is less than in a Brinell test, indentation size is reduced. This requires more elaborate techniques for measuring indentation dimensions, and also means that the surface of the material must be suitably prepared, for example by polishing it. For these reasons, the Vickers hardness test is not as widely used as the Brinell test. However, the variations in load allowed by the Vickers test enable assessment of the hardness of materials with widely varying plastic flow resistance. The DPH also relates to flow stress as given by Eq. (1.23).

Microhardness testing can also be done. In these tests, indentation dimensions are comparable to microstructural ones. Thus, microhardness testing is useful for assessing the relative hardnesses of various phases or microconstituents in two- or multiphase alloys and can also be used to monitor hardness gradients that may exist, e.g., near the surface of a surface-hardened part. To provide the small indentations required for microhardness studies necessitates correspondingly low loads. As a corollary, the indentation dimensions are measured at the optical microscopic level, and this, of course, means the surface of the material must be prepared appropriately. For these reasons, microhardness assessments are not as often used industrially as other hardness tests. However, microhardness testing is widely employed in research studies.

A hardness test used commonly in the United States is the Rockwell hardness test. Various combinations of load and indenter can be used in the Rockwell series of tests, and different combinations are designated by different subscripts to the Rockwell hardness number. Thus, when the test is carried out with a 150-kg load and a diamond indenter, the resulting hardness number is called the Rockwell C (R_C) hardness. If the applied load is 100 kg and the indenter is a 1/16-in diameter hardened steel ball, a Rockwell B (R_B) hardness number is obtained. Depending on the material evaluated, other combinations of load and indenter are available.

In contrast to the other hardnesses described, the Rockwell hardness number has no units, and does not relate unambiguously to material yield strength. However, the test is easy to perform and rapidly accomplished. As a consequence, it is used widely in industrial applications, particularly in quality control situations.

Figure 1.18

Torsion of a solid bar. Application of a twisting moment (M_t) causes angular displacement, i.e., rotation, of the bar about its axis. The external moment is balanced by the internal shear stress developed in response to it.

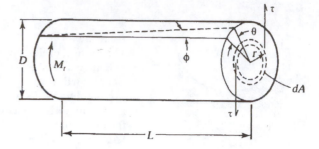

In this text, we occasionally discuss material hardness in lieu of material flow strength when the latter are not, or only sparsely, available. The brief discussion here is intended to provide the necessary background for relating the two properties.

F. The Torsion Test

A torsion test is especially useful for studying material flow at large plastic strains. As discussed later, this capability derives from the stress state developed in such a test. A schematic of a torsion test is shown in Fig. 1.18. A twisting moment, M_t (having dimensions of force $\times$ distance, e.g., N·m) is applied and this results in relative angular displacement of points along the circumference (and the interior) of the bar. The relative displacement varies linearly with axial separation distance (L) between the planes considered (Fig. 1.18). The external moment gives rise to internal shear stresses that resist it.

The shear strain varies linearly with radial position within the bar, increasing from zero at the bar center to a maximum on the circumference. Referring to Fig. 1.18, we see that the shear strain γ is given by

$$\gamma = \tan \phi = \frac{r\theta}{L} = r\theta' \tag{1.24}$$

where r represents radial position, θ the displacement angle, and L the axial separation distance. The parameter θ/L is frequently expressed as θ', the angle of twist per unit length.

For elastic deformation, the shear stress also varies linearly with radius. By balancing the external twisting moment with that arising from the internal shear stress (Prob. 1.14), we find the latter is given by

$$\tau = \frac{32M_t r}{\pi D^4} \tag{1.25}$$

During the test, the twisting moment and angle of twist (θ) are measured. The shear stress and strain can then be calculated from Eqs. (1.24) and (1.25) (usually evaluated at the sample circumference, $r = D/2$) and a $\tau - \gamma$ curve can be constructed. Since, for elastic deformation, $\tau = G\gamma$, accurate measurement of M_t and θ allows determination of the shear modulus.

During plastic deformation, shear strain still varies linearly with radius and is given by Eq. (1.24). However, the shear stress no longer does; thus, Eq. (1.25) does not hold for plastic shear strains. However, analyses have been developed that

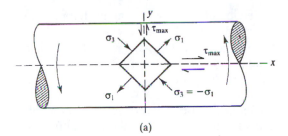

(a)

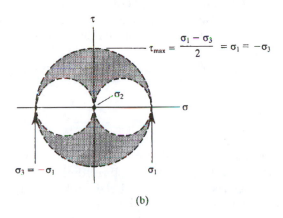

(b)

Figure 1.19
(a) The stress state developed in torsion is a biaxial one, with principal stresses related by $\sigma_1 = -\sigma_3$ with $\sigma_2 = 0$. With respect to the system of axes of the figure, maximum shear stresses are developed in the y-x plane. (b) Mohr's circle representation of the torsional stress state. Large shear stresses relative to the normal stress (i.e., $\tau_{max} = (\sigma_1 - \sigma_3)/2 = \sigma_1$) are found in torsion, and the test therefore is useful for facilitating plastic flow.

allow calculation of a $\tau - \gamma$ curve from M_t vs. θ data obtained during torsional plastic deformation. By further manipulation, the procedure also allows for construction of an equivalent $\sigma_T - \varepsilon_T$ curve. For example, high-temperature flow behavior during hot-working is usually obtained from torsional plastic torque-twist curves.

The usefulness of the torsion test for studying large plastic strain behavior comes from the stress state developed in torsion. This is illustrated in Fig. 1.19a, which shows that a state of biaxial tension, with principal stresses related through $\sigma_3 = -\sigma_1$, is developed in torsion; the corresponding Mohr's circle representation is given in Fig. 1.19b. The stress-state characteristic of torsion leads to development of large maximum shearing stresses relative to the principal normal stresses. Indeed, if one were to characterize the tendency for plastic flow relative to fracture by the ratio of τ_{max} to the mean pressure [$= (\sigma_1 + \sigma_2 + \sigma_3)/3$], this ratio is infinite for torsion testing. By contrast, it is only 1.5 and 0.75 for simple ($\sigma_1 > 0$; $\sigma_2 = \sigma_3 = 0$) and balanced biaxial tension ($\sigma_1 = \sigma_2 > 0$; $\sigma_3 = 0$), respectively. This characteristic of the torsion test makes its employment worthwhile in many situations, despite the difficulty in its execution.

This ends our overview of plastic deformation. We again emphasize that an understanding of the topics reviewed here is requisite to a good appreciation of material properties and behavior.

1.4 FRACTURE

The modulus, E, is a material property used in design against elastic deflection. For example, the properties of a spring are determined by the spring geometry and the modulus of the spring material. The deflection of an aircraft wing in flight is likewise controlled by the wing architecture and the modulus of the wing material.

A material's yield strength, σ_y, is the property used in design against permanent deformation. Knowledge of the stress state (coupled with appropriate safety factors) permits design against permanent deformation of a material, provided its yield strength is known.

Elastic deflection greater than that intended constitutes a form of material failure. So, too, does permanent deformation of a structural material. Material fracture is another form of failure. In fact, it can legitimately be thought of as the "ultimate" failure. We have described tests that provide material properties (E and σ_y) useful in design against deflection and plastic yielding. In this section we briefly summarize measurement of an additional property, the material fracture toughness (K_c), used in design against fracture. The treatment here is introductory; the topic is discussed at length in Chaps. 9 and 10.

A. Fracture Toughness

Modern design against fracture assumes that most materials contain flaws—cracks or pores—either within their interior or on their surface. Design, therefore, seeks to prevent the catastrophic propagation of such flaws under load; i.e., it seeks to prevent material fracture. Large flaws are more deleterious in this respect than are small ones. To use this design philosophy, some idea of the size of these preexisting flaws must be known. This is obtained through engineering experience—"know how," if you will. For example, it might be known what size flaws or cracks might go undetected by inspection prior to placing a component into service. Given an idea of this size permits estimation of the stress required to propagate a crack through the formula

$$\sigma_F = \frac{K_c}{(\alpha \pi c)^{1/2}} \tag{1.26}$$

In Eq. (1.26), σ_F is the stress required to propagate a surface crack of length c (or an interior crack of length $2c$), α is a constant on the order of unity and dependent on the precise crack shape, and K_c (units of N/m$^{3/2}$ or Pa·(m)$^{1/2}$) is a material property called the *fracture toughness*.

Equation (1.26) indicates that for a given K_c, the fracture propagation stress decreases with increasing crack size. There is a physical reason for this behavior; a crack concentrates stress in front of it. That is why, for example, when we break a glass rod we first file its surface to produce a notch. The notch concentrates the stress at its tip insuring (most often) that the crack propagates at the location of the notch.

The parameter, K_c, is not truly a material property for it depends on the dimensions (size) of the material being fractured. That is, fracture toughness is both material and geometry dependent. However, as described in Chap. 9, if the sample dimensions are made sufficiently large, then the fracture toughness measured becomes independent of sample size. For this situation, the fracture toughness is designated K_{Ic} and is called the *plane strain fracture toughness*. This is the fracture toughness usually used in design, since it represents conservative design; i.e., K_{Ic} represents a lower limit on fracture toughness.

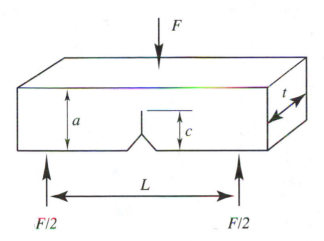

Figure 1.20
Fracture toughness can be determined in a three-point bend test, as illustrated here. A crack is developed from a preexisting notch. The depth of this crack (including the original notch depth) is the value of *c* used in Eq. (1.26) to determine fracture toughness. The beam is subjected to bending until the crack propagates. The stress used in Eq. (1.26) is obtained from measurement of the force (*F*), and the resulting moment, that causes catastrophic crack propagation.

Fracture toughness can be measured in a number of ways. Several are described in Chap. 9, and Fig. 1.20 shows one of them. In this test, a sharp crack is initiated at the tip of a notch in the bar subjected to bending. The sharp crack is formed by subjecting the relatively blunt notch to an environment that causes a sharpened crack to emanate from it. This can be accomplished, for example, by fatiguing the beam or by subjecting it to thermal shock. The beam with the sharp crack is then bent until the crack propagates catastrophically. Mechanics provides a value of the bending stress at which this occurs, and analysis also provides a value of the constant α appropriate to the beam geometry. K_{Ic} is then calculated from Eq. (1.26).

Provided the fracture toughness determined in this way is a plane strain fracture toughness, it can be used for design purposes since values of α appropriate to most typical loading and sample geometries are known.

Fracture toughness testing of ceramics poses problems. Producing a sharp notch is not the least of them, since a crack generated from a notch in a ceramic is likely to propagate catastrophically upon its formation. However, hardness testing has been found to provide a reliable way of reasonably estimating the fracture toughness of many ceramics. This is the case when secondary cracks emanate from the notch indentation, as indicated in the schematic of Fig. 1.21. Correlation of the lengths of indentation cracks with measured values of K_{Ic} indicate that ceramic fracture toughnesses correlate with these lengths as

$$K_{Ic} = \alpha_0 \left(\frac{E}{H}\right)^{1/2} \left(\frac{P}{d'^{3/2}}\right) \tag{1.27}$$

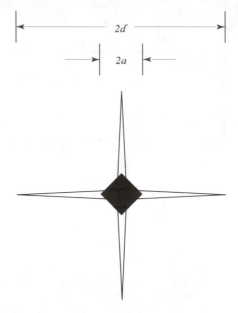

Figure 1.21
Secondary cracks developed during hardness testing can be used to assess the fracture toughness of brittle materials according to Eq. (1.27). Illustrated here is a schematic of such secondary cracks emanating from a diamond pyramid indenter (used in a Vicker's hardness test). The indentation length is 2a. Secondary cracks emanate from the indentation corners and have a length 2d′ as measured on the sample surface.

where E is the material modulus, H its hardness (which can be determined from the same test), P the indentation load, and $2d'$ the secondary crack length (Fig. 1.21). The value of toughness in Eq. (1.27) is expressed in $MN/m^{3/2}$ if the constant α_0 is taken as 0.016 and the other terms in the equation are expressed in fundamental SI units (i.e., N and m). This value of α_0 is accurate to about 25% as determined by correlation of indentation fracture toughnesses with those obtained through more fundamental means. Although the correlation is, therefore, only approximate, it is used widely for preliminary assessment of ceramic fracture toughness since the hardness-derived toughness is relatively easily obtained.

Material fracture can also take place when preexisting cracks are not present or, if they are present, they are very small in size and not likely to cause fracture. This is usually the situation in a tension test, and it can also be the case for creep fracture, fatigue fracture, and fracture due to embrittling environments. These forms of fracture are discussed in Chaps. 9–13. However, a brief overview of them is provided here. We first consider tensile fracture.

B. Tensile Fracture

The microscopic processes accompanying crack nucleation and propagation in a material loaded in tension are described in Chaps. 9 and 10. Here we summarize them in the context of their manifestation in a tensile test.

Ductile fracture is characterized by a finite %R.A. in a tensile test and by the formation of a neck prior to fracture. Ductile fracture is observed in most polycrystalline metals at ordinary temperatures, in metallic and many ceramic single crystals at low temperatures, and is also seen in certain polycrystalline ceramics at relatively higher temperatures. Fracture is frequently instigated by microscopically heterogeneous plastic deformation. This can be caused by inherently heterogeneous plastic flow or by that arising from the presence of internal boundaries (e.g., grain or interphase boundaries). The plastic strain gradient accompanying such micro-scopic flow results in internal stress concentrations that serve first to nucleate cracks or voids (e.g., by inclusion fracture or decohesion) and then to promote their growth. Void growth takes place by plastic deformation and, when the intervoid spacing becomes small enough, the voids link up. In a tensile test this produces an internal penny-shaped crack, with final tensile separation taking place by shear fail-ure of the "tube" surrounding the crack. High values of %R.A. are promoted, there-fore, by (1) delaying crack nucleation (e.g., by having a low density of potential void nucleation sites) or (2) having the voids grow to a large extent before they link up. In this sense, %R.A. is a measure of a material's fracture resistance as well as its ductility.

When a brittle material is tested in tension, its R.A. is zero or is limited to a few percent, and necking does not occur. Thus, a brittle solid manifests no, or very lit-tle, capacity for macroscopic permanent deformation. In some brittle solids, crack propagation does not involve any microscopic plastic deformation either; glass at room temperature is an example.[5] For these materials, preexisting surface or interior cracks—no matter how small they might be—serve as the crack nuclei. As noted, stress is concentrated at the tips of such flaws and when the stress concentration is such that Eq. (1.26) is satisfied, catastrophic crack propagation takes place.

In some "brittle" solids in which preexisting flaws are minimal in size (as might be the situation for a tensile test) microscopic plastic deformation precedes crack nucleation, and may also accompany crack propagation. Such solids include some polycrystalline metals at low temperature and some polycrystalline nonmetals at low and intermediate temperatures. Thus, even though macroscopic plastic defor-mation is limited, plastic deformation does take place on a microscopic level and materials behaving in this way are, therefore, "tougher"—i.e., they have higher K_{Ic} values—than inherently brittle materials such as glass.

Inherently brittle materials can also fracture in compression. This is unex-pected on the basis of previous comments that fracture is facilitated by a high value of the mean pressure. It is the presence of preexisting cracks in a brittle solid that leads to their compressive fracture. Some of the preexisting cracks in a brittle solid are oriented with respect to the applied load in a way that favors their growth. How-ever, the compressive fracture stress is typically much greater (about a factor of ten greater) than the tensile one. Current ideas on this type of fracture, which is prone to large experimental variations in measured fracture strengths, invoke statistical analysis of crack size, shape, and orientation.

A different type of fracture occurs at high temperatures. This creep fracture, as it is called, is discussed in Chap. 11, but an overview of it is presented in the fol-lowing section.

C. Creep Fracture

Materials at high temperature can flow plastically and eventually fracture under conditions of a constant applied stress. The continuing permanent deformation with time at a fixed stress is called creep, and the resulting failure is termed creep

---

[5]Glass fibers can be made so that they contain very few preexisting flaws and the flaws that are present are small in size. This accounts for the usefulness of glass fibers in composites.

fracture. Creep deformation and fracture both relate to thermally assisted deformation processes that provide for and/or assist creep strain, void nucleation, and the subsequent growth of voids which is a precursor to creep fracture.

Some materials at high temperature neck down to a point before separating into two parts. Thus %R.A. = 100 for these *viscous* solids, as they are called. More limited high-temperature R.A.s, however, are far more common. Void nucleation in these solids is thermally assisted, and takes place at regions of microscopic heterogeneous deformation such as grain boundaries and/or particle-matrix interfaces. It should be noted, though, that thermally assisted deformation to some extent relaxes strain gradients associated with heterogeneous deformation; this kind of relaxation does not happen at ordinary temperatures. Thus, strains at which voids nucleate at high temperatures are often greater than they are at low temperatures.

Nucleated voids often grow by the same processes responsible for creep deformation. Creep fracture takes place when the voided volume fraction attains some critical value or when the intervoid spacing becomes small enough that voids link up rather than continue to grow individually.

The total creep fracture strain is composed of two parts: that due to the creep deformation itself and that resulting from void growth. The latter strain can be appreciable if voids are distributed homogeneously throughout the material and their link up occurs within grains, leading to *transgranular* creep fracture. However, frequently the voids whose growth leads to fracture are situated on grain boundaries. *Intergranular* creep fracture (ICF) is the term applied to this fracture mode, and the strain due to void growth in this circumstance is limited. Indeed, %R.A. for ICF is close to zero, and a macroscopic ICF fracture surface bears a distinct resemblance to a low-temperature brittle fracture surface, even though considerable permanent deformation may precede ICF.

Stress state influences creep fracture, just as it does other kinds of fracture. One particularly important factor differentiates low- from high-temperature tensile fracture. Without an applied external tensile force, voids spontaneously shrink at high temperatures and eventually disappear. This *sintering* does not happen at low temperature because the atomic motion responsible for it takes place so slowly at low temperatures. Void shrinkage is accelerated by compressive stresses. Thus, void growth leading to creep fracture does not occur during compressive loading and, in fact, a tensile stress exceeding the inherent "void shrinkage stress" (called the *sintering limit*) must be applied for void growth to occur.

High-temperature deformation is considered in Chap. 7, where mechanisms leading to creep are described and microstructural features promoting creep resistance are presented. In Chap. 11, processes leading to creep fracture are identified, and mechanisms of void growth are discussed.

D. Fatigue Fracture

Fatigue refers to material failure under a time-varying stress that would not result in fracture under an equivalent static stress. For example, if a material were loaded at low temperature to a stress below its tensile strength and the stress were held constant thereafter, the material could withstand this load indefinitely without fracture. However, if it were loaded initially to this same stress and then the stress were re-

reduced to an extent that the material cannot withstand the maximum applied stress or by the crack tip having grown to a sufficient depth that the maximum stress-crack depth combination exceeds that permitted by the material's fracture toughness.

Fatigue fracture surfaces often mirror the processes just described. For example, visual inspection of a fatigue fracture surface is often sufficient to delineate the respective areas of fast and slow crack growth. The latter are often smooth and burnished as a result of the abrasion of the separated mating surfaces during the slow crack growth period. Moreover, if the material is exposed to a corrosive environment, corrosion products (e.g., rust, if the material is a steel) often develop on the slow crack growth surfaces. At the microscopic level, particularly for high-cycle fatigue, the slow crack growth surfaces are sometimes characterized by *fatigue striae*. These are ridges or undulations on the fracture surface, with the spacing between these features being the slow-crack advance distance per cycle. The presence of striae is often used to identify a failure as being due to cyclic loading. However, it is important to note that striations can be obscured by environmental and other factors and, therefore, lack of a clear-cut identification of them does not necessarily mean that failure was not fatigue in origin. The other side of the coin is that striations are found in other cases of intermittent crack growth (e.g., as in some environmentally assisted fracture). Thus, the presence of surface striae does not, per se, identify the failure mechanism as that of fatigue.

The response of materials to cyclic loading is different from that to monotonic loading. The tensile flow curve serves as an example of monotonic loading and, at least at low temperatures, work hardening is a ubiquitous characteristic of such a curve. In contrast, metals may either harden or soften during cyclic deformation. If the stress required to maintain a specified strain range decreases with the number of cycles, the material is said to cyclically work soften. Generally, "hard" alloys of a given metal tend to cyclically soften, whereas softer alloys of the same material cyclically work harden. In some cases, the "steady-state" cyclical flow stress is independent of the initial state of the material. For example, annealed and cold-worked alloys of the same material may eventually require the same cyclical stress range to produce the same cyclical strain, even though the cold-worked alloy initially required a much greater stress range. When this occurs, an equivalent steady-state deformation microstructure—observable at the electron microscope level—characterizes the alloy.

Brittle solids, such as glass and many polycrystalline ceramics at room temperature, are not considered subject to fatigue. This is because if fatigue cracks do form in them, they grow to a size sufficient to induce brittle fracture in relatively few cycles. Polymers are subject to fatigue, and in a phenomenological sense their cyclical response is comparable to that of metals. But the microscopic manifestations of fatigue in the two material classes are often quite different. As one example, the viscoelastic feature of many polymers leads to hysteretic heating, and an increase in temperature of the material, during cyclical loading. As a result, polymer fatigue lives are sensitive to the frequency of cyclical stressing/straining, whereas this is seldom the case for metals. Certain microscopic features of polymer deformation, particularly at temperatures near their glass transition temperature, also make their response to cyclical loading differ from that of metals. Further discussion of metallic and polymeric fatigue is presented in Chap. 12.

E. Embrittlement

Environmental effects can lead to unexpected and/or premature material failure. Materials exposed simultaneously to a static stress and a "hostile" environment may experience delayed failure (*static fatigue*). Such environments include corrosive fluids and liquid or solid metals; the respective phenomena are called stress corrosion cracking (SCC) and liquid/solid metal embrittlement (LME/SME). Static fatigue involves slow, often intermittent, crack growth, somewhat akin to that taking place in ordinary fatigue. Crack advance is facilitated by the environment, which may cause material dissolution at the crack tip or adsorption of the environmental species there, both effects facilitating crack advance. As with ordinary fatigue, final (fast) fracture takes place when the crack has grown to the extent that either tensile overload failure happens or the material's fracture propagation stress is exceeded.

Undesirable "tramp" elements (H, O, S, and P are examples) can also embrittle metals. Hydrogen is especially harmful to the ductility and toughness of many high-strength metals. Hydrogen embrittlement is often a special case of SCC. Hydrogen formed via an electrochemical cathode reaction either enters into the metal in the vicinity of the crack tip or becomes adsorbed on it, with crack advance promoted thereby. In other circumstances, hydrogen or other tramp elements are incorporated within the material during processing. This kind of embrittlement is manifested by a reduction in material ductility and toughness, often unaccompanied by any diminishment in material strength.

Nuclear radiation can also be considered a "hostile environment." Bombardment by high-energy neutrons can create excess vacancies and self-interstitials in a material, often with an adverse effect on malleability and fracture resistance. At certain temperatures, the excess vacancies collapse into voids and this can lead to material swelling. Nuclear transmutation products often cause the same effect; e.g., gaseous fission by-products can result in formation of internally pressurized gas bubbles. Because of the technological importance of embrittlement, it is discussed in Chap. 13.

1.5
SUMMARY

In this chapter an overview of the topics that constitute this book's emphases—deformation and fracture—has been presented. Some topics—elastic deformation and fracture—were presented in a preview format, as the background for treating them in depth is adequately covered in individual chapters. However, some areas, those relating to permanent deformation, have been treated in more detail. This was done because the background material necessary for a full appreciation of topics discussed subsequently is not presented in similar chapters treating permanent deformation.

While a variety of material properties were introduced, three of them—the elastic modulus, E, the yield strength, σ_y, and the fracture toughness, K_{Ic}—are central to low-temperature mechanical design. The modulus is a key property when design against elastic deflection is paramount. The yield strength is the material parameter used to design against unwanted permanent deformation. And the fracture toughness is used to design against crack propagation resulting in material fracture. These three properties vary among the materials classes (ceramics, metals, and polymers) and these differences are elaborated on in later chapters dealing with elasticity, plastic flow, and fracture.

Several concepts related to plastic deformation were also reviewed. The tensile test was described at some length, and material characteristics such as the yield strength (the stress at which plastic flow initiates in a tensile test) and work hardening were emphasized. The tensile geometrical instability known as necking was discussed, and the relationship of this instability to a material's work-hardening characteristics was brought forth. The concepts of engineering and true stress and strain were introduced. While the results of a tensile test are most often described in terms of engineering stress and strain, which are based on initial sample dimensions, true stress and strain (defined on instantaneous sample dimensions) more accurately portray the plastic-flow properties of a solid.

Plastic-deformation behavior is frequently represented by means of a tensile flow curve, which is appropriate for uniaxial loading. However, in actual service many materials are subjected to combined shear and normal stresses. There exists one set of orthogonal axes for which the stress state is described solely in terms of principal, or normal, stresses. Yielding criteria under multiaxial loading conditions are expressed in terms of these stresses. One criterion, the Tresca condition, states that yielding takes place when the algebraic difference between the maximum and minimum principal stresses exceeds the tensile yield strength, whereas the other, the von Mises, is a more complicated function of the differences between the principal stress components. Both criteria correspond to plastic flow being the result of shear stresses. In contrast, fracture is often promoted by a large sum of the principal normal stresses. Since fracture and flow are often competitive, stress state plays an important role in determining whether a material plastically deforms or fractures, and this point is amplified in later chapters.

Having performed the necessary preliminaries, we now direct our attention to the primary purposes of this book. The following chapter discusses the several kinds of elastic deformation, and the atomic and molecular characteristics responsible for elastic deformation.

REFERENCES

Ashby, Michael F., and David R. H. Jones: *Engineering Materials 1—An Introduction to their Properties and Applications,* Pergamon Press, Oxford, 1980.

Caddell, Robert M.: *Deformation and Fracture of Solids,* Prentice-Hall, Englewood Cliffs, N.J., 1980.

Dieter, George E.: *Mechanical Metallurgy,* 3rd ed., McGraw-Hill, New York, 1986.

Felbeck, David K., and Anthony G. Atkins: *Strength and Fracture of Engineering Solids,* Prentice-Hall, Englewood Cliffs, N.J., 1984.

Hertzberg, Richard W.: *Deformation and Fracture Mechanics of Engineering Materials,* 4th ed., Wiley, New York, 1996.

1.1 The tensile stress-strain behavior of pure aluminum is shown in the graphs following. Determine the following properties for pure Al. (a) Young's modulus, (b) 0.2% offset yield strength, (c) tensile strength, and (d) engineering fracture strain.

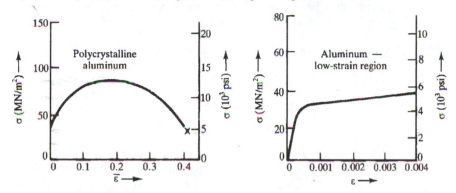

1.2 Following fracture of a steel bar tested in tension and having an initial gage length of 2 in and diameter of 0.505 in, the final gage length and diameter are found to be 2.48 in and 0.3 in, respectively. Calculate (a) percentage elongation, (b) %R.A., and (c) true fracture strain.

1.3 Using the definition of true and engineering strain, show that ε_T and ε_E are related through Eq. (1.11), provided deformation along the gage length is uniform.

1.4 Using the data for aluminum given in Prob. 1.1, plot the true stress-true strain curve of Al over the stress-strain range for which this is possible. Compare the engineering and true strain at the necking point.

1.5 We can estimate the relationship between ε_f and ε_{Eu} in the following way. Let the total elongation be written as $\delta l = \delta l_n + \delta l_u$, where δl_n is the elongation in the necked region of the tensile bar and δl_u the sample elongation in the remainder of the bar.
  a If α represents the effective fraction of the sample length that experiences necking, show that $\varepsilon_f = \alpha \varepsilon_{fn} + (1 - \alpha)\varepsilon_{Eu}$, where ε_{fn} is the effective fracture strain in the necked portion of the bar.
  b Schematically plot ε_f vs. α. How does α depend on sample gage length? Using your result, schematically plot ε_f vs. sample gage length and comment on the result.

1.6 a Assuming Eq. (1.13) holds, how are ε_{Eu} and n related?
  b Develop an expression for the true stress at necking in terms of the material's tensile strength and strain hardening exponent.

1.7 a Using Eq. (1.13) and the criterion for necking, show that $\varepsilon_T = n$ at necking.
  b What is the implication from (a) relative to a material's tendency for neck development vis-à-vis its work-hardening characteristics?

1.8 The tensile plastic deformation work to fracture is equal to the integral of force over the extension during tensile loading.

a Show that when this work is expressed on a unit-volume basis it is equal to the integral of the true stress-true strain curve. (State any assumptions made in this derivation.)

b If true stress and true strain are related through $\sigma_T = K(\varepsilon_T)^n$, determine the plastic deformation work (i) up to necking and (ii) through fracture in terms of the necking and true fracture strains. What material properties lead to a large plastic deformation work to fracture?

1.9 The engineering stress-strain curve for annealed polycrystalline Cu is shown below.

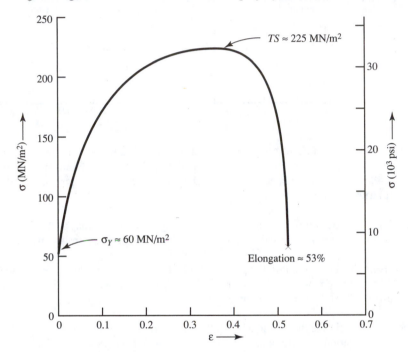

a Is annealed polycrystalline copper a material that work hardens significantly or a material that exhibits a low work-hardening rate? Explain your answer.

b Suppose this Cu were loaded to a stress of 150 MN/m², and then the load were removed. How much strain would be recovered on unloading? (For Cu, $E = 11.1 \times 10^{10}$ N/m².)

c Suppose the unloaded sample were left on a table in the laboratory. A student comes into the laboratory and measures the yield strength of this material. What would the student quote for the value of this yield strength?

1.10 Wire drawing is a process used widely for production of rods, wires, etc. A schematic of wire drawing is shown below. A cylindrical piece of material is drawn through a die by a force, F. The cross-sectional area of the wire is reduced from A_0 to A_f in the process. The drawing stress, p, is defined on the basis of the final cross-sectional area; i.e., $p = F/A_f$. A simple analysis of the process, which does not consider frictional effects between the work piece and the die, can be conducted for wire drawing.

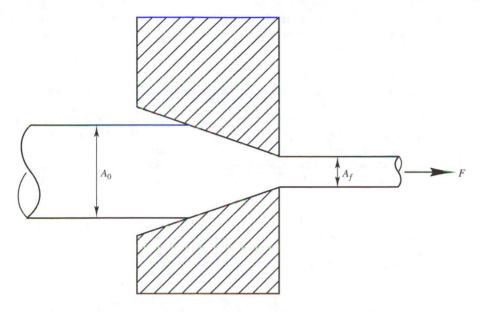

a It can be shown that the drawing stress is equal to the plastic work per unit volume done on the material during drawing. This work is the integral of the true stress–true strain curve though the drawing strain. Consider first a material that does not undergo work hardening; instead the material flows at the initial yield strength regardless of the amount of plastic strain it has experienced. Derive a simple relationship among the drawing stress, the yield strength, and the area reduction ratio, A_0/A_f.

b Such a material cannot be drawn beyond a certain reduction ratio. This ratio corresponds to the drawing stress exceeding the yield strength. What is the maximum area reduction ratio for this nonwork-hardening material?

c Consider now a work-hardening material whose true stress–true strain behavior is described by Eq. (1.13). On the stipulation in (a) (relative to the value of the drawing stress) determine a relation among p, K, n and the area reduction ratio.

d What is the maximum area reduction ratio for such a work-hardening material? (Assume the same "drawability" criterion as in (b).)

1.11 The yield strength of a eutectoid steel can be increased significantly through extensive wire drawing. If the true wire drawing strain is designated as ε, the yield strength of the drawn steel can be expressed as

$$\sigma_y \ (MN/m^2) = 140{,}000 + 700{,}000\varepsilon$$

What do you expect to find for the true strain at necking in a tensile test of a wire processed to a drawing strain of 2?

1.12 Consider the forged Cu disk of Example Problem 1.1.

a A tensile sample is cut from the forged disk. Plot the engineering stress–engineering strain curve for this sample. (Hint: Recall the condition for the necking instability, Eq. (1.12b). Noting that the term $-dA/A$ is also equivalent to $d\varepsilon_T$, show that the necking condition can also be written as $d\sigma_T/d\varepsilon_T = \sigma_T$. Compare the values of the terms on both sides of this equation in the forged disk.)

b A small disk is cut from the original forged disk and forged further to a reduction in height of 50%, also under frictionless conditions. Plot the forging pressure as a function of reduction in height.

1.13 a Draw a Mohr's circle diagram for uniaxial compression, $\sigma_1 = \sigma_2 = 0$, $\sigma_3 = -p$. If this situation corresponds to a hardness test, how would p be related to the material's tensile yield strength?

b The constraint present in a hardness test produces stresses $\sigma_1 = \sigma_2 \neq 0$. If the pressure required to cause plastic flow in a hardness test is three times the material's yield strength in tension, what are the values of σ_1 and σ_2? (Calculate these stresses using both the Tresca and von Mises yielding conditions.)

c Using your answer(s) to (b), sketch a Mohr's circle diagram appropriate for a hardness test.

1.14 Equation (1.25) can be derived by setting the external twisting moment, M_t, equal to the internal resisting one. With reference to Fig. 1.18, the latter is given by

$$\int_0^{r=D/2} \tau(r) r \, dA$$

where $dA = 2\pi r \, dr$ and $\tau(r)$ varies linearly with r for elastic deformation (i.e., $\tau = G\gamma = Gr\theta'$). By equating M_t with the internal moment, show that τ and M_t are related through Eq. (1.25).

1.15 Refer to Fig. 1.15. Calculate the ratio of τ_{max} to the mean pressure (= one-third of the sum of the principal normal stresses), with the latter expressed in terms of the magnitude of the maximum normal stress, for (a) uniaxial tension, (b) uniaxial compression, (c) biaxial tension, (d) tension with balanced biaxial compression and $\sigma_2 = \sigma_3 = -\sigma_1/2$, and (e) hydrostatic tension. Tabulate the ratios in order of increasing value, and compare them with the similar ratio for torsion. Comment on your results in terms of the effect that the stress state has on the tendency of material to flow as opposed to fracture.

1.16 Flaws, such as holes, concentrate stress in their vicinities. When a thin rectangular sheet containing a circular hole of radius a is loaded in simple tension, a state of biaxial stress is developed in the vicinity of the hole. The stress state is best described in cylindrical coordinates r and θ; thus, σ_{rr} is the stress acting along a radial direction and on a plane whose normal is in the same direction, whereas $\sigma_{\theta\theta}$ is the stress orthogonal to σ_{rr}. Both of these stresses vary with r and θ; their variation with r is given by

$$\sigma_{\theta\theta} = \sigma\left[1 + \frac{1}{2}\left(\frac{a}{r}\right)^2 + \frac{3}{2}\left(\frac{a}{r}\right)^4\right]$$

$$\sigma_{rr} = \frac{3\sigma}{2}\left[\left(\frac{a}{r}\right)^2 - \left(\frac{a}{r}\right)^4\right]$$

where σ is the nominal tensile stress.

a Plot σ_{rr} and $\sigma_{\theta\theta}$ vs. (r/a). How far away from such a hole must one be for the stress $\sigma_{\theta\theta}$ to be only 10% greater than the nominal stress?

b What do you expect the effect such a hole has on the tendency of the material to fracture rather than flow? Give your reasoning.

1.17 A steel having a yield strength of 1450 MN/m^2 and a fracture toughness of 55 MN/m$^{3/2}$ is placed into service. Quality control indicates that brittle inclusions, which fractured during the processing of the steel, are present in the steel and that the maximum dimension of these inclusions is about 20 μm. This dimension can be considered to be the length of an existing interior crack in the steel. A thick plate of this steel is placed into service.

    a Show that the steel will yield plastically before cracks associated with the inclusions propagate through the material. (Assume the parameter, α, of Eq. (1.26) has a value of 1.05.)

    b What size would the inclusions have to be in order to have the material fracture before it yielded plastically?

CHAPTER 2

Elastic Behavior

2.1
INTRODUCTION

In this chapter we initiate our discussions of the relationship between mechanical properties and structure by considering elastic properties of materials. Elastic properties are important in several aspects of material behavior. The most commonly encountered of these is the stiffness of a structural member (e.g., a component of a bridge) designed to undergo only elastic deformation. However, as will be shown in subsequent discussions of permanent deformation, the stresses required to initiate low-temperature plastic flow (and some types of high-temperature flow) also relate to elastic properties, since permanent deformation involves aspects of atomic bond stretching and distortion. Finally, material fracture is related to elastic properties in that some of the driving force for fracture is related to the elastic energy release accompanying it.

We discuss various types of elastic deformation, but their elastic nature is their principal, and in some cases sole, commonalty. Linear elasticity, which is intrinsically related to atomic, molecular, or ionic bonding forces is discussed first. The elastic behavior of crystalline materials is almost entirely of this type. So is that of amorphous solids at low temperature. The linearly elastic behavior of crystalline solids is inherently anisotropic; that is, the strain response to an applied stress depends on the orientation of the stress with respect to the crystallographic axes of the solid. In a polycrystalline aggregate, the anisotropy of individual grains is masked by the large number of randomly oriented grains within the aggregate and thus most polycrystals are elastically isotropic. On the other hand, extensive permanent deformation of polycrystals, as well as certain heat treatments, can produce a textured structure in which the individual crystallites have certain axes aligned preferentially in a particular direction. These polycrystals are elastically anisotropic.

44

The low-temperature elastic response of long-chain polymers is also linearly elastic. However, at moderate temperatures long-chain polymers also display a form of time-dependent elasticity; such behavior is termed viscoelasticity. The degree to which viscoelasticity is manifested in these polymers is determined by their molecular architecture, i.e., by the degree and frequency of interchain linking and the configuration and size of molecular side groups. In rubbers, interchain linking occurs to such an extent that these elastomers display large reversible extensions. This rubber elasticity, however, is not associated with primary bond stretching or distortion but, instead, the elastic deformation reduces the system entropy. Rubber elasticity is discussed on this basis, which allows constitutive equations to be developed that agree reasonably well with experiment.

Finally, damping of both organic and inorganic materials is considered. Damping is a time-dependent phenomenon in which the responsive strain lags behind a cyclically applied stress, and this leads to mechanical hysteresis.

2.2
RANGE OF ELASTIC MODULI

What are typical elastic moduli values of engineering materials? Moduli values vary widely among the several materials classes—ceramics, metals, and polymers—as well as within a specific material class. A bar chart illustrating this is provided in Fig. 2.1 which shows that moduli of engineering solids vary over six orders of magnitude although the materials with the lowest moduli—foamed polymers—are porous solids so their moduli do not reflect the inherent stiffness of their solid component. Nonetheless, the range in moduli is striking. The physical basis for this variation among and within materials classes is discussed later. However, a few comments providing an overview of this variation are made here.

Diamond has the highest modulus of all solids, about 1000 GN/m^2. The strong covalent bonding among carbon atoms in diamond is responsible for its high stiffness. Among the ceramics, other covalent bonded and polar covalent (a mixture of ionic and covalent bonding) solids have the highest moduli after that of diamond. Examples include alumina (Al_2O_3), silicon carbide (SiC), and silicon nitride (Si_3N_4). The high stiffnesses of these materials (as well as their high strengths, Chap. 4) combined with their low densities make these materials potentially attractive for high-temperature structural use in spite of their low fracture toughnesses.

Moduli of ionic solids, such as the alkali halides, are not as impressive as those of covalent solids and this is a reflection of the lesser strength of ionic bonds compared to covalent ones. Graphite, a form of carbon, has a modulus almost two orders of magnitude less than that of diamond. This is a consequence of the structure of graphite which is layered; carbon bonding within layers is very strong but interlayer bonding is much weaker. This interlayer bonding is the cause of the low modulus of ordinary graphite. When graphite is processed to fiber form and the fibers are stressed along the fiber axes, a very high modulus is found. In this situation, intralayer bonds are stretched and the high modulus is a consequence. Ice has a low modulus because the bonding in ice is of the "hydrogen" variety (a strong version of a van der Waals bond). Thus, we see that modulus relates to the nature of the chemical bonds between/among atoms, ions, or molecules in a solid. For ceramics, this bonding is strongest for covalent solids, less so for polar covalent solids, even less for ionic solids, and fairly low for hydrogen-bonded ceramics.

As a consequence of the relatively strong metallic bond, moduli of metals are relatively high although not as high as those of covalent solids. Moduli of metals vary by about a factor of thirty; lead is the least stiff metal and osmium (Os) the stiffest. Among the useful engineering metals (Os is not one), the refractory metals—tungsten, molybdenum, chromium, etc.—display the highest Young's moduli.

The highest moduli values of engineering polymers (ca. 10 GN/m^2) are only about as high as those of low modulus metals. The chemistry and molecular architecture of polymers are the cause of this behavior. Polymers come in two architectures; network structures (in which case they are called thermosets) and chain structures (termed thermoplastics). Chemical bonding along the chains and within the networks is covalent. The bonding between chains (in a chain polymer) and branches (in a network structure) is generally a weak van der Waals bond. During

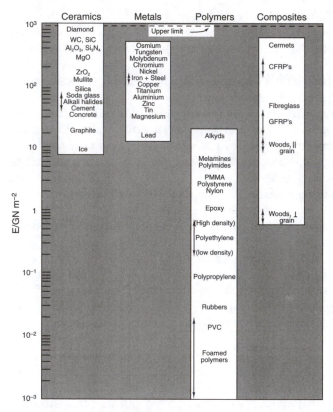

Figure 2.1

A bar chart illustrating elastic moduli values of the primary material classes (ceramics, metals, and polymers) and of composites (a hybrid of materials from the different primary classes). Although there is considerable variation in elastic moduli within a given material class, ceramics as a whole have the highest elastic moduli and polymers the lowest. Moduli of composites are intermediate to those of their constituents. It is noteworthy that elastic moduli of engineering solids span about six orders of magnitude. (*From Michael F. Ashby and David R. H. Jones, Engineering Materials I—An Introduction to their Properties and Applications, Pergamon Press, Oxford, 1980.*)

low-temperature[1] linear elastic deformation of a polymer, small (and recoverable) interchain or interbranch displacements occur and these displacements, rendered against a van der Waals bond, account for the low moduli of polymers. Some thermoplastics are drawn into filaments where they are used in this form in textiles and carpeting. This processing aligns the chains so that the filament modulus, since it now reflects stretching of primary bonds along the chain, is much greater than the modulus of the corresponding bulk polymer. The situation is comparable to the behavior of graphite in fiber and bulk form.

Moduli of composites are intermediate to those of their constituents. Thus, carbon fiber reinforced plastics (CFRPs) have moduli intermediate to graphite fibers and the plastic matrix containing them. A similar situation holds for glass fiber reinforced plastics (GFRPs). The strengths of composite materials also are intermediate to their constituent strengths (Chap. 6). In distinction, composite fracture toughnesses frequently (and considerably) exceed the toughnesses of their components (Chap. 9). This is one reason these composites are attractive as structural materials.

2.3
ADDITIONAL ELASTIC PROPERTIES

Before proceeding to discuss linear elasticity at the atomic level, elastic properties in addition to those introduced previously must be introduced. As mentioned in Chap. 1, when a uniaxial tensile force is applied to a material, the material extends in the direction of this force. Concurrent with this longitudinal extension there is a lateral or transverse contraction in the directions normal to the applied stress (Fig. 2.2). For permanent deformation, the accompanying contractions are such that the

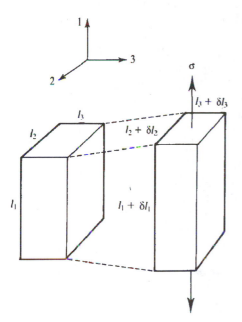

Figure 2.2

Application of a uniaxial stress leads to longitudinal extension along the tensile axis ($\varepsilon_1 = \delta l_1/l_1$) and to transverse contractions along the two perpendicular axes ($\varepsilon_2 = \varepsilon_3 < 0$). For linear elastic deformation, the longitudinal and lateral strains are related through Poisson's ratio ν; $\varepsilon_2 = \varepsilon_3 = -\nu\varepsilon_1$.

[1]Higher-temperature polymer elasticity is discussed in Sect. 2.7.

volume of the material is conserved. This is not so during elastic deformation. During elastic tensile deformation the volume increases, and during elastic compressive deformation the volume decreases, slightly. In the terms of Fig. 2.2, the lateral contractions, ε_2 and ε_3, are equal and related to the primary (longitudinal) extension by

$$\varepsilon_2 = \varepsilon_3 = -\nu\varepsilon_1 \tag{2.1}$$

In Eq. (2.1), ν is Poisson's ratio and is a material elastic property; the negative sign in Eq. (2.1) indicates that the sample dimensions normal to the primary extension decrease (increase) as the axial length of the sample increases (decreases). For metals, the value of ν is often on the order of ⅓.

The change in volume associated with the small strains of linear elastic deformation can be obtained by differentiating the expression for the volume ($V = l_1 l_2 l_3$) and keeping terms only to first order. The result is

$$\frac{\Delta V}{V} = \frac{\delta l_1}{l_1} + \frac{\delta l_2}{l_2} + \frac{\delta l_3}{l_3} = \varepsilon_1 + \varepsilon_2 + \varepsilon_3 \tag{2.2}$$

For uniaxial deformation, Eqs. (2.1) and (2.2) yield $\Delta V/V = \varepsilon(1 - 2\nu)$. Given that $\nu \cong$ ⅓, an elastic uniaxial strain of 0.5% would produce a volume change of ca. 0.2%. Since linear elastic strains are typically smaller than this, the volume change during this type of deformation is usually quite small.

Equations (2.1) and (2.2) also indicate that the elastic volume change decreases as ν increases. For an incompressible material, such as a plastically deforming metal for which the volume change is zero, the ratio of lateral to uniaxial strain is $-$½. Such a value does not imply that ν, an elastic property, has a value of 0.5 for a metal during plastic deformation. However, the value of ν observed during elastic deformation nonetheless gives a clue as to the degree to which a material is characterized by linear elasticity. Table 2.1 lists values of ν for some selected solids. As can be seen, long-chain polymers typically have values of ν greater than metals. Hence, and as noted in the previous section, these materials differ substantially from other linear elastic materials.

The principle of superposition of strain allows the linear elastic response of a material subject to multiaxial stress states to be predicted. In Fig. 2.3, for example, a linearly elastic solid is subjected to three principal normal stresses. The tensile strain along axis "1" is obtained by setting it equal to the sum of the strains expected on the basis of each stress acting individually. Thus,

$$\varepsilon_1 = \frac{\sigma_1}{E} - \frac{\nu\sigma_2}{E} - \frac{\nu\sigma_3}{E} \tag{2.3}$$

with corresponding expressions for ε_2 and ε_3.

If the principal stresses are algebraically equal then, depending on whether the stresses are tensile or compressive, the material is in a state of *hydrostatic* tension or compression. For this situation, using Eqs. (2.2) and (2.3) (with $\sigma_1 = \sigma_2 = \sigma_3$), the change in volume is found as

$$\frac{\Delta V}{V} = \frac{3\sigma}{E}(1 - 2\nu) = \frac{\sigma}{K} \tag{2.4}$$

where K is the *bulk modulus* of the material. K and E are thus related through Poisson's ratio;

Material class	Material	ν
Metallic crystalline solid	Ag	0.38
	Al	0.34
	Au	0.42
	Cu	0.34
	α-Fe	0.29
	Ir	0.26
	Ni	0.31
	W	0.29
Covalent crystalline solid	Ge	0.28
	Si	0.27
	Al_2O_3	0.23
	TiC	0.19
Covalent ionic solid	MgO	0.19
Covalent glass	Silica glass	0.20
Network polymer	Bakelite	0.20
	Ebonite (hard rubber)	0.39
Elastomer	Natural rubber	0.49
Chain polymer	Polystyrene	0.33
	Polyethylene	0.40
van der Waals solid	Argon*	0.25

*At 0 K

Table 2.1
Room-temperature Poisson's ratio for selected solids

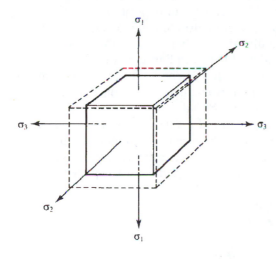

Figure 2.3
A triaxial stress applied to a cube (solid lines) produces a "distorted" cube (dotted lines). The elastic strain along any axis is obtained by addition of the strain resulting from each individual stress component (e.g., $\varepsilon_1 = (\sigma_1/E) - (\nu\sigma_2/E) - (\nu\sigma_3/E)$, etc.).

$$K = \frac{E}{3(1 - 2\nu)} \qquad (2.5)$$

For a metal with $\nu \cong \frac{1}{3}$, the bulk and Young's moduli are approximately numerically equal. It can also be shown that the shear modulus, G, and Young's modulus are related (Prob. 2.2) by

$$G = \frac{E}{2(1 + \nu)} \qquad (2.6)$$

Thus, only two of the four elastic constants (E, G, ν, K) of linear elasticity are independent. This is true for *isotropic* materials of the type considered so far. In an isotropic material the response to, for example, a tensile stress is independent of the orientation of the tensile axis with respect to the material. Isotropy is, therefore, implicit in our discussion of yielding under multiaxial stress states (Chap. 1). Isotropic behavior is observed in most polycrystalline materials, in which the inherent anisotropic behavior of individual grains is masked by their number and random orientation in space, and in noncrystalline materials such as inorganic glasses. However, in single crystals the chemical bonding responsible for linear elastic behavior is intrinsically anisotropic; that is, the bonding forces depend on crystallographic direction.

Certain deformation and/or recrystallization processes produce textured materials. Texturing is associated with preferential alignment of crystallographic planes and/or directions in a particular geometrical direction(s), and this can lead to anisotropic behavior. Aspects of anisotropic linear elasticity are discussed in Sect. 2.5.

2.4
BASIS FOR LINEAR ELASTICITY

As mentioned, linear elastic behavior is a macroscopic manifestation of atomic bonding. The bulk modulus, for example, is directly related to the external force required to compress or extend interatomic distances in opposition to the internal forces that seek to establish an equilibrium, or undistorted, interatomic distance. A shear modulus similarly represents a distortion or bending of atomic bonds, and Young's modulus measures a combination of bond bending and extension/compression.[2] The elastic constants of crystalline materials can be described in terms of these effects, as can those of amorphous inorganic glasses at temperatures where their elastic behavior is only of the linear elastic kind.

In this section, we describe the physics of linear elasticity. Before treating the matter in depth, it is worthwhile to describe the phenomenon more simply. The elastic modulus can be considered analogous to an atomic level "spring constant." For a macroscopic spring, the spring constant, S, is the ratio of an applied force to the spring displacement; thus, S has units of N/m. At the atomic level, S can be converted to an effective modulus by dividing by an appropriate length; we take this length to be the interatomic spacing, r_0, so

$$E = \frac{S}{r_0} \qquad (2.7)$$

From Eq. (2.7), we see that moduli depend upon both the "spring constant" and the interatomic spacing, although they depend much more so on S than on r_0. This is because interatomic spacings do not vary all that much among materials. For

---

[2]The lateral contraction in a tension test produces bond distortion as well as stretching.

example, atomic volumes (which scale with r_0^3) of the elements vary by about a factor of eight or so, and usually much less than this. Thus, r_0 only varies typically by about a factor of two among solids.[3] S varies by much more than a factor of two, both within a specific materials class and among the classes. Covalent solids are "stiff;" their S values range from about 20 to 200 N/m. Metals and ionic solids are somewhat less stiff, having values of S in the range from 15 to 100 N/m. Intrachain spring constants in polymers are high since they are a manifestation of covalent bonding within the chains. However, interchain bonding is of the weak van der Waals type (S = 0.5 to 2 N/m), and it is this bonding that is responsible for the previously noted low moduli of long-chain polymers.

The preceding discussion is a good qualitative description of the relationship between atomic bonding and the elastic modulus. However, the relationship between macroscopic elastic properties and internal bonding can be developed in a more quantitative way. To do this we consider the variation of the internal potential energy of a simple cubic array of atoms with the interatomic spacing; a two-dimensional representation of such an array is shown in Fig. 2.4. The total potential energy, U, of this system of atoms can be written as

$$U = \frac{1}{2}\sum_{i,j} z u_{ij} \tag{2.8}$$

In Eq. (2.8), u_{ij} is the interaction potential energy between any two near-neighbor atoms, and the sum is taken over all the atoms in the crystal. The coordination number, z (= number of near neighbors), represents the number of interactions each atom has (z = 6 for the simple cubic structure of Fig. 2.4), and the factor of ½ is necessary to avoid duplicate counting of interactions. Equation (2.8) explicitly assumes that interactions between second and third near neighbors are unimportant in determining system potential energy.

The variation of U with interatomic spacing is shown in Fig. 2.5a. The "zero point" energy (as $r \rightarrow \infty$) is arbitrarily selected. Physically, this state corresponds to

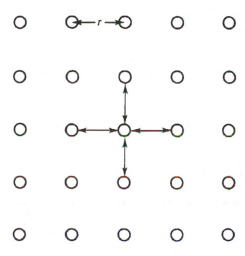

Figure 2.4
A two-dimensional representation of a simple cubic array of atoms with near-neighbor separation distance r. The crystal potential energy can be considered a sum of attractive and repulsive energies between near-neighbor atoms, as suggested by the arrows between the central atom and its four near-neighbors in this (100) plane.

[3]A typical value of r_0 is on the order of 0.2 nm.

the crystal being an ordered gas for which no potential energy interactions exist as a result of the large interatomic spacings. The equilibrium spacing (r_0 in Fig. 2.5a) corresponds to a minimum in the potential energy; the internal energy, U_0, at this atomic spacing is the material's cohesive energy and is related to the heat of sublimation. That U_0 is algebraically less than zero (the energy of the "gas") indicates that the crystal is more stable than the gas at the low temperatures we are considering. The shape of the U-r curve is related to competitive attractive and repulsive forces. The attractive force is longer range than the repulsive one, and for $r > r_0$ the long-range force dominates the interatomic interaction and draws the atoms closer together with a concurrent reduction in crystal energy. On the other hand, for spacings less than r_0, the repulsive force, which arises primarily from electron-electron overlap of the inner electron shells of the individual atoms, dominates. Hence, the minimum in potential energy at $r = r_0$ corresponds to a balance between these competing forces.

Equilibrium can be discussed further in terms of these forces. Figure 2.5b shows the variation of the attractive force, the repulsive force, and the sum of the two with interatomic spacing. Force is related to the potential energy through its derivative; i.e., $F = -\partial U/\partial r$. Since U in Eq. (2.8) can also be represented as a sum of attractive (U_{att}) and repulsive (U_{rep}) energies, this allows for the respective forces to be plotted as they are in Fig. 2.5b. Note that equilibrium is attained when $F_{att} + F_{rep} = 0$; i.e., the attractive and repulsive forces are equal and opposite at $r = r_0$. That the force at equilibrium is zero is associated with the minimum in energy since

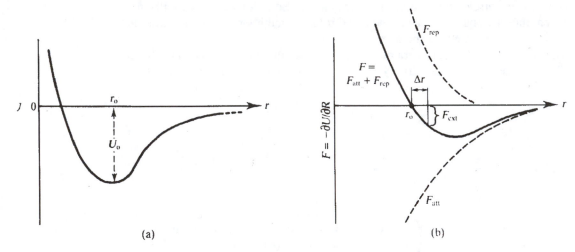

(a)   (b)

Figure 2.5
(a) The crystal potential energy of a simple cubic array of atoms as a function of the interatomic distance r. A minimum in potential energy ($-U_0$) is found at $r = r_0$, and this defines equilibrium. Increases or decreases in r result in a higher energy and a configuration that is unstable in the absence of external forces. In (b) the long-range attractive (F_{att}) and short-range repulsive (F_{rep}) forces and their sum are shown as functions of r. In the absence of an external force, equilibrium is found at $r = r_0$ where the forces balance ($F_{att} + F_{rep} = 0$). If an external tensile force (F_{ext}) is applied so as to increase the interatomic separation from r_0 to ($r_0 + \Delta r$) this force must be equal and opposite to the net attractive force, i.e. $F_{ext} + F_{att} + F_{rep} = 0$. Thus, elastic behavior is related to atomic level forces.

$F = -\partial U/\partial r$. Thus, in the absence of *external* forces, the crystal assumes a simple cubic lattice with an equilibrium interatomic spacing of r_0.

When an external force is applied, for example, along a [100] axis, the interatomic spacing is altered. It becomes greater along [100] and smaller along the two orthogonal directions ([010] and [001]). Both the axial extension and the lateral contractions increase the lattice potential energy, and the cause is the applied force. For small elastic extensions (Δr), it is found that the force increases linearly with $\Delta r/r_0$, i.e., with the strain. The slope of the external force $-\Delta r/r_0$ relation (or, alternatively, the second derivative of the energy-distance curve) is related to Young's modulus in the [100] direction. In effect, the external force required to produce the extension Δr is equal in magnitude to the *net* attractive force—at the separation $r_0 + \Delta r$—that would exist in the absence of the external force. Alternatively, the new equilibrium interatomic spacing ($r_0 + \Delta r$) is defined by the condition $F_{att} + F_{rep} + F_{ext} = 0$, where F_{att} and F_{rep} have their previous meaning (but are evaluated at $r_0 + \Delta r$) and F_{ext} is the applied external force.

Directly relating Young's modulus in the [100] direction to the force and energy curves is difficult, because one dimension of the crystal increases while the other two decrease. Further discussion of the modulus-potential energy relationship is facilitated by considering hydrostatic compressive or tensile forces. In Fig. 2.6a, the crystal potential energy is plotted vs. the volume per atom (Ω); in the simple-cubic structure, the atomic volume is r^3 and is obtained by dividing the crystal volume by the number of atoms within it. When a hydrostatic tensile stress is applied, the atomic volume increases from Ω_0 to $\Omega_0 + \Delta\Omega$ and the potential energy is also increased. The force-energy relationship can be converted to a force-volume relationship that takes the form $F = -A(\partial U/\partial\Omega)$, where A is the cross-sectional area normal to the applied force. Thus, the hydrostatic stress required to increase the volume is $\sigma = -(\partial U/\partial\Omega)$. In the vicinity of the equilibrium spacing (or volume) both the U-r and U-Ω relationships are nearly symmetrical about $r = r_0$ and

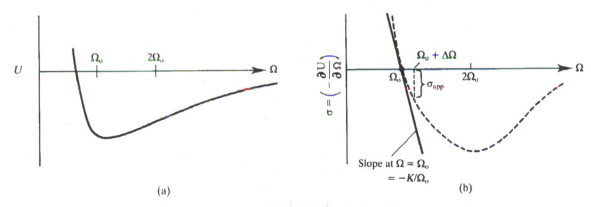

(a)   (b)

Figure 2.6
(a) Potential energy vs. volume per atom for the atomic array of Fig. 2.4. The equilibrium volume is Ω_0 ($= r_0{}^3$). (b) The "internal" hydrostatic stress as a function of volume per atom for this solid. Equilibrium at $\Omega = \Omega_0$ is associated with $\sigma = 0$. To increase the volume to ($\Omega_0 + \Delta\Omega$) requires an applied stress, σ_{app}. For small $\Delta\Omega$, the $\sigma - \Omega$ curve is linear in the vicinity of Ω_0, and the slope $(\partial\sigma/\partial\Omega)_{\Omega_0} = -K/\Omega_0$, where K is the bulk modulus. Since $\sigma = -(\partial U/\partial\Omega)$, we also have $K = \Omega_0(\partial^2 U/\partial\Omega^2)_{\Omega_0}$.

$\Omega = \Omega_0$. This means that the slope (e.g., $\partial U/\partial \Omega$) increases linearly with $\Delta\Omega$ ($= \Omega - \Omega_0$) in the vicinity of Ω_0 and that σ increases likewise (Fig. 2.6b). Thus, the bulk modulus (cf. Eq. (2.4)) in the vicinity of $\Omega = \Omega_0$ is obtained as

$$K = -\Omega_0\left(\frac{\partial \sigma}{\partial \Omega}\right)_{\Omega_0} = +\Omega_0\left(\frac{\partial^2 U}{\partial \Omega^2}\right)_{\Omega_0} \tag{2.9}$$

that is, the bulk modulus is related to the curvature of the U-Ω relationship. Since the derivative of the U-Ω curve increases linearly with increases in Ω, this means that $\partial^2 U/\partial\Omega^2$ is a constant near $\Omega = \Omega_0$ and linear elasticity is observed. At larger strains this is no longer true and linear elasticity is not observed even though the elastic behavior manifested remains a result of stretching or compression of atomic bonds. This kind of behavior is found in tensile deformation of whiskers and in hydrostatic compression at large values of applied pressure.

Modulus values relate to bond energy as just described. They also depend on the equilibrium interatomic spacing and nature of the chemical bonding as discussed earlier. In general the potential energy between two atoms, ions, or molecules can be expressed by an equation of the form

$$u_{ij} = -Ar^{-n} + Br^{-m} \tag{2.10}$$

The first term on the right-hand side of Eq. (2.10) represents the longer-range attractive energy and the second term the corresponding repulsive energy. The long-range/short-range nature of the two forces is satisfied if $m > n$; in fact, this is a condition for crystal stability (Prob. 2.4). Minimum total potential energy is achieved when u_{ij} is a minimum. This allows B to be expressed in terms of A and r_0 by setting $\partial u_{ij}/\partial r = 0$ and substituting the result in Eq. (2.10). On doing so, Eq. (2.10) can be written as

$$u_{ij} = -Ar^{-n}\left[1 - \frac{n}{m}\left(\frac{r}{r_0}\right)^{-(m-n)}\right] \tag{2.11}$$

For ionic solids, the attractive force is well represented by a simple Coulombic interaction between oppositely charged ions; i.e., $F \sim r^{-2}$ and $u \sim r^{-1}$. Hence, $n = 1$ for ionic materials and m is empirically determined as a number usually greater than 6. Thus, at equilibrium, the cohesive energy is determined primarily by the attractive energy and can (within typical accuracy of 15 percent or better) be so approximated. Equation (2.11) can be written in terms of atomic volume ($\Omega \sim r^3$), and determination of the bulk modulus for ionic materials via Eq. (2.9) (neglecting the repulsive energy) indicates that $K \sim (\Omega_0)^{-4/3}$ or $K \sim r_0^{-4}$. Bulk moduli of ionic materials with a given structure do, in fact, exhibit a dependence on r_0 of this type. Although all interactions between atoms are fundamentally Coulombic in nature, it is nevertheless surprising that K also varies with r_0^{-4} for the alkali metals and for covalent-bonded elements and compounds having the diamond cubic structure (Fig. 2.7). (There are other theoretical reasons, though, for expecting this behavior in diamond cubic structure materials.) On the other hand, the bulk moduli of transition metal carbides (e.g., TiC, TaC, etc.) do not exhibit an r_0^{-4} dependence even though the bonding in these materials is a mixture of primary bond types.

Hydrostatic normal stresses, by application of which bulk moduli are measured, are associated only with bond stretching or compression and not bond distortion.

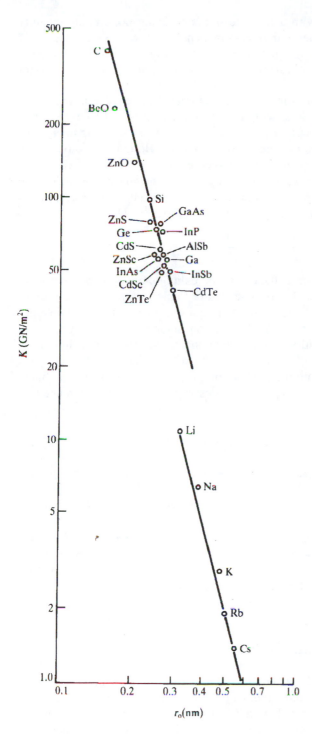

Figure 2.7
Bulk moduli of the alkali metals
and tetrahedral covalently
bonded crystals as a function
of their interatomic spacing.
The slope of −4 observed on
these logarithmic coordinates
shows $K \sim (r_0)^{-4}$. The generally
higher moduli of the covalent
solids is indicative of their
inherently stronger bonding.
(*Data obtained from J. J. Gilman,*
Micromechanics of Flow in
Solids, McGraw-Hill, New York,
***1969, Chap. 2, pp. 29–41.*)**

Shear stresses are associated with bond distortion, and shear moduli consequently do not vary in as direct a manner with interatomic spacing as do bulk moduli.

Although bulk (and Young's) moduli are fundamentally related to the curvature of the bonding energy–atomic volume (interatomic separation) curves, they also relate to the bond energy. This is because materials having a high bond energy also typically manifest large curvatures in the energy–atomic volume (separation) curve. Further, bond energy also correlates with the melting temperature. To express this relation in units of energy, we write $U_0 \sim kT_m$, where U_0 is the bonding energy, T_m is the material's absolute melting temperature, and k is the Boltzmann constant. To obtain a dimensionally correct correlation between E and kT_m requires dividing the latter by a volume; we take this as the atomic volume. On this basis, E should scale with kT_m/Ω. Empirical correlation indicates that the appropriate scaling constant is about 100 (when SI units are used; i.e., kT_m in J and Ω in m^3). Thus,

$$E \cong \frac{100kT_m}{\Omega} \tag{2.12}$$

Figures 2.5 and 2.6 only depict the crystal potential energy situation pertinent to 0 K where potential energy is the sole contributor to the system energy. With increasing temperature, atoms vibrate about their equilibrium positions and this motion is caused by thermal energy. While this additional lattice energy is small relative to U_0, the atomic vibrations do lead to an increase in the equilibrium interatomic separation (i.e., the material thermally expands). The modulus decreases concurrent with the increased atomic separation. This decrease is essentially linear with temperature, and an approximate equation describing the modulus-temperature relationship is

$$\frac{E}{E_0} = \left[1 - a\left(\frac{T}{T_m}\right) \right] \tag{2.13}$$

where E is the modulus at temperature T and E_0 the modulus at 0 K. The proportionality constant a for most crystalline solids is on the order of 0.5. Thus, for such a "typical" material, the modulus decreases by about 50% as the temperature increases from 0 K to the material's melting point.

We have mentioned that the forces between atoms, molecules, or ions in crystals depend on the distances between them. Thus, they also vary with crystallographic direction so it should not be surprising that crystalline moduli are anisotropic. This is discussed in the following section.

2.5
ANISOTROPIC LINEAR ELASTICITY

Anisotropic linear elasticity is expected in a single crystal since, as noted, the forces between atoms depend on crystallographic direction. This is true even if the forces are central in nature, i.e., they act along a line connecting the atom or ion centers. Noncentral forces, which are as much the rule as the exception, produce even greater anisotropy.

Consider a single crystal in the shape of a cube (Fig. 2.8). If the axes of the crystalline form are orthogonal (i.e., if the crystal is cubic, tetragonal, or orthorhombic), the macroscopic 1, 2, and 3 axes are chosen parallel to the unit cell axes in Fig. 2.8. If the crystal does not display this type of symmetry, the macroscopic axes are chosen so as to have a specific relationship with the unit cell axes.

Anisotropic linear elasticity can be described by either a linear relationship between each of the individual stresses of Fig. 2.8 and the resulting strains, or by a corresponding relation between the strains and the stresses (Prob. 2.7). Describing anisotropic linear elasticity with the former approach leads to the following constitutive equations:

$$
\begin{aligned}
\sigma_{11} &= C_{11}\varepsilon_{11} + C_{12}\varepsilon_{22} + C_{13}\varepsilon_{33} + C_{14}\gamma_{23} + C_{15}\gamma_{13} + C_{16}\gamma_{12} \\
\sigma_{22} &= C_{21}\varepsilon_{11} + C_{22}\varepsilon_{22} + C_{23}\varepsilon_{33} + C_{24}\gamma_{23} + C_{25}\gamma_{13} + C_{26}\gamma_{12} \\
\sigma_{33} &= C_{31}\varepsilon_{11} + C_{32}\varepsilon_{22} + C_{33}\varepsilon_{33} + C_{34}\gamma_{23} + C_{35}\gamma_{13} + C_{36}\gamma_{12} \\
\tau_{23} &= C_{41}\varepsilon_{11} + C_{42}\varepsilon_{22} + C_{43}\varepsilon_{33} + C_{44}\gamma_{23} + C_{45}\gamma_{13} + C_{46}\gamma_{12} \\
\tau_{13} &= C_{51}\varepsilon_{11} + C_{52}\varepsilon_{22} + C_{53}\varepsilon_{33} + C_{54}\gamma_{23} + C_{55}\gamma_{13} + C_{56}\gamma_{12} \\
\tau_{12} &= C_{61}\varepsilon_{11} + C_{62}\varepsilon_{22} + C_{63}\varepsilon_{33} + C_{64}\gamma_{23} + C_{65}\gamma_{13} + C_{66}\gamma_{12}
\end{aligned}
\qquad (2.14)
$$

As written in this form, 36 independent stiffness constants (i.e., the number of different C_{ij} coefficients in Eq. (2.14)) are required to describe anisotropic behavior. Considerations of mechanical equilibrium show, however, that the *off-diagonal* coefficients (i.e., C_{ij}; $i \neq j$) are related by $C_{ij} = C_{ji}$; e.g., $C_{14} = C_{41}$, etc. This reduces the number of independent coefficients to 21: the 6 diagonal terms (C_{11}, C_{22}, etc.) plus 15 off-diagonal ones. Twenty-one coefficients are, in fact, required to describe linear elasticity in triclinic crystals, this crystal form having the lowest symmetry.

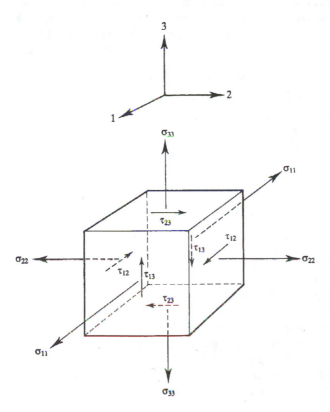

Figure 2.8
A single crystal in the form of a cube subjected to a combination of normal (σ_1, σ_2, σ_3) and shear (τ_{12}, τ_{23}, τ_{13}) stresses. The coordinate axes are chosen parallel to <100> directions if the crystal system is orthogonal (i.e., cubic, tetragonal, or orthorhombic) or to have a specified orientation with respect to the unit cell if it is not.

For reasons described below, as crystal symmetry increases, the number of independent elastic constants decreases. For example, nine independent elastic constants are required to describe linear elasticity in orthorhombic crystals and six are necessary for tetragonal crystals.

Cubic crystals require but three independent elastic constants to describe their behavior. To illustrate why, assume that the crystal of Fig. 2.8 is cubic with <100> directions parallel to the 1, 2, and 3 axes. It is clear that a stress applied along the 1 axis (i.e., the [100] direction) would produce a response equivalent to a stress applied along the 2 axis (the [010] direction). Likewise, a similar response would be obtained if the stress were along the 3 or [001] axis; this leads to the conclusion that $C_{11} = C_{22} = C_{33}$. Similarly, shear stresses applied along the family of {001} planes would result in identical shear responses, from which we conclude that $C_{44} = C_{55} = C_{66}$. Further considerations like these reduce the complicated series of Eqs. (2.14) to a much simpler one for a cubic material:

$$\sigma_{11} = C_{11}\varepsilon_{11} + C_{12}\varepsilon_{22} + C_{12}\varepsilon_{33}$$
$$\sigma_{22} = C_{12}\varepsilon_{11} + C_{11}\varepsilon_{22} + C_{12}\varepsilon_{33}$$
$$\sigma_{33} = C_{12}\varepsilon_{11} + C_{12}\varepsilon_{22} + C_{11}\varepsilon_{33}$$
$$\tau_{23} = C_{44}\gamma_{23} \tag{2.15}$$
$$\tau_{13} = C_{44}\gamma_{13}$$
$$\tau_{12} = C_{44}\gamma_{12}$$

Hence, the number of independent elastic constants for a cubic material is three: C_{11}, C_{12}, and C_{44}. Cubic materials are *not* inherently isotropic, since isotropic materials require only two independent elastic constants. If the material of Fig. 2.8 were isotropic (for which case selection of the axes with respect to the crystal is arbitrary), the linear elastic stress-strain response would be described by

$$\sigma_{11} = (\Gamma + 2G)\varepsilon_{11} + \Gamma\varepsilon_{22} + \Gamma\varepsilon_{33}$$
$$\sigma_{22} = \Gamma\varepsilon_{11} + (\Gamma + 2G)\varepsilon_{22} + \Gamma\varepsilon_{33}$$
$$\sigma_{33} = \Gamma\varepsilon_{11} + \Gamma\varepsilon_{22} + (\Gamma + 2G)\varepsilon_{33} \tag{2.16}$$
$$\tau_{ij} = G\gamma_{ij} \ (i,j = 1,2; \ 1,3; \ 2,3)$$

where

$$\Gamma = \frac{E\nu}{(1 + \nu)(1 - 2\nu)} \tag{2.17}$$

Thus, for a cubic material to display isotropy, we must have

$$\Gamma + 2G = C_{11}; \qquad \Gamma = C_{12}; \qquad G = C_{44} \tag{2.18}$$

or $(C_{11} - C_{12})/2C_{44} = 1$. This ratio $(C_{11} - C_{12})/2C_{44}$, is called the anisotropy ratio and values of it for several materials are shown in Table 2.2. Only for tungsten is the ratio essentially unity and tungsten, therefore, is elastically isotropic. Some of the other cubic materials show considerable anisotropy in their elastic behavior. The anisotropy ratio has some physical significance. The constant C_{44} represents resistance to shear on {100} planes in <100> type directions (cf. Fig. 2.8). On the other hand, the parameter $(C_{11} - C_{12})/2$ represents resistance to shearing on {010} planes in <110> directions. Since it would not be expected that the shear resistance would be the same for both of these cases, it is not surprising that anisotropy ratios are seldom unity.

Material class	Material	C_{11} (10^{10} N/m²)	C_{12} (10^{10} N/m²)	C_{44} (10^{10} N/m²)	Anisotropy ratio $(C_{11} - C_{12})/2C_{44}$
Metals	Ag	12.4	9.3	4.6	0.34
	Al	10.8	6.1	2.9	0.81
	Au	18.6	15.7	4.2	0.35
	Cu	16.8	12.1	7.5	0.31
	α-Fe	23.7	14.1	11.6	0.41
	Mo	46.0	17.6	11.0	1.29
	Na	0.73	0.63	0.42	0.12
	Ni	24.7	14.7	12.5	0.40
	Pb	5.0	4.2	1.5	0.27
	W	50.1	19.8	15.1	1.00
Covalent solids	Si	16.6	6.4	8.0	0.64
	Diamond	107.6	12.5	57.6	0.83
	TiC	51.2	11.0	17.7	1.14
Ionic solids	LiF	11.2	4.6	6.3	0.52
	MgO	29.1	9.0	15.5	0.65
	NaCl	4.9	1.3	1.3	1.38

Table 2.2
Stiffness coefficients for selected cubic materials

The set of Eqs. (2.14) linearly relate the normal and shear stresses to the normal and shear strains. Because these relationships are linear, we can also express normal and shear strains in terms of the corresponding stresses. For example, we have in analogy to Eq. (2.14)

$$\varepsilon_{11} = S_{11}\sigma_{11} + S_{12}\sigma_{22} + S_{13}\sigma_{33} + S_{14}\tau_{23} + S_{15}\tau_{13} + S_{16}\tau_{12} \qquad (2.19)$$

with analogous expressions for the other (both normal and shear) strains. The coefficients S_{ij} of Eq. (2.19) and their analogs are termed *compliances*. For cubic materials, there are but three independent compliances just as there are three independent stiffnesses. The independent compliances—S_{11}, S_{12} and S_{44}—are related to the three independent stiffnesses by (Prob. 2.8)

$$C_{11} = \frac{S_{11} + S_{12}}{(S_{11} - S_{12})(S_{11} + 2S_{12})} \qquad (2.20a)$$

$$C_{12} = \frac{-S_{12}}{(S_{11} - S_{12})(S_{11} + 2S_{12})} \qquad (2.20b)$$

$$C_{44} = \frac{1}{S_{44}} \qquad (2.20c)$$

For a cubic crystal the modulus in an arbitrary [hkl] direction can be expressed in terms of stiffnesses or compliances and the orientation of the [hkl] direction to the three <100> directions. It is more convenient to use compliances for this purpose, and on doing so we find

$$\frac{1}{E_{[hkl]}} = S_{11} - 2(S_{11} - S_{12} - \tfrac{1}{2}S_{44})(\alpha^2\beta^2 + \alpha^2\gamma^2 + \beta^2\gamma^2) \qquad (2.21)$$

where α, β, and γ are the direction cosines of the [hkl] direction and the [100], [010], and [001] directions, respectively.

Material class	Material*	$E_{polycrystal}$ (10^9 N/m^2)	$E_{<111>}$ (10^9 N/m^2)	$E_{<100>}$ (10^9 N/m^2)	$E_{<100>}/E_{<111>}$	Anisotropy ratio†
Metals	Al	70	76	64	0.84	0.81
	Au	78	117	43	0.37	0.35
	Cu	121	192	67	0.35	0.31
	α-Fe	209	276	129	0.47	0.41
	W	411	411	411	1.00	1.00
Covalent solids	Diamond	—	1200	1050	0.88	0.83
	TiC	—	429	476	1.11	1.14
Ionic solids	MgO	310	343	247	0.72	0.65
	NaCl	37	32	44	1.38	1.38

*For the materials listed $E_{<111>} = E_{max}$ and $E_{<100>} = E_{min}$ except for TiC and NaCl, for which the reverse applies.
†Note: $E_{<100>}/E_{<111>}$ should scale with the anisotropy ratio (Table 2.2).

Table 2.3
Maximum and minimum moduli for selected cubic materials*

Thus, we see that cubic crystals, because they are not isotropic, exhibit Young's moduli that depend on the direction of the applied stress relative to the crystal axes. The formulation of Eq. (2.21) can be written in another, sometimes more convenient, form. In particular, the elastic modulus in any particular crystallographic direction can be obtained if it is known for two other specific directions. For most cubic materials $E_{<111>}$ (i.e., the modulus measured with the stress axis along the <111> direction) is larger than $E_{<100>}$, and these represent extremum values of the modulus. The modulus along an arbitrary [hkl] direction is obtained from $E_{<111>}$ and $E_{<100>}$ as

$$\frac{1}{E_{[hkl]}} = \frac{1}{E_{<100>}} - 3\left(\frac{1}{E_{<100>}} - \frac{1}{E_{<111>}}\right)(\alpha^2\beta^2 + \alpha^2\gamma^2 + \beta^2\gamma^2) \quad (2.22)$$

In the absence of texture, a polycrystalline cubic material is, as mentioned, elastically isotropic as a result of the averaging out over a large number of grains of the anisotropy inherent to each grain. Table 2.3 lists the values of E_{max} (usually $E_{<111>}$) and E_{min} (usually $E_{<100>}$) and the polycrystalline moduli for a number of materials. Polycrystalline moduli are usually close to an arithmetic mean of the extremum moduli. For polycrystalline cubic materials, elastic anisotropy is considered in engineering design only in the case of pronounced crystallographic texturing. It is much more of a concern in, for example, hexagonal materials, for which moderate texturing can produce appreciable anisotropy as a result of the inherently greater single-crystal anisotropy. In the remainder of this text, deformation and fracture are discussed primarily in terms of isotropic materials since the mathematical description of these phenomena is considerably simplified by doing so. In these treatments a cautionary note—that isotropic behavior need not always be the case—is in order.

EXAMPLE PROBLEM 2.1. For iron, $C_{11} = 237$ GN/m^2, $C_{12} = 141$ GN/m^2, and $C_{44} = 116$ N/m^2.

a. Use Eqs. (2.20) to determine the respective compliances for Fe.
b. Determine $E_{[100]}$, $E_{[110]}$, and $E_{[111]}$ for Fe.

Solution. a. From Eq. (2.20c) we have $S_{44} = 1/C_{44}$, so $S_{44} = 0.862 \times 10^{-11}$ m²/N. Taking the ratios of Eqs. (2.20a) and (2.20b)

$$\frac{S_{11} + S_{12}}{-S_{12}} = \frac{C_{11}}{C_{12}} = 1.681 \quad \text{or} \quad \frac{S_{11}}{S_{12}} = -2.681$$

Substitute back into (a slightly rearranged) Eq. (2.20b) to solve for S_{12};

$$C_{12} = 141 \times 10^9 \frac{N}{m^2} = \frac{-1}{S_{12}\left(\dfrac{S_{11}}{S_{12}} - 1\right)\left(\dfrac{S_{11}}{S_{12}} + 2\right)}$$

Substituting in the values for C_{12} and S_{11}/S_{12}, we get

$$S_{12} = -0.283 \times 10^{-11} \text{ m}^2/\text{N}$$

Using the previously determined value of $S_{11}/S_{12}(= -2.681)$, we have

$$S_{11} = 0.757 \times 10^{-11} \text{ m}^2/\text{N}$$

b. Use Eq. (2.21). Here α is the cosine of the angle between [hkl] (the direction in which the modulus is to be calculated) and [100]; β and γ are the comparable cosines for the directions [010] and [001]. Thus,

for [100]; $\alpha = 1$, $\beta = \gamma = 0$ and $\alpha^2\beta^2 + \alpha^2\gamma^2 + \beta^2\gamma^2 = 0$

for [110]; $\alpha = \beta = 1/\sqrt{2}$, $\gamma = 0$ and $\alpha^2\beta^2 + \alpha^2\gamma^2 + \beta^2\gamma^2 = 1/4$

for [111]; $\alpha = \beta = \gamma = 1/\sqrt{3}$ and $\alpha^2\beta^2 + \alpha^2\gamma^2 + \beta^2\gamma^2 = 1/3$

Now evaluate Eq. (2.21) using the values of the several S_{ij}s determined from part (a) and the just obtained geometrical relationships.

$$\frac{1}{E_{[hkl]}} = S_{11} - 2(S_{11} - S_2 - \tfrac{1}{2}S_{44})(\alpha^2\beta^2 + \alpha^2\gamma^2 + \beta^2\gamma^2)$$

$$\frac{1}{E_{[100]}} = 0.757 \times 10^{-11} \text{ m}^2/\text{N} \quad \text{or} \quad E_{[100]} = 132 \text{ GN/m}^2$$

$$\frac{1}{E_{[110]}} = 10^{-11}\frac{m^2}{N}\{0.757 - 2(0.757 + 0.283 - 0.431)\}\tfrac{1}{4} = 0.453 \times 10^{-11} \text{ m}^2/\text{N}$$

$$\text{or } E_{[110]} = 221 \text{ GN/m}^2$$

$$\frac{1}{E_{[111]}} = 10^{-11}\frac{m^2}{N}\{0.757 - 2(0.757 + 0.283 - 0.431)\}\tfrac{1}{3} = 3.51 \times 10^{-11} \text{ m}^2/\text{N}$$

$$\text{or } E_{[111]} = 285 \text{ GN/m}^2$$

2.6
RUBBER ELASTICITY

Rubber and linear elasticity represent the extremes of elastic behavior. Elastomeric materials (rubbers) demonstrate macroscopic reversible extensions that can be as great as 1000 percent. This is in contrast to linear elastic deformations, which are typically on the order of a few tenths of a percent at most. Rubber elasticity is exhibited only by certain polymerics, and elastomeric behavior is apparent in them only over a certain temperature range.[4] At low temperatures, for example, an elastomer displays brittle, or glass-like, behavior. Furthermore, the physics of rubber elasticity differ fundamentally from that of linear elasticity. Linear elasticity is a result of primary bond stretching and distortion. This type of distortion occurs in elastomers only at very large strains. As we will see, ordinary rubber elasticity is a mechanical response dictated by entropic, rather than potential energetic, considerations.

Elastomeric behavior is found only in noncrystalline, long-chain polymers. For such a polymer to display rubber elasticity, the backbone of the chain must be very long and have many kinks and bends. The chains must also be cross-linked, by primary chemical bonds, every several hundred atoms along their length (the degree of cross-linking determines the elastomer stiffness) and, in addition, the chains must be in constant motion; i.e., they must have considerable kinetic energy over the temperature range of elastomeric behavior.

In most cases the kinking or chain bending in elastomers is related to carbon-carbon double bonds along the chain; these produce a helical pattern to the chain. The effect the carbon-carbon double bond has on chain kinking can be illustrated by considering two isomers of natural rubber. (Isomers have the same composition, but display different structures. In a sense, structural isomers are to polymers what allotropic forms are to crystalline materials.) The bonding and geometrical arrangement observed in gutta-percha (Fig. 2.9a) produce a relatively straight chain and gutta-percha does not display elastomeric behavior. However, in the cis-isomer

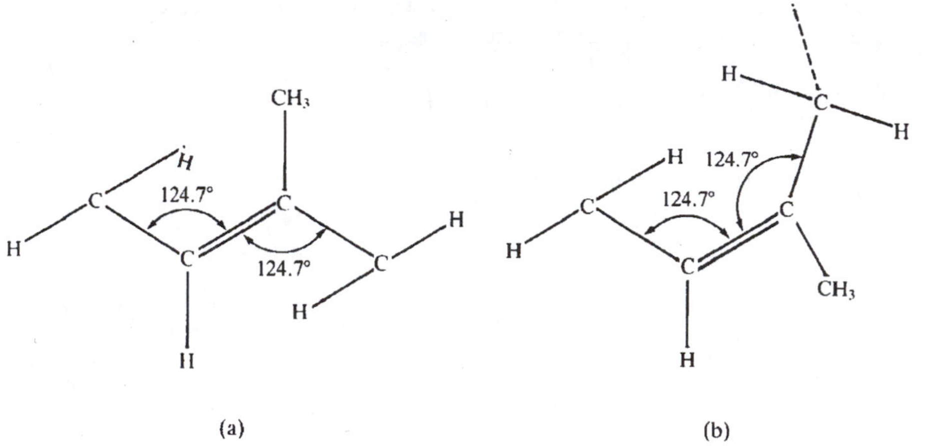

(a)   (b)

Figure 2.9
The structure of the monomers in two isomeric forms of natural rubber $(-CH_2 - CH = C(CH_3) - CH_2 -)$. (a) The *trans*-isomeric form in gutta-percha; H and CH_3 side-groups alternate along the chain backbone, resulting in a relatively straight chain that is not associated with elastomeric behavior. (b) The *cis*-isomeric form of natural rubber. Highly coiled chains are formed as a result of the H and CH_3 side-groups being on the same side of the chain. The invariant 124.7° angle associated with the carbon-carbon double bond gives rise to a helical chain configuration and to elastomeric behavior.

[4]All noncrystalline polymers exhibit "rubbery" behavior over a certain temperature range, as is discussed in the following section. However, only elastomers—the materials discussed in this section—demonstrate large reversible extensions over the specified temperature range.

of isoprene (Fig. 2.9*b*), the arrangement of the CH_3 "molecule" along the chain is such that considerable kinking is effected and *cis*-isoprene is a rubbery material.

Chain cross-linking is also required for elastomeric behavior to be displayed. Cross-linking can be effected by the addition of sulfur to the rubber, as in vulcanization. The carbon-carbon double bond is broken and replaced by a single C-C bond at the cross-linking points. The additional bonding electron is joined to a sulfur atom, which mates in a similar manner with a carbon atom on an adjacent chain, thus producing a primary bond cross-link. The cross-links provide attachment points so that on load removal the rubber "springs back" to its original dimensions.

From the viewpoint of a mechanical model the structure of an elastomer can be considered a highly coiled skeleton of primary bonds (including the cross-links) immersed in a viscous-like medium. During application of a tensile load the coils are "unwound" to an extent that depends on the applied force. The resulting rubber modulus is small, and this is analogous to the elongation or compression of a spring for which the forces of extension or compression are much less than those required for elastic distortion of primary atomic bonds. Associated with the spring-like response of the skeleton is a viscous elastic flow that represents sliding of interchain segments; the segments of chains that are displaced with respect to each other are those located between the chemical cross-links. The stress associated with the extension of an elastomer is shown in Fig. 2.10. Only at extensions where the chains are approximately fully extended does the force begin to stretch primary bonds, and this is associated with the rapidly increasing stress at an extension ratio of ca. 4.0 in

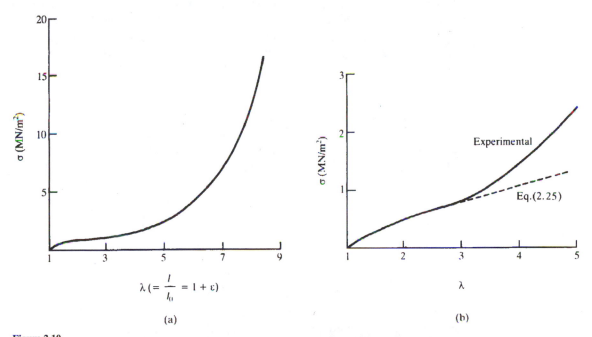

Figure 2.10
(a) The stress-strain curve of natural rubber (strain is represented by the extension ratio $\lambda (= l/l_0)$; $\lambda = 1 + \varepsilon$). For extension ratios less than ca. 3.0–4.0, the rubber strain results from uncoiling of elastomer chains. At larger strains, stretching of primary bonds also occurs, and this results in higher stresses being required to continue deformation. In (b) the low-strain region of the stress-strain curve is expanded. The constitutive Eq. (2.25) adequately describes the rubber stress-strain behavior for $\lambda < 3.0$; i.e., it applies over the strain range for which the assumptions underlying its derivation are valid.

Fig. 2.10. When the load on the elastomer is released, the cross-linking atoms act to restore the original dimensions. Without these links, this restoration would not take place. For example, in their absence, natural rubber is a gummy, viscous liquid that flows under its own weight. As mentioned, the "pinning point" density is related to the number of cross-linking atoms. The greater their number, the stiffer the rubber and also the lesser the strain at which primary bond stretching commences. Ebonite, for example, is a heavily cross-linked elastomer for which the extent of cross-linking is so high that the structure resembles a three-dimensional skeleton rather than a group of long-chain polymers cross-linked occasionally. As a consequence, ebonite is a "hard," tough elastomer whose mechanical characteristics (high strength and low extensibility) represent a mechanical extremum of an elastomer.

Since elastomeric extension is not associated with primary bond stretching, it is not surprising that rubber moduli are small. Moreover, as noted, the physics of rubber elasticity differ fundamentally from those of linear elasticity. In rubber elasticity the entropy of the material is increased as the chains are extended, and this is the primary, almost sole, contribution to the free energy increase of the rubber as it is stretched. In rubber elasticity, the system potential energy is unchanged because primary bond stretching does not occur. The opposite is the case for linear elasticity; the potential energy increase arising from bond stretching is the primary contribution to the free energy increase of a linearly elastic material under stress.

The mathematical description of rubber elasticity is akin to that describing compression of an ideal gas, for which the potential energy is also independent of volume. This independence results from the absence of intermolecular interactions in an ideal gas owing to the large spacings between molecules. For a gas, the independence of potential energy on volume (V) leads to a constitutive equation for its compression,

$$p = T\left(\frac{\partial p}{\partial T}\right)_V \qquad (2.23)$$

where p is the gas pressure and T the absolute temperature. (Substitution of the ideal gas law into Eq. (2.23) serves to verify its correctness for this case.) We note that the pressure required to compress a gas increases with temperature. This is a result of the increased *kinetic* energy of the gas which is manifested by the pressure that must be overcome during gas compression.

Because the physics of rubber elasticity is the same as for a gas, the constitutive equation for an elastomer is analogous to Eq. (2.23). The alteration is that F (the force) substitutes for p and l (the elastomer length) for the volume V. Thus,

$$F = T\left(\frac{\partial F}{\partial T}\right)_l \qquad (2.24)$$

As for a gas, the stiffness of a rubber, as measured by the force required to extend it, increases with temperature. In contrast, linear elastic moduli decrease slightly with temperature (cf. Eq. (2.13)). That rubber becomes stiffer with increasing temperature can be verified by a simple experiment. If a thick rubber band is stretched and then slightly heated, an increase in the force required to maintain the extension is observed. A corollary is that the thermal expansion coefficient of a rubber under stress is negative. That is, the increase in temperature is manifested by a tendency on the part of the rubber to contract; thus, to keep the rubber length fixed the applied force must be increased. Both of these observations are consistent with the essential kinetic nature of rubber elasticity. That is, as temperature is increased, the kinetic energy of the polymer chains is also increased. This results in an increased tendency for the chains to reduce their end-to-end length, which increases the rubber entropy. In fact, the $(\partial F/\partial T)$ term of Eq. (2.24) can be shown by thermodynamic arguments to be equal to $-(\partial S/\partial l)_T$, where S is the rubber entropy. This entropy can be calculated in terms of the extension ratio, l/l_0, of the long-chain polymer which relates directly to the macroscopic extension ratio. Following such a procedure permits a constitutive equation for elastomeric tensile deformation to be developed;

$$\sigma = \left(\frac{\rho RT}{M_c}\right)\left(\lambda - \frac{1}{\lambda^2}\right) \tag{2.25}$$

In Eq. (2.25), λ $(= l/l_0)$ is the extension ratio, ρ the rubber density (kg/m^3), R the gas constant (J/mol K), and M_c the molecular weight of the rubber *between* the pinning points (units of kg/mol). Hence, the rubber stiffness $(\rho RT/M_c)$ increases with temperature and with decreases in M_c; that is, as the distance between cross-links is decreased, M_c decreases and rubber stiffness increases.

Equation (2.25) reasonably describes the tensile behavior of a rubber, at least at strains less than those at which primary bond stretching commences (Fig. 2.10b). At the onset of this, the rubber stiffness, as measured by the instantaneous slope of the stress-extension ratio curve, increases. In this region, primary bonds are stretched and some cross-links may be broken. If the latter occurs, the rubber exhibits a decreased stiffness on subsequent reloading. Moreover, on unloading, the rubber length does not instantaneously resume its original unstretched value. (This can be demonstrated by snapping a rubber band, and quickly measuring the length of the "fractured" rubber in comparison to the original length.) However, provided not too many cross-links have been broken, the rubber gradually contracts to very near its original dimensions. This is a result of the chain molecules attempting to attain their relaxed configuration, and this retarded elasticity (viscoelasticity) is also observed commonly in polymers that do not exhibit elastomeric behavior. Polymer elasticity is discussed next.

2.7
POLYMER ELASTICITY AND VISCOELASTICITY

Polymer moduli are both temperature and time dependent (i.e., polymers are strain rate sensitive). Moduli of three different polymers (polystyrene, an epoxy, and polyisobutylene, (an elastomer)), as they depend on temperature and time, are shown in Figs. 2.11(a)–(c). Polystyrene (PS) illustrates the time and temperature dependence of the mechanical response of a long-chain polymer. The polystyrene modulus at low temperatures ($<100°C$; $T/T_g < 1$ where T_g is the glass transition temperature) is much less than that of a typical inorganic solid. The reason, mentioned previously, is that this modulus reflects the "spring constant" of the weak van der Waals bonds between the polymer chains. The modulus of PS decreases with temperature during this "glassy" region, where polymers are relatively brittle and, thus, behave in a "glass-like" manner. The cause for this temperature sensitivity is similar to that causing the comparable decreases in modulus of linear elastic

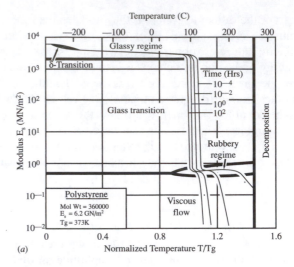

(a)

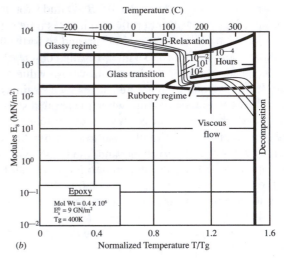

(b)

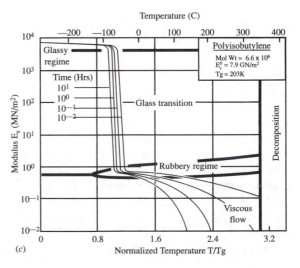

(c)

Figure 2.11
Modulus-temperature diagrams for: (a) an amorphous chain polymer (polystyrene, PS); (b) a cross-linked polymer (an epoxy); and (c) a typical elastomer (polyisobutylene, PIB). All three polymers exhibit linear elastic behavior below their glass transition temperature (T_g), although their low-temperature moduli are much less than for inorganic solids. At and around T_g, polymer moduli decrease rapidly with temperature and become time dependent as well. At somewhat higher temperatures, rubber behavior is manifested in all of the polymers, although large reversible extensions are found only in PIB. And at even higher temperatures, viscous (permanent) deformation characterizes the response of a polymer to an applied stress; then the "moduli" shown on the diagram bear no relation to either a linear elastic modulus or a viscoelastic modulus. (*From Lorna J. Gibson and Michael F. Ashby, Cellular Solids—Structure and Properties, Pergamon, Oxford, 1988.*)

materials; that is, thermal expansion increases the separation between polymer chains and reduces the bonding between them. In addition, for long-chain polymers small recoverable elastic displacements between segments of adjoining chains take place during the glassy region and these displacements are facilitated by increased temperature. The displacements are polymer specific, involving different molecular groups for each polymer. Polymer chemists assign "names" to these elastic effects or "transitions"; thus, for PS, the transition is termed a "δ" transition, whereas a comparable transition in the epoxy (Fig. 2.11(b)) is called a "β" transition. We note that all polymer classes demonstrate a "glassy regime," which extends from 0 K to about $T = T_g$. The temperature dependence of the modulus during the glassy regime is expressed by an equation having the form of Eq. (2.13). The difference is that T_g substitutes for the melting temperature in the case of polymers.[5] The constant a in the equation is about the same (i.e., $a \cong 0.5$) for both polymers and inorganic solids.

At and around T_g polymer moduli decrease precipitously with increasing temperature, by over three orders of magnitude for the long-chain polymer and the elastomer and by about an order of magnitude for the epoxy, over a fairly narrow temperature range. Further, the modulus is time sensitive, as indicated by the lines in Fig. 2.11 marked 10^{-4} hr, 10^{-2} hr, and so forth. This is indicative of viscoelastic behavior, and we return to a more detailed description of viscoelasticity later. However, here we note that the moduli, whose values are shown in Figs. 2.11, are obtained by division of an applied stress by a strain as indicated in Figs. 2.12. Figure 2.12a illustrates the situation when the applied stress is maintained constant in such a "modulus" determination. If the strain experienced by the polymer were time independent (e.g., as it usually is at low temperatures), then the modulus is independent of time. If, on the other hand, the strain experienced by the material increases with time (Fig. 2.12a), then the modulus is time dependent; the longer the time at which the strain is measured, the lower the modulus. This modulus, termed a relaxation modulus, can also be obtained in a stress relaxation test (Fig. 2.12b). In this test, a specified strain is applied to the material and the stress required to maintain this strain measured. For a viscoelastic material, the measured stress decreases with time and so, too, does the relaxation modulus. We see that the moduli of the polymers whose behavior is illustrated in Figs. 2.11 are quite sensitive functions of time during the transition from glassy to viscoelastic behavior ("the glass transition") that takes place near T_g. In terms of characterizing the polymer behavior, we say that polymer goes from being "glass-like" to "leathery-like" (with the polymer becoming far more flexible) during this transition.

We note, too, that all polymer classes exhibit a "rubbery" regime in which modulus increases with temperature. This regime is much more extensive for the elastomer, polyisobutylene (PIB), than for the other two classes of polymers. Further, the rubbery regime encompasses room temperature for PIB, so this material is an elastomer at room temperature. For PS, rubbery behavior begins at about 100°C; for the epoxy, it starts at about 150°C. In contrast to PIB, however, neither PS nor the epoxy are elastomers. That is, even though they display rubbery characteristics

---

[5]Although a bit oversimplified, T_g can be taken as a "melting temperature" in the sense that T_g can be viewed as that temperature at which "melting" of the van der Waals bonds between the polymer chains occurs.

and although their moduli increase with temperature in the rubbery regime, neither material displays the extensive reversible extensions of a true elastomer. For PS, the cause is the absence of cross-links between the polymer chains. If PS were suitably cross-linked it would display elastomeric behavior in the rubbery regime and would be classified as a "high temperature" elastomer. The epoxy, on the other hand, cannot be modified to render it an elastomer. The short "segments" that constitute the backbone of this thermoset do not allow for the extensive chain "uncoiling" characteristic of rubber elasticity.

At higher temperatures still, "viscous flow" describes the polymers' mechanical response. The moduli values of Fig. 2.11 lose meaning here, since viscous flow is permanent in nature and use of a "modulus" to characterize such behavior is inappropriate. (Hence, moduli "values" in the viscous region represent only a ratio of a stress to a (time-dependent) permanent strain.) Viscous flow is put to good use in long-chain polymers; for example, they are formed into useful shapes when they are viscous. Long-chain polymers, thus, are inherently recyclable. During their reclamation, they are heated to a temperature in the viscous region and shaped into a suitable part (e.g., plastic knives and forks). Viscous flow in an epoxy is limited relative to that of a long-chain polymer. In a practical sense this prevents recycling of thermosets for they are prone to decompose before they flow to an extent permitting them to be reformed. Similar considerations apply to conventional chemically cross-linked elastomers. Because of the inability of conventional elastomers to be recycled, other types of elastomers—thermoplastic elastomers (TPEs)—have been

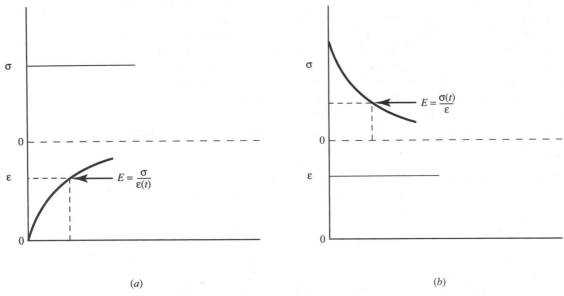

(a) (b)

Figure 2.12
Schematic of methods used to determine the "modulus" of a viscoelastic solid. In (a) a finite stress is applied at $t = 0$ and held constant. The strain is then measured and a modulus defined as the ratio of the applied stress to the strain. Since the strain is a function of time, so is the modulus. (b) Shows how the modulus can be measured by a stress-relaxation test. In the test, a specified strain is applied and the stress required to maintain this strain is monitored as a function of time. This stress decreases with time as does the "modulus," which is again determined by the ratio of the stress to the strain.

developed. These materials are thermoplastic "alloys." The major component of the chain in a TPE is a thermoplastic (e.g., polyisobutylene or polybutadiene) possessing the characteristics (i.e., highly coiled and in constant motion) of a chain in a conventional elastomer. However, the chains are not chemically cross-linked in a TPE. The "elastomer-like" portions of the chain are separated by segments of another thermoplastic having a glass transition temperature higher than the intended temperature use of the elastomer. These stiff, or glassy, segments segregate into domains as indicated in Fig. 2.13, which illustrates the structure of a styrene-butadiene TPE. Here the butadiene is the rubbery phase and styrene the glassy one. When stretched, the chains in a TPE uncoil, just as they do in a conventional elastomer. The glassy constituent of the chain remains rigid, however, and this prevents

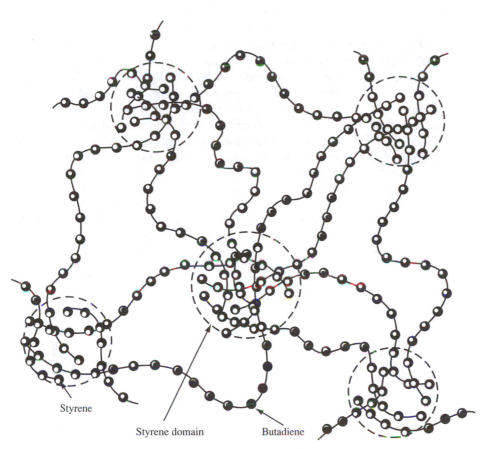

Styrene

Styrene domain   Butadiene

Figure 2.13
Schematic of a specific thermoplastic elastomer (SB, styrene-butadiene). This elastomer is a "block" copolymer (or "block" polymer alloy). Chains of butadiene—a rubbery material—are terminated by blocks of styrene; styrene is a glassy material over the intended range of elastomeric behavior. The blocks of styrene form domains which effectively serve as pinning points for the butadiene elements when the elastomer is extended. (*From* The Science and Engineering of Materials, 3rd ed., by D. R. Askeland. © 1994. Reprinted with permission of Brooks/Cole Publishing, a division of Thomas Learning. *PWS, Boston, 1994.*)

permanent interchain displacement; that is, the glassy phase acts as a "physical," as opposed to chemical, cross-link. Thermoplastic polymers can be recycled by heating them above the viscous flow temperature of the glassy phase. Because TPE properties are comparable to those of ordinary elastomers and because of their ability to be recycled, they are being increasingly used.

Discussion of viscous behavior (and of TPEs) represents a diversion. We now return to viscoelastic deformation. As noted, viscoelasticity refers to time-dependent elastic deformation. It is observed in a number of metallic and non-metallic materials, but is most important in polymers. The time dependence of viscoelastic behavior is frequently related to atomic displacements that can also cause permanent deformation. However, during viscoelastic behavior these displacements disappear on load removal.

A mechanical model that phenomenologically describes viscoelasticity, and which can also be related to structural considerations, is shown in Fig. 2.14. This *Voigt* model consists of two components arranged in parallel. One is a linear elastic member for which stress is related linearly to strain. The other is a viscous element (or dashpot) for which stress is related linearly to *strain rate* (cf. Eq. (1.14)).

We consider the response of this model when a tensile stress (σ) is instantaneously applied at $t = 0$, held there for a time t, and then removed (Fig. 2.15). In the Voigt model, the strains experienced by both elements are equal, but the stresses on them are not. The stress is partitioned in a time-varying manner between the elastic member and the dashpot. Since the viscous member requires a finite time for it to experience strain, it does not extend on initial application of the load, but rather momentarily supports the whole of this load. (At $t = 0$ the stress on the elastic member must be zero; otherwise it would experience a finite strain.) The initial stress on the dashpot imparts to it a finite strain rate, and the initial slope of the strain-time curve of Fig. 2.15 can be related to the dashpot viscosity (Example Prob. 2.2). As the dashpot elongates, some of the load is transferred to the spring.

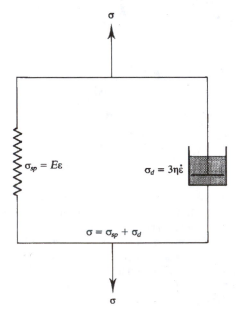

Figure 2.14
A Voigt model can be used to mimic a viscoelastic material. The model consists of a linear elastic and a viscoplastic member in parallel. For the former, strain and stress are linearly related; for the latter, strain rate varies linearly with stress. While the strains experienced by the spring and dashpot are equal, the stresses are not and vary with time as well.

With increasing time, this load transference continues, and is manifested by a con-
tinuously decreasing strain rate ($\dot{\varepsilon} = d\varepsilon/dt$, the instantaneous slope of the strain-time
curve). At very long times, the transfer is complete and the spring carries all, and
the dashpot none, of the applied stress. This limiting strain (Fig. 2.15) is determined
as σ divided by the elastic component modulus.

On removal of the stress, the process is reversed. Since the dashpot cannot instantaneously return to its initial length, the strain does not instantaneously return to
zero (Fig. 2.15). The finite positive strain indicates that the spring experiences a
tensile force, whereas the negative strain rate ($\dot{\varepsilon} < 0$) shows that the dashpot is in
compression. The algebraic sum of the stresses on the two components is equal to
zero, the new value of the "applied" stress. As the process continues, the magnitudes of the compressive and tensile stresses and the strain rate decrease, and the total strain is eventually recovered. Since the strain is recovered fully, it is elastic
strain; since the strain is a function of time, it is a viscoelastic strain.

Long-chain polymers in the vicinity of their glass transition temperature are capable of viscoelastic behavior, as discussed previously. The viscous flow component is involved with stress-assisted interchain sliding, which relates to chain
molecular architecture. Various molecular side-groups and branches along the
chains produce a geometrical (steric) hindrance to relative chain displacements. Although the applied stress is the force causing such displacements, they are aided by
raised temperature, which produces vibration and motion of the chains and their
side-groups. In effect, steric hindrance is periodically removed when, so to speak,
one of the side-groups moves "out of the way" as a result of thermal vibrations. As
temperature is important to this process, in terms of the model of Fig. 2.14, the
dashpot viscosity decreases with increasing temperature. Thus, at higher temperatures, the initial strain rate increases, and the time to complete the load transfer from
the dashpot to the spring decreases.

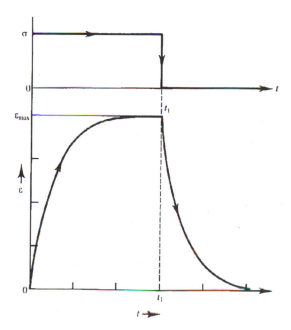

Figure 2.15
**The strain-time relationship
for a Voigt model when a constant stress is applied at $t = 0$
and held for a time t_1. The
strain-time behavior is nonlinear. However, the strain is
elastic since it is recovered
fully when the load is removed.**

The mechanism responsible for strain recovery on load removal is similar to that discussed for elastomers. On unloading, the polymer chains tend to return to their original configuration. However, in the absence of cross-linking common to elastomers, the springback does not occur more or less instantaneously but rather by the same viscous processes responsible for the original material extension.

While reasonable and physically descriptive, the processes just described only partially represent those occurring in long-chain polymers for $T \geq T_g$. In the absence of cross-linking, viscous deformation is frequently so extensive that the polymer is incapable of returning to its original dimensions on load removal (recall the viscous flow region for polystyrene in Fig. 2.11a). A more accurate model for simulating mechanical behavior under these conditions is illustrated in Fig. 2.16. In this "standard linear solid" a spring and a dashpot in series are connected to a Voigt model. The second spring simulates the limited linear elasticity that occurs during loading. The permanent deformation observed after load removal is simulated by the additional dashpot. The strain response as a function of time for a standard linear solid under the stress-time conditions shown in Fig. 2.15 is given in Fig. 2.17. The initial linear elastic response (ε_s) on both loading and unloading is due to the spring in series. The strain occurring for $t > 0$ is part viscoelastic and part viscoplastic. The viscous strain found at long times (and at a constant strain rate) is indicative of continued chain displacement with viscosity characteristic of the second dashpot. On unloading, linear elastic strain is immediately recovered and the viscoelastic strain is sluggishly recovered. Some permanent viscoplastic strain remains, however.

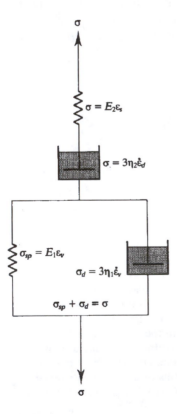

$\sigma = E_2 \varepsilon_s$

$\sigma = 3\eta_2 \dot{\varepsilon}_d$

$\sigma_{sp} = E_1 \varepsilon_v$

$\sigma_d = 3\eta_1 \dot{\varepsilon}_v$

$\sigma_{sp} + \sigma_d = \sigma$

Figure 2.16

A standard linear solid consists of a Voigt model in series with additional spring and viscous elements. The total strain is the sum of the Voigt element strain and the strains of the added spring and dashpot. The stress on both of these additional elements is equal to the applied stress; their strains, however, are not equal.

EXAMPLE PROBLEM 2.2. a. The strain-time behavior of the Voigt model (Fig. 2.15) can be described analytically. Let a stress, σ, be applied at time $t = 0$ and held constant thereafter. The stress is partitioned between the elements such that $\sigma_{sp} + \sigma_d = \sigma$ ($\sigma_{sp,d}$ = stress carried by spring and dashpot, respectively), with the strain and the strain rate for the dashpot and spring being equal. Realizing that $d\sigma/dt = 0$, determine the time variation of σ_{sp}, σ_d, $\dot{\varepsilon}$, and ε.
b. Assume the Voigt element has been extended to its asymptotic strain. Remove the stress and repeat the exercise of part (a).

Solution. a Since stress is constant, $d\sigma/dt = 0 = d\sigma_{sp}/dt + d\sigma_d/dt$. Further, $\sigma_{sp} = E_1\varepsilon$ and $\sigma_d = 3\eta_1\dot{\varepsilon}$. Therefore we have the following equation for the variation in strain rate with time;

$$E_1\dot{\varepsilon} + 3\eta_1\left(\frac{d\dot{\varepsilon}}{dt}\right) = 0 \qquad (a)$$

Integrating this equation from $\dot{\varepsilon} = \dot{\varepsilon}_0$ to $\dot{\varepsilon}$, where $\dot{\varepsilon}_0$ is the initial strain rate we get

$$\dot{\varepsilon} = (\dot{\varepsilon}_0)\exp[-E_1t/3\eta_1] \qquad (b)$$

What is $\dot{\varepsilon}_0$? At $t = 0$, the dashpot carries all of the applied stress; thus, $\dot{\varepsilon}_0 = \sigma/3\eta_1$ and

$$\dot{\varepsilon} = \left(\frac{\sigma}{3\eta_1}\right)\exp[-E_1t/3\eta_1] \qquad (c)$$

Integrating (c) from $t = 0$ to $t = t$ gives the strain as a function of time;

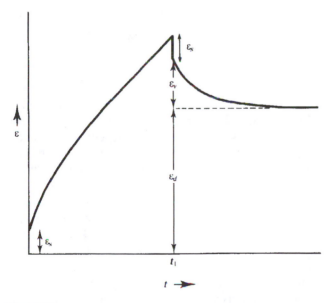

Figure 2.17
The strain-time behavior of a standard linear solid for a constant stress applied at $t = 0$ and held to $t = t_1$ when it is removed. The initial strain (and that likewise recovered on unloading) is that of the spring in series. Time-dependent strain for $t > 0$ is due to the viscoelastic Voigt element and the second viscoplastic dashpot. The former strain (ε_v) is recovered on unloading; the latter (ε_d) is not.

$$\varepsilon = \left(\frac{\sigma}{E_1}\right)(1 - \exp[-E_1 t/3\eta_1]) \qquad (d)$$

The stress carried by the dashpot is $3\eta_1$ multiplied by $\dot{\varepsilon}$; the stress carried by the spring is E_1 multiplied by ε. Thus,

$$\sigma_{sp} = \sigma(1 - \exp[-E_1 t/3\eta_1]) \text{ and } \sigma_d = \sigma \exp[-E_1 t/3\eta_1] \qquad (e)$$

Note that all of the stress is initially borne by the dashpot and none of it by the spring. At very long times, the opposite holds. In addition, $\dot{\varepsilon}_0$ is determined solely by the dashpot viscosity and the stress. The strain rate decreases with time, eventually approaching zero. The initial strain is zero and the limiting strain—corresponding to the spring carrying all of the stress—is σ/E_1.

b The governing differential equation for $\dot{\varepsilon}$ remains as in part (a). Thus, only the initial conditions are altered, along with the new stipulation that $\sigma_d + \sigma_{sp} = 0$. Let t' be the time measured from when the load is released. The equation for $\dot{\varepsilon}$ is the same as in part (a); however, the strain rate is now negative. The new strain is the negative of Eq. (d) above, with the addition of the initial strain ($= \sigma/E_1$) to it. So for unloading

$$\dot{\varepsilon} = -\left(\frac{\sigma}{3\eta_1}\right)\exp[-E_1 t'/3\eta_1]; \quad \varepsilon = \left(\frac{\sigma}{E_1}\right)\exp[-E_1 t'/3\eta_1]$$

The dashpot is in compression (its strain rate is negative). The spring remains in tension so long as the strain is finite. Thus,

$$\sigma_{sp} = \sigma \exp[-E_1 t'/3\eta_1]; \sigma_d = -\sigma \exp[-E_1 t'/3\eta_1]$$

Note that the total stress is zero, the strain rate is negative and that it (and the strain) gradually approach zero.

2.8
MECHANICAL DAMPING

Mechanical damping, related to time-dependent elasticity, can lead to mechanical hysteresis. Under the action of a cyclically varying stress, this hysteresis is manifested by a strain that "lags behind" an applied stress. Time-dependent elastic behavior of this type gives rise to the term *damping*, for when a material is subjected to, for example, an impulse stress, the resulting strain vibration is eventually dampened out. Such a damping capacity is, in part, responsible for the use of cast iron in lathe beds, because cast iron quickly dampens stress pulses. On the other hand, the resonating tone of a struck choir bell is a direct result of the low damping capacity of the bell material.

The hysteretic behavior of a viscoelastic material can be modeled using the standard linear solid model without the dashpot in series. Eliminating the dashpot removes the viscoplastic component of deformation, as illustrated in Fig. 2.18, which represents the strain-time variation as a result of a stress applied at $t = 0$ and held there for a time ($t = t_1$) sufficient to produce the equilibrium Voigt component extension. If elastic modulus is defined as the instantaneous ratio of stress to strain, it can be seen that two limiting values of the modulus can be defined. The *unrelaxed modulus*, E_u, calculated at $t = 0$, is defined on the basis of the initial linear elastic strain. The unrelaxed modulus is greater than the *relaxed modulus*, E_r, which

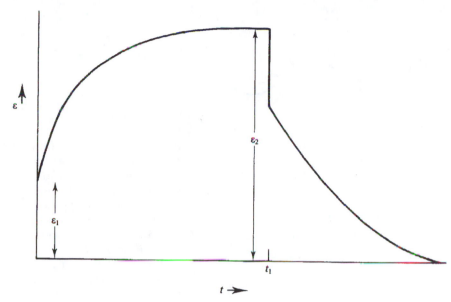

Figure 2.18
The strain-time response to a constant stress, applied at $t = 0$ and held to $t = t_1$ and then removed, for a standard linear solid without the second viscous element. All of the strain is elastic; some of it is viscoelastic. If the elastic modulus is defined as the stress divided by the (instantaneous) strain, the modulus lies between two extremes: the unrelaxed modulus $E_u = \sigma/\varepsilon_1$ and the relaxed modulus $E_r = \sigma/\varepsilon_2$. On unloading, the strain returns to zero since it is elastic.

is based on the equilibrium strain. The unloading curve, after the stress is removed (Fig. 2.18), consists of immediate recovery of the linear elastic strain and time-dependent recovery of the viscoelastic strain. The time required to reduce the strain to zero is the same as that needed to produce the equilibrium extension during loading.

The stress-strain behavior, under a cyclically varying stress such as that illustrated in Fig. 1.22, of a material of the type considered depends on the frequency of the applied stress. If this frequency (ν) is so great that ν^{-1} (which has dimensions of time) is much less than the "time constant" of the viscoelastic medium, the material does not respond viscoelastically to the stress. In such a case the stress and strain are "in phase" (Fig. 2.19a) and a stress-strain plot (Fig. 2.19b) during cyclic loading is characteristic of a typical linear elastic material with a modulus equal to the unrelaxed modulus. If, on the other hand, a time-varying stress of very low frequency is applied, then ν^{-1} is much greater than the time required to effect the full viscoelastic component (i.e., ν^{-1} is much greater than t_1, Fig. 2.18). In this case, too, the strain is in phase with the stress and a plot of stress vs. strain will be linear (although, of course, this does not here reflect linear elasticity) with a lower modulus, E_r (Fig. 2.19b).

In the intermediate case, the applied frequency is comparable to the inverse of the viscoelastic time constant. When this occurs, the viscoelastic deformation is unable to "keep up"—i.e., to be in phase—with the stress, and the stress and strain

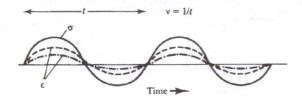

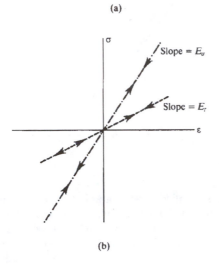

(a)

(b)

Figure 2.19
(a) The variation of strain with time for the time-varying stress shown. In the cases depicted, one strain (—•—•) corresponds to an applied stress frequency so great that the strain is characteristic of the unrelaxed modulus ($\varepsilon = \sigma/E_u$). The other case (---) corresponds to a very low frequency so that $\varepsilon = \sigma/E_r$. (b) A cross-plot of σ vs. ε for these two cases. The stress-strain curves are linear with moduli of E_u and E_r, respectively.

vary with time as shown in Fig. 2.20*a*. When the stress-strain curve is plotted (Fig. 2.20*b*), we see that it is path-dependent; the stress-strain curves on loading and unloading differ. The "average" modulus as defined, for example, by the ratio $\sigma_{max}/\varepsilon_{max}$ in Fig. 2.20*b* lies between E_r and E_u since some, but not all, of the potential viscoelastic strain is manifested at the intermediate frequency. The area enclosed by the unloading and loading curves (the cross-hatched region in Fig. 2.20*b*) represents an irreversible (hysteretic) energy loss per stress cycle. Analogous phenomena are observed in magnetic materials when the internal magnetic field does not respond instantaneously to an applied external field, and in electrical capacitors subjected to time-varying electrical fields. In all cases, the irreversible energy loss heats the material unless some external means to cool it are provided. That irreversible *permanent* deformation is accompanied by internal heating is easily ascertained. Hammering thin a metal is accompanied by a discernible increase in its temperature. That hysteretic *elastic* deformation also leads to heat generation can be demonstrated by touching the (warm) sidewall of an automobile tire after a drive of moderate length. In this case, the tire deformation is assuredly elastic. The above ambient temperature of the sidewall is caused by the viscoelastic response of some components of the tire.

Polymers and, surprisingly, metals and other crystalline materials display elastic mechanical hysteresis. In polymers, this behavior is due to viscoelasticity of the type discussed previously. Elastic hysteresis can be considerable in polymers capable of large viscoelastic deformations and in which, therefore, E_r and E_u differ appreciably. Indeed the energy loss per cycle can be correlated with the ratio E_r/E_u

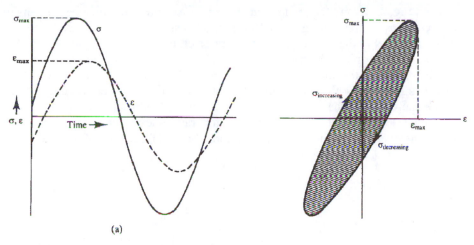

Figure 2.20
(a) The strain response to a cyclical stress when the applied stress frequency is comparable to the "natural" frequency of the viscoelastic material. In this case the strain "lags" behind the stress and the ratio of $\sigma_{max}/\varepsilon_{max}$ lies between E_r and E_u. (b) A crossplot of stress vs. strain from (a). The stress-strain behavior is different on loading ($\sigma_{increasing}$) and unloading ($\sigma_{decreasing}$). The area enclosed by the σ-ε curve represents hysteretic elastic work performed each cycle, and is manifested by heat generation.

(Prob. 2.21); the lower this ratio, the greater the energy loss per cycle. Metals and their alloys can be considered similarly. However, their characteristic values of E_r/E_u are sufficiently close to unity that only small *fractional* energy losses occur during stress cycling. (The fractional energy loss is the hysteresis energy loss per cycle divided by the elastic energy stored in the material during deformation; it thus represents a ratio of irreversible to recoverable elastic energy.) Because the energy loss for a metal is generally much less than in a polymer, this behavior in metals is distinguished from similar behavior in polymers. *Anelastic* deformation is the term used to describe the time-varying, hysteretic elastic behavior of metals.

It is worthwhile to examine briefly the structural and atomistic features responsible for metallic anelasticity since, according to previous discussions, metal elasticity is fundamentally linear and elastic hysteresis should not be expected. In fact, in a polycrystalline material, atoms within grain interiors do indeed undergo only linear elastic behavior. However, atoms at grain boundaries are less regularly arranged and can experience peripheral atomic displacements, essentially viscous in nature, in response to an applied stress. That is, atoms in grain boundaries are capable of moving relative to one another in much the same way that polymer chains are in a long-chain polymer. At high temperatures this grain-boundary sliding can be extensive and lead to viscoplastic strain. At ordinary temperatures, however, the boundary atomic displacements are fractions of an atomic diameter, and are viscoelastic displacements because they are recovered on load removal.

Transition metal, body-centered cubic interstitial solid solutions also demonstrate mechanical damping. This is caused by the stress-assisted preferential occupancy of one set of interstitial sites. The unstressed body-centered cubic unit cell is shown in Fig. 2.21*a* and the (exaggerated) response of the cell lattice parameters to a uniaxial tensile stress along a <001> direction is shown in Fig. 2.21*b*. The unit-cell axis parallel to the stress direction is preferred for interstitial atom occupancy as a result of the greater interstitial volume along this axis. Whether such atoms avail themselves of this more spacious environment depends on the frequency of the applied stress. If it is high enough that the atoms have insufficient time to reposition themselves along the preferred axis, anelasticity is not observed, and vice versa. Thus, the onset of anelasticity at a critical (temperature-dependent) stress frequency yields fundamental information regarding the vibration frequency of the interstitial atoms.

2.9
SUMMARY

In this chapter the elastic behavior of a variety of materials has been discussed. Linear elasticity results from forces between atoms in the condensed state. Linear elastic moduli are related to the slope of the interatomic force–separation (strain)

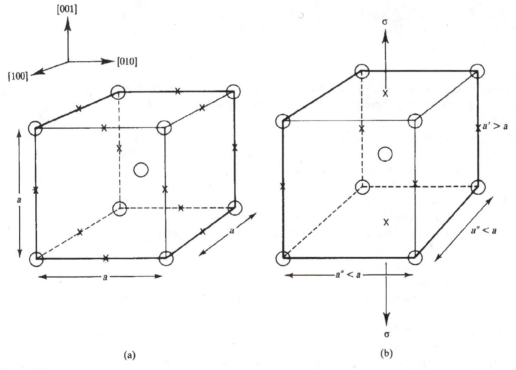

(a)   (b)

Figure 2.21
(a) An unstressed body-centered cubic unit cell. The crosses denote octahedral sites along unit cell edges. These are occupied by interstitial atoms, as are other octahedral sites (not designated) at the center of the cell faces. In the unstressed state, all sites have equal probability of being occupied by interstitial atoms. (b) Under stress, the sites along [001] (the direction of the applied tensile stress) become preferred occupancy sites as a result of the increased interstitial volume there. Thus, interstitial atoms tend to jump into these sites from sites along adjacent [100] and [010] axes. If the stress applied is cyclical, and of frequency comparable to the natural vibration frequency of interstitial atoms, mechanical hysteresis results.

relationship. Alternatively, they are related to the curvature of the potential energy–interatomic separation curve. Values of elastic moduli mirror the material bond strength. Thus, metals and inorganic solids—in which atoms/ions are bonded by a primary chemical bond—have relatively high moduli. Moduli of polymers, in contrast, are fairly low as a result of the weak van der Waals bond between polymer chains and/or segments.

The interatomic separation in crystals depends on crystallographic direction and, as a result, elastic constants vary with crystallographic direction. The extent to which such anisotropy is displayed depends on crystal symmetry. Cubic materials, for example, have three independent elastic constants, one more than the two elastic constants needed to describe elastically isotropic materials. Triclinic crystals, members of the least-symmetrical crystal system, require 21 independent elastic constants to describe fully their elastic behavior. Although single crystals are inherently anisotropic, in polycrystalline aggregates the mixture of them usually results in isotropic behavior. This is due to the "averaging" of the specific elastic responses of the large number of individual anisotropic grains in them.

Rubber elasticity differs fundamentally from linear elasticity. The stress required to extend an elastomer arises from entropic, rather than potential energy, considerations. The large reversible extensions displayed by elastomers usually result from the presence of primary bond cross-links between elastomer chain molecules. The cross-links cause the elastomer to snap almost instantaneously back to its original dimensions on removal of an applied stress. The rubber stiffness, which increases with temperature, also increases as the cross-linking frequency is increased.

Most metals and inorganic solids display linear elastic behavior up to their melting points, although their moduli do decrease with temperature. Polymers become capable of extensive viscoelastic (i.e., time-dependent elastic) deformation at their glass transition temperature. This temperature is somewhat lower than would be the equilibrium melting temperature of the polymer in crystalline form. During polymer viscoelastic deformation, material extension is accompanied by time-dependent interchain sliding which is recovered on load removal. Owing to structural differences that exist between a long-chain polymer and an elastomer, this recovery is sluggish or retarded for the polymer rather than instantaneous as it is for an elastomer. Viscoelasticity and its companion in crystalline materials—anelasticity—can lead to mechanical hysteresis manifested by strain not being in phase with a cyclically applied stress. The resulting energy loss per cycle leads to heat dissipation, and the frequency and temperature dependence of the energy loss can be related to the internal structure of viscoelastic and anelastic materials.

Elastic behavior is important in itself. It is also important in material fracture. However, many other aspects of deformation and fracture are related to processes taking place during permanent deformation and this topic forms the basis for the discussions of the next several chapters.

REFERENCES

Ashby, Michael F., and David R. H. Jones: *Engineering Materials 1—An Introduction to their Properties and Applications,* Pergamon Press, Oxford, 1980.

Dieter, George E.: *Mechanical Metallurgy,* 3rd ed., McGraw-Hill, New York, 1986.

Hertzberg, Richard W.: *Deformation and Fracture Mechanics of Engineering Materials*, 4th
ed., Wiley, New York, 1996.

Meyers, Marc André, and Krishan Kumar Chawla; *Mechanical Metallurgy—Principles and
Applications*, Prentice-Hall, Englewood Cliffs, NJ, 1984.

PROBLEMS

2.1 Show that Eq. (2.2), relating the fractional change in volume to the principal tensile
strains, is valid for the small-strain condition. First do this by determining δV using
$l_1 = l_{01} + \delta l_{01}$, etc., and considering only terms to first order in $\delta l_i/l_i$. Repeat the exercise by setting $V = l_1 l_2 l_3$, taking the logarithms of both sides of the equation and differentiating the resultant expression. (Note: For small δx, $dx \cong \delta x$.)

2.2 A state of biaxial stress involving a tensile stress and a compressive stress of equal
magnitude is equivalent to a state of pure shear:

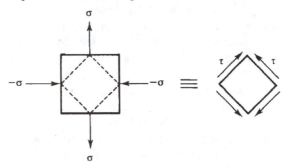

a Using the generalized equation for axial strain (Eq. (2.3)), show that for an isotropic
elastic material $G = E/2(1 + \nu)$.

b Show that a state of pure shear leads to no volume change during linear elastic deformation.

2.3 Plane stress and plane strain are important concepts, particularly in the realm of fracture. Plane stress is defined by two finite principal stress components, one principal
stress component being zero. Plane strain is defined analogously. Use Eq. (2.3) and
its analogs to show that plane stress conditions lead to three principal strain components and that plane strain conditions result in three principal stress components.

2.4 Show that the condition $m > n$ must be satisfied in Eq. (2.10) for it to describe an
equilibrium situation. (Note: Equilibrium can be obtained only if the interaction energy, u_{ij}, is a minimum.)

2.5 Using Eqs. (2.9) and (2.11), show that $K \sim (r_0)^{-4}$ for an ionic solid.

2.6 The bulk modulus is defined by Eq. (2.9). Consider an ionic solid for which $n = 2$ and
$m = 8$. Estimate the fractional change in volume required to alter the bulk modulus by
10% when this solid is subjected to hydrostatic compression.

2.7 a Linear elastic strains can be related to stress by equations having the form of Eq.
(2.19). Substitute the expressions for the σ_{ij} from Eqs. (2.14) into the above equations

to determine relationships among the respective compliance (S_{ij}) and stiffness (C_{ij}) co-efficients. (Note: your result is consistent with $[S][E] = [1]$ where $[S]$ and $[E]$ are the compliance and stiffness matrices and $[1]$ is the unit matrix.) How many independent compliance coefficients are there? Give your reasoning.

b For iron, the elastic compliances are $S_{11} = 0.757 \times 10^{-11}$, $S_{12} = -0.283 \times 10^{-11}$ and $S_{44} = 0.862 \times 10^{-11}$ (all in units of m²/N). What is the bulk modulus of Fe?

2.8 For a cubic material, there are only three independent compliances: S_{11}, S_{12}, and S_{14}. Using the results from Prob. 2.7, show that

$$C_{11} = \frac{S_{11} + S_{12}}{(S_{11} - S_{12})(S_{11} + 2S_{12})}$$

$$C_{12} = \frac{-S_{12}}{(S_{11} - S_{12})(S_{11} + 2S_{12})}$$

$$C_{44} = \frac{1}{S_{44}}$$

2.9 For iron, $C_{11} = 237$ GN/m², $C_{12} = 141$ GN/m², and $C_{44} = 116$ GN/m².

a Determine $E_{[100]}$, $E_{[110]}$, and $E_{[111]}$ of Fe.

b Suppose a polycrystalline Fe wire is composed of grains with either [100], [110], or [111] crystal directions lying along the wire axis. If all three orientations are present in equal amounts, what is E for the wire? (State your assumptions.)

c The polycrystalline elastic modulus of Fe is 209 GN/m². How does this compare with the modulus estimated in part (b)?

d By changing the processing conditions, a different kind of texture can be produced in Fe wire; 70% of the grains are oriented so that a <100> direction is aligned with the wire axis and the remaining 30% are evenly split between <110> and <111> directions lying along this axis. What is the modulus along the wire axis for this case?

e Is E transverse to the wire axis different from the modulus along the axis for the situations pertaining to parts (b) and (d)? Explain your reasoning.

2.10 Consider simple tension and shear tests, and use the relationships among G, E, and ν appropriate for isotropic materials to show that Eqs. (2.16) follow from Eqs. (2.15).

2.11 a Sketch a (001) plane in a face-centered cubic material and an arbitrary vector within it making an angle θ with the [100] direction. Plot the Young's modulus for copper as a function of θ for directions between [110] and [100].

b Sketch a (110) plane in Cu and a vector in the plane making an angle α with the [Ī10] direction. Plot E vs. α for directions between [Ī10] and [001].

2.12 Crystallographic texture can develop in rolled Cu sheet. Consider two different textures in rolled sheet. Texture A has {100} planes parallel to the plane of the sheet and <001> directions parallel to the rolling direction. Texture B has {110} planes parallel to the plane of the sheet and <001> directions parallel to the rolling direction.

If tensile samples are obtained by cutting at different angles, θ, to the rolling direction in the plane of the sheet ($\theta = 0$ corresponds to the rolling direction) and one sample is obtained with the tensile axis perpendicular to the plane of the sheet, estimate the values of the modulus for the sample perpendicular to the sheet and for samples cut at $\theta = 0$, 45°, and 90°. Do the exercise for both textures. (Refer to Table 2.3 for data.)

2.13 Compare the value of the anisotropy ratio ($C_{11} - C_{12}/2C_{44}$) to the values of E_{max}/E_{min} for materials whose properties are given in Tables 2.2 and 2.3.

2.14 Schematically sketch how sulfur, with two bonding electrons, can lead to cross-linking between polymer chains in natural rubber.

2.15 a Show that Eq. (2.23) is valid for an ideal gas.
b The equation of state for a van der Waals gas is

$$\left(p + \frac{a}{V^2}\right)(V - b) = RT$$

where the coefficients a and b represent the *small* interaction energy between gas molecules. (That is $a/V^2 \ll p$ and $b \ll V$.) Does a van der Waals gas satisfy Eq. (2.23)? If not, is the deviation from it a large or a small one?

2.16 a Using Fig. 2.10, determine the instantaneous elastic modulus (the slope of the stress-extension ratio curve) at an extension ratio of eight ($\varepsilon = 7$). How does this value of "E" compare with linear elastic moduli of crystalline materials (Table 2.3)?
b For a natural rubber $\rho \cong 1200$ kg/m³. Calculate a density-compensated modulus (E/ρ) of rubber at $\varepsilon = 7$ and compare with E/ρ for some of the crystalline materials listed in Table 2.3. Data: Take $M_c = 5000$; $\rho_{Al} = 2700$ kg/m³; $\rho_{Fe} = 7870$ kg/m³; $\rho_W = 19{,}300$ kg/m³; $\rho_{MgO} = 3580$ kg/m³; $\rho_{NaCl} = 2230$ kg/m³.

2.17 Consider the thermoplastic elastomer illustrated in Fig. 2.13. How could you manipulate the modulus of this elastomer by altering the relative amounts of styrene and butadiene in it?

2.18 a The additional spring and dashpot in series with the Voigt model of Fig. 2.14 constitute a standard linear solid (Fig. 2.16). The spring and dashpot in series alone are referred to as a Maxwell model. If a stress σ is applied at time $t = 0$ and held constant, sketch the strain-time response expected for a Maxwell model.
b Release the load in part (a) some time after its application. Sketch strain vs. time following this unloading.
c Show that the addition of the strains drawn in (a) and (b) to those schematized in Fig. 2.15 leads to the response illustrated in Fig. 2.17.

2.19 a Apply a strain ε to the Maxwell model (cf. Prob. 2.18). Determine how the stress required to keep this strain constant varies with time. (The result is an example of a stress-relaxation process.) (Hint: Since they are in series, the stress on the dashpot and on the spring are equal. The applied strain is the sum of the strains of the spring and the dashpot.)
b During relaxation, stress varies with time as $\sigma \sim \exp(-t/\tau)$. What is τ in terms of the properties of the elements in the Maxwell model? How does this "time constant" compare with that for strain relaxation of the Voigt model (Example Prob. 2.2)?

2.20 Refer to Fig. 2.20b. Schematically plot $\sigma_{max}/\varepsilon_{max}$ as a function of the applied frequency. (σ_{max} is the maximum stress observed during the hysteresis cycle and ε_{max} is the maximum strain; σ_{max} typically is not found when $\varepsilon = \varepsilon_{max}$, cf. Fig. 2.20b.)

2.21 a The standard linear solid without the second viscous element can be regarded as a spring (s) in series with a Voigt element (v). Let a stress $\sigma = \sigma_0 \sin 2\pi\nu t$ be applied to such a model. Determine the strain response ($\varepsilon = \varepsilon_s + \varepsilon_v$) as a function of time.

b How does ε_{max} compare to ε_1 and ε_2 of Fig. 2.18? (Consider a large range of frequencies.)

c The energy loss per cycle is the area enclosed by the hysteresis loop shown in Fig. 2.20*b*. Determine the energy loss as a function of ν divided by the "natural" frequency of the Voigt model. (Note: The latter is equal to the inverse of the Voigt-model time constant, see Example Prob. 2.2 and Prob. 2.19.) Correlate the energy loss with the ratio, E_r/E_u.

2.22 The microstructure of certain ceramics consists of a crystalline phase surrounded by a glassy one. A two-dimensional model which mimics this arrangement is sketched below.

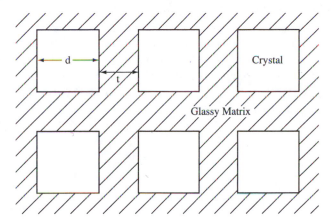

Over a selected temperature range, the glassy phase deforms viscously while the crystalline solid deforms only via linear elasticity. Derive a constitutive equation which describes the strain rate response of this material when it is subjected to a stress σ at $t = 0$ and is held constant thereafter. Your answer should be expressed in terms of the stress, the time, the elastic modulus of the crystalline phase, the viscosity of the glassy phase *and* the volume fraction of the crystalline phase ($= d^2/(d + t)^2$ in this two-dimensional model). Schematically sketch the strain-time response of the material, illustrating the key concepts of your answer. (Hint: This is a problem similar to those immediately preceding. However, there is one important qualification. While the "force" in a parallel element is partitioned according to the equal strain condition, the force partitioning ratio depends on the respective phase (element) volume fractions. You may wish to consider reading the opening sections of Chap. 6 for further clarification.)

2.23 a The anelastic energy loss of a metal is some small fraction of the total elastic energy stored ($= E\varepsilon^2/2$) during cyclical elastic straining. For grain-boundary anelasticity, the fraction is related to the ratio of the number of atoms located within grain boundaries compared to the number of atoms in grain interiors. Assume a metal composed of grains having the shape of cubes. What is the ratio of the number of grain boundaries to interior atoms for grain sizes of 0.1 μm, 1 μm and 100 μm? (You may wish to assume a particular crystal structure to facilitate calculations, but it is not necessary to do so to obtain reasonable estimates.)

b On the basis of the above results, do you expect anelastic grain-boundary energy loss to be a strong function of grain size?

2.24 In order for carbon atoms in iron to cause anelasticity, they must be able to move interatomic distances in response to a cyclical stress. Estimate the temperature dependence of the stress frequency below which this kind of anelastic behavior will not be found. To do so, assume that the time (t) required for an atomic jump is $t = x^2/D$, where x is the jump distance between interstitial octahedral sites in body-centered cubic iron, and D is the diffusion coefficient of carbon in bcc iron. Data: lattice parameter of bcc iron = 0.290 nm; $D = 7 \times 10^{-7} \exp(-10,060/T)$ m^2/sec.

CHAPTER 3

Dislocations

3.1
INTRODUCTION

The stress required to cause permanent deformation is a concept important to material mechanical behavior. In crystalline materials, this plastic flow is related to the presence, and response to an applied stress, of certain crystallographic defects called *dislocations*. For example, the presence of dislocations explains the low stresses required to cause flow in crystals containing them compared to the much higher stresses necessary for this in crystals not containing dislocations.

In this chapter the properties of dislocations are described, and their role in plastic deformation is emphasized. The topic of dislocations is sufficiently important and their behavior is so intriguing that entire books have been devoted to them. In this chapter we not only describe the characteristics of dislocations that are directly germane to topics in following chapters, but also emphasize some of their other features. This is warranted in view of the key role dislocations play in crystal plasticity. Moreover, treatment of additional topics helps to develop a better "feel" for the characteristics of dislocations, thereby facilitating a deeper understanding of plastic deformation processes. Thus, we discuss not only the structure and crystallography of dislocations, but also their energies and associated stress fields, since these stress fields influence plastic flow behavior. A plastically deforming material is, moreover, a dynamic system. The number and morphology of dislocations changes within a material as plastic strain increases, and this is manifested by a flow stress that varies with accumulated plastic strain. It is the purpose of this chapter to illustrate how dislocations are responsible for plastic flow, and then to describe the changes taking place in a crystal as flow proceeds. In doing so it establishes the basis for describing the work-hardening behavior of crystals. Additionally, a good understanding of the concepts presented in this chapter is requisite for appreciation of the means by which resistance to plastic flow is increased by the presence, for example, of impurity atoms and second-phase particles. These and other facets of plastic flow are discussed in chapters following this one.

3.2
THE YIELD STRENGTH OF A PERFECT CRYSTAL

The yield strength of a "perfect" crystal is considerably greater than the stresses commonly associated with plastic flow initiation in most crystalline solids. For example, a reasonably sized pure single crystal of tungsten, the most refractory of the metals, can be bent permanently by hand. The dichotomy between "theory" and experiment is resolved by the dislocation concept. But we first consider the problem of just how strong a perfect crystal should be.

To estimate this, we assume that atomic sliding on slip planes is the cause of plastic flow. Such an ideal shearing process is illustrated schematically in Fig. 3.1a. For the atoms in the upper row (plane) to slide over those in the lower one, strong interatomic forces must be overcome by the applied stress. The variation in crystal

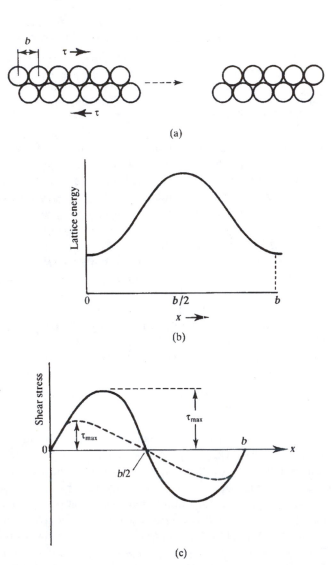

Figure 3.1
The process of slip in a perfect crystal. In (a), the shear stress acts to displace atoms in the upper row (plane) by the interatomic separation distance, b, with respect to the atoms in the lower plane. In (b) the system energy is plotted schematically vs. the displacement across the planes. Lattice energy is a maximum at the position $x = b/2$. In (c) the stress required to effect the displacement, proportional to the derivative of the energy-distance curve, is shown. The maximum stress is the theoretical shear strength. The solid curve is appropriate to the sinusoidal variation of energy with position shown in (b). The dotted line, which results in a lesser value of τ_{max}, more accurately represents the situation in a real material.

energy with relative atomic displacement across the rows is shown in Fig. 3.1b. When the atoms in the upper row have been displaced by one-half of their transit distance (b/2), the crystal energy is at a maximum and atoms in the upper row are, in the absence of an applied stress, as likely to return to their initial position as to complete the slip transit. The internal restoring stress, which the applied stress must overcome, is related to the derivative of the energy-distance curve (cf. Chap. 2). The shear stress required to produce the atomic displacements described is indicated in Fig. 3.1c. The negative stress at displacements greater than b/2 indicates that for such displacements the lattice force acts to complete the slip process. The applied stress required to overcome the lattice resistance to shear is indicated by τ_{max}; it occurs at a displacement approximately equal to b/4. This stress represents the theoretical shear strength and it can be estimated by assuming that the stress-displacement curve varies in a sinusoidal manner, i.e.,

$$\tau = \tau_{max} \sin\left(\frac{2\pi x}{b}\right) \tag{3.1}$$

At low (elastic) values of shear strain, Eq. (3.1) must satisfy $\tau = G\gamma$ or, alternatively, $d\tau/d\gamma = G$. On differentiating Eq. (3.1),

$$\frac{d\tau}{dx} = \frac{2\pi}{b} \tau_{max} \cos\left(\frac{2\pi x}{b}\right) \tag{3.2a}$$

and

$$\left(\frac{d\tau}{dx}\right)_{x=0} = \frac{2\pi}{b} \tau_{max} \tag{3.2b}$$

For small displacements, $\gamma = x/a$ (where a is the slip plane spacing) and on using $d\tau/dx = (d\tau/d\gamma)(d\gamma/dx)$, we obtain

$$\left(\frac{d\tau}{d\gamma}\right)_{x=0} = \frac{2\pi a}{b} \tau_{max} \tag{3.3}$$

On setting Eq. (3.3) equal to G,

$$\tau_{max} = \frac{G}{2\pi} \frac{b}{a} \tag{3.4}$$

In Eq. (3.4), b (the slip distance) and a (the spacing between slip planes) are comparable. Thus, the theoretical shear strength is estimated to be on the order of $G/2\pi$. This estimate is somewhat high, since the asymmetry of the force–atomic separation curve is pronounced at the large strains occurring in the slip process (e.g., $\tan \gamma \cong 0.5$ at $x = b/2$). A more realistic stress-displacement curve, which takes into account this asymmetry, is shown in Fig. 3.1c. Use of this more refined approach lowers the estimate of the theoretical strength to $\tau_{max} \cong G/30$. Even this is much higher than shear yield strengths observed for single crystals of pure materials. For example, $G/30$ for aluminum is ca. 9×10^8 N/m^2, whereas the observed shear yield strength is 7.8×10^5 N/m^2. Values of observed and theoretical shear strengths for other materials are compared in Table 3.1. In all cases the discrepancy between the two values is so great that there is no doubt that atomic shear, by which plastic deformation occurs in crystalline solids, does not take place by the process just described. Instead, it occurs by the movement of structural defects called dislocations,

Material	τ_{th} $(= G/30)$ $(10^9$ N/m$^2)$	τ_{exp} $(10^6$ N/m$^2)$	τ_{exp}/τ_{th}	τ_f $(10^6$ N/m$^2)$*
Ag	1.0	0.37	0.00037	20
Al	0.9	0.78	0.00087	30
Cu	1.4	0.49	0.00035	51
Ni	2.6	3.2	0.0070	121
α-Fe	2.6	27.5	0.011	150

*Overestimate, based on narrow dislocation width, cf. Eq. (3.5).

Table 3.1
Values of the theoretical, experimental, and frictional yield strengths for several materials

the existence of which was postulated in the 1930s to explain the discrepancies between experimentally observed and theoretically calculated shear strengths. Further credence to the dislocation explanation of plastic deformation, as well as to the essential correctness of the theoretical shear strength calculations, was obtained when whiskers—dislocation-free crystals—were manufactured. The yield strengths of whiskers are much greater than those of ordinary materials, and within a factor of 10 of their theoretical shear strengths.

3.3
THE EDGE DISLOCATION

An appreciation of the atomic arrangements in the vicinity of the defects causing plastic flow is necessary to understand how they facilitate atomic slip. Thus, we begin our discussions on dislocations by describing these arrangements in the vicinity of a line defect called the edge dislocation, and then show how slip is caused by motion of this defect.

A. Slip by Edge Dislocation Motion

A sketch of an edge dislocation is shown in Fig. 3.2.[1] The defect can be considered an additional partial plane (commonly called an extra "half" plane) of atoms inserted into the upper portion of the crystal and terminating on a {100} plane. The means by which atomic displacements are achieved by edge dislocation motion are shown in Fig. 3.3. Here a shear stress is applied on the top and bottom faces of the crystal so as to produce shearing forces in a <100> direction. Inspection of the atomic arrangement in the vicinity of the partial plane termination shows that, in the absence of an applied stress, the atom (atom A) at the termination is equally attracted to atoms B and C in the plane below it. Application of the stress alters this condition so that atom A is attracted preferentially to atom C. As a result of this,

---

[1]The simple cubic structure is used here to illustrate. Although the structure is a hypothetical one, it serves to describe the physics of dislocation structure and motion, and is used in preference to other structures for which the atomic arrangements in the vicinity of a dislocation are not so easily visualized.

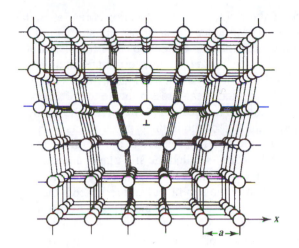

Figure 3.2

**A schematic of an edge disloca-
tion, represented by a partial
atomic plane in a simple cubic
structure. The "core" of the
dislocation is localized at the
partial plane termination.
Atomic positions are distorted
in the dislocation core region.**
(*Adapted from A. G. Guy and
J. J. Hren, Elementary Physical
Metallurgy, 3rd ed., Addison-
Wesley, Reading, Mass., 1974.
Reprinted with permission.*)

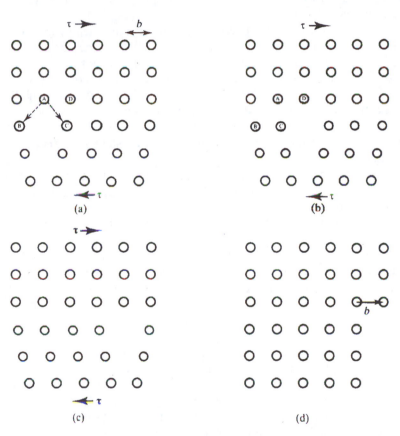

Figure 3.3

**Motion of an edge dislocation in response to a stress. (a) The stress blases atom A to
lie closer to atom C than to B. (b) As a consequence the location of the core moves to
atom D. (c) The process repeats and the dislocation moves to the right until (d) it
exits the crystal effecting a displacement, b, of the top part of the crystal with respect
to the bottom portion. (For clarity atoms in only one plane of the crystal are shown.)**

minor atomic rearrangements occur in the vicinity of the dislocation core (i.e., in the region of the partial plane termination) to produce the arrangement shown in Fig. 3.3*b*. The new atomic positions are such that the dislocation has translated one atomic position to the right; that is, the extra half plane now terminates at atom D. We emphasize that the additional partial plane has not moved as such. Rather, the small displacements resulting from the applied shear stress relocate the position of the partial plane termination. With continued application of the stress, the dislocation continues to move in the direction indicated, and in the process atoms lying on the plane of the termination are displaced to the right relative to atoms on the plane below it. The plane containing the dislocation is called the slip plane, and the slip direction is the direction of the dislocation motion. The magnitude of the slip displacement is *b*, the interatomic separation in the slip plane, and represents the displacement of atoms on this plane relative to those on the plane below it. Continued motion of the dislocation from left to right (Fig. 3.3*c*), until it emerges from the side of the crystal (Fig. 3.3*d*), produces a permanent offset of magnitude *b* of the top part of the crystal with respect to the bottom one. That is, plastic deformation has resulted from the motion of the dislocation.

Comparison of the means by which permanent deformation is achieved by dislocation motion and by perfect slip is worthwhile. For the latter, every atom in the slip plane moves simultaneously with respect to atoms in the plane below it. This concurrent motion produces a displacement (*b*) of the two portions of the crystal, and the force required for this is reflected in the theoretical shear strength. If slip occurs by edge-dislocation motion, the displacement *b* is effected only over one column of atoms at a time. It is only when the dislocation has moved the entire length of the slip plane that displacement of the two halves of the crystal by the distance *b* is achieved. Indeed, if the crystal is composed of *N* atomic columns, each time the dislocation moves a distance *b* in the slip direction, translation of the upper part of the crystal with respect to the lower part by an amount *b/N* takes place. It is this incremental nature of slip by dislocation motion that makes the stress required to produce it so much less than the theoretical strength associated with simultaneous slip of all atoms.

That the stress required to induce dislocation motion should be much less than the theoretical strength can be appreciated by closer inspection of Fig. 3.3. Indeed, intuition suggests that the dislocation motion stress should be vanishingly small, since atom A is attracted equally to atoms B and C and it would be expected that only a trivial "push" would be required to get the dislocation to "move over." However, the attractive and repulsive forces between atoms depend differently on interatomic separation (Chap. 2), and because of this the energy of the crystal increases somewhat as the partial plane termination moves from atom A to atom D. This variation in energy is illustrated schematically in Fig. 3.4 as is the stress, proportional to the derivative of the energy-distance curve, required to move the dislocation. However, this lattice *frictional stress* is much less than the theoretical strength. Peierls and Nabarro have calculated the frictional stress; it is given by

$$\tau_f = G \exp\left[\frac{-2\pi a}{(1 - v)b}\right] \tag{3.5}$$

where *G* is the shear modulus, *v* Poisson's ratio, *a* the separation distance between slip planes, and *b* the slip distance. As given by Eq. (3.5), the frictional stress is

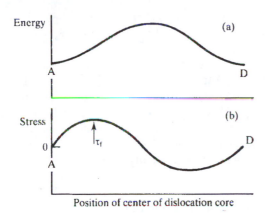

Energy

(a)

A

D

Stress

0

τ_f

(b)

D

A

Position of center of dislocation core

Figure 3.4
During edge dislocation motion (e.g., from its core being located below atom A to below atom D in Fig. 3.3), the energy of the crystal increases. The energy increase is much less than that shown in Fig. 3.1b and the associated stress required for dislocation motion (τ_f, the frictional stress) is much less than the theoretical shear strength.

low when a is large and b is small; hence, slip should occur most readily on close-packed atomic planes, which are characterized by the greatest separation distance, and in close-packed directions for which the atomic slip distance is a minimum. In most materials, slip does occur prevalently on close-packed planes and, but for a few exceptions, in close-packed atomic directions. (Although there are reasons other than the Peierls stress for favoring such slip directions.) Comparison of values of τ_f calculated through Eq. (3.5) with experimentally observed shear strengths (Table 3.1) indicate the former is greater. The reason is that the Peierls-Nabarro treatment deals with a *narrow* dislocation, i.e., one for which the atomic distortion at the partial plane termination is only found in the immediate vicinity of the termination. When the distortion is spread out over a number of atomic distances, as it is in fcc metals and—to a lesser extent—in bcc metals,[2] the frictional stress is decreased because the atomic displacements associated with dislocation motion become even less than b.

The frictional stress can be alternatively expressed in terms of the width, w, of the dislocation:

$$\tau_f = G \exp\left[\frac{-2\pi w}{b}\right] \tag{3.6}$$

As indicated by Eq. (3.6), the frictional stress decreases as w increases. The concept of dislocation width can be illustrated with reference to Fig. 3.5, which shows dislocations in two-dimensional "bubble rafts." Such a raft is a close-packed array of spherical bubbles, and the interaction among them simulates to a degree the forces between atoms in certain solids. The two bubble rafts illustrated contain dislocations. In the raft composed of large ("soft") bubbles, the dislocation core is clearly evident and the atomic distortion (i.e., the deviation from ideal crystal packing) is localized in the vicinity of the partial plane termination. Such a dislocation is considered narrow; i.e., it has a small value of w. In contrast, the dislocation in the raft composed of small ("hard") bubbles is more difficult to place; the atomic distortion is spread over many atomic planes, resulting in a wide dislocation. As noted, dislo-

---

[2]Equation (3.5) describes well the frictional stress for covalent solids, though. Because of the nature of their bonding, dislocations are narrow in covalent crystals.

cations in metals are typically "wide." Reconciliation of their observed shear yield strengths to the friction stress as given by Eq. (3.6) can be made assuming reasonable dislocation widths in them (Prob. 3.2).

An alternative means of characterizing dislocation width is to superimpose a grid, in the form of a perfect lattice, on the bottom of Fig. 3.2 and to plot the displacement in the slip direction of the slip-plane atoms in the dislocated structure with respect to atoms in the perfect lattice. This is done in Fig. 3.6a. The displacement is $-b/2$ on the left-hand side of the crystal and $+b/2$ on the right. The slip-plane distance over which the displacement changes from $-b/2$ to $+b/2$ is a measure of the dislocation width; this is shown schematically in Fig. 3.6b for both wide and narrow dislocations.

As mentioned, τ_f decreases as dislocation width increases as a result of the smaller relative atomic displacements required for motion of a wide dislocation. The width of a dislocation is also temperature-sensitive, decreasing with decreasing temperature, and this leads to an increased frictional stress at low temperatures. The

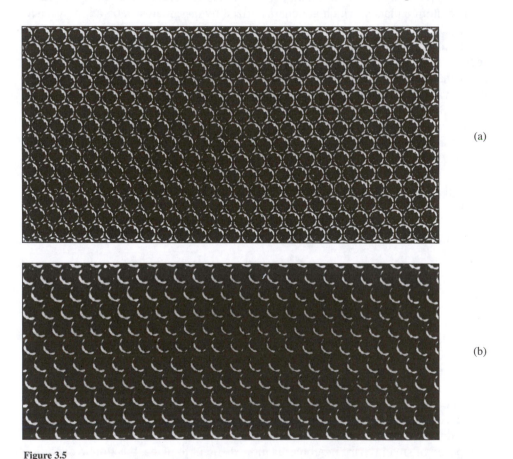

(a)

(b)

Figure 3.5
Dislocations in close-packed, two-dimensional soap bubble arrays. In (a) the large bubbles (1.9 mm diameter) are "soft" and give rise to a narrow dislocation, the core of which is easily identified. In (b) the smaller (0.76 mm diameter) and "harder" bubbles produce a "wide" dislocation; the core position of this dislocation is difficult to identify specifically. (*After W. C. Bragg and J. F. Nye, Proc. Roy. Soc. (London), A190, 474, 1947.*)

temperature variation of dislocation width, and hence τ_f, is sensitive to the nature of atomic bonding and crystal structure. For example, covalent bonded solids demonstrate a relatively strong temperature dependence of the frictional stress, whereas it is nearly temperature-independent for most face-centered cubic metals. A comparison of the dislocation width, and magnitude and temperature sensitivity of τ_f for several classes of materials, is given in Table 3.2. As indicated there, covalent solids have large frictional stresses and their temperature sensitivity of this stress is moderately high. Face-centered cubic metals have both a low friction stress and a low-temperature sensitivity of it. Ionic solids and body-centered cubic metals are intermediate cases, although the temperature sensitivity of τ_f can be large for the body-centered cubic transition metals. This sensitivity affects the low-temperature fracture behavior of bcc transition metals as is discussed in Chap. 10.

Material class	Dislocation width	Frictional stress	Yield strength temperature sensitivity
FCC metals	Wide	Very small	Small
BCC metals	Narrow	Moderate	Large (for transition metals)
Ionic solids	Narrow	Moderate	Moderate to large
Covalent solids	Very narrow	Large	Moderate to large

Source: Revised from R. W. Hertzberg, *Deformation and Fracture Mechanics of Engineering Materials,* Wiley, New York, 1976.

Table 3.2
Relationship among frictional (Peierls) stress, dislocation width, and temperature sensitivity of yield strength for various material classes

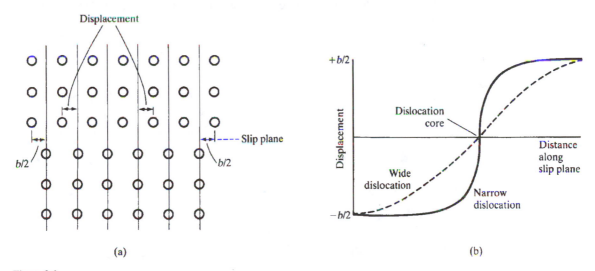

(a) (b)

Figure 3.6
A way of defining dislocation width. In (a) a grid, with spacing equal to the interatomic spacing, is superimposed on the atomic configuration associated with an edge dislocation. Above the slip plane, the atoms on the left side of the crystal are displaced by $-b/2$, and those on the right side by $+b/2$, with respect to the grid. In (b) the relative displacement of atomic columns with respect to the grid is plotted as a function of distance along the slip plane. The distance over which the displacement changes from $-b/2$ to $+b/2$ is a measure of dislocation width. This is illustrated for both a narrow and a wide dislocation.

B. Climb of Edge Dislocations

Edge dislocation motion of the type illustrated in Fig. 3.3 is called glide, and is referred to as conservative motion. This dislocation can also move by climb, a form of nonconservative motion. Nonconservative motion requires addition to the dislocation core of atoms or vacancies, as illustrated in Fig. 3.7. In Fig. 3.7*a,* vacancies interchange positions with atoms at the partial plane termination and, in doing so, the dislocation "climbs" by one atomic layer at the interchange position and a "jog" is introduced into the dislocation at this point (Fig. 3.7*a*). If the opposite process occurs, i.e., if a vacancy below the partial plane termination exchanges places with an adjacent lattice atom, a jog of the "opposite sign" forms (Fig. 3.7*b*) and climb in the opposite sense takes place. If repetitive processes of the type shown in Fig. 3.7 occur, a net motion of the line in a direction *normal* to the slip plane occurs. Although we shall see that a dislocation tries to minimize its total length (i.e., a dislocation does not like to be "jogged"), nevertheless there is always a certain number of jogs on each dislocation and, in the absence of an applied stress, the number of "positive" and "negative" jogs is about the same (Fig. 3.8). For a dislocation to undergo a net motion by climb requires, therefore, a preferential flow of either vacancies or lattice atoms to the dislocation core. An applied stress can alter these respective flow rates, thereby providing a driving force for dislocation climb. Such a mechanism allows, for example, a dislocation that has encountered an obstacle on its slip plane to climb to an adjacent plane, thereby circumventing the obstacle. Since climb involves diffusive motion of atoms, temperatures at which diffusion can occur are required to effect it. The activation energy associated with climb is typically a vacancy-diffusion activation energy, and thus climb occurs only at moderate to elevated temperatures. Climb is important for some mechanisms of high-temperature flow (creep). Further discussion of climb as it relates to creep is presented in Chap. 7.

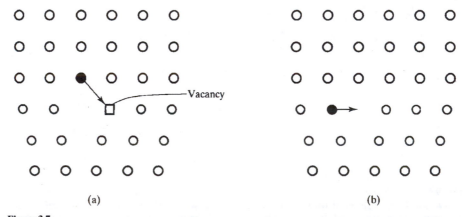

(a)   (b)

Figure 3.7
(a) A dislocation can climb by interchange of atomic sites along its core with an adjacent vacancy. "Reverse" climb is shown in (b). Here, an adjacent atom moves into the dislocation core leaving behind a lattice vacancy.

C. Topological Considerations

Topological ideas give rise to the concept of a dislocation as a line defect. As indicated in Figs. 3.2 and 3.5, the atomic packing in the vicinity of an edge dislocation is distorted in approximately a cylindrical volume; the axis of the cylinder is the termination of the partial plane and the cylinder diameter is approximately the dislocation width. Although this distortion constitutes a finite volume, the long axis of the cylinder is much greater than its diameter, and thus a dislocation is logically referred to as a line defect.

The character of the edge dislocation of Fig. 3.2 can be clarified further by application of an atomic circuit, called a Burgers circuit, which encloses the axis of the line. In a crystal containing no dislocations, a Burgers circuit would start at one atomic position and by traversing a path of the type shown in Fig. 3.9a would return

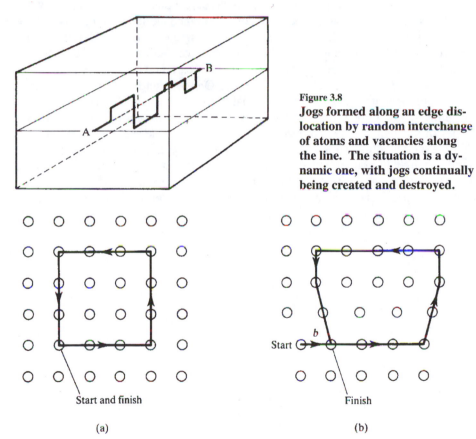

Figure 3.8
Jogs formed along an edge dislocation by random interchange of atoms and vacancies along the line. The situation is a dynamic one, with jogs continually being created and destroyed.

(a) (b)

Figure 3.9
Determination of the slip vector of an edge dislocation by application of a Burgers circuit. In (a) a circuit, obtained by traversing equal (and reversed) distances in two mutually perpendicular directions, closes on itself in a "perfect" crystal. In (b) a similar circuit applied around an edge dislocation does not close. The closure fault is the Burgers vector, $\bar{b}$; $\bar{b}$ defines the dislocation slip direction and is normal to the edge dislocation line. The plane containing both $\bar{b}$ and the line defines the slip plane.

to this same position. When executing a similar path enclosing an edge dislocation (Fig. 3.9b), the circuit fails to close on itself. The closure defect has magnitude and direction; it is called the Burgers vector. The magnitude of it is b, the slip displacement, and the direction of $\bar{b}$ is the slip direction; that is, it is the direction in which relative atomic displacements take place during slip. For an edge dislocation, the Burgers vector is normal to the dislocation line and the two directions define the slip plane. Thus, the slip plane contains both $\bar{b}$ and the dislocation line or, alternatively, the vector cross-product of $\bar{b}$ and a vector parallel to the dislocation line is normal to the slip plane.

D. Motion of Edge Dislocations Containing Jogs

The motion in response to a shear stress of a dislocation containing jogs provides further insight into the ways dislocations move in response to forces. As shown in Fig. 3.10a, the dislocation line lies in its original plane along the segments AB and EF. The line lies in the plane above this plane along the segment CD; thus BC and DE are jogs. We assume no further nonconservative motion occurs. The Burgers vector is the same along all segments of the line; i.e., $\bar{b}$ is the same for a circuit applied around BC as it is for one enclosing AB. On the other hand, the slip planes of the jog components BC and DE are normal to the slip planes of AB, CD, and EF. Since the stress applied in Fig. 3.10a does not produce a shearing stress in the slip planes of the jogs, they do not move in response to it. The segments AB, CD, and EF do experience this stress but, being pinned at the jog locations, they no longer remain straight but rather bow out in loop-like fashion from the pinning points (Fig. 3.10b). This being so, the atomic distortions along the line are no longer characteristic of those associated with the presence of a partial atomic plane. Instead, these configurations are characteristic of screw or mixed dislocations. These types of dislocations are considered in the following section.

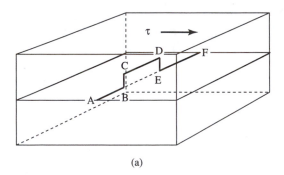

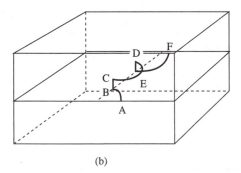

(a)   (b)

Figure 3.10
(a) The jogged edge dislocation (AF) moves in response to a shear stress as shown in (b). The portion of the crystal below CD but above BE is considered a row of vacancies and the jogs BC and DE experience no net stress in their slip plane (which is normal to the slip plane of AB, CD, and EF). Thus, as shown in (b), BC and DE remain fixed whereas the other portions of the line bow out in response to the stress.

3.4
SCREW AND MIXED DISLOCATIONS

A. Slip by Screw Dislocation Motion

A *screw dislocation* is another type of line defect the motion of which causes plastic flow in crystals. The atomic arrangements in the vicinity of a screw dislocation are illustrated in Fig. 3.11*a*. The distorted volume is again approximately a cylinder with AB the cylinder axis and the dislocation width the cylinder diameter. For the screw dislocation, the width corresponds to the diameter of the helical distortion of the structure containing atoms lying on the plane of AB and in the parallel plane immediately below it (Fig. 3.11*a*). Motion of a screw dislocation occurs if the top and bottom faces of the crystal are subjected to the shear stresses illustrated in Fig. 3.12. As with edge dislocation motion, only small atomic rearrangements along the line are required to move the dislocation to the left (Fig. 3.12). (Note that for a screw dislocation the motion of the line is normal to the direction of the applied stress; an edge dislocation line moves in the same direction as the stress.) With continued movement of the screw dislocation to the left, further displacement of the upper half of the crystal with respect to the bottom half is achieved, and when the dislocation reaches the crystal edge, the crystal halves are displaced relative to each other by magnitude *b* and in the *direction* of the applied stress (Fig. 3.12*b*). Thus, plastic deformation results from the motion of the screw dislocation.

The terminology "screw" derives from the helical nature of the structural distortion in the vicinity of the dislocation core (Fig. 3.11). This can be illustrated by taking a Burgers circuit around a screw dislocation (Fig. 3.13). The closure fault $\bar{\mathbf{b}}$ is parallel to the dislocation line. Taking repeated Burgers circuits results in the tracing of a helical path around the dislocation core, the path resembling that of the recesses along the axis of a screw. The Burgers vector for a screw, as well as for an edge, dislocation is parallel to the slip direction, and the vector magnitude represents the slip distance in both situations. As noted, however, the Burgers vector for a screw dislocation is parallel to the line, and thus, unlike for an edge dislocation, the Burgers vector and the screw dislocation line do not define a unique slip plane. Hence, although the sense of the stress illustrated in Fig. 3.12 defines clearly a specific slip plane in that example, this is not always the case for screw dislocations. Slip in face-centered cubic metals provides an example. In these materials, slip occurs invariably on {111} planes; these are the most densely packed planes and the separation distance between them is greatest. In face-centered cubic metals, slip directions are of the close-packed <110> type (the Burgers vector magnitude is $a/\sqrt{2} = 2r$, where *a* is the lattice parameter and *r* the atomic radius). In fcc metals, the {111} family of planes contains common slip directions. For example the (111) and (1$\bar{1}$1) planes have in common the direction [$\bar{1}$01] (Fig. 3.14). Thus, if a screw dislocation traveling on a (111) plane in a fcc metal, and having a Burgers vector $(a/2)[\bar{1}01]$, encounters an obstacle on this plane, it can circumvent it by *cross-slipping* onto a (1$\bar{1}$1) plane (Fig. 3.15). Once the obstacle has been surmounted, the dislocation can then return, by an additional cross-slip process, to a (111) plane coplanar with the initial glide plane. Hence, a screw dislocation is able to overcome obstacles to slip by conservative motion involving cross-slip. This is in contrast to the climb process required of edge dislocations for this purpose.

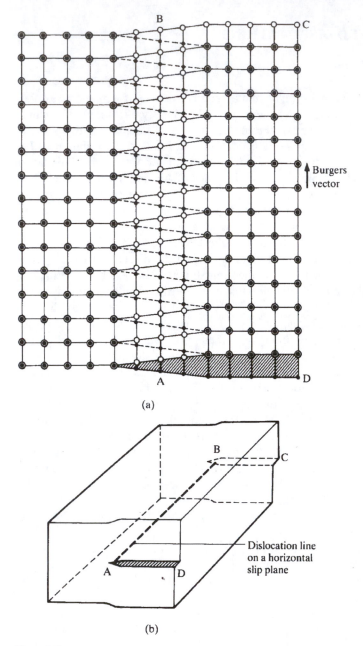

(a)

(b)

Figure 3.11
(a) The atomic arrangements in the vicinity of a screw dislocation, resulting from the crystal distortion shown in (b). The arrangement along AB is illustrated by super-position of atoms in the plane below AB (solid circles) and those in the plane above it (open circles). The resultant atomic pattern is akin to a helix centered along AB. The dislocation width here is about four interatomic spacings. (*Adapted from W. T. Read, Jr., Dislocations in Crystals, McGraw-Hill, New York, 1953.*)

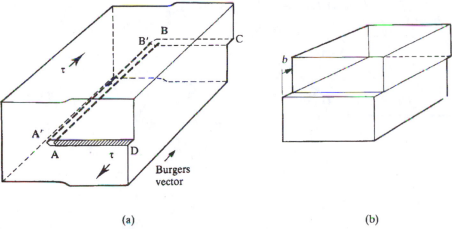

(a)   (b)

Figure 3.12
Motion of the screw dislocation of Fig. 3.11 in response to a stress. The open-circle atoms of Fig. 3.11 move preferentially in the direction of the stress applied to the top crystal surface, and the solid-circle atoms move in the opposite direction. Only small atomic displacements are necessary for the dislocation line to move in the direction *normal* **to the applied stress (e.g., to the position A'B'). (b) When the dislocation has reached the crystal edge, slip of the top half of the crystal with respect to the lower half, and in the direction of the applied stress, is accomplished.**

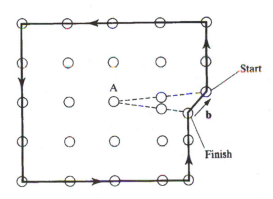

Figure 3.13
A Burgers circuit taken around a screw dislocation (e.g., around the front face of the crystal of Fig. 3.11b) produces a closure fault parallel to the line. The vector is in the slip direction but, unlike for an edge dislocation, the vector and the dislocation line do not define a unique plane.

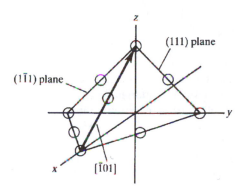

Figure 3.14
The atomic arrangements on the (111) and (1̄11) close-packed planes in the face-centered cubic structure. Both planes contain three nonparallel close-packed directions of the <110> type and the [1̄01] direction is common to both.

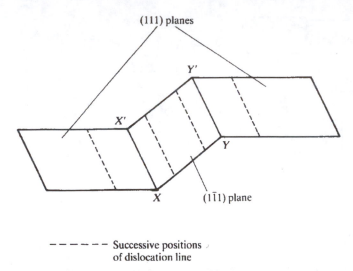

Figure 3.15
Schematic of a screw dislocation cross-slipping in a fcc lattice. The dislocation line (with $\bar{b} = a/2[\bar{1}01]$) originally is moving on a (111) plane when it encounters an obstacle along the line XX'. It then cross-slips to a $(1\bar{1}1)$ plane (which also contains a $[\bar{1}01]$ slip direction) and then glides on it until, at YY', it cross-slips again onto a (111) plane parallel to the original slip plane.

B. Slip by Mixed Dislocation Motion

An example of a mixed dislocation is shown in Fig. 3.16. The atomic arrangements along the line are, at point A, similar to those of a screw dislocation and, at point B, similar to those of an edge dislocation. At other portions of the line, the atomic arrangements are intermediate to those of the "pure" edge or "pure" screw ones. A Burgers circuit enclosing the line at A produces a closure fault parallel to the line; a similar circuit around B yields a Burgers vector normal to the line. However, the magnitude and direction of $\bar{b}$ is the same irrespective of the location of the circuit. Thus, a circuit constructed around the line at an arbitrary position yields a closure fault making an angle between 0 and 90° to the line. We see, therefore, that the Burgers vector of a dislocation line is invariant. Any variation in dislocation character along the line is only structural, as defined by the angle between the Burgers vector and a tangent drawn to the line at a specific location; e.g., if the line is normal to $\bar{b}$, the dislocation structure there is edge-like.

As for edge and screw dislocations, the Burgers vector defines the slip direction of a mixed dislocation. Thus, on application of a stress (Fig. 3.17a), the mixed dislocation expands in a loop-like fashion; i.e., at all locations the motion of the line is normal to the line. When the loop reaches the crystal boundaries (Fig. 3.17b), a permanent offset of the crystal portions is accomplished. The slip magnitude is b, and the slip direction has the sense of the Burgers vector and is consistent with the direction of the applied stress.

The atomic arrangements around the expanding loop of Fig. 3.10 can now be examined further. To do this, a mirror image of the crystal shown in Fig. 3.16 is

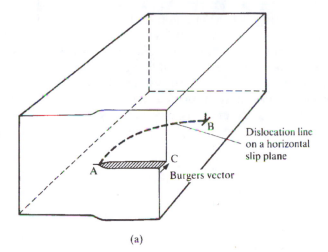

(a)

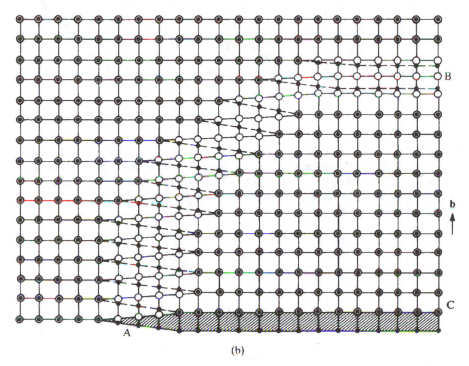

(b)

Figure 3.16
(a) A mixed dislocation (line AB) in a crystal. (b) The atomic arrangements along the line are such that atoms in the plane below the line are represented by solid circles and those above by open ones. In the vicinity of A, the atomic arrangement is like that of a screw dislocation; $\bar{b}$ is parallel to the line there. Near B the atomic arrangements are those of an edge dislocation, and $\bar{b}$ is normal to the line here. (*Adapted from W. T. Read, Jr., Dislocations in Crystals, McGraw-Hill, New York, 1953.*)

attached to itself (Fig. 3.18). The atomic distortion along the line ABA′ is akin to that found along the segment CD of Fig. 3.10*b*. At the jog-anchored points (C and D) the line is parallel to the Burgers vector and thus is screw-like. Conversely, at the "tip" of the loop the line is normal to $\bar{\mathbf{b}}$ and is edge-like. Along the rest of the line, the dislocation is mixed. The dislocation loop of Fig. 3.10 serves to illustrate that in actual crystals most dislocations are neither pure edge nor pure screw in character. That is, since the resistance to dislocation motion varies along a slip

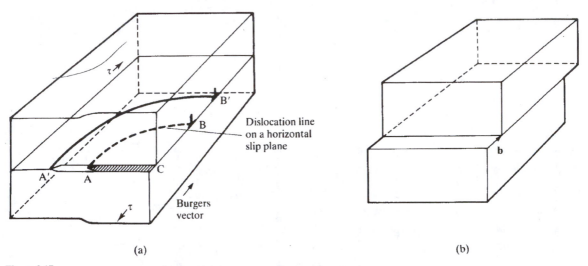

(a)   (b)

Figure 3.17
Motion of a mixed dislocation in response to a stress. (a) The dislocation expands in a loop-like manner (from AB to A′B′) producing atomic displacements across the slip plane. The line moves in the direction of the stress at the edge locations and normal to it at the screw locations. (b) After the dislocation has swept to the side and rear faces of the crystal, the upper part of the crystal has been displaced with respect to the lower part in the slip direction.

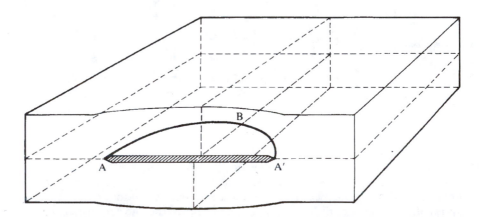

Figure 3.18
A semicircular dislocation produced by taking a mirror image of the crystal of Fig. 3.16 and joining the crystals along the plane containing point B. The atomic distortion along ABA′ is the same as that along the curved dislocation CD of Fig. 3.10*b*.

plane, curved dislocations result. Since the Burgers vector of a dislocation line is invariant, a curved dislocation can be neither entirely pure edge nor entirely pure screw in character.

3.5
TWINNING

Although dislocation glide is the dominant mechanism of plastic deformation in crystals, twinning is another way by which permanent shape changes can be effected. The atomic displacements associated with twinning in a body-centered cubic metal are illustrated in Fig. 3.19. As shown there, atoms above the twin plane move in a coordinated manner to produce a shape change and, at the same time, a mirror image of the crystal across the twin plane results. As with slip in perfect crystals, twinning unassisted by dislocation motion requires the cooperative and simultaneous motion of a number of atoms; thus, the theoretical twinning stress is high. For this reason, twinning is believed associated with dislocation motion, and a mechanism by which twinning can take place by edge dislocation motion is shown in Fig. 3.20. In contrast to slip, which can take place by the uncoordinated and independent glide of numerous dislocations, twinning by dislocation motion requires cooperative dislocation displacements.

When is twinning important? Since twinning and slip can be considered competitive, that mechanism requiring the lowest stress to effect it should be observed. For fcc metals, the stress required for flow via slip is almost always less than the

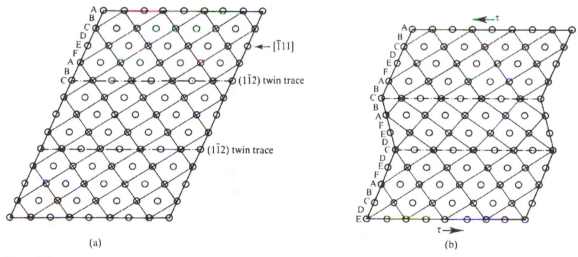

(a)   (b)

Figure 3.19

Twinning in the bcc structure. (a) The untwinned structure. The plane of the paper contains a (110) plane and shows the ABCDEFABC . . . stacking of (1$\bar{1}$2) planes that project normally out of the paper. The [$\bar{1}$11] direction lies in the plane of the paper and within a (1$\bar{1}$2) plane. (Only half of the intersections of the (1$\bar{1}$2) planes are shown.) (b) Twinning occurs by cooperative displacement of atoms in the [$\bar{1}$11] direction. Twinning results in a permanent shape change (i.e., plastic deformation). The process is called twinning because, subsequent to deformation, a mirror image of the crystal is found across the twin plane boundary.

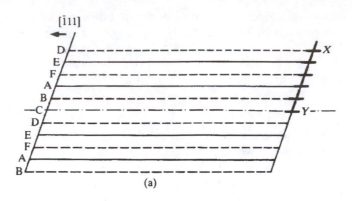

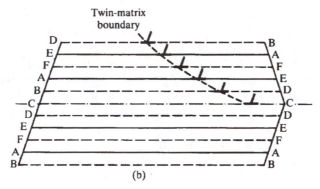

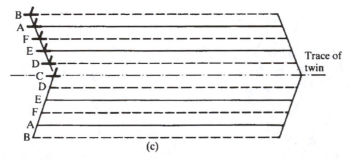

Figure 3.20
Mechanical twinning in the bcc lattice by edge-dislocation motion. In (a) the stacking in the untwinned crystal is schematized similarly to that of Fig. 3.19a. (b) Motion of dislocations in the [Ī11] direction produces the twin misorientation. In (c) the process is complete. (*Adapted with permission from Derek Hull, Introduction to Dislocations, Pergamon Press, Oxford, 1965.*)

twinning stress and so twinning is seldom observed in these materials.[3] Body-centered cubic transition metals are more prone to exhibit twinning. This is because their yield strengths associated with slip are strongly temperature dependent (Chap. 4). Thus, with decreasing temperature, twinning becomes more likely in them. In

---

[3]As with all generalizations, there are exceptions. In this case, we note that twinning is observed in certain fcc metals at low temperatures. Further, certain alloying elements in Cu promote twinning at room to moderately elevated temperatures.

fact, for certain bcc transition metals (Fe being among the most important of these) twinning becomes the dominant low-temperature plastic deformation mode.

Deformation twinning is important in hexagonal close-packed metals. As indicated in Table 3.3 (see page 116), this material class has relatively few slip systems (see also Sect. 3.7A). As is discussed in Chap. 4, it is a requirement for plasticity in polycrystals that a material exhibit a certain critical number of slip systems. Hexagonal close-packed metals, for the most part, do not display this critical number. However, many of them (e.g., Mg, Zn, Ti, and Zr) exhibit moderate to extensive plasticity in polycrystalline form. This is so because the twinning deformation of which they are capable relaxes the slip system requirement. The twin plane in hcp metals is the pyramidal $\{10\bar{1}1\}$ plane. Deformation twinning across this plane can provide either extension or compression depending on the c/a ratio of the metal. Twinning can also serve to facilitate slip. For example, the reorientation of crystal planes during twinning can place them in orientations in which the applied stress is more efficient for causing slip.

As implied by Fig. 3.19, twinning produces a more heterogeneous deformation pattern than does slip. This can be seen in Fig. 3.21, which is a photograph of zirconium (Zr) that has deformed by mechanical twinning. The region within the twin boundaries has experienced a shear analogous to that shown in Fig. 3.19; the regions without the boundaries have not undergone such a shear. Thus, there is a plastic deformation gradient across the boundaries and also at the intersection of twins with grain boundaries. Unless this incompatibility is removed by additional plastic deformation, it leads to initiation of fracture at the boundaries. This happens at low temperatures in bcc transition metals that twin. The effect has a profound impact on the fracture behavior of bcc transition metals, as is discussed further in Chap. 10.

Figure 3.21
Deformation twins in polycrystalline zirconium. The twins are the lenticular-shaped areas in four of the grains shown. The twin thicknesses here are about 1 μm.
(Adapted from Robert E. Reed-Hill, Physical Metallurgy Principles, 2nd ed., van Nostrand, New York, 1973.)

3.6
PROPERTIES OF DISLOCATIONS

In addition to their topological characteristics that cause plastic flow via their glide, dislocations have a number of other features. An understanding of them is required to appreciate fully the role of dislocation motion in plastic deformation. In this section, we discuss some of these characteristics. Topics considered include dislocation stress fields and energy, the shape of a dislocation as it controlled by energetic considerations, and dislocation velocity as it depends on stress. The latter is important in discussion of dislocation dynamics, i.e., the relationship among strain rate, stress, and dislocation population and mobility.

A. Dislocation Stress Fields

The atomic distortions in the vicinity of dislocations produce internal stress fields. Since dislocation motion is in response to the total stress acting on the dislocation, which includes components arising from the stress fields of other dislocations as well as from external forces, the stress fields associated with dislocations are discussed.

A complete description of the atomic displacements and resulting stress field within the dislocation core is difficult, and depends on accurate knowledge of interatomic potentials. However, the stress field at distances well removed from the core can be determined by elastic analysis, since the long-range displacement (i.e., the Burgers vector) of atoms at opposite sides of the dislocation is specified. Moreover, description of the core is not all that critical for our purpose; many, if not most, effects of dislocation behavior are due to the long-range stresses and not to effects associated with the core. Since the relative displacements of the atomic planes persist over distances large in comparison to the atomic size, the stress field of a dislocation is also a long-range one and the application of elasticity theory to estimate it is well accepted.

The stress field of a screw dislocation is most easily characterized, since the atomic displacements around it (Fig. 3.11) give rise to only shear stresses. The shear stress associated with a screw dislocation is given by

$$\tau = \frac{Gb}{2\pi r} \qquad (3.7)$$

where r is the radial distance from the dislocation core. The shear stress acts in the slip direction, i.e., along the axis of the dislocation. This stress must be considered part of the total stress experienced by another dislocation in the vicinity of a screw dislocation.

The stress field of an edge dislocation is more complicated. Both shear and dilational stresses (i.e., tensile or compressive stresses due to changes in local atomic volume) are present in the vicinity of an edge dislocation. The nature of the dilational strain can be visualized by reconsidering Fig. 3.2. The atoms in the region above the slip plane are compressed owing to the presence of the additional atom plane; that is, the local atomic volume is decreased in this region. Conversely, the atoms below the slip plane experience tensile forces, as reflected by the increase in interatomic separation in this region. That shear stresses also result from the pres-

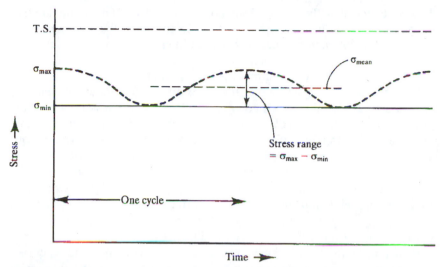

Figure 1.22
A time-varying tensile stress that might result in fatigue fracture. The maximum stress is less than the material tensile strength. The number of stress cycles required to cause fatigue fracture depends on the material, the stress range ($\sigma_{max} - \sigma_{min}$), and the mean stress ($\sigma_{max} + \sigma_{min}$)/2. The higher the stress range and the higher the mean stress, the fewer the number of cycles required to cause fatigue failure.

duced somewhat, followed by an increase to the initial stress, and the process repeated (Fig. 1.22), the material might eventually fracture. How many stress cycles this would take depends on the material and the range and mean (average) value of the applied stress (Fig. 1.22); the higher the stress range and the greater the mean stress, the shorter the material's *fatigue life*.

Fatigue lives can vary from tens (low-cycle fatigue) to many millions (high-cycle fatigue) of cycles. Short lives are associated with high mean stresses and concomitantly large strain ranges. In some cases the strain ranges are large enough that the material deforms plastically at a macroscopic level. When this happens, fatigue lives correlate better with the plastic strain range than they do with the stress range or mean stress. During high-cycle fatigue the material deforms only elastically in a macroscopic sense. Fatigue lives then correlate equally well with elastic strain range ($\Delta\varepsilon_{el}$) and the stress range, since the two are related through $\Delta\varepsilon_{el} = \Delta\sigma/E$, where E is the material's elastic modulus.

Plastic deformation also takes place during high-cycle fatigue, albeit on a microscopic, rather than macroscopic, scale. Local stress concentrations or microstructural inhomogeneities promote this kind of flow, and the repetitious nature of the loading eventually leads to nucleation of a crack in regions that deform plastically. After a certain period the developed crack grows slowly and in a direction normal to the stress axis (for uniaxial loading). The rate of crack advance per cycle (the crack growth rate) correlates in many cases with the stress range, the mean stress having a decidedly secondary effect on the slow crack growth rate. After the crack has grown to a critical length, it advances rapidly and final fracture takes place. The latter is caused either by having the cross-sectional area

ence of an edge dislocation is evident from the relative atomic displacements across the slip plane. With reference to the coordinate system of Fig. 3.22a, the dilational and shear stresses are given by

$$\sigma_x = \frac{-Gb}{2\pi(1 - v)r} \sin \theta \, \{ 2 + \cos 2\theta \} \qquad (3.8a)$$

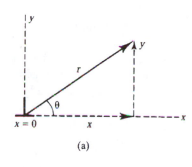

(a)

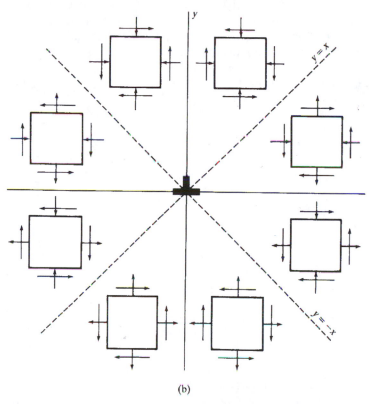

(b)

Figure 3.22
(a) Coordinate system used to describe the stress state of an edge dislocation located at the origin of it. The Cartesian coordinates are x, y; cylindrical coordinates are r and θ ($y = r \sin \theta, x = r \cos \theta$). (b) The sense of the shear (τ_{xy}) and dilational (σ_x, σ_y) stress components of an edge dislocation. (*Part (b) adapted from W. T. Read, Jr., Dislocations in Crystals, McGraw-Hill, New York, 1953.*)

$$\sigma_y = \frac{Gb}{2\pi(1 - \nu)r} \sin\theta \cos 2\theta \qquad (3.8b)$$

$$\tau_{xy} = \tau_{yx} = \frac{Gb}{2\pi(1 - \nu)r} \cos\theta \cos 2\theta \qquad (3.8c)$$

and the signs of these stresses are illustrated in Fig. 3.22b. There are no shear stresses acting in the direction parallel to the dislocation line. There is, however, a third dilational stress, σ_z (normal to the plane of Fig. 3.22), given by

$$\sigma_z = \frac{-Gb\nu}{\pi(1 - \nu)r} \sin\theta \qquad (3.9)$$

For a mixed dislocation whose Burgers vector makes an angle α with the dislocation line, the stress fields are equal to those associated with a screw dislocation (with b being replaced by $b \cos\alpha$) plus those due to an edge dislocation (with b being replaced by $b \sin\alpha$).

B. Dislocation Energies

An elastic energy is associated with the dislocation stress fields. The energy of a dislocation is important in aspects of plastic flow as dislocations try to reduce their energy by minimizing their length as well as by developing arrays that minimize the total system dislocation energy. When a dislocation bypasses an obstacle and this is, as it usually is, accompanied by a bowing of the dislocation (i.e., an increase in its length), the plastic flow resistance of the material increases as a result of the energy increase accompanying the increase in line length.

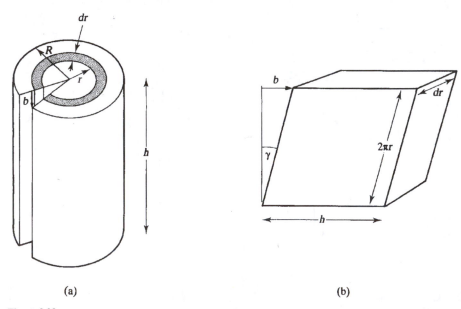

(a)   (b)

Figure 3.23
(a) A cylindrical crystal of radius R contains one screw dislocation whose line lies along the cylinder axis. (b) "Unrolling" the cylindrical element of radius r and thickness dr shows that the shear strain is equal to $b/2\pi r$.

Evaluation of the elastic strain energy of a screw dislocation can be accomplished by considering a single crystal containing a screw dislocation lying along the crystal central axis (Fig. 3.23a). By unrolling the cylinder (Fig. 3.23b), the shear strain at the radial position, r, is obtained:

$$\gamma = \frac{b}{2\pi r} \tag{3.10}$$

The strain energy per unit volume is $G\gamma^2/2$;[4] thus for the volume element located at position r in Fig. 3.23a, the incremental elastic strain energy, dU'_s, is

$$dU'_s = (2\pi rl \, dr)\left(\frac{Gb^2}{8\pi^2 r^2}\right) = \frac{Gb^2 l}{4\pi} \frac{dr}{r} \tag{3.11}$$

On integrating Eq. (3.11) from the dislocation core radius, r_0, to some arbitrary distance, r_1, we obtain for the dislocation energy (neglecting the core energy)

$$U_s = \frac{U'_s}{l} = \frac{Gb^2}{4\pi} \ln\left(\frac{r_1}{r_0}\right) \tag{3.12}$$

Note that U_s is the dislocation strain energy *per unit length* of dislocation line. Approximate calculations indicate the core energy is but a small fraction of the total strain energy; thus, it is usually neglected. The calculations reflect the fact that most of the dislocation energy is associated with the large number of atoms well removed from the core that undergo small displacements rather than with the few core atoms that experience large displacements. The outer limit of integration, r_1, is taken as the crystal radius if the crystal contains but a single dislocation. In a real material containing many dislocations, r_1 is that distance beyond which the stress field of the dislocation effectively vanishes as the result of overlap with stress fields of other dislocations; thus, r_1 is typically of microscopic dimensions. Because of this complexity, estimation of the coefficient of Gb^2 in Eq. (3.12) presents difficulties. Therefore, the energy per unit length of a screw dislocation is approximated by taking the coefficient as ½, an estimate certainly valid within an order of magnitude. Further, the essential physics of dislocation energy—that the energy is proportional to shear modulus and to the Burgers vector squared—is preserved in so doing. Hence

$$U_s \cong \frac{1}{2} Gb^2 \tag{3.13}$$

Since dislocation energy is minimized by its having a small Burgers vector, this accounts for the observation that slip directions in most materials are close-packed ones. Dislocations with larger slip distances are energetically unfavorable in comparison.

Calculations on the elastic energy of an edge dislocation proceed in much the same way. With the same approximations, the energy per unit length, U_e, for an edge dislocation is given as

$$U_e \cong \frac{Gb^2}{2(1 - \nu)} \tag{3.14}$$

---

[4]This is obtained by integrating the elastic shear stress with respect to elastic shear strain; i.e., $\int \tau \, d\gamma = G\gamma^2/2$ when $\tau = G\gamma$.

where ν is Poisson's ratio. Comparison of Eqs. (3.13) and (3.14) shows that an edge dislocation has a somewhat greater strain energy than a screw dislocation.

Since dislocation energy is expressed per unit line length of dislocation, the units of Eqs. (3.13) and (3.14) are those of force. This force, called the line tension, acts so as to minimize total dislocation line length. Hence, it is also the force required to change the shape of a dislocation from a straight to a curved line, or to expand a linear dislocation into a loop of the type shown in Fig. 3.10b.

C. Forces Between Dislocations

The elastic stress fields of dislocations give rise to forces among them. These affect the external stress required for plastic deformation. In this section we discuss the nature and magnitude of interdislocation forces.

Calculation of the force between two dislocations proceeds as follows. The total elastic strain energy of such a system is equal to the sum of the individual dislocation energies (e.g., Eqs. (3.13) and (3.14)) plus the interaction energy arising from overlap of the individual dislocation stress fields. In a Cartesian coordinate system, the force in the x direction, F_x, between the dislocations (the force acts equal and opposite on the individual dislocations) is obtained as the gradient of the interaction energy with respect to distance between the dislocations. That is, $F_x = \partial U/\partial x$, where U is the interaction energy; similar expressions hold for F_y and F_z.[5]

Several examples of the forces between parallel dislocations are shown in Fig. 3.24. If, as in Fig. 3.24a, the dislocations are screw dislocations, the force (per unit dislocation length) between the dislocations lies on a line connecting their centers and its magnitude is

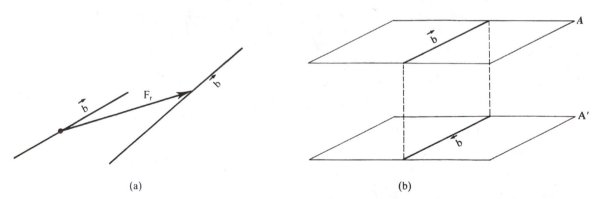

(a)   (b)

Figure 3.24
(a) A repulsive force, acting along a line between the dislocation cores, exists between two screw dislocations with parallel Burgers vectors (the force would be attractive for anti-parallel slip vectors). (b) If the dislocation slip planes are parallel (A and A'), dislocations with anti-parallel Burgers vectors glide to a position where they are directly over one another.

---

[5]The sign convention used here is that a negative force signifies attraction between the dislocations, and vice versa.

$$F_s = \frac{Gb^2}{2\pi r} \qquad (3.15)$$

where r is the center-to-center distance between the dislocations. If the dislocations have parallel Burgers vectors, F_s is repulsive; if the dislocations are of opposite sign (i.e., their Burgers vectors are anti-parallel) they attract each other. It is important to note that F_s varies inversely with dislocation separation; that is, it is greater when dislocations are close together than when they are widely separated. Provided a component of the interaction force exists on the dislocation slip plane and in the slip direction, it can contribute to glide of the dislocations. For example, if the slip planes of the dislocations of Fig. 3.24a contain the line joining their centers, the dislocations either glide away from each other (if the force is repulsive) or come together and *annihilate* (if the force is attractive.) On the other hand, if the glide planes are parallel to each other, screw dislocations of opposite sign will glide along these planes until they lie directly above and below each other (Fig. 3.24b).

In terms of the coordinates of Fig. 3.25a, the forces between two parallel edge dislocations are

$$F_x = \pm\, b\tau_{xy} \qquad (3.16a)$$

$$F_y = \mp\, b\sigma_x \qquad (3.16b)$$

where σ_x and τ_{xy} are given by Eqs. (3.8a) and (3.8c) and the $\pm$ signs depend on whether the dislocations are of the same or opposite sign (i.e., on whether their Burgers vectors are parallel or antiparallel). The x component of force gives rise to a force in the slip plane (Fig. 3.25b) for both parallel and anti-parallel dislocations. Note that if the dislocations are of the same sign, an equilibrium configuration is obtained when they lie directly above each other in the xy plane. This kind of interaction among numerous edge dislocations of the same sign can produce the array of Fig. 3.25c. Such an array is called a tilt boundary, as there is a slight misorientation of the glide planes to either side of the boundary. A tilt boundary is a special case of a low-angle or subboundary, and such boundaries are observed commonly.

If the two edge dislocations of Fig. 3.25 have opposite Burgers vectors, a metastable array of dislocations is obtained when the x and y coordinates of the dislocation separation are equal (Fig. 3.25d). As with dislocations of the same sign, an array of numerous dislocations of opposite sign can produce a subboundary; similar low-misorientation boundaries can also be produced by arrays of screw dislocations.

Although the y component of the force between parallel edge dislocations does not promote dislocation glide, it does facilitate climb. For the situation of Fig. 3.25a, for example, the force F_y aids climb so as to increase the vertical spacing between dislocations. When the dislocations are of opposite sign, climb is in the opposite direction and a combined climb-glide process can result in annihilation of the two edge dislocations (Prob. 3.8).

Finally, we note that there is no force between edge and screw dislocations having lines parallel to each other. In toto, however, it is clear that forces between and among dislocations can be appreciable and must be taken into account when discussing dislocation motion. This topic is revisited in Sect. 3.8C where dislocation arrays in materials having large numbers of dislocations are considered.

The total force per unit length acting on a dislocation is given by an expression analogous to Eq. (3.16a); i.e., the force is given by

$$F = \tau b \tag{3.17}$$

where τ is the sum of all internal and external shear stresses acting to induce dislocation motion. The force acting on the line is instantaneously normal to it. Thus, the motion of the screw dislocation of Fig. 3.12 is explained; that is, although the shear stress is applied in a direction parallel to the line, the force promoting glide is normal to it.

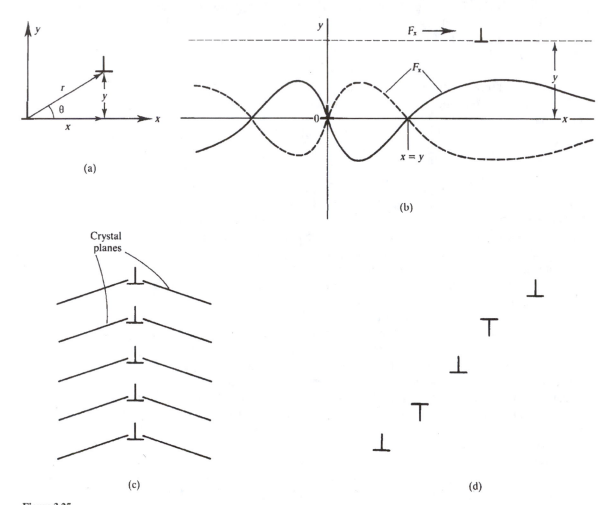

(a)

(b)

(c)

(d)

Figure 3.25
(a) The coordinate system used to define the force between two parallel edge dislocations in terms of Eqs. (3.16). (b) The F_x component (which acts to promote glide) between the parallel dislocations of (a). The solid line applies when the dislocations have the same sign (parallel Burgers vectors), and the dotted one when they have anti-parallel slip vectors. (c) A metastable array obtained from parallel dislocations with the same Burgers vectors. This results because $F_x = 0$ at $x = 0$ (see (b)). A slip plane misorientation exists across the sub-(tilt) boundary formed. (d) A different metastable array results for dislocations with alternating anti-parallel Burgers vectors. This array comes about because $F_x = 0$ at $x = y$ for such dislocations.

D. Kinks in Dislocations

As mentioned, a dislocation tends to minimize its total length because its strain energy is minimized by doing so. Conversely, when dislocations glide they encounter obstacles to motion along the slip plane, and this can lead to the presence of curved dislocations. The periodically varying energy that results in the Peierls-Nabarro stress is useful for discussing the competition between these factors. In Fig. 3.26, the plane of the drawing is the slip plane. The dotted lines, referred to as valleys, represent low-energy positions for the dislocation, and the dashed lines (peaks) represent energy maxima between the valleys (cf. Fig. 3.4). We consider a dislocation line with a kink in it. The portions AB and CD of the line in Fig. 3.26 lie in valleys; the kink portion traverses the energy peaks. The shape of the kink region is defined by the angle θ of Fig. 3.26, and the value of θ is determined by the difference between the peak and valley energies. If this energy difference is substantial, the dislocation length in the high-energy regions is as small as possible and the kink is a sharp one. Such a kink maximizes total dislocation length, but this is compensated for by having but a small fraction of the length in the high-energy regions. On the

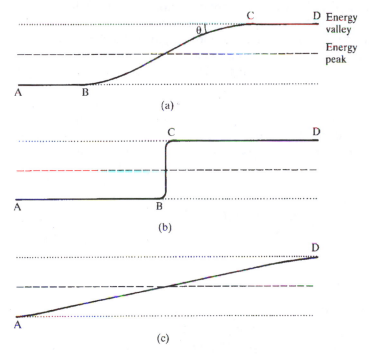

Figure 3.26
(a) A view of a kinked dislocation (the slip plane is the plane of the drawing). The portions AB and CD of the line lie in energy "valleys." The BC portion of the line lies in a relatively high-energy lattice region. The shape of the kinked portion of the curve is defined by the angle θ. (b) When the energy of the "peaks" is high relative to that of the valleys, a sharp kink ($\theta = 90°$) is formed, even though the total length of line between A and D is maximized in so doing. (c) If the energy difference between peaks and valleys is small, the dislocation minimizes its total length and becomes very nearly a straight line ($\theta \to 0°$).

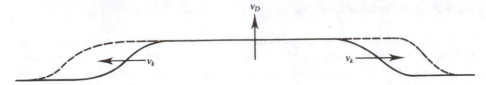

Figure 3.27
**Velocity of a dislocation normal to itself is achieved by lateral spreading of disloca-
tion kinks with velocity v_k; i.e., a net translation of the line is accomplished by the
spreading. Continued motion of the dislocation by this mechanism requires that
kinks be repeatedly nucleated along the dislocation line.**

other hand, if the energy difference between the peaks and the valleys is low, θ becomes very small (Fig. 3.26c), and the kink is so "diffuse" that the dislocation line
is essentially straight.

Kinks of the type shown in Fig. 3.26 permit dislocation motion by associated
kink motion (Fig. 3.27). Here the dislocation moves normal to itself with velocity
v_D, but this motion is achieved by lateral motion of the kinks with velocity v_k, which
is generally greater than v_D. (Kink lateral motion is not associated with the overcoming of energy barriers.) Dislocation motion resulting from kink motion requires
an equilibrium number of kinks per unit dislocation length, and thus v_D is limited by
the kink nucleation frequency. In summary, motion of dislocation lines containing
kinks (i.e., lines for which $\theta > 0°$, cf. Fig. 3.26) can be achieved by the lateral
spreading of kinks.

Kinks should not be confused with jogs. A kink on a dislocation line is merely
a special example of a curved dislocation. Or, to put it another way, the kink portion and the rest of the dislocation line not only have the same Burgers vector, but
also lie within the same plane. Further, as noted, kinks facilitate dislocation motion.
Jogs generally do the opposite. The reason is that, notwithstanding that the Burgers
vector of a dislocation containing a jog is the same along the whole of the dislocation line (just as it is for a dislocation containing a kink), a dislocation with a jog
moves in and out of a slip plane. This generally restricts dislocation motion as already described in Sect. 3.3D. The effect is elaborated on further in Sect. 3.8.

E. Dislocation Velocities

The velocity of a dislocation depends strongly on the value of the applied shear
stress to which it is subjected. Dislocation velocities can be determined to reasonable accuracy by suitable techniques, but ascertaining the stress causing dislocations
to move is more difficult. In effect, the dislocation density[6] must be sufficiently low
to insure that interactions among dislocations can be neglected and, therefore, that
the applied stress is the one causing dislocation motion. A number of carefully ex-

---

[6]Dislocation density is defined as the length of dislocation line per unit volume; it has units of m/m^3.
Alternatively, dislocation density can be expressed as the numerically equivalent $number/m^2$. The latter
is visualized physically as the number of dislocations intersecting an arbitrary crystal plane per unit area
of this plane.

ecuted experiments have related dislocation velocity to applied stress. The results
of some of these are shown in Fig. 3.28. As noted in Fig. 3.28a, the limiting velocity of a dislocation is the material sound velocity. For stress levels well below those
producing dislocation velocities of this magnitude, the relation between dislocation
velocity (v_D) and applied stress can be represented by the empirical equation

$$\frac{v_D}{v_0} = \left(\frac{\tau}{\tau_0}\right)^P \tag{3.18}$$

where τ_0 and P are experimentally determined material constants and v_0 has the significance of being the dislocation velocity at the stress τ_0. The strong dependence
of velocity on stress can, as we shall see, cause yield point phenomena in certain
materials. The greater dislocation velocities at higher temperatures in silicon iron
(Fig. 3.28b) are related to previously mentioned factors such as an increase in dislocation width with increasing temperature, thermally assisted formation of dislocation kinks, and increased atomic vibrations at higher temperatures, all of which
promote readier dislocation motion.

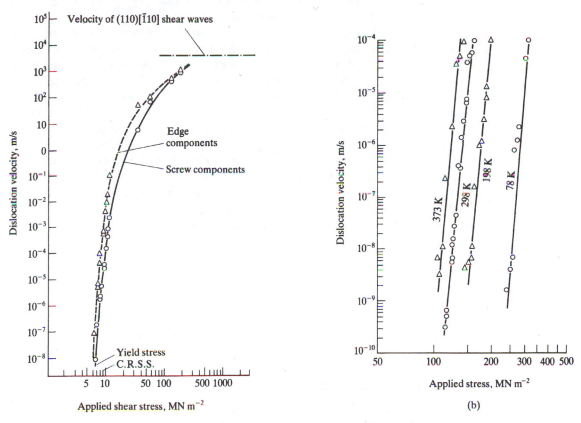

Figure 3.28
**The relation between dislocation velocity and stress for (a) screw and edge dislocations in lithium fluo-
ride (*from W. G. Johnston and J. J. Gilman, J. Appl. Phys., 30, 129, 1959*) and (b) edge dislocations in
an Fe-3.25% Si alloy (bcc structure)(*from D. F. Stein and J. M. Low, J. Appl. Phys., 31, 362, 1960*).
The limiting velocity is the material sound velocity (= 3.6 × 10³ m/s for LiF; not shown for the iron).
Dislocation velocity is a strong function of stress.**

3.7

DISLOCATION GEOMETRY AND CRYSTAL STRUCTURE

A. Slip Systems

As mentioned, slip is expected to occur in close-packed atomic directions. Indeed, this is observed for plastic flow of metals and a number of nonmetallics. Further, and also as mentioned, slip can, and does, happen in nonclose-packed directions in certain ionic solids. Also from earlier discussion, slip planes should be those with the densest atomic packing, for they are characterized by the greatest separation between parallel crystal planes. This "rule" is generally observed, but not to the same extent as the one regarding slip directions.

Slip in face-centered cubic metals is consistent with the "rules" on slip directions and planes; that is slip takes place on close-packed {111} planes and in <110> type directions. Body-centered cubic metals slip in <111> close-packed directions characteristic of this structure and also slip preferentially on the closest-packed {110} planes. However, slip is also observed frequently on {112} as well as on other planes in bcc metals. Regardless of the slip plane, the slip direction in hexagonal close-packed metals is the close-packed basal plane direction. In the ideal hcp structure (based on stacking of close-packed layers of spherical atoms and for which $c/a = (8/3)^{1/2} = 1.633$), slip should occur on the basal or {0001} plane. Basal plane slip is also expected when $c/a > 1.633$, since the separation distance between basal planes is increased for this circumstance. Conversely, if $c/a < 1.633$, the basal planes are less widely separated and additional slip planes may become active. Thus, prism and pyramidal slip (Table 3.3) occur in some hcp metals.

The number of combinations of slip planes and directions provided by the crystallography of a crystal structure determines the number of its *slip systems*. Thus, the face-centered cubic structure has 12 slip systems that results from there being four geometrically distinct (although crystallographically equivalent) planes of the {111} type, each of which contains three distinct <110> type directions. As demonstrated in Table 3.3 for the common metallic structures, the number of slip systems

Crystal structure	Slip plane	Slip direction	Number of nonparallel planes	Slip directions per plane	Number of slip systems
Face-centered cubic	{111}	<1$\bar{1}$0>	4	3	12 = (4 × 3)
Body-centered cubic*	{110}	<$\bar{1}$11>	6	2	12 = (6 × 2)
	{112}	<11$\bar{1}$>	12	1	12 = (12 × 1)
	{123}	<11$\bar{1}$>	24	1	24 = (24 × 1)
Hexagonal close-packed†	{0001}	<11$\bar{2}$0>	1	3	3 = (1 × 3)
	{10$\bar{1}$0}	<11$\bar{2}$0>	3	1	3 = (3 × 1)
	{10$\bar{1}$1}	<11$\bar{2}$0>	6	1	6 = (6 × 1)

*Slip other than the {110}<$\bar{1}$11> type in bcc metals is system- and temperature-specific. That is, at higher temperatures secondary slip systems shown may come into play, depending on the material. See also comments in Table 4.1.

†Slip in hcp metals is system-specific, depending on the *c/a* ratio. Basal plane ({0001}) slip is favored by high *c/a* ratios. See also comments in Table 4.1.

Table 3.3
Slip systems in metallic crystal structures

is obtained by multiplication of the number of (nonparallel) geometrically distinct slip planes by the number of nonparallel slip directions contained in each of them.

B. Partial Dislocations

Slip in face-centered cubic materials can be complicated by the slip path preferred by atoms during shear displacements of atomic planes. A portion of a {111} glide plane in the face-centered cubic structure is shown in Fig. 3.29, along with the Burgers vector corresponding to a full atomic displacement ($\mathbf{b} = (a/2)[\bar{1}01]$, with magnitude $b = 2r = a/\sqrt{2}$). A somewhat easier path for atomic gliding is provided if slip occurs by the two partial slip steps shown in Fig. 3.29. That this process of slip is physically plausible can be verified by placing a billiard ball on a previously racked cue held secure. If the upper ball is glided from positions corresponding to the complete slip distance of Fig. 3.29a, it is observed that the transit occurs more readily by segmental motion equivalent to steps X and Y as this motion is not accompanied by the "up and over an adjacent ball" process that the complete slip displacement is.

The vector components of segments X and Y are $(a/6)[\bar{2}11]$ and $(a/6)[\bar{1}\bar{1}2]$, respectively; the components add vectorially to $(a/2)[\bar{1}01]$. If slip occurs via partial steps, the separate segments are considered *partial dislocations,* since their vectors are less than the complete slip vector. Whether slip in fcc metals is associated with a full slip step or with two partial ones depends on energy considerations. The dissociation of a dislocation into two partials is *favored* on strain energy grounds because the total dislocation energy is reduced by the splitting. That is, the sum of Gb^2 for the two partials is $Ga^2/3$ $(=(Ga^2/36)(4 + 1 + 1) + (Ga^2/36)(1 + 1 + 4))$, whereas Gb^2 for the full dislocation is $Ga^2/2$ $(=(Ga^2/4)(1 + 0 + 1))$. Thus, on strain energy considerations, the *dislocation reaction*

$$\frac{a}{2}[\bar{1}01] = \frac{a}{6}[\bar{2}11] + \frac{a}{6}[\bar{1}\bar{1}2] \tag{3.19}$$

is expected.

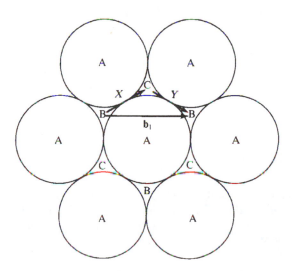

Figure 3.29
A portion of the atoms in a (111) plane in a fcc crystal. Unit slip in the B plane occurs with displacement $(a/2)[\bar{1}01]$. Instead of slip occurring by unit displacement it can take place in two segmental steps (X and Y). The displacement X results in atoms in plane B temporarily occupying a C stacking sequence in the fcc lattice.

Following dissociation, the two partial dislocations repel each other. This can be realized by considering the initial complete dislocation as an edge dislocation; thus, the X segment of the partial has two components; X_1, a screw component, and X_2, an edge component (Fig. 3.30a). The second partial dislocation is considered similarly. The force between the partials is the sum of the repulsive force between the like edge components and the attractive force between the unlike screw components. The former dominates as a result of the larger Burgers components involved. Thus, the partial dislocations move away from each other as shown in Fig. 3.30b.

This somewhat abstract discussion of partial dislocations can be related to slip crystallography. If the dislocation (an edge dislocation) of Figs. 3.29 and 3.30 is moving to the right in the plane of the paper, then the region to the right of segment Y is unslipped, while the region to the left of segment X has undergone a complete slip step; (that is, the displacement across the slip plane is $(a/2)[\bar{1}01]$). In the region between the partials, *partial slip* has taken place; the atoms directly above the slip plane are displaced with respect to the atoms in the plane below by the partial dislocation vector. This disrupts the ABCA/BCABC atomic stacking characteristic of the fcc structure. In the region between the partials, the stacking is ABCA/CABCA where, in the notation used, the A plane prior to the / is considered the slip plane. The stacking of planes in the sequence ACA is called a stacking fault, and since the atomic packing within it is no longer characteristic of the fcc structure, the stacking fault has an associated energy (stacking fault energy, SFE, with units of J/m^2). As the partials separate, more stacking fault is created and the equilibrium spacing of the fault (w, Fig. 3.30b) is obtained when the net repulsive force (per unit length) between the partials equals the SFE (Example Prob. 3.1). Thus, partial dislocations are separated widely if the SFE is low; conversely, if the SFE is high, dislocation dissociation into partials will not occur and the slip dislocation will be a unit dislocation, i.e., of the type $(a/2)[\bar{1}01]$. In three dimensions, a stacking fault can be viewed as a ribbon of faulted material bonded on two sides by partial dislocations and having a thickness of several atomic diameters. Two views of a stacking fault according to this scheme are shown in Fig. 3.31.

The SFE magnitude controls the ease of cross-slip in fcc metals. As mentioned, cross-slip of screw dislocations can occur in these metals. However, if as a result of a low SFE, a screw dislocation dissociates into partials, cross-slip requires recombination of the partials because the latter contain edge components that cannot

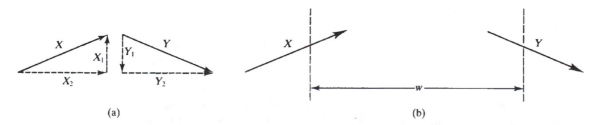

(a)   (b)

Figure 3.30
(a) Resolution of the X and Y partial slip vectors into screw (X_1,Y_1) and edge (X_2,Y_2) components. The total slip vector $= X_2 + Y_2$ is assumed to be that of an edge dislocation. (b) Although the screw components of the partials attract, the repulsion between the edge components is greater, and the partial dislocations separate. Their equilibrium separation (w) depends on the stacking-fault energy.

cross-slip. The more widely separated the partials, the more difficult the recombination process. Thus, fcc materials with low SFEs cross-slip with difficulty and vice versa. A reduced tendency to cross-slip, by which obstacles to dislocation motion can be bypassed, generally results in greater work hardening. For example, brass and stainless steel, with low SFEs, work-harden rapidly.

Analogous considerations apply to stacking faults and partial dislocations in hcp metals, particularly when basal slip is dominant. Body-centered cubic materials can also have stacking faults, but their crystallography is more complex than for fcc or hcp metals, and stacking fault energies of bcc metals are also generally higher.

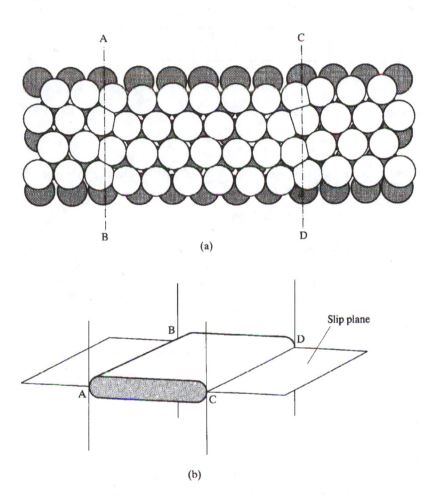

(a)

(b)

Figure 3.31
Two views of a stacking fault in a fcc crystal. (a) A view normal to the fault; the positions of the partial slip vectors are designated by AB and CD, and the atomic distortion along these lines is evident. Between AB and CD the atomic stacking is faulted (e.g., ACA rather than ABC). (b) A three-dimensional perspective showing that the fault is a narrow ribbon of thickness several atomic diameters and bonded by the partial dislocations AB and CD. (*Part (a) adapted from W. G. Moffatt, G W. Pearsall, and J. Wulff, The Structure and Properties of Materials, Vol. I, Wiley, New York, 1964.*)

EXAMPLE PROBLEM 3.1. Refer to Fig. 3.30a. Take the components X_1 and Y_1 of the partials as screw and take the components X_2 and Y_2 as edge. Consider the net force between the partials as that due to repulsion between the edge components and attraction between the screw components. As the partials separate, there is an additional contribution to the crystal energy. It increases by SFEl, where SFE is the stacking fault energy and l the fault separation distance. Thus, the "chemical" force resisting separation is SFE (dimensions of energy per unit area or force per unit length). Determine the equilibrium separation of the partials in terms of SFE and material properties.

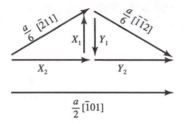

Solution. Figure 3.30a is elaborated on above. The vectors $\bar{X}$ and $\bar{Y}$ are (cf. Eq. (3.19))

$$\bar{X} = \frac{a}{6}[\bar{2}11] \quad \text{and} \quad \bar{Y} = \frac{a}{6}[\bar{1}\,\bar{1}2]$$

The edge vectors, $\bar{X}_2$ and $\bar{Y}_2$, are equal and given by $(a/4)[\bar{1}01]$. The screw components have opposite sign. By vector algebra;

$$\bar{X}_1 = \frac{a}{6}[\bar{2}11] - \frac{a}{4}[\bar{1}01] = \frac{a}{12}[\bar{1}2\bar{1}]$$

We use Eqs. (3.16a), (3.8c), and (3.7) to solve for the respective forces, i.e., $F_x = \pm b\tau_{xy}$ with (since $\theta = 0$) $\tau_{xy} = Gb/[2\pi(1-v)r]$ for an edge dislocation and $\tau = Gb/2\pi r$ for a screw dislocation (r = slip plane separation of the partials). (Note the dimensions of the "force" are force per unit length.) To calculate the repulsive force between the edge components, we use the square of the magnitude of their Burgers vectors in the equations noted; i.e., $b_e^2 = (a^2/16)(1 + 1) = a^2/8$. For the screw components; $b_s^2 = (a^2/144)(1 + 4 + 1) = a^2/24$. The magnitude of the net force (it is repulsive) is obtained by subtracting the force between the screw components from that between the edge ones. Thus, using, Eqs. (3.8c) and (3.7)

$$F = \frac{Ga^2}{16\pi(1-v)r} - \frac{Ga^2}{48\pi r} = \left(\frac{Ga^2}{16\pi r}\right)\left[\frac{(2+v)}{3(1-v)}\right]$$

The term in square brackets (the one containing v) on the right-hand side of the above equation has a value = ⅞, if v is taken as ⅓. In order to simplify the arithmetic, we take this term as equal to unity.

As the partials move apart (thereby reducing dislocation elastic strain energy) a stacking fault—which increases system energy—is formed. Per unit length of dislocation line, the stacking fault energy is (SFE)r (units of J/m). The associated force (units of N/m) resisting the formation of the fault is, therefore, SFE. Equating the elastic energy and the "chemical" forces,

$$\frac{Ga^2}{16\pi r} = \text{SFE} \quad \text{or} \quad r = \frac{Ga^2}{16\pi(\text{SFE})}$$

Note that, as expected, if SFE is high, the spacing of the partials is low and vice versa.

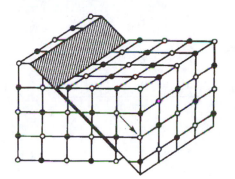

Figure 3.32
The slip displacement in NaCl. The open circles represent the Na^+, and the closed the Cl^-, ions. As in fcc metals the slip direction is of the <110> type, but owing to considerations of Coulombic energy, slip in NaCl occurs on {110} planes rather than on the {111} slip planes of fcc metals. (*Adapted from J. J. Gilman, Micromechanics of Flow in Solids, McGraw-Hill, New York, 1969.*)

C. Dislocations in Nonmetallic Materials

Discussion to this point has focused on dislocations in typical metallic crystals because dislocation geometry in them is relatively simple. However, dislocations are present and are also responsible for plastic flow in crystalline nonmetallics. For example, layered materials (e.g., talc and graphite) plastically deform by dislocation glide. In them, the atomic bonding within certain crystal planes is much greater than between them, and slip occurs by relative displacement of the strongly bonded planes. Partial dislocations can also be found in layered materials.

The ionic material sodium chloride has the fcc structure; i.e., it consists of interpenetrating fcc lattices of Na^+ and Cl^- ions. As in fcc metals, slip occurs in <110> type directions but slip planes are of the {110} type (Fig. 3.32). This plane is selected in preference to the {111} planes of fcc metals because, in doing so, repulsive forces between like ions are minimized during dislocation glide (Prob. 3.13). Similar considerations apply to CsCl, which has simple-cubic symmetry with two interpenetrating ion lattices such that Cl^- ions occupy the corners, and Cs^+ ions the center, of a cubic unit cell. Although ions touch along <111> directions in CsCl, the slip direction is of the <001> type, consistent with the simple-cubic structure. On the other hand, slip planes are of the {110} variety as a result of considerations similar to those applying to NaCl.

D. Slip and Dislocations in Ordered Structures

Slip in ordered structures, common to many intermetallic compounds, is also more complicated than in most metals. A two-dimensional portrayal of a simple ordered structure is shown in Fig. 3.33a. Note that if the crystal is displaced by one interatomic spacing across the slip plane (Fig. 3.33b), undesirable A-A and B-B near neighbors are formed with a concurrent increase in crystal energy. Such an arrangement would happen if a simple edge dislocation with Burgers vector $\bar{b}$ passed through the lattice, although a second dislocation would restore the proper bonding across the slip plane. Because of this, dislocations in structures of this type are often double ("superlattice") dislocations with, as shown in Fig. 3.33c, a Burgers vector equal to $2\bar{b}$. Propagation of the superdislocation as a unit is not accompanied by the formation of "poor" bonds across the interface, yet the elastic strain energy ($\cong 2Gb^2$) of a superdislocation is greater than that of two widely separated single

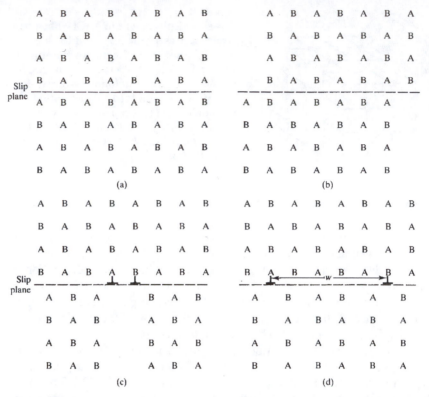

Figure 3.33
(a) A two-dimensional schematic of an ordered AB crystal. A and B atoms are pre-ferred near neighbors. (b) Slip (e.g., by the passage of a simple edge dislocation) by the unit distance, *b*, along the slip plane produces undesirable atomic bonding across the plane. (c) Slip by movement of a "superlattice" dislocation (with Burgers vector = 2b̄) does not produce the undesirable bonding. (d) Because of the elastic repulsion between the individual dislocations in the superdislocation, it splits. In analogy with partial dislocations, the equilibrium separation of the unit dislocations depends on the value of the anti-phase boundary energy.

dislocations ($\cong Gb^2$). Thus, the situation is similar to that for partial dislocations, with the dislocations establishing the structure shown in Fig. 3.33*d*. As a result of their elastic repulsion, the individual dislocations are separated and improper bonding across the slip plane occurs between them. This anti-phase boundary (APB) has an associated energy, and the equilibrium separation of the unit dislocations is dictated by the APB energy (Prob. 3.14); if this energy is high, the separation distance is small and vice versa.

In crystals such as Cu_3Au, with A_3B stoichiometry and possessing the face-centered cubic structure, the superlattice dislocation is composed of two unit $(a/2)[110]$ dislocations. Dissociation of the unit dislocations can occur so as to produce a dislocation arrangement schematically illustrated in Fig. 3.34. The region across the slip plane over which the APB exists is bounded on both sides by individual stacking faults, associated with the dissociation of the unit dislocations into partials.

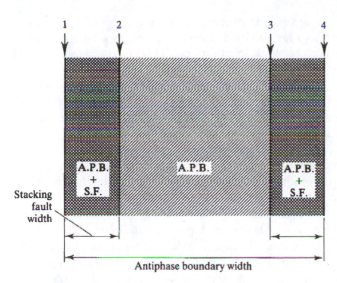

Figure 3.34
Schematic view looking down on a slip plane in an ordered A$_3$B fcc crystal. The unit dislocations of the superlattice dislocation have split into partials (1 and 2; 3 and 4). Thus, both anti-phase and stacking fault boundaries exist across the slip plane. (*Adapted with permission from M. J. Marcinkowski, N. Brown, and R. M. Fisher, Acta Metall., 9, 129, 1961.*)

3.8
INTERSECTION OF MOVING DISLOCATIONS

Dislocation topology is altered by interactions between moving dislocations. In this section, we consider several features of such interactions. We first discuss how dislocation topology is altered by dislocation intersections. These take place during plastic flow, especially when it is accomplished by dislocation glide on multiple slip systems. These intersections generally give rise to a reduction in dislocation velocity, thus promoting work hardening. The effect is mitigated to a degree by the generation of dislocation arrangements suitable for dislocation multiplication. Such multiplication, observed commonly in plastically deforming materials, allows for continued plastic flow in a material of high dislocation density even though the mobilities of the dislocations are reduced in such a circumstance. The section concludes with a description of how dislocation arrangement is altered during plastic deformation. In particular, the dislocation pattern becomes heterogeneous, and this pattern emerges in response to the material's attempt to minimize total dislocation strain energy.

A. Jogs on Dislocations

An important dislocation interaction involves intersection of two dislocations. Dislocation intersection generally requires an increase in lattice energy, since intersection often introduces jogs in one or both of the dislocations, thus increasing dislocation line length. Further, jogs created by dislocation intersections are frequently less mobile than other portions of the dislocation line. This effect can lead to work hardening.

Dislocation intersection can be visualized with the aid of Fig. 3.35a. Here the moving edge dislocation CD intersects the edge dislocation AB whose slip plane differs from, and is normal to, CD's slip plane. Following dislocation intersection, a portion of the plane containing AB is displaced in the direction of the Burgers vector of CD. In the process, a jog PP' forms on AB (Fig. 3.35b). The energy increase accompanying jog formation is approximately $Gb^3/2$; that is, the dislocation line with energy per unit length $Gb^2/2$ is increased in length by an amount b. The Burgers vector of the jog is the same as that of the rest of AB. And, since $\bar{b}$ is normal to AB, the jog is edge-like. However, the slip plane of the jog is not the same as that of AB. The formation of the jog in AB does not, per se, reduce the mobility of it, provided a component of the stress results in a force acting on the slip plane of the jog and in the direction of its Burgers vector. However, if no such stress is present, the jog is immobile. Moreover, if the shear stress on the jog is less than that on the remainder of AB, the jog moves at a lesser velocity (cf. Eq. (3.18)) and hence restrains motion of the remainder of the line.

The intersection of a moving edge dislocation (EF, Fig. 3.36a) with a screw dislocation (GH) is illustrated in Fig. 3.36. The helical atomic arrangement around GH is schematized by a like configuration of the lattice planes of the figure. After intersection has occurred, the point E on the line (now E') is displaced by one atomic

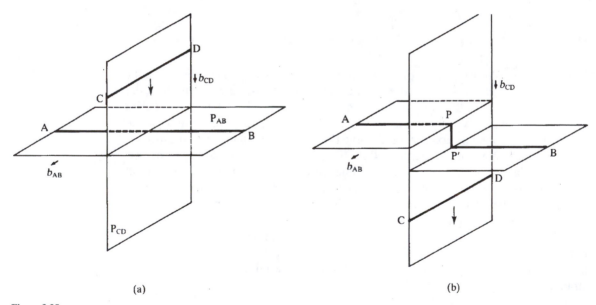

(a)   (b)

Figure 3.35
Intersection of two edge dislocations. (a) The moving dislocation CD intersects dislocation AB. (b) After it does, and as a result of the atomic displacements on the slip plane of CD, AB has a jog PP' on it. PP' has the same Burgers vector as the rest of AB and is, therefore, an edge jog. However, PP's slip plane is *not* the slip plane of AB. (*Adapted from W. T. Read, Jr., Dislocations in Crystals, McGraw-Hill, New York, 1953.*)

plane with respect to point F (Fig. 3.36b); that is, a jog is created on dislocation EF.[7]
Considerations similar to the case just described also apply here. The jog in the
edge dislocation is edge-like and has a slip plane different from that of the unjogged
portion of the line. As such, the velocity of the jog (PP′) may be less than the unre-
strained velocity of the original line (EF), causing a net reduction in dislocation mobility.

The intersection of a moving screw dislocation with another screw dislocation
is illustrated in Fig. 3.37. After intersection, the line arrangement is similar to that
shown in Fig. 3.36. The jog formed is edge-like, and continued motion of the screw
dislocation requires this jog to move in a direction *normal* to its slip plane. Thus,
climb is necessary for jog motion if the atomic structure in the jog vicinity is to be
preserved; alternatively the jog can move if it generates a row of vacancies or self-
interstitials (which of the two depends on the slip direction) in the jog plane (Fig.
3.37b). It can be seen, therefore, that the formation of a jog on a screw dislocation
such as that just described results in a substantial reduction in dislocation mobility.
Moreover, since most dislocations are mixed dislocations, jog formation on them
leads to relatively immobile jogs.

(a)

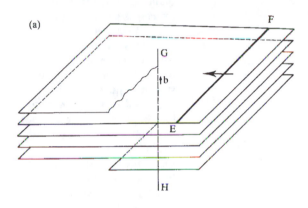

(b)

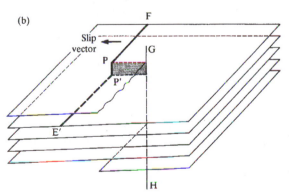

Figure 3.36
**Intersection of a moving
edge dislocation (EF) with a
screw dislocation (GH). (a)
The situation prior to the intersection and, (b), subsequent to it. Point E is now
displaced to E′, which lies on
the plane below F. Thus, the
jog PP′ is created; PP′ is an
edge jog because its Burgers
vector is the same as that of
EF.** (*Adapted from W. T.
Read, Jr., Dislocations in
Crystals, McGraw-Hill, New
York, 1953.*)

[7]Consideration of Figs. 3.35 and 3.36 confirms a rule governing dislocation intersections which states
that "when one dislocation cuts another, each acquires a jog equal to the component normal to its own
slip plane of the other dislocation's Burgers vector."

The reduced mobility of a jogged dislocation depends on the jog height. If this height is but the interatomic distance, b, the mobility reduction is marked for edge jogs on screw dislocations, and jog motion requires the high stresses needed to generate vacancies or interstitials. On the other hand, consecutive intersections can lead to a composite jog with a height of several Burgers vectors or greater. This is illustrated in Fig. 3.38 where the jog segment PP′ is assumed sufficiently long so that the stress fields of the components AP and P′B do not interact significantly.

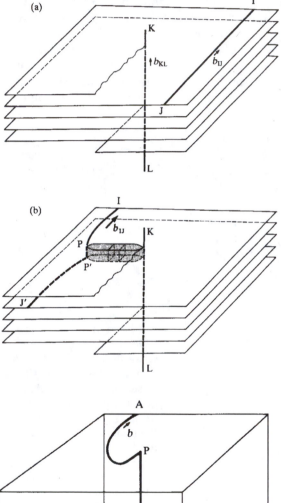

Figure 3.37
Intersection of a moving screw dislocation (IJ) with another screw dislocation (KL). The schematic of the before (a) and after (b) process is similar to that of Fig. 3.36 and the jog (PP′) formed is an edge jog. Continued motion of the screw dislocation in the direction shown requires motion of the jog in a direction *normal* to its slip plane. In the absence of climb, this leaves behind a "void" (a row of vacancies— the shaded area) in the crystal. (*Adapted from W. T. Read, Jr., Dislocations in Crystals, McGraw-Hill, New York, 1953.*)

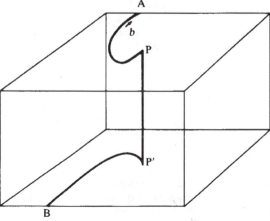

Figure 3.38
A composite jog PP′ of length sufficient that interaction between the stress fields of components AP and P′B of this screw dislocation is minimal. Under the action of an applied stress, the segments pinned at P and P′ continue to produce slip by their bowing in the slip planes.

Hence, AP and P'B can move independently of each other and, as shown in Fig. 3.38, continued slip does *not* necessitate nonconservative processes; that is, dislocation motion of segments AP and P'B continues even while the jog connecting them remains immobile.

B. Dislocation Multiplication

Immobilization of dislocation line segments can lead to dislocation *multiplication,* a process observed ubiquitously during plastic flow in crystals. A well-annealed metal can have dislocation densities as low as ca. 10^{10} m/m^3; yet, if the metal is strained sufficiently, the dislocation density can increase to as much as 10^{15}–10^{16} m/m^3. It may appear contradictory that while dislocations are responsible for plastic flow an increase in their density causes work hardening, but this is not so. In effect, the reduction in dislocation mobility concurrent with their greater numbers more than compensates for their greater numbers.

A particular mechanism for dislocation multiplication, the Frank-Read source, is shown in Fig. 3.39. Within the crystal (Fig. 3.39a) the dislocation line segments AB and CD considered immobile in the presence of the shear stress that acts to extend the segment BC on its slip plane. (The immobility of AB and CD may arise from intersection processes and/or as a result of the sense of the applied stress.) The stress causes BC to bow out (Fig. 3.39b). The stress required to expand the segment into a circular arc with radius of curvature r is shown in Fig. 3.40 for an infinitesimal portion, δs, of such an arc. The force on the line is $\tau b \, \delta s = \tau r b \, \delta \theta$. The line tension (energy per unit length) opposing the dislocation bowing is $2T \sin(\delta\theta/2) \cong T \, \delta\theta$. Taking $T = Gb^2/2$, the required stress is

$$\tau = \frac{Gb}{2r} \tag{3.20}$$

The smallest radius of curvature, obtained when the loop is in the form of a semicircle (Fig. 3.39c), corresponds to the maximum stress required to effect the process shown in Fig. 3.39 and is given by $r = l/2$, where l is the distance between the pinned segments. After the loop assumes a semicircular shape, it expands spontaneously (Fig. 3.39d). As the points X and X' approach they annihilate each other (they have opposite character),[8] leading to the situation of Fig. 3.39e. As a result of dislocation line tension, the cusps in the dislocation subsequently straighten producing the configuration of Fig. 3.39f. The net result is the generation of a dislocation loop. Under the action of the applied stress, this loop continues to expand, producing slip across the slip plane and, simultaneously, the generation of a new loop begins. Each additionally generated loop produces further slip, and it appears that the process could continue indefinitely. In practice, the expanding loops encounter either obstacles to slip or a crystal grain boundary that is impenetrable to

---

[8]Although the Burgers vector is the same along all portions of the loop, this is a consequence of taking Burgers circuits along the line while traversing it in one direction. If an observer were to look (horizontally) into the plane of the paper of Fig. 3.39 and take separate clockwise circuits around the portion of the lines containing X and X', the Burgers vectors found would be the opposite of each other.

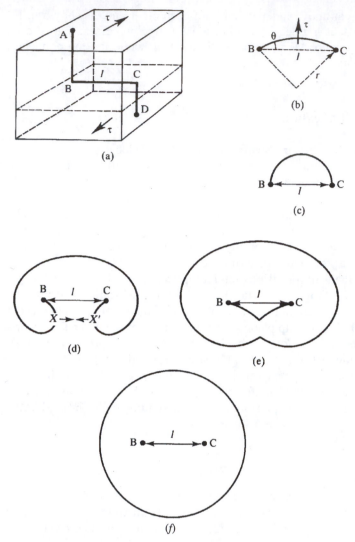

Figure 3.39
(a) A segmented dislocation (ABCD) in a crystal. The line is an edge dislocation for which only the component BC is driven to move when the shear stress shown is applied. (b)–(f) Successive shapes of the segment BC as viewed on its slip plane. The dislocation initially bows out in a circular arc (b), eventually attaining the shape of a semicircle (c). A maximum shear stress is required to reach this shape; beyond this point, the shape is unstable and winds about itself (d). The portions of the line near X and X' meet and annihilate (e). Finally (f), the cusps on the line are removed by the dislocation line tension. The result is a net production of a dislocation loop; i.e., dislocation multiplication has occurred.

dislocations. The resulting restricted motion of the lines produces a "back stress" opposing that causing loop generation; as a result, every potential loop generator generates only a finite number of loops. However, that dislocation generation by this, and similar, means takes place is not in question. A particularly clear and striking example of a Frank-Read source in silicon is shown in Fig. 3.41.

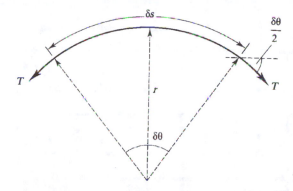

Figure 3.40
Calculation of the stress
required to produce a curved
dislocation with radius of
curvature r. The arc length
δs ($= r\,\delta\theta$) **is acted on by the**
force $\tau b \delta s$. This is balanced by
the force due to the line tension;
$2T \sin(\delta\theta/2) \cong T\,\delta\theta$. **Thus, with**
$T = Gb^2/2$, **we have $\tau = Gb/2r$.**

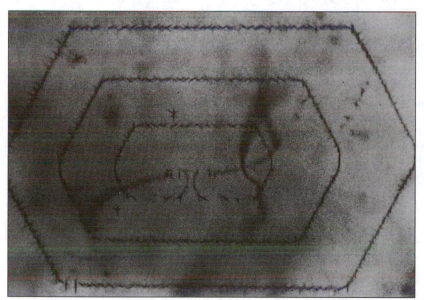

Figure 3.41
A Frank-Read source in silicon. The dislocation loops have been delineated by chem-
ical etching. In silicon, the loops are not circular; the anisotropic bonding of Si cre-
ates "loops" composed of approximately straight segments. (*From W. G. Dash,*
Dislocations and Mechanical Properties of Crystals, ed. J. C. Fisher, Wiley, New York,
***1957.*)**

C. Dislocation Arrangements at High Dislocation Densities

Previous discussion of dislocation intersections is couched in terms of random
processes. As a corollary, it is likewise considered that the distribution of dislocations with respect to their location and orientation within the crystal is random. This
description is reasonable at low dislocation densities where interactions among individual dislocations are relatively small. However, it does not apply when dislocation densities have increased to the extent that these interactions are significant.
With high dislocation densities, interdislocation interactions affect the pattern and
morphology that the dislocations assume. The patterns arise because—for a specific dislocation density—the crystal attempts to minimize the internal elastic en-

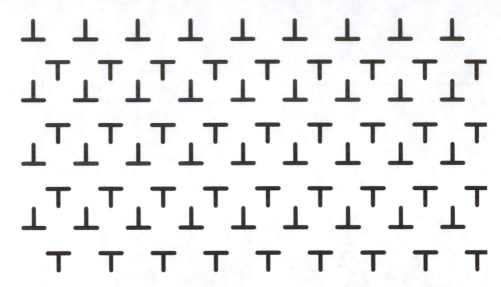

Figure 3.42
A schematic of a low-energy dislocation structure appropriate to slip taking place on one slip system. (For which situation, dislocation intersections and jog formation do not happen.) The dislocation array is composed of rows of dislocations of the same Burgers vector alternating with rows of dislocations of the opposite Burgers vector. For a specified dislocation density, the array minimizes the internal strain energy due to the presence of the dislocations (also recall Fig. 3.25(d)).

ergy due to the presence of the dislocations. Since total dislocation energy can be reduced by such interactions (cf. Sect. 3.6C), lower-energy dislocation configurations (i.e., structures) are a consequence.

A simple example of such a low-energy structure, appropriate to a single crystal deforming by single slip, is shown in Fig. 3.42. The dislocation population illustrated consists of equal numbers of edge dislocations of opposite sign. The dislocation pattern in this "Taylor lattice," as it is known, is an extension of the array previously depicted in Fig. 3.25d. The array depicted in Fig. 3.42 corresponds to the lowest possible energy for a specified total dislocation population. The spacing between individual dislocations decreases with increasing dislocation density, and is proportional to $\rho^{-1/2}$ where ρ is this density.

When slip occurs on multiple slip systems, intersections among dislocations are commonplace and a different type of dislocation pattern emerges. It is called a "cellular" structure, and its evolvement with plastic strain in a typical metal that develops a cell structure is shown in Figs 3.43a and b. Figure 3.43a is a transmission electron micrograph[9] of lightly deformed iron that has not yet developed a well-defined cellular structure. The "cell" structure can be seen in the more heavily deformed iron (Fig. 3.43b). The structure consists of boundaries in which dislocations are preferentially stored and cell interiors which are relatively dislocation free. This

[9]Transmission electron microscopy (TEM) is widely used to characterize dislocation density and morphology. The lattice distortion associated with dislocations gives rise to diffraction effects so that, under appropriate conditions, dislocations appear as lines (e.g., Figs. 3.43).

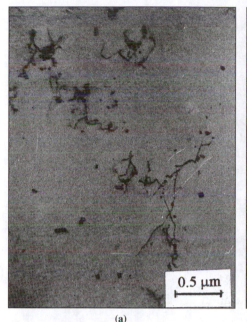

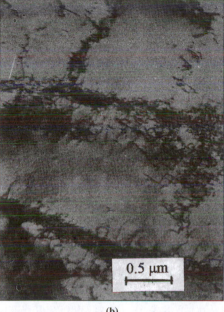

(a)   (b)

Figure 3.43
Dislocation arrangements in Fe plastically strained (a) 1% and (b) 9%. In these
transmission electron micrographs, dislocations appear as lines. (a) At the low strain,
dislocations are randomly distributed. (Some evidence of dislocation "tangles,"
though, can be seen.) (b) At the higher strain, a cellular dislocation pattern emerges.
The cell boundaries have a much higher dislocation density than the cell interiors.
(After A. S. Keh and S. Weissman, Electron Microscopy and Strength of Crystals, eds.
G. Thomas and J. Washburn, Wiley, New York, 1963, p. 242.)

dislocation arrangement minimizes total strain energy. With continued plastic
strain—and a concurrent increase in dislocation density—cell size decreases. An
empirical relationship between the cell size (D) and the applied shear stress (τ) has
been found;

$$D = \frac{KGb}{(\tau - \tau_0)} \cong \frac{KGb}{\tau} \qquad (3.21)$$

where K is a constant with a value close to ten and τ_0 is the friction stress. The approximation on the far right side of Eq. (3.21) can be used when the material is
work-hardened to a strength level well in excess of the friction stress. Another characteristic feature of cells is the "mesh length" of dislocations contained in the cell
walls (boundaries). This length is the spacing between dislocation intersections in
the walls and, clearly, this is much less than the cell size.[10] As does the cell size,
mesh length decreases with increases in dislocation density and plastic strain. The
mesh length plays a role in the work-hardening behavior of materials manifesting
cellular structures, as is discussed further in Chap. 5.

---

[10]The structural details in the cell walls cannot be seen in Fig. 3.43*b*.

3.9
DISLOCATION DENSITY AND MACROSCOPIC STRAIN

The value of macroscopic shear strain is related to the number of dislocations within the crystal and the degree to which each of them has slipped. This is illustrated in Fig. 3.44 in which the crystal considered plastically flows via edge dislocation motion. The macroscopic shear strain is $\gamma = \Delta/h$, where h is the crystal height and Δ is the displacement of the top crystal plane relative to the bottom one. This displacement is the sum of the displacements of the individual dislocations, i.e., $\Delta = \Sigma\Delta_i$, where Δ_i represents an individual dislocation displacement. The latter is given as $(x_i/L)b$ where x_i is the distance from the left side of the crystal of dislocation i, and L is the slip plane length. Hence, x_i/L is the fractional displacement of the dislocation on the slip plane. (When this is unity, the dislocation has produced a slip offset equal to the Burgers vector.) Thus,

$$\gamma = \frac{b}{Lh} \sum x_i = \frac{b}{Lh}N\bar{x} \tag{3.22}$$

where N is the total number of dislocations and $\bar{x}$ is the average dislocation displacement on the slip planes. The dislocation density ρ is given as Nz/Lzh ($= N/Lh$), where z is the crystal width; thus

$$\gamma = \rho b\bar{x} \tag{3.23}$$

Equation (3.23) relates shear strain to the dislocation population and the extent to which the "average" dislocation has slipped. Differentiation of this equation with respect to time leads to a relationship between the shear strain rate ($\dot{\gamma}$) and the av-

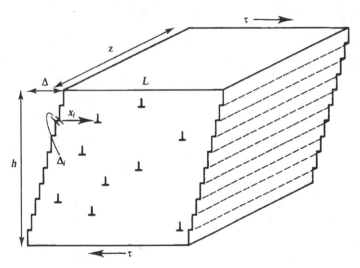

Figure 3.44
The relationship between macroscopic shear strain and dislocation transit can be deduced from this figure. The shear strain is $\gamma = \Sigma\Delta_i/h$, where Δ_i is the offset resulting from each of the N dislocations (here all assumed edge dislocations). Each dislocation produces a relative offset $(x_i/L)b$ and if $\bar{x}$ is the average slip distance of the dislocations on each plane, $\gamma = N\bar{x}b/Lh$. (*Adapted from J. J. Gilman, Micromechanics of Flow in Solids, McGraw-Hill, New York, 1969.*)

$$\dot{\gamma} = \rho b \bar{v} \qquad (3.24)$$

Confusion sometimes arises as to whether Eqs. (3.23) and (3.24) apply only to mobile dislocations or whether they refer to the total dislocation population, but it shouldn't. The equations are appropriate to the total dislocation density, provided $\bar{x}$ and $\bar{v}$ are measured relative to this population. The same result is obtained if the *mobile* dislocation density is considered, provided $\bar{v}$ and $\bar{x}$ apply only to these dislocations. That is, $\rho_{tot}\bar{v}_{tot} = \rho_{mob}\bar{v}_{mob}$ (Prob. 3.18), where the subscripts refer to the total and mobile dislocation population, respectively.

Equation (3.23) is derived for a particular dislocation arrangement. However, it is a specific example of the general rule that can be used to calculate the displacement, Δ, caused by a moving dislocation. This rule is that if a dislocation on a slip plane of area A has produced slip over an area A' of the plane, then the shear displacement is

$$\Delta = b\left(\frac{A'}{A}\right) = ba \qquad (3.25)$$

where a is the fractional area of the glide plane having slipped. As written, Eq. (3.25) is more general than Eq. (3.23), for Eq. (3.25) can be applied to curved dislocations, dislocation loops, etc. That Eq. (3.25) can be reduced to Eq. (3.23) for the case illustrated in Fig. 3.43 is left as an exercise (Prob. 3.19). It should be noted that use of Eq. (3.25) to define displacements resulting from dislocation motion does not, except for a change in the proportionality constant that we neglect, alter the relationship between shear strain rate and dislocation density and velocity given by Eq. (3.24).

As noted, Eq. (3.24) was derived assuming a constant dislocation density. However, dislocation density increases with strain while the fraction of dislocations that are mobile simultaneously decreases. This behavior is illustrated schematically in Figs. 3.45a and b. The population of mobile dislocations, obtained by multiplication of total dislocation density by the mobile fraction, is shown as a function of strain in Fig. 3.45c. As indicated, this population actually increases at low strains and "strain softening" might therefore be expected. In fact, in a number of non-metallic materials, characterized by low initial dislocation densities, the rapid dislocation multiplication at low strains is not compensated for by a marked reduction in dislocation mobility, and strain softening—in the form of a yield point—is observed. On the other hand, if the initial dislocation density is fairly high, dislocations become immobilized quickly and work hardening is concurrent with the initiation of plastic deformation. This is the situation for most metals and their alloys.

3.10
SUMMARY

A number of concepts relative to the role of dislocations in plastic flow has been introduced in this chapter. Here we summarize the most important of these, empha-

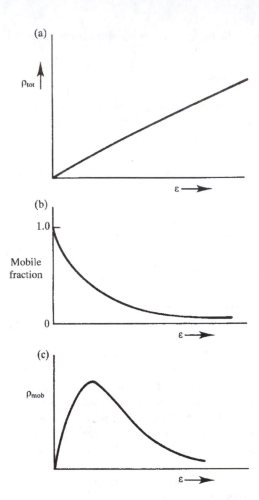

Figure 3.45
Schematic of the dependence of (a) total dislocation density, (b) fraction of the dislocation population that is mobile, and (c) mobile dislocation population (obtained by multiplication of the functions in (a) and (b)), as a function of plastic strain. For the situation shown (characteristic of a material with a low initial dislocation density and rapid dislocation multiplication with strain), the mobile dislocation density peaks at a finite strain. The initial increase in the mobile population is associated with work softening. In materials with a reasonable initial dislocation density and/or a low multiplication coefficient, ρ_{mob} decreases continuously with strain and work hardening occurs at the onset of plastic deformation.

sizing those used in subsequent chapters that relate dislocation behavior to microstructural features.

Dislocations are structural line defects that account for plastic deformation by slip in crystals. This is because the atomic distortions accompanying dislocation motion are considerably less than those happening during "perfect slip." In the latter, atomic planes glide over one another in a manner requiring correlated and simultaneous motion of all atoms in the slip plane. Screw and edge dislocations represent extreme examples of these line defects, and they are characterized by different atomic arrangements in the vicinity of the dislocation core. While the slip direction is the Burgers vector for both kinds of dislocations, this vector is normal to an edge dislocation line but parallel to a screw dislocation line. Thus, the slip plane of an edge dislocation is defined by these nonparallel vectors, but a screw dislocation slip plane is not so specified. As a result, a screw dislocation can "cross-slip" from one slip plane to another. An edge dislocation can move out of its slip plane only by displacements normal to it; this "climb" involves addition of vacancies or interstitials to the dislocation core. A mixed dislocation, structurally intermediate to an edge and a screw dislocation, is most common. The Burgers vector of a mixed dislocation—while still defining its slip direction—makes an angle between 0 and 90° to the dislocation line. As for an edge dislocation, therefore, the slip plane for a mixed dislocation is defined specifically, and cross-slip of a mixed dislocation is restricted to those portions of the line having a screw character. The force per unit length on the dislocation promoting slip is given by τb, where τ is the shear stress on the slip plane; τ may include contributions from internal, as well as external, stresses.

An elastic energy results from the atomic structural irregularities associated with line defects. This strain energy can be approximated as $Gb^2/2$ per unit length of line, where G is the material shear modulus and b the Burgers vector. This elastic energy gives rise to dislocation stress fields and results in forces among dislocations. Such forces can either aid or hinder dislocation motion. In some cases the interaction can lead to dislocation annihilation; in others it can produce metastable dislocation arrays resembling small-angle crystal boundaries. When present in large quantities, dislocations arrange themselves in a manner so as to minimize their total strain energy. Dislocation cell structures—common to cold-worked metals—are an example of this effect. Dislocations lying within cell boundaries are arranged so as to approximately cancel out their long-range stress fields, thereby producing a low-energy structure.

Since dislocation energy is proportional to the square of its Burgers vector, it is not surprising that dislocations with the smallest such vectors (the interatomic spacing) are most frequently found. Thus, slip directions are usually close-packed atomic directions. Additionally, the friction stress resisting dislocation motion is minimized for slip planes having a large interatomic spacing. Thus, slip on a close-packed atomic plane is favored, and the number of geometrically distinct combinations of slip planes and directions, which is dictated by crystal structure, determines the number of slip systems of a material. The rules regarding slip planes and directions are general ones, with the slip direction requirement being almost universally observed. On the other hand, specific factors relating to atomic packing and bonding can lead to slip on other than close-packed planes, as is often observed in ionic materials. In some crystal structures, dissociation of the unit dislocation (i.e., that dislocation with Burgers vector magnitude equal to the interatomic spacing) into partial dislocations occurs. A full slip displacement requires the passage of the two partial dislocations across the slip plane. The area between the partials has slipped incompletely and this region is characterized by improper atomic stacking (a stacking fault). Cross-slip of a dissociated screw dislocation requires recombination of the partials, and if they are widely separated this (and cross-slip) is difficult. If the energy associated with the improper atomic stacking (the stacking fault energy) is low, the partials are widely separated and vice versa.

Intersections between moving dislocations usually result in the formation of jogs on dislocations. Not only does this increase dislocation length (and energy), but jogs are generally less mobile than the unjogged portions of the dislocation line. Hence, dislocation intersections can lead to reduced dislocation mobility and material work hardening. On the other hand, immobilization of dislocation line segments can lead to dislocation multiplication; for example, by the Frank-Read mechanism.

Shear strains and strain rates relate to the dislocation density. The strain rate also increases with dislocation velocity which, for mobile dislocations, is a strong function of stress. Increasing plastic strain is accompanied by both an increase in dislocation density and a reduction in the fraction of mobile dislocations. If the latter decreases more rapidly than the population increases, work hardening is concurrent with the onset of plastic deformation. If not, work softening, manifested by a yield point, is found.

REFERENCES

Friedel, J.: *Dislocations*, Pergamon Press, New York, 1964.

Gilman, John J.: *Micromechanics of Flow in Solids*, McGraw-Hill, New York, 1969.

Hertzberg, R. W.: *Deformation and Fracture Mechanics of Engineering Materials*, 4th ed., Wiley, New York, 1996.

Hirth, J. P., and J. Loethe: *Theory of Dislocations*, McGraw-Hill, New York, 1968.

Hull, Derek: *Introduction to Dislocations,* Pergamon Press, Oxford, 1965.

Meyers, Mark André, and Krishan Kumar Chawla: *Mechanical Metallurgy—Principles and Applications*, Prentice-Hall, Englewood Cliffs, NJ, 1984.

Weertman, J., and J. R. Weertman: *Elementary Dislocation Theory*, MacMillan, New York, 1964.

PROBLEMS

3.1 The asymmetry of the bond energy leads to a lattice frictional stress for a dislocation. The physics that lead to the frictional stress can be simply illustrated. Consider the atoms (1), (2), and (3) in the figure below. Atom (1) can be thought of as the atom at the terminus of an edge dislocation core. As the dislocation moves from left to right, the geometrical arrangement of the atoms changes as approximately shown in the sketch.

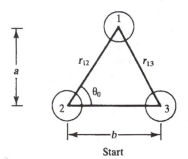
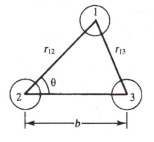
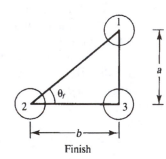

Assume the lattice energy that changes during dislocation motion is the sum of u_{12} and u_{13}. Set up an expression for this energy as a function of the angle θ, and A, B, m, and n of Eq. (2.10). Plot this energy as a function of θ from $\theta = \theta_0$ to $\theta = \theta_f$. To simplify the calculation, let $A = B$, $n = 2$, and $m = 4$ and express the energy in units of A/B^2. Schematically plot F ($\sim dU/d\theta$) vs. θ to display the existence of a friction force. (Note: Your results will not accurately simulate the "real" force, nor will the position at which the force is a maximum be correct. This is a result of the simplified geometry, and of only considering two interatomic interactions. Nonetheless, the exercise is useful for demonstrating how a friction stress arises.)

3.2 The frictional stresses listed in Table 3.1 are calculated on the basis of a narrow dislocation (Eq. (3.5)). Assume that the experimental values of yield strength listed in this table correspond to the formulation based on dislocation width (Eq. (3.6)). Determine the value of dislocation width necessary to reconcile the frictional strength to the experimentally observed strengths. Express your answer in terms of the ratio w/b. Comment on your results.

Data: All of the materials listed in Table 3.1 are fcc, except for α-Fe which is bcc. Note that the spacing, a, in Eq. (3.5) is the spacing between parallel close-packed planes; likewise, b is the interatomic distance (i.e., the spacing between atoms in a close-packed direction). Refer to Table 2.1 for Poisson ratio values. Lattice parameters; Ag (0.409 nm), Al (0.405 nm), Cu (0.362 nm), Ni (0.352 nm), α-Fe (0.287 nm).

3.3 Sketch a dislocation loop in the form of a circle and specify the Burgers vector of the loop. Denote the portions of the loop that are edge and those that are screw. Given that edge and screw dislocations have somewhat different energies per unit length, would you expect dislocation loops generally to be circular in shape? If not, show how the loop you have drawn will change its shape because of energy differences along the circumference.

3.4 Refer to Fig. 3.19. Determine the crystal shear strain caused by the twinning deformation shown. What is the "local" shear strain (that is, the shear strain across the twin boundary)? Why are the two strains different? (Hint: What fraction of the crystal has twinned?)

3.5 Determine the mean pressure (approximated as $(\sigma_x + \sigma_y + \sigma_z)/3$), expressed in units of Gb/r, for an edge dislocation. At what value of θ is the mean pressure a maximum? For $r = 10b$, what is this maximum pressure (in units of G)? Is this a large or a small stress?

3.6 a Equation 3.12 expresses the dislocation line energy per unit length in terms of the dislocation core radius (r_0) and some outer radius (r_1). The outer radius is the distance at which the stress field of the dislocation considered effectively vanishes owing to overlap with stress fields from other dislocations. Let this outer distance be approximated as one-half the mean dislocation spacing. Let $r_0 = 3b$, $b = 0.250$ nm, and assume $\nu = \frac{1}{3}$. Calculate the dislocation energy per unit length for dislocation densities of 10^{10}, 10^{12}, and 10^{14} per square meter. How do these energies compare with the approximate Eq. (3.13)? (Hint: How is dislocation spacing related to the dislocation density?)

b The total dislocation energy per unit volume is estimated as $Gb^2\rho$. Show that this parameter has units of J/m^3. Estimate $Gb^2\rho$ for copper with $\rho = 10^{10}/m^2$, $10^{12}/m^2$, and $10^{14}/m^2$. How do these energies compare with the latent heat of fusion for Cu ($= 1840$ MJ/m^3)? Data for Cu: $b = 0.256$ nm, $G = 4.1 \times 10^{10}$N/m^2.

c Based on internal energy considerations, what dislocation density in Cu is needed to make the crystalline form of Cu unstable with respect to the liquid form? What would the dislocation spacing be at this density? Comment on your answer.

3.7 Illustrate how dislocation climb can increase the spacing between edge dislocations in the tilt boundary of Fig. 3.25c. What is the stress driving the climb process?

3.8 For the situation of Fig. 3.25a, but with dislocations of opposite Burgers vector, show how a combined climb-glide process can lead to dislocation annihilation. (In the exercise, ensure that dislocation motion is always consistent with the forces between the dislocations.)

3.9 Consider dislocations advancing by the lateral spreading of kinks. Let the advanced portion of the line be a distance δ beyond the portion of the line that has not advanced, and let the kinks be separated by a distance λ along the line (i.e., the kink density per unit length is λ^{-1}). Determine a relationship between the dislocation velocity, v_D, and the kink velocity, v_k.

3.10 Calculate, as a function of the *c/a* ratio, the distance between parallel basal, pyramidal, and prismatic planes in the hexagonal close-packed crystal structure. Assuming slip takes place on the most widely separated planes, determine which planes are active as a function of *c/a*.

3.11 Sketch a close-packed plane in a face-centered cubic structure—call it plane A in the ABCABC stacking of this structure. Place on top of it another close-packed plane (B), and show that when slip occurs by two partial slip processes the intermediate stage corresponds to atoms of the B plane assuming C positions in the stacking sequence. That is, show that a stacking fault is generated by partial slip.

3.12 Consider partial slip in the basal plane of the hcp structure. Describe the crystallography associated with a basal plane stacking fault in hcp materials. (That is, how is the ABAB stacking sequence altered by partial slip?)

3.13 Slip in NaCl-structure ionic crystals occurs most commonly on {110} planes and in <1$\bar{1}$0> directions.
a Sketch the (001) plane showing the ions in a square region containing 16 unit cell faces: (i) indicate the (110) slip plane trace and the [$\bar{1}$10] slip direction; (ii) remove half of the anions on a (1$\bar{1}$0) plane trace and half of the cations on a neighboring [(2$\bar{2}$0)] plane trace. Show that the result is an edge dislocation having the correct Burgers vector.
b How are the anions and cations arrayed along the edge dislocation line (i.e., along the termination of the half plane)?

3.14 Consider a "super" edge dislocation (Fig. 3.33). By procedures similar to those used in Example Prob. 3.1, calculate the equilibrium anti-phase boundary width, *w*, in terms of the anti-phase boundary energy (APBE) and material properties.

3.15 Could an extended jog (Fig. 3.38) be considered equivalent to a sharp kink?

3.16 Show that the jog configurations resulting from the dislocation intersections shown in Figs. 3.35–3.37 are consistent with the rule "when one dislocation cuts another, each acquires a jog equal to the component normal to its own slip plane of the other dislocation's Burgers vector."

3.17 The stress required to operate a Frank-Read source (Eq. (3.20)) was determined by means of a line tension balance. This stress can also be approximated using a simpler approach. The energy increase accompanying the dislocation "extrusion" can be approximated as that due to the increase in dislocation line length (i.e., the change in line length relative to the initial line length) associated with operation of the source.
a Calculate the increment in dislocation line energy when the dislocation is in the form of a semi-circle.
b The associated force (not force per unit length) is the gradient of the energy change. Estimate the distance over which the increase in energy takes place, and then determine this force.

c Estimate the stress required to operate the Frank-Read source and compare the value you obtain with that given by Eq. (3.20).

3.18 Let the total dislocation density, ρ_{tot}, be the sum of the mobile (ρ_{mob}) and the immobile (ρ_{im}) density. Using the definition of an average velocity for the total population, show that $\rho_{tot}\,\bar{v}_{tot} = \rho_{mob}\,\bar{v}_{mob}$. (Note: The velocity of the immobile dislocations is zero.)

3.19 Use Fig. 3.44 to show that Eq. (3.23) is a specific application of Eq. (3.25).

3.20 The relationship $\gamma = \rho b \bar{x}$ can be used to determine the variation of strain rate with dislocation velocity when dislocation density changes with strain. Show that, for this situation,

$$\gamma = \frac{\rho b \bar{v}}{\left[1 - \dfrac{d\ln\rho}{d\ln\gamma}\right]}$$

(Hint: Recall the chain rule for differentiation; $\dfrac{dy}{dt} = \dfrac{dy}{dx}\dfrac{dx}{dt}$.)

3.21 A 1-cm³ cube of NaCl is subjected to a shear stress such that slip takes place on a $(1\bar{1}0)$ plane and in a [110] direction. The crystal is oriented so that the slip planes are parallel to the top and bottom crystal surfaces. For NaCl, the lattice parameter, a, is 0.544 nm; the slip plane spacing is $a/\sqrt{2}$.
a An edge dislocation with $\bar{b} = (a/2)[110]$ passes halfway through the crystal on every sixth plane. What is the shear strain?
b What is the dislocation density after deformation of the type just described?

Plastic Deformation in Single and Polycrystalline Materials

4.1
INTRODUCTION

In this chapter plastic flow in crystalline materials is discussed in the context of crystallographic and other considerations. Plastic-flow initiation in single crystals is described in terms of a critical stress, the critical resolved shear stress. This stress depends on temperature, strain rate, and material purity; it increases with increasing strain rate and impurity content and with decreasing temperature.

The work hardening of single crystals is described in terms of dislocation interactions taking place during plastic straining. Provided that multiple slip, associated with dislocation intersections that may immobilize dislocations, does not occur, the work-hardening rate of single crystals is low. Multiple slip, however, markedly increases this rate.

The deformation behavior of polycrystals is inherently related to slip processes occurring within individual crystals of the polycrystalline aggregate. However, since some grains are oriented more favorably than others for effecting dislocation glide, plastic flow in polycrystals differs substantially from that in single crystals. The strain displacements across grain boundaries separating individual grains must be matched, and this constraint imparts restrictions that lead to the greater strengths of polycrystals relative to single crystals. The relation between poly- and single-crystal deformation can be rationalized provided these constraints are taken into account.

The grain boundary constraint also leads to the presence of strain gradients within individual grains; that is, the dislocation density in the vicinity of grain boundaries is greater than in grain interiors. The additional dislocations are not found in single crystals deformed to the same strain, and these dislocations are termed "geometrically necessary" dislocations. Since the stress required for plastic flow increases with dislocation density, the geometrically necessary dislocations impart an additional component to the flow stress of polycrystals compared to single crystals.

140

4.2
INITIATION OF PLASTIC FLOW IN SINGLE CRYSTALS

As described in Chap. 3, crystalline slip occurs in specific crystallographic directions and on particular atomic planes. The several combinations of geometrically distinct slip planes and directions account for the number of slip systems, which is determined by crystal structure (cf. Table 3.3).

If a single crystal of a specific material is subjected to a tensile (or compressive) stress, plastic flow can be accomplished by slip on one or more of the possible slip systems. Which of the slip systems is activated can be determined through consideration of Fig. 4.1. For simplicity, this figure considers only one of the potential slip systems in a sample subjected to an applied tensile force, F. The transverse cross-sectional area of the crystal is A_0 and hence, the tensile stress is given as $\sigma = F/A_0$. That the shear stress promoting slip is less than σ is shown through geometry. In Fig. 4.1, the slip plane normal makes an angle ϕ with respect to the tensile direction, and λ is the corresponding angle between the slip direction and the tensile axis. In general the three directions (the tensile axis, the slip plane normal, and the slip direction) are not coplanar so that $\phi + \lambda \neq 90°$; indeed the minimum possible value of $(\phi + \lambda)$ is $90°$, attained only when these directions are coplanar. The resolved force in the slip direction is $F \cos \lambda$. The slip plane area, A_s, is greater than A_0, and the areas are related by $A_s = A_0/\cos \phi$. Thus, the resolved shear stress, τ_{RSS}, acting on the slip plane and in the slip direction is given by

$$\tau_{RSS} = \frac{F \cos \lambda}{A_s} = \frac{F}{A_0} \cos \phi \cos \lambda = \frac{\sigma}{m} \qquad (4.1)$$

where $m \ (= (\cos \phi \cos \lambda)^{-1})$ is determined by the orientation of the slip system with respect to the tensile axis.

Since the various slip systems are usually oriented differently with respect to the tensile axis, for a given σ each will have a different value of τ_{RSS}, and one of

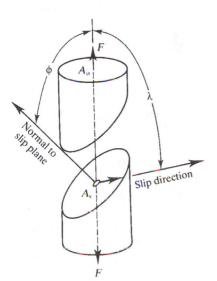

Figure 4.1
Resolution of an applied stress ($\sigma = F/A_0$) into a shear stress acting in the slip direction on a slip plane in a single crystal. The slip plane area A_s is given by $A_0/\cos \phi$ where ϕ is the angle between the tensile axis and the slip-plane normal. The resolved force (F_s) in the slip direction is $F \cos \lambda$, where λ is the angle between the tensile axis and the slip direction. Thus, the resolved shear stress, $\tau_{RSS} = F_s/A_s = \sigma \cos \phi \cos \lambda$. Slip occurs when the resolved shear stress attains a critical value, τ_{CRSS}, characteristic of the material.

142

CHAPTER 4
Plastic
Deformation in
Single and
Polycrystalline
Materials

these (the one with the minimum value of m or the maximum value of $\cos \phi \cos \lambda$) will experience the greatest resolved shear stress. This "most-favored" slip system is the one in which plastic deformation initiates.

Plastic flow initiates when τ_{RSS} reaches some critical value, characteristic of the material. This stress is designated τ_{CRSS}, and is related to the tensile yield stress (σ_y) through

$$\sigma_y = m\tau_{CRSS} \tag{4.2}$$

Thus, $\sigma_y > \tau_{CRSS}$. The essential correctness of Eq. (4.2) (Schmid's law) has been verified by tensile testing single crystals of a material oriented so that different values of $(\cos \phi \cos \lambda)_{max}$ are generated in the most-favored slip system. Measured values of σ_y differ for these crystals. However, when they are divided by the different values of m for each orientation, it is found that, for the most part, τ_{CRSS} ($= \sigma_y/m$) is invariant for a given material. Exceptions are found in the body-centered cubic transition metals. The contradiction, however, is explained by the existence of "anomalous" slip in these materials. That is, calculated values of m are based on the assumption of {110}<111> slip in the bcc structure and, for certain crystallographic orientations, slip in systems other than these takes place. Hence the rule—that in a given material slip is initiated when the resolved shear stress attains a critical value appropriate to the material—generally holds.

The value of τ_{CRSS} depends on test conditions, such as temperature and strain rate, as well as structural features such as the material's initial dislocation density and purity. Decreases in temperature and purity and increases in strain rate and dislocation density increase τ_{CRSS}. The role that strain rate and temperature play is illustrated schematically in Fig. 4.2. As shown there, three temperature regions can be identified. At the highest temperatures (Region III where typically $T \geq 0.7T_m$ with T_m being the material's absolute melting temperature), τ_{CRSS} decreases rapidly with increasing temperature. This is a result of the important role diffusive processes play in permanent deformation at high temperatures and, thus, τ_{CRSS} is also a function of strain rate in Region III. (Further discussion of Region III is provided in Chap. 7.) In the low-temperature Region I (for which typically $T \leq 0.25T_m$), τ_{CRSS} again is a function of temperature and strain rate, increasing with decreasing temperature and increasing $\dot{\varepsilon}$. For temperatures $\leq 0.7T_m$, therefore, the critical stress can be represented by

$$\tau_{CRSS} = \tau_a + \tau^* \tag{4.3}$$

where τ_a is the athermal (i.e., temperature-independent) component of the stress and τ^* the thermally dependent term. In effect, τ^* approaches zero at the temperature marking the transition from Region I to Region II.

The two components of τ_{CRSS} relate to microstructural features. The athermal component arises from the stress required to move dislocations in the presence of long-range internal stress fields, i.e., stresses felt present over distances large in comparison to atomic dimensions. Examples of such stresses are those due to the presence of other dislocations. On the other hand, τ^* represents the resistance to dislocation motion presented by "short-range" barriers, existing over distances sufficiently small so that thermal energy is useful for surmounting them. Such obstacles include those coming from the lattice-energy variation responsible for the

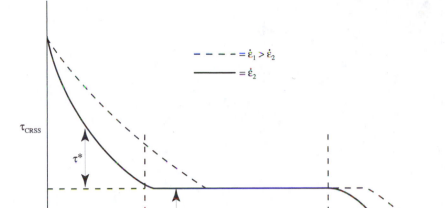

Figure 4.2

Schematic illustrating the variation of τ_{CRSS} with temperature and strain rate ($\dot{\varepsilon}$). At high temperatures (Region III), τ_{CRSS} is a strong function of both variables. At intermediate temperatures, τ_{CRSS} is independent of strain rate and temperature, and is given by τ_a. At lower temperatures, τ_{CRSS} again increases with decreases in temperature and increases in $\dot{\varepsilon}$. Thus, for example, at temperature T_1, τ_{CRSS} can be considered to be the sum of the athermal stress τ_a and a thermally dependent stress τ^*. At the transition from Region I to Region II, τ^* effectively becomes zero.

Peierl's stress and the resistance to dislocation glide provided by some impurity atoms. Thermal activation of dislocation kinks, for example, is one mechanism by which temperature aids in reducing the stress necessary to overcome short-range obstacles. Since the efficacy of thermal activation increases with temperature, τ^* decreases with increases in temperature. The temperature at which τ^* goes to zero is essentially that temperature at which thermal activation renders short-range obstacles ineffective for restricting dislocation motion.

The temperature variation of τ_{CRSS} is shown for a number of materials in Fig. 4.3. It is interesting that this measure of plastic deformation resistance is relatively low in fcc metals, not only in comparison to the nonmetals but also to the bcc transition metals. Also clear is that impurity atoms increase τ_{CRSS}, in some cases considerably; this *solid solution strengthening* is described further in Chap. 5. Finally, although not all of the materials whose behavior is illustrated in Fig. 4.3 demonstrate all of the features of the schematic Fig. 4.2, the latter nonetheless remains a generally useful scheme for describing the temperature variation of τ_{CRSS}. Section 4.5 further discusses yield strength levels and the temperature variation of this strength as they vary among the material classes.

144

CHAPTER 4
Plastic
Deformation in
Single and
Polycrystalline
Materials

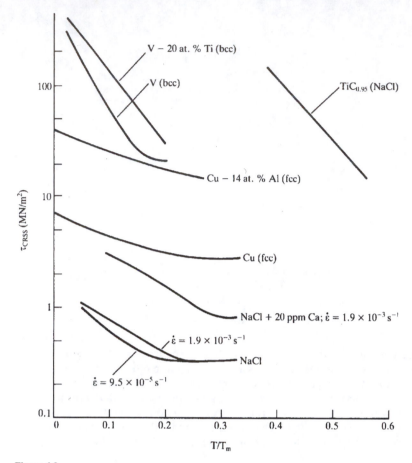

Figure 4.3
The temperature variation of τ_{CRSS} for materials with different structures and bonding characteristics. The body-centered cubic metals have higher values, and a stronger temperature dependence, of τ_{CRSS} than do face-centered cubic metals. The intrinsic strengths of ionic crystals (e.g., NaCl) are generally low. The covalently bonded material TiC, which has the same crystal structure as NaCl, possesses an inherently great resistance to plastic flow. Both the metals (e.g., V–20 at. % Ti vs. V, and Cu–15 at. % Al vs. Cu) and the nonmetals (e.g., NaCl plus 20 ppm Ca vs. NaCl) show solid-solution hardening and a strain-rate dependence of τ_{CRSS}. (*Data for V and V-Ti from E. Pink and R. J. Arsenault, Metal Sci. J., 6, 1, 1972; for Cu and Cu-Al from T. J. Koppenaal and M. E. Fine, Trans. TMS-AIME, 224, 347, 1963; for NaCl and NaCl-20 ppm Ca from J. Hesse, Phys. Stat. Sol., 9, 209, 1965 and Reinstoffe problemme, ed. E. Rexner, III, Akademic-Verlag, Berlin, 1967, p. 413; for TiC from W. S. Williams, J. Appl. Phys., 35, 1329, 1964.*)

EXAMPLE PROBLEM 4.1. Hexagonal close-packed zinc slips by basal plane slip. A zinc single crystal is oriented so that the normal to its slip plane makes an angle of 60° with the tensile axis. If the three slip directions have angles of 38°, 45°, and 84° with respect to this axis, and the critical resolved shear stress for Zn is 2.3 MN/m², determine the tensile stress at which plastic deformation commences.

Solution. The angle ϕ is 60°. The slip direction that will first be activated is the one having the greatest value of cos λ; i.e., it is the direction making an angle of 38° with the tensile axis. Thus, the tensile stress at which slip commences is

$$\sigma = \frac{\tau_{CRSS}}{\cos\phi\cos\lambda} = \frac{\tau_{CRSS}}{\cos 60°\cos 38°} = \frac{2.3\ \text{MN/m}^2}{(0.5)(0.788)} = 5.84\ \text{MN/m}^2$$

EXAMPLE PROBLEM 4.2. A single crystal having a simple cubic structure (slip planes {100}, slip directions <100>) is oriented such that the tensile axis is parallel to the [010] crystal axis. Make a list of the slip systems in this crystal and calculate the Schmid factor for this loading geometry.

Solution. A sketch of the situation, indicating the tensile axis as well as the slip planes and directions, is given below. The following table summarizes the results.

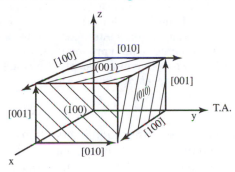

Slip Plane	ϕ (°)–cos ϕ	Slip Direction	λ (°)–cos λ	Schmid Factor
(100)	90–0.00	[010]	0–1.00	0
		[001]	90–0.00	0
(010)	0–1.00	[100]	90–0.00	0
		[001]	90–0.00	0
(001)	90–0.00	[100]	90–0.00	0
		[010]	0–1.00	0

Note that none of the possible slip systems experiences a shear stress for this loading condition. Thus, plastic deformation will not result from an applied stress. When the applied stress becomes sufficiently high, the material will fracture instead.

Problem 4.3 continues this exercise for a situation involving loading in a direction that does result in shear stresses on active slip systems. Problems 4.1, 4.2, 4.4, and 4.6–4.9 also deal with problems of this nature.

4.3
STRESS-STRAIN BEHAVIOR OF SINGLE CRYSTALS

A general description of the stress-strain behavior of single crystals following yielding is provided in the schematic Fig. 4.4. The crystal work-hardening rate after yielding is initially low (Stage I, Fig. 4.4, has a low value of $d\tau/d\gamma$). This "easy glide" is associated with single slip on the slip system having the maximum value

CHAPTER 4
Plastic
Deformation in
Single and
Polycrystalline
Materials

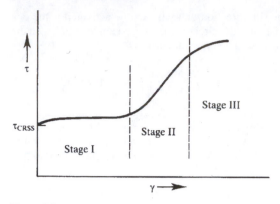

Figure 4.4
Schematic of the shear stress–shear strain curve of a single crystal. The low work-hardening rate immediately following yielding is observed when one slip system is active. For this circumstance, dislocation interactions impeding their motion are minor. Stage II (linear) hardening is quite pronounced, and associated with multiple slip and significant dislocation-motion impediments as a result of intersection processes. The reduced work-hardening characteristic of Stage III is termed saturation or "exhaustion" hardening.

of $\cos \phi \cos \lambda$. Since this is so, dislocation intersections that give rise to jog formation and dislocation immobility are absent, and the strong work hardening resulting from such interactions does not occur. The work hardening observed during easy glide results from the overlap of dislocation stress fields among dislocations gliding on parallel planes; the resulting force between these dislocations is manifested by a shear stress that increases with plastic strain.

The transition from Stage I to Stage II behavior is almost invariably associated with the onset of multiple slip, and the strong work hardening resulting from interactions among dislocations on nonparallel planes. Indeed, the slope of the τ–γ curve in Stage II, the "linear hardening" region, is large, on the order of $G/300$. Stage III, "exhaustion hardening," is characterized by a reduction in the work-hardening rate in comparison to Stage II.

The above description is in accord with many experiments. For example, the strain extent of Stage I decreases with temperature, and this is consistent with the easier onset of multiple slip at higher temperatures. Likewise, the extent of Stage II is reduced as temperature is raised and this is consistent with recovery processes operating more effectively at higher temperatures. Moreover, in face-centered cubic metals, the transition from Stage II to Stage III correlates with stacking-fault energy (SFE). Cross-slip occurs more readily in materials with high SFE, and thus the transition from Stage II to Stage III occurs at lower stress levels in materials having a high SFE.

While Fig. 4.4 is conceptually useful as a general description of single crystal plasticity, it applies most specifically to fcc metals. There are differences between the crystallography of slip among the metallic (as well as nonmetallic) materials, and this affects the details of their single crystal plastic behavior. Here we describe some of these differences, with special emphasis on the different behavior of fcc and bcc transition metals.

Although slip directions in bcc metals are close-packed <111> directions, the slip plane is not always of the {110} variety. Other planes, in particular the {112} and {123} ones, have comparable atomic packing densities to {110} planes while also containing <111> type directions. As noted before, these slip planes become active in the bcc metals at moderate temperatures; exclusive $<\bar{1}11>\{110\}$ slip is restricted to low temperatures. Thus, a dislocation in a bcc metal has a choice of glide planes. Further, SFEs in bcc metals are high and partial dislocations are seldom observed in them. This permits easy cross-slip for screw dislocations. The combination of these factors just mentioned often leads to corrugated slip plane surfaces (Fig. 4.5a) (a result of what is termed dislocation *pencil glide*) in bcc metals. Pencil glide and the corrugated pattern can result from the motion of both edge and screw dislocations (Figs. 4.5b and c). This type of glide also gives rise to wavy slip steps on surfaces of deformed bcc crystals. Figure 4.5d illustrates this schematically and a photograph of wavy slip bands in bcc iron is shown in Fig. 4.6.[1] Pencil glide impacts Stage I behavior of bcc metals in at least two ways. First, τ_{CRSS} need not be characteristic of $<\bar{1}11>\{110\}$ slip if pencil glide operates. Second, the non-planar path of dislocations moving in pencil glide results in different Stage I work hardening than that found in fcc metals.

The difference in behavior of the bcc transition metals (as exemplified by V in Fig. 4.3) and the fcc metals (of which Cu is typical) can be amplified by examining the temperature dependence of their flow behavior. Figure 4.7a illustrates this behavior for Cu; Fig. 4.7b does the same for the bcc transition metal Nb, both metals being oriented initially for single slip. For Nb, we see that τ_{CRSS} is strongly temperature dependent; for example, it increases by about a factor of three—from ca. 20 MN/m^2 to about 60 MN/m^2—as temperature decreases from 273 K to 201 K.[2] Further, the easy glide region decreases with temperature effectively vanishing at 423 K; since Nb melts at 2740 K, this temperature is about $0.17T_m$ for Nb. We also note that the work-hardening rate of Nb in the Stage II region of single crystal plasticity is very nearly temperature independent over the temperature range during which Stage II behavior is evident.

Copper (Fig. 4.7a) behaves differently. To begin with, τ_{CRSS} for it is very nearly temperature independent between 93 K and 623 K ($0.07 - 0.46T_m$). In common with Nb, though, the deformation extent of Stage I decreases with increasing temperature. And, while not obvious from Fig. 4.7a, the work-hardening rate of Cu in Stage II is more strongly temperature dependent than is the comparable rate of Nb. Since Stage II single crystal work-hardening rates mimic those of polycrystals, we shall see (Sect. 4.5) that this difference in behavior is reflected in the temperature dependence of the tensile elongations of the bcc transition and fcc metals.

---

[1] The surface pattern shown in Fig. 4.6 does not reflect individual dislocations. Rather, the pattern is a result of "bands" of slip planes that have exited the iron surface.

[2] The anomalous flow behavior of Nb for $T \le 175$ K results from twinning becoming an important plastic deformation mode at such temperatures.

148

CHAPTER 4
Plastic
Deformation in
Single and
Polycrystalline
Materials

That the onset of multiple slip is associated with the Stage I–Stage II transition can be demonstrated by considering the stress-strain behavior of single crystals oriented so that multiple slip is favored from the onset of plastic deformation. As noted, the τ–γ curve of Fig. 4.4 schematizes the stress-strain response of single crystals for which only one slip system operates initially. However, certain crystal orientations promote slip in two or more systems (Prob. 4.7) at the onset of plastic

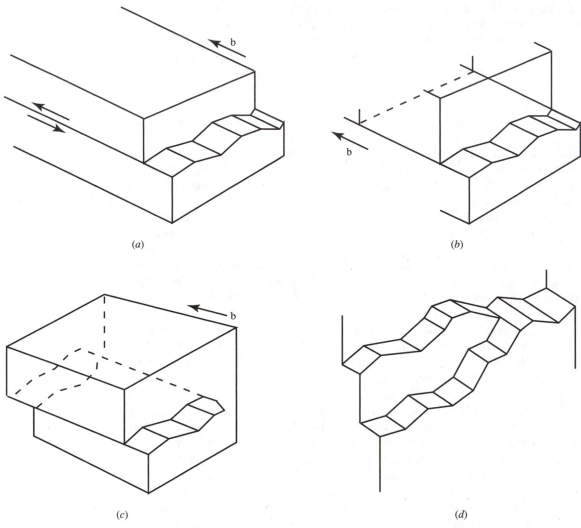

(a)

(b)

(c)

(d)

Figure 4.5
(a)–(c) Pencil glide leads to corrugated slip surfaces. (a) Since slip takes place on several planes during pencil glide, the slip transit is not planar. (b) Shows how the corrugated surface can be obtained by edge dislocation motion. (c) Does the same for screw dislocation motion. (d) The surface manifestation of pencil glide is the presence of wavy slip steps on the crystal surface. These intersect periodically, giving rise to a mosaic pattern.

deformation. Figure 4.8 shows the stress-strain response of such crystals in comparison to when only single slip is initially favored. While τ_{CRSS} is the same for all crystals, easy glide is observed only for the crystal oriented for single slip. For the other crystals, the work-hardening response even in the early stages of plastic deformation is characteristic of Stage II hardening.

Multiple slip in crystals initially oriented for single slip is facilitated by constraints common to tension or compression testing. The effect of such constraints is shown in Fig. 4.9. Figure 4.9a illustrates how sample shape changes in response to single slip in the absence of grip constraints common to a tensile test. Without these, the axis of the crystal translates away from the original tensile axis. As shown in Fig. 4.9b, the grips cause the tensile and the original long crystal axis to remain aligned. This is accomplished by a rotation of the crystal slip planes; the rotation is such that the angle between the slip direction and tensile axis decreases with strain. Lattice rotation also takes place in compression. However, as indicated in Fig. 4.10, in compression the rotation causes the angle between the loading axis and the slip plane normal to decrease with strain. In either case, however, the rotation results in an alteration of the m value of the primary slip system, and in most

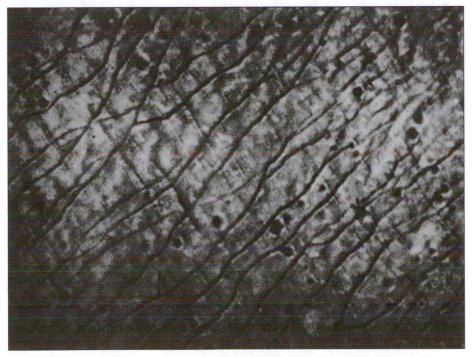

Figure 4.6
The surface of a deformed Fe single crystal exhibiting a wavy slip pattern as a result of pencil glide. (*After H. W. Paxton, M. A. Adams, and T. B. Massalski, Phil. Mag., 42, 257, 1952.*)

CHAPTER 4
Plastic
Deformation in
Single and
Polycrystalline
Materials

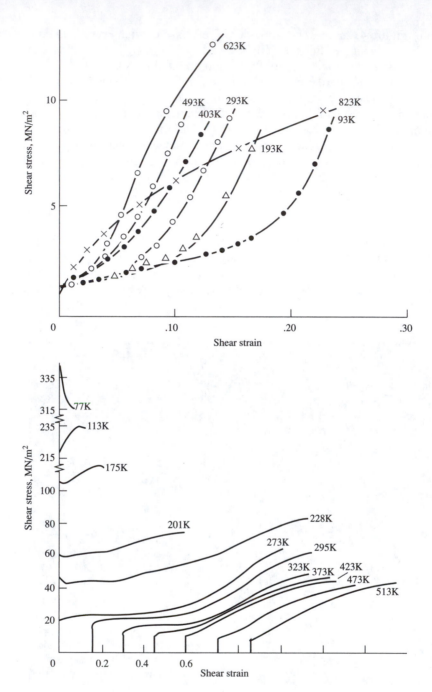

Figure 4.7

Illustration of different temperature dependencies of the flow behavior of fcc metal and bcc transition metal single crystals. (a) Flow curves for fcc Cu as a function of temperature. τ_{CRSS} is very nearly temperature independent over a broad range of temperature. (b) In contrast, τ_{CRSS} is strongly temperature dependent for the bcc transition metal, Nb. (Both the Cu and Nb crystals were initially oriented for single slip.)

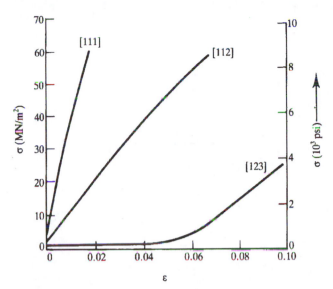

Figure 4.8

The tensile stress-strain behavior of copper single crystals oriented differently with respect to the tensile axis ([123], for example, signifies that the tensile axis is parallel to the (123) plane normal). Although τ_{CRSS} is the same for all crystals, only the [123] crystal that flows initially by single slip displays an easy-glide region. Duplex slip occurs at yielding for the [112] crystal, and six systems are initially active for the [111] crystal (see also Fig. 4.12b). Thus, multiple slip, with attendant significant work hardening, occurs immediately following yielding in the [112] and [111] crystals and the hardening is most pronounced for the latter, which has the most active slip systems. (*After J. Diehl, Z. Metallk.*, **47, 331, 1956.**)

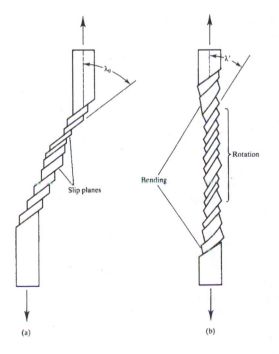

Figure 4.9

Constraints on sample deformation during tensile testing. (a) If the sample grips did not geometrically restrain a single crystal, it would assume the shape shown. (b) The grips provide for axial alignment of the crystal and tensile axes. This is accomplished by a rotation such that the angles between the tensile axis and slip plane normal (ϕ) and the slip direction (λ) are changed from their original values. Concurrently, other slip systems are having their values of these angles changed. When the Schmid factor on a secondary system equals that on the primary one, duplex slip initiates. Near the specimen grips, some bending of the crystal may take place. (*Adapted from R. W. Hertzberg, Deformation and Fracture Mechanics of Engineering Materials, copyright 1975, with permission of John Wiley & Sons, Inc.*)

CHAPTER 4
Plastic
Deformation in
Single and
Polycrystalline
Materials

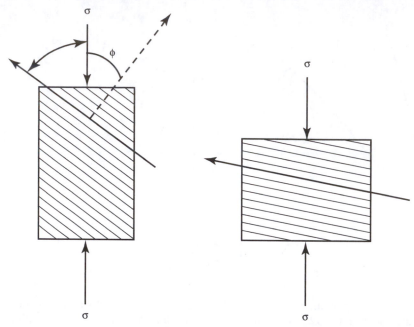

Figure 4.10

Change in crystal orientation during compression. (a) The initial orientation of the slip plane and direction with respect to the compression axis. (b) During compression the slip direction rotates away from the compression axis, while the slip plane normal rotates toward it. Even though the respective rotations are different for compression and tension, the result—that other slip systems eventually become active during single crystal plastic deformation—is the same for both situations.

cases, m is increased.[3] Simultaneous with the rotation of the primary slip plane, other slip systems become more favorably oriented for plastic flow in both tensile and compressive loading. When the resolved shear stress on such a secondary system equals that on the primary one, slip in both systems takes place. Hence, by their nature, tensile and compression tests eventually effect multiple slip at a strain corresponding to the Stage I to Stage II transition.

The rotation of a single crystal during tensile testing can be described with the aid of the stereographic projection. The construction of such a projection is shown in Fig. 4.11. In Fig. 4.11*a*, a cubic single crystal is located within an enclosing sphere. The normals [*hkl*] to the various (*hkl*) planes of the crystal are drawn to intersect the sphere surface. The stereographic projection is obtained by drawing a tangent plane to the sphere and placing a "light source" diametrically opposite the tangent point (Fig. 4.11*b*). The projection of the sphere onto the plane is a circle. The location of the (*hkl*) plane normal on the projection is obtained by extending the

---

[3]Certain initial crystal orientations may actually lead to a decrease in m, with a concurrent increase in the value of the resolved shear stress. In the absence of work hardening, this causes a decrease in the force necessary to continue plastic flow; the effect is called "geometrical softening."

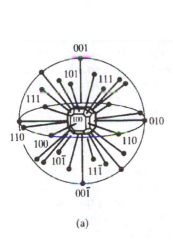

(a)

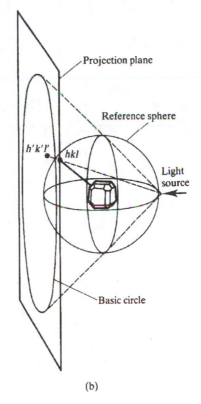

(b)

Figure 4.11
Construction of the stereographic projection of a cubic crystal. (a) First, normals to the various (*hkl*) planes are drawn, and projected onto the surface of a surrounding sphere. (b) Next, a tangent plane is placed along the sphere and a "light source" placed diametrically opposite the plane. The stereographic projection of the crystal is a circle obtained by projecting the "mid circle" of the sphere onto the plane. The position of the (*hkl*) plane normal on the projection is obtained by drawing a line from the light source through the intersection of the direction [*hkl*] with the sphere, and continuing on to the stereographic circle.

line defined by the light source and the intersection of the direction [*hkl*] with the sphere onto the projection plane; e.g., the direction [*hkl*] is denoted by [*h'k'l'*] in Fig. 4.11*b*.[4]

The [001] stereographic projection of a cubic crystal is shown in Fig. 4.12*a;* the [001] designation indicates the crystal is oriented such that [001] projects onto the center of the projection. As can be seen in Fig. 4.12*a*, this projection can be divided into three-sided elements ("triangles"). The apexes of each triangle represent the projection of normals of each of the {001}, {110}, and {111} planes. Because of

---

[4]The stereographic projection appears cumbersome at first glance. However, it is a very useful tool for specifying single crystal orientations. Moreover, facility with the projection is acquired with only moderate practice.

154

CHAPTER 4
Plastic
Deformation in
Single and
Polycrystalline
Materials

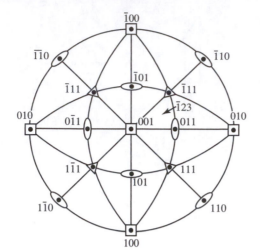

(a)

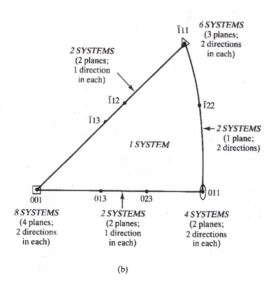

(b)

Figure 4.12
(a) The [001] stereographic pro-jection of a cubic crystal; [001] means that the crystal is oriented so that the [001] direction projects onto the center of the stereo-graphic circle. The projection can be divided into three-sided ele-ments ("triangles") bounded by projections of {100}, {110}, and {111} plane normals. The "stan-dard" triangle is usually taken as the one defined by the [001], [011], and [$\bar{1}$11] intersections. The posi-tion of the [$\bar{1}$23] projection is shown within this triangle. (b) Expansion of the standard trian-gle indicating the location of vari-ous plane normals on the perimeter of the triangle. While the standard triangle is the same for all cubic crystals, the occur-rence of multiple slip depends on the specific crystal structure. In the example above, appropriate for a fcc metal, single slip is ob-tained for crystals whose orienta-tions are located within the standard triangle. Duplex slip oc-curs for crystals with orientations lying along the triangle perimeter, and even more slip systems oper-ate when [001], [011], and [$\bar{1}$11] directions are parallel to the stress axis.

cubic symmetry, each triangle contains within it, or on its boundaries, one of all pos-sible (*hkl*) projections. Thus, the orientation of a single crystal can be specified by the location of its projection within or on the perimeter of a triangle; e.g., the crys-tal plane normal [$\bar{1}$23] is indicated in Fig. 4.12*a*. Because of crystal plane multi-plicity—and the resulting "redundancy" of the triangles in a cubic projection—one "standard" triangle can be used to describe the orientation of a single crystal. The standard triangle commonly chosen is the one defined by the projection of the (001), (011), and ($\bar{1}$11) plane normals (Fig. 4.12*b*). As shown in Fig. 4.12*b,* an fcc crystal oriented such that the projection of its normal lies within the triangle is ori-ented so that slip initially occurs on one slip system. A crystal with orientation such

that the normal projects onto the circumference of the triangle has two equally favored slip systems initially, a crystal with [011] parallel to the tensile axis has four
favored systems, a [$\bar{1}$11] crystal has six, and a [001] crystal has eight.[5]

Consider the rotation of an fcc crystal during tensile straining, such a crystal
having the initial orientation P within the standard triangle (Fig. 4.13). During deformation, the crystal rotates with respect to the tensile axis so that P moves toward
the slip direction, [$\bar{1}$01] in Fig. 4.13. The rotation path is along the arc of a great circle as projected onto the projection. When the crystal orientation reaches the
boundary of the triangle, slip on an alternate system [($\bar{1}\bar{1}$1),[011]] comes into play
because the Schmid factor on it is now the same as on the primary slip system. Provided the crystal does not "overshoot" into the secondary triangle shown in Fig.
4.13, slip now occurs on both the secondary (conjugate) system and the (111)[$\bar{1}$01]
system. Moreover, if the extent of further slip is the same in both systems, continued crystal rotation causes the crystal orientation to move along the boundary line
of the two triangles and towards the direction [$\bar{1}$12], the vector sum of the two slip
directions. Thus, analogously to the fact that the direction of crystal rotation during
tensile slip is in the slip direction when one slip system operates, the crystal rotation
is in the direction of the *net* slip displacement during duplex slip. Upon the orientation attaining the [$\bar{1}$12] direction, further rotation of the crystal does not take place
for the case considered. Similar considerations apply during compression; the difference being that rotation during duplex slip in compression corresponds to the
crystal rotating so that the vector sum of the slip plane normals approaches the compression axis. In either case, however, crystal rotation during single-crystal plasticity is a means for activating multiple slip and initiating the Stage I to Stage II
transition.

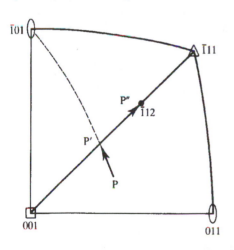

Figure 4.13
**Rotation of a crystal, oriented initially
for single slip, in a tensile test. The crystal, with projection normal P, slips initially on (111) and in the [$\bar{1}$01] direction.
The tensile axis rotates relative to the
crystal and towards the slip direction
along the path PP′. When the orientation reaches the boundary of the standard triangle, duplex slip involving the
conjugate slip system (($\bar{1}\bar{1}$1);[011]) initiates. Further rotation is towards the
sum of the slip directions on both active
planes. Thus, if the extent of slip is the
same on both, the tensile axis and [$\bar{1}$12]
direction eventually become parallel
(path PP″ is followed). Further rotation
does not take place after this happens.**

[5]These are determined for fcc crystals by assuming {111} <$\bar{1}$10> slip. For a body-centered cubic crystal, the stereographic projection is the same as for an fcc material. On the other hand, the deductions as
to which orientations give rise to different degrees of multiple slip differ for bcc and fcc structures since
their slip crystallographies differ.

156

CHAPTER 4
Plastic
Deformation in
Single and
Polycrystalline
Materials

4.4
PLASTIC FLOW IN POLYCRYSTALS

Although the basic slip mechanisms are the same in polycrystals and single crystals, their stress-strain behavior differs substantially. Even as the grips of a testing machine provide constraints on single crystal deformation, so do adjacent grains within a polycrystal provide similar constraints in polycrystalline deformation. The deformation response of grains within a polycrystal are thereby altered in comparison to the response that would be found if each grain were tested as a single crystal. More specifically, the displacements across grain boundaries must be matched, so as to permit the grains to deform in concert. In the absence of such cooperative displacements, voids or cracks would appear at the grain boundaries. In a physical sense, therefore, neighboring grains restrain the plastic flow of each other and, in so doing, provide a polycrystal with an intrinsically greater resistance to plastic flow than that of a single crystal.

The constraints can be described in a more quantitative, yet still simple, way by considering tensile (or compressive) deformation of a bicrystal (Fig. 4.14). Each crystal can be considered to have six strain components; three tensile (ε_x, ε_y, ε_z) and three shear (γ_{xy}, γ_{xz}, γ_{yz}) components. With the coordinates of Fig. 4.14, the following conditions must be satisfied at the grain boundary in order to provide material continuity across it:

$$\varepsilon_x^A = \varepsilon_x^B \tag{4.4a}$$

$$\varepsilon_z^A = \varepsilon_z^B \tag{4.4b}$$

$$\gamma_{xz}^A = \gamma_{xz}^B \tag{4.4c}$$

where the superscripts A and B designate the individual crystals of the bicrystal. Since one grain has a higher value of $\cos \phi \cos \lambda$ than the other, the constraints described by Eqs. (4.4) restrict the deformation of this more favorably oriented

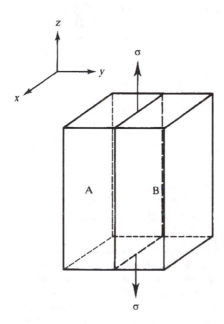

Figure 4.14
Tensile deformation of a bicrystal. The crystals A and B are of the same material, but oriented differently with respect to the tensile axis. Dilational and shear strains must be matched along the interface (the *xz* plane) between the crystals. This constraint increases the flow stress of a bicrystal in comparison to that of a single crystal.

grain and result in a higher yield stress (and a greater work-hardening response) of the bicrystal. In a polycrystalline aggregate, the grain boundary constraints are more restrictive than those for a bicrystal and, thus, the level of the stress-strain curve for a polycrystal is correspondingly higher. Examples illustrating these points are given in Fig. 4.15.

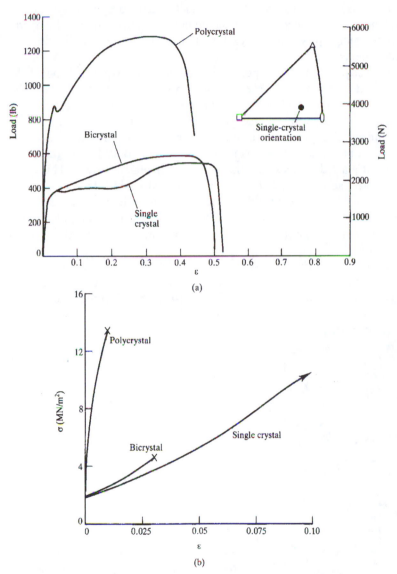

Figure 4.15
Room-temperature tensile (force/stress)-strain curves of single crystals, bicrystals, and polycrystals of (a) niobium and (b) NaCl. The higher stress levels and lower ductilities of polycrystals compared to single crystals are due to the restraints that adjacent crystals in a polycrystal place on the deformation of each other. The level of the flow stress of a bicrystal depends on the relative misorientation of the two crystals comprising it. (*Data from C. S. Pande and Y. T. Chou, Treat. Matls. Sc. Tech., ed., H. Herman, 8, 43, 1975; for NaCl from R. J. Stokes, Proc. Br. Cer. Soc., 6, 189, 1966.*)

158

CHAPTER 4
Plastic
Deformation in
Single and
Polycrystalline
Materials

As mentioned, the more restrictive analogs to Eqs. (4.4) must be satisfied in order for plastic flow to occur in polycrystals. Each grain in a polycrystal has three shear and tensile components of strain as described above for a bicrystal. However, only five of these are independent, because the dilational strains are related through the constant-volume condition ($\varepsilon_x + \varepsilon_y + \varepsilon_z = 0$) of plastic deformation. It can be shown that five independent slip systems are required to meet the boundary compatibility requirements (these arise from the five independent components of strain). That is, the matching of displacements across the boundary necessitates the operation of five independent slip systems, at least in the vicinity of the grain boundaries, in each crystal of the aggregate.

The above description can be experimentally verified. A photograph of the surface of a strained bicrystal of Fe-3% Si is shown in Fig. 4.16. Secondary slip lines emanate from the bicrystal boundary, whereas in regions removed from this boundary slip on only one system is found. Comparable studies on polycrystals indicate that multiple slip (on the five independent slip systems) takes place in the vicinity of grain boundaries whereas near grain centers only two to three slip systems operate (at least this is so near the initiation of plastic flow). This additional slip/dislocation activity leads to a gradient in dislocation density between grain boundaries and grain centers; the dislocation density is higher near the boundaries. These additional dislocations yield a greater hardness near the boundaries and careful studies—using microhardness techniques—confirm this. Section 4.7 further discusses this matter.

The term *independent slip systems* needs clarification. The number of independent slip systems is related to, but not equal to, the number of geometrical slip systems. An independent slip system is one for which slip displacements on it can-

Figure 4.16
Surface slip traces on a Fe-3% Si bicrystal. The grain boundary is the vertical line in the middle of the picture. Primary slip traces in each of the crystals run from lower left to upper right and are found in grain interiors as well as near the grain boundary. Traces of secondary slip (running from upper left to lower right), which maintains strain compatibility across the boundary, are restricted to the boundary region. (*After J. P. Hirth, Metall. Trans., 3, 3047, 1972.*)

not be duplicated by a combination of displacements on other slip systems. The concept can be illustrated by consideration of basal slip in the hexagonal close-packed structure. As shown in Fig. 4.17, there are three geometrically distinct slip systems (the three nonparallel close-packed directions ($\bar{a}_1$, $\bar{a}_2$, $\bar{a}_3$) within the basal plane) in this structure. However, an arbitrary displacement in the (positive) $\bar{a}_3$ direction can be duplicated by a combination of equal slip in the (negative) $\bar{a}_1$ and $\bar{a}_2$ directions. Thus, the number of independent slip systems for hexagonal basal plane slip is two and not three. As noted in Table 4.1, the number of independent slip systems is most always less than the number of geometrical slip systems. This table also shows that the maximum number of independent slip systems is five; thus, a slip displacement in a sixth system can always be duplicated by a combination of slip on five other slip systems.

The crystallographic nature of some materials (cf. Table 4.1) cannot provide for five independent slip systems. When this is so, extensive polycrystalline plasticity in them is precluded and they fracture in a macroscopically brittle manner. Thus, several of the hexagonal close-packed metals and ceramics listed in Table 4.1 are brittle in polycrystalline form even though they may be capable of extensive plastic flow in single crystal form. Polycrystalline ductility is sometimes provided by the activation of secondary slip systems (Table 4.1) that come into play at elevated temperatures. For example, polycrystalline NaCl is brittle at low temperatures, but at temperatures greater than about half of the absolute melting temperature, ductility is effected by the advent of slip on {001} type planes. Observations such as this indicate clearly that flow in polycrystals requires slip on more than one slip system; this accounts for the absence of an easy glide region during polycrystalline flow as well as, in part, the higher tensile flow stresses of polycrystals compared to single crystals.

The higher flow stresses of polycrystals are also due to geometrical considerations relating to the differing orientations of the individual crystals of the aggregate with respect to the tensile axis. Each crystal has its own characteristic value of the Schmid factor ($1/m$) and those with the lowest tend to deform last in a tensile test. It is reasonable that the tensile yield strength (σ_y) relates to τ_{CRSS} by geometrical relations existing between τ and σ. Thus, in analogy to Eq. (4.1), we can write

$$\sigma_y = \overline{m}\,\tau_{\text{CRSS}} \qquad (4.5)$$

where $\overline{m}$ is a suitable average for the polycrystal. In evaluating $\overline{m}$, care must be taken that it reflects the greater constraints provided by the least-favorably oriented grains; that is, $\overline{m}$ is not a geometrical average of the m values of all of the (presumably) randomly oriented crystals but is, rather, somewhat higher than this.

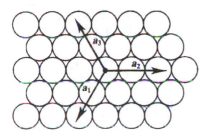

Figure 4.17
The basal plane in a hexagonal close-packed crystal; the three nonparallel slip directions are noted. Although there are three geometrically distinct slip systems, only two are independent in that an arbitrary displacement along the $\bar{a}_3$ slip direction can be duplicated by a suitable combination of slip in the $\bar{a}_1$ and $\bar{a}_2$ directions.

160

CHAPTER 4
Plastic
Deformation in
Single and
Polycrystalline
Materials

The procedural details for calculating $\overline{m}$ on this basis are complicated, although the basic scheme is relatively straightforward. For a *specific* grain orientation, there are a number of combinations of five slip systems that can produce a specified plastic strain. The combination that produces the specified strain and that involves the least amount of *total* shear strain is selected. An average value of m for these five systems is then calculated. Then an average value of m for the aggregate is determined by averaging such m values over all possible grain orientations. With this procedure, $\overline{m}$ is found to be 3.06 for the fcc structure and 2.75 for the bcc structure. These $\overline{m}$ values are significantly higher than those obtained by averaging the lowest m value (i.e., the m value corresponding to single slip for all possible grain orientations). The latter procedure, for example, leads to a $\overline{m}$ value of 2.25 for the fcc structure.

An additional strengthening effect associated with polycrystals comes from grain size; smaller grain size materials have higher values of σ_y than coarse-grained materials. This phenomenon is further discussed in Chap. 5.

Material class	Primary system	Number of geometrical systems	Number of independent systems	Secondary system	Number of geometrical systems	Number of independent systems	T/T_m for activation of secondary slip
Face-centered cubic metals	$\{111\}<1\bar{1}0>$	12	5	—	—	—	—
Body-centered cubic metals	$\{110\}<\bar{1}11>$	12	5	$\{112\}<11\bar{1}>$* $\{123\}<11\bar{1}>$	12 24	5 5	—
Hexagonal close-packed metals†	$\{0001\}<11\bar{2}0>$ $\{1\bar{1}00\}<11\bar{2}0>$	3 3	2 2	$\{1\bar{1}01\}<11\bar{2}0>$	6	4	—
Rock salt (fcc) (e.g. MgO, LiF, NaCl)	$\{110\}<1\bar{1}0>$	6	2	$\{001\}<1\bar{1}0>$	6	3	0.5
Diamond cubic (fcc) (e.g. Si, Ge, diamond)	$\{111\}<1\bar{1}0>$	12	5	—	—	—	—
TiC (fcc)	$\{111\}<\bar{1}10>$	12	5	—	—	—	—
CsCl (simple cubic)	$\{110\}<001>$	6	3	None for T/T_m ≤ 0.6	—	—	—
Al_2O_3 (hexagonal)	$\{0001\}<11\bar{2}0>$	3	2	$\{11\bar{2}0\}<1\bar{1}01>$ $\{1\bar{1}02\}<1\bar{1}01>$	3 3	2 2	0.8
BeO (hexagonal)	$\{0001\}<11\bar{2}0>$	3	2	$\{1\bar{1}00\}<11\bar{2}0>$ $\{10\bar{1}0\}<0001>$	3 1	2 1	0.5

*Slip in the bcc metals invariably involves glide in the <111> type of direction. However, active slip planes are very system specific. Slip may be restricted to $\{110\}$ kinds of planes or it may involve $\{112\}$, $\{123\}$ and other planes as well.

†Primary and secondary slip systems in hcp are system specific, depending often on the c/a ratio. When this ratio is greater than the ideal one, slip planes are usually the basal $\{0001\}$ planes. The basal plane is the primary slip plane for Be, Cd, Zn, and Mg. The $\{1\bar{1}00\}$ plane is the primary slip plane for Ti and Zr. Secondary slip occurs on $\{10\bar{1}1\}$ for Cd, Co, Zn, Mg, Zr, and Ti. Secondary slip can also take place on the $\{11\bar{2}2\}<11\bar{2}3>$ system (five independent systems) as it does in Cd, Zn, Mg, Zr, and Be.

Taken from the following sources: For hexagonal close-packed metals: R. G. Partridge, *Metall. Rev.,* **12,** 118, 1967. For rock salt structures: A. Kelly and G. W. Groves, *Crystallography and Crystal Defects,* Longman, London, 1970, p. 175; J. J. Gilman, *Micromechanics of Flow in Solids,* McGraw-Hill, New York, 1969. For TiC: G. E. Holloy and R. E. Smallman, *J. Appl. Phys.,* **37,** 818, 1966. For Al_2O_3: J. B. Wachtman, Jr. and L. H. Maxwell, *J. Am. Cer. Soc.,* **40,** 377, 1957; D. J. Gooch and G. W. Groves, *J. Am. Cer. Soc.,* **55,** 105, 1972; D. J. Gooch and G. W. Groves, *Phil. Mag.,* **28,** 623, 1973; M. V. Klassen-Neklyudova, V. G. Govorkov, A. A. Urosovkaya, N. N. Voinova and E. P. Kozlovskaya, *Phys. Stat. Sol.,* **39,** 679, 1970; J. D. Snow and A. H. Heuer, *J. Am. Cer. Soc.,* **50,** 153, 1973. For BeO: G. G. Bentle and K. T. Miller, *J. Appl. Phys.,* **38,** 4248, 1967; D. T. Livey, *High Temperature Oxides,* ed. A. M. Alper, Part 3, Academic Press, New York, 1970, p. 1.

Table 4.1
Slip characteristics in various classes of materials and the effect of crystal structure

The temperature and strain-rate dependence of σ_y for polycrystals (Fig. 4.18) parallels that of τ_{CRSS} for single crystals. In some materials, such as NaCl, additional slip systems are activated at high temperatures and this leads to a greater temperature dependence for σ_y than for τ_{CRSS}. For metals, σ_y for the face-centered cubic metals does not depend greatly on temperature (Fig. 4.18), but the tensile strength does increase significantly at low temperatures. On the other hand, body-centered cubic metals have a large low-temperature temperature dependence of σ_y. This is caused by factors relating to chemistry and bond strength, and which include the strong temperature dependence of the Peierls stress of these metals, the advent of additional slip systems at higher temperatures, and the presence of interstitial atoms (such as carbon and nitrogen) that are very effective hardeners of these metals at low temperatures. Regardless of the source of the temperature dependence of σ_y, it can lead to premature fracture at low temperatures when σ_y becomes so great that the material fractures before undergoing macroscopic plastic flow (Chaps. 9 and 10).

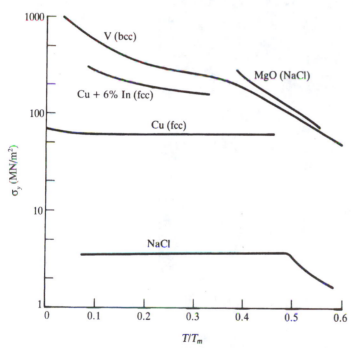

Figure 4.18
The temperature variation of the polycrystalline yield strength (σ_y) closely parallels that of τ_{CRSS} of single crystals. Body-centered cubic metals (e.g., V) have intrinsically higher σ_y values than do face-centered cubic metals (e.g., Cu). Further, σ_y is more temperature sensitive for bcc metals than for fcc ones. The high strength of MgO is related to its strong (polar covalent) bonding. MgO is much stronger than ionically bonded NaCl, which has the same crystal structure. (*Data for Cu and V from H. Conrad, High Strength Materials, ed. V. F. Zackay, Wiley, New York, 1965, p. 436; for Cu-6 percent In from M. M. Hutchinson and R. T. Pascoe, Metal. Sc. J., 6, 189, 1966; for MgO from T. G. Langdon and J. A. Pask, J. Am. Cer. Soc., 50, 365, 1967.*)

162

CHAPTER 4
Plastic
Deformation in
Single and
Polycrystalline
Materials

4.5
PLASTIC-FLOW BEHAVIOR AND MATERIAL CLASS

Although Fig. 4.18 provides examples of how yield strength relates to material class, further discussion of this topic is warranted. The bar chart of Fig. 4.19 illustrates room-temperature yield strengths of a number of materials; the chart is classified on the basis of material class. As with elastic moduli (Fig. 2.1), we see that yield strengths of engineering solids vary by about six orders of magnitude. Ceramics are the strongest of the material classes. Their yield strengths shown in Fig. 4.19 were estimated from hardness tests (cf. Sect. 1.3E) because ceramics are brittle, and it is difficult to test them in tension for they often crack when gripped to do so. Further, many ceramics contain preexisting flaws which cause them to fracture prematurely. Thus, even though ceramics are the strongest of the engineered materials in terms of resistance to plastic flow, their strengths shown in Fig. 4.19 flatter

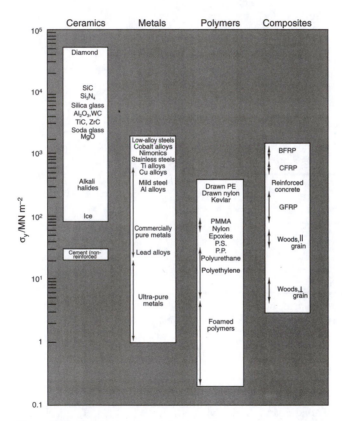

Figure 4.19
Bar chart of yield strengths classified on the basis of material class. Ceramics are the strongest of the material classes, although they are prone to fracture at stress levels below those shown in this figure. Polymers are the least strong of the material classes, and strengths of metals are intermediate to those of polymers and ceramics. Composite strengths are an average of those of their constituents. (*Adapted from Michael F. Ashby and David R. H. Jones, Engineering Materials I: An Introduction to their Properties and Applications, Pergamon Press, Oxford, 1980.*)

them too much; that is, ceramics almost always fracture at stress levels less than their yield strengths. Because of this, and as is discussed in Chaps. 9 and 10, much current effort is directed at improving the fracture resistance of ceramics.

As a class, engineered plastics are the least resistant to plastic deformation. The softest of them—foamed polymers—are very soft indeed, but this also reflects their porous nature (Chap. 14). The strongest plastics—heavily drawn thermoplastic filaments—are about as strong as Al alloys. As mentioned in Chap. 3 (and also Chap. 8), filament strengths derive from the alignment of the polymer chains along the filament axis. Thus, an applied stress is resisted by strong covalent bonds rather than by weak interchain bonding as is the situation for bulk polymers. Since extensive deformation precedes chain alignment, these high polymer strengths are only found in fine filaments. In this form, they constitute the backbone of the textile and carpet industry. More is written about the flow behavior of polymers in Chap. 8.

The strengths of several of the strongest composites of Fig. 4.19 (boron-, carbon-, and glass fiber-reinforced composites; BFRP, etc.) reflect primarily the strengths of their reinforcing filaments. As discussed in Chap. 6, composite strengths are a suitable average of their constituent strengths. In the case of BFRP etc., the composite matrix strength (i.e., that of the polymer) is much less than that of the reinforcement; thus, the matrix contributes little to strength. The matrices do serve, however, to "glue" the reinforcement together. As discussed further in Chaps. 9 and 10, composite materials display surprisingly high fracture toughnesses.

On the average, the strengths of metals are intermediate to those of ceramics and polymers. Ultra-pure metals are quite soft. We have noted previously that a pure W single crystal can be bent by hand. Most high-strength engineering metals are alloys; that is, they contain more than one element. As discussed in Chap. 5, there are a number of causes for the higher strengths of alloys compared to those of the elements on which they are based.

The malleability of materials relates to their strengths. Typically, a strong material is not very malleable and vice versa. Thus, ceramics are strong and brittle, and thermoplastics are relatively weak and malleable.[6] Metals and their alloys are generally malleable, moderately strong, and fairly fracture resistant. These characteristics account for their widespread use as structural materials. Their malleability allows them to be formed in a variety of shapes, and their strengths and fracture resistances permit them to be used safely in demanding stress environments. The malleabilities of specific metal classes depend on temperature just as do their strengths. Thus, it is worthwhile to conclude this section by comparing tensile ductilities of the bcc and fcc metals as they vary with temperature.

Figure 4.18 illustrates in general how the yield strengths of fcc and bcc metals vary with temperature. Specifically, σ_y is a strong function of temperature for the bcc metals and a fairly weak one for fcc metals. The temperature dependence of the work-hardening behavior of these metal classes also differs. We have mentioned that the T.S. of Cu increases significantly with decreasing temperature even though its σ_y is hardly affected by such changes. The increased difference between

---

[6]Provided, of course $T \geq T_g$ for the thermoplastic. Further, thermosetting polymers (e.g., epoxies) are not malleable. Once set, they cannot be reformed. Attempts to do so by heating them are pointless; thermosets decompose before becoming capable of permanent deformation.

164

CHAPTER 4
Plastic
Deformation in
Single and
Polycrystalline
Materials

the yield and tensile strength with decreasing temperature indicates that the work-hardening rates of the fcc metals increase with such temperature alterations. The work-hardening rates of the bcc metals are only mildly temperature dependent (Fig. 4.7b). The different temperature dependencies of strength and work-hardening rates of the fcc and bcc metals impacts their tensile ductilities as shown schematically in Fig. 4.20. Figure 4.20a plots the true stress-strain curves for a fcc metal at a "high" and a "low" temperature. Superimposed on the figure are graphs of $d\sigma_T/d\varepsilon_T$ at these temperatures. Tensile instability takes place when $\sigma_T = d\sigma_T/d\varepsilon_T$ (Eq. (1.12b)). Thus, the necking strain (defined by the intersection of the slope of the stress-strain curve with the stress-strain curve) decreases with increasing temperature for fcc metals. On this measure, these materials become *both* less formable and less strong with increasing temperature. More intuitively satisfactory results are found for the bcc metals; $d\sigma_T/d\varepsilon_T$ (assumed temperature independent) and σ_T are schematically plotted vs. ε_T for bcc metals in Fig. 4.20b. We see that necking commences in these materials at lower strains as temperature is decreased. In closing, though, we note that the necking strain only represents a material's resistance to neck development. While this factor is important in metal forming operations, it is not the best measure of a material's ductility or malleability. As noted in Chap. 1, %R.A. assesses this property much better.

4.6
GEOMETRICALLY NECESSARY DISLOCATIONS

During plastic flow, the dislocation density increases and the dislocation mobility decreases as straining progresses. Many of the dislocations accumulated during plastic flow result from multiplication processes. Since dislocation encounters leading to multiplication are "chance" encounters, dislocations accumulated by such processes are called *statistically stored* dislocations; their corresponding density is designated ρ_s.

Some plastic deformation is accompanied by internal plastic strain gradients. When such gradients are present, *geometrically necessary* dislocations are accumulated in addition to the statistically formed ones. The concept of geometrically necessary dislocations can be illustrated by considering plastic bending of a single crystal (Fig. 4.21). Here the slip planes and directions of the crystal are taken parallel to the axis of bending. Prior to deformation, the crystal is of length l and thickness t (Fig. 4.21a). On bending the bar to a radius of curvature r (Fig. 4.21b) the upper portion of the crystal undergoes tensile deformation; i.e., its length is increased from l ($= r\theta$) to $l + \delta l$ ($= (r + t/2)\theta$), with δl being positive and of magnitude $t\theta/2$. Conversely the inner circumference undergoes compression with a negative length change $t\theta/2$. Thus, a strain gradient accompanies the bending and the magnitude of the strain gradient is the strain difference between the two surfaces ($= 2\delta l/l$) divided by the distance (t) over which the gradient exists, i.e.

$$\text{Strain gradient} = 2\frac{\delta l}{lt} = \frac{\theta}{l} = r^{-1} \qquad (4.6)$$

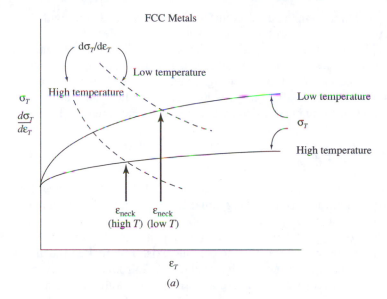

(a)

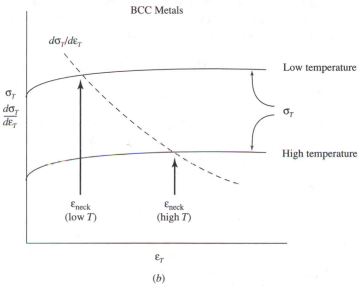

(b)

Figure 4.20
**Schematic true stress–true strain curves for (a) a "typical" fcc metal and (b) a "typi-
cal" bcc transition metal. Plots of $d\sigma_T/d\varepsilon_T$, the work-hardening rates of the respective metal classes, are also shown in the figure. For a fcc metal, σ_y is relatively
insensitive to temperature, but the work-hardening rate increases with decreasing
temperature. Since tensile geometrical instability occurs when $d\sigma_T/d\varepsilon_T = \sigma_T$, these
features lead to a lesser necking strain with increasing temperature for fcc metals.
For the bcc metals, the yield stress is highly temperature sensitive but the work-hardening rate is not. Thus, for these metals the tensile necking strain increases with increasing temperature.**

166

CHAPTER 4

Plastic
Deformation in
Single and
Polycrystalline
Materials

The number of crystal planes on the tension surface is the surface length divided by the interatomic spacing b (b is also the Burgers vector magnitude). Likewise the number of atomic planes on the compressed surface is $(l - \delta l)/b$. This difference in the number of atomic planes between the surfaces is accommodated by the introduction of edge dislocations (all of the same sign) into the crystal, as illustrated in Fig. 4.21c. These are geometrically necessary dislocations. Their number is $2\delta l/b$ and their density ρ_G is this number divided by the crystal surface area, lt. Hence

$$\rho_G = 2\frac{\delta l}{lbt} = (rb)^{-1} = \frac{\text{strain gradient}}{b} \tag{4.7}$$

Note that if there were no strain gradient, no geometrical dislocations would be present. Moreover, we emphasize that the geometrical dislocations present are in addition to those stored statistically so that the total dislocation density is $(\rho_s + \rho_G)$.

The precise density of geometrical dislocations depends on the orientation of the slip plane and direction with respect to the bending axis. Figure 4.22 illustrates a situation where bending can be accomplished without the introduction of geometrical dislocations. This is because the slip plane normals are parallel to the axis of bending, and the slip directions are perpendicular to this axis. Ordinary dislocation glide accommodates the shape change for this situation. Thus, for the general case, the geometrically necessary dislocation density is expressed as

$$\rho_G = \alpha \frac{\text{strain gradient}}{b} \tag{4.8}$$

where α is a constant of order unity.

As inferred from previous discussion, deformation of polycrystals also produces strain gradients. An alternative way of viewing the constraints that individual crystals place on each other during polycrystalline deformation is provided in Fig. 4.23. The figure shows that geometrically necessary dislocations can provide compatibility of displacements between adjacent grains. Figure 4.23b shows the

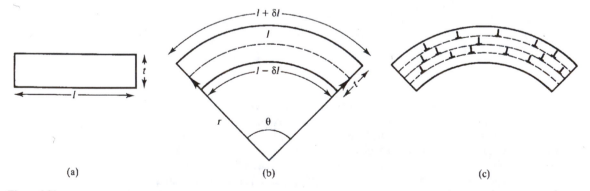

(a) (b) (c)

Figure 4.21
(a) Plastic bending of a bar of length l and thickness t to a radius of curvature r as in (b) produces a tensile strain on the outer, and a compressive strain on the inner, bar surface. Thus, there are a greater number of atomic planes ([= $(l + \delta l)/b$]; b = interatomic spacing) on the outer surface than on the inner one [= $(l - \delta l)/b$]. (c) This strain gradient is accommodated by introduction of $2\delta l/b$ geometrically necessary dislocations into the crystal.

deformation that each grain would experience in the absence of constraints, i.e., the strain it would undergo if deformed as a single crystal. Without the constraints, either voids or grain "overlap" would ensue, and this does not happen. In addition to previously discussed aspects of multiple slip, voids or overlaps among grains can be eliminated by geometrically necessary dislocations that accommodate the strain gradients that exist between individual grains.[7] The dislocation arrangements required to produce compatibility are illustrated in Figs. 4.23c and 4.23d.

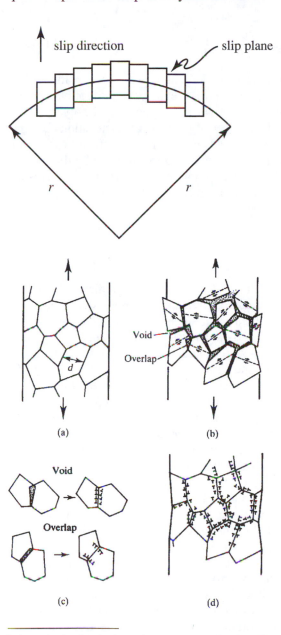

Figure 4.22
The single crystal shown here can be bent without the need of introducing geometrical dislocations into it. This is because the slip direction is normal to the axis of bending and the slip plane normal is parallel to this axis. Thus, the change in shape is accomplished solely by dislocation motion and no plastic strain gradient exists between the sample surfaces.

Figure 4.23
(a) Schematic of tensile plastic deformation of a polycrystalline solid with grain size *d*. (b) If the grains deformed as unconstrained single crystals, voids between, or overlaps of, the grains would occur. The voids or overlaps can be eliminated by introducing geometrically necessary dislocations ((c) and (d)). (*After M. F. Ashby, Phil. Mag.,* **21,** *399, 1970.*)

---

[7]Alternatively, the plastic strain gradients can be viewed as arising from the different plastic strains in the grain boundary vicinity and the grain centers.

168

CHAPTER 4
Plastic
Deformation in
Single and
Polycrystalline
Materials

Although the dislocation arrangements are different across each boundary, the geometrically necessary dislocation density can be estimated by considering an "average" grain. The amount of overlap (or void displacement) between two adjacent grains is proportional to $\bar{\varepsilon}d$, where $\bar{\varepsilon}$ is the average strain and d the grain diameter; that is, the displacement δl is proportional to $\bar{\varepsilon}$ multiplied by the "gage length" with the latter taken as d for a polycrystal. Dividing by the Burgers vector gives the number of such dislocations and further division by the grain area ($\cong d^2$) determines ρ_G as

$$\rho_G \cong \frac{\bar{\varepsilon}}{bd} = \frac{\bar{\varepsilon}}{4bd} \tag{4.9}$$

where the factor of 4 arises from more specific geometrical considerations than have been employed here. Note that ρ_G measures the dislocation density over and above that obtained (ρ_s) at an equivalent strain in a hypothetical polycrystal of sufficiently large grain size so that the number of geometrical dislocations is minimal. This increase in dislocation density leads to a grain-size effect during work hardening; that is, fine grain-sized materials work-harden more rapidly than coarse grain-sized ones. The following chapter, dealing with strengthening mechanisms, further considers grain size effects on yielding and plastic flow.

4.7
SUMMARY

In this chapter certain features of the plastic-flow behavior of single and polycrystalline materials were presented and compared. Tensile or compressive plasticity in single crystals initiates when the resolved shear stress on the slip system oriented most favorably with respect to the stress axis reaches a critical value. This stress, τ_{CRSS}, is characteristic of the material, and in general is both a function of temperature and strain rate. The temperature and strain-rate behavior of τ_{CRSS} can be divided conveniently into three regions. At the highest temperatures (Region III), diffusive aspects of permanent deformation are important and τ_{CRSS} decreases with increasing temperature and decreasing strain rate. At intermediate temperatures (Region II), this critical stress is approximately independent of both temperature and strain rate. At lower temperatures still (Region I), τ_{CRSS} again increases with decreases in temperature and increases in strain rate. The thermal dependence of the stress results from thermal activation aiding in overcoming short-range internal stress fields that obstruct dislocation motion. Thus, at low and intermediate temperatures, τ_{CRSS} consists of an athermal (temperature-independent) component and a temperature-dependent component, the latter effectively vanishing at the temperature marking the transition from Region I to Region II.

The work-hardening behavior of single crystals can be divided into three stages. Provided only one slip system is initially active, the work-hardening rate immediately following yielding is low; this Stage I is referred to as easy glide. After a certain plastic strain, multiple slip comes into play and the work-hardening rate increases dramatically, consistent with what is expected from dislocation intersections and immobilization arising from slip on intersecting slip planes. At a still higher stress level, which corresponds to Stage III of the stress-strain curve, the hardening rate decreases.

As a result of constraints arising from testing machine geometry, crystal rotation occurs during tensile and compression testing of single crystals oriented initially for single slip. The rotation eventually activates slip on other systems, and the onset of multiple slip marks the transition from Stage I to Stage II hardening. The extent and progression of rotation can be followed using the stereographic projection.

Although slip mechanisms are the same in poly- and single crystals, the former plastically yield and flow at considerably higher stress levels. This is a direct result of constraints that neighboring, and differently oriented, grains within the polycrystal place on each other. These constraints necessitate the operation of five independent slip systems for polycrystalline plastic flow and, as a result, an easy glide region is not observed in the tensile flow curves of polycrystals. Instead, the work-hardening behavior of polycrystals more closely resembles that for Stage II–Stage III hardening of single crystals. Additionally, certain crystal systems do not provide the necessary five independent slip systems required for polycrystalline flow. Thus, polycrystals of materials with these structures are incapable of such flow and fail in a brittle manner even though single crystals of them may exhibit extensive plastic flow.

The stress (σ_y) required to initiate polycrystalline plasticity, and the temperature dependence of this stress, vary among the material classes. Ceramics are inherently resistant to plastic flow, so much so that in service they usually fracture before macroscopic plastic flow is observed in them. That fracture occurs preferentially indicates that the fracture propagation stress is less than σ_y for these materials. There are several reasons why ceramic yield strengths are high. One of them is that their Peierl's stresses are high, and this provides an inherently high flow stress. In addition, many ceramics do not display the five independent slip systems necessary for polycrystalline plasticity.

The yield and flow behavior of face-centered cubic metals and body-centered cubic metals also differ. The bcc metals display a strong temperature dependence of σ_y, especially at low temperatures. On the other hand, their work-hardening rates are not very temperature sensitive. Face-centered cubic metals behave oppositely. Their yield strengths are only mildly temperature sensitive. However, their work-hardening rates increase significantly with decreasing temperature.

The level of the flow curve of a polycrystal in comparison to a single crystal can be expressed in terms of an orientation (Schmid) factor of the "average" polycrystal grain. This description is proper provided that the slip system requirements discussed above are considered in evaluating the average orientation factor. Grain size also plays a role in polycrystalline deformation. Strain gradients within a polycrystal give rise to the generation of "geometrically necessary" dislocations as deformation proceeds. Thus, the dislocation density of a polycrystal is greater than that of a single crystal at the same strain and this feature raises the flow stress of a polycrystal. The smaller the polycrystalline grain size, the greater the geometrically necessary dislocation density; thus, this additional strengthening is most pronounced in materials of fine grain size.

170

CHAPTER 4
Plastic
Deformation in
Single and
Polycrystalline
Materials

REFERENCES

Arsenault, R. J.: "Low Temperature Deformation of BCC Metals and Their Solid Solution Alloys," *Treat. Mat. Sc. Tech.*, ed. R. J. Arsenault, **6**, 1, 1975.

Evans, A. G., and T. G. Langdon: "Structural Ceramics," *Prog. Matls. Sc.*, **21**, 171, 1976.

Evans, K. R.: "Solid Solution Strengthening of Face-Centered Cubic Alloys," *Treat. Mat. Sc. Tech.*, ed. H. Herman, **4**, 113, 1974.

Hertzberg, Richard W.: *Deformation and Fracture Mechanics of Engineering Materials,* 4th ed., Wiley, New York, 1996.

PROBLEMS

4.1 A zinc single crystal ($c/a = 1.856$) is loaded in compression in a direction at an angle of 7° to the normal of the slip plane (the basal plane). The basal plane and the twin plane normals are coplanar with the compression axis. For Zn, $\tau_{CRSS} = 2.3$ MN/m^2 for slip and the slip direction ([11$\bar{2}$0]) makes an angle of 31° with the compression axis. Assuming that there is a critical resolved shear stress for twinning of 8.0 MN/m^2 and that the twinning direction is [10$\bar{1}$1] on the (10$\bar{1}$2) plane, determine the value of compressive stress at which plastic deformation begins. State whether this deformation is a result of slip or twinning.

4.2 The following data were obtained during tensile tests on magnesium single crystals at room temperature.

ϕ (°)	λ (°)	Tensile Yield Strength (psi)
82	11	402
63	36	155
52	41	121
35	55	100
27	63	148
14	77	258

Confirm the validity of Schmid's law for Mg and determine τ_{CRSS} for this material.

4.3 Consider the single crystal having a simple cubic structure as discussed in Example Prob. 4.2. Repeat this problem for a situation where the tensile axis is parallel to the [011] crystal axis.

4.4 A square plate-shaped sample is prepared from a single crystal which has the simple cubic structure. (See Example Prob. 4.2 for slip planes and directions.) The crystal axes are oriented with respect to the principal stress directions as shown. That is, σ_1 is normal to the (1$\bar{1}$0) plane and σ_2 is normal to the (110) plane.
a Derive expressions for the plane stress yield surface for this single crystal. Plot the resulting yield surface.
b The result differs from the Tresca and von Mises yield surfaces. Physically explain why this is so.

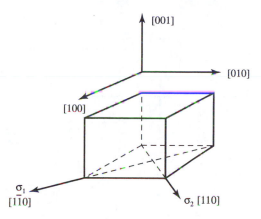

4.5 Convert the tensile stress-strain curve of the copper single crystal with a [123] orientation (cf. Fig. 4.8) into a shear stress–shear strain curve.

4.6 Consider a face-centered cubic material with a tensile stress parallel to <001>, <011>, and <111> types of directions. How many initial slip systems operate for these respective cases?

4.7 Refer to Fig. 4.12*b*. The sides of the unit triangles connecting [001] and [$\bar{1}$11] contain directions of the form [$\bar{1}$1*n*] ($n \geq 1$); the other sides contain directions of the form [0*n′m′*] ($m′ \geq n′$; $m′,n′ \geq 0$) and [$\bar{1}n′n′$] ($n′ \geq 1$), respectively. Show that two slip systems initially operate in fcc crystals when the tensile axis is parallel to planes with normals situated on any of the three sides of the [001] stereographic triangle.

4.8 a Assume that a bcc crystal slips on {110}<111>. For a tensile axis parallel to the [123] crystal axis, calculate the Schmid factor for all possible slip systems. (Hint: Use of a stereographic projection is recommended.)
b Is this crystal oriented for single slip or for multiple slip?
c What is the conjugate slip system for tensile deformation if the initial tensile axis is parallel to the [123] crystal axis?
d What is the conjugate slip system for compressive deformation if the initial compression axis is parallel to the [123] crystal axis?

4.9 A single crystal in the form of a cube has two possible slip planes, (1) and (2), as shown in the sketch below. Each of these planes has two possible slip directions, (*a*) and (*b*), as is also shown.

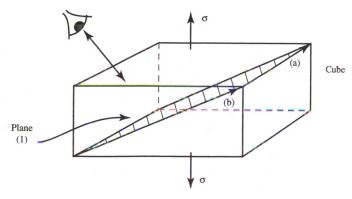

172

CHAPTER 4
Plastic
Deformation in
Single and
Polycrystalline
Materials

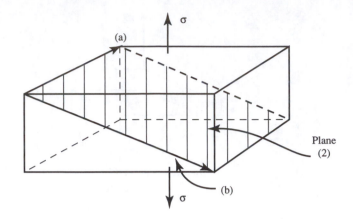

a If a stress is applied to this crystal in the direction noted, specify which particular (e.g., plane (1), direction (a)) slip system(s) will be active at the beginning of plastic deformation.

b If the critical resolved shear stress is 7 MN/m^2, calculate the tensile stress at which plastic deformation initiates.

c Looking normal to slip plane (1), four distinct types of dislocations—*A*, *B*, *C*, and *D*—are observed as shown below.

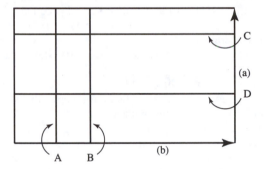

Dislocations *A* and *B* are parallel to slip direction (*a*) whereas *C* and *D* are parallel to slip direction (*b*). Dislocations *A* and *C* are observed to glide during plastic deformation; dislocations *B* and *D* do not. Construct a table—with headings as shown below—that lists characteristics of the four dislocation types. Explain your reasoning. (Dislocation "character" refers to the dislocation as edge, screw, or mixed.)

Dislocation Type (e.g., A)	Slip Direction ((*a*) or (*b*))	Dislocation Character

d Suppose the tensile axis were parallel to direction (*b*) in slip plane (1). Briefly describe the tensile stress-strain behavior of the single crystal for this condition.

4.10 a The [$\bar{1}$23] direction falls within the standard stereographic triangle in a [001] projection. Show that only one slip system operates when the tensile axis is aligned with this direction during a tensile test of a fcc metal single crystal. Identify the specific plane and direction on which slip initiates.

b If τ_{CRSS} is 5 MN/m^2, what is the value of the tensile stress required to initiate slip if the material is tested as described in part (a)?

c As discussed in the text, the slip direction rotates towards the tensile axis during single slip. Show that for the crystal orientation of part (a), the tensile axis intersects the $11x$ side of the standard triangle at $\bar{1}12$. (Hint: This can be done in more than one way. An easy way is to show that the initial slip direction, the direction $[\bar{1}23]$, and $[\bar{1}12]$ are all in the same plane. This can be accomplished employing dot products and simple geometry.)

d As the slip direction rotates, the tensile stress required to continue plastic deformation increases as a result of the slip plane/direction orientation changes. Assuming that the shear stress required to plastically deform the material does not change (i.e., τ remains at 5 MN/m²), what will be the value of the tensile stress at the transition from single to duplex slip?

4.11 Hexagonal close-packed magnesium slips by basal plane slip at room temperature. Polycrystalline Mg has a yield strength of about 100 MN/m².
  a Estimate the critical resolved shear stress for Mg. State any assumptions.
  b What are the highest and lowest yield strength values that could be found for a single crystal of Mg?
  c Will the work-hardening rate of single crystal Mg be higher, lower, or the same as that of a single crystal of an hcp metal that exhibits both basal and prismatic slip? Give your reasoning.

4.12 Show that grain overlap and voids (Fig. 4.23) can be eliminated by introducing discs of vacancies and plates of atoms, respectively. Sketch the appearance of such vacancy and interstitial "loops" in two dimensions. Show how they can be schematically represented by edge dislocations.

4.13 Statistical dislocations are both generated and removed during cold work. One reasonable expression describing this behavior is due to Kocks. The expression relates the statistical dislocation density, ρ_s, to strain, ε, through

$$\frac{d\rho_s}{d\varepsilon} = C_1(\rho_s)^{1/2} - C_2\rho_s$$

Integrate the above equation from $\rho_s = \rho_0$ (the initial dislocation density) to $\rho_s = \rho_s$, and obtain ρ_s as a function of strain and the constants C_1 and C_2. Plot dislocation density as a function of strain for materials having the values of C_1 and C_2 shown in the table below. (Take $\rho_0 = 10^{10}$/m².)

Material	C_1 (m^{-1})	C_2
Nickel	2.3×10^7	1.40
Silver	3.0×10^7	2.10
Copper	4.2×10^7	2.90
Iron	4.4×10^6	0.01

4.14 A 4-cm-long wedge-shaped metal strip is cold-rolled, as illustrated in the sketch following. The thickness of the thin part of the wedge is 0.4 cm; that of the initial thick end is 0.8 cm. The roll spacing is set at 0.4 cm; thus, the thin edge of the wedge is not deformed but the thick end has its thickness reduced by 50% during rolling.

174

CHAPTER 4
Plastic
Deformation in
Single and
Polycrystalline
Materials

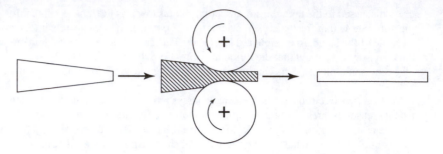

a Would the rolled wedge contain geometrically necessary dislocations? Explain.
b If the wedge does contain geometrical dislocations, calculate their approximate density. Does the number you calculate correspond to a high dislocation density or a low one?

CHAPTER 5

Strengthening of Crystalline Materials

5.1
INTRODUCTION

In this chapter we consider important ways by which the flow strength of crystalline solids is increased by restricting dislocation motion. Various types of "obstacles," either alone or in combination, can be present in a crystalline material to produce such an effect. The most commonly employed obstacles are other dislocations, internal boundaries (such as grain, subgrain, or cell boundaries), solute atoms, and second-phase particles.

Additional dislocations generated during plastic deformation give rise to work hardening. Material strength increases owing to the decrease in dislocation mobility concurrent with an increase in dislocation population. A grain boundary is a particularly effective strengthening agent because grain boundaries cannot be penetrated by a moving dislocation. However, effective grain-size strengthening can be obtained only when the grain boundary "density" is high; i.e., when the grain size is small, usually on the order of 5 μm or less. Subboundaries and cell boundaries, the latter often formed during plastic deformation, provide lesser restraints to dislocation motion than grain boundaries. (Subboundaries and cell boundaries are capable of being penetrated by moving dislocations.) Nonetheless, the fine cell sizes that can be produced by extensive cold deformation of some materials can be used to harden them appreciably. Solid solution impurity atoms are generally considered "weak" hardeners. This is true for substitutional elements that produce spherically symmetrical distortions within the crystal. However, certain interstitial elements in the body-centered cubic transition metals and divalent cation impurities in monovalent ionic solids, produce nonspherically symmetric distortions and this leads to appreciable strengthening. In some circumstances, second-phase particles provide exceptional strengthening. In general, the strengths of such precipitation- or dispersion-hardened alloys are limited by the fineness of the particle dispersion that an be produced in the matrix.

Most high-strength alloys are hardened by more than one of the mechanisms noted above. In such cases, the total hardening can be approximated as the sum of the strength contributions resulting from the separate hardening mechanisms. Several examples of technologically important materials and the processes by which high strength is achieved in them are provided near the end of this chapter.

Although the basic physics of strengthening is well known, accurate quantification of the strength provided by specific dislocation obstacles is difficult. Since the additional strength is determined by the resistance to dislocation motion provided by the obstacles, we begin our discussions on strengthening by considering this factor.

5.2
GENERAL DESCRIPTION OF STRENGTHENING

As mentioned, the strength of a crystalline material is increased by obstacles that restrict dislocation motion within it. Consider a slip plane containing an array of such obstacles (Fig. 5.1), the nature of the obstacles remaining unspecified for now. As shown in Fig. 5.1, the obstacles are characterized by a mean spacing, L', within the slip plane. In response to an applied stress, the dislocations at first bend between the obstacles. That is, as shown in Fig. 5.1, they are "extruded" between them until, at a critical extrusion angle, ϕ_c, the dislocation breaks through the obstacles and progresses on the slip plane until it encounters other obstacles. The stress required to produce continued motion of the dislocation through the obstacle array is the macroscopic yield or flow stress.

The obstacle "strength" essentially determines this flow stress. "Strong" obstacles resist dislocation penetration. This is reflected by ϕ_c approaching zero (Fig. 5.2a). The spacing between obstacles is also important. For strong obstacles, the effective spacing is the mean spacing between obstacles on the slip plane (L, Fig. 5.2a). If, on the other hand, the obstacles are weak, ϕ_c is very large, approaching 180° (Fig. 5.2b), and this means that the dislocation line is much straighter for

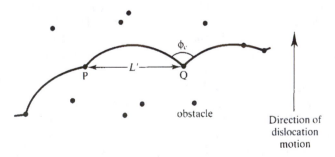

Figure 5.1
A dislocation held up by a random array of slip-plane obstacles. The stress required to overcome the obstacles depends on the effective spacing (L') between the obstacles along the dislocation line and the angle (ϕ_c) to which the dislocation bends before it breaks through them. (*After L. M. Brown and R. K. Ham, Strengthening Mechanisms in Crystals, ed. A. Kelly and R. B. Nicholson, Wiley, New York, 1971, p. 10.*)

the situation of weak obstacles. As a corollary, the "effective" obstacle spacing, L', is much greater than L. In the limit of very weak obstacles, L' is the average distance between obstacles found when a random line is drawn across the slip plane; i.e., L' is the average spacing between the intersections that the obstacles make with such a line.

The stress required to bend the dislocation to the angle ϕ_c is calculated in a manner analogous to that used to determine the stress required to operate a Frank-Read source (Sect. 3.8B). An appropriate line tension balance (taking the dislocation line tension as $Gb^2/2$) at the critical angle shows that the necessary shear stress for continued dislocation motion is

$$\tau \cong \frac{Gb}{L'} \cos \frac{\phi_c}{2} \tag{5.1}$$

where L' is the effective spacing.

As noted, for strong obstacles $\phi_c \cong 0$ and $L' \cong L$. Thus, the maximum strength associated with a given obstacle array is

$$\tau_{max} = \frac{Gb}{L} \tag{5.2}$$

With decreasing obstacle strength, ϕ_c increases. An overestimate of the operative stress for strong obstacles is obtained by utilizing Eq. (5.1) with $L' = L$, i.e.,

$$\tau(\text{strong obstacles}) \cong \frac{Gb}{L} \cos \frac{\phi_c}{2} \tag{5.3}$$

Friedel has considered the case of weak obstacles and has shown that as $\phi_c \rightarrow \pi$, L' becomes approximately equal to $L/(\cos(\phi_c/2))^{1/2}$. Thus, the approximate strength provided by weak obstacles is given by

$$\tau(\text{weak obstacles}) \cong \frac{Gb}{L} \left(\cos \frac{\phi_c}{2} \right)^{3/2} \tag{5.4}$$

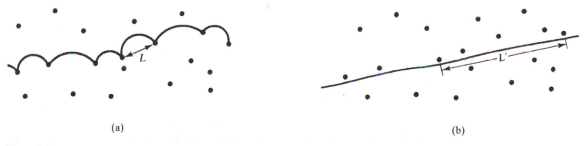

(a)   (b)

Figure 5.2
(a) Dislocation topology appropriate to a "strong" obstacle. Here, ϕ_c is small and the effective obstacle spacing (L') is approximately equal to the mean center-to-center spacing of the obstacles in the plane. (b) When the obstacles are "weak," the dislocation line is nearly straight (i.e., $\phi_c \cong 180°$), and the spacing between the obstacles is large in comparison to the mean obstacle center-to-center distance such that L' is approximately equal to the distance between obstacle intersections that a random line makes on the slip plane. (Adapted from R. L. Fleischer, The Strengthening of Metals, ed. D. Peckner, Reinhold, New York, 1964, p. 93.)

Figure 5.3 shows the variation of the increase in flow stress due to obstacles as a function of ϕ_c as predicted by Eqs. (5.3) and (5.4). Both equations predict the same maximum strength ($= Gb/L$) at $\phi_c = 0$, but the Friedel relation (Eq. (5.4)) shows that weak obstacles are much weaker than predicted by Eq. (5.3). Shown also in Fig. 5.3 are the results of computational simulations of dislocations moving through an obstacle array. In the simulation, the mean spacing L and critical angle are specified and the stress required to move the dislocations through the array is calculated. The results show that the maximum strengthening is somewhat less than predicted by Eq. (5.2), although the difference is well within the accuracy of most calculations of this nature. The simulation also indicates that the Friedel relation is quite accurate for large values of ϕ_c.

It is clear, therefore, that a good approximation of the strength provided by an obstacle is obtained if both the obstacle spacing and the angle ϕ_c are known. Unfortunately, it is difficult to determine ϕ_c accurately for most obstacles. Instead, obstacles are classified generally as either strong ($\phi_c \cong 0°$) or weak ($\phi_c \cong 180°$). If the obstacle is strong, the added strength is always on the order of Gb/L. For weak obstacles, however, the strengthening is small and depends critically on the value of ϕ_c. Hence, in discussing weak obstacles, such as solid-solution elements, estimates

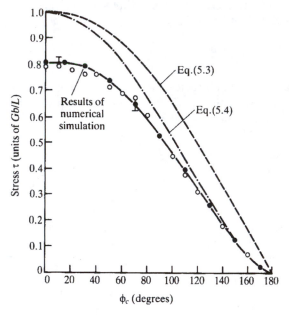

Figure 5.3
The shear stress (units of Gb/L) required to overcome slip-plane obstacles as a function of ϕ_c. The uppermost curve (Eq. (5.3)) holds for strong obstacles. The intermediate curve (Eq. (5.4)) describes weak obstacles; at large values of ϕ_c, it is an excellent approximation to the results of computational simulations. The "strong-hardening" approximation of Eq. (5.3) overestimates the required stress compared to computational results (even when $\phi_c \to 0$, for which Eq. (5.3) is expected to be most valid). However, Eq. (5.3) remains an excellent approximation to the strength provided by strong obstacles. (After L. M. Brown and R. K. Ham, Strengthening Mechanisms in Crystals, ed. A. Kelly and R. B. Nicholson, Wiley, New York, 1971, p. 10.)

of their strength contribution are obtained by approaches different than those described here.

179

SECTION 5.3
Work Hardening

5.3
WORK HARDENING

The increase in dislocation density accompanying continued plastic deformation is responsible for work hardening; that is, dislocations themselves are obstacles to dislocation motion. Depending on the type of interaction that takes place among moving dislocations, dislocations can be either soft or hard obstacles.

Dislocation interactions during easy glide of single-crystal plasticity provide an example of soft obstacles. Stage I hardening can be considered to arise from the stress field interaction of dislocations moving on parallel slip planes (Prob. 5.4; cf. also Fig. 3.42). In this description, the obstacles of Fig. 5.1 correspond to locations of interaction energy troughs or hills[1] on active slip planes. The resulting increase in flow strength depends on the spacing between active slip planes and the overall dislocation density. Small interplanar spacings and larger dislocation densities result in greater strengthening. However, in all cases the hardening is relatively modest.

Dislocation intersections produce hard obstacles. This is manifested by the high work-hardening rates of polycrystals and single crystals during Stage II deformation. Flow in both instances involves multiple slip and, as discussed in Chap. 3, the consequential dislocation intersections can lead to the creation of immobile jogs on the dislocations. These jogs are "hard" obstacles, and they are circumvented by dislocation extrusion between them in a manner analogous to the operation of a Frank-Read source. The flow stress can be described in terms of the approximate Eq. (5.2). That is, assuming all dislocations represent obstacles, then, if the average dislocation density is ρ, the obstacle spacing is obtained using the relationship (Prob. 5.3) $L^2\rho$ = constant. From Eq. (5.2) the shear flow stress is then given by

$$\tau = \tau_0 + \alpha Gb(\rho)^{1/2} \tag{5.5}$$

where τ_0 is the intrinsic strength of a material having a dislocation density low enough so that dislocation interactions are inconsequential, and the empirical constant α represents the correction factor necessitated by the approximate applicability of Eq. (5.2). For body-centered cubic metals, α is on the order of 0.4; for face-centered cubic metals it is typically 0.2. The essential correctness of Eq. (5.5) has been verified by a number of studies. Results of investigations on Cu are shown in Fig. 5.4, in which the square root dependence of flow strength on dislocation density is observed over a range of several orders of magnitude in dislocation density.

In some respects, it is surprising that the flow stress can be uniquely correlated with dislocation density for, as described in Chap. 3, dislocation morphology is often altered as plastic deformation proceeds. At small strains, the dislocation

---

[1]Recall that the force required for dislocation motion is proportional to the gradient of the interaction energy. Thus, both energy troughs and hills restrict dislocation motion.

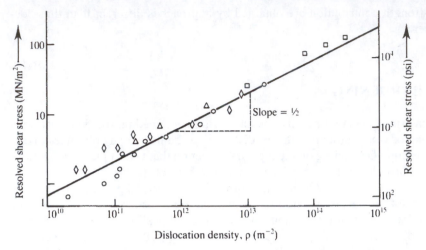

Figure 5.4
Critical resolved shear stress as a function of dislocation density for Cu single crystals and polycrystals. The observed slope of ½ on the logarithmic coordinates verifies that Eq. (5.5) describes the flow strength of work-hardened materials as it relates to dislocation density. □, polycrystalline Cu; ○, single-crystal Cu—one slip system; ◊, single-crystal Cu—two slip systems; △, single-crystal Cu—six slip systems. (*After H. Weidersich, J. Metals, 16, 425, 1964.*)

arrangement can be considered a random array of intersecting lines in three dimensions. However, with increasing plastic strain this arrangement evolves into the heterogeneous dislocation distribution described as a cellular structure (Sect. 3.8C). Such a structure can be described as heavily dislocated in the cell boundaries, with cell interiors having fairly low dislocation densities (cf. Fig. 3.43*b*). As deformation continues cell size is reduced, and for some body-centered cubic metals it continues to decrease even at true strains on the order of five. For most face-centered cubic metals, however, a limiting cell size is attained at some strain beyond which no appreciable further reduction in cell size accrues.

In spite of the different dislocation configuration associated with cells, the relation between flow stress and dislocation density is still given by Eq. (5.5) although when it is applied to cellular structures there is a somewhat different physical basis to the relationship. The pertinent physics are contained in Kuhlmann-Wilsdorf's theory of work hardening. In this theory, plastic flow is accompanied by extrusion of dislocations from the cell walls through something analogous to a Frank-Read mechanism. The structure of these walls consists of "links" of dislocation lines meeting at nodes. Extrusion takes place along the longer links in the walls; these links (as well as the cell size) become smaller as deformation proceeds. The relationship between flow stress and dislocation density comes directly from Eq. (3.21), which relates cell size (*D*) to shear flow stress.

$$D = \frac{K'Gb}{(\tau - \tau_0)} \qquad (3.21)$$

On rearranging Eq. (3.21) we obtain

$$\tau = \tau_0 + \frac{K'Gb}{D} \qquad (5.6)$$

The cell size and link spacing (l) scale, and the link spacing varies with dislocation density (ρ) as $l \sim \rho^{-1/2}$. Thus, the flow stress is expected to depend on the square root of dislocation density in materials with cellular structures. There is one difference, though, between the otherwise equivalent expressions of Eqs. (5.5) and (5.6). The proportionality constant between the flow stress and dislocation density is less for cellular structures. (More precisely, the value of the constant α in Eq. (5.5) decreases with strain.) The reason is that the dislocation line tension is less for dislocations within cells. In particular, the more precise expression for dislocation line tension (Eq. (3.12)) contains a "cut off" length in its logarithmic term. This length scales in proportion to the dislocation link length, and this leads to a gradual decrease in dislocation line tension with strain and a like decrease in α.

5.4
BOUNDARY STRENGTHENING

Internal boundaries also present obstacles to dislocation motion. On geometrical considerations alone, boundaries are expected to provide stronger obstacles to dislocation motion than line defects (i.e., dislocations) or point defects (e.g., solute atoms), since the intersection of a surface with a slip plane is a line rather than a point. Hence, a surface such as a boundary impedes dislocation motion along its entire slip plane length, and this provides a greater resistance to slip than isolated obstacles on the slip plane.[2]

Grain boundaries are particularly effective obstacles to dislocation motion, as crystallographic factors do not permit the passage of a dislocation from one grain to an adjacent one through a grain boundary. Instead the effect of grain boundaries on yield strength is related to the activation of simultaneous slip in adjacent grains in a polycrystal. A number of models have been put forth to describe how grain boundaries harden a material in this manner. Many of them are physically plausible. Here, one of them is described.

Prior to macroscopic yielding, microscopic yielding can occur in a grain having slip systems oriented favorably with respect to the stress axis. As a result, dislocations within such a grain pile up against the grain boundary separating it from an adjacent grain not so favorably oriented (Fig. 5.5). Macroscopic yielding occurs when dislocation motion (or sources) is (are) activated in such adjacent grains. This can be effected by the emission of dislocations from the grain boundary or, alternatively, the dislocation pileup at the boundary in the deforming grain can produce a stress concentration sufficient to activate slip in the nondeforming grains.

The process can be described semiquantitatively. Suppose τ^* is the stress required to activate dislocation motion in the unfavorably oriented grain, and these dislocations are located a distance r from the boundary (Fig. 5.5). The stress concentration due to the dislocation pileup in the active grain increases with the

---

[2]Because of this factor, strengthening by grain boundaries cannot be discussed in terms of Sect. 5.2 which applies to discrete obstacles to dislocation motion.

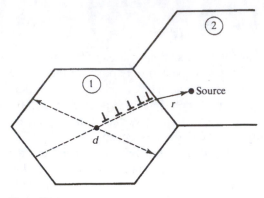

Figure 5.5
**Microyielding in a grain (Grain 1) favorably oriented for slip may precede macro-
scopic yielding. Macroscopic flow requires dislocation activity in all grains (e.g.,
Grain 2), and this may be induced by the internal stress caused by the dislocation
pileup at the boundary in Grain 1. This stress may cause dislocation emission from
the boundary or may activate a dislocation source (at point r) in Grain 2. The magnitude of the stress concentration depends on the number of dislocations in the pileup,
and increases with the grain diameter, d.**

number of dislocations that come up against the impenetrable boundary. In turn, this
number increases as the mean slip distance (the grain size, d) in the active grain increases. Analysis shows that dislocation activation in the second grain occurs when

$$(\tau_{app} - \tau_0)\left(\frac{d}{4r}\right)^{1/2} = \tau^* \tag{5.7}$$

In Eq. (5.7), τ_{app} is the applied shear stress at which this activation occurs and τ_0 is
the intrinsic stress resisting dislocation motion in the deforming grain. The parameter $(d/4r)^{1/2}$ represents the stress concentration arising from the pileup; this increases with the number of dislocations in it (i.e., with d). Rearrangement of Eq.
(5.7) allows the applied shear stress to be expressed in terms of the grain diameter;

$$\tau_{app} = \tau_0 + 2\tau^* r^{1/2} d^{-1/2} = \tau_0 + k_y' d^{-1/2} \tag{5.8}$$

In terms of tensile yield strength, Eq. (5.8)'s analog is

$$\sigma_y = \sigma_0 + k_y d^{-1/2} \tag{5.9}$$

Equation (5.9) is known as the Hall-Petch equation and its prediction that the yield
strength of a polycrystal increases linearly with $d^{-1/2}$ has been substantiated in a
number of materials. It should be noted that the essential physics of grain-size
strengthening in the model just described is contained in Eq. (5.7); that is, if the
grain size is large, a greater stress concentration is developed in the adjacent grain,
and thus the applied stress needed to activate flow in this grain is relatively low, and
vice versa.

The above arguments have physical appeal, yet there have been few observations of dislocation pile-ups at boundaries that would confirm the model based on
them. This led Li to consider that the grain size effect on yield strength is caused by
dislocation emission from grain boundaries (more specifically, from ledges on such

boundaries). He argues that a certain dislocation activity (active dislocation line length) must be achieved for plastic flow to occur. He defines a parameter, μ, that characterizes the ability of a grain boundary to emit dislocations; physically, μ corresponds to a total length of dislocation line emitted per unit area of grain boundary. The corresponding dislocation density at yielding, ρ_c, is related to μ by $\rho_c = 3\mu/d$ where d is the grain size. This value of dislocation density is substituted in Eq. (5.5) which relates shear yield strength to dislocation density to obtain

$$\tau = \tau_0 + \alpha Gb\left(\frac{3\mu}{d}\right)^{1/2} \tag{5.10}$$

Thus, a Hall-Petch relationship is obtained; k_y' is now equal to $\alpha Gb(3\mu)^{1/2}$.

Equations (5.8) and (5.9) (and Eq. (5.10) and its tensile equivalent) are coupled through the average value of the orientation factor $m = (\cos \phi \cos \lambda)^{-1}$ (Chap. 4). Thus, for a given r and τ^* (or G and α for the Li model), the proportionality constant k_y should increase with increasing m. Table 5.1 lists experimentally determined values of k_y; larger values of k_y are associated with higher values of m (e.g., the hcp metals with a limited number of slip systems). Likewise, higher values of k_y correlate with increasing inherent strength (τ^* and G; e.g., some of the body-centered cubic transition metals).

The degree to which grain-size hardening can be fruitfully utilized depends on the material's Hall-Petch coefficient and the degree of grain-size refinement possible in the material. Engineered ceramics typically have finer grain sizes than metals and fine-grained ceramics are both stronger and more fracture resistant than their coarse-grained counterparts. To appreciably grain-size harden a metal typically necessitates having a grain size below 5 μm.[3] Special precautions during processing, such as introducing fine dispersoids that do not dissolve in the metal but are

Material	Crystal structure	$k_y(MN/m^{3/2})$
Low-carbon steel	bcc	0.307
Armco iron	bcc	0.583
Molybdenum	bcc	1.768
Zinc	hcp	0.220
Magnesium	hcp	0.279
Titanium	hcp	0.403
Copper	fcc	0.112
Aluminum	fcc	0.068

Source: Adapted from J. D. Embury, *Strengthening Methods in Crystals*, ed. A. Kelly and R. B. Nicholson, Wiley, New York, 1971. Original data from: R. Armstrong et al., *Phil. Mag.*, **7**, 45, 1962; E. Anderson et al., *Trans TMS-AIME*, **242**, 115, 1968; A. A. Johnson, *Phil. Mag.*, **4**, 194, 1959; F. E. Hauser et al., *Trans TMS-AIME*, **206**, 889, 1956; R. W. Guard, *WADC Tech. Report 55-RL-1339*, 1955; F. Feltham and J. E. Meakin, *Phil. Mag.*, **2**, 105, 1959; R. P. Carreker and W. R. Hibbard, *Trans. TMS-AIME*, **209**, 1157, 1957.

Table 5.1
Values of k_y for several materials

---

[3]This is a generalization, of course; there are exceptions to it.

effective in restricting grain growth during processing, are required to achieve such grain sizes. Recent research has led to the development of nanocrystalline materials which typically have grain sizes on the order of 10 nm or so. One (of several) technique used to produce nanocrystals involves vapor deposition of nanocrystalline powders. Careful processing of these powders permits them to be consolidated into a fully dense structure, thereby permitting assessment of their mechanical characteristics.

A distinguishing feature of consolidated nanocrystals is the structural difference between them and conventional materials. For example, if crystal diameters are on the order of 10 nm and if the thickness of the boundaries between them is taken as 1 nm (about 4 atomic diameters), then about 30% of the atoms of the solid are situated at grain boundaries; this is a much higher fraction than in conventional polycrystals (Prob. 5.10). Further, nanocrystalline grain interiors are thought to be essentially dislocation free. In spite of these features, so different from those of ordinary polycrystals, the strengths of nanocrystals increase with decreasing grain size, similar to the behavior of polycrystals.[4] Because of their fine grain sizes, therefore, nanocrystalline material yield strengths are greater than those of ordinary polycrystals. Recent work has attempted to correlate the mechanical behavior of nanocrystals with their structure. For example, nanocrystals might be reasonably viewed as a "composite" containing a noncrystalline phase (the grain boundaries) and a crystalline one. Dislocation activity might account for some of the plastic strain of which nanocrystals are capable because, even though not found in grain interiors, dislocations might be emitted from grain boundaries during plastic deformation. Other recent work has suggested parallels between the deformation characteristics of nanocrystalline and glassy metals (behavior of glassy metals is discussed in Chap. 8).

As noted in Chap. 4, geometrical dislocations, of density ρ_G ($\cong \varepsilon/bd$), are generated during polycrystalline deformation. These provide additional strengthening. If ρ_G is considerably greater than ρ_s, a feature expected at relatively small strains, the work-hardening strength increment is proportional to $(\rho_G)^{1/2}$. Thus, the flow stress of a polycrystal at relatively low plastic strains can be written as

$$\sigma_{\text{flow}} = \sigma_0 + \left[k_y + \alpha G b \left(\frac{\varepsilon}{b} \right)^{1/2} \right] d^{-1/2} = \sigma_0 + k_y'' d^{-1/2} \qquad (5.11)$$

Equation (5.11) predicts that the proportionality constant between flow stress and $d^{-1/2}$ ($= k_y''$) should increase with strain, and this is observed experimentally at low to moderate strains. However, at larger strains k_y'' approaches a saturation value. This reflects that the geometrical dislocations—most prevalently situated near grain boundaries—are not as effective strengthening agents as cell boundaries are at these larger strains.

As mentioned previously, grain boundaries are remarkably effective barriers to dislocation motion. However, it is only materials with fine grain sizes (Prob. 5.8) for which grain boundaries appreciably increase material strength. This is because it is only when d has such a magnitude that the obstacle spacing becomes small enough

---

[4]There is evidence that nanocrystalline yield strengths follow a Hall-Petch relationship, although their values of k_y are not necessarily the same as those for ordinary polycrystals of the same material.

to take advantage of the intrinsically high "strength" of the boundary. Cell boundaries, discussed previously, and subboundaries are not completely impenetrable by dislocations. While cell boundary hardening is more properly considered a special case of work hardening, it and subboundary hardening are often empirically described by an equation analogous to Eq. (5.9); i.e.,

$$\sigma_{\text{flow}} = \sigma_0 + k_y''' (d')^{-1/2} \tag{5.12}$$

where d' is the subboundary or cell boundary size, and the constant k_y''' is typically $\frac{1}{2}$ to $\frac{1}{3}$ of the constant k_y of Eq. (5.9). The lower value of k_y''' reflects the lesser resistance to dislocation penetration provided by subboundaries or cell boundaries compared to grain boundaries. On the other hand, subboundary and cell boundary dimensions are always less than the grain size, and this feature of theirs can provide for substantial strengthening. For example, it is used to advantage in the production of heavily drawn high-strength pearlitic steel wires (Sect. 5.9A).

Although it is more appropriate to think of cell boundary strengthening in terms of dislocation strengthening, subboundary strengthening can also be considered a special case of boundary hardening. Subboundaries can be produced by low-temperature annealing of cold-worked materials with a cellular structure. During annealing, some of the dislocations in the cell walls are annihilated while others rearrange themselves into the more "orderly" arrangements that differentiate a subboundary from a cell boundary (Fig. 5.6). Subboundaries are more resistant to dislocation penetration than a cell boundary and this is reflected in a higher value of k_y''' in Eq. (5.12) for subboundaries.

The effect that grain boundaries have on the yield strength and work-hardening behavior of polycrystals can be summarized. Yield strength follows the Hall-Petch

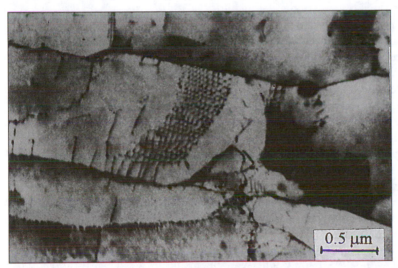

Figure 5.6
A subboundary produced in iron after deformation at 25°C followed by annealing for 16 hours at 600°C. A subboundary is more resistant to dislocation penetration than a cell boundary (cf. Fig. 3.43b). (*After A. S. Keh and S. Weissman, Electron Microscopy and Strength of Crystals, ed. G. Thomas and J. Washburn, Wiley, New York, 1963, p. 231.*)

relation, Eq. (5.9). Following yielding, geometrical dislocations—whose density increases with decreases in grain size—form, and this produces a grain-size effect on work-hardening behavior at low to moderate plastic strains. At higher strains, the effect of geometrical dislocations on the flow stress is swamped compared to that arising from other work-hardening effects.

EXAMPLE PROBLEM 5.1. Using Eq. (5.11), determine the strain at which geometrical dislocations contribute more to grain-size hardening than does inherent grain-size hardening (as represented by the term $k_y d^{-1/2}$).

Solution. The two terms in the brackets of Eq. (5.11) represent the separate effects of inherent grain-size strengthening and geometrical dislocations. Equating the terms permits determination of the strain above which geometrical dislocations contribute most to strength. Doing this we find the "critical" strain is

$$\varepsilon \geq \frac{k_y^2}{\alpha^2 G^2 b}$$

Note that the term α in the above equation pertains to a tensile, rather than shear, flow stress. Thus, the α term above is about twice that of the α term in Eq. (5.5) (using Mohr's circle to define the relation between shear and tensile stress) or can also be estimated as $\overline{m}$ (cf. Sect. 4.4) times the α of Eq. (5.5). Numerical application of the above relations are reserved for Prob. 5.7.

5.5
SOLID-SOLUTION STRENGTHENING

Solute atoms increase the yield strength of crystalline materials. This is a result of the interactions that take place between a moving dislocation and solute atoms. Since the dislocation stress field is a long-range one, solute atoms located both on the slip plane and above or below it interact with dislocations. However, since the maximum interaction is experienced with those atoms lying in close proximity to the glide plane, it is only these that need be considered in estimating the increase in material yield strength arising from the presence of the solute atoms.

Introduction of a substitutional solute atom into a crystal produces a lattice dilation that typically gives rise to a spherically symmetric stress field around the solute. If the solute atom has an atomic size less than the host (solvent) atom, the dilation is negative (i.e., the atomic volume is locally decreased in the solute atom's vicinity) and vice versa. The resultant stress field interacts with that of a dislocation, giving rise to a solute atom–dislocation interaction energy. Consider, for example, an edge dislocation moving on a slip plane as it encounters a smaller solute atom lying directly above this plane (Fig. 5.7). In this case, the interaction energy is negative, because the small atom reduces the dislocation energy resulting from the compressed volume lying above the dislocation core. Hence, the dislocation is attracted to the solute atom. Conversely, if the small atom were lying directly below the glide plane, the situation would be reversed and a positive interaction energy would result; i.e., the dislocation would be repelled by the solute atom. For such an edge dislocation, therefore, it might be expected that, on the average, there would

Direction of dislocation motion

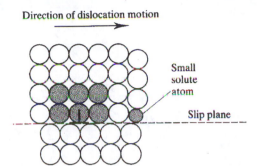

Small
solute
atom

Slip plane

Figure 5.7
**An edge dislocation moving on a slip plane containing a solute atom of atomic size
less than the solvent. When the dislocation core reaches the solute, the compressive
strain energy (shaded region) is relieved somewhat. This leads to an attractive interaction energy between the solute and the dislocation. If the solute atom were positioned below the glide plane, a repulsive energy would result. Thus, a small solute
atom can, depending on its position relative to the glide plane, attract or repel an
edge dislocation. A similar result is obtained when the solute atom has a size larger
than the solvent. As explained in the text, on the average both relatively larger and
smaller solute atoms interact attractively with an edge dislocation and an increase in
flow stress results.**

be as many repulsive as attractive encounters. However, since a dislocation is flexible, it spends more time at (or, more precisely, a greater fraction of its length is located at) the locations of negative interaction energy. Thus, on the average, the
interaction energy between an edge dislocation and a small solute atom is negative.
Analogous considerations show that this is true also for large solute atoms.

As indicated by Eq. (3.7), the stress field of a screw dislocation is solely shear
in nature; i.e., while atomic bonds are distorted in the region of a screw dislocation,
there is no dilational strain. As a result, unlike the situation for an edge dislocation,
there is no "size effect" interaction between a screw dislocation and a solute atom.[5]

Some point defects produce a tetragonal distortion of the lattice. Comparison
of tetragonal and spherical distortions is given in Fig. 5.8. For the latter, the princi-
pal components of strain (ε_1, ε_2, ε_3) are algebraically equal. For a tetragonal distortion, ε_1 is positive (or negative) whereas the other principal strain components (ε_2,
ε_3) are negative (or positive) and equal in magnitude. Examples of tetragonal defects are interstitials in body-centered cubic metals (Fig. 5.9a) and divalent substitutional ions in a monovalent ionic crystal (Fig. 5.9b). Self-interstitials, which can
be produced by high-energy neutron bombardment, also give rise to a tetragonal
distortion, as does the collapse of excess vacancies (which can also be generated by
irradiation) into a vacancy loop. The interaction between screw dislocations and
tetragonal defects is substantial and is a direct result of the relatively large shear
strain associated with the most common tetragonal defects. In fact, this energy is so

[5]More refined calculations show there is a "second-order" volume change associated with a moving
screw dislocation. This gives rise to a size interaction energy, but its magnitude is small compared to
that for an edge dislocation and, therefore, is neglected in our discussion.

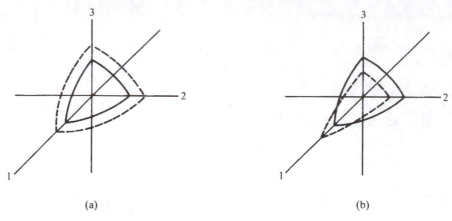

(a)   (b)

Figure 5.8
**Comparison of the strains associated with (a) spherical and (b) tetragonal distortions,
(——, original shape; – – –, distorted shape). A spherical distortion produces a
dilation for which $\varepsilon_1 = \varepsilon_2 = \varepsilon_3$. For the tetragonal distortion, an octant of an
ellipsoid of revolution is formed; i.e., $\varepsilon_1 > 0$, $\varepsilon_2 = \varepsilon_3 < 0$ (or $\varepsilon_1 < 0$, $\varepsilon_2 = \varepsilon_3 > 0$).**

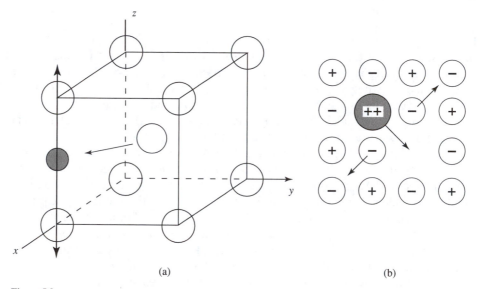

(a)   (b)

Figure 5.9
**Examples of tetragonal distortions caused by (a) an interstitial atom in a bcc lattice
and (b) a divalent ion plus positive ion vacancy in a monovalent ionic solid. In (a) the
interstitial atom does not fit without distortion in the structure (the drawing is not to
scale). This results in displacement of the solvent atoms along the z axis; i.e., a tetragonal distortion results. In (b) the divalent ion and the associated vacancy "relax" by
approaching each other. This also produces a tetragonal distortion. (The arrows indicate the directions in which atoms/ions move in response to the presence of the dis-
torting agent.)**

large that, unlike the case for typical substitutional atoms, tetragonal defects are considered "hard" obstacles to dislocation glide.

The modulus of a solute atom relative to that of the solvent also plays a role in solid-solution strengthening. Since the self-energy of a dislocation is proportional to the modulus (cf. Eq. (3.13)), a substitutional atom "softer" than the solvent attracts a dislocation. If the solute is "hard" relative to the solvent, there is a repulsion between the dislocation and the solute atom. In contrast to the size-effect interaction, the modulus effect (be it attractive or repulsive) does not depend on whether the solute atom lies above or below the glide plane.

Both the size and modulus effects produce a dislocation–solute atom interaction energy that is elastic in nature, and the interplay between the two effects for an edge dislocation is described in Fig. 5.10. Figure 5.10a corresponds to the "soft" atom case for which the modulus and size interaction energies are both negative and the two effects reinforce each other. For the case of a hard atom (Fig. 5.10b), the modulus interaction energy is positive and the two energies tend to cancel each other. Hence, we arrive at the intuitively surprising conclusion that, other things being equal, a "soft" atom will strengthen a crystal more than a "hard" one. Further consideration of Figs. 5.10a and b provides explanation of this apparent anomaly. The resulting force (Figs. 5.10c and d) to move the dislocation through the lattice is proportional to the derivative of the energy-distance profiles of Figs. 5.10a and b. As the dislocation approaches the solute atom, it is attracted to it, as indicated by the negative forces to the left of the position $x = 0$. However, a positive force, F_{max}, is required to tear the dislocation away from the solute atom and this force relates to the associated shear stress by $\tau_{max} = F_{max}/bL'$. Figure 5.11 illustrates the situation for a hard atom in which the modulus interaction energy has greater magnitude than the size interaction energy. The resulting net positive interaction energy (Fig. 5.11a) gives rise to an initial repulsion between the edge dislocation and the solute. Nonetheless, the value of F_{max} of Fig. 5.11b relates to the yield stress in the same manner as the values of F_{max} of Fig. 5.10. That is, dislocation motion is retarded irrespective of the algebraic sign of the interaction energy.

The magnitude of the size-effect interaction energy between an edge dislocation and a spherically symmetric distortion is estimated as follows. If a solvent atom of radius r is replaced by a solute atom having radius $r(1 + \varepsilon_b)$, the local volume change δV is given to first order as $\delta V = 4\pi r^3 \varepsilon_b$ (Prob. 5.11). The interaction energy between the edge dislocation and the atom is obtained by multiplying δV by the sum of the tensile and compressive stress fields of the edge dislocation. This scales with $(\sigma_x + \sigma_y + \sigma_z)/3$, where σ_x, etc., have been defined for edge dislocations in Sect. 3.5A. The subsequent mathematical development determines the size interaction energy, U_D^E, as

$$U_D^E = \frac{4(1 + v)Gbr^3 \, \varepsilon_b \sin \theta}{3(1 - v)R} \tag{5.13}$$

where R is the distance between the dislocation core and the solute atom and θ is the angle between the slip direction and a line connecting the dislocation core and the solute atom (cf. Fig. 3.21). (An estimate of the maximum interaction energy can be obtained by substituting b for R in Eq. (5.13) with $\sin \theta = 1.0$.) The parameter ε_b is a measure of the relative size difference between the solute and solvent atoms. In a

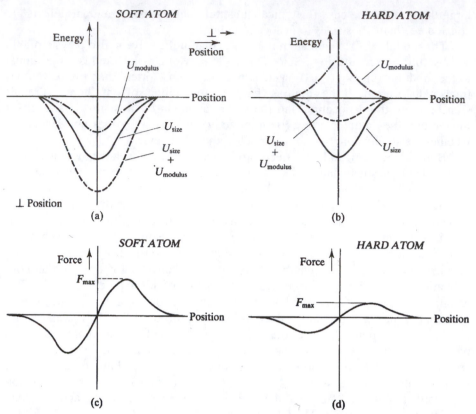

Figure 5.10
Interaction energy vs. relative slip plane position of an edge dislocation and a solute atom for (a) a "soft" solute and (b) a "hard" solute. For the soft solute, the size and interaction energies reinforce. This is not so for the hard atom. (In the case illustrated, the size interaction is taken as the more important one, and this results in a net attraction between the dislocation and the solute.) The interaction force-position relationship for these respective cases is shown in (c) and (d). The force (proportional to the derivative of the energy-position curve) is negative (attractive) as the dislocation approaches the solute atom for both situations. The force necessary to continue dislocation motion (F_{max}) is that required to "tear" the dislocation away from the solute. As a result of the "adding" of the size and modulus effects, this force is greater for the soft atom.

binary alloy, ε_b can be estimated as the fractional change in lattice parameter per unit concentration of solute atom, i.e.,

$$\varepsilon_b = \frac{1}{a}\frac{da}{dc} \tag{5.14}$$

where a is the lattice parameter and c the solute concentration (expressed as an atomic fraction). As discussed previously, U_D^E is always taken as negative irrespective of the sign of ε_b.

Estimation of the modulus interaction energy proceeds by considering the variation of the dislocation energy as it approaches a solute atom. The details of the cal-

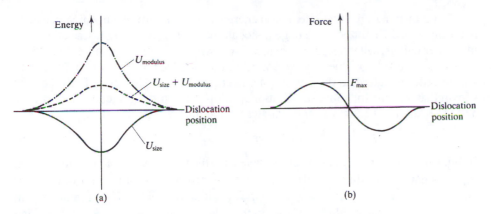

Figure 5.11
(a) The dislocation–solute atom interaction energy vs. position for a hard atom for which the modulus interaction energy has greater magnitude than the size interaction energy. The net positive energy results in the force-position curve shown in (b). Here, the force (F_{max}) required to continue dislocation motion is that necessary to *push* the dislocation by (rather than, as for the case of a negative interaction energy, the force required to pull the dislocation away from) the solute atom. Strengthening results regardless of whether the solute atom–dislocation interaction is attractive or repulsive.

culation are similar to those employed to calculate the size interaction energy. In contrast to the latter, however, the modulus effect involves both screw and edge dislocations. Results approximate the modulus interaction energies as

$$U_G^S = \frac{G\varepsilon_G' b^2 r^3}{6\pi R^2} \tag{5.15}$$

and

$$U_G^E = \frac{U_G^S}{(1 - \nu)} \tag{5.16}$$

where U_G^S and U_G^E are the modulus interaction energies for screw and edge dislocations and ε_G' is defined by

$$\varepsilon_G' = \frac{\varepsilon_G}{\left(1 + \frac{1}{2}|\varepsilon_G|\right)} \tag{5.17}$$

The parameter ε_G of Eq. (5.17) is analogous to ε_b. That is, ε_G represents the fractional change in shear modulus per unit solute concentration;

$$\varepsilon_G = \frac{1}{G}\frac{dG}{dc} \tag{5.18}$$

In contrast to the size-effect interaction energy, the modulus interaction energy can be either positive or negative depending on the sign of ε_G.

Combination of the two interaction energies into one form from which τ_y can be determined is difficult because edge dislocations are subject to both size and modulus interactions, whereas screw dislocations experience only the modulus effect. Hence, one must know the relative contributions of edge and screw dislocations to slip in order to calculate the appropriate average energy. However, Fleischer has shown that, at least for some substitutional solid-solution strengtheners, this energy and the resulting shear stress correlate with a parameter, ε_s, defined as

$$\varepsilon_s = |\varepsilon_G' - \beta\varepsilon_b| \tag{5.19}$$

In Eq. (5.19), ε_b is always taken as positive as is the parameter β. For "soft" atoms, ε_G' is negative and thus the size and modulus effects reinforce each other. For "hard" atoms, ε_G' is positive and the opposite situation holds. The parameter β in Eq. (5.19) is an empirical one related to the relative importance of screw and edge dislocations during plastic flow and, therefore, also depends on the relative importance of the size effect. As shown in Fig. 5.12, the increase in yield stress per unit solute concentration for copper single crystals containing various solutes correlates well with ε_s provided the value of β is taken as 3.

Solid-solution strengthening can also be discussed in terms of concepts introduced in Sec. 5.2 and Figs. 5.10 and 5.11. If L' is the effective obstacle spacing, the increase in flow stress associated with the solute atoms is

$$\tau = \frac{F_{\max}}{bL'} \tag{5.20}$$

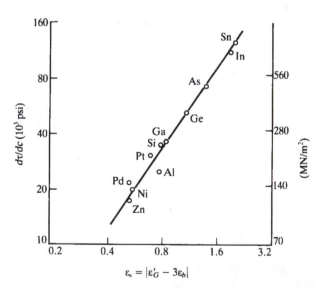

Figure 5.12
The increase in shear yield stress per unit solute concentration for copper alloy single crystals vs. the parameter ε_s. The correlation is excellent when β is taken as 3; this value of β indicates that screw dislocation motion is more important than edge dislocation motion in the flow behavior of copper and its alloys. (_From Acta Metall._, 11, R. L. Fleischer, Copyright 1963, _203_, used with permission from Elsevier Science.)

Equation (5.20) is analogous to that used in developing Eq. (5.1), for which F_{max} is given as $Gb^2 \cos(\phi_c/2)$. For hard obstacles, F_{max} is on the order of Gb^2. Approximate estimates of F_{max} for tetragonal distortions are in the range of $Gb^2/5$ to $Gb^2/10$; i.e., these qualify as "hard" obstacles. In contrast, F_{max} for the spherically symmetric distortions associated with most substitutional solutes is on the order of $Gb^2/120$; these solutes are, therefore, "soft" obstacles. The effective spacing also relates to obstacle strength. For tetragonal distortions, L' is the mean free path (L) between solute atoms associated with the slip plane and is given as $b/(2c)^{1/2}$, where c is the impurity concentration. The effective spacing is much greater for "soft" obstacles for which $L' \cong b/[c(\pi - \phi)]^{1/2}$, where $(\pi - \phi)$ is typically on the order of 1°. Hence, for tetragonal defects (hard obstacles) the solid solution strengthening is given as

$$\tau_{TET} \cong \gamma Gb\left(\frac{c^{1/2}}{b}\right) = \gamma Gc^{1/2} \tag{5.21}$$

where γ is a proportionality constant less than, but of order of magnitude of, unity. Equation (5.21) reflects the strong hardening arising from tetragonal defects and is analogous to Eq. (5.5) describing dislocation hardening—i.e., the strength is proportional to the mean slip plane distance between strengthening agents and the proportionality constant is on the order of Gb.

Hardening due to spherical distortions is much less than for tetragonal distortions. The flow stress still varies with $c^{1/2}$, but the proportionality constant is far less than G. Using the relationship between τ_y and ε_s displayed in Fig. 5.12, along with other considerations, Fleischer estimates the yield strength increment due to "conventional" substitutional solid-solution strengthening to be

$$\tau_y = \frac{G\varepsilon_s^{3/2} c^{1/2}}{700} \tag{5.22}$$

Reasonable estimates of ε_s show that $\varepsilon_s^{3/2}/700$ is much less than unity (Example Prob. 5.2).

The $c^{1/2}$ dependence of solid-solution hardening is observed in a number of materials strengthened by atoms that cause either spherically symmetric or tetragonal distortions (Fig. 5.13). For the materials whose behavior is illustrated in Fig. 5.13, the $c^{1/2}$ dependence of the yield strength is well established. However, in some other materials, τ_y correlates with c^n where, depending on the material, n is less than ½ (but typically no less than ⅓) or greater than ½ (but typically no greater than unity). Likewise, the dependence of τ_y on ε_s reflected in Eq. (5.22) is basically an empirical one and may vary from material to material. Nonetheless, the predictions of Eqs. (5.21) and (5.22) can be viewed as good, useful approximations and are additionally attractive in view of the physically plausible principles underlying their formulations.

For most crystalline solids, the elastic interaction energy (including both size and modulus effects) dominates solid-solution strengthening. However, there are other causes of strengthening, and these are primarily chemical or electrical in nature. In metals, for example, addition of a solute atom with valence different from the solvent locally alters the electronic charge distribution and energy, and this can lead to interactions with dislocations. In metals, such interactions are generally minor. For ionic materials, however, addition of a divalent ion to a monovalent crystal not only produces a tetragonal distortion but may also lead to significant electrical interactions between the impurity and the charged ions that comprise a dislocation in an ionic material. It is thought that this may lead to substantial strengthening, but this strengthening (also proportional to $\sqrt{c}$) is expected to be less than that provided by the tetragonal distortion. For materials containing stacking faults, solute atoms may be preferentially attracted to or repelled from the faults, thus effectively lowering the SFE. Additional "chemical strengthening" like this can

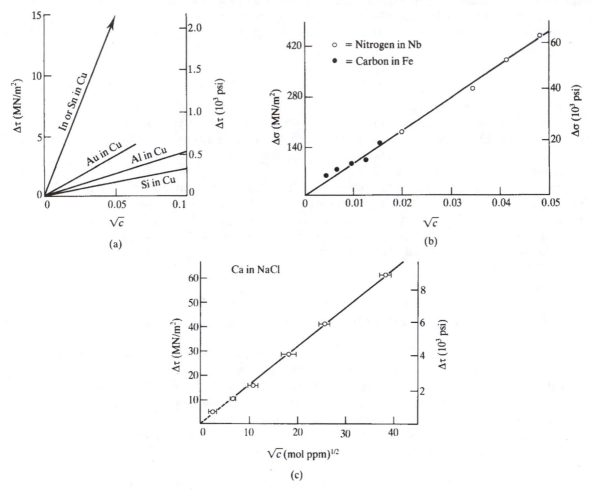

(a)

(b)

(c)

Figure 5.13
The $c^{1/2}$ dependence of solid-solution hardening is verified for (a) the increase in critical resolved shear stress for a Cu alloy single crystal hardened by substitutional atoms (spherical distortions), (b) the tensile yield stress increase due to interstitials in Fe and Nb (tetragonal distortions), and (c) the increase in critical resolved shear stress due to the presence of divalent Ca ions in NaCl (tetragonal distortions). Tetragonal distortions harden materials to a much greater degree than do spherical ones. (*Data for (a) from R. L. Fleischer, Acta Metall., 11, 203, 1963; J. O. Linde and S. Edwards, Arkiv Fysik, 8, 511, 1954; and T. J. Koppenal and M. E. Fine, Trans. TMS-AIME, 224, 347, 1962. For (b) from C. Wert, Trans. TMS-AIME, 188, 1242, 1950 and P. R. V. Evans, J. Less Common Metals, 4, 78, 1962. For (c) from A. G. Evans and T. Langdon, Prog. Matls. Sc., 21, 171, 1976.*)

be provided by long- or short-range ordering, and such effects are touched on in the section following.

In summary, solid-solution strengthening can be caused by several types of dislocation–solute atom interactions. Many of these are difficult to quantify accurately. However, the mechanisms discussed here provide for reasonable estimates of the magnitude of solid-solution strengthening and its variation with solute concentration.

As described in Sect. 4.2, lattice vibrations impart a temperature dependence to the flow stress. The temperature-dependent component, τ^*, can be described in terms of mechanisms discussed above, as most solute atoms are "short-range" obstacles. To illustrate how thermal fluctuations can reduce the stress necessary for dislocation glide, a schematic force-obstacle separation spacing sketch is shown in Fig. 5.14. At 0 K, in bypassing the obstacle the external stress does work given by the integral of the force-distance curve from $x = a$ (the outer "limit" of the interaction distance) to $x = 0$ (when the dislocation and solute atom lie directly above (or below) each other). This work is the area under the curve in Fig. 5.14a. At a finite temperature, thermal vibrations can produce an energy on the order of kT to help the obstacle overcome the barrier. Thus, the work done by the external stress is reduced by kT and this is reflected in a reduction in the applied stress ($\tau_{app} = F_{app}/bL'$) required for bypassing the obstacle (Fig. 5.14b). When the temperature is high enough that kT is equal to the external work required at 0 K, the obstacle presents no further barrier to glide; thermal energy supplies the necessary work for overcoming barriers. If the force-distance profile of Fig. 5.14a is known, the temperature dependence of the flow stress can be determined; for certain defects τ^* can be expressed as

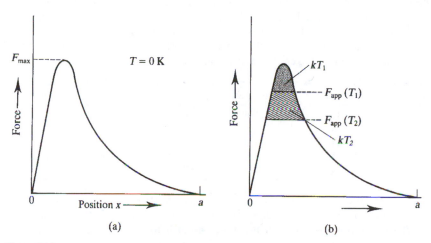

(a) (b)

Figure 5.14

A schematic force-distance curve for a dislocation approaching an obstacle. (a) At 0 K, the external force to overcome the barrier is F_{max}, and this force does work given by the area under the force-distance curve. (b) At finite temperatures (T_1 and $T_2 > T_1$), thermal energy aids the applied force, reducing the work required of it by approximately kT (the respective shaded areas). Thus, the required external force is less than F_{max}. The higher the temperature, the greater the reduction in the applied force.

$$\left(\frac{\tau^*}{\tau_0^*}\right)^{1/2} = 1 - \left(\frac{T}{T_c}\right)^{1/2} \qquad (5.23)$$

where τ_0^* is the thermal component of the yield stress at 0 K and T_c is the above-mentioned critical temperature. The relationship between τ^* and T predicted by Eq. (5.23) has been verified experimentally for a number of materials. As with many other relations we have presented, Eq. (5.23) is only an approximation; in this case this is so because the temperature dependence of the flow stress depends on the detailed shape of the force-distance curve.

EXAMPLE PROBLEM 5.2. **a.** The elements listed below are being considered as substitutional solid-solution strengtheners for nickel. Calculate the values of ε_s for them, and list the elements in order of decreasing ability to solid-solution strengthen nickel.

Element	Atomic radius (nm)	G (GN/m²)
Ni	0.125	86
Al	0.143	23
Fe	0.124	81
Cu	0.128	45
W	0.137	159
Nb	0.143	45

b. Calculate values of the parameter $(\varepsilon_s)^{3/2}/700$ for the above elements. Are these solid-solution atoms "hard" or "soft" obstacles? Explain.

Solution. We first calculate (estimate) the separate terms that go into ε_s and then determine their sum. The results are shown in the table below. There are "assumptions" made in developing this table. First, we assume that the lattice parameter increases linearly with composition and that it scales with the "average" atomic radius (considering both solute and solvent) of the solution. This assumption is used in calculating ε_b. We also assume (used in calculating ε_G) that the shear modulus increases linearly with composition. Finally, we have taken $\beta = 3$ in calculating ε_s.

Element	ε_b ($1/a \cdot da/dc$)	ε_G ($1/G \cdot dG/dc$)	ε_G'	$\varepsilon_s = |\varepsilon_G' - \beta\varepsilon_b|$	$\varepsilon_s^{3/2}/700$
Al	0.144	−0.709	−0.523	0.955	0.0013
Fe	−0.008	−0.058	−0.056	0.080	0.000032
Cu	0.024	−0.048	−0.047	0.119	0.000059
W	0.096	0.849	0.596	0.308	0.00024
Nb	0.144	−0.477	−0.385	0.817	0.0011

Listed in order of decreasing ability to strengthen (that is, first element listed is the most potent hardener) we have: Al, Nb, W, Cu, and Fe. Note that the value of the parameter $\varepsilon_s^{3/2}/700$ is low even for the most potent hardening elements. Thus, these solid-solution strengthening atoms are "weak" obstacles.

5.6
PARTICLE HARDENING

Small particles of a second phase dispersed in a matrix can markedly increase the matrix yield strength even when the dispersoid volume fraction is low (ca. 1–10 percent). This is so because an aggregate of solute atoms (one way of viewing such a particle) resists dislocation penetration to a considerably greater degree than does an isolated solute atom. Particle sizes in the range of several to thousands of atoms in diameter are effective in promoting particle hardening. Particles with sizes greater than this are still effective hardeners on a per particle basis; however, the concentration of particles is smaller with larger particle sizes and this results in an interparticle spacing too great to provide extensive hardening (cf. Eqs. (5.1) and (5.2)).

The degree of strengthening provided by a particle dispersion depends on a number of factors. These include the particle size and volume fraction (L, the mean particle separation distance, is defined if particle size and volume fraction are specified), particle shape, and the nature of the boundary between the particle and the matrix. The effect of particle size and volume fraction on strength is discussed later where we consider particles that are spherical in shape. Nonspherical particles (e.g., plates and needles) are common to many precipitation-hardened alloys. They also can be more effective as strengtheners than spherical particles; however, their more complicated geometry makes them more difficult to treat and here we are concerned with the fundamentals, and not the details, of particle strengthening.

The nature of the interface between the particles and the matrix depends on various factors including particle size and the manner in which particles are introduced into the matrix. Small particles produced by precipitation from supersaturated solid solutions most often have interphase boundaries that are coherent with the matrix (Fig. 5.15a). Atomic bonds match up across such an ordered interface even though, as shown in Fig. 5.15a, the interatomic spacings in the particle and matrix may differ. The consequential distortion gives rise to a coherency strain; dislocations interact with the stress field of a coherent particle similar to the way they interact with solute atoms having a different size from the solvent. During precipitation, coherency may eventually become lost when the particle attains a certain size.

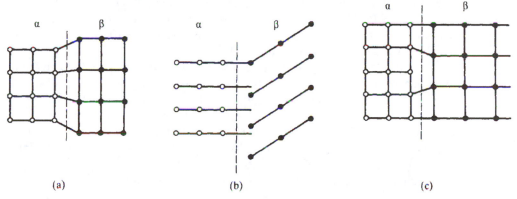

(a)   (b)   (c)

Figure 5.15

Three types of interphase boundaries (IPBs). In (a), a coherent or ordered IPB exists between α and β phases. The atoms match up, one to one, along such a boundary. Owing to different lattice parameters of the phases, a coherency strain energy is associated with this type of boundary. In (b) a fully disordered IPB is shown. Here, there are no coherency strains. A dislocation can penetrate an ordered IPB, but not a disordered one. (c) An intermediate type of IPB (a partially ordered one). Here, coherency strains are partially relieved by the periodic introduction of dislocations along the boundary.

Essentially, the energy associated with the strained interface becomes too high and it is replaced by a disordered interface, the structure of which resembles that of a high-angle grain boundary (Fig. 5.15b). Similarly disordered interphase boundaries result when the particle dispersion is introduced into the matrix "artificially," as is commonly done in powder metallurgy and ceramic processes. An intermediate structure (a partly ordered boundary, Fig. 5.15c) can also be found at particle-matrix interfaces. In this case, the volume difference (and the resulting strain) between the particle and the matrix is accommodated partially by interphase boundary dislocations.

Coherent boundaries and small particles generally result in dislocations passing through or "cutting" the particle (i.e., $\phi_c > 0°$). Conversely, dislocations tend to bow between large particles or between particles with disordered interphase boundaries ($\phi_c = 0°$). The contribution to strength depends on whether dislocation "cutting" or "bowing" is the predominant slip mechanism. In this section we first discuss the hardening expected for the situation of dislocation cutting. This is followed by a description of strengthening for dislocation bowing. Finally, we discuss how factors such as particle size, volume fraction, etc., determine whether the cutting or bowing process dominates slip.

A. Deforming Particles

There are a number of ways by which a matrix can be hardened by a dispersion of particles through which dislocations pass. Several are discussed in this section. All are likely to be manifested to one degree or another, although the extent to which they are is very system specific; that is, the nature of deformable particles varies significantly from one material system to another. Even within a given matrix material—Al, for example—the dominant hardening mechanism depends on the solute atom that is the basis for the resulting precipitate. Further, there are a large number of metallic elements amenable to precipitation hardening.[6] These include Al, Cu, Co, Fe, Ni, and Pb (Al and Ni are the two most important commercially). The nature of the precipitate responsible for hardening is system specific and so, therefore, is the dominant hardening mechanism.

Most treatments dealing with dislocation cutting of particles apply to precipitation hardenable alloys. A more specific description of such alloys is provided later. However, we note here that the precipitation process involves stages. In the early stages, the particles are small. As precipitation proceeds, particle size and volume fraction increase until the equilibrium volume fraction of the precipitate is formed. The particles coarsen; i.e., their size and spacing increase—both during precipitation and following its completion. Most of the hardening mechanisms treated in this section are appropriate to the early stages of precipitation.

1. Coherency hardening

Precipitate coherency hardening is analogous to the size effect in solid solution hardening. A precipitate ordinarily has an atomic volume different than the matrix

---

[6]Ceramics (e.g., MgO) can also be precipitation hardened.

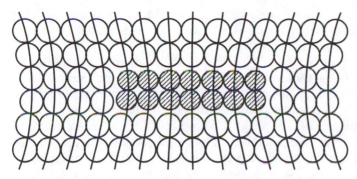

Figure 5.16
Schema of a coherent precipitate (shaded atoms) in a matrix. Here, the precipitate has a lattice parameter less than that of the matrix. The atomic "match" across the IPB leads to an internal stress field that interacts with moving dislocations.

from which it forms. If the precipitate is coherent, this leads to an internal lattice strain (Fig. 5.16). The associated stress field interacts with dislocations, either attracting or repulsing them; either situation, of course, results in an increase in yield strength. Although the basic physics of coherency hardening and size-effect solid-solution strengthening are similar, the particle "problem" is more difficult to treat. In part this is because of the finite particle size and the consequently stronger interaction between it and a dislocation compared to the comparable interaction between a solute atom and a dislocation. Thus, for example, the assumption of a straight dislocation line is generally not tenable for coherency hardening, but the degree to which the dislocation bends is also difficult to ascertain. Even so, the problem has been treated. An approximate expression for the increase in resolved shear stress during early stage precipitation is

$$\tau_{coh} \cong 7|\varepsilon_{coh}|^{3/2} G\left(\frac{rf}{b}\right)^{1/2} \tag{5.24}$$

where r is the precipitate radius, f its volume fraction, and the parameter ε is analogous to the size parameter of solid-solution strengthening. For the particle case, ε_{coh} depends on the difference between the particle (a_p) and matrix (a_m) lattice parameters (Fig. 5.16) and is defined as

$$\varepsilon_{coh} = \frac{a_p - a_m}{a_m} \tag{5.25}$$

We note that when the particle lattice parameter is greater than that of the matrix, the particle is in a state of compression whereas when $a_m > a_p$, the particle is in tension. However, both situations give rise to material strengthening since the increment in strength depends on the absolute value of ε_{coh}.

Equation (5.24) should be compared to Eq. (5.22), the expression for the strengthening due to solid-solution atoms. The two equations are similar. Both contain a "strengthening factor" (ε) raised to the ³⁄₂ power. Further, the parameter rf/b of Eq. (5.24) is analogous to the solute concentration of Eq. (5.22). That is, f is a particle "concentration"; the multiplying factor r/b effectively scales the precipitate size relative to that of an atom. The only significant difference between Eqs. (5.24) and (5.22) is the value of their numerical proportionality constants; this constant is much greater for particle hardening.

Comparison of the predictions of Eq. (5.24) with experimental results for systems (e.g., a number of Cu-base alloys) believed strengthened primarily by coherency hardening has not been satisfactory. While the strength increases para-metrically follow Eq. (5.24), the observed strengths are less than predicted using the numerical constant of Eq. (5.24). Although this equation is meant to describe hard-ening in the early stages of precipitation (i.e., when the particle size and volume fraction are small), treatments which attempt to consider later stage precipitation also overestimate the strength. Notwithstanding the lack of quantitative success in predicting the magnitude of coherency hardening, the ideas put forth here do reflect its essential physics.

2. Modulus hardening

When a dislocation enters a precipitate having a shear modulus different from that of the matrix, the dislocation line tension ($\cong Gb^2/2$) is altered. The physics here are again analogous to solid-solution strengthening. We initiate our discussion of modulus hardening with a simplified treatment. This does not yield a correct an-swer, but it has the advantage of outlining the approach taken in more advanced treatments.

A view looking down on the slip plane of a (presumed straight) dislocation entering a particle with shear modulus G_p (the corresponding matrix modulus is G_m) is provided in Fig. 5.17. The line length of dislocation affected is the particle di-ameter ($2r$), and the associated change in dislocation line tension is $(G_p - G_m)b^2/2$.

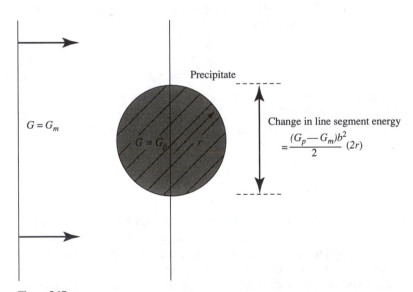

Figure 5.17
View looking down on a slip plane as a straight dislocation enters a particle having radius r on the plane. The dislocation line tension is changed when it enters the par-ticle. The maximum dislocation length affected is the particle diameter; the line ten-sion change takes place gradually over a distance equal to r.

(We assume the matrix and precipitate Burgers vectors are the same.) Thus, the maximum change in dislocation self-energy (when the dislocation is halfway through the particle) is $b^2r(G_p - G_m)$. The distance over which this energy changes is the particle radius. Thus, the force, proportional to the gradient in energy, is on the order of

$$F = b^2(G_p - G_m) = Gb^2\varepsilon_{Gp} \tag{5.26}$$

where, on the far right side of Eq. (5.26) G is taken as the matrix shear modulus and the parameter $\varepsilon_{Gp} = (G_p - G_m)/G_m$ is akin to the parameter (ε_G') used to characterize modulus hardening in solid solution strengthening.

While the above approach is physically appealing, it is too simple. For example, the dislocation bends both before, and on, entering the particle. This increases the affected dislocation line length. Moreover, the change in energy takes place over a distance less than r. The latter can be taken into account by substituting b for r to yield a revised Eq. (5.26);

$$F = br(G_p - G_m) = Gbr\varepsilon_{Gp} \tag{5.27}$$

The associated increase in shear stress (τ_{Gp}) is obtained by dividing F by bL', or

$$\tau_{Gp} = \frac{Gr\varepsilon_{Gp}}{L'} \cong \frac{1}{2}Gf\varepsilon_{Gp} \tag{5.28}$$

where, on the far right side of Eq. (5.28), we have substituted $f \cong 2r/L'$ (f = particle volume fraction), appropriate for a straight dislocation. (The magnitude of ε_{Gp} is to be used in Eq. (5.28).)

Even this improved analysis does not fully describe modulus hardening. The details of the correct treatment are complex. Here we only provide the result, appropriate to the early stages of precipitation;

$$\tau_{Gp} = 0.01G\varepsilon_{Gp}^{3/2}\left(\frac{fr}{b}\right)^{1/2} \tag{5.29}$$

(The magnitude of ε_{Gp} is used in Eq. (5.29), of course.) The proportionality constant in Eq. (5.29) is approximate and is accurate to within about a factor of two. Note the similar forms of the equations describing coherency and modulus strengthening as well as their similarity to Eq. (5.22) which describes solid-solution strengthening. Coherency and modulus hardening both vary with a "mismatch" parameter (ε) to the ½ power and with $(fr/b)^{1/2}$. The values of the numerical proportionality constants in the equations are different, though.

3. Chemical strengthening

When a dislocation passes through a particle, an additional particle-matrix interface is formed (Fig. 5.18). Since there is a surface energy associated with such an interface, work must be done by the process. This type of strengthening is called chemical strengthening. We approach it as we did modulus hardening. That is, we first describe the basic physics and derive a simplified yet plausible expression for the associated hardening. Then the results of more refined treatments are provided.

As Fig. 5.18 shows, interphase surface area is created both when the dislocation enters the particle and when it exits it. After complete passage of the dislocation through the particle, the interphase area has increased by approximately $2\pi r b$, where r is the particle radius (Fig. 5.19). If r is large in comparison to b, approximately half of the energy increase (total energy increase $= 2\pi r \gamma_s b$, where γ_s is the particle-matrix interphase surface energy) occurs as the dislocation enters the particle and half of it when it exits the particle. Thus, the maximum force required to push the dislocation through the particle (the maximum value of dU/dx of Fig. 5.19) is approximated by

$$F_{max} = \frac{\pi r \gamma_s b}{b} = \pi r \gamma_s \qquad (5.30)$$

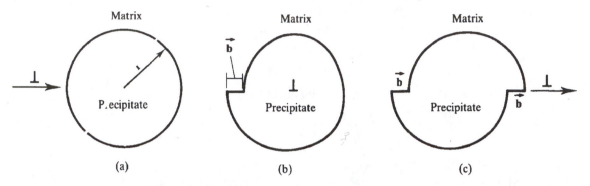

(a)   (b)   (c)

Figure 5.18
View of an edge dislocation penetrating a particle. (a) The dislocation is approaching the particle. (b) It is within the particle and an offset, *b*, of a portion of the upper part of the particle with respect to the lower part accompanies the dislocation entry. (c) A similar offset is effected when the dislocation exits the particle. The complete transit is accompanied by creation of matrix-precipitate surface area of approximate magnitude $2\pi r b$.

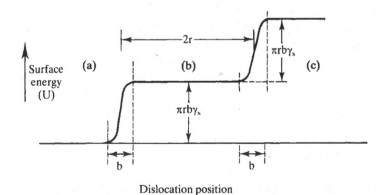

Dislocation position

Figure 5.19
Surface energy as a function of dislocation position for the situation of Fig. 5.18. When the dislocation enters and exits the particle, an increment of surface energy of $\pi r b \gamma_s$ is effected. The energy increase occurs over a distance about equal to the slip vector, *b*; thus, the force required to produce the increase (proportional to the slope of the energy-distance curve) is approximately $\pi r \gamma_s$.

The associated increase in flow stress is

$$\tau_{chem} = \frac{F_{max}}{bL'} = \frac{\pi r \gamma_s}{bL'} \tag{5.31}$$

where L' is the effective particle spacing. If the dislocation remains straight, we can substitute $2r/L' = f$, and thus

$$\tau_{chem} \cong \frac{\pi f \gamma_s}{2b} \tag{5.32}$$

This simplified treatment suffers from shortcomings similar to those of the simplified treatment of modulus hardening. A more detailed analysis leads to the following expression for τ_{chem};

$$\tau_{chem} = 2G\left(\frac{\gamma_s}{Gr}\right)^{3/2}\left(\frac{fr}{b}\right)^{1/2} = 2G(\varepsilon_{ch})^{3/2}\left(\frac{fr}{b}\right)^{1/2} \tag{5.33}$$

The numerical coefficient of Eq. (5.33) is, as with the other similar coefficients in this section, approximate. We see that chemical strengthening depends on parameters similar to those describing coherency and modulus hardening. Specifically, the strength increase varies with $(fr/b)^{1/2}$, and there is another parameter—γ_s/Gr—for which strength increases with the $\frac{3}{2}$ power. In this sense, γ_s/Gr is analogous to both ε_{Gp} and ε_{coh}, and that is why we have defined γ_s/Gr as ε_{ch} in Eq. (5.33).

There are other reasons for designating γ_s/Gr with an ε. This dimensionless term is the ratio of a factor causing hardening to a "baseline" factor (as are the previously defined ε's.) For coherency hardening, the ratio is the lattice parameter difference between the precipitate and matrix divided by a "baseline" (i.e., matrix) lattice parameter. For modulus hardening it is a corresponding ratio involving the respective precipitate and matrix shear moduli, and for chemical hardening the ratio is γ_s/r divided by the matrix shear modulus. The term γ_s/r has units of N/m² and, therefore, can be considered a surface energy "stress" having the same units as G.

Chemical strengthening is thought not to generally have an important role in precipitation hardening. The possible exception is when particles are very fine, as in the earliest stages of precipitation. That is, for such a circumstance the $r^{-3/2}$ term in $\varepsilon_{ch}^{3/2}$ dominates the $r^{1/2}$ term in $(fr/b)^{1/2}$.

4. Order strengthening

There are other ways surface energy can contribute to particle hardening. Two prominent cases are stacking fault strengthening and order strengthening. If the stacking-fault energies of the particle and the matrix differ, dislocation motion is impeded because the equilibrium separation of the partial dislocations is different in the matrix and particle. Stacking fault strengthening is not discussed further here. The physics of it is similar to those of order hardening which we do discuss. Our choice is made because the slip crystallography in the order hardening case can be discussed in a simple manner.[7]

---

[7]The paper of Ardell, noted in the references, provides an extended discussion of stacking fault strengthening.

Figure 5.20 illustrates a (presumed straight) edge dislocation passing through an ordered particle. As it does, A—A and B—B bonds form across the slip plane. These represent a higher energy state relative to the A—B bonds that existed across this plane prior to the dislocation transit. The associated energy increase (units of J/m^2) is called an antiphase boundary energy (APBE). The problem can be treated similarly to the simplified way we handled chemical strengthening. The only change is that the surface energy increase due to the APBE takes place gradually as the dislocation passes through the particle rather than when it enters (and leaves) it. The simplified analysis (Prob. 5.21) predicts that the increased shear stress due to order hardening is given by $\tau_{ord} = \pi(ABPE)f/2b$.

This simplified treatment suffers the shortcomings of our previous simplified treatments. It also does not consider the effect of a second dislocation following the first one. The second dislocation is attracted to the particle for when it traverses it, the antiphase boundary formed by the passage of the first dislocation is removed. The interplay between the two dislocations is complicated. On elastic energy considerations, the two dislocations repel each other and would wish to be widely separated. Were it not for this repulsion, though, the dislocations would desire to travel in pairs (as "superdislocations") since, in doing so, no antiphase boundary is formed during their passage through the particle. The actual situation is that if the APBE is low, the elastic repulsion between the dislocations keeps them far enough apart so that hardening is realistically analyzed on the basis of a single dislocation passage. In the more general case, though, both factors must be taken into account. When they are, it is found that the separation distance of the two dislocations depends on the ratio (APBE)/Gb. This is a ratio of two forces; one (APBE) is a force that binds the dislocations while the other (Gb) represents a force that separates them. When the ratio is low the dislocations travel separately (Fig. 5.21a); when it is high they travel in pairs (Fig. 5.21b). Consistent with other treatments of this section we define a ratio (ε_{ord}) to characterize order hardening; i.e., $\varepsilon_{ord} = (APBE/Gb)$.

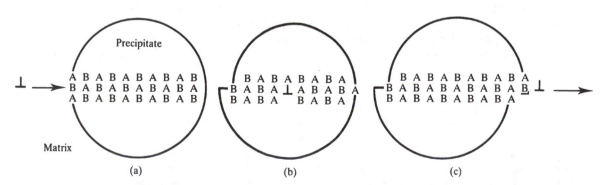

Figure 5.20
A view of an edge dislocation penetrating an ordered particle. (The particle crystal structure is cubic and its composition is AB.) In (a) the dislocation has not yet entered the particle. In (b) it is partially through it. Slip in the particle is accompanied by the formation of an antiphase boundary (A—A and B—B bonds) across the slip plane. After the dislocation exits the particle, the antiphase boundary occupies the whole of the slip plane area of the particle; the associated energy increase is ca. πr^2(APBE). The increase in energy is roughly linear with dislocation position in the particle.

More detailed considerations lead to the following expression for the increase in shear stress due to order hardening in the early stages of precipitation and for the situation when the dislocations are widely separated (i.e., low APBE or, equivalently, low ε_{ord});

$$\tau_{ord} \cong 0.7G(\varepsilon_{ord})^{3/2}\left(\frac{fr}{b}\right)^{1/2} \quad \text{(low } \varepsilon_{ord}\text{; early stage precipitation)} \quad (5.34a)$$

When the dislocations are not widely separated, the increase in strength is less and is given by

$$\tau_{ord} \cong 0.7G\left[\varepsilon_{ord}^{3/2}\left(\frac{fr}{b}\right)^{1/2} - 0.7\varepsilon_{ord}f\right] \quad \text{(high } \varepsilon_{ord}\text{; early stage precipitation)} \quad (5.34b)$$

As with other particle strengthening mechanisms discussed, the dislocation shape assumed during particle shearing is different in the early and late stages of precipitation. This alters the extent of hardening. For order hardening, the equations analogous to Eqs. (5.34) and which hold for late stage precipitation are

$$\tau_{ord} \cong 0.44G\varepsilon_{ord}f^{1/2} \quad \text{(low } \varepsilon_{ord}\text{; late stage precipitation)} \quad (5.35a)$$

and

$$\tau_{ord} \cong 0.44G\varepsilon_{ord}[f^{1/2} - 0.92f] \quad \text{(high } \varepsilon_{ord}\text{; late stage precipitation)} \quad (5.35b)$$

Note that during the later stages of precipitation, particle size does not come into play; only ε_{ord} and precipitate volume fraction impact strength. Further, a saturation in strength is predicted. (It is also predicted in late stage precipitation for the other hardening mechanisms discussed in this section.)

Order hardening is important in the strengthening of certain nickel-base superalloys used at high temperature and also in some precipitation-hardened stainless

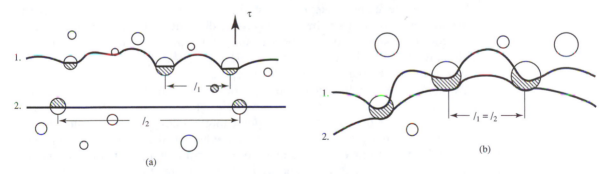

(a)   (b)

Figure 5.21
View looking down on a slip plane for the situation corresponding to (a) a low APBE (or, equivalently, a low ε_{ord}) and (b) a high APBE (high ε_{ord}). (a) When the APBE is low the dislocations are widely separated. The leading dislocation line shape corresponds to that of a hard obstacle situation and the effective obstacle spacing is the mean center-to-center spacing of obstacles on the slip plane. The trailing dislocation is straight, corresponding to a weak obstacle situation for which the effective obstacle spacing is much greater. (b) When the APBE is high, the second dislocation closely trails the leading dislocation which bends significantly during its passage through the particle.

steels. Equations (5.34) and (5.35) describe well the strengths of the nickel-base superalloys.

EXAMPLE PROBLEM 5.3. Strengthening can result from the change in dislocation self-energy as it enters a particle. We have discussed the effect in terms of the change in dislocation energy arising from the different shear moduli of the particle and the matrix. Now we consider the change resulting from a difference in the Burgers vector between the particle and the matrix. Let b_m and b_p be the matrix and precipitate Burgers vectors, respectively. Show that the magnitude of the energy difference per unit length of dislocation line is about $Gb|\Delta b|$ for small $|\Delta b| = |b_m - b_p|$. (Assume G of the matrix and precipitate are the same.)

Solution. The energy per unit length of the dislocation lines is about $Gb^2/2$. Thus, this energy change (per unit line length) when the dislocation enters the particle is $(G/2)(b_p^2 - b_m^2)$. We can write this as

$$\Delta E = \frac{1}{2}G(b_p - b_m)(b_p + b_m) = \frac{1}{2}G|\Delta b|(b_p + b_m)$$

Since the Burgers vectors of the particle and matrix are not that much different, we can approximate $(b_m + b_p)$ as $2b_m = 2b$. Therefore

$$\Delta E = Gb|\Delta b|$$

Estimation of the magnitude of strengthening expected in this instance is left as an exercise (Prob. 5.17).

---

5. Summary

In this section we have described strengthening for the situation where dislocations shear (cut through) particles in a matrix. Application to "real" material behavior is difficult for there is a large number of matrix and precipitate combinations in commercial alloys. Each combination exhibits several strengthening mechanisms, one of which might be dominant.

We can be even more skeptical. Here is how an "exact" analysis would proceed. First, the interaction energy between a passing dislocation and the particle is calculated for all possible hardening mechanisms, and then these energies are summed. This sum, of course, depends on the slip plane distance separating the particle and the dislocation. The derivative of the energy-distance curve is calculated and its maximum value correlates with the value of the applied shear stress necessary to cause dislocation passage through the particle. This, too, is not a trivial exercise. Some interactions (e.g., coherency strengthening) are long range in nature whereas others (e.g., chemical hardening) only are felt when the dislocation and the particle are proximate. Sorting out such a mess for a system in which a number of interactions are possible is so difficult that we might be tempted to say our treatment here is nothing but an academic exercise. But that is a bit too harsh an assessment. While specifications are difficult, generalizations can be made. Thus, some summarizing statements are made in this section.

One is that during the early stages of precipitation the strength increase depends on two terms. The first is a dimensionless parameter (or parameters, if more than one strengthening mechanism is important) designated ε that contains within it the physics of strengthening. For example, ε depends on the fractional change in matrix

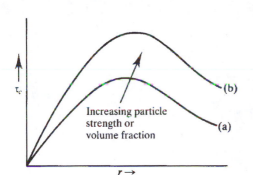

Figure 5.22
Schematic of the stress required for dislocations to shear precipitates as a function of precipitate size. The stress increases approximately with $r^{1/2}$ at small r. However, for a fixed particle volume fraction, this stress may decrease at larger values of r owing to an increase in particle spacing. The overall level of the $\tau_c - r$ curve is raised by increases in either inherent particle "strength" or particle volume fraction.

and precipitate lattice parameter for coherency strengthening and so on. The second dimensionless parameter which affects strength is fr/b. The precipitate volume fraction, f, is analogous to the solute concentration in solid-solution strengthening. For particle strengthening, f is "amplified" by the ratio r/b which is a measure of the precipitate size relative to that of an atom. Further, in most situations the strength increase varies with $\varepsilon^{3/2}(fr/b)^{1/2}$; that is, the increase in strength has the same form for particle hardening as it does for solid-solution strengthening. However, the numerical proportionality constants in the equations for particle strengthening are greater.

We have discussed here late stage precipitation hardening only for order hardening. However, all of the particle hardening mechanisms are predicted to yield "plateaus" in strength in the later stages of precipitation. (In some cases, a maximum strength is found at an intermediate precipitation stage.) The overall scheme is sketched in Fig. 5.22 which is a plot of the incremental strength as a function of particle radius. At a fixed f, strength initially increases with particle size (e.g., curve (a) of Fig. 5.22) but then it either plateaus or, as is suggested in the figure, a maximum is found. Curve (b) of Fig. 5.22 simulates either an increase in particle volume fraction or particle "strength" (the latter exemplified, for example, by an increased numerical constant in the strength equation). The schematic of Fig. 5.22 is verified experimentally in several alloy systems for which f is approximately constant and, therefore, for which r can be varied independently. The results for several systems are shown in Fig. 5.23.

In many commercial alloys, both f and r increase concurrently during the early stages of precipitation. After precipitation is complete, particle size continues to increase by particle coarsening. Since f is now fixed, these (now larger) particles become more widely separated as they coarsen. At a certain particle separation, dislocations find it easier to bow between, rather than cut, particles. Typical particle sizes at which this transition occurs are in the range of tens of nanometers. We next address this phenomenon of dislocation bowing.

B. Nondeforming Particles

Dislocation bowing, as illustrated in Fig. 5.24, occurs when the spacing between particles exceeds a certain size or when the particle-matrix interface is disordered.

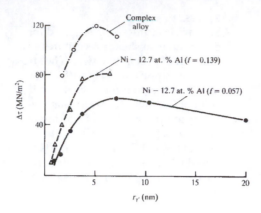

Figure 5.23
Precipitation hardening in Ni-Al alloys as related to γ′ precipitate size. The data are in accord with the schematic of Fig. 5.22. That is, the hardening increases with particle volume fraction (compare $f = 0.139$ with $f = 0.057$) and particle strength (the "complex" alloy vs. the others). Additionally, a maximum in strength occurs at a critical particle size. (Data from L. M. Brown and R. H. Ham, Strengthening Mechanisms in Crystals, ed. A. Kelly and R. B. Nicholson, Wiley, New York, 1971, p. 10.)

The increase in shear strength associated with dislocation bowing is determined from Eq. (5.2) as

$$\tau_B \cong \frac{Gb}{(L - 2r)} \tag{5.36}$$

where the mean spacing between particles (L in Eq. (5.2)) is replaced by $(L - 2r)$ to account for the finite particle sizes. For a fixed volume fraction of particles, L increases concurrently with r; that is, $(L - 2r)$ increases as the dispersion becomes coarser and thus τ_B is greatest for a fine dispersion. This is shown schematically in Fig. 5.25, where τ_B is plotted versus r for two different values of f. As indicated by Eq. (5.36), particle "strength," per se, does not influence τ_B. That is, once the particle is "hard" enough that bowing, rather than cutting, is the slip mechanism, further increases in the resistance of the particle to dislocation penetration do not affect τ_B. On the other hand, the maximum possible particle hardening is frequently related to particle "strength," for this dictates both the particle size and the stress at which the transition from dislocation cutting to bowing occurs.

C. The Transition from Cutting to Bowing and the Maximum Particle Hardening

Figure 5.26 illustrates schematically the interaction among particle size, volume fraction, and "strength" for the cutting and bowing processes. For particles of A with volume fraction f_1, maximum strength (τ_1) is obtained at the particle radius r_{c1} where the curves (d) and (a) of the bowing and cutting stresses intersect. If the particles are made harder (say with particles of B as shown in Fig. 5.26), the cutting stress will be higher and the maximum stress (τ_2) is obtained at the intersection of curves (b) and (d), which obtains at a particle radius $r_{c2} < r_{c1}$. Increasing the volume fraction of A will increase the levels of both the cutting (curve (c)) and bowing

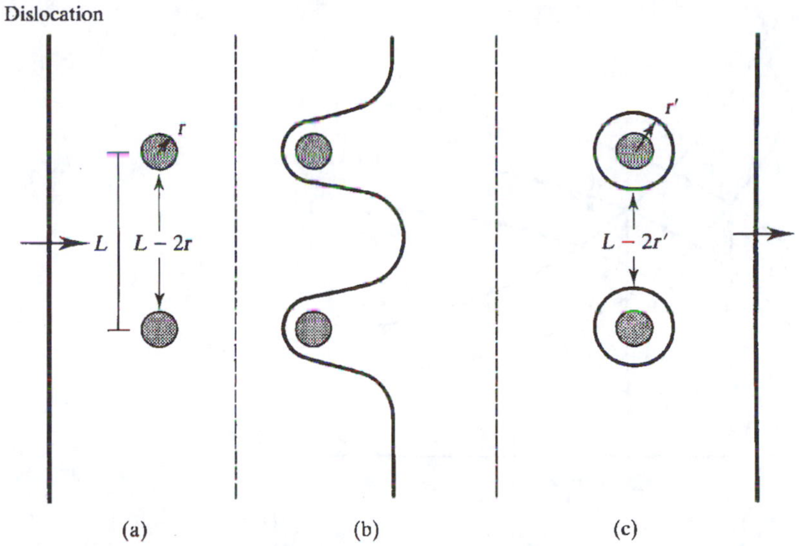

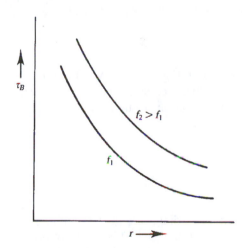

(a)   (b)   (c)

Figure 5.24
(a) A view looking down on a slip plane as a dislocation approaches nondeforming particles. (b) The dislocation bows around the particles ($\phi_c = 0$); the stress required to effect the bypassing is inversely proportional to the interparticle spacing ($L - 2r$), where r is the slip plane particle radius. (c) Dislocation loops encircle the particles after the bypass operation. A subsequent dislocation would have to be extruded between the loops. Thus, the effective particle spacing for the second dislocation is reduced to ($L - 2r'$), and the bypassing stress for this dislocation will be greater than for the first one.

Figure 5.25
The bowing stress–particle size relationship. For a fixed particle volume fraction, τ_B decreases with increasing r as this is accompanied by an increase in particle spacing. Increasing f increases the level of the stress as a result of a finer particle spacing. The level of τ_B is unaffected by particle strength. That is, once a particle is "strong" enough to resist cutting, any further increase in its resistance to dislocation penetration has no effect on τ_B which depends only on matrix properties and effective particle spacing.

(curve (e)) stresses and result in a greater maximum strength. The particle radius, r_{c3}, at which the intersection of the cutting and bowing stresses occurs for this situation, may be either greater or less than r_{c1}, depending on the shape of the $\tau_c - r$ curve.

The scenario illustrated in Fig. 5.26, in which maximum strength is obtained at the cutting-bowing transition, conforms approximately to experimental observations. In some systems, however, maximum strength is found when the particles are

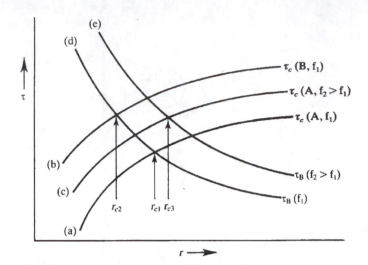

Figure 5.26
The competition between cutting and bowing as schematized by τ–r curves. If parti-cles of A of volume fraction f_1 are dispersed in a matrix, particles are sheared for $r < r_{c1}$ and are bypassed for $r > r_{c1}$. Maximum strength is obtained at $r = r_{c1}$, where the cutting and bowing stresses are equal. If inherently "harder" particles of B of the same volume fraction are present, the level of the τ_c curve is increased but that of the τ_B one is not. Maximum hardening, greater than that for A particles, is found at $r_{c2} < r_{c1}$. Increasing the volume fraction of A raises the level of both τ_B and τ_c and increases the maximum strength obtained. The latter is found at r_{c3}, which may be either less than or greater than r_{c1} depending on the shape of the $\tau_c - r$ curves.

Curves: (a) τ_c (A; $f = f_1$); (b) τ_c (B; $f = f_1$); (c) τ_c (A; $f_2 > f_1$); (d) τ_B ($f = f_1$); (e) τ_B ($f = f_2$).

still being sheared. That this is possible is shown in Fig. 5.27, in which the inter-section of the τ_c and τ_B curves takes place at a particle radius larger than that asso-ciated with the maximum cutting stress. This situation occurs, for example, in several precipitation-hardened Al-Cu alloys.

5.7
STRAIN-GRADIENT HARDENING

When a material is deformed so that plastic strain gradients are present within it, it "work hardens" more than if such gradients are not present. This has been previ-ously discussed in terms of the grain-size effect on work hardening (Sect. 5.4 and Eq. (5.11)) and in bending of a single crystal (Sect. 4.6). The cause of the added strength is the geometrical dislocations that accommodate the strain gradient. These interfere with dislocation flow just as (although not necessarily in the same manner as) statistical dislocations do. Inspection of the expression for the geometrical dis-location density (ρ_G; Eq. (4.7)) shows that strain gradients must exist over distances on the order of micrometers or less (Probs. 5.29–5.32) if values of ρ_G high enough to significantly increase flow stress are to be found. Thus, for a bent bar to be

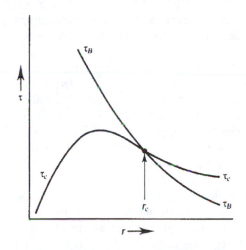

Figure 5.27
If the intersection of the cutting and bowing stresses occurs at a particle size greater than that at which τ_c is a maximum, maximum strength is obtained while the particles are still being sheared. This is in contrast to the situation of Fig. 5.26, where it is assumed the intersection of the respective stresses takes place at a stress less than the maximum possible cutting stress.

demonstrably stronger than a bar strained in tension, the bar would have to be quite thin. Microelectronic components, for example, might be prone to this effect.

Figure 5.28 amplifies the point just made. This figure shows the tensile stress-strain curves (Fig. 5.28a) of fine Cu wires; these can be compared to the torque-twist curves of the same wires tested in torsion (Fig. 5.28b). Outside of (a relatively small number of) geometrical dislocations formed due to the strain gradients existing across grain boundaries, a tensile test gives rise to uniform deformation and a

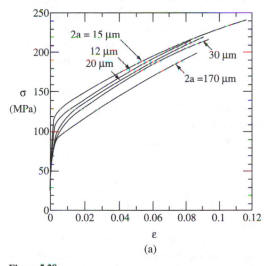

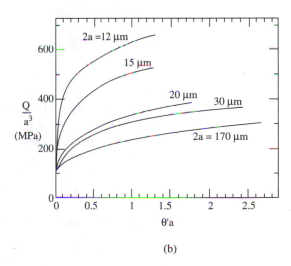

(a)

(b)

Figure 5.28
**(a) Tensile true stress-true strain curves for fine copper wires (diameter ($2a$) in the range of 12–170 μm). There is a negligible effect of wire diameter on tensile behavior (except for the wire having the largest diameter; this wire likely has a coarser grain size than the others). (b) Torsional response of these same wires. The "normalized" torque (defined as the twisting moment (Q) divided by a^3) is plotted vs. normalized twist (twist per unit length (θ' (Eq. (1.37)) multiplied by a)). In distinction to tensile behavior, torsional behavior depends greatly on wire diameter. The finer wires manifest significantly greater flow stresses. This is a result of the greater dislocation densities in the wires tested in torsion. (*Reprinted from N. A. Fleck, G. M. Muller, M. F. Ashby and J. W. Hutchinson, "Strain Gradient Plasticity: Theory and Experiment," 475, copyright 1994, with permission from Elsevier Science.) Acta Metall. et Mater., 42.*

preponderance of statistical dislocations. As a result, the tensile stress-strain curves are, to a good approximation, size independent. In contrast, the torque-twist curves are dependent on the wire size. In a torsion test, the plastic strain varies linearly from the wire axis to its perimeter (Sect. 1.3F). The resulting strain gradient is large if the wires are fine enough. Geometrical dislocations form to accommodate such a gradient. Thus, fine wires twisted to the same extent as large-diameter wires of the same material require significantly higher torques to deform them. The behavior illustrated in Fig. 5.28 is thus compelling evidence that, provided the characteristic distance over which the plastic strain gradient exists is small enough, appreciable hardening by geometrical dislocations can take place.

Other mechanical tests also may give rise to large plastic strain gradients and commensurably high geometrical dislocation densities. If the indentor impression in a microhardness test is on the order of micrometers, the hardness measured is indentation size dependent; the finer the indentation, the greater the hardness because the strain gradient between the material plastically deformed by the indentation and the elastic region that surrounds it is greater with a fine indentation size. Similarly, plastic zones in front of a crack tip (Chaps. 9 and 10) are associated with plastic strain gradients. If the zone sizes are sufficiently small, appreciable geometrical dislocation densities can be generated in them.

Strain gradients also can come about as a result of microstructural features; witness the grain-size effect on work hardening. However, microstructurally related geometrical dislocations display their greatest effect in the work-hardening behavior of some particulate-strengthened materials.

The work hardening of materials containing particles depends strongly on whether dislocations shear or bow around particles. The work-hardening rates of materials for the first situation do not differ much from those of the same material not containing particles. Or, to put it another way, the stress required to move a dislocation through a particle does not depend on the number of dislocations that have previously passed through it.[8] For example, Fig. 5.29 shows the flow behavior of pure Cu and that of a precipitation-hardened Cu alloy in which slip occurs by particle shearing. The initial yield strength of the precipitation-hardened alloy is much greater than that of pure Cu, but the work-hardening rate (gaged by the slope of the stress-strain curves) of the two materials is about the same.

The work-hardening rates of materials containing nondeforming particles, however, are larger. The flow curve of the Cu-BeO alloy, also shown in Fig. 5.29, demonstrates this. The beryllium oxide particles—about which dislocations bow— do not impressively increase the matrix yield strength, particularly in comparison to the precipitation-hardened alloy. However, the work hardening rate of the Cu-BeO material is impressive. That such strong work-hardening might be expected can be deduced from Fig. 5.24. The effective obstacle spacing for a second dislocation to be extruded between the particles shown there is reduced from $L - 2r$ to $L - 2r'$, where $2r'$ is the diameter of the dislocation loop formed by the passage of the first dislocation. Thus, the stress required to generate the second loop is increased. Further, the internal stress field developed by successive dislocation looping would be substantial. Indeed, calculations indicate that only a few such loops would produce an internal stress high enough to fracture the particles enclosed by the loops. Such

---

[8]For an interesting exception, see Prob. 5.23.

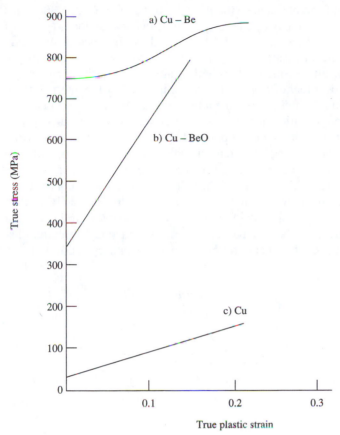

Figure 5.29
True stress–true strain behavior for pure Cu (curve c), a precipitation-hardened Cu-Be alloy in which dislocations shear particles (curve a), and a dispersion-hardened Cu-BeO alloy for which dislocations bow around particles. While the flow stress of the Cu-Be alloy is much greater than that of pure Cu, the work-hardening rate of the two materials is comparable. Although the yield strength of the Cu-BeO alloy is quite a bit less than that of Cu-Be, the work-hardening rate of Cu-BeO is much greater. This is a consequence of geometrical dislocation generation during plastic flow of Cu-BeO. (*Adapted from A. Kelly, Proc. Roy. Soc. (London), A282, 63, 1964.*)

premature fracture, though, does not happen. This is because only at very small strains do the dislocations that bow between the particles leave a loop configuration in their wake.

Understanding of the plastic-strain-induced dislocation configuration in materials containing nondeforming particles is facilitated by further consideration of geometrically necessary dislocations. A shear loop (Fig. 5.24) is one means of accommodating the plastic strain gradient between the deforming matrix and the nondeforming particle. In effect, the loop provides a region of elastic matrix between the particle and the plastic matrix. There are alternative dislocation arrays that can accomplish this same purpose, and which do not result in the large internal stresses associated with shear loops. Since, in essence, geometrically necessary dislocations "move" matrix material from one side of the particle to the other (Fig. 5.30a), this accommodation can be provided by a series of prismatic dislocation loops (plates of vacancies or self-interstitial atoms, Fig. 5.30c) as well as by shear loops (Fig. 5.30b). In actuality, the dislocation array associated with dislocation bowing changes from that of shear to prismatic loops with strain. This transition, which occurs at shear strains on the order of 1 percent, is accomplished by the internal dislocation stress fields, which promote the prismatic array at the higher strains. The dislocation configuration transition can be seen in internally oxidized Cu-Al$_2$O$_3$ alloys in which the alumina particles are not penetrated by moving dislocations (Fig. 5.31). The dislocation maneuvering by which prismatic, rather than shear, loops are formed relates to cross-slip processes during bowing (these hardly alter the bowing stress) as illustrated in Fig. 5.32. At even higher plastic strains (typically about 25 percent) the prismatic loop configuration is superseded by dislocation activation on other than the primary slip system. Such flow also accommodates a strain gradient (cf. Fig. 4.16), and is favored when the particle-matrix interface is disordered (disordered interfaces are especially suitable for emission of secondary dislocations).

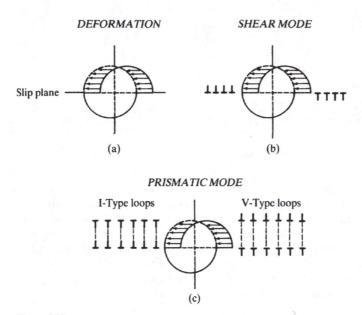

Figure 5.30
A view looking end-on at a slip plane containing a nondeformable particle. (a) If the dislocation had passed through the particle, relative displacements within it would have occurred. (The dotted circle represents the original particle position; the solid circle that following particle slip.) (b) The displacement in (a) can be accommodated by shear loops (see also Fig. 5.24c, which is a view looking down on a slip plane containing one shear loop). (c) The accommodation can also be provided by an array of prismatic loops (plates of vacancies and self-interstitials that "shift" matrix material from one side of the particle to the other). Shear loops accommodate the displacement at small strains; at larger strains, prismatic loops are favored owing to the lower dislocation energy associated with them. (*From M. F. Ashby, Strengthening Mechanisms in Crystals, ed. A. Kelly and R. B. Nicholson, Wiley, New York, 1971, p. 137.*)

To summarize, plastic strain gradients are accommodated by shear loops at low

strains and for small particles having ordered particle-matrix boundaries. Prismatic loops, formed by cross-slip, do the same at intermediate strains and are favored also by small particles, small particle volume fractions, and coherent—or at least "strong"—interphase boundaries. (A "strong" boundary is one for which secondary dislocation emission is difficult.) Multiple slip is the accommodation mechanism at large plastic strain and for large particles, high particle volume fractions, and disordered or "weak" interphase boundaries.

The density, ρ_G, of the geometrically necessitated dislocations can be estimated for the prismatic loop configuration of Fig. 5.30b. The volume "defect" associated with the slip process is $\cong V_p\gamma$, where γ is the shear strain and V_p the particle volume. Each loop has a diameter on the order of the particle radius and the loop

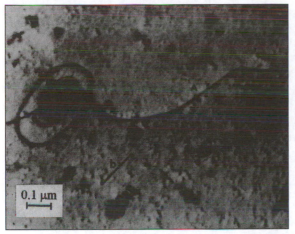 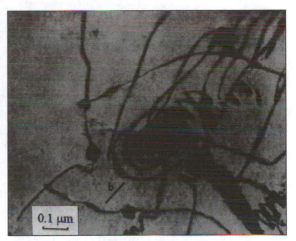

Figure 5.31
(a) Shear loops, formed at small plastic strains, around Al$_2$O$_3$ particles in a Cu-30% Zn crystal. (b) At larger strains, strain accommodation is provided by an array of prismatic loops. (*After P. B. Hirsch and F. J. Humphries, Physics of Strength and Plasticity, ed. A. Argon, M. I. T. Press, Cambridge, 1969.*)

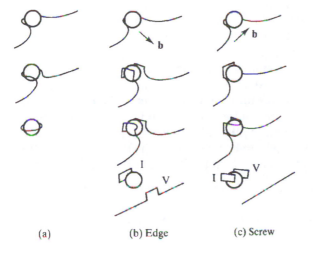

(a)   (b) Edge   (c) Screw

Figure 5.32
(a) Strain accommodation for a nondeforming particle as provided by shear loops (cf. also Fig. 5.24c). (b) Particle bypassing involving cross-slip by an edge dislocation generates prismatic loops. (c) The similar process for a screw dislocation. (*After M. F. Ashby, Strengthening Mechanisms in Crystals, ed. A. Kelly and R. B. Nicholson, Wiley, New York, 1971, p. 137.*)

"thickness" is b; thus the volume of each loop is on the order of r^2b. The number of loops per particle, n_T, is thus,

$$n_T \cong \frac{V_p \gamma}{r^2 b} \cong \frac{r\gamma}{b} \qquad (5.37)$$

where we have taken $V_p \cong r^3$ in developing the far right side of Eq. (5.37). Multiplication of n_T by N_v, the number of particles per unit volume ($N_v r^3 \cong f$), gives the loop density per unit volume, N_T:

$$N_T = \frac{f\gamma}{br^2} \qquad (5.38)$$

The geometrically necessary dislocation density is then obtained by multiplying N_T by the line length per loop ($\cong r$); thus,

$$\rho_G \cong \frac{f\gamma}{br} = \frac{4f\gamma}{br} \qquad (5.39)$$

where the factor of 4 comes from a more detailed geometrical treatment than used here. The quantity f/r can be identified with the inverse of a mean slip distance, λ_G, associated with the geometrical dislocations. Thus,

$$\rho_G \cong \frac{4\gamma}{b\lambda_G} \qquad (5.40)$$

According to Eq. (5.40), ρ_G increases linearly with plastic strain and inversely with λ_G, the latter being defined by the particle dispersion. An equation analogous to Eq. (5.40) can be written in differential form for statistical dislocations:

$$d\rho_s = \left(\frac{4}{b\lambda_s}\right) d\gamma \qquad (5.41)$$

For statistical dislocations, λ_s decreases with increasing strain and thus ρ_s increases more rapidly with plastic strain than does ρ_G.[9] Since the work-hardening behavior is related to total dislocation density, whichever is the greater of ρ_s or ρ_G contributes most to the associated increase in flow stress. Values of ρ_s for copper single- and polycrystals are shown in Fig. 5.33 and compared there to values of ρ_G calculated from Eq. (5.40). It is clear that ρ_G contributes most to work hardening at low, and ρ_s at high, strains. The transition strain increases as λ_G becomes finer; when λ_G is small enough, the transition strain can be relatively high.

The exceptional work-hardening rates of alloys containing nondeforming particles (e.g., Cu-BeO, Fig. 5.29) buttresses the above arguments. High work-hardening rates are also observed in Al alloys containing nondeforming particles. Figure 5.34 shows stress-strain curves for Al alloys containing both particles that shear and those that don't. The latter manifest considerably lesser yield strengths, yet their work-hardening rates are much greater. (The work-hardening rates of the alloys containing deforming particles are comparable to those of pure Al.) Thus, while Al alloys containing nondeforming particles are not initially highly resistant to plastic flow, plastic strain and the dispersion act in a synergistic way so as to im-

---

[9]This is so during the early stages of plastic strain. At higher strains, dislocation annihilation also occurs; this factor is not taken into account in Eq. (5.41).

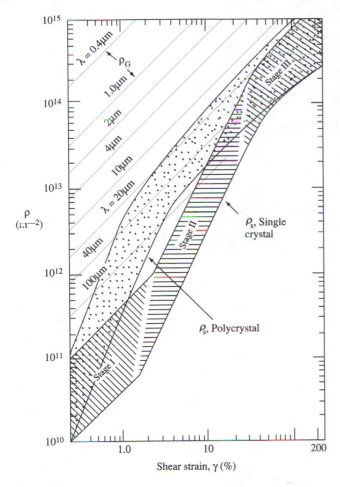

Figure 5.33
Variation of ρ_s (experimental data for single crystals; inferred from stress-strain curves for polycrystals) and ρ_G (calculated) with shear strain. (For the single crystals, the stages noted correspond to the deformation stages shown in Fig. 4.4.) At small strains, ρ_G can exceed ρ_s and contribute substantially to work hardening if the geometrical slip distance (λ_G) is small enough. At intermediate and larger strains, ρ_s dominates ρ_G except when fine geometrical slip distances characterize the material. *(Reprinted from Acta Metall. et Mater., 42, N. A. Fleck, G. M. Muller, M. F. Ashby and J. W. Hutchinson, "Strain Gradient Plasticity: Theory and Experiment," 475, Copyright 1994, with permission from Elsevier Science.)*

part considerable strength. Moreover, the dislocation-particle array developed by plastic strain in them is often quite resistant to recovery processes. These facets of the behavior of alloys containing nondeforming particles are utilized commercially, and are described further in Sect. 5.9. However, it is worth concluding this section by noting that one important application of strain gradient hardening is found in Al beverage containers. The alloy utilized for such contains a nondeforming dispersed phase. This accounts for its high work hardening, which is a requirement for preventing plastic instability in forming operations.

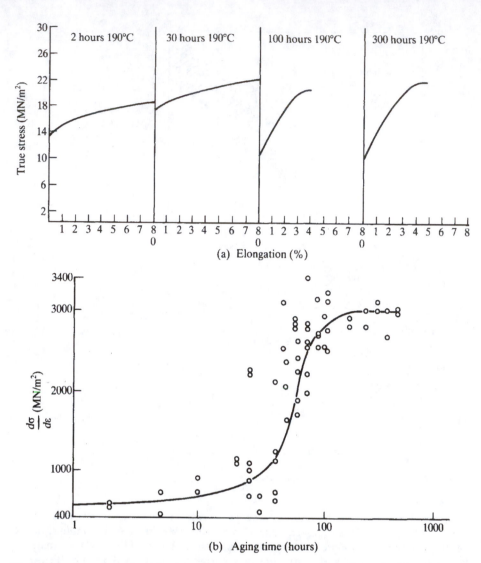

Figure 5.34
(a) Room-temperature stress-strain curves for Al-1.6 wt.% Cu alloys aged for various times at 190°C. For aging times of 30 hours or less, plastic deformation takes place by particle shearing; for aging times of 100 hours or more, it occurs by dislocation bowing. Dislocation bowing is associated with lower yield strengths. However, work-hardening rates are much greater when bowing is the slip mechanism. This is illustrated further in (b) where the work-hardening rate (at 2 percent strain) is plotted vs. aging time. The transition from dislocation cutting to bowing is accompanied by a substantial increase in this rate. (*From Prog. Matls. Sc., 10, "Precipitation Hardening," A Kelly and R. B. Nicholson, 151, Copyright 1963, with permission from Elsevier Science.*)

5.8
DEFORMATION OF TWO-PHASE AGGREGATES

The deformation of two-phase aggregates is discussed in this section. An aggregate differs from the two-phase alloys discussed previously in one major respect; in aggregates, the volume fractions of both phases are comparable. This distinguishes them from particle-strengthened alloys, for which the volume fraction of the dispersed phase is typically on the order of only several percent. Additionally, the microstructural scale of aggregates can be much greater than for particle-hardened alloys.

Because of the small volume fraction of the particles in a particle-hardened alloy, their strengths can be discussed solely in terms of the hardening provided to the matrix by the particles; the applied stress carried by the particles is negligible. This is not the case in aggregates. In elastic-plastic aggregates, for example, the elastic phase supports a portion of an applied force. When the elastic phase is in the form of a fiber, with its long axis aligned parallel to the direction of an applied stress, the load-carrying capability of the elastic phase is significant and serves to reduce the load carried by the weaker matrix. Such materials are fiber-reinforced composites, and they are discussed further in Chap. 6. When the dispersed elastic phase is in the approximate form of a sphere, however, it is not as effective as a fiber in carrying load. The elastic phase in an elastic-plastic aggregate also can provide constraints, arising from geometry and unrelated to microstructure, that effectively increase the plastic shear resistance of the plastic phase. This constraint is more important the higher the elastic-phase volume fraction of the aggregate (or, alternatively, the higher the ratio of elastic-phase particle radius to its interparticle spacing); such an effect is essentially absent in particle-hardened alloys. A macroscopic example of such a constraint is found in thin brazed joints (Fig. 5.35a). The tensile strength of a brazed joint depends on the ratio of joint thickness to joint diameter and not on the joint thickness per se. That is, thick joints can be made as strong as thin ones provided they have a sufficiently large diameter. As the thickness-to-diameter ratio of a brazed joint approaches zero, the joint strength approaches that of the base metal (Fig. 5.35c; bulk braze strengths are typically less than the metals they join). The "strength" increase arises from the biaxial stress state developed at the joint-base metal interface (Fig. 5.35b). In effect, the base metal restricts the softer braze from plastically deforming, and this gives rise to a secondary tensile stress in the braze. This secondary stress reduces the shear stress causing plastic flow (recall Mohr's circle, Chap. 1), and the magnitude of the secondary stress increases as the thickness-diameter ratio decreases. In elastic-plastic aggregates, similar constraints may operate, and they depend on the elastic phase volume fraction but not on its size.

However, microscopic effects must also be considered in discussion of elastic-plastic aggregate deformation. These effects arise at the onset of plastic-phase deformation, when geometrical dislocations are generated to accommodate the absence of plastic deformation in the elastic phase. In most aggregates, the elastic-phase volume fraction is sufficiently high, and the interphase boundary sufficiently weak, that such dislocations are secondary ones, emitted from the particle-matrix interface. With continued deformation, the internal stress field associated with the secondary dislocations increases and this stress, acting in concert with the applied stress, eventually becomes large enough to fracture the elastic particle (or

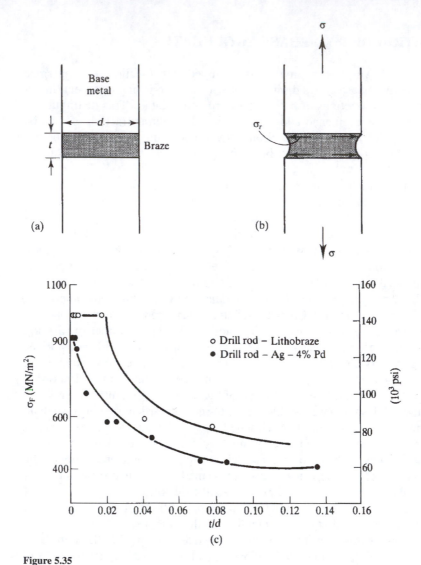

Figure 5.35
(a) The geometry of a brazed joint is characterized by its thickness (t) to diameter (d)
ratio. (b) When subjected to a tensile stress, the weaker braze is restrained from
plastically deforming. This is manifested by the development of a secondary stress
(σ_r) at the braze-base metal interface. The lower t/d is, the greater the ratio of σ_r to
σ and the greater the braze flow resistance. (c) Tensile strengths at liquid-nitrogen
temperature of brazed joints for two different brazes joining pieces of drill rod
(hardened steel). The drill rod tensile strength is ca. 960 MN/m². This level of
strength is attained with brazed joints having small t/d values. (This strength
exceeds by far the tensile strength of the braze alone, $\cong$ 400 MN/m² for the Ag-4
**percent Pd braze). (*Part (c) from H. J. Saxton, A. J. West, and C. R. Barrett, Metall.*
Trans., 2, 999, 1971.)**

to decohere it from the matrix). This type of fracture initiation is also observed in elastic-plastic alloys containing only a few volume percent of elastic phase. In aggregates, the fracture initiation strain is intermediate to that of the particle and the matrix not containing the particles. Further aspects of this type of fracture are discussed in Chaps. 9 and 10.

221

SECTION 5.9
Strength,
Microstructure
and Processing:
Case Studies

Whether the deformation behavior of elastic-plastic aggregates is better described by a continuum-mechanics approach or an approach incorporating the stress fields of secondary dislocations depends on microstructural scale. Fine microstructures are better described by the latter; coarse ones by the former. However, it does seem clear that "microscopic" aspects remain pertinent even in "coarse" dispersions (i.e., ones for which particle sizes are as large as 50 μm). The cutting tool WC-Co, which consists of a dispersion of elastic tungsten carbide particles in a cobalt matrix, is a "classic" example of an elastic-plastic aggregate. In spite of the limited aggregate ductility, this material finds widespread use for demanding cutting and drilling operations as a result of the high hardness and excellent wear resistance of WC.

Discussion of plastic-plastic aggregate deformation also must consider microstructural scale. In plastic-plastic aggregates, both phases are capable of at least moderate plastic deformation. If the size of each phase is "large," the aggregate stress-strain behavior is approximated by a suitable average of the stress-strain behavior of each of the phases (Prob. 5.34). However, when the microstructural scale is fine, geometrical dislocations influence aggregate plastic-flow behavior. Further, their contribution to strength should be greater than in a comparable single-phase material. This is so since the deformation incompatibility (which is removed by the geometrical dislocations) across a boundary between phases with different plastic-flow behavior is greater than that across a grain boundary in a single-phase material. Patented steel (piano) wire, a heavily drawn eutectoid steel, is one example of a high-strength plastic-plastic aggregate. It, and other examples of processing used to develop high-strength materials, are discussed in the following section.

5.9
STRENGTH, MICROSTRUCTURE, AND PROCESSING: CASE STUDIES

A number of important alloys serve to illustrate the principles discussed in this chapter. The structure and properties of several of them are described in this section. Of these materials, a number were developed and utilized long before an understanding of their behavior was had. However, synthesis of more recently developed materials incorporated a good understanding of fundamental strengthening principles. In many industrially important materials, hardening is achieved by several strengthening mechanisms. To a reasonable engineering approximation, the incremental strengths provided by the separate mechanisms can be considered additive.

A. Patented Steel Wire

High strength is achieved in eutectoid steel by extensive cold drawing of it into wire form. Indeed, the yield strengths of such steels (T.S. $\cong 4140$ MN/m^2 = 600,000 psi or 0.02 E, where E is the elastic modulus) approach the theoretical strength. Eutectoid steel is a two-phase mixture of body-centered cubic ferrite (containing $\cong 0.02$ wt.% C in solid solution) and the compound Fe_3C (cementite). The mixture is formed by the eutectoid decomposition of fcc austenite to pearlite (= ferrite + Fe_3C). Ferrite is the major phase ($\cong 89$ percent by volume) in this pearlite. While cementite is considered brittle in "massive" form (e.g., as it occurs in white cast irons or hypereutectoid steels), in pearlite it is capable of being cold drawn to essentially the same extent as the ferrite to which it is coupled. During wire drawing, the microstructural scale of the phases decreases commensurate with the wire diameter. The cell structure formed during drawing controls the strength of the steel wire, and the key to their high strengths lies in the very fine ferrite cell size generated. This size is considerably finer than that found in comparably deformed single-phase alloys (Fig. 5.36). In patented steel, the cell size scales approximately with the wire diameter d. Since d is related to the initial wire diameter (d_0) and the true drawing strain (ε_T) through

$$\varepsilon_T = 2 \ln\left(\frac{d_0}{d}\right) \tag{5.42}$$

the strength of the wire, assumed proportional to (cell diameter)$^{-1}$ (cf. Eq. (5.6)), varies with d_0 and ε_T as[10]

$$\sigma \sim \frac{\exp(\varepsilon_T/2)}{d_0} \tag{5.43}$$

For a given drawing strain, strength can be increased somewhat by starting with a finer eutectoid product that can be obtained by transforming the austenite at greater

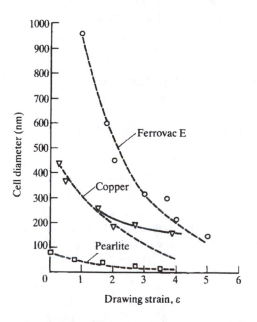

Drawing strain, ε

Figure 5.36
Cell diameter as a function of true drawing strain for several materials (Ferrovac E is nominally pure bcc Fe). The cell size in ferrite contained in pearlite is much less than that developed in the single-phase materials. This is a cause of the high strength of patented steel wire. (*From J. D. Embury, Strengthening Mechanisms in Crystals, ed. A. Kelly and R. B. Nicholson, Wiley, New York, 1971, p. 331.*)

---

[10]If strength is assumed to vary as $d^{-1/2}$, the equation analogous to Eq. (5.43) is $\sigma \sim \exp(\varepsilon_T/4)/d_0^{1/2}$.

223

SECTION 5.9
Strength,
Microstructure,
and Processing:
Case Studies

undercoolings. It is unfortunate that the high steel strengths realizable by this type of processing are limited to steels in wire form. However, the drawing strains required to yield very fine cell sizes are so great that, for all practical purposes, the final product is restricted to fine dimensions.

B. Steel Martensites

Most commercial steels contain, as does patented steel wire, the two phases of ferrite and cementite. These phases are generally distributed heterogeneously in microconstituents. One microconstituent is ferrite and the other is pearlite which contains both ferrite and cementite. The properties of these kinds of steel depend primarily on the amount of cementite (or equivalently, for hypoeutectoid steels, the amount of pearlite) which increases with carbon content. Higher carbon contents provide higher strengths but lesser ductilities.

Mechanical properties of these ferritic-pearlitic steels are sufficient to permit their widespread structural use. However, greatest steel strengths are achieved in martensitic steels. Indeed, such steels represent the most technologically important application of solid-solution strengthening. In martensitic steels, this comes about through carbon atoms placed in interstitial sites of a variant of the body-centered cubic form of iron; the interstitial carbon produces a tetragonal distortion in the crystal lattice. Although effective strengtheners on a per unit concentration basis, tetragonal distortions generally do not lead to exceptional strengths because the energy associated with the distortion precludes the responsible agents from dissolving into the lattice in amounts sufficient to permit their hardening capabilities to be utilized to the fullest. Thus, ordinary body-centered cubic iron can dissolve only 0.1 at.% carbon at most, and this ferrite demonstrates a relatively low strength.

The allotropic transformation in iron, which involves the decomposition of the high-temperature face-centered cubic form of iron (austenite) to ferrite, provides a mechanism for incorporating substantial amounts of interstitial carbon in ferrite. Austenite can dissolve fairly large amounts of such carbon (=12 at.%, although typical steels contain 5 at.% C or less). Under ordinary cooling conditions, austenite containing high amounts of carbon "disposes" of the excess via the eutectoid reaction product of cementite in pearlite. This transformation involves carbon diffusion. Thus, it can be suppressed under appropriate conditions such as rapid cooling or by the introduction of alloying elements into the steel that retard diffusion. Under these circumstances, a diffusionless $\gamma \rightarrow \alpha'$ transformation (the martensite transformation) initiates at an appropriate undercooling (the martensite start temperature, M_s) and the reaction is essentially complete at a lower temperature (M_f, the martensite finish temperature). During this transformation, carbon dissolved in the austenite is carried over as interstitial carbon in the α' (martensite). Martensite can be considered a highly supersaturated solid solution of carbon in ferrite.[11]

---

[11]Martensite actually has a tetragonal crystal structure that can be viewed as a distorted form of cubic ferrite. The tetragonality arises from the preferential occupancy by carbon of one of the three equivalent interstitial sites in the bcc lattice. The extent of the tetragonality is measured by the c/a ratio of the tetragonal unit cell; in martensites, c/a increases from 1.0 to 1.06 as the carbon content increases from 0 to ca. 1.4 wt.% C.

The supersaturation of "virgin" martensite is so great that it is highly metastable; carbide precipitation in it occurs in short times after its formation, even at temperatures below room temperature. Thus, carefully conceived experiments must be done to characterize the structure of virgin martensite and to estimate its strength. Such studies show that, in addition to the high interstitial carbon level, virgin martensite is characterized by (1) a fine martensite grain size—microstructural scale is reduced further by the presence of fine twins in many steels; and (2) a high dislocation density resulting from the shear nature of the martensite transformation which is accompanied by a change in atomic volume.[12] Although both factors contribute to the strength of virgin martensite, the greatest contribution is made by interstitial carbon. This strengthening is augmented by the fine twin structure, and the strength increase due to both factors varies as

$$\sigma \sim \left(\frac{c}{d_T}\right)^{1/2} \tag{5.44}$$

where c is the atomic carbon concentration and d_T the martensite twin spacing.[13] Equation (5.44) holds only for carbon contents less than ca. 0.4 wt.% ($\cong$ 2 at.%). At concentrations in excess of this, plastic flow occurs increasingly by twinning rather than slip and the twin strength of virgin martensite depends much less on carbon content than the "slip" strength does. On the other hand, the concentration range for which Eq. (5.44) is approximately valid includes carbon contents of many important martensitic steels.

As noted, martensites held at room temperature are invariably slightly "tempered," i.e., carbides precipitate from the supersaturated ferrite. This precipitation further strengthens martensite, for the particle hardening is greater than the concurrent loss in strength due to depletion of interstitial carbon during precipitation. Both virgin martensite and room-temperature "tempered" martensite are exceedingly hard and brittle. Their brittleness prevents them from being used as structural materials. However, additional tempering is accompanied by a strength reduction (albeit a high strength level remains) and an accompanying increase in ductility and fracture toughness. These tempered martensites find extensive use as critical structural materials; aircraft landing gears, automotive drive shafts, and high-strength bolts are examples.

Tempering is accomplished typically by heating as-formed martensite to temperatures in the region of 400°C for times on the order of 1 hour. The heat treatment is accompanied by further carbide precipitation (in a coarser form than at room temperature) and some recovery of the heavily dislocated martensite. This recovery, along with the further reduction in interstitial carbon, decreases strength to a greater extent than the carbide precipitation increases it. However, tempered martensite strengths are still high; they typically range from 1200 to 1800 MN/m². Such tempered martensites are characterized by reasonable toughness as well as high strength.

---

[12]In a sense these dislocations can be thought of as geometrically necessary. They help to accommodate the volume change accompanying martensite formation.
[13]Some investigators correlate martensite strength with $c^{1/3}$. Within the accuracy of experimental results, there is not much to choose between the two formulations. Our use of Eq. (5.44) maintains consistency with our previous treatment of solid-solution strengthening.

C. Ausformed Steels

225

SECTION 5.9
Strength,
Microstructure,
and Processing:
Case Studies

Steel microstructures resembling those of tempered martensites can be produced without recourse to a quench-heat treatment cycle by a process called ausforming. Ausforming is accomplished by cooling an austenite structure to a temperature above M_s, warm working it at this temperature, and then cooling to below M_f. Steels for which ausforming can be applied have isothermal transformation diagrams similar to that shown in Fig. 5.37, which indicates that the warm working must be completed before the austenite diffusionally transforms to pearlite or bainite. During ausforming, the high dislocation density generated is inherited by the product martensite. More importantly, the working is accompanied by precipitation of alloy carbides. Thus, an ausformed steel typically contains strong carbide-forming elements such as molybdenum, and their final structures are heavily dislocated with a fine precipitate dispersion; i.e., they resemble a cold-worked alloy containing nondeformable precipitates. Since carbide precipitation is accompanied by depletion of carbon from the austenite, the martensite product contains less interstitial carbon than the original austenite. Hence, the structures of ausformed steels are similar to those of tempered martensites, and the properties of the two structures are comparable.

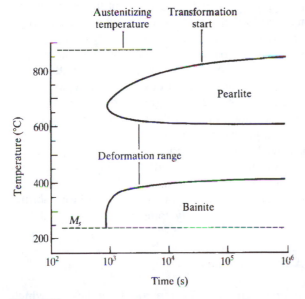

Figure 5.37
A TTT diagram for a steel susceptible to ausforming. Forming is carried out in the temperature "bay" between temperatures where the pearlite and bainite transformations occur at a reasonable rate. Such a temperature "window" is necessary in ausforming in order for the warm-worked austenite to convert wholly to martensite on cooling and not to one of the diffusional transformation products.

D. Microalloyed Steels

Microalloyed steels are not as strong as quenched-and-tempered martensites or ausformed steels. Yet microalloyed steels play a critical role in the transportation industry. They are, to a large extent, responsible for vehicle structural weight reduction, substituting for ferritic-pearlitic steels in considerably thinner section thicknesses because of the much higher strengths of microalloyed steels. To achieve this economically requires a readily processed material whose cost is not too high. Microalloyed steels, which contain only about 0.1 wt.% of expensive alloying elements such as niobium and titanium, fulfill these requirements.

Other than manganese (an element ubiquitous in steels), microalloyed steels contain minor amounts of strong carbide-forming elements (e.g., Nb) and carbon (carbon contents on the order of 0.1 wt.%). In microalloyed steels, the carbide-forming element reacts with carbon to form fine carbide particles in what is essentially a ferrite matrix. The volume fraction of the carbides is relatively low, on the order of several tenths of a percent. However, the fine carbide size provides two beneficial effects. One comes from the carbides present within the ferrite grains; these function as nondeformable dispersoids and strengthen the material. Beyond that, some of the carbides are located at ferrite grain boundaries; these particles restrict grain growth of the ferrite formed by austenite decomposition. Thus, microalloyed steels have finer grain sizes than conventional steels and demonstrate correspondingly greater grain-size strengthening.

The combined effect of both factors is shown in Fig. 5.38, which compares the yield strength of a microalloyed steel to a conventional steel. The figure plots yield strength as a function of $d^{-1/2}$ (d = grain size), and the linearity of the relationships validates the Hall-Petch relation for both steels. Further, and as would be expected, the slope of the lines in the figure are the same for both steels [$= k_y$ for ferrite (cf. Eq. (5.9))]. At the same grain size, the microalloyed steel strengths are about 100 MPa higher than those of the conventional steel. This increment in strength is provided by the intragranular carbides. The maximum strengths of the conventional steel correspond to a grain size of ca. 8 μm, the finest grain size achievable in it. Ferrite grain size can be refined—to a value as low as about 3 μm—in the microalloyed steel. The associated additional grain-size hardening is substantial. For example, the microalloyed steel has a yield strength of about 475 MPa for a grain size of 8 μm. The additional grain-size refinement provided by the intergranular carbides increases this strength by about another 75 MN/m^2, to 550 MN/m^2. The high strengths of microalloyed steels are accompanied by good ductility and formability. The latter is critical to fabrication of automotive body components.

E. Precipitation-Hardened Aluminum Alloys

Owing to their low densities, aluminum and its alloys, particularly those in which high strength can be realized, are especially important in the aerospace industry where a high strength-to-density ratio is required. Although a number of Al and other element-based precipitation-hardenable alloys have been developed, the general characteristics of precipitation hardening can be discussed with reference to the original Al-Cu precipitation-hardenable alloys.

227

SECTION 5.9
Strength,
Microstructure,
and Processing:
Case Studies

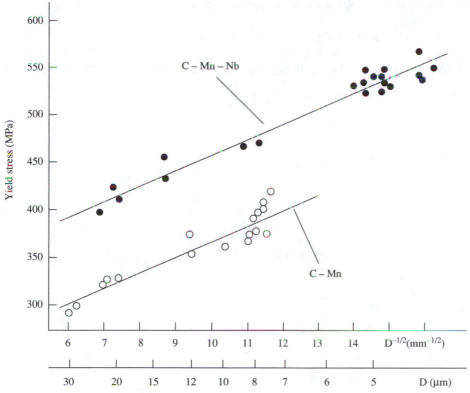

Figure 5.38
**Variation of yield stress with $d^{-1/2}$ (d = grain size) for a microalloyed steel (C-Mn-Nb)
and a plain carbon steel (C-Mn) of the same carbon content. The plain carbon steel
contains only C and Mn. The finest grain size achievable in this steel is about 8 μm.
The yield strength of the microalloyed steel is higher. The difference in the strength
levels of the two steels reflects added strength coming from intragranular carbides
present in the microalloyed steel. The finer grain size achievable in the microalloyed
steel provides additional strengthening. Note that the slopes of the curves (= k_y) of
the yield strength–$d^{-1/2}$ relationship are the same for both steels. (*After A. B. LeBon
and L. N. de Saint-Martin, Microalloying '75, Union Carbide, New York, 1977, p. 90.*)**

The Al-Cu phase diagram is shown in Fig. 5.39a. Precipitation hardening is accomplished by first heating the alloy into a single-phase region (solutionizing, Fig.
5.39b). Rapid cooling to room temperature prevents the diffusion-controlled precipitation of Cu-rich particles. The strength of such a "solutionized and quenched"
alloy is a result of solid-solution hardening of Al by Cu. Strength can be increased
by a subsequent precipitation heat treatment. The amount, nature, and size of the
precipitate, and thus the strength of the alloy, depend on the time and temperature of
the heat treatment. In Al-Cu alloys, as well as in many other precipitation-hardened
alloys, precipitates other than the equilibrium one (θ = $CuAl_2$ in Al-Cu) are formed
during precipitation. In Al-Cu, three metastable precipitates—GP-I, GP-II, and
θ'—in order of increasing stability, can be produced in addition to θ. The more

metastable the precipitate, the lower the maximum temperature at which it can form
(Fig. 5.39a).

The GP-I structure (Guinier-Preston zones of the first kind, named after the investigators who first elucidated their structure) consists of platelike arrays of Cu
atoms, several atoms thick and ca. 25 atoms in diameter, oriented parallel to {100}

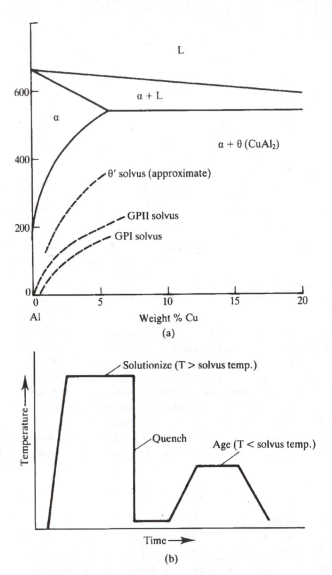

Figure 5.39
(a) The Al-rich end of the Al-Cu phase diagram. In addition to the equilibrium
θ (= $CuAl_2$) phase, metastable phases (GP-I, GP-II, and θ') can precipitate from
solid solution below their metastable solvus temperatures (designated here by dotted
lines). (b) Precipitation hardening is accomplished by first solutionizing an alloy containing less than 5.65 wt.% Cu at a temperature above the equilibrium solvus.
Quenching rapidly from this temperature produces a metastable supersaturated solid
solution. Fine precipitates are formed from this by a subsequent aging treatment at a
temperature below (usually well below) the equilibrium solvus.

planes of the Al matrix. GP-I zones, which are sheared by dislocations, provide moderate strengthening. GP-II zones are thicker ($\cong$ 10 atoms) and larger in diameter ($\cong$ 75 atoms) than GP-I zones. Moreover, GP-II zones contain Al in an approximate stoichiometric ratio to Cu. GP-II zones also are sheared by dislocations; however, they give rise to strengthening greater than GP-I zones do.

229

SECTION 5.9
Strength,
Microstructure,
and Processing:
Case Studies

The θ' precipitate has a composition the same as or close to that of θ. However, the structure of θ' is distorted in comparison to θ so that a coherent or partially coherent boundary between θ' and the matrix is formed. In contrast to GP-I and GP-II zones, dislocations bow around both θ' and θ particles. The hardness variation with aging time at a temperature at which all four precipitates form sequentially[14] (i.e., GP-I $\rightarrow$ GP-II $\rightarrow$ θ' $\rightarrow$ θ) is shown in Fig. 5.40. Here, room-temperature hardness is plotted as a function of precipitation time at various temperatures. The increase in strength at short aging times is related to increases in size and volume fraction of the GP zones. Maximum strength is obtained near the GP-II $\rightarrow$ θ' transition. At longer aging times, the structure coarsens and the strength decreases as the spacing between the particles increases and dislocation bowing supplants cutting as the dominant slip mode.

As shown previously (Fig. 5.34), yield strengths of alloys containing θ' and θ are not remarkable, but their work-hardening rates are. The high work-hardening rates of alloys containing nondeformable precipitates are utilized in Al and other dispersion-hardened alloys. In some of these, a fine dispersoid is introduced artificially rather than, as in precipitation-hardened alloys, from solid-state reactions. Many of the techniques to produce such structures involve powder metallurgy which generally does not yield an extremely fine particle dispersion in the matrix.[15] On the other hand, when mechanical deformation is carried out properly, high strengths can be achieved. Dispersion-hardened alloys are also more useful at elevated temperatures than are precipitation-hardened alloys. Precipitate particles dissolve to some extent at high temperatures and, more importantly, the particle dispersion coarsens, resulting in lower strengths.[16] A careful choice of dispersoid minimizes coarsening at high temperature. Dispersions of nitrides, carbides, and, especially, oxides that have limited solubility in the matrix are useful in this regard. Moreover, the particle-dislocation structure generated by deformation is resistant to recovery and recrystallization. Thus, as illustrated in Fig. 5.41, appropriately processed dispersion-hardened alloys display superior high-temperature properties; these alloys are finding increased use as high-temperature structural materials.

High work-hardening rates are useful in materials subjected to extensive forming to produce their final shape because, as mentioned previously, a high work-hardening rate delays localized necking during forming. Brass, which work hardens rapidly, is subjected to extensive deformation during its processing into cartridge shells, for example. So are the Al alloy beverage containers. Certain steels—TRIP steels—also have excellent formability. The acronym TRIP stands for "Transformation Induced Plasticity." In brief, TRIP steels are metastable austenitic steels

---

[14] Owing to the metastability of the other precipitates, the θ phase eventually forms in all alloys if the heat treatment time is long enough.

[15] Recently developed processes, such as mechanical alloying in which very fine dispersions are produced by "alloying" in a high-energy mill, are changing this picture.

[16] Certain nickel-base alloys are resistant to such coarsening, and because of this they are used widely as high-temperature precipitation-hardened alloys.

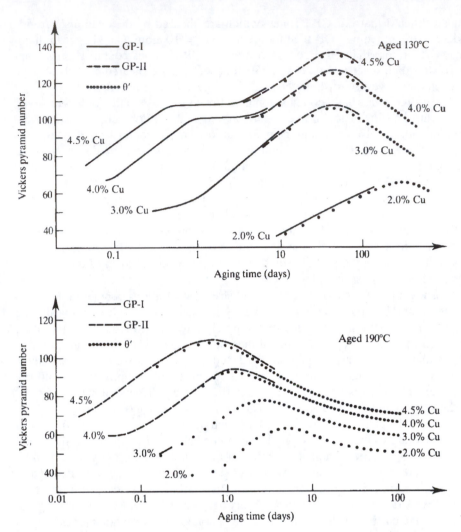

Figure 5.40
Room-temperature hardness vs. aging time at two different temperatures for several Al-Cu alloys. In most of the alloys, maximum strength is associated with both the GP-II and θ′ phases; the former deforms by shear whereas dislocations typically bow around θ′ particles. Highest strengths are achieved with higher solute contents that yield greater precipitate volume fractions. Higher strengths are typically achieved at lower aging temperatures (albeit only by aging for longer times) owing to a greater precipitate volume fraction and, especially, a finer precipitate size achieved with low-temperature aging. Overaging is associated with particle coarsening. (*After J. Silcock, T. Heal, and H. Hardy, J. Inst. Metals, 82, 239, 1953–54.*)

which, when plastically deformed, undergo a gradual strain-induced martensitic transformation, and the martensite so-formed is stronger than its parent austenite. This leads to an increase in flow stress with strain over and above that due to conventional work-hardening effects. TRIP steels find use as beverage containers, primarily in Europe.

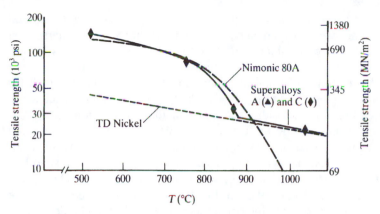

Figure 5.41

Tensile strength-temperature relationships for several high-temperature alloys. TD (thoria-dispersed) nickel is a dispersion-hardened Ni base alloy. Its low-temperature strength is less than, but its high-temperature strength is greater than, that of Nimonic 80A, a precipitation-hardened Ni alloy. Superalloys A and C contain artificial dispersoids and precipitates. The latter provide low-temperature strength equivalent to that of Nimonic 80A, and the former provide elevated temperature strength parallel to that of TD Ni. (*From J. S. Benjamin, Metall. Trans., 1, 2945, 1970.*)

5.10
SUMMARY

In this chapter, the strengthening provided by various obstacles to dislocation motion has been discussed. The hardening associated with an obstacle array is approximated by Eq. (5.1), which relates the incremental strength to the effective obstacle spacing and the angle through which the dislocation bends before it passes through the obstacle. For weak obstacles, dislocations bend only slightly before breaking through the obstacles and little strengthening per obstacle is provided, and vice versa. When the obstacles become strong enough, dislocations do not pass through them. Rather, dislocations loop around obstacles, leaving a dislocation array in its wake; the array is composed of shear loops at small strains and prismatic loops and secondary dislocations at progressively higher strains.

Dislocations pass through substitutional atoms in approximately the form of a straight line, and the hardening per solute atom is weak. Certain impurity atoms can give rise to a tetragonal distortion of the lattice, producing greater strengthening. Likewise, some precipitate particles, although sheared by dislocations, impart considerable strength to the matrix. Dislocation arrays produced during deformation are likewise strong obstacles, and moving dislocations can penetrate such arrays only with difficulty. Grain boundaries and certain large and/or incoherent particles are nondeformable; i.e., dislocations cannot pass through them. These obstacles produce large strengthening effects per obstacle, yet in most microstructures they cannot be produced in concentrations sufficient to generate great additional strengths. The dislocation structure developed by deformation in materials containing nondeforming particles contributes substantially to their work hardening, though.

Grain boundary hardening is not represented by Eq. (5.1) because a grain boundary cannot be penetrated by a dislocation and the grain boundary intersection with a slip plane is a line rather than a point (the latter forming the basis for Eq. (5.1)). Dislocations in polycrystals are "hard" obstacles; thus, their contribution to strength is given by a version of Eq. (5.2) where the spacing between obstacles is related to the dislocation density through $L \cong \rho^{-1/2}$. The square root dependence of strength on dislocation density holds even when dislocations assume a cellular structure, as they are prone to do with increasing plastic strain. Solid-solution strengthening varies with the square root of the solute concentration, and also scales with a parameter, ε, that contains within it the factors (e.g., modulus and size) that lead to hardening. However, the numerical constant in the solid-solution hardening equation is small (except for tetragonal distortions), indicative of the generally weak hardening associated with solid-solution strengthening. Precipitation hardening depends on microstructural features, and the hardening response bears resemblance to solid-solution hardening. That is, irrespective of the specific hardening mechanism (coherency strengthening, modulus hardening, etc.), the strength provided by particles through which dislocations pass varies with a parameter we have designated ε. Further, the increase in strength scales with a modification of the particle "concentration"; the modified factor is the particle volume fraction multiplied by the ratio of particle radius to Burgers vector. Although the phenomenology of particle and solid-solution strengthening is similar, particles are more effective strengthening agents. Strengthening provided by particles that are bypassed by dislocations is given by Eq. (5.2) modified to take into account the finite particle size.

Hardening mechanism	Nature of obstacle	Strong (S) or Weak (W)	Hardening law
Work hardening	Other dislocations	S	$\Delta\tau = \alpha G b \rho^{1/2}$; see [1]
Grain size	Grain boundaries	S	$\Delta\tau = k_y' d^{-1/2}$; see [2]
Solid solution	Solute atoms	W (see [3])	$\Delta\tau = G\varepsilon_s^{3/2} c^{1/2}/700$ (see [4])
Deforming particles	Small, coherent particles	W (see [5])	$\Delta\tau = CG\varepsilon^{3/2}(fr/b)^{1/2}$ (see [6])
Nondeforming particles	Large particles, incoherent particles	S (see [7])	$\Delta\tau = Gb/(L - 2r)$

Notes:

[1] α equals about 0.2 for fcc metals, about 0.4 for bcc metals.

[2] k_y' scales with inherent flow stress and/or shear modulus; therefore k_y' is generally greater for bcc metals than for fcc metals.

[3] Exception to weak hardening occurs for interstitials in bcc metals; the shear distortion interacts with screw dislocations leading to strong hardening.

[4] Equation apropos to substitutional atoms; parameter ε_s is empirical, reflecting a combination of size and modulus hardening.

[5] Coherent particles can be "strong" in optimally aged materials.

[6] Constant C depends on specific mechanism of hardening; parameter ε relates to hardening mechanism(s). Equation shown applies to early stage precipitation. Late stage precipitation results in saturation hardening.

[7] Highly overaged alloys can represent "weak" hardening.

Symbols: G = shear modulus; b = Burgers vector; ρ = dislocation density; d = grain size; c = solute atom concentration (atomic fraction); f = precipitate volume fraction; r = precipitate radius; L = spacing between precipitates on slip plane.

Table 5.2
Summary of Hardening Mechanisms in Crystalline Materials

Aggregates differ from particle-strengthened alloys in that the volume fractions of the phases are comparable in an aggregate. Further, in many aggregates the microstructural scale is coarse enough for phase interactions to be neglected in discussing their mechanical behavior. For this situation, the deformation response is approximated by a suitable averaging of the behavior of the individual phases. If the microstructural scale is fine, however, phase interactions involving microscopic concepts become important. In a real sense, aggregates are a special case of a composite material, the topic of the following chapter.

But before proceeding to discuss composite materials, let us concede that the material covered and treated in this chapter has been a "real dose." Perhaps Table 5.2 will help in this regard. It lists the common strengthening mechanisms, whether they represent hard or soft obstacles, and the constitutive equation describing the effect they have on the shear yield strength of a material. Such a summary has a tendency to be a bit oversimplified, so don't disregard the footnotes that accompany this table!

REFERENCES

Ardell, A. J.: "Precipitation Hardening," *Metall. Trans. A,* **16A**, 2131, 1985.

Ashby, M. F.: "The Deformation of Plastically Non-Homogeneous Alloys," in *Strengthening Methods in Crystals,* ed. A. Kelly and R. B. Nicholson, Wiley, New York, 1971, p. 137.

Brown, L. M., and R. K. Ham: "Dislocation-Particle Interactions," in *Strengthening Methods in Crystals,* ed. A. Kelly and R. B. Nicholson, Wiley, New York, 1971, p. 10.

Embury, J. D.: "Strengthening by Dislocation Substructures," in *Strengthening Methods in Crystals,* ed. A. Kelly and R. B. Nicholson, Wiley, New York, 1971, p. 331.

Evans, K. R.: "Solid Solution Strengthening of Face-Centered Cubic Alloys," *Treat. Matls. Sc. and Tech.,* ed. H. Herman, **4**, 113, 1974.

Fleck, N. A., G. M. Muller, M. F. Ashby, and J. W. Hutchinson: "Strain Gradient Plasticity: Theory and Experiment," *Acta Metall. Mater.* **42**, 475, 1994.

Fleischer, R. L.: "Solid Solution Hardening," in *The Strengthening of Metals,* ed. D. M. Peckner, Reinhold, New York, 1964, p. 93.

Hansen, N., and D. Kuhlmann-Wilsdorf: "Low Energy Dislocation Structures due to Unidirectional Deformation at Low Temperatures," in *Low Energy Dislocation Structures,* ed. M. N. Bassin, W. A. Jesser, D. Kuhlmann-Wilsdorf and H. G. F. Wilsdorf, Elsevier Sequoia, Lausanne, 1986, p. 41.

Kelly, A.: *Strong Solids,* Clarendon Press, Oxford, 1966.

Kuhlmann-Wilsdorf, Doris: "Theory of Work Hardening 1934–1984," *Metall. Trans. A,* **16A**, 2091, 1985.

PROBLEMS

5.1 Derive Eq. (5.1) by means of a line tension balance of the kind discussed in Chap. 3.

5.2 Consider Stages I, II, and III of single crystal deformation (Fig. 4.4). Classify the dislocation interactions in each of the stages as being of the "soft" or "hard" obstacle variety. Give your reasoning.

5.3 a Show that Eq. (5.5) follows directly from Eq. (5.2) by using the relationship between dislocation density and spacing.

b Using Eq. (5.5), and assuming that dislocations are distributed randomly through-out the crystal, what can you say about the critical angle, ϕ_c? (Assume $\alpha = 0.3$.)

5.4 Two edge dislocations are present in a crystal as shown in the sketch below.
a Plot the glide force on dislocation 2 resulting from the presence of dislocation 1 as a function of x_1. Take $x_2 = 25b$.

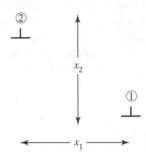

For the remainder of this problem do not assume any special value for x_2. Construct derivations for the general case.
b Derive general expressions for the value of x_1 corresponding to the maximum glide force on dislocation 2 arising from the presence of dislocation 1.
c Derive a general expression for the stress necessary to push dislocation 2 past dislocation 1 (i.e., from $x_1 = -\infty$ to $+\infty$).
d Make a simple model of a dislocation array in a crystal and use it to derive an expression for the increase in yield strength as a function of dislocation density.
e Compare your results from part (d) to Eq. (5.5) and the discussion in Sect. 5.3.

5.5 Use reasonable estimates of τ^* and r in Eq. (5.8) and the relationship $\tau = \sigma/\bar{m}$ to develop an order of magnitude estimate for k_y in Eq. (5.9). Compare your estimated value with those listed in Table 5.1.

5.6 The following data were obtained for a carbon steel and an aluminum alloy.

Carbon Steel		Aluminum alloy	
d (μm)	σ_y (MN/m^2)	d (μm)	σ_y (MN/m^2)
406	93	42	223
106	129	16	225
75	145	11	225
43	158	8.5	226
30	189	5.0	231
16	233	3.1	238

a Show that the yield strengths of this steel and aluminum alloy obey the Hall-Petch relationship. Determine σ_0 and k_y for each material.
b Certain microalloyed steels contain small additions of vanadium or niobium that permit the grain size to be reduced to about 2 μm if the processing of the steel is carefully controlled. Likewise, advanced aluminum alloys containing special types of particles can be processed to yield a grain size of about 2 μm. Suppose we reduce the grain size of steel and aluminum from 150 μm to 2 μm by such processing. Would a substantial increase in the strengths of these materials result? Comment on your answer.

5.7 a Using the results of Example Prob. 5.1, calculate—for the materials listed in Table 5.1—the strain beyond which geometrical dislocations contribute more to grain-size hardening than does the inherent grain-size effect on yield strength. (Additional information: see Chap. 2 for elastic properties; G can be calculated assuming isotropic elastic behavior with $\nu = \frac{1}{3}$; let $b = 0.250$ nm.)

b Do your results imply that geometrical dislocations are the primary strengthening mechanism for strains greater than those calculated in part (a)? Explain.

5.8 a Calculate $k_y d^{-1/2}$ for the materials listed in Table 5.1. Consider grain sizes of 0.1 μm, 1 μm, and 10 μm.
b Compare the $k_y d^{-1/2}$ values with typical values of σ_0 at $T/T_m = 0.2$. (See Chap. 4 for data—e.g., Figs. 4.3 and 4.18, and relate τ_{CRSS} to σ_0; if σ_0 is not known, estimate it on the basis of bonding and crystal structure—e.g., Figs. 4.3 and 7.17.)

5.9 Schematically plot flow stress vs. $d^{-1/2}$ for grain-size strengthening (Eq. (5.9)) and cell strengthening (Eq. (5.12)) and demonstrate a transition from grain-size to cell-size control of this stress.

5.10 Consider grains having diameters of 10, 100, and 1000 nm. Let the grain boundary thickness be four atomic diameters. Considering grains as cubes, determine the fraction of atoms that are situated in grain boundaries for the three grain diameters considered. (You will have to assume a reasonable value for the atomic diameter to obtain a numerical result.)

5.11 Show that if an atom of radius $r(1 + \varepsilon_b)$ is substituted for one of radius r in a crystal, the local volume change is about $4\pi r^3 \varepsilon_b$. (Note: The assumptions are that the larger atom is incompressible and the volume change is relatively small; i.e., you can neglect terms in ε_b^2 and higher order.)

5.12 When copper is added to body-centered cubic iron, the lattice parameter change is minimal; i.e., the atomic size of Cu is almost the same as that of Fe. The shear modulus of bcc iron increases by about 2.5% when 10 at.% Cu is dissolved in it. (For Fe, $G = 8.1 \times 10^{10}$ N/m^2.)
a Use the Fleischer model to predict the solid-solution strengthening of bcc Fe by Cu. Plot your results as a graph of $\Delta\tau_y$ vs. at.% Cu in solid solution; the concentration scale should cover the range of Cu solubility in bcc Fe at room temperature as well as compositions corresponding to supersaturated alloys.
b Do you believe Cu could be used to effectively solid solution strengthen iron and steel?

5.13 Use the equations for σ_{xx}, σ_{yy}, σ_{zz}, and τ_{xy} provided in Chap. 3 to derive Eq. (5.13). Note that the energy is given by $p \, \delta V$ where $p = (1/3)(\sigma_1 + \sigma_2 + \sigma_3)$, and σ_1, σ_2, and σ_3 are the principal stresses.

5.14 a Show that for "hard" obstacles (for which L' is the mean free path, L) L is related to the obstacle concentration by $L = b/(2c)^{1/2}$.
b Show that for "soft" obstacles, $L' = b/[c(\pi - \phi)]^{1/2}$. (Hint: Recall Figs. 5.2a and b.)

5.15 Estimate τ^* at 0 K for pure copper and vanadium (Fig. 4.3). At a critical temperature, T_c, thermal energy is able to overcome the obstacles causing the stress, τ^*. The volume, v, associated with the obstacles is related to τ^* and T_c by $\tau^* v \cong kT_c$. Calculate v for copper and vanadium. Recall that $v = ba_s$ where a_s is the slip plane area over which the barrier's presence is felt. What do your results say about a_s in Cu vis-à-vis a_s in V?

5.16 We approximate the force per unit length a dislocation experiences as it approaches a short-range obstacle as shown in the sketch following.

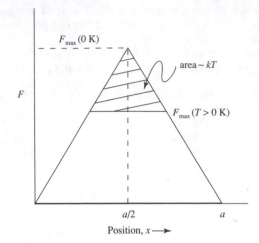

a How is τ_0^* (cf. Eq. (5.23))related to F_{max}?

b As indicated in the sketch, the force per unit length required to move the dislocation past the obstacle is reduced at finite temperatures, and at T_c (also Eq. (5.23)), short-range obstacles are ineffective in restricting dislocation motion. Derive an expression for T_c in terms of τ_0^*, k, b, a, and l (l = slip plane separation distance of the obstacles).

c Consider a Cu-14 at.% Al alloy. The obstacle slip plane spacing is approximately $l = 4r/3c$ where r is the atomic radius and c the impurity concentration (expressed in atomic fraction). T_c for this alloy is about 400 K. Taking $a = 10b$, estimate τ_0^*. (For Cu, $b = 0.256$ nm.) Refer to Fig. 4.3. Is your answer reasonable in terms of the value of τ_0^* you obtained?

d F_{max} is reduced at finite temperatures. Using the force-distance profile shown in the above figure, derive an expression for the temperature dependence of τ^* in terms of τ_0^*, T and T_c. How does this equation compare to Eq. (5.23)?

5.17 Refer to Example Prob. 5.3. What is the value of the shear stress increase associated with the different Burgers vectors of the matrix and the particle? Assume the disloca-tion remains straight in its passage through the particle. Compare the magnitude of this stress to those arising from chemical and order hardening. (To be consistent, use the values of the chemical- and order-hardening stresses that would be found were the dislocation to remain straight in its passage through the particle.) (Take $\gamma_s = 0.2$ J/m^2, $f = 0.03$, $r = 10$ nm, $\Delta b/b = 0.10$, and $G = 50$ GN/m^2.)

5.18 Using principles of quantitative microscopy, show that for a straight dislocation, L' is related to particle volume fraction (f) and particle radius (r) through $f = 2r/L'$.

5.19 Spinodal decomposition, which occurs in a number of alloys, is an alternative to pre-cipitation. The phase diagram of an "ideal" system exhibiting spinodal decomposition is shown in the sketch following. Spinodal decomposition may occur in preference to precipitation for alloy compositions between c_1 and c_2 at the indicated decomposition temperature.

   During spinodal decomposition, the composition "difference" across the boundary between the α_1 and α_2 phases is not sharp, but diffuse, as indicated in the sketch below. The composition is predicted to vary with distance as;

$$c = \frac{1}{2}(c_{max} + c_{min}) + \frac{1}{2}(c_{max} - c_{min}) \sin \frac{2\pi x}{\lambda}$$

Here c_{max} can be thought of as the maximum concentration of B in the α_2 phase and c_{min} can be viewed as the minimum B concentration in α_1; λ is a measure of the scale over which decomposition occurs.

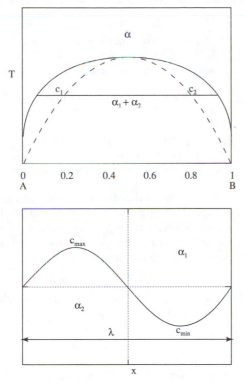

A number of mechanisms can strengthen a spinodally decomposed alloy. We consider one: the variation of the dislocation self-energy ($Gb^2/2$ per unit length) as it moves through the lattice. Since the composition is periodic, the shear modulus and Burgers vector are likewise periodic.

a Consider spinodally decomposed alloys in two different systems, A-B and A-C, having identical phase diagrams. The shear modulus of A is 40 GN/m^2; that of both B and C is 50 GN/m^2. The respective atomic radii are: 0.13 nm (for A), 0.14 nm (for B), 0.12 nm (for C). Based solely on dislocation self-energy considerations, will "dislocation energy hardening" be greater in the A-B or the A-C system?

b Assume that spinodal decomposition has occurred (in both the A-B and A-C systems) such that the concentration of A varies between 0.3 and 0.7 over the distance, λ. Assume G and b vary linearly with alloy composition. Set up an expression for the dislocation self-energy as it moves through the lattice. Recalling that the force is the gradient of the energy, determine F_{max} (the maximum force per unit length) for both systems. (Your answer will involve the distance, λ.) (Hint: You are interested in determining the increase in the force relative to that of a homogeneous alloy of the same overall composition. Thus, the change in energy you calculate is the change in dislocation self-energy relative to that of an alloy of uniform composition.)

c What is the increase in shear yield stress (for both systems) associated with the compositional inhomogeneity and resulting from the variation in dislocation self-energy?

d Values of λ for spinodal decomposition are typically small; $\lambda = 10b$ would not be uncommon. Using this value of λ, calculate the values of the increase in shear yield stress coming from the hardening mechanism considered. Are these large or small increases compared to other strengthening mechanisms discussed in this chapter?

5.20 a Consider chemical hardening described by Eq. (5.33). Consider particles for which $r \gg b$ does not hold. By sketching curves similar to Fig. 5.19, show that small particles do not chemically harden a material to the extent that large ones do.
b How will dislocation width come into play when determining the magnitude of the chemical strengthening?

5.21 Show that a simplified treatment of order hardening predicts $\tau_{ord} = \pi APBE f/2b$. In this treatment, assume (a) a straight dislocation and (b) the increase in antiphase boundary area is linear with dislocation position in the particle.

5.22 a Schematically plot τ_{ord} (units of G) as a function of particle radius using Eqs. (5.34a) and (5.34b). Discuss the different situations pertaining to a low and a high APBE.
b For a Ni base-alloy ((APBE) = 0.1 J/m², r = 50 nm, b = 0.250 nm, f = 0.10, and $G = 86 \times 10^3$ MN/m²), what is the ratio of τ_{ord} calculated from Eq. (5.34a) to that from Eq. (5.34b)? Which equation do you believe is most appropriate in this situation?

5.23 Consider a small (e.g., $r = 10b$) particle that causes strengthening as a result of particle-matrix surface area created by dislocations passing through the particle. Briefly describe how τ_{chem} varies with the passage of successive dislocations. (Hint: "Look down" on the slip plane and consider the particle-matrix area created as a function of the number of dislocations cutting the particle.)

5.24 An aluminum alloy is strengthened by ordered, coherent particles and used at a temperature of 200°C. At this temperature, the average particle radius (r) increases (coarsens) with time (t) according to

$$r^3 = r_0^3 + Kt$$

where r_0 is the initial particle radius and K is a constant ($= 10^{-28}$ m³/s). Assume that the particle shearing stress is that of order hardening (Eq. (5.34a)); the particle bowing stress is given by Eq. (5.36).

Calculate, as a function of time (0 through 1000 hours) at 200°C, how the strength of the alloy varies. (Data: $G = 25 \times 10^9$ N/m²; $b = 0.29$ nm; APBE = 50 mJ/m²; $f = 0.05$; $r_0 = 10$ nm.)

5.25 An aluminum-lithium alloy contains 10 at.% Li. At 800 K, all of the Li is in solid solution, a situation that can be maintained on quenching to room temperature. On aging the quenched alloy at 450 K, the supersaturated solid solution decomposes into an Al-rich matrix containing Al_3Li, a metastable precipitate phase. The maximum volume fraction of Al_3Li that forms is $f \cong 0.22$. At this stage of aging, the Li content of the matrix is reduced to 7 at.%.
a Estimate the increase in critical resolved shear stress of aluminum due to the presence of lithium in solid solution in the quenched alloy. Take $\varepsilon_s = 1.2$.
b Assuming that the particle shearing stress arises from order hardening, calculate the maximum change (i.e., the change in late stage precipitation associated with particle shearing) in critical resolved shear stress that could be obtained in the alloy by the precipitation sequence described. (Take APBE = 50 mJ/m².)
c During precipitation, the solid-solution hardening decreases as a result of the reduction in matrix Li content. Estimate the maximum change in τ_{CRSS} of the alloy in comparison to pure Al, taking into account both precipitation and matrix solid-solution strengthening effects.
d In part (c) you found that the increase in τ_{CRSS} due to precipitation is much larger than the decrease in it due to diminished solid-solution hardening. Provide an exam-

ple in which τ_{CRSS} decreases on precipitation because the decrease in strength accompanying solute depletion is greater than the concurrent increase due to precipitation.

5.26 Voids can be generated in metals. For example, they are common products of radiation damage. Do you think voids can "harden" a metal? Explain. If your answer is "yes," what factors mitigate the hardening relative to the potential of using voids to improve the mechanical properties of a metal?

5.27 Consider a precipitation-hardened system having the following characteristics: the lattice parameter of the precipitate is 5% greater than that of the matrix; G of the matrix is 40 GN/m² while G of the precipitate is 60 GN/m²; the matrix Burgers vector is 0.250 nm; precipitate-matrix surface energy = 0.25 J/m²; APBE of precipitate = 0.1 J/m². Consider coherency, modulus, chemical, and order hardening. Determine the relative increases in strength accompanying each mechanism. Consider particles having $r = 10b$ and particles with $r = 100b$.

5.28 Compare Eq. (5.41) with that derived in Prob. 4.13, which also described the statistical dislocation density as a function of strain. What are the similarities and differences between the two expressions?

5.29 a Are the work-hardening rates of overaged Al-Cu alloys consistent with those expected on the basis of geometrical dislocation hardening? (Hint: Recall that $\Delta\tau = \alpha Gb\rho^{1/2}$ with $\rho = \rho_G$; ρ_G is given by an expression similar to Eq. (5.40).)
b If geometrical dislocations are responsible for work hardening of overaged Al precipitation hardenable alloys, how would you expect work-hardening rates to change with continued overaging?

5.30 The statistical dislocation density (ρ_s) in polycrystalline Cu is estimated to vary with deformation strain as

$$\rho_s \ (\text{m}^{-2}) = 2.1 \times 10^{14}[1 - 0.99 \exp(-1.45\varepsilon)]^2$$

a The following data apply to a commercially pure Cu alloy cold-worked to different strain levels as indicated

Deformation strain	Yield strength (MN/m²)
0.058	195
0.116	205
0.232	250
0.463	310
0.928	345
1.39	365

Determine the expected values of ρ_s at the above-noted strains, and determine whether or not the increase in flow strength with strain is approximately consistent with work-hardening theory.
b Consider dispersing 3 vol.% of BeO in this copper, with the BeO particles having a radius of 20 nm. Assume that the work-hardening rate of the material, $d\sigma/d\varepsilon$, is given by the product of $d\sigma/d\rho$ and $d\rho/d\varepsilon$. Obtain an expression for this work-hardening rate considering both statistical and geometrical dislocations. How do your "predictions" compare with the behavior of the Cu-BeO alloy whose behavior is illustrated in Fig. 5.29?

5.31 In this problem we estimate geometrical dislocation densities (ρ_G) generated by the following processes: (1) torsion testing of a 1-mm-diameter wire; (2) bending of a

1-mm thick beam; (3) hardness testing; (4) plastic deformation in front of a crack tip in a fracture toughness test; (5) plastic deformation in a particle-strengthened alloy for which slip takes place by dislocation bowing; (6) tensile deformation of a polycrystal. These dislocation densities will be compared to the corresponding statistical disloca-tion densities (ρ_s). For the metal we treat, ρ_s is given by

$$\rho_s \ (m^{-2}) = 10^{14}[1 - 0.99 \exp(-\varepsilon)]^2$$

where ε is the strain in a tensile test. Other material data are: $G = 6 \times 10^{10} \ N/m^2$; $b = 2.5 \times 10^{-10}$ m; α (of Eq. (5.5)) $= 0.2$; $k_y = 0.15 \ MN/m^{3/2}$; $\sigma_y = 500 \ MN/m^2$; $K_{Ic} = 25 \ MN/m^{3/2}$. (In this problem you need not differentiate between shear and tensile strains; any resulting "error" will cause no more than a factor of $\sqrt{2}$ in the calculated values of ρ_G.)

a Let the torsion bar be subject to a true shear strain of one. Calculate ρ_G. Assuming the average plastic strain in the bar is one-half the maximum strain, calculate ρ_s. Estimate the additional hardening due to the geometrical dislocations.

b Repeat part (a) for the bent bar. Assume a maximum tensile strain of 0.5 on the bar's surface.

c The material's measured Vicker's hardness is 1400 MN/m². Vicker's hardness tests employ a diamond indenter, and the hardness is calculated from

$$\text{Hardness (N/m}^2) = \frac{1.854 \ P}{(2a)^2}$$

where P is the load ($= 2 \times 10^5$ N in the present situation) and $2a$ is the diagonal (in meters) of the indentation. This diagonal length can be taken as a measure of the size of the plastically deformed region beneath the indentation (cf. Fig. 1.17). The average plastic strain in the indented region can be taken as 5%. We further assume that the plastic strain varies linearly between the indenter surface and the boundary of the elas-tic/plastic region beneath the indenter. Estimate ρ_G in this plastically deformed region. Estimate the error in measured hardness measured due to the geometrical dislocations generated during the test. (Recall that hardness scales with flow stress and, thus, hard-ness varies with dislocation density in the same manner as flow stress.)

d The stress concentration in front of a crack tip produces a plastic zone there. The ra-dius (r) of this zone is approximately

$$r = \frac{K_{Ic}^2}{2\pi\sigma_y^2}$$

where σ_y is the average flow stress of the material. The maximum plastic strain in this zone is the fracture strain; we conveniently take this as one. Estimate ρ_G in the plastic zone, and compare its value to that of ρ_s in the zone. Estimate the effective increase in flow stress in the plastic zone due to the geometrical dislocations. Note that the in-crease in flow stress reduces plastic zone size. That is, the value of r used in the above calculation is overestimated. Iterate the procedure until you arrive at a "stable" answer.

e Assume nondeforming second-phase particles are dispersed in our material. The particles are of radius 10^{-8} m, and their volume fraction is 0.02. Estimate the value of ρ_G when the alloy is strained 10% in tension. How much of an increase in flow stress is associated with the geometrical dislocations?

f The grain size of the material is 5 μm. Repeat part (e) for the situation when the ma-terial is deformed 30% in tension.

5.32 Consider Eq. (5.41). Assuming that λ_s varies as $\rho_s^{-1/2}$, integrate this equation to ob-tain ρ_s as a function of plastic shear strain. Schematically plot ρ_s and ρ_G vs. shear strain and compare with behavior illustrated in Fig. 5.33.

5.33 Assume that a shear loop is left in the wake of each dislocation that bows around a nondeforming particle. Derive an approximate relationship for the density of these geometrical dislocations as a function of shear strain, the Burgers vector, and the mean spacing between the particles on the slip plane. Show that ρ_G is proportional to the strain gradient between the plastically deforming matrix and the elastic particle.

5.34 Consider a two-phase material containing a volume fraction V_α of phase α and volume fraction $(1 - V_\alpha)$ of phase β. Assume that the flow behavior of each phase in the material is unaffected by the presence of the other. Also assume that the two phases undergo equivalent tensile strains (as if they were parallel slabs with the long axes of the slabs aligned with the tensile axis). Show that the tensile flow behavior of the mixture is represented by

$$\sigma_c(\varepsilon) = V_\alpha \sigma_\alpha(\varepsilon) + (1 - V_\alpha)\sigma_\beta(\varepsilon)$$

where σ_c is the mixture flow stress at strain ε, and $\sigma_\alpha(\varepsilon)$ and $\sigma_\beta(\varepsilon)$ are the tensile flow stresses of α and β at this same strain.

5.35 a Plot the cell size of Ferrovac E, copper, and ferrite in pearlite (Fig. 5.36) vs. exp $(\varepsilon_T/2)$ to see whether the assumption (Eq. (5.42)) of the relation between strain and cell size holds. (Alternatively, you may wish to plot $\ln(d_0/d)$ vs. $\varepsilon_T/2$ where d_0 is some arbitrary initial cell size (e.g., $d_0 = d$ at $\varepsilon_T = 0$) and d is the instantaneous cell size.)
b Can you provide a physically plausible explanation of why the ferrite cell size should be finer in pearlite than it is in single-phase ferrite drawn to the same strain?

5.36 The strengths of "virgin" (i.e., as quenched) and quenched-and-tempered martensite are dealt with here.
a Consider a 0.40 wt.% C (= 1.83 at.% C) quenched martensitic steel. The dislocation density in the quenched martensite is $10^{16}/m^2$. What is the yield strength of the as-quenched martensite? (Consider solid-solution strengthening by carbon, dislocation strengthening, and the friction stress. The latter can be estimated by extrapolating the yield strength–grain size relationship of Fig. 5.38 to an "infinite" grain size. Take α (Eq. (5.5)) = 0.4. Equation (5.5) pertains to a shear yield stress, so you must convert this to a tensile yield strength.
b The steel is now tempered. During tempering the dislocation density is reduced to $5 \times 10^{14}/m^2$, and the martensitic matrix carbon content is reduced to 0.20 wt.% C (= 0.92 at.% C). This reduction in carbon content is accompanied by precipitation of iron carbide particles that are not sheared by dislocations. Estimate the yield strength of this quenched-and-tempered martensite. (Consider solid-solution strengthening, carbide particle strengthening, dislocation strengthening, and the friction stress. Hint: You must determine the carbide volume fraction. While the densities of iron and the carbide are different, the difference is sufficiently small so that you can equate phase weight fractions to phase volume fractions.)
c Are the values of yield strength you found in parts (a) and (b) reasonable? Explain. Data for this steel: $G = 8.1 \times 10^{10}$ N/m²; $b = 2.48 \times 10^{-10}$ m.

5.37 Plot schematic free energy-composition diagrams at several temperatures for (a) fcc Al alloyed with Cu, (b) the phase $CuAl_2$, and (c) one of the metastable phases in the Al-Cu system. Show that metastability is associated with an increased solubility of Cu in Al (i.e., show that the solvus lines are displaced to higher copper contents for the metastable phase), and that the metastable phase dissolves at a lower temperature than the equilibrium one.

5.38 Is it necessary that a precipitate be metastable in order for it to be an effective precipitation hardening agent? Rationalize your answer.

5.39 Cu alloy C82500 has the approximate composition 98 wt.% Cu-2 wt.% Be. C82500 can be precipitation hardened. (See the Cu-Be alloy phase diagram shown below.) The precipitate phase (denoted as γ_2 in the phase diagram) is CuBe, which has the CsCl structure which is a simple cubic array with Cu atoms at the unit cell corners and a Be atom in the cell center.

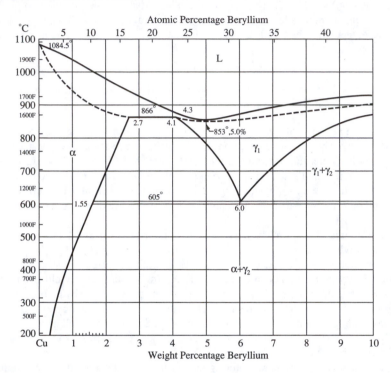

a Describe the steps you would take to precipitation harden C82500. Explicitly state the temperature range that could be used for each treatment. Assume the precipitate volume fraction increases linearly with aging time until the equilibrium volume fraction of precipitate is formed. The purpose of this part of the problem is to calculate the time (t_c) required to precipitate the equilibrium volume fraction. Use the following information and assumptions.

i Assume the precipitate density is constant ($=10^{23}/\text{m}^3$) from the beginning of aging and the initial particle radius is negligibly small.

ii Assume particle growth rate is governed by Be diffusion, for which case the particle radius r varies with aging time, t, as $r^2 \cong Dt$, where $D = D_0 \exp(-Q/RT)$ ($D_0 = 3 \times 10^{-5}$ m²/s; $Q = 165$ kJ/mol). Consider an aging temperature of 643 K.

b Plot precipitate radius as a function of aging time for $t \leq t_c$.

c Assume particle coarsening follows attainment of the equilibrium precipitate volume fraction. During coarsening, precipitate size increases according to

$$r^3 = r^3(t_c) + KDt$$

where t is the time after completion of precipitation, $r(t_c)$ is the particle size at t_c and $K = 2 \times 10^{-9}$ m. Plot precipitate radius vs. time during coarsening on the same plot you used in part (b).

d Plot the bowing and particle shearing stresses as functions of aging time. Take $L' = r(\pi/f)^{1/2}$ for dislocation bowing. Assume the particle shearing stress is due to coherency hardening with $\varepsilon_{coh} = 0.01$. Take $G = 40 \times 10^9$ N/m² and $b = 0.26$ nm.

e Define aging times corresponding to underaging, peak aging, and overaging at 643 K. Does the peak aging time correspond to the transition between precipitation and coarsening? Does it correspond to the transition between particle shearing and bowing?

f The diagram below shows experimental results for aging of C82500 at 643 K. Compare your calculated curve to this one. Specifically relate (i) time to peak hardness, (ii) strength increase at peak hardness, and (iii) shape of the curve for $t < t_c$. Discuss factors that may contribute to differences between experimental and calculated results. (That is, which assumptions used in the calculations are reasonable and which might not be.)

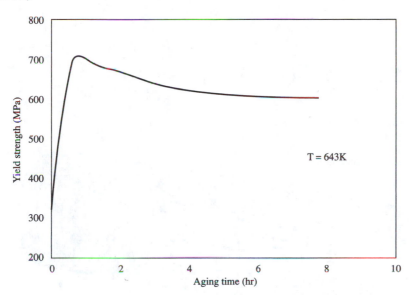

5.40 Consider a microalloyed steel whose structure consists of a ferrite matrix containing 0.2% by volume of NbC particles. The carbide particles are spherical and quite small; their average radius (r) is 10 nm. The NbC particles serve two functions (cf. Fig. 5.38). One is that they restrict ferrite grain growth during processing, providing additional grain-size strengthening. (The grain size (d) of the microalloyed steel varies with r and f as $d = r/f$.) Some carbide particles are also situated within grains. Dislocations bow around them during plastic deformation, providing further strengthening. The effective spacing between particles within the grain is $L' = r(\pi/f)^{1/2}$.

a The yield strength of ferrite with a grain size of 50 μm is 260 MN/m². How much of a strength increase results from grain refinement provided by the carbides?

b How much of a strength increase is due to the particles situated within the grains?

c What is the yield strength of this microalloyed steel? Compare with Fig. 5.38. In your comparison consider not only the total strength of the steel, but also the respective contributions to strength from grain-size refinement and particle hardening.

d Suppose this steel is processed improperly such that the carbide particle size is increased by a factor of two (but f remains equal to 0.002). What is the yield strength of this improperly processed material?

Data for steel: $k_y = 0.60$ MN/m³/²; $G = 8.1 \times 10^{10}$ N/m²; $b = 2.48 \times 10^{-10}$ m.

CHAPTER 6

Composite Materials

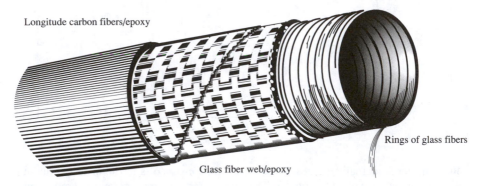

Longitude carbon fibers/epoxy

Rings of glass fibers

Glass fiber web/epoxy

Frontispiece
An advanced vaulting pole used by "elite" vaulters. The pole is hollow and employs three different types of fiber to provide strength and resistance to twisting. Two of the fibers are in the form of composites—the glass fiber web/epoxy and the carbon fiber/epoxy. (The "pole" is hollow. The design is comparable to that which utilizes materials efficiently in the shape of I-beams. That is, solid material is placed preferentially on the surface where the greatest stresses are experienced.) (*From K. E. Easterling, Advanced Materials for Sports Equipment, Chapman and Hall, London, England, 1993, p. 2.*)

6.1
INTRODUCTION

The frontispiece of this chapter illustrates the construction of a modern vaulting pole. It is a hybrid structure, composed of glass fibers and two composites; one composite is a carbon fiber/epoxy mixture and the other is a glass fiber web/epoxy. Sporting goods constitute a significant fraction of the composite materials market.

So does the aerospace industry where, as with sporting goods, high strength and high stiffness coupled with a low density are paramount factors in materials selection. The primary structural material for the Voyager aircraft, for example, was a carbon fiber composite. The Voyager's 40,000-kilometer nonstop flight around the world was possible only because of its light weight.

It is easier to correlate the improved performance of the Voyager to the materials from which it was made than to similarly link the athletic performance of amateurs to the implements they choose to wield. Agility, coordination, and athleticism count more for weekend warriors than any fancy tool they may use in their games. But it should come as no surprise that we ordinary folks plunk down the cash in what is usually a quixotic quest to make us seem better athletes than we actually are. But in the hands of the skilled, the tools of the trade are important. Figure 6.1 is a graph of the winning pole vault heights in the Olympic games; the figure spans a century. Prior to 1900, vaulting poles were made of hickory wood, and in 1896 the gold medal vault height was a trifling 3.3 m. The bamboo pole was introduced in 1904. It can be seen that it had an immediate impact. Bamboo remained the vaulting pole material of choice until about 1952. The continued improved performance between 1904 and 1952 resulted from improvements in technique and greater athletic ability of the vaulters. Aluminum was used as a pole vault material in the 1950s. It resulted in slightly improved performance, but its time in the sun was short. Glass fiber reinforced poles, lighter and stiffer than aluminum, supplanted aluminum in 1964. Use of this new material immediately resulted in improved performance. Further, its implementation permitted vaulters to alter their vaulting technique and this has led to further increases in winning pole vault heights.

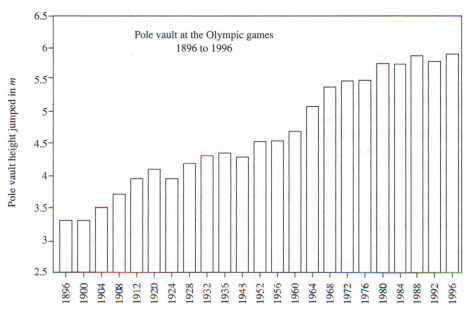

Figure 6.1
Winning pole vault heights at the Olympics from 1896 to 1996. Increased winning heights often correlate with improved vaulting materials, as discussed in the text. *(From F. H. (Sam) Froes, MRS Bulletin, vol. 23, No. 3, 1998, p. 32.)*

What makes a good material for a vaulting pole? First it must be light enough to be carried by a human being running at a rapid pace. Second, it must resist twisting. (Twisting of the pole reduces its "lift" efficiency. The glass fiber rings provide twisting resistance to the pole displayed in the frontispiece.) But, most important, the pole must store energy. A vaulting pole is an elastic spring, and a measure of its potential in this regard is the maximum elastic energy per unit volume that can be stored in it; this energy is $\sigma^2/2E$ where σ is the material strength and E its modulus. On this basis, steel would make a dandy vaulting pole. But we must also take into account weight. When this is done, the pole material *performance index* is $\sigma^2/2E\rho$ where ρ is the material density. Problem 6.1 lets the reader check for herself the performance indices of a number of possible vaulting pole materials.

Now we must get more technical. Unlike the situation for the two-phase materials considered in Chap. 5, the strengths of composite materials are usually not defined by microscopic interactions between phases. Rather, composite strengths, as well as their other mechanical characteristics, are an appropriate average of the properties of the materials comprising the composite. *Fiber composite materials* are especially useful composites because they contain a reinforcing phase in which high tensile strengths are realized when they are processed to fine filaments, wires, or whiskers (hereinafter called fibers). Such fibers typically have diameters ranging from ca. 1 μm to 0.025 cm. The inherently high fiber strengths are effectively utilized in composites by incorporating the fibers into a suitable matrix. In such fiber composites, the matrix "glues" the fibers together and also transfers stress to them. In many cases, the matrix also protects the fiber from deleterious interactions with the environment, such as oxidation or corrosion. Thus, many fiber-reinforced materials consist of a high-strength material imbedded in a much less strong matrix that serves the purposes described above.

Strong, yet brittle, materials can also be reinforced if materials of reasonable ductility are incorporated within them. In this case the term "reinforced" is something of a misnomer, for the function of the added phase is to toughen the matrix; i.e., this phase serves to resist brittle crack propagation rather than to improve material strength.

As mentioned, in most cases composite strengths are unrelated to microscopic effects, because the phase "scale" of most composites is large compared to that at which such effects become important. However, in some cases both the size and separation of the reinforcing phase is on the order of micrometers or less. When this is so, description of composite mechanical behavior must also consider aspects of deformation discussed in Chap. 5.

Fiber composites come in various forms. For example, the fibers can be essentially continuous; that is, their lengths are as long as the dimension of the component of which they are a part. Discontinuous fibers (e.g., glass fibers) can provide significant composite strengthening although, as discussed later, the strengths of these types of fiber composites are always less (but often only slightly less) than those containing continuous fibers. Fiber orientation can also be varied. Some composites have the fiber axes aligned along one direction only. This produces excellent reinforcement in this direction, but yields inferior properties in other directions. This anisotropy can be partially reduced by producing plate-like arrangements in which the fiber axis directions differ in different plates of the arrangement.

The reinforcing phase need not be in the shape of a fiber to impart desired properties. Plate-like reinforcements provide strengthening comparable to that of fibers and, in fact, plate-like reinforcements generally result in greater composite fracture toughnesses. Reinforcements also come in particulate shape. While not as efficient as fibers in improving stiffness, strength, and toughness, particle-reinforced composites are usually less costly and also display isotropic properties.

Different types of materials can be used as composite matrices. Thus, we have *polymer matrix composites* (PMCs), *metal matrix composites* (MMCs) and *ceramic matrix composites* (CMCs). Polymer matrix composites are most common. Examples include ordinary "fiber glass" and many composites used as sporting goods. "Graphite" golf club shafts, for example, are not graphite but a PMC reinforced with graphite fibers. The primary drawback to PMCs is the low maximum temperature at which they can be used. Higher temperatures call for MMCs or CMCs. Metal matrix composites are typically reinforced by strong fibers or particles. Silicon carbide, for example, can be used in either form to reinforce metals. While MMC strengths exceed those of their matrix, their toughnesses usually do not. Ceramic matrix composites are primarily employed because they are more fracture tough than the matrix on which they are based. While CMCs are discussed in this chapter, discussion of toughening in them is deferred until Chap. 10.

There are a large number of potential reinforcement materials. Glass fibers, as noted, are commonly used for this purpose, particularly in PMCs. Ceramic materials—having melting temperatures much higher than the temperatures at which glass fibers soften—are used in MMCs and also in CMCs. Some discussion of the various types of fibers employed in modern composites is provided in Sect. 6.11.

6.2
BASIC PRINCIPLES OF REINFORCEMENT

The appropriate "average" of the individual phase properties to be used in describing composite tensile behavior can be elucidated with reference to Fig. 6.2. Although this figure illustrates a plate-like composite, the results that follow are equally applicable to fiber composites having similar phase arrangements. The two-phase material of Fig. 6.2 consists of N slabs (lamellae) of α and β phases of thickness l_α and l_β, respectively. Thus, the volume fractions (V_α, V_β) of the phases are

$$V_\alpha = \frac{l_\alpha}{(l_\alpha + l_\beta)} \qquad V_\beta = \frac{l_\beta}{(l_\alpha + l_\beta)} \tag{6.1}$$

In Fig. 6.2a, a tensile force F is applied normal to the broad faces (dimensions $L \times L$) of the phases. In this arrangement the stress borne by each of the phases ($= F/L^2$) is the same, but the strains they experience are different. Provided l is not so small in comparison to L that the "brazed joint" effect (cf. Chap. 5) must be considered, we can assume that the strains in both phases (ε_α, ε_β) are equivalent to those that would be found in tension tests of them individually. Thus, the elongation ($= \Delta l_\alpha$) in one slab of the phase α is

$$\Delta l_\alpha = \varepsilon_\alpha l_\alpha \tag{6.2}$$

with a corresponding expression for Δl_β. The total elongation of the composite, Δl_c, is obtained as

$$\Delta l_c = N\Delta l_\alpha + N\Delta l_\beta \tag{6.3}$$

and the composite strain (ε_c) is

$$\varepsilon_c = \frac{\Delta l_c}{N(l_\alpha + l_\beta)} = \frac{\Delta l_\alpha + \Delta l_\beta}{l_\alpha + l_\beta} \tag{6.4}$$

Substitution of Eqs. (6.1) and (6.2) into Eq. (6.4) gives

$$\varepsilon_c = V_\alpha \varepsilon_\alpha + V_\beta \varepsilon_\beta \tag{6.5}$$

That is, the composite strain is a volumetric weighted average of the strains of the individual phases.

If the deformation in both phases is elastic (i.e., $\varepsilon_\alpha = \sigma_\alpha/E_\alpha$, etc.), Eq. (6.5) becomes

$$\varepsilon_c = \sigma\left(\frac{V_\alpha}{E_\alpha} + \frac{V_\beta}{E_\beta}\right) \tag{6.6}$$

and the composite modulus E_c ($= \sigma/\varepsilon_c$) is found from

$$\frac{1}{E_c} = \frac{V_\alpha}{E_\alpha} + \frac{V_\beta}{E_\beta} \tag{6.7}$$

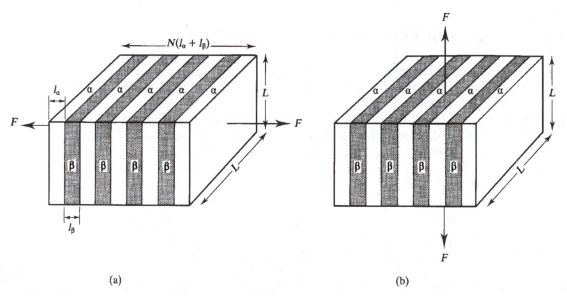

(a)   (b)

Figure 6.2
Lamellar arrangements of two phases (α and β) having different properties. In (a) a tensile force is applied in a direction such that the stress borne by both phases is the same, but they experience different strains. In (b), the applied force results in the phases experiencing the same strain but different stresses. This is the arrangement that best utilizes the properties of the stronger phase.

or

$$E_c = \frac{E_\alpha E_\beta}{(V_\alpha E_\beta + V_\beta E_\alpha)} \qquad (6.8)$$

Equation (6.5) is one form of a volume-fraction rule, and is appropriate for a phase geometry in which both phases experience equal stress. However, in fiber composites in which fibers are aligned parallel to the tensile axis the arrangement of the phases is analogous to that illustrated in Fig. 6.2b. Here the strains in both phases are equal (and the same as the composite strain), but the external force is partitioned unequally between the phases. The division of force can be determined by realizing that the sum of the forces $(F_\alpha + F_\beta)$ borne by the individual phases is equal to the total force F. Each α lamellae carries a stress σ_α; thus F_α is obtained as

$$F_\alpha = \sigma_\alpha N l_\alpha L \qquad (6.9)$$

with a corresponding expression for F_β. Hence

$$F = NL(\sigma_\alpha l_\alpha + \sigma_\beta l_\beta) \qquad (6.10)$$

Division of Eq. (6.10) by the cross-sectional area $(= NL(l_\alpha + l_\beta))$ transverse to the applied force defines the composite stress:

$$\sigma_c = \frac{\sigma_\alpha l_\alpha}{(l_\alpha + l_\beta)} + \frac{\sigma_\beta l_\beta}{(l_\alpha + l_\beta)} = \sigma_\alpha V_\alpha + \sigma_\beta V_\beta \qquad (6.11)$$

The composite modulus for the equal strain condition is obtained by substituting into Eq. (6.11) appropriate terms; i.e., $\sigma_c = E_c \varepsilon$ with corresponding terms for σ_α and σ_β, keeping in minds that the strains of the two phases are equal and equal also to the composite strain. On doing this, we find

$$E_c = V_\alpha E_\alpha + V_\beta E_\beta \qquad (6.12)$$

Equation (6.11) (and also Eq. (6.12)) are another form of a volume-fraction rule. It can be shown that the moduli of *any* two-phase material lies between those given by Eqs. (6.8) and (6.12); Eq. (6.8) represents a lower bound, and Eq. (6.12) an upper bound, on the modulus. The variation of elastic modulus with volume fraction of α phase (presumed stiffer in this instance) is shown in Fig. 6.3 for a specific ratio (= 3) of the α and β phase moduli. It is clear that the equal-strain condition provides more effective reinforcement (the upper-bound modulus) than does the equal-stress condition (the lower-bound modulus). As discussed in the following section, the same applies to composite strength.

That the equal-strain condition is most useful for reinforcement can be illustrated by further consideration of elastic deformation for this situation. The ratio of the forces carried by the α and β phases in the elastic region is $V_\alpha E_\alpha / V_\beta E_\beta$ (cf. Eq. (6.9) and Prob. 6.3). This ratio is plotted in Fig. 6.4 as a function of V_α (α is again presumed the stiffer phase) for several values of E_α / E_β and compared to the equal stress condition (for which $F_\alpha / F_\beta = 1$). It can be seen that, provided E_α is sufficiently large compared to E_β, the α phase is utilized more efficiently in the equal-strain condition. However, as we shall see, for this to be so a certain volume fraction of the stronger phase must be present in the composite. This is the situation for most useful composite materials. Before proceeding to discuss reinforcement

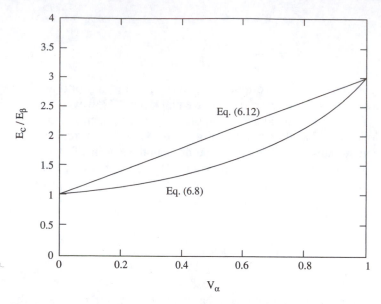

Figure 6.3
A graph of the upper- (Eq. (6.12)) and lower-bound modulus (Eq. (6.8)) of a composite as a function of the volume fraction of the stiffer (α) phase in it. The ordinate is the ratio of the composite modulus to that of the phase (β) having the lesser modulus, and is constructed for a specific value ($E_\alpha/E_\beta=3$) of the ratio of the modulus of the two phases. The equal-strain condition, which leads to the upper-bound modulus, clearly more effectively utilizes the greater stiffness of the α phase.

with fibers—which often gives rise to the equal-strain condition—some discussion of particle-reinforced composites is called for. They, too, are useful engineering materials.

6.3
PARTICLE REINFORCEMENT

Moduli of particle-reinforced materials generally lie between the values predicted by Eqs. (6.8) and (6.12). Empirical equations are often used to represent this situation. Thus, moduli of particle-reinforced composites can be expressed by a variant of Eq. (6.12); i.e.,

$$E_c = V_m E_m + K_c V_p E_p \qquad (6.13)$$

where $E_{p,m}$ and $V_{p,m}$ are the particle and matrix modulus and volume fraction, respectively. In Eq. (6.13), K_c is empirically determined. Its value is less than one, reflecting that particle-reinforced composites are not characterized by the equal-strain condition. An equation similar to Eq. (6.13) can be used to describe the tensile strengths of particle-reinforced composites. The equation is a variant of the equal-strain-condition equation applying to composite strength (Eq. (6.11)). Thus,

$$(T.S.)_c = V_m (T.S.)_m + K_s V_p (T.S.)_p \qquad (6.14)$$

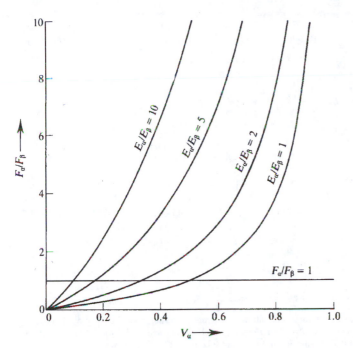

Figure 6.4
The ratio of the force carried by the (strong) α phase to that carried by the β phase for the geometries of Fig. 6.2. For the equal-stress condition (Fig. 6.2a), $F_\alpha/F_\beta = 1$, and the force ratio is independent of V_α, the α-phase volume fraction. For the equal-strain condition, F_α/F_β increases with both V_α and the modulus ratio (E_α/E_β). The results shown here hold for linear elastic deformation.

where $(T.S.)_{p,m}$ are the respective particle and matrix tensile strengths and K_s is an empirically determined constant. (We note that K_s and K_c generally do not have the same value.) Equations (6.13) and (6.14) generally well describe composite modulus and strength, but usually only over a limited range of particle volume fraction.

6.4
REINFORCEMENT WITH ALIGNED CONTINUOUS FIBERS

As mentioned, various phase arrangements (fiber, particulate, and plate) are found in composite materials. However, the most widely utilized phase geometry is one in which the reinforcing, strong phase is in the form of a fiber. In part this is due to the extraordinary strengths attainable in many materials in fibrous form. Table 6.1 (see also Table 8.5) lists the properties of a number of nonmetallic and metallic fibers. (Further discussion of the structure and processing of these fibers is given in Sect. 6.11.) Nonmetallic fibers demonstrate not only high strengths but also very high strength-to-density ratios; these characteristics render them attractive in aerospace applications. Further, the strengths realizable in a specific fiber material have, as a result of improved production techniques, increased substantially in the last several decades. It is expected that such quality improvements will continue.

Material class	Material	E (GN/m^2)	T.S. (GN/m^2)	ρ (Mg/m^3)	E/ρ (MNm/kg)	T.S./ρ (MNm/kg)
Metals	Be	315	1.3	1.8	175	0.72
	Pearlitic steel	210	4.2	7.9	27	0.53
	Stainless steel	203	2.1	7.9	26	0.27
	Mo	343	2.1	10.3	33	0.20
	β-Ti	119	2.3	4.6	26	0.50
	W	350	3.9	19.3	18	0.20
Ceramics	Al_2O_3	380–480	1.4–2.4	3.9–4.0	95–123	0.35–0.62
	Al_2O_3 whiskers	300–1500	2–20	3.3–3.9	77–455	0.51–6.1
	B	386–400	3.1–7.0	2.6	148–154	1.2–2.7
	BN	90	1.4	1.9	47	0.74
	Graphite whiskers	700	20	2.2	318	9.1
	Graphite	390–490	1.5–4.8	1.95–2.2	177–251	0.68–2.5
	E Glass	72–76	3.5	2.55	28–30	1.4
	S Glass	72	6	2.5	29	2.4
	SiC	380–400	2.4–3.9	2.7–3.4	112–148	0.71–1.4
	SiC whiskers	400–700	3–20	3.2	125–219	0.94–6.3
	Si_3N_4	380	5–7	3.2–3.8	100–119	1.3–2.2
Polymer	Kevlar	133	2.8–3.6	1.4–1.5	89–95	1.9–2.6

Notes: Variations in properties for a given material result from different processing conditions employed to manufacture them. *Data from*: (1) *Modern Composite Materials*, ed. L. J. Broutman and R. H. Krock, Addison-Wesley, Reading, Mass., 1967, articles of P. T. B. Shaeffer (p. 197), J. A. Roberts (p. 228), F. E. Wawner, Jr., (p. 244). (2) J. D. Embury, in *Strengthening Methods in Crystals*, ed. A. Kelly and R. B. Nicholson, Wiley, New York, 1971, p. 331. (3) *Metal Matrix Composites: Processing and Interfaces*, ed. R. K. Everett and R. J. Arsenault, *Treat. Matls. Sc. and Tech.*, Academic Press, San Diego, 1991, articles by W. C. Harrigan, Jr. (p. 1) and R. B. Bhagat (p. 43). (4) D. Hull, *Introduction to Composite Materials*, Cambridge University Press, Cambridge, England, 1981.

Table 6.1
Properties of Selected Fibers and Whiskers

Useful fiber composites require not only high-strength fibers but a technique to incorporate them into a (usually much weaker) matrix. Incorporation of fibers into polymeric matrices (typically epoxy or polyester resins, but thermoplastic-matrix composites are now under development) is comparatively inexpensive vis-à-vis incorporation into metallic or ceramic matrices. Polymeric matrix composites, however, are limited in their temperature use owing to degradation of properties and/or polymer decomposition, whereas this is not as true for nonmetallic matrices. Nonetheless, the economics of production is such that current use of PMCs in recreational (e.g., golf clubs) and aerospace applications far exceeds that of MMCs and CMCs.

The deformation behavior of aligned fiber-reinforced composites is frequently adequately described by a VFR appropriate to the equal-strain condition (Eq. (6.11)). For fiber composites we alter the subscripts of Eq. (6.11) so that α, the "hard" phase, corresponds to the fiber (f) and the softer phase corresponds to the matrix (m). Thus, for a fiber-reinforced material with a tensile force applied parallel to the long fiber axis (Fig. 6.5):

$$\sigma_c = V_f\sigma_f + V_m\sigma_m \tag{6.15}$$

Implicit in Eq. (6.15) is that the stress-strain behavior of the composite constituents (the fiber and the matrix) are ascertained by testing them individually.

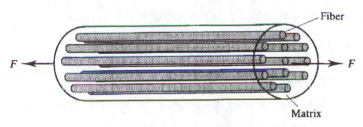

Fiber

F ← → F

Matrix

Figure 6.5
A fiber composite consisting of fibers imbedded in a matrix. When the force is applied in the direction shown, the equal-strain condition holds.

The uniaxial stress-strain response of a fiber composite can be divided into several stages. In the first (Stage I), the strain is small and both the fiber and matrix deform elastically. If this strain is linearly elastic, we have

$$\sigma_c = E_c \varepsilon_c = \varepsilon_c [V_f E_f + V_m E_m] \qquad \text{(Stage I)} \qquad (6.16)$$

In some fiber-reinforced materials (particularly those with a metal matrix), the matrix deforms permanently at a strain at which the fiber remains elastic. This constitutes the onset of Stage II deformation, for which

$$\sigma_c = V_f E_f \varepsilon_c + V_m \sigma_m(\varepsilon_c) \qquad \text{(Stage II)} \qquad (6.17)$$

where, as mentioned, $\sigma_m(\varepsilon_c)$ is assumed to be the stress carried by the matrix (at strain ε_c) as determined from a tensile test of the matrix. The Stage II "modulus," E'_c, is defined as the instantaneous slope of the composite stress-strain curve during Stage II deformation. That is,

$$E'_c = \frac{d\sigma_c}{d\varepsilon_c} = V_f E_f + V_m \left(\frac{d\sigma_m}{d\varepsilon_c}\right) \qquad \text{(Stage II)} \qquad (6.18)$$

In most cases the second term of Eq. (6.18) is much less than the first, so that

$$E'_c \cong V_f E_f \qquad (6.19)$$

Although $d\sigma_m/d\varepsilon_c$ is presumed to be the slope of the stress-strain curve of the matrix tested by itself, this is not always the case during Stage II. This is due to the incompatibility of the lateral matrix and fiber deformation strains during this stage. In the absence of fiber-matrix interactions, the lateral strain is $-\nu\varepsilon_c$ for the fiber (where ν is Poisson's ratio of the fiber) and $-\varepsilon_c/2$ for the plastically deforming matrix. Compatibility of lateral deformation is provided by introduction of internal stresses. These affect the stress state of both fiber and matrix. The magnitude of the internal stress depends directly on the incompatibility that would exist in the unconstrained state; i.e., the stress is proportional to $(½ - \nu)$. Thus, the matrix work-hardening coefficient in Stage II is that of a constrained matrix, and is invariably greater than that for the unconstrained matrix. The adequacy of the approximation of Eq. (6.19) depends on the value of $V_f E_f$ relative to the second term of Eq. (6.18). Provided V_f is sufficiently large, Eq. (6.19) remains a reasonable approximation for the secondary modulus.

Many high-strength fibers do not deform permanently prior to fracture. Thus, the tensile strengths of composites containing them are usually found in Stage II. On the other hand, metallic fibers usually deform plastically before fracture, and composites containing permanently deforming fibers exhibit a third stage in their tensile curve. During this Stage III, the VFR is expressed as

$$\sigma_c(\varepsilon_c) = V_f\sigma_f(\varepsilon_c) + V_m\sigma_m(\varepsilon_c) \tag{6.20}$$

where $\sigma_f(\varepsilon_c)$ and $\sigma_m(\varepsilon_c)$ are the respective fiber and matrix flow stresses at the composite strain ε_c. During Stage III deformation the lateral strains in the constituents are comparable (the effective plastic Poisson's ratio for each is 1/2), and thus the constraints present in Stage II deformation are absent. This is true at least up to and near the composite tensile strength. As the necking strain of one of the phases is approached, the other phase acts on it in a manner so as to delay necking. Thus, the uniform or necking strain of the composite lies between the necking strains of the composite constituents. For example, if $V_f(T.S.)_f$ is much greater than $V_m(T.S.)_m$, composite necking strain is very nearly the fiber necking strain. When the converse holds, the composite necking strain is approximately that of the matrix. When neither of the above conditions apply, the composite necking strain is intermediate to that of the matrix and the fiber.

The stages of composite behavior just described are illustrated schematically in Figs. 6.6a and b. Figure 6.6a is appropriate when only the first two stages of composite deformation are observed, and Fig. 6.6b illustrates composite stress-strain behavior manifesting all three stages. The fracture event illustrated in Fig. 6.6a is

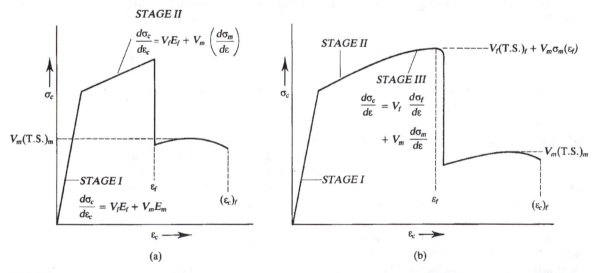

(a) (b)

Figure 6.6
Composite stress-strain curves predicted by the volume fraction rule. Stages I and II are shown in both (a) and (b). In Stage I, the fiber and matrix deform elastically; the composite modulus is the volume fraction-weighted modulus of the phases. In Stage II the matrix deforms plastically and the fiber elastically. Thus, the "secondary" modulus, i.e., the slope of the stress-strain curve, is reduced. In (b), a third stage, in which both the matrix and the fiber deform plastically, is represented. The fiber failure strain (ε_f) is taken as less than that of the matrix. Matrix failure is not necessarily concurrent with fiber failure. Thus, a secondary tensile strength ($V_m(T.S.)_m$) is observed.

appropriate when all of the fibers fail at the same strain. This is not always the case; composite failure under these circumstances is discussed further in Sect. 6.7. As is also shown in Fig. 6.6a, composite failure in a tension test is not necessarily concurrent with fiber fracture; that is, although for the composite whose behavior is illustrated in Fig. 6.6a composite tensile strength is simultaneous with fiber fracture, the matrix remains able to sustain load and to deform subsequent to fiber fracture. Thus, a "secondary" tensile strength $(= V_m(\text{T.S.})_m)$ is observed.

Provided composite tensile strength is coincident with fiber failure, this strength is expressed as

$$(\text{T.S.})_c = V_f(\text{T.S.})_f + V_m\sigma_m(\varepsilon_f) \tag{6.21}$$

where $(\text{T.S.})_f$ is the fiber tensile strength and $\sigma_m(\varepsilon_f)$ is the matrix flow stress at the fiber failure strain ε_f. The VFR applied to composite tensile strength, as given by Eq. (6.21), is shown by the curve marked (a) in Fig. 6.7 which plots $(\text{T.S.})_c$ vs. V_f. On the other hand, if V_f is low, the secondary tensile strength may be greater than that given by Eq. (6.21). This possibility is illustrated in Fig. 6.8, which shows composite stress-strain curves for both "low" and "high" values of V_f. The secondary tensile strength is plotted against V_f as the curve marked (b) in Fig. 6.7. As can be seen, the secondary strength exceeds that predicted by Eq. (6.21) if V_f is less than some critical value V_c. Thus, composite tensile strength is given as

$$(\text{T.S.})_c = V_m(\text{T.S.})_m \qquad V_f \leq V_c \tag{6.22a}$$

or

$$(\text{T.S.})_c = V_f(\text{T.S.})_f + V_m\sigma_m(\varepsilon_f) \qquad V_f \geq V_c \tag{6.22b}$$

The critical volume fraction is found by equating Eqs. (6.22a) and (6.22b); thus,

$$V_c = \frac{[(\text{T.S.})_m - \sigma_m(\varepsilon_f)]}{[(\text{T.S.})_f + (\text{T.S.})_m - \sigma_m(\varepsilon_f)]} \tag{6.23}$$

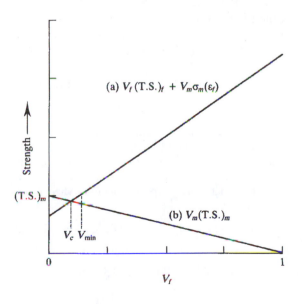

Figure 6.7
Composite tensile strength as predicted by the volume fraction rule (curve (a)) and the "secondary" tensile strength (curve (b)) as a function of fiber volume fraction. The observed tensile strength is the greater of these two strengths. For $V_f < V_c$, the secondary tensile strength exceeds the VFR strength and composite strength decreases with V_f. For composite tensile strength to exceed that of the matrix, a minimum volume fraction (V_{min}) of fibers must be present.

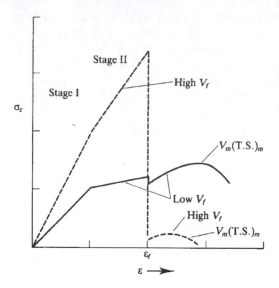

Figure 6.8
Composite stress-strain curves (Stages I and II) for two different fiber volume fractions. The curve marked "low V_f" is appropriate for $V_f < V_c$, and composite tensile strength is the secondary tensile strength. For the curve marked "high V_f" ($V_f > V_c$), composite tensile strength is given by the VFR.

As can also be seen in Fig. 6.7, composite tensile strength exceeds that of the matrix only if the value of Eq. (6.21) exceeds (T.S.)$_m$. Thus, a certain *minimum* fiber volume fraction, V_{min}, must be had if composite tensile strength is to exceed that of the matrix. The value of V_{min} (Prob. 6.5) is

$$V_{min} = \frac{[(\text{T.S.})_m - \sigma_m(\varepsilon_f)]}{[(\text{T.S.})_f - \sigma_m(\varepsilon_f)]} \tag{6.24}$$

For typical fiber-matrix combinations, both V_c and V_{min} are quite low (Prob. 6.6), and these volume fractions are greatly exceeded in most fiber composites. Thus, Eq. (6.21) provides a reasonable approximation to the tensile strengths of aligned fiber composites containing continuous fibers.

The discussion of this section pertains to fiber composites for which the applied stress is parallel to the long axis of the fibers. When this is not the situation, composite strengths are less—sometimes considerably less—than indicated by the treatment of this section. Aspects of such "off-axis" loading are discussed in Sect. 6.6. Before we do that, though, we consider the tensile behavior of fiber composites comprised of discontinuous fibers having their axes aligned parallel with the stress axis. And even before that, some example problems!

EXAMPLE PROBLEM 6.1. A sheet of graphite-fiber reinforced epoxy contains 30 vol.% aligned, continuous graphite fibers; for the fibers, $E = 490 \times 10^9$ N/m² and $E = 5 \times 10^9$ N/m² for the epoxy. What is the modulus of the composite measured with an applied stress parallel to the fiber axis? What is the composite modulus when the applied stress is perpendicular to the fiber axis?

Solution. When the stress is parallel to the axis of the fiber, we have an equal-strain condition. Call this modulus $E_\parallel$.

$$E_\parallel = V_f E_f + V_m E_m = (0.3)490 \text{ GPa} + (0.7)(5 \text{ GPa}) = 150.5 \text{ GPa}$$

When the stress is perpendicular to the fiber axis, we have an equal-stress condition. Call this modulus $E_\perp$.

$$E_\perp = \frac{E_f E_m}{(V_f E_m + V_m E_f)} = \frac{(490)(5)}{(0.3 \times 5 + 0.7 \times 490)} = 7.1 \text{ GPa}$$

EXAMPLE PROBLEM 6.2. Consider a composite containing 30 vol.% of aligned, continuous fibers having a tensile strength of 3200 MPa imbedded in a matrix with a tensile strength of 100 MPa.
a. If the matrix and the fiber both fail at the same strain, what is the composite tensile strength?
b. If the failure strain of the matrix is 50% of the failure strain of the fiber, what is the composite tensile strength?

Solution. **a.** With the tensile axis parallel to the fiber axis, and with the fiber and matrix failing at the same strain, the composite tensile strength is the volume weighted average of the strengths of the composite constituents. (Also see sketch of stress-strain behavior below.)

$$(\text{T.S.})_c = V_f(\text{T.S.})_f + V_m(\text{T.S.})_m = 0.3(3200 \text{ MPa}) + 0.7(100 \text{ MPa}) = 1030 \text{ MPa}$$

b. With the matrix failing before the fiber does, the composite tensile strength is now only that resulting from the fiber. (Sketch of stress-strain behavior is provided below.)

$$(\text{T.S.})_c = V_f(\text{T.S.})_f = 0.3(3200 \text{ MPa}) = 960 \text{ MPa}$$

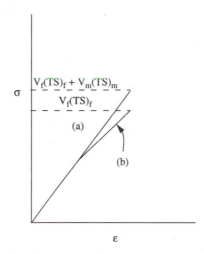

6.5
REINFORCEMENT WITH DISCONTINUOUS FIBERS

The equal-strain condition, for which the VFR equations of the previous section hold, applies to composites containing continuous fibers. This situation does not hold if the fibers are discontinuous. This can be demonstrated by considering Fig. 6.9. Figure 6.9a illustrates a composite containing discontinuous fibers, and a region of this material containing but one fiber is shown in Fig. 6.9b. The stress-strain state at the fiber end cannot correspond to the equal tensile strain condition. This can be visualized by regarding a volume of matrix near the fiber end and lying along an extension of the fiber axis. This strained matrix region cannot "instantaneously" transfer tensile load, in an amount equivalent to that carried by continuous fibers, to the fiber at its end. Instead, tensile force is transmitted from the matrix to the fiber by means of shear stresses that develop along the fiber-matrix interface (Fig. 6.9c).[1] The tensile force borne by the matrix deforms it and the matrix is displaced relative to the fiber along the interface. The relative displacement is zero at the fiber mid-point and a maximum at its ends. This results in an interfacial shear stress, greatest at the fiber ends, that transmits tensile load to the fiber. A force balance (Fig. 6.9d) allows a relationship between the interfacial shear stress and fiber tensile stress to be developed.

$$\tau_m (\pi d_f)\, dx = \left(\frac{\pi d_f^2}{4} \right) d\sigma_f \qquad (6.25a)$$

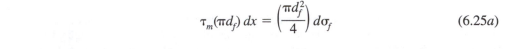

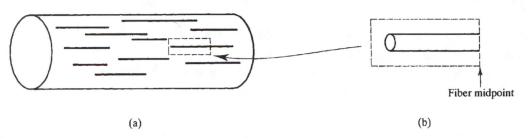

(a)   (b)

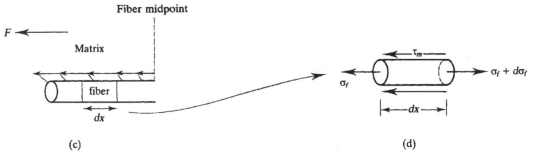

(c)   (d)

Figure 6.9
(a) A schematic of a matrix containing discontinuous fibers. (b) The geometry of one fiber is shown in the cross-hatched region. At the fiber end, tensile load cannot be transferred instantaneously from the matrix to the fiber. (c) Tensile load transfer is accomplished by development of a shear stress at the fiber-matrix interface owing to the relative displacement of the fiber and matrix along this interface. The displacement is proportional to the arrows shown, and is zero at the fiber mid-point and a maximum at the fiber end. (d) A small increment of length dx of a fiber; the incremental fiber tensile stress ($d\sigma_f$) is obtained by a force balance; i.e., $(\pi d_f^2/4)\, d\sigma_f = \tau_m (\pi d_f\, dx)$, where τ_m is the interfacial shear stress.

[1]The discussion here can be physically visualized by a "hands on" experiment. Grasp the index finger of one of your hands with the other. The grasping hand can be considered the matrix; the index finger, the fiber. Note that when you attempt to extract your index finger from the hand grasping it, you sense a tensile force on your finger. This force develops by transmission of the sliding friction force between your index finger and your grasping hand.

or

$$\frac{d\sigma_f}{dx} = \frac{4\tau_m}{d_f} \qquad (6.25b)$$

259

SECTION 6.5
Reinforcement
with
Discontinuous
Fibers

where σ_f is the (position specific) tensile stress carried by the fiber, τ_m the shear stress (also position specific), d_f the fiber diameter, and x the distance from the fiber end.

The resulting variation of τ_m and σ_f with distance along the fiber is illustrated schematically in Fig. 6.10 for the situation where the fiber and matrix deform elastically. The fiber tensile stress is a maximum at the fiber mid-point and zero at the fiber ends.

The elastic analysis shows that after only a very small strain, the magnitude of the shear stress developed at the fiber end becomes large. So large, in fact, that either fiber-matrix delamination (which occurs typically in PMCs) takes place or, if the fiber-matrix bond is strong (as it is in many MMCs), plastic shear of the matrix occurs. The latter happens even when the bulk of the material is, on the basis of the applied load, expected only to elastically deform. We first consider the case in which the matrix deforms plastically along the interface. This allows the mechanism of stress transfer in a discontinuous fiber composite to be illustrated without undue mathematical complexity. At the same time, the analysis displays the basics of the stress-transfer process irrespective of the nature of the matrix.

If the matrix is considered ideally plastic (i.e., it does not work harden), the variation of the interfacial shear stress with distance from the fiber end is as illustrated in Fig. 6.11a. Only over a small portion of the interface does the value of τ_m differ from either the matrix shear yield strength, τ_{my}, or from zero. We neglect this intermediate region in our analysis. The variation of tensile fiber stress along the fiber length is shown in Fig. 6.11b. As noted, this is a maximum at the fiber mid-point, and the maximum fiber stress attained depends on fiber length as shown in Fig. 6.12. Provided this length exceeds a critical value, the fiber mid-point and regions adjacent experience a tensile stress corresponding to the equal-strain condition. That is, the fiber tensile stress at these positions is the same as would be found for a continuous fiber. (For example, for an elastic fiber in a composite strained to

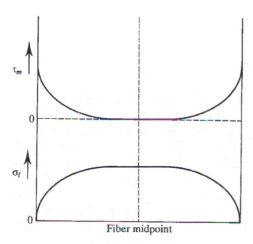

Figure 6.10

Variation of τ_m and σ_f with position along the fiber when the matrix (and fiber) deform elastically; τ_m is zero at the fiber mid-point and a maximum at its end. The reverse holds for σ_f.

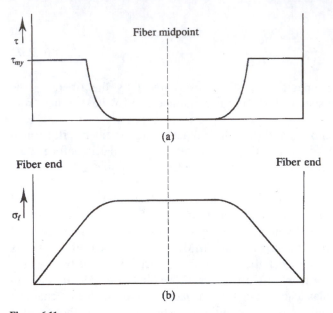

Figure 6.11
When the matrix plastic yield stress is exceeded, τ_m attains a constant value τ_{my} (assuming the matrix does not work harden) near the fiber end. Curves (a) and (b) show the variation of τ_m and σ_f for this situation. For the most part, τ_m is constant ($= \tau_{my}$) or zero; this means that σ_f increases approximately linearly from the fiber end and reaches a fixed value at the position where τ_m goes to zero.

ε_c, the stress carried by the fiber at its midpoint is $E_f\varepsilon_c$.) This critical length, l_c, can be ascertained via inspection of Fig. 6.12; it is determined by setting the stress at the mid-point ($x = l_c/2$) equal to $\sigma_f(\varepsilon_c)$, i.e.,

$$\sigma_f(\varepsilon_c) = 2\frac{\tau_{my}l_c}{d_f} \qquad (6.26a)$$

or

$$\frac{l_c}{d_f} = \frac{\sigma_f(\varepsilon_c)}{2\tau_{my}} \qquad (6.26b)$$

The ratio, l_c/d_f, is called the "critical aspect ratio" and, according to Eq. (6.26b), it increases with composite strain (and stress), since $\sigma_f(\varepsilon_c)$ does likewise. Thus, as shown in Fig. 6.13, a fiber of a certain length may have its mid-point stressed to the "equal-strain" condition at a low composite stress, but this will not be so at a higher stress level. Hence, for the mid-point of a fiber to be stressed to the equal-strain condition at composite fracture, its length must be at least $d_f(\text{T.S.})_f/2\tau_{my}$.

So what is the average tensile stress experienced by a fiber in a composite containing discontinuous fibers? We first consider the case in which the fiber lengths exceed l_c. In a given transverse cross-sectional area, some fibers (those whose ends

261

SECTION 6.5
Reinforcement
with
Discontinuous
Fibers

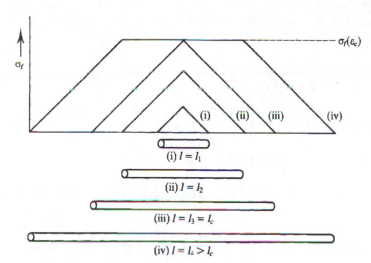

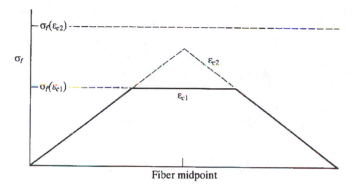

Figure 6.12
Whether the fiber mid-point carries the stress it would if it were a continuous fiber depends on its length. If the fiber length (e.g., l_1, l_2) is less than a critical length l_c, then σ_f does not reach this value. When the fiber length is greater than l_c (e.g., l_4), it does. In this diagram, $l_3 = l_c$.

Figure 6.13
Fiber tensile stress as a function of position along the fiber for two different strains. At the lower strain (ε_{c1}), the fiber mid-point and adjacent regions carry the same stress as would a continuous fiber. At the higher strain (ε_{c2}), this is not the case. Thus, the fiber critical length is a function of ε_c, and increases with it. (If the fiber were elastic, the increase in l_c with ε_c would be linear.)

are greater than a distance $l_c/2$ from the given plane) will be carrying the stress $\sigma_f(\varepsilon_c)$. However, those whose ends are within $\pm l_c/2$ from the plane will be experiencing a stress less than this. Thus, the average fiber stress is less than $\sigma_f(\varepsilon_c)$. It can be shown that the appropriate average stress is the same as the average stress borne by one fiber if the transverse section is displaced axially along its length (Fig. 6.14a). The fraction of the fiber length carrying stress $\sigma_f(\varepsilon_c)$ is $(l - l_c)/l$.

The remaining fraction (l_c/l) bears an average stress $\sigma_f(\varepsilon_c)/2$. Thus, the average stress $(\overline{\sigma}_f)$ carried by the fiber is

$$\overline{\sigma}_f = \sigma_f(\varepsilon_c)\left[1 - \left(\frac{l_c}{l}\right)\right] + \frac{1}{2}\sigma_f(\varepsilon_c)\left(\frac{l_c}{l}\right) = \sigma_f(\varepsilon_c)\left[1 - \left(\frac{l_c}{2l}\right)\right] \quad l \geq l_c \quad (6.27)$$

For fibers with $l < l_c$, the average stress (cf. Fig. 6.14b) is $\sigma_{max}/2$ with $\sigma_{max} = 2\tau_{my}l/d_f$. On substituting for $2\tau_{my}$ in Eq. (6.26b), we obtain

$$\overline{\sigma}_f = \frac{1}{2}\sigma_f(\varepsilon_c)\left(\frac{l}{l_c}\right) \quad l \leq l_c \quad (6.28)$$

The VFR for continuous fiber composites is thus modified for discontinuous ones by substituting $\overline{\sigma}_f$ for σ_f. Hence,

$$\sigma_c(\varepsilon_c) = V_f\sigma_f(\varepsilon_c)\left[1 - \left(\frac{l_c}{2l}\right)\right] + V_m\sigma_m(\varepsilon_c) \quad l \geq l_c \quad (6.29a)$$

and

$$\sigma_c(\varepsilon_c) = V_f\sigma_f(\varepsilon_c)\left(\frac{l}{2l_c}\right) + V_m\sigma_m(\varepsilon_c) \quad l \leq l_c \quad (6.29b)$$

To determine how effective discontinuous fibers are vis-à-vis continuous ones for reinforcement requires knowledge of l_c. Evaluation of the ratio l_c/d_f at $(T.S.)_f$ (Prob. 6.8) shows that it is typically only 10–20, and most useful fibers have aspect ratios considerably greater than this. For example, a fiber in a composite with a length such that $l/l_c = 10$ would exhibit an average fracture strength only 5% less than if the fiber were continuous. Thus, appreciable reinforcement can be provided by discontinuous fibers provided their lengths are much greater than the (usually) small critical lengths.

The above discussion applies strictly to composites, such as MMCs, in which the shear stress developed at the interface results in matrix plastic flow. In most

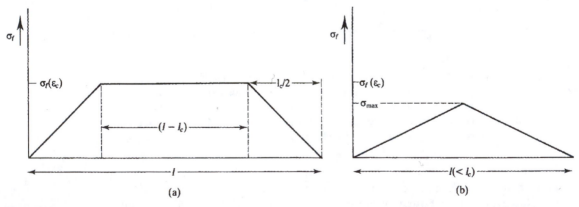

Figure 6.14
Method used to compute the average fiber stress $(\overline{\sigma}_f)$ in a composite containing discontinuous fibers. In (a), $l > l_c$, and a fraction of the fiber carries the same load as a continuous fiber; the remaining fraction carries, on average, half of this stress. In (b), $l < l_c$ and the average stress is $\sigma_{max}/2$, where σ_{max} is less than the stress carried by a continuous fiber.

resin-based composites the interfacial stress is sufficient to cause delamination because the resin is not capable of extensive plastic deformation. Subsequent to delamination at the fiber end, the matrix flows by the fiber, and tensile stress is transferred to it by friction. The friction stress is given by μP where μ is the friction coefficient between the matrix and the fiber, and P is an internal pressure resulting from shrinkage of the resin during curing and/or incompatible lateral deformation. Thus, for resin composites, τ_m of Eqs. (6.25) and (6.26) is replaced by μP and

$$\frac{l_c}{d_f} = \frac{\sigma_f(\varepsilon_c)}{2\mu P} \tag{6.30}$$

Values of μP for resin-based composites are typically an order of magnitude less than τ_{my} for MMCs, and, thus, values of l_c are about an order of magnitude higher than for MMCs. However, as evidenced by their widespread use, this does not detract much from the usefulness of resin-based PMCs.

From the standpoint of strength, it is desirable for a composite to contain fibers having lengths much greater than l_c. Toughness is another matter. Consider a composite containing fibers having lengths $< l_c$. These fibers provide some strength, but much less than if their lengths were considerably in excess of l_c. However, during composite fracture, and as a direct result of their short lengths, the short fibers do not fracture. Instead they are pulled out of the matrix and the resulting fracture surface has an irregular appearance; fibers protrude from each fracture surface and corresponding "holes" are observed on their mating surface. The work associated with fiber pull-out provides an added component to the fracture work. This component is greater than for composites containing fibers of lengths much greater than l_c. In the latter (assuming that the fibers fracture prior to the matrix), many of the fibers will fracture in the fracture plane and only those whose ends are $\pm l_c/2$ from the fracture plane will contribute to the added "pull-out" work. Thus, we see that for composites, as well as for most engineering materials, there are tradeoffs between strength and toughness. Further discussion of toughening mechanisms in composites is presented in Chaps. 9 and 10.

6.6
FIBER ORIENTATION EFFECTS

Previous discussion has focused on fiber composites for which the tensile axis is parallel to the fiber axes. When this is not the situation, composite strengths can be reduced considerably because other failure modes come into play. These are illustrated in Fig. 6.15 which considers a tensile force applied at an angle θ to the fiber axes. The situation can be investigated experimentally by taking a plate of an aligned fiber composite and machining from it tensile bars oriented differently within the plate.

Fracture of an aligned fiber composite is associated with fiber fracture since the loading geometry results in the equal-strain condition. We term this type of fracture longitudinal fracture and designate the corresponding composite tensile strength by $\sigma_\parallel^*$. ($\sigma_\parallel^*$ has the value predicted by Eq. (6.29a)). For small misorientation angles, composite tensile strength is expected to be greater than $\sigma_\parallel^*$. This is due

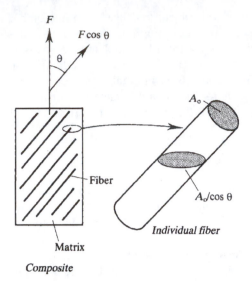

Figure 6.15
The effect of a small fiber misorienta-
tion on fiber tensile fracture. The
stress required to effect longitudinal
tensile (i.e., fiber) fracture is increased
by the factor $(\cos \theta)^{-2}$ as a result of
the reduced force ($F \cos \theta$) and in-
creased cross-sectional area ($A_0/\cos \theta$)
of the fiber.

to the change in the loading geometry. The resolved force acting along the fiber length is $F \cos \theta$, and the cross-sectional area on which the force acts is $A_0/\cos \theta$, where A_0 is the nominal fiber cross-sectional area. Thus, if the mode of composite failure remains fiber fracture, composite tensile strength is predicted to be

$$(\text{T.S.})_c \text{ (longitudinal fracture)} = \frac{\sigma_{\parallel}^*}{\cos^2 \theta} \tag{6.31}$$

Increasing misorientation can lead to composite failure by matrix shear. Here the shear force is $F \cos \theta$. However, the matrix area on which the shear force acts is $A_m/\sin \theta$ where A_m is the nominal cross-sectional area of the matrix in the "aligned" (i.e., $\theta = 0$) condition. Thus, the expected composite tensile stress for failure via matrix shear is

$$(\text{T.S.})_c \text{ (shear failure)} = \frac{\tau_{my}}{\sin \theta \cos \theta} \tag{6.32}$$

When the misalignment of the fiber with respect to the tensile axis becomes even greater, another and different composite failure mode is expected. For example, if the composite is loaded at $\theta = 90°$, we have the equal-stress condition. In this situation the fiber does carry load, but does so nowhere near as effectively as in the equal-strain condition. The failure mode in this situation is likely to be matrix tensile failure. We designate the composite failure stress in this transverse loading condition as $\sigma_{\perp}^*$. Transverse (matrix) failure can also be the dominant fracture mode for $\theta < 90°$. When this is so, though, the composite fracture stress is greater than $\sigma_{\perp}^*$ by the factor $\sin^2 \theta$; one factor of $\sin \theta$ comes from the component of the applied force that loads the matrix in tension and the other factor comes from the effective matrix area that is so loaded. Thus, we have the following expression for composite transverse fracture stress;

$$(\text{T.S.})_c \text{ (transverse fracture)} = \frac{\sigma_{\perp}^*}{\sin^2 \theta} \tag{6.33}$$

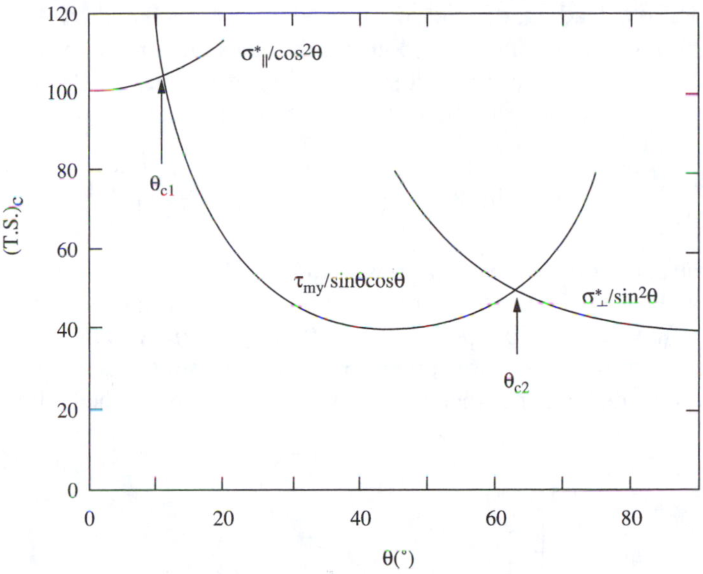

Figure 6.16
The variation of composite tensile strength with misorientation angle (θ) as predicted by Eqs. (6.31)–(6.33). For $\theta < \theta_{c1}$, tensile strength increases with θ and fracture occurs by "longitudinal" fracture. When $\theta > \theta_{c1}$, but is less than θ_{c2}, composite failure occurs by matrix shear (Eq. (6.32)). And when $\theta > \theta_{c2}$, composite fracture takes place by transverse fracture. For most fiber composites, the angle θ_{c1} is typically several degrees. (The graph is constructed for relative values of strength such that $\sigma_{\parallel}^* = 100$, $\sigma_{\perp}^* = 40$ and $\tau_{my} = 20$.)

Composite failure takes place when the applied stress attains a value equal to the *least* of the strengths predicted by Eqs. (6.31)–(6.33). The failure mode depends on the angle θ as shown in Fig. 6.16. We see that composite tensile strength decreases rapidly with increasing θ following the transition in fracture mode from longitudinal fracture to matrix shear. The angle (θ_{c1}) at which this transition takes place is found by equating Eqs. (6.31) and (6.32);

$$\theta_{c1} = \tan^{-1}\left[\frac{\tau_{my}}{\sigma_{\parallel}^*}\right] \tag{6.34}$$

Values of θ_{c1} are typically small ($< 10°$), so that only small misorientations are sufficient to negate the high strength capabilities of aligned fiber composites.

While the above treatment is useful for delineating the various ways that composites may fail in off-axis loading, a more accurate failure criterion has been developed by Tsai and Hill. In general form, this criterion is expressed as

$$\left(\frac{\sigma_{\parallel}}{\sigma_{\parallel}^*}\right)^2 - \left(\frac{\sigma_{\parallel}\sigma_{\perp}}{\sigma_{\parallel}^*}\right)^2 + \left(\frac{\sigma_{\perp}}{\sigma_{\perp}^*}\right)^2 + \left(\frac{\tau}{\tau_{my}}\right)^2 = 1 \tag{6.35}$$

In Eq. (6.35), $\sigma_{\parallel}$ is the resolved stress along the fiber axis, $\sigma_{\perp}$ is the resolved stress normal to the fiber axis, and τ is the resolved shear stress causing matrix shear (cf. Eq. (6.32)). For off-axis loading, $\sigma_{\parallel} = \sigma \cos^2 \theta$, $\tau = \sigma \sin \theta \cos \theta$ and $\sigma_{\perp} = \sigma \sin^2 \theta$ with σ being the nominal tensile stress. Substituting these relationships into Eq. (6.35), we can solve for the stress (σ) that satisfies this equation; this stress is the composite tensile stress and the Tsai-Hill criterion predicts it to be

$$(T.S.)_c = \left[\frac{\cos^4 \theta}{(\sigma_{\parallel}^*)^2} + \cos^2 \theta \sin^2 \theta \left(\frac{1}{(\tau_{my})^2} - \frac{1}{(\sigma_{\parallel}^*)^2} \right) + \frac{\sin^4 \theta}{\sigma_{\perp}^{*2}} \right]^{-1/2} \quad (6.36)$$

Comparison of the Tsai-Hill failure criterion with those of Eqs. (6.31)–(6.33) is given in Figs. 6.17a and 6.17b. Figure 6.17a provides results for off-axis failure of a continuous graphite fiber ($V_f = 0.50$)/epoxy composite; Fig. 6.17b does the same for a short-glass fiber ($V_f = 0.15$) reinforced thermoplastic. In both situations, the Tsai-Hill failure criterion fits the experimental data better, although for the carbon

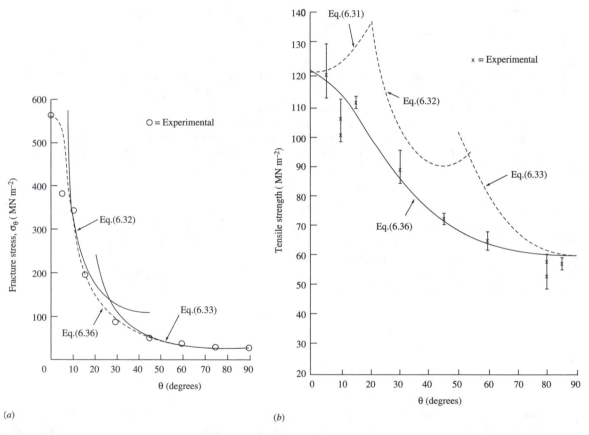

(a)   (b)

Figure 6.17
Off-axis tensile strength as a function of misorientation angle for (a) an aligned 50 vol.% carbon fiber/ epoxy resin composite and (b) an aligned 15 vol.% short-glass fiber/polymethylmethacrylate (PMMA) composite. Tensile strengths are compared both with the Tsai-Hill criterion (Eq. (6.36)) and those predicted from Eqs. (6.31)–(6.33). The Tsai-Hill criterion is more accurate. ((a) From **J. H. Sinclair and C. C. Chamis, Proc. 34th SPI/RP Annual Tech. Conf., paper 22-A, Society of the Plastics Industry, New York, 1979.** (b) From **F. Ramsteiner and R. Theysohn, Composites, 10, 111, 1979.**)

fiber composite there is not much to choose between the two criteria.[2] However, for the short-glass fiber-reinforced thermoplastic (for which the ratio $\sigma_\perp^*/\sigma_\parallel^*$ is much greater than it is for the graphite fiber composite), the Tsai-Hill criterion clearly better describes off-axis tensile strength. Regardless of these "details," the important result is that tensile strengths of aligned fiber composites decrease rapidly when they are loaded off-axis. This causes problems in composite design when the material is to be subjected to other than uniaxial loading.

There are several ways to circumvent this "anisotropy problem" in fiber composites. For example, if isotropy in two dimensions is called for (e.g., as for in-plane loading), the fibers can be distributed randomly within the plane. And, if three-dimensional isotropy is called for, the fibers can be distributed randomly within the matrix. Both kinds of fiber distribution lead to a reduction in composite modulus and strength. Composite modulus for these situations can be expressed as

$$E_c = KV_fE_f + V_mE_m \qquad (6.37)$$

where K is an empirically determined "reinforcement factor." An equation analogous to Eq. (6.37) can be written for composite strength. Clearly, K has a value less than unity. For example, if the fibers are distributed randomly within a plane, in-plane strength can be approximated with $K = \frac{3}{8}$. When the fibers are distributed randomly in three dimensions, K is about equal to 0.20.

Better utilization of composites, though, is obtained by "cross-plying" aligned fiber composites so as to produce "sandwich" structures. An example of a two-ply 0–90° composite is schematically shown in Fig. 6.18. (Engineered structures like this typically contain more than two plies.) Other ply arrangements (e.g., 0–45° or 0–45–90°) are also available. These sandwich structures are particularly useful for in-plane loading. To analyze their behavior in detail requires relatively advanced mechanics. Mechanical analyses of this nature have enabled cross-plied composites to be fruitfully used in many applications.

To this point, we have treated fibers as if they all had the same strength. In fact they do not and, in many cases, are prone to fracture over a rather wide range of stress. Engineering design, therefore, utilizes an appropriate "average" fiber fracture stress. This is sufficient in many cases and, when so, concepts relating to the vari-

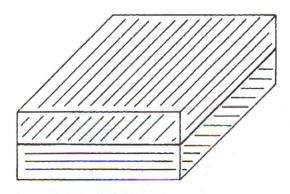

Figure 6.18
Schematic of a 0–90° "cross-ply" fiber composite structure. Such configurations are useful for biaxial loading of composites. Other configurations (e.g., 45–45°) can be similarly employed.

---

[2]We also note that the Tsai-Hill criterion does not produce "cusps" in the fracture strength associated with transitions in failure mode.

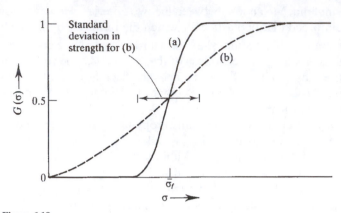

Figure 6.19
Method for describing fiber strength variability. The ordinate here ($G(\sigma)$) is the fraction of fibers that fail at a stress less than or equal to σ. Thus, an average, or expected, fiber tensile strength is that stress at which half of the fibers fail (i.e., the stress at which $G(\sigma) = 0.5$). For curve (a), the fibers do not have a wide variability in strength, whereas they do for curve (b). The variability can be described by the standard deviation in fiber strength.

ability in fiber fracture strength do not have to be incorporated into design. However, in other situations fiber strength variability is important. Thus, in the following section we consider some aspects of the statistics of composite failure as they devolve on fiber fracture strength variations.

6.7
STATISTICAL FAILURE OF COMPOSITES

As noted, it has been previously and tacitly assumed that all fibers in a composite fail at the same stress, i.e., at $(T.S.)_f$. This is demonstrably not true for most of the inorganic nonmetallic fibers listed in Table 6.1. Their strengths are affected markedly by the presence of defects within them or on their surface. The most common defects are small surface flaws or cracks that appreciably reduce strength. Even high-strength metallic fibers display variations in tensile strength, although these variations are nowhere near as large as they are for nonmetallic fibers.

The tensile strengths listed in Table 6.1 are average values obtained from a series of tests. A schematic illustration of the results from such a test series is given in Fig. 6.19, where the fraction of fibers of length l that have failed at a given stress is plotted against this stress level. The ordinate is labeled $G(\sigma)$; $G(\sigma)$ represents the fraction of fibers that fail at a stress less than or equal to σ. An "expectation" value of the fiber strength ($\overline{\sigma}_f$) can be defined as that stress for which 50 percent of the fibers fracture; i.e., $G(\overline{\sigma}_f) = [1 - G(\overline{\sigma}_f)] = 0.50$. Different fibers may have the same value of $\overline{\sigma}_f$, but exhibit significant differences in their strength variations. Thus, curve (a) in Fig. 6.19 is for a material that exhibits little variation in strength from fiber to fiber, whereas curve (b) is for a material manifesting intrinsically large strength variations. The strength variation is defined by the standard deviation in

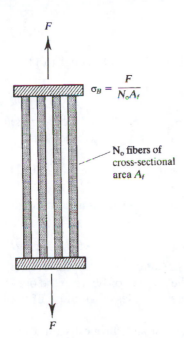

Figure 6.20
Schematic of a tensile test on a fiber bundle. The nominal bundle stress (σ_B) is defined as $F/N_0 A_f$ where N_0 is the total number of fibers of cross-sectional area A_f in the bundle. This nominal stress is used to characterize bundle stress even after some of the fibers in it have broken.

fiber strength (cf. Fig. 6.19); this parameter is closely related to another term, called the fiber coefficient of variation.

Related to $G(\sigma)$ is the *fraction* of fibers that fail between the stress levels σ and $\sigma + d\sigma$. This function, $g(\sigma)\, d\sigma$, is defined by $G(\sigma + d\sigma) - G(\sigma) = dG(\sigma)$ or

$$dG(\sigma) = g(\sigma)\, d\sigma \tag{6.38a}$$

and

$$G(\sigma) = \int_0^\sigma g(\sigma)\, d\sigma \tag{6.38b}$$

These ideas can be further developed by considering the tensile behavior of a *bundle* of fibers of number N_0 and length l for which the expected fiber strength and standard deviation are determined by the function $G(\sigma)$ (or, alternatively, $g(\sigma)$). Such a bundle subjected to an increasing tensile force F is illustrated in Fig. 6.20 and resembles a "composite" for which V_f is equal to one. On initial loading, the tensile force is divided equally among the fibers so that the force on each is F/N_0 and the corresponding stress (σ_f) is $F/N_0 A_f$, where A_f is the transverse cross-sectional area of the fiber. The nominal tensile stress on the bundle, σ_B, is defined by

$$\sigma_B = \frac{F}{N_0 A_f} = \frac{N_0}{N_0}\sigma_f \tag{6.39}$$

Equation (6.39) applies up to the stress at which fibers within the bundle begin to fracture. Above this stress level, the N_0 term in the numerator of the right-hand side of Eq. (6.39) must be replaced by N, the number of unbroken fibers. (N_0 in the denominator remains unchanged because bundle stress is based on the original bundle cross-sectional area ($N_0 A_f$)). Thus,

$$\sigma_B = \frac{N}{N_0} \sigma_f \qquad (6.40)$$

and since N/N_0 is the fraction of unbroken fibers it is identically equal to $[1 - G(\sigma_f)]$. Hence,

$$\sigma_B = \sigma_f[1 - G(\sigma_f)] \qquad (6.41)$$

Figure 6.21 schematically plots σ_f, $[1 - G(\sigma_f)]$, and σ_B (the product of these terms) as a function of σ_f. Maximum strength is obtained at the stress, σ_{fmax},[3] such that

$$\left(\frac{d\sigma_B}{d\sigma_f}\right)_{\sigma_{fmax}} = [1 - G(\sigma_{fmax})] - \sigma_{fmax}g(\sigma_{fmax}) = 0 \qquad (6.42)$$

If the function $g(\sigma)$ is known, σ_{fmax} can be solved for and the bundle tensile strength is given by

$$\sigma_B = \sigma_{fmax}[1 - G(\sigma_{fmax})] \qquad (6.43)$$

Since the bundle contains fibers having a distribution in strength, bundle strengths also exhibit such variations. However, if the number of fibers in the bundle is reasonably large, bundle strength variations are much less than those of the fibers comprising the bundle.

Bundles are utilized in a number of applications; stranded ropes, yarns, and steel wire are all bundles. At first glance it is surprising that the fibers in these, and other bundles, are of finite length and less than the bundle length. However, the stranding allows tensile stress to be transferred from one strand to another as a result

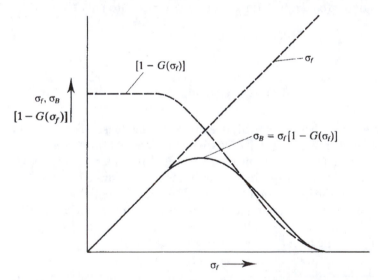

Figure 6.21
The bundle stress is obtained by multiplying σ_f by the fraction of unbroken fibers $[1 - G(\sigma_f)]$. The variation of σ_f, $[1 - G(\sigma_f)]$, and their product, σ_B, is shown as a function of σ_f. Bundle strength corresponds to the product $\sigma_f[1 - G(\sigma_f)]$ being a maximum.

---

[3]Note σ_{fmax} is not equal to σ_f, the expected fiber strength. Problems 6.11–6.14 illustrate this point.

of the friction force generated when the strands attempt to slide by one another. This feature means that stranded wires and ropes are not bundles in the precise sense of the term we have used in the above analysis. Indeed, the stress transfer process in a stranded rope or wire bears some resemblance to that in a fiber composite. Stress transfer in a bundle is, however, nowhere near as efficient as it is in a composite. As a result, fiber-reinforced materials utilize fiber strengths better than do bundles or stranded wire/rope.

This can be shown by applying bundle theory to fiber composites, as illustrated in Fig. 6.22. So that we can neglect the complications associated with finite-length fibers, Fig. 6.22 is applied to composites containing continuous fibers of length l, the sample length. When a fiber in such a composite fails at a random location along the gage length, it is not, as it is in a bundle, precluded from carrying tensile load at other positions along the gage length. Instead, the matrix reloads the fiber to its expected stress level at a distance $\pm l_c/2$ from the original fiber fracture location. Subsequent to the first fiber fracture, the unbroken fibers in the transverse cross-section where this break occurred carry the load previously borne by the now-fractured fiber. And if the fibers all have the same strength, the composite fractures at this cross-section. On the other hand, if the distribution in fiber strengths is large, the remaining fibers are able to withstand the additional stress caused by the fracture of the first fiber, and the next fiber fracture takes place at a different location along the gage length. For this situation, composite fracture takes place by a statistical accumulation of fiber breaks along the gage length and final fracture (or tensile strength) corresponds to the failure of one additional fiber "link" such that the remaining unbroken fibers at that specific cross-section are unable to absorb the incremental load without themselves fracturing. We see that, in distinction to bundles,

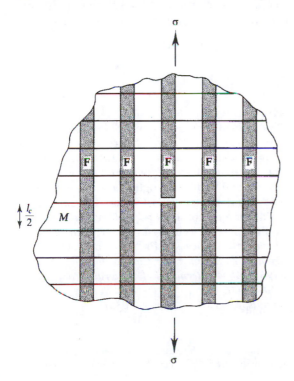

Figure 6.22
Bundle theory applied to composites containing fibers of variable strength. In contrast to a fiber in a bundle, one in a composite can be reloaded after fracture. In terms of carrying load, a broken fiber is ineffective only over a length $l_c/2$ above and below the plane of its fracture. Thus, the composite is divided into links of length $l_c/2$, and the fiber length employed in this "bundle" treatment is $l_c/2$. (M = matrix, F = fiber.)

fibers may be broken at more than one location along their gage length during this "statistical" composite fracture.

Composite tensile behavior in this scenario can be described by considering each fiber as consisting of a series of links of length $l_c/2$. Composite tensile strength (neglecting the contribution from the matrix) is given by an equation analogous to Eq. (6.43) with one important difference: the function $G(\sigma)$ is replaced by one that describes the variation in strength of fibers having length $l_c/2$ rather than l. Calling this function $F(\sigma)$,[4] the expected composite strength is

$$\sigma_c = \sigma_{fmax}[1 - F(\sigma_{fmax})] \qquad (6.44)$$

where σ_{fmax} is the stress at which the product $\sigma_f[1 - F(\sigma_f)]$ is a maximum.[5] Composite strengths will always exceed those of bundles for the following simple reason. A fiber in a bundle fails at its weakest link, and thereafter is incapable of bearing load. Provided $l > l_c/2$, this is not so in a composite, and a broken fiber contributes to composite stress-carrying capability all along its length except in the vicinity of the break.

How much stronger a composite is than a bundle depends on the number of links in each fiber, i.e., on the ratio l/l_c. If there are a great number of links (i.e., if $l/l_c \gg 1$), composite strength can far exceed that of the corresponding bundle. Likewise, composite strength relative to bundle strength is greatest for fibers with intrinsically large strength variations. In a bundle, the weak links fail at a low stress level and partially negate the higher-strength capacities of the remaining fibers. This is not so if the fiber is imbedded in an efficient stress-transferring matrix. These points are illustrated in Fig. 6.23, where the ratio of composite strength to average fiber strength is plotted versus the fiber coefficient of variation for various values of l/l_c. High values of l/l_c enhance composite strength, and the effect is most pronounced for fibers with intrinsically large strength variations.

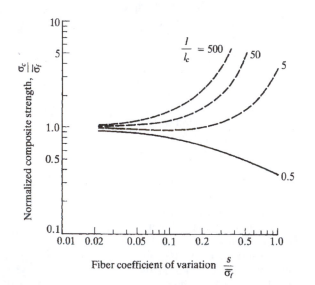

Figure 6.23
The ratio of composite strength to average fiber strength as a function of fiber coefficient of variation (a measure of fiber strength variability) for different values of l/l_c. For $l < l_c$, we have a bundle and bundle strength decreases with increasing fiber strength variation. The opposite holds for a composite ($l > l_c$) as the weak links in the fibers are rendered relatively harmless through the process of fiber reloading. (Adapted from B. W. Rosen, Fiber Composite Materials, (1965), ASM International, Materials Park, OH 44073-0002 (formerly American Society for Metals, Metals Park, OH 44703).)

---

[4]If $G(\sigma)$ for fibers of length l is known, $F(\sigma)$ for fibers of length $l_c/2$ can be determined (Prob. 6.12).
[5]This value of stress is not the same as the value of stress for which the function $\sigma_f[1 - G(\sigma_f)]$ is maximum.

EXAMPLE PROBLEM 6.3. Consider a fiber population for which the function $g(\sigma)$ is constant between a lower stress σ_1 and an upper stress σ_2, but is zero for stresses less than σ_1 and greater than σ_2.
a. Determine $G(\sigma_f)$.
b. Determine the fiber stress, σ_{fmax}, at which the function $\sigma_f[1 - G(\sigma_f)]$ is a maximum.
c. Calculate the bundle strength.

Solution. **a.** $G(\sigma_f) = \int_0^{\sigma_f} g(\sigma)\,d\sigma = 0$; $\sigma_f \leq \sigma_1$: $G(\sigma_f) = \int_{\sigma_1}^{\sigma_f} A\,d\sigma = A(\sigma_f - \sigma_1)$;

$\sigma_1 \leq \sigma_f \leq \sigma_2$

We also must have $G(\sigma_2) = 1$, since all fibers have failed at this stress. Thus, for the constant A, we obtain $A = 1/(\sigma_2 - \sigma_1)$. In summary:

$$G(\sigma_f) = 0; \sigma_f \leq \sigma_1: G(\sigma_f) = \frac{\sigma_f - \sigma_1}{\sigma_2 - \sigma_1}; \sigma_1 \leq \sigma_f \leq \sigma_2: G(\sigma_f) = 1; \sigma_f \geq \sigma_2$$

The function $G(\sigma_f)$ is plotted below.

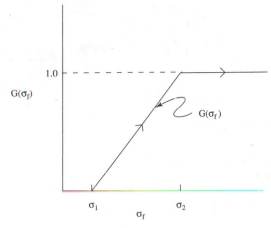

b. To obtain the stress carried by the bundle, multiply σ_f by $[1 - G(\sigma_f)]$.

$$\sigma_B = \sigma_f; \sigma_f \leq \sigma_1: \sigma_B = \sigma_f\left[\frac{\sigma_2 - \sigma_f}{\sigma_2 - \sigma_1}\right]; \sigma_1 \leq \sigma_f \leq \sigma_2: \sigma_B = 0; \sigma_f \geq \sigma_2$$

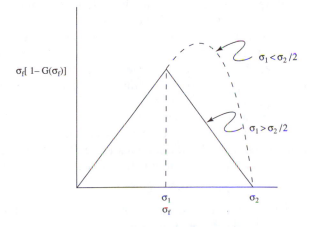

The above functions are plotted vs. σ_f in the sketch below. Note that the maximum in the function depends on the value of σ_1 relative to σ_2. If $\sigma_1 \geq \sigma_2/2$, the maximum occurs at $\sigma_f = \sigma_1$; if $\sigma_1 \le \sigma_2/2$, the maximum occurs at a stress intermediate to σ_1 and σ_2. The intermediate stress has the value $\sigma_2/2$, as shown by differentiating σ_B in the intermediate stress range and setting it equal to zero; i.e.,

$$\frac{d(\sigma_f[1 - G(\sigma_f)])}{d\sigma_f} = \frac{-\sigma_f}{(\sigma_2 - \sigma_1)} + \frac{(\sigma_2 - \sigma_f)}{(\sigma_2 - \sigma_1)} = \frac{(-2\sigma_f + \sigma_2)}{(\sigma_2 - \sigma_1)} = 0$$

or

$$\sigma_f = \frac{1}{2}\sigma_2$$

c. To determine the bundle strength, substitute the values above for σ_f in the expression for the bundle stress. Thus,

$$\overline{\sigma}_B = \sigma_1; \quad \sigma_1 \ge \sigma_2/2$$

$$\overline{\sigma}_B = \frac{1}{2}\sigma_2\left[\frac{\frac{1}{2}\sigma_2}{(\sigma_2 - \sigma_1)}\right] = \frac{\sigma_2^2}{4(\sigma_2 - \sigma_1)}; \quad \sigma_1 \le \sigma_2/2$$

Note that when $\sigma_1 = \sigma_2/2$, $\sigma_B = \sigma_1$ for both situations.

6.8
STRAIN-RATE EFFECTS

Many composites, particularly MMCs, are intended for use in the temperature regime where they are strain-rate sensitive, and thus subject to time-dependent deformation. The strain-rate sensitivity of the matrix in such composites can be described by the empirical Eq. (1.14)

$$\sigma_m = K_m(\dot{\varepsilon}_m)^m \tag{1.14}$$

(the m subscript here indicates "matrix," and should not be confused with the exponent, m, which describes the strain-rate sensitivity).

Potential high-temperature composites include many anticipated to be comprised of fibers displaying linear elastic behavior and a matrix that is expected to creep under high-temperature conditions. Thus, the fibers in the composites are expected to enhance not only strength but creep resistance. Microscopic mechanisms of creep are considered in Chap. 7. However, the strain-rate sensitivity of a composite, as well as its creep behavior, can be discussed here using phenomenological and other concepts presented previously.

We consider first a simple example—a composite consisting of continuous fibers aligned along the tensile axis. The fibers deform elastically, whereas the matrix that surrounds them has a stress-strain rate response described by Eq. (1.14). The volume fraction rule gives the stress-strain-strain rate behavior of the composite;

$$\sigma_c = V_m K_m(\dot{\varepsilon}_m)^m + V_f E_f \varepsilon_f \tag{6.45}$$

We note that because of the equal-strain condition, the strain and the strain rate are the same for the fiber and the matrix and, thus, also for the composite. As a corollary, because the strain on the matrix is a function of time, so is the fiber strain, ε_f.

That is, $\varepsilon_f = \int \dot{\varepsilon}\, dt$ where t is the time duration of the applied stress and $\dot{\varepsilon}$ is the instantaneous strain rate.

We apply a stress (σ_c) at time $t = 0$ and hold this stress constant thereafter. The composite strain rate is then measured. The situation is similar to the viscoelastic deformation treated in Chap. 2, with but two differences. First, in the current situation, the relationship between stress and strain rate for the "viscous component" (the matrix) is not linear; that is, the exponent m in Eq. (6.45) is not generally equal to one. Second, the volume fractions of the fiber and matrix are variable; in Chap. 2 they were (implicitly) taken as equal. Because the strains on the fiber and matrix are the same and also because a fiber elastic strain cannot be instantaneously attained on application of the stress, the initial strain on the composite is zero; the fiber carries none of the applied stress and the matrix carries all of it. Thus, the second term on the right-hand side of Eq. (6.45) is zero at $t = 0$, and the initial composite strain rate is

$$\dot{\varepsilon}_c(t = 0) = \left(\frac{\sigma_c}{V_m K_m}\right)^{1/m} \tag{6.46}$$

With increasing time, load is transferred to the fiber and the strain rate decreases. It eventually attains a value of zero at which point all of the load is carried by the fiber phase and none of it by the matrix. The asymptotic composite strain is

$$\varepsilon_c(t = \infty) = \sigma_c / V_f E_f \tag{6.47}$$

There are clear parallels to this situation and the Voigt model described in Chap. 2 (Prob. 6.15).

When the fibers are discontinuous the situation is more complicated. Consider Fig. 6.24 which illustrates an elastic fiber imbedded in a matrix that flows past the fiber at a relative shear strain rate $\dot{\gamma}$. The magnitude of $\dot{\gamma}$ varies along the fiber length; it is zero at the fiber midpoint ($z = 0$) and a maximum at the fiber ends ($z = l/2$). We assume that this variation in strain rate is linear with position along the fiber length and also that it increases linearly with the matrix tensile strain rate, $\dot{\varepsilon}_m$; i.e.,

$$\dot{\gamma} = \beta' z \dot{\varepsilon}_m \tag{6.48}$$

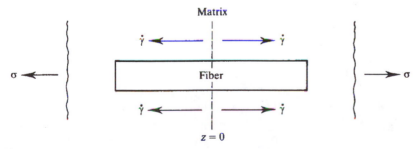

Figure 6.24
An elastic fiber imbedded in a matrix deforming at a tensile strain rate, $\dot{\varepsilon}_m$. The tensile strain rate results in a relative shear strain rate, $\dot{\gamma}$, at the fiber-matrix interface; $\dot{\gamma}$ is zero at the fiber mid-point and a maximum at the fiber ends.

In Eq. (6.48), the constant β' has dimensions of $(\text{length})^{-1}$. To keep our analysis consistent with previous treatments of load transfer to discontinuous fibers, we take $\beta' = \beta/d_f$, where d_f is the fiber diameter. The shear stress–shear strain rate behavior of the matrix is described by an equation analogous to Eq. (1.14), i.e., $\tau = K_\tau(\dot{\gamma})^m$. Thus, the shear stress at the fiber-matrix interface is

$$\tau = K_\tau(\beta\dot{\varepsilon})^m \left(\frac{z}{d_f}\right)^m \tag{6.49}$$

The variation of τ with position along fiber length is illustrated in Fig. 6.25. The figure shows that the stress increases to higher values the greater the strain-rate sensitivity of the matrix. It also shows that when the matrix flow behavior is not strain-rate sensitive (i.e., when $m = 0$) the shear stress is constant as would be expected on the basis of our earlier treatment (Sect. 6.5). Although not indicated in Fig. 6.25, the higher the strain rate the greater the shear stress developed at the interface (Eq. (6.49)).

The interface shear stress transfers tensile load to the fiber, just as the matrix shear stress does in the nonstrain-rate sensitive case. The variation of the fiber tensile stress with position along the fiber is illustrated schematically in Fig. 6.26a. Note that the fiber mid-point (and regions removed from this position) can be strained to the strain it would experience were it fully exploited. That is, at the mid-point the tensile stress on the fiber is $E_f\varepsilon_c$ where ε_c is the composite strain. However, because the matrix creeps, ε_c is a function of time ($\varepsilon_c = \dot{\varepsilon}_m t$, where t is the time of the test). Thus, with increasing time the region of the fiber strained to the maximum possible extent becomes less and less. Indeed, for very long times, this condition is not even fulfilled at the fiber mid-point. The variation of fiber tensile stress with position along the fiber length for such long times is shown in Fig. 6.26b. For this situation, the average fiber tensile stress is obtained by averaging σ_f along the fiber length. When this is done, the asymptotic average fiber tensile stress is found as

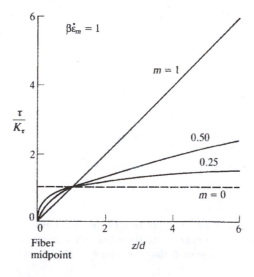

Figure 6.25
Shear stress developed at the fiber-matrix interface as a function of position along the fiber. (The strain rate is fixed in this instance.) The shear stress increases more rapidly as the strain-rate sensitivity of the matrix increases.

$$\bar{\sigma}_f = \left[\frac{K_m(\dot{\bar{\varepsilon}})^m}{m+2}\right](2)^{2-m}\left(\frac{l}{d_f}\right)^{m+1} \qquad (6.50)$$

The average stress increases with time, as indicated in Fig. 6.27 (the asymptotic stress in this figure is given by Eq. (6.50)).

The composite creep rate is found by rearranging Eq. (6.45);

$$\dot{\varepsilon}_c = \left[\frac{\sigma_c - V_f\bar{\sigma}_f(t)}{V_m K_m}\right]^{1/m} \qquad (6.51)$$

The stress $\sigma_f(t)$ in Eq. (6.51) increases with time (Fig. 6.27) until the value of it is given by Eq. (6.50). This corresponds to the onset of "steady-state" creep. The steady-state creep rate is reduced by increasing V_f and l/d_f since both factors result in an increase in stress carried by the fibers (Prob. 6.16). During steady-state creep,

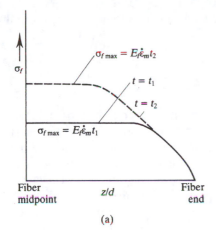

(a)

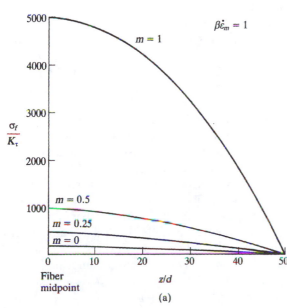

(a)

Figure 6.26
Variation of σ_f as a function of position along the fiber length for "short" and "long" creep times. (a) At short times, the fiber mid-point and positions adjacent experience a stress ($= E_f\varepsilon_m t$) corresponding to the equal strain condition. With increasing time, the region that suffers this stress becomes less and less. (b) Eventually the fiber mid-point is stressed to a level less than that of the equal strain condition. How much less depends on the matrix strain-rate sensitivity; increasing m values lead to greater fiber stresses. So do increases in strain rate (although this factor is not illustrated in this figure).

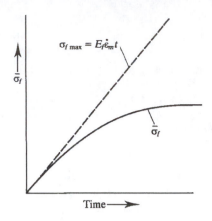

$$\sigma_{f\,max} = E_f \dot{\varepsilon}_m t$$

$\overline{\sigma}_f$

Time→

Figure 6.27
Variation of average tensile stress carried by the fiber ($\overline{\sigma}_f$) as a function of creep time. For short times, $\overline{\sigma}_f = E_f \dot{\varepsilon}_m t$, indicating that all of the fiber is strained to the equal-strain condition. This situation does not hold at longer creep times and $\overline{\sigma}_f$, therefore, decreases. Its limiting value is given by Eq. (6.50).

the two-phase mixture can be viewed as an aggregate in which the elastic fibers carry a certain fraction of the applied load, and a matrix that flows by these fibers.

6.9
MICROSCOPIC EFFECTS

To this point, discussion of composites has been couched in macroscopic terms. In effect, we have assumed that the presence of the fibers does not alter the matrix flow behavior and vice versa. For this to be so, an interfiber spacing large in comparison to those of other flow obstacles in the matrix is required. If, say, the matrix is a metal whose strength is defined by its grain size, this requirement translates into the grain size being less than the interfiber spacing. If, on the other hand, each matrix grain contains a large number of fibers, these act as additional obstacles to matrix flow, and any constitutive equation describing composite mechanical behavior will have to take this microscopic feature into account.

Most composite processing yields phase dispersions on a scale large enough that microscopic interactions between the phases need not be considered in describing composite properties. However, some fabrication procedures produce *in situ* composites in which typical interphase spacings are small (on the order of μm or less). Examples are aligned two-phase composites produced by directional eutectic solidification or directional eutectoid decomposition (Fig. 6.28).

That microscopic effects must be considered in describing mechanical behavior of *in situ* composites can be appreciated by realizing that one of the phases is stronger and more resistant to plastic flow than the other. Plastic deformation of the weaker phase is constrained by the stronger phase and the interface between the phases, and these constraints become more effective as the interphase spacing is reduced. If the stronger phase is in the form of a fiber it further serves as an obstacle to matrix plastic flow as well as a load-carrying member. (And a fiber is more effective in restricting matrix flow than is a spherical particle because it is more difficult to effect cross-slip around a fiber.) In spite of these complexities, a "modified" VFR can be used to describe the stress-strain behavior of *in situ* composites. The important modification is the incorporation into the VFR of a matrix flow stress that takes into account the effect of the presence of the other phase.

In situ composites produced via eutectic/eutectoid decomposition are potentially attractive high-temperature materials, as they are chemically stable and relatively resistant to phase coarsening that, for example, degrades the high-temperature properties of precipitation-hardened alloys. However, thermal cycling of *in situ* composites of these types produces an internally generated and likewise fluctuating stress field. This comes about as a result of the different thermal expansion coefficients of the phases. In extreme cases, the effect can lead to dimensional instability (warpage).

Other types of *in situ* composites utilize the beneficial effect of the greater work hardening of two-phase materials compared to single-phase ones. Thus, for example, pearlitic steel (Fig. 5.36) manifests extraordinary strengths when deformation processed to high strains. Other two-phase metals demonstrate similar behavior. For example, mixtures of Ag and Cu (Fig. 6.29) continue to work harden at strains well beyond those at which their individual constituents "saturation" harden. While the deformation processing produces a phase arrangement aligned along the deformation direction and, therefore, a structure that resembles a composite, the strengths of these *deformation processed composite materials* (DPCMs) derive from their high work-hardening capacity and not from the phase morphology. Similar high-strain generated strengths are observed in a number of other two-phase

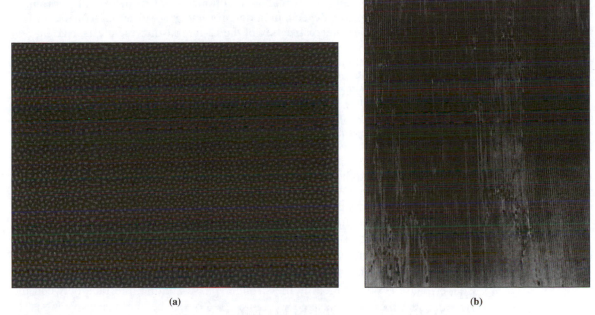

(a)   (b)

Figure 6.28
(a) A transverse section of a directionally solidified Cr fiber-NiAl matrix eutectic. The Cr fibers (ca. 1 μm diameter) extend in and out of the plane of the photograph. (*From J. L. Walter and H. E. Cline, Metall. Trans., 1, 1221, 1970.*) **(b) A longitudinal section of a Co-Co$_2$Si lamellar *in situ* composite produced by directional eutectoid decomposition. A transverse section of this plate-like array would appear similar. Typical spacings for directionally decomposed eutectoids are an order of magnitude less than they are for eutectics.** (*Reprinted from Acta Metall., 23, J. D. Livingston, "Properties of Aligned Co-Si Eutectoid," 54, Copyright 1975, with permission from Elsevier Science.*)

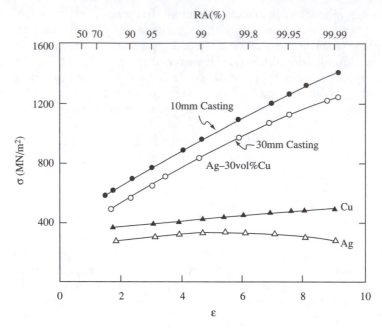

Figure 6.29
Tensile strengths of Ag-30 vol.% Cu alloys as a function of true deformation strain. The strengths are high and depend on initial processing as well as on deformation strain. The initially finer microstructure of the 10-mm casting manifests itself in higher strengths at equivalent strains. The strengths of the composite components as a function of deformation strain are also shown. Composite strengths are clearly much in excess of their components, and this is indicative of the significant work hardening accompanying deformation of the two-phase material. (***From Acta Metall., 23, G. Frommeyer and G. Wasserman, "Microstructure and Anomalous Mechanical Properties of In Situ-Produced Silver-Copper Composite Wires," 1353, Copyright 1975, with permission from Elsevier Science.***)

metal mixtures (e.g., Cu/Fe and Cu/Nb). As is evident from Fig. 6.29, though, considerable processing strain is required before particularly high strengths are developed in DPCMs. Thus, these materials—as is patented steel wire—are restricted to fine wire and/or sheet form. The thermal stability of DPCMs also leaves much to be desired. Their high stored energies of cold work render them susceptible to recrystallization and/or coarsening at relatively low temperatures.

6.10
REINFORCEMENT OF BRITTLE MATRICES

In most PMCs and MMCs the high strengths of fibers are utilized by incorporating the fibers in a relatively weak matrix. Generally, the fibers fail at a lesser strain than the matrix, with the result that the matrix also serves as a crack arrester whereby the fracture toughness of the composite is increased.

In many CMCs the toughness of intrinsically brittle materials is increased by incorporation of *ductile* fibers within them. In the absence of such fibers, fracture of

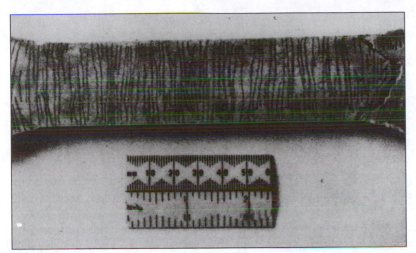

Figure 6.30
Photograph of cracked Portland cement containing 1.5 vol.% of aligned, chopped fibers ($d = 150$ μm, $l = 2.5$ cm). The cracks, too small to be seen at this magnification, have been outlined. The multiple cracking displayed is not observed in monolithic cement, and is a manifestation of the toughening provided by the fibers. (*After A. Kelly, Metall. Trans., 3, 2313, 1972.*)

a monolithic sample of a brittle material occurs catastrophically and at a relatively low strain. The resultant fracture surface is macroscopically planar and material failure is restricted to the first (and last) fracture location (i.e., secondary cracks do not form). When brittle materials are reinforced with ductile fibers, numerous fracture events may occur prior to final composite fracture. This is illustrated in Fig. 6.30 for Portland cement containing 1.5% by volume of chopped metallic fibers. Such a cement exhibits a considerably greater failure strain and a concurrent increase in fracture work compared to the "unreinforced" material.

To see how a fracture of the type shown in Fig. 6.30 develops, we investigate the stress state that exists after the initial fracture of the matrix. This is illustrated in Fig. 6.31. It is clear that the unbroken fibers in the vicinity of the fracture plane sustain more stress than do the fibers removed from it. In the regions away from the fracture plane, the tensile stress is apportioned between the fibers and the matrix according to the VFR,

$$\sigma_c = V_f \sigma_{f0} + V_m \sigma_m \tag{6.52}$$

where σ_{f0} is the stress carried by the fibers. At the fracture plane, the matrix no longer sustains load so the stress on the fibers is increased to σ_f' such that

$$\sigma_f' = \sigma_{f0} + \sigma_m \left(\frac{V_m}{V_f} \right) \tag{6.53}$$

If σ_f' is less than the fiber tensile strength, composite fracture is not concurrent with the first matrix fracture; rather a series of fracture planes develops along the sample length. The separation of the planes is about equal to the matrix critical length. That is, even as the matrix can reload a broken fiber within a distance $\pm l_c/2$ from the ends of it, so can the fiber transmit tensile stress to a fractured matrix. The matrix critical length (l_{cm}) is

$$l_{cm} = \left[\frac{V_m \sigma_m}{V_f(2\tau_i)} \right] d_f \tag{6.54}$$

where d_f is the fiber diameter and τ_i the interfacial stress. The similarity of Eq. (6.54) to Eq. (6.26b) is obvious; the additional fiber stress ($V_m\sigma_m/V_f$) deduced from Eq. (6.53) substitutes for σ_f of Eq. (6.26b).

During tensile fracture of reinforced brittle materials, the matrix is, in effect, held together by the fibers thus delaying final fracture. The engineering usefulness of brittle materials is increased as a result. Not only is their toughness increased by reinforcement, but so are such properties as resistance to impact, fatigue, and thermal shock.

Our discussion here pertains to tensile loading of a brittle material reinforced by a more ductile phase. This is the situation for many CMCs. Further aspects of

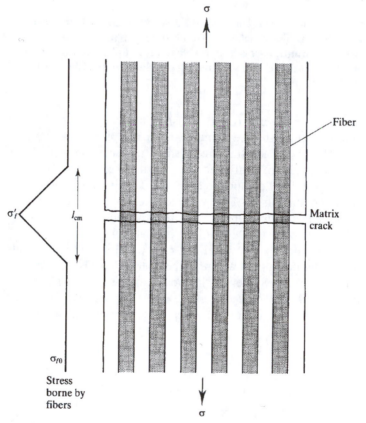

Figure 6.31
The situation in a ductile fiber-brittle matrix composite following matrix fracture. The unbroken fibers in the cracked region carry a stress (σ_f') greater than the fibers in the unbroken matrix. If σ_f' is less than the fiber tensile strength, composite fracture is not coincident with formation of the first matrix crack. That is, the matrix above and below the fracture surface reloads the fibers, and this can lead to multiple matrix cracking (e.g., Fig. 6.30).

toughening in CMCs is presented in Chap. 10 where the focus is on the work associated with crack propagation and the material property of fracture toughness.

6.11
MODERN COMPOSITE MATERIALS

A. Fibers

The properties of a number of fibers commonly used as reinforcement in composites are given in Table 6.1. An extensive description of the methods used to manufacture these fibers is outside our scope. Nonetheless, a brief description is worthwhile in order to illustrate the creativity of the processes employed in fiber manufacture.

The most common fiber material is glass. It is extensively employed in PMCs. Glass fibers can be made by "blowing" molten glass through a suitable orifice and into a quenching medium (e.g., a coolant gas). Fibers produced this way are macroscopically of short length although they possess a sufficiently high aspect ratio to result in substantial strengthening when they are used in composites. It is of interest that glass fibers typically have low glass transition and softening temperatures. These features permit their manufacture at moderate temperatures and, since the resultant fibers are used as low-temperature reinforcements, do not negatively impact their low-temperature use.

Glass filaments can be produced in essentially continuous form. In this form, they can be "wrapped" around a mandrel in various orientations. The filaments are "soaked" in an epoxy resin prior to wrapping. Thus, the wrapping and setting operations are combined. Rocket motor cases, for example, have been produced by such processing for more than 30 years.

Graphite fibers are among the strongest and stiffest fibers. Further, they can be used for both high- and low-temperature reinforcement. To display high strengths and moduli, the layer (basal) planes of the graphite must be aligned along the fiber axis. This is because the bonding between these planes is not nearly as strong as that within the planes. In practice, the alignment process is not perfect; there are many "alignment" faults and defects within the fiber as illustrated in Fig. 6.32.

Kevlar is a polymeric fiber that competes surprisingly well with inorganic fibers and filaments. Aromatic rings constitute the chain backbone of Kevlar, and this feature produces a relatively rigid chain. Kevlar fibers are produced by extrusion and spinning, followed by drawing. The drawing produces molecular alignment along the fiber axis similar to that found in more conventional polymeric filaments and fibers (Chap. 8).

Some of the other inorganic fibers listed in Table 6.1 are formed by a vapor deposition technique. Examples are boron and silicon carbide in which the fiber component(s) are deposited from a vapor phase onto a suitable substrate. Tungsten serves as a substrate for boron, and carbon does the same for silicon carbide. Processing improvements and better quality control continue to result in improved properties of the many fibers and filaments available for use in composites.

B. Matrices

As discussed in Sect. 6.1, composite materials can be classified on the basis of the nature of their matrices. Thus, we have ceramic matrix composites (CMCs), metal matrix composites (MMCs) and polymer matrix composites (PMCs). The latter are the least expensive. As a result they are not only used in structural applications (where cost is often a secondary concern), but are widely utilized in more mundane applications. Examples, some of them cited previously, include fiberglass panels and graphite-reinforced fishing rods, golf club shafts and squash rackets (all erroneously referred to as graphite poles, clubs, and rackets). Maximum use temperatures of PMCs are relatively low, as their matrices are prone to softening or decomposition at moderate temperatures. Elevated temperature use necessitates employment of MMCs or CMCs. Metal matrix composites can be used to relatively high homologous temperatures. They are also superior to CMCs in terms of toughness and ductility. While CMC development is relatively recent, these materials are of intense current interest owing to the high temperatures at which they can be used and to their inertness in many environments. Most CMCs use the reinforcing phase as a means of imparting toughness to the matrix, rather than to strengthen it as is the usual situation for PMCs and MMCs.

Many PMCs are strengthened with glass fibers or filaments. Their most common matrices are epoxy resins, although recent work has focused on employing thermoplastics in that capacity. Polymer matrix composites are relatively inexpensive owing to their low setting/forming temperatures. As noted, axially symmetric structures can be made by filament winding. However, fiberglass is more widely used in less sophisticated applications. The fiberglass can be set in place during (or partially set prior to) fabrication.

Fiber
axis

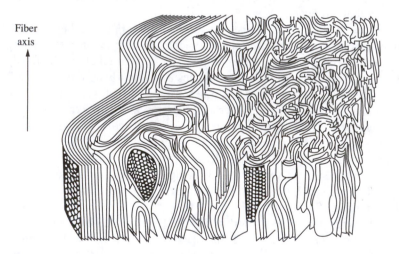

Figure 6.32
Schematic representation of the structure of carbon fibers. The sheets represent the graphite basal planes. Although the planes are aligned with the fiber axis, they contain many structural "faults." (*From S. C. Bennett, Ph.D. thesis, University of Leeds, 1976, as presented in "An Introduction to Composite Materials," Derek Hull, Cambridge University Press, Cambridge, England, 1981.*)

Reinforcing fibers are often chemically quite distinct from the matrices in which they are imbedded. This is particularly true for PMCs, and we have previously noted that polymeric matrices are not often well-bonded to the fibers reinforcing them. This is reflected in high critical aspect ratios (cf. Eq. (6.30)), and is also manifested by the appearance of PMC tensile fracture surfaces. Examples of these are provided in Figs. 6.33. The "pull-out" fiber lengths displayed there scale with the fiber ineffective length, and the "shaving brush" appearance of the fracture surfaces reflects the relatively high critical aspect ratios that characterize PMCs.

While glass fibers/filaments—because of their relatively low cost—are widely used to reinforce polymers, they are seldom employed in MMCs. There are several reasons. One is that the high cost of MMCs makes it only marginally more expensive to use a higher performance and more costly reinforcement in them. Another is that many MMCs are intended for high-temperature use. The relatively low softening temperature of most glass fibers/filaments mitigates their effective employment at high temperatures.

Most MMC development has, as might be expected, focused on light metals as matrix materials. Their low densities, in conjunction with high specific strengths and moduli (i.e., strength or modulus divided by density), make them particularly attractive for aerospace applications where performance is often a more compelling design requirement than is cost. The point is illustrated in Fig. 6.34, where specific

(a)

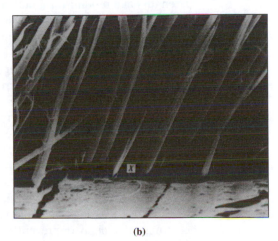

(b)

Figure 6.33
Scanning electron micrographs of the tensile fracture surfaces of (a) a glass fiber-polyester resin composite and (b) a Kevlar (aromatic polyamide)-epoxy resin composite. Extensive fiber pullout gives rise to the "brush-like" appearance of the surfaces, and is characteristic of composites containing fibers with long ineffective lengths. The diameter of the Kevlar fibers is ca. 5 μm. (*From D. Hull, An Introduction to Composite Materials, Cambridge University Press, Cambridge, England, 1981.*)

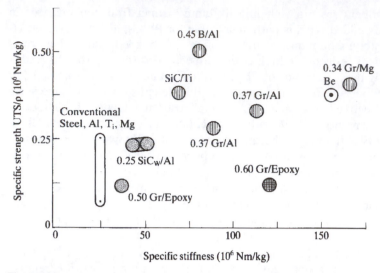

Figure 6.34
Specific strength–specific stiffness comparison of various composites and conventional metallic materials. Excepting the metal beryllium, most of the composites whose properties are displayed exhibit specific strength-stiffness combinations superior to the metals. The number in front of the composite represents the volume fraction of the reinforcement (e.g., 0.45B/Al indicates an aluminum matrix containing 45 vol.% of boron; 0.37 Gr/Al indicates an aluminum matrix containing 37 vol.% graphite). (*Adapted from B. J. Maclean and M. S. Misra, Mechanical Behavior of Metal-Matrix Composites, ed. J. E. Hack and M. F. Amateau, TMS-AIME, Warrendale, Pa., p. 301, 1983.*)

strengths and moduli for a number of advanced composites are shown. The composites whose properties are displayed in Fig. 6.34 offer clear advantages relative to conventional metals when high strength coupled with low density is a design requirement.

In distinction to PMCs, fiber-matrix bonds in MMCs are often chemical in nature, and this results in shorter critical aspect ratios. For a matrix well bonded to a fiber, the shear stress at the fiber-matrix interface can be taken as the matrix shear flow stress. This assumption, in fact, formed the basis for much of the analysis of this chapter. The fracture surface of an SiC-reinforced titanium composite is shown in Fig. 6.35. Comparison to Fig. 6.33 clearly demonstrates the relatively short fiber critical length of this MMC. Fiber-matrix bonding in MMCs can be enhanced by fiber surface treatment prior to dispersing the fibers in the matrix. The treatment usually involves producing a thin layer of a different chemical constituent on the fiber surface, the layer facilitating bonding to the matrix. In addition, such surface coatings often delay or prevent chemical interactions between the fiber and the matrix. These can occur at high temperatures and result in diminished composite performance.

Fiber-reinforced MMCs are expensive owing to the "cottage industry" manner in which they are made. A typical manufacturing process, for example, might involve laying fibers between sheets of matrix, followed by consolidation of the array by hot pressing. Consolidation conditions must be carefully chosen so as not

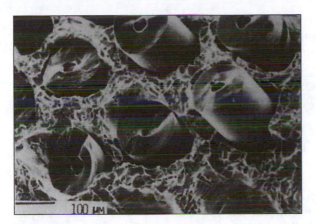

Figure 6.35
Fracture surface of an SiC-reinforced titanium composite. Fiber pullout is much less
extensive than in PMCs (e.g., Fig. 6.33), indicating superior fiber-matrix bonding.
The dark central core of the SiC fiber is a substrate on which SiC is deposited dur-
ing manufacture of the fiber. (*W. D. Brewer and J. Unman. Reprinted with permission*
from Mechanical Behavior of Metal-Matrix Composites, ed. J. E. Hack and M. F. Am-
ateau, TMS-AIME, Warrendale, Pa., p. 39, 1983.)

to damage the fibers during the process. Particle-reinforced MMCs are not as expensive as fiber-reinforced ones. A typical such composite—SiC particle-
reinforced aluminum, for example—might be produced by codeposition of powders
of both phases followed by extrusion to yield a fully dense product.

For reasons noted previously, CMCs are of current interest. Carbon-carbon
composites, in which high-strength carbon fibers are imbedded in a graphite matrix,
are among the most prominent CMCs. Carbon-carbon composites display excellent
high-temperature mechanical properties, although the graphite in them must be protected from oxidation. As noted, reinforcements in CMCs are used as much (or
more) for toughening as for strengthening. Fibers and plates are especially effective
shapes for providing toughness. Since the ceramic matrix usually fails prior to the
reinforcement, the unbroken fibers (plates) reload the matrix following this failure
(Sect. 6.10). The unbroken constituent also provides a crack-closing force during
crack extension, thereby increasing fracture toughness. Dispersed ductile particles,
while not as effective as fibers or plates in this regard, also can be used to increase
ceramic fracture toughness. In this instance the increased toughness comes from the
plastic work associated with the ductile-phase fracture during crack extension. The
topic is further considered in Chap. 10.

6.12
SUMMARY

In this chapter the strengths of composite materials were discussed in terms of the
properties of the composite constituents, their volume fractions, their shapes, and
their arrangements. Reinforcement shape and arrangement dictate the extent to

which the composite constituents experience the same stress or the same strain under an applied load. If they suffer the same stress, they experience different strains and vice versa. The latter situation is better for effectively utilizing the high strengths of certain fibrous materials.

Composites come in different forms; they can have polymer, metal, or ceramic matrices. Polymer matrix composites are the most widely used. They are inexpensive compared to composites having other types of matrices. However, PMCs suffer from a low maximum use temperature; their matrices decompose or soften at relatively low temperatures. High-temperature use requires a MMC or a CMC. Metal matrix composites are strengthened by a reinforcing phase, while CMCs are toughened by a (generally) more ductile phase.

Reinforcements come in different shapes. Particles can be used as reinforcement. However, they are not as efficient in doing so as are fibers. Nonetheless, particle-reinforced composites find use because they are typically less expensive. When the dispersed phase is in the form of a particle, the equal-strain condition does not apply. However, empirical equations describing the variation of composite modulus and strength with particle volume fraction and matrix and particle properties are available.

For a composite in which the fibers are arranged so that the tensile axis is parallel to the fiber axis, the volume fraction rule (VFR—Eqs. (6.12) and (6.21)) is appropriate for describing composite modulus and strength. This VFR states that the stress carried by the composite at a specified strain is a volume fraction–weighted average of the stresses borne by the fiber and the matrix at this same strain. The VFR is but an approximation if the fibers are of finite length; yet it is a good one provided the fiber length is long in comparison to the fiber critical length. The latter is the length over which shear stresses, developed by relative matrix-fiber displacement at a fiber end, load the fiber to a tensile strain equal to that of the matrix. The VFR disregards phase interactions, which may arise if the microstructural scale of the composite is fine. This effect can often be disregarded, since structural scale in composites is typically relatively coarse. When it is not, as for *in situ* composites, the VFR must be modified to take into account additional strengthening due to phase interactions.

When aligned fiber composites are tested "off-axis," their strengths are (except for very small misorientations) less than when the tensile axis is parallel to the fiber axis. This behavior can arise from different composite failure modes for off-axis loading; for example, matrix shear or matrix tensile failure. Models that predict off-axis strengths as a function of the angle between the applied stress and the fiber axis are available. These models are couched in terms of the composite tensile strength in the "parallel" (i.e., stress parallel to the fiber axis) and "perpendicular" (stress normal to the fiber axis) orientations. The Tsai-Hill model, in particular, well describes off-axis strengths of aligned fiber composites.

Because the strengths of aligned fiber composites decrease dramatically when they are subjected to off-axis loading, measures must be taken when composites are subjected, for example, to biaxial loading. Composite structures, consisting of plies of aligned fibers oriented in different directions, can be used for biaxial loading conditions. When isotropic properties are required, composites containing randomly oriented discontinuous fibers can be employed.

When fibers having an intrinsically large variability in fracture strengths are used in composites, statistical considerations must be incorporated into the VFR for composite strength. Provided the distribution in strengths for fibers of length used in the composite is known, and provided, too, that the fiber critical length is known, values of composite strength for a composite failing via a statistical accumulation of fiber fractures can be predicted. The *effective* fiber strength in such a composite is much greater than that of an individual fiber. This is a result of the capacity of the matrix to reload a fiber after it is broken. Likewise, the variation in composite strengths is much less than the corresponding variation in fiber strengths, particularly if the composite contains fibers of length much greater than the critical length.

High-temperature composite use entails consideration of matrix creep and strain-rate effects. In a tension (or creep) test at a temperature at which such effects are important, the plastically flowing matrix transmits load to the fibers; however, in contrast to temperatures at which strain-rate effects are unimportant, the stress-transfer is time dependent. At low strains (or times), a goodly portion of the fiber length experiences the same tensile strain as the matrix; i.e., the fiber elastically creeps along with the plastic matrix. However, even though the fiber carries a higher stress with increasing strain, a smaller fraction of its length is strained to the same extent as the matrix. At even longer times, the fiber carries a load that does not increase further with additional composite strain. At this deformation stage, the composite contains load-carrying "rigid" fibers imbedded in a matrix that creeps around them.

Ductile fibers imbedded in a brittle matrix "toughen" it, thereby increasing its fracture toughness. Multiple cracking, in layers, of the matrix may precede tensile fracture provided the fiber is able to transmit tensile load to the matrix above and below the fracture plane. If so, the crack plane spacing is related to the matrix in-effective length over which reloading occurs.

REFERENCES

Easterling, K. E.: *Advanced Materials for Sports*, Chapman and Hall, London, England, 1993.

Everett, R. K. and Arsenault, R. J., eds.: *Metal Matrix Composites; Mechanisms and Properties*, Academic Press, San Diego, 1991.

Holister, G. S. and Thomas, C.: *Fiber Reinforced Materials*, Elsevier, London, England, 1986.

Hull, D.: *An Introduction to Composite Materials*, Cambridge University Press, Cambridge, England, 1981.

Kelly, A.: *Strong Solids* (Chaps. V and VI), Oxford University Press, Oxford, England, 1966.

Piggott, M. R.: *Load Bearing Fibre Composites*, Pergamon Press, Oxford, England, 1980.

PROBLEMS

6.1 a Consider the following materials as potential materials for vaulting poles. Calculate the values of the performance index, $\sigma_f^2/2E\rho$, for these materials and select the best vaulting pole material on this basis.

Material	ρ (kg/m³)	E (GPa)	σ_f (MPa)
High-strength Al alloy	2710	69	500
High-strength steel	7850	207	1590
CFRP	1600	145	1240
GFRP	2100	80	1020
Hickory	600	15	100

b In part (a), you found that high-strength steel—which is not a suitable vaulting pole—has a performance index comparable to some materials that have been so used. Why is steel unsuitable as a vaulting pole?

6.2 a Would selection of a material for a diving board involve the same performance index as that of a vaulting pole? Explain.

b Suppose a diving board were constructed of a CFRP having the properties listed in Prob. 6.1. Let the diver have a mass of 70 kg and consider a 3-m-long board with a thickness and width of 2 cm and 30 cm, respectively. If the board stored the maximum possible elastic energy when the diver was launched from it, how high would it "catapult" the diver? (Hint: Conversation of energy.) Comment on your result.

6.3 Use Eq. (6.9) to show that the ratio of the forces carried by the α and β phases during composite elastic deformation is $F_\alpha/F_\beta = V_\alpha E_\alpha/V_\beta E_\beta$.

6.4 The "secondary" (Stage II) modulus during composite tensile deformation is given by Eq. (6.18). For MMCs, compare typical values of the first and second terms on the right-hand side of Eq. (6.18). (Take $V_f = 0.3$ and values of E_f for several of the fibers whose properties are listed in Table 6.1. Estimate $d\sigma_m/d\varepsilon$ for some typical metals from data of the kind provided in Chap. 5.)

6.5 a Do the calculations leading to the expression for V_c (Eq. (6.23)). Use Eqs. (6.21) and (6.22) as starting points.

b Derive Eq. (6.24), the expression for V_{min}. Use logic similar to that employed in part (a).

6.6 Estimate V_{min} for a number of fibers listed in Table 6.1. Do the calculations for an Al matrix with $\sigma_y = 70$ MN/m² and T.S. = 100 MN/m².

6.7 Consider the graphite-epoxy composite whose constituents have moduli as given in Example Prob. 6.1. Three sheets of this material are glued together. In two of the sheets the fibers are aligned parallel to each other; in the third sheet the fiber axes are aligned perpendicular to the fiber axes in the other two sheets. (See sketch below.)

Fibers

Sheets glued on top of each other

a Estimate the modulus of the three-sheet combination for the situation where the axis of the applied stress is parallel to the fiber axes in two of the three sheets.
b Estimate the modulus of the three-sheet combination for the situation where the axis of the applied stress is parallel to the fiber axes in one of the three sheets.

6.8 Calculate critical aspect ratios for several fibers listed in Table 6.1. Assume an aluminum matrix with $\tau_{my} = 70$ MPa.

6.9 Use Eq. (6.34) to calculate the critical misorientation angle, θ_{c1}, for the same material combinations of Prob. 6.6. Are these critical angles large or small?

6.10 Consider the graphite-epoxy composite of Prob. 6.7. Using the Tsai-Hill criterion, estimate the tensile strength of this composite when the tensile axis makes an angle of 45° to the fiber axis. Assume τ_{my} of the matrix is 50 MPa.

6.11 Assume two different types of fibers with different functions $g(\sigma)$ describing their strengths. For the first kind of fiber, $g(\sigma)$ is a constant between two stresses, σ_1 and σ_2, and zero at all other stress levels (i.e., no fibers fail at a stress below σ_1 and all will have failed at a stress σ_2; this is the same situation as in Example Prob. 6.3). For the second kind of fiber, $g(\sigma)$ increases linearly between σ_1 and σ_2, and is zero at all other stresses.
a Calculate the fiber stress at which the function $\sigma_f[1 - G(\sigma_f)]$ is a maximum.
b Calculate the bundle strength.
c Calculate the average fiber breaking stress ($\overline{\sigma}_f$) obtained by a series of tensile tests on the two kinds of fibers. Compare the $\overline{\sigma}_f$ values to the bundle strengths calculated in part (b).

6.12 Let $g(\sigma)$ $d\sigma$ be the fraction of fibers of length l that fail between the stresses σ and $\sigma + d\sigma$, and

$$G(\sigma) = \int_0^\sigma g(\sigma) \, d\sigma$$

The function $f(\sigma)$ $d\sigma$ represents the fraction of fibers of length $l_c/2$ that fail between σ and $\sigma + d\sigma$, with

$$F(\sigma) = \int_0^\sigma f(\sigma) \, d\sigma$$

The latter function can be obtained through knowledge of $g(\sigma)$ and $G(\sigma)$ as follows. Divide a fiber into a series of links $l_c/2$. Recall that for a fiber to fail between σ and $\sigma + d\sigma$ all of the links of length $l_c/2$ must *not* have failed at the stress σ, and one of the links must fail between σ and $\sigma + d\sigma$. Use concepts of probability to relate $f(\sigma)$ and $F(\sigma)$ to the functions $g(\sigma)$ and $G(\sigma)$.

6.13 Consider composites containing fibers of length $l = 50l_c$. For fibers having the strength distributions, $g(\sigma)$ specified in Prob. 6.11, calculate (a) the stress at which the function $\sigma_f[1 - F(\sigma_f)]$ is a maximum, and (b) the expected composite strength. (You must use the results of Prob. 6.12 here.) Compare these stresses (see Prob. 6.11) with (i) the average strength for fibers of length l, (ii) the stress at which the function $\sigma_f[1 - G(\sigma_f)]$ is a maximum, and (iii) the bundle strength for fibers of length l.

6.14 The distribution in strengths of brittle fibers of length l and diameter d often can be described by a Weibull distribution:

$$P_s(l) = \exp\left[-\left(\frac{\sigma}{\sigma_0}\right)^m\right]$$

Here $P_s(l)$ is the probability that a fiber (of length l) "survives" (without fracture) an imposed tensile stress, σ. (Thus, P_s is analogous to the function $[1 - G(\sigma_f)]$.)

a Consider a material for which $m = 5$. Suppose a series of tensile tests were conducted on fibers whose strength distribution is described by $P_s(l)$. What would be the "expected" fiber failure strength (i.e., the stress at which 50% of the fibers would have fractured)? Your answer will be expressed in terms of σ_0.

b Consider a bundle of these fibers. What is the expected value of the bundle strength? (Again, your answer is expressed in terms of σ_0. You can do this part of the problem either graphically or analytically. If you choose the latter approach, you might find the "arithmetic" simpler if you substitute $\Sigma = \sigma/\sigma_0$ before attempting differentiation.) How does the value of the bundle strength compare to the "expected" fiber strength?

c Now consider a composite containing these same fibers with a critical length equal to l_c. The survival probability for fibers of length $l_c/2$ is given by

$$P_s\left(\frac{l_c}{2}\right) = \exp\left[-\left(\frac{l_c}{2l}\right)\left(\frac{\sigma}{\sigma_0}\right)^m\right]$$

$(P_s(l_c/2)$ is analogous to the function $[1 - F(\sigma_f)]$.) Determine the expected composite strength (in units of σ_0) for fibers having lengths between $l = l_c/2$ and $l = 100\ l_c$. (Hint: When $l = l_c/2$, your results should reduce to that for a bundle.) Graphically portray these results.

d From what you have learned in this problem, can you determine how bundle and composite strength should vary with the parameter, m? If so, describe this relationship. What are the implications with respect to composite "design" using brittle fibers?

6.15 Consider the composite whose creep behavior is described by Eqs. (6.45)–(6.47). Compare the equations for the initial strain rate and asymptotic strain to those that apply to the Voigt model. What are the similarities and differences of the Voigt model and that exemplified by Eqs. (6.45)–(6.47)?

6.16 **a** Consider Eqs. (6.50) and (6.51) applied to a material having a strain-rate sensitivity, m, equal to 0.5. Derive an expression for the ratio of the stress carried by the fibers relative to that borne by the matrix as a function of (l/d_f) and V_f. Plot this function for different values of (l/d_f) (ranging from 10 to 100) and a fixed ($= 0.3$) value of V_f.

b Now let V_f vary from 0.1 to 0.7. Plot the above ratio as a function of V_f taking a value of (l/d_f) of 20.

c Use your results from parts (a) and (b) to schematically plot the composite creep rate, first as a function of (l/d_f) for a fixed V_f and, second, as a function of V_f for a fixed value of (l/d_f).

6.17 Derive Eq. (6.54), the expression for the matrix critical length in a brittle matrix–ductile fiber composite. (Hint: Use similar reasoning to that used to calculate the fiber critical length in a brittle fiber–ductile matrix composite.)

6.18 For a brittle matrix reinforced with ductile fibers, multiple matrix cracking occurs prior to final composite fracture (cf. Fig. 6.30). How is this spacing related to the matrix critical length? How do you expect the spacing to be affected by variations in the matrix fracture strength along the sample gage length?

High-Temperature Deformation of Crystalline Materials

7.1
INTRODUCTION

Use of materials at elevated temperatures involves additional considerations to those important at lower temperatures. At low temperatures, permanent deformation is prevented by assuring that the material is not subjected to an effective stress exceeding its yield strength. At high temperatures, on the other hand, permanent deformation can be effected over a *period of time* at a stress level well below the material's tensile yield strength. This time-dependent deformation is called creep and is observed in both crystalline and noncrystalline solids. In this chapter, our discussion concerns creep in crystalline solids. (Deformation in noncrystalline solids is discussed in Chap. 8.) Several mechanisms contributing to creep of crystalline materials involve dislocation motion. Others, however, involve only diffusional flow of atoms. In distinction to dislocation creep, these latter creep mechanisms can be observed in both crystalline and noncrystalline materials.

Whether time-dependent (creep) or time-independent (yielding) flow needs to be considered in design depends on the operating stress and temperature. Low stress levels and high temperatures (typically ≥ 0.4–$0.5 \, T_m$ where T_m is the material's absolute temperature) usually call for consideration of creep. Aids in determining which type of deformation dominates at different stress/temperature combinations are provided by "deformation maps." These maps, having axes of stress and temperature, are divided into regions that delineate the dominant deformation mode (elasticity, time-independent plasticity, time-dependent plasticity).

Although deformation maps are useful, they are unable, per se, to predict the high-temperature engineering life of a material. Instead, this is estimated through engineering approximation. Since the intended life of most components often far exceeds the duration of any reasonable laboratory test, service-life estimates are based on extrapolation of data obtained at higher temperatures and/or stress levels

294

CHAPTER 7
High-Temperature
Deformation of
Crystalline
Materials

than those intended for service. One such extrapolation procedure, and its basis, are described in this chapter. Likewise, engineering development of materials suitable for high-temperature use involves synthesis of microstructures resistant to creep deformation, and such considerations are considered briefly here.

Closely related to creep occurring in fine-grained materials is a phenomenon called superplasticity. Superplastic materials exhibit a strong strain-rate dependence of the flow stress over a limited temperature range and in this range they are capable of extensive uniform plastic deformation; e.g., in a tension test, necking is delayed until very large strains. Delayed necking in superplasticity is related to the strain-rate dependence of the flow stress (strain-rate hardening, cf. Chap. 1). Superplasticity is useful for forming certain materials, for the stress required to plastically deform superplastic materials is low, and, as a result of the attendant high plasticity, shapes that cannot be produced by conventional hot-working processes can be produced by superplastic forming.

7.2
PHENOMENOLOGICAL DESCRIPTION OF CREEP

The essential difference between time-independent and time-dependent plasticity is illustrated in Fig. 7.1. For the stress-temperature combination applying to the curve marked (a), permanent deformation is time-independent. That is, subsequent to the application of a constant stress at time $t = 0$, essentially instantaneous elastic and plastic strain of magnitude ε_E and ε_p is obtained and, irrespective of the time duration of the stress, the material strains no further. On load removal, the elastic strain

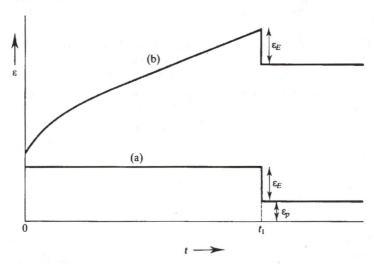

Figure 7.1
The difference between time-independent (curve a) and time-dependent (curve b) deformation. Strain is plotted for a constant stress, applied at $t = 0$ and held until time t_1 at which it is removed. For (a), the initial (elastic plus plastic) strain is the only strain suffered by the material. The elastic strain (ε_E) is recovered on unloading. In (b), strain increases with time. On unloading, only the initial elastic strain is recovered; the initial (time-independent) and time-dependent plastic strains are not.

is recovered and the permanent strain remains. This behavior is in contrast to curve (b) of Fig. 7.1, which is appropriate for a material subjected to a stress-temperature combination at which time-dependent plasticity occurs. Here, if the stress level is greater than the yield strength, the initial strain may include some time-independent plastic strain. But even if it does not, permanent strain, the extent of which increases with time, is observed subsequent to application of the stress. Upon load removal, only the elastic component of strain is recovered; both the initial and time-dependent (creep) permanent strain remain.

Figure 7.2a illustrates schematically the creep behavior of a material subjected to a constant true stress as the behavior depends on stress and temperature. Both stress and, particularly, temperature increase the creep strain rate ($\dot{\varepsilon} = d\varepsilon/dt$).

The creep curves illustrated in Fig. 7.2a can conveniently be divided into three stages. In Stage I (transient creep) $\dot{\varepsilon}$ decreases with time and strain.[1] Transmission-microscopic examination of materials subjected to transient creep strains at relatively high stress levels shows that the structure evolves with increasing strain in a manner somewhat analogous to that observed during work hardening at lower temperatures. For example, the dislocation density increases and, in many materials, a dislocation subgrain structure is formed with a cell size that decreases with strain. These structural changes are consistent with a decreasing creep rate during Stage I creep.

During Stage II creep, similar observations show that a "steady-state" microstructure obtains, and this is consistent with the constant-creep-rate characteristic of steady-state creep. The invariant microstructure is indicative that recovery

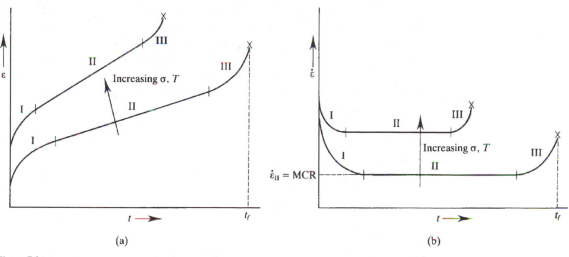

(a)   (b)

Figure 7.2
Strain (a) and strain rate (b) vs. time in a constant-stress creep test. Creep behavior can be divided into three stages. In Stage I (transient creep), the strain rate decreases until it attains a steady-state, minimum value (Stage II). Tertiary creep (Stage III), characterized by an increasing strain rate, precedes fracture at t_f. Increasing stress and/or temperature raises the overall level of the creep curve and also results in higher creep rates.

---

[1]In certain metal alloys, the creep rate increases during transient creep. Nonetheless, these materials also display a constant Stage II creep rate. These "special" materials are discussed further in Sect. 7.3C.

296

CHAPTER 7
High-Temperature
Deformation of
Crystalline
Materials

effects are concurrent with deformation during Stage II. If that were not the case, the dislocation density would increase and a subgrain structure would become progressively finer with increasing strain. Hence, hardening mechanisms effective at low temperature are not as effective at higher temperatures, and Stage II creep can be viewed as a regime in which the work-hardening capacity of the material is balanced by recovery or "softening" effects. On a microscopic level, a steady-state structure is also consistent with the view that nonconservative dislocation motion renders obstacles to glide less effective at elevated temperatures, and also with the notion that dislocations can be removed from the material by recovery. Mechanisms by which these may take place are discussed in the following section. For the present we simply note that nonconservative dislocation motion requires the presence of a reasonable concentration of mobile vacancies, and thus creep deformation is significant only at elevated temperatures.

Subsequent to Stage II creep, tertiary (Stage III) creep is observed. (The relative strain extent of Stage III creep vis-à-vis the other creep stages depends on the applied stress level and generally increases with increasing stress.) During tertiary creep, the creep rate exceeds that of Stage II and also increases continuously. This is shown in a plot of *creep rate* versus strain or time (Fig. 7.2b), which also shows that Stage II is characterized by a constant, minimum creep rate (MCR = $\dot{\varepsilon}_{II}$).[2] The accelerating creep deformation of Stage III eventually leads to material fracture (point X, Fig. 7.2a) and is related to several factors. Tests conducted under constant *true* tensile stress conditions, for example, show that the transition from Stage II to Stage III creep can be correlated with microscopic changes in the material. Such alterations include the onset of recrystallization, the coarsening of second-phase particles and/or the formation of internal cracks or voids. The latter are precursors to fracture that occurs at the time t_f (the time to rupture or, more precisely, the time to fracture).[3] High-temperature fracture is considered in Chap. 11.

A material's engineering creep resistance is frequently characterized by one or both of the parameters $\dot{\varepsilon}_{II}$ and t_f. However, scientific discussion of creep is couched almost solely in terms of $\dot{\varepsilon}_{II}$. Provided Stage II creep constitutes a significant fraction of the material's total creep strain, $\dot{\varepsilon}_{II}$ and t_f are inversely related. In a practical sense, $\dot{\varepsilon}_{II}$ and t_f represent extremum design criteria. For example, if a component is intended for short-time elevated temperature use (e.g., a "one-shot" rocket engine component), creep deformation may be tolerable but fracture is not. In such a case, t_f is an appropriate design parameter. On the other hand, many components are designed for high-temperature operation of hundreds to thousands of hours (e.g., a jet engine turbine blade) or, in some cases, many years (e.g., boiler tubing). In these situations, structural integrity, as reflected in a low permanent strain, is required and $\dot{\varepsilon}_{II}$ is the critical design parameter.

As mentioned and as indicated in Fig. 7.2, creep rate increases with both stress and, especially, temperature. For many materials the steady-state creep rate can be correlated to these variables by an equation of the form[4]

---

[2]For the "special" alloys mentioned in Footnote 1, the steady-state creep rate is not the minimum one.
[3]As we shall see in Chap. 11, the term "rupture" is a misnomer here. However, it is widely used in the high-temperature engineering community. In our discussion, we designate the fracture or rupture time by the symbol t_f rather than t_r as is more common in the high-temperature materials community.
[4]Creep mechanisms that determine the basis of the empirical Eq. (7.1) are discussed in Sect. 7.3.

$$\dot{\varepsilon}_{\mathrm{II}} = A\sigma^{m'} \exp\left[\frac{-Q_C}{RT}\right] \tag{7.1}$$

where A and m' are material constants and Q_c is the creep activation energy. In general, Eq. (7.1) can be applied only over a limited stress/temperature range as the constants A, m' and Q_c can vary with stress and temperature. Variations in these parameters are related to changes in creep mechanism as discussed in the following section. However, provided that $T \geq 0.5\, T_m$ (the temperature regime for which creep processes are most important), the activation energy Q_c is often essentially equal to the activation energy for self-diffusion (Fig. 7.3). This correlation gives additional credence to the view that vacancy motion is indispensable to creep processes, and we now turn our attention to describing several of these.

7.3
CREEP MECHANISMS

In this section, several mechanisms of creep are discussed to illustrate the basis of Eq. (7.1). Depending on temperature and applied stress, dislocation glide, dislocation climb, or diffusional-flow mechanisms may dominate creep deformation. Some of the mechanisms described, particularly those involving dislocations, are speculative in that they cannot be (or at least have not yet been) verified by direct microstructural examination. Nonetheless, processes similar to the ones envisaged

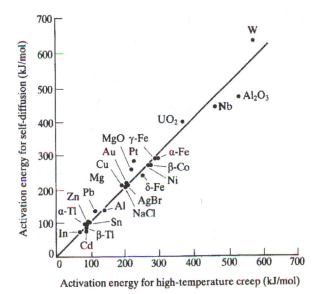

Figure 7.3

High-temperature creep activation energies correlate closely with self-diffusion activation energies. The correlation indicates that creep deformation is associated with diffusional flow. (*Reprinted from O. D. Sherby and P. M. Burke, Prog. Matls. Sc., 13, 325, Copyright 1967, with permission from Elsevier Science.*)

298

CHAPTER 7
High-Temperature
Deformation of
Crystalline
Materials

surely occur during creep deformation. Beyond that, description of the processes allows correlation of Eq. (7.1) with microstructure and the "forces" of temperature and stress.

A. Dislocation Glide at Low Temperature

Even at low temperatures, thermal activation affects lattice resistance to dislocation glide. This was illustrated in Sect. 5.4, in which the temperature variation of the flow stress was related to the temperature-assisted force required to overcome obstacles to dislocation motion. Although overcoming such barriers is not associated with diffusional flow and, therefore, lies outside the main thrust of this chapter, discussion of this process serves to illustrate the role that temperature and stress play in determining creep rates.

The role of temperature in such "dislocation glide" creep can be illustrated with reference to Figs. 7.4a and b, which schematically portray the lattice energy before and after a dislocation bypasses an obstacle on its slip plane. The situation of Fig. 7.4a applies in the absence of an applied stress. In this circumstance, the dislocation is equally "satisfied" on either side of the barrier. The activation energy U_0 is the integral of the force-distance curve of Fig. 5.14a up to the position $x = 0$ and represents the energy required for a dislocation to bypass the obstacle; such an energy can be supplied thermally (the magnitude of the thermal energy is on the order of kT). An applied stress tends to drive the dislocation past the barrier, and alters the energy-distance profile (Fig. 7.4b). Now the dislocation is in a lower energy state after passing the obstacle; the energy difference is noted as δU. Principles of kinetics allow determination of the creep rate.

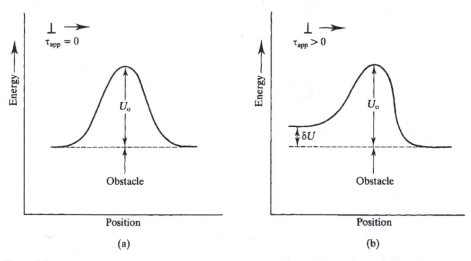

Figure 7.4
(a) Schematic of energy variation with dislocation position as a dislocation approaches a slip-plane obstacle. (b) Part of the work (δU) required to overcome the barrier (U_0) is provided by the applied stress and the remainder by thermal energy. The dislocation glide-creep rate can be deduced from kinetic principles as discussed in the text.

The rate at which dislocations bypass the obstacle in the forward (stress-aided) direction is given by

$$\text{Forward rate} \sim \exp\left(-\frac{(U_0 - \delta U)}{kT}\right) \qquad (7.2)$$

Likewise, the reverse reaction rate (i.e., dislocations can move in a direction *opposite* to that favored by the applied stress, but at a lesser rate) is proportional to $\exp(-U_0/kT)$. The net rate (which is proportional to the strain rate) scales with the difference between these terms. Thus, the dislocation glide creep rate, $\dot{\varepsilon}_{DG}$, can be written as

$$\dot{\varepsilon}_{DG} = \dot{\varepsilon}_0 \exp\left(-\frac{U_0}{kT}\right)\left[\exp\left(\frac{\delta U}{kT}\right) - 1\right] \qquad (7.3)$$

At the low temperatures of concern, $\exp(\delta U/kT)$ is much greater than unity (except at very low stress levels); thus,

$$\dot{\varepsilon}_{DG} \cong \dot{\varepsilon}_0 \exp\left(-\frac{U_0}{kT}\right)\exp\left(\frac{\delta U}{kT}\right) \qquad (7.4)$$

In Eqs. (7.3) and (7.4), $\dot{\varepsilon}_0$ is a material parameter relating to the frequency of atomic vibrations and has units of s^{-1}.

The energy δU is related to, and increases with, the applied stress; e.g., $\delta U = 0$ if $\tau = 0$. In essence, δU is the work done by this stress in approaching the obstacle and this work reduces the thermal energy required to overcome the barrier. If L is the effective spacing between slip plane obstacles, then $\delta U = \int L\tau b\, dx$, where the integral is taken from $x = \infty$ to the position at which $\tau b = F_{\text{app}}$ (cf. Fig. 5.14). The integral can be approximated as $b\tau a_s$ where a_s is an appropriate area on the slip plane. On doing so the strain rate is obtained as

$$\dot{\varepsilon}_{DG} = \dot{\varepsilon}_0 \exp\left(-\frac{U_0}{kT}\right)\exp\left(\frac{\tau b a_s}{kT}\right) \qquad (7.5)$$

Although Eq. (7.5) is derived on a model for which plastic strain does not devolve on atomic diffusion, it bears similarity to creep equations involving such diffusion. In particular, the creep rate is controlled by an intrinsic activation energy ($= U_0$ in Eq. (7.5)). Moreover, the rate depends also on the ratio of a stress-assisted energy ($= \tau b a_s$ in Eq. (7.5)) to thermal energy. The numerator in this ratio has dimensions of energy; it can be viewed as a product of stress and volume. As this ratio increases, the creep rate does also. We shall see that all creep-rate equations, irrespective of mechanism, contain similar terms. However, the significance of the activation energy and the stress-volume terms depend on the specific creep mechanism.

B. Diffusional Flow Creep Mechanisms

i. NABARRO-HERRING CREEP.
Dislocation glide creep does not involve atomic diffusion. Nabarro-Herring (NH) creep is the opposite in that NH creep is accomplished solely by diffusional mass transport. Nabarro-Herring creep dominates creep behavior at much lower stress levels and higher temperatures than those

300

CHAPTER 7
High-Temperature
Deformation of
Crystalline
Materials

at which creep is controlled solely by dislocation glide. Since it does not involve dislocations, NH creep is also observed in amorphous materials. However, discussion of NH creep is facilitated by considering first how it is accomplished in a crystalline material subjected to the stress state shown in Fig. 7.5.

The grain illustrated in Fig. 7.5 may be considered either an isolated single crystal or an individual grain within a polycrystal. As indicated, the lateral sides of the crystal are subjected to a compressive stress, and the horizontal sides to a tensile stress. The stresses alter the atomic volume in these regions; it is increased in regions experiencing a tensile stress and decreased in the volume under compression. As a result, the activation energy for vacancy formation is altered by $\pm\sigma\Omega$, where Ω is the atomic volume and the $\pm$ signs refer to compressive and tensile regions, respectively. Thus, the fractional vacancy concentration in the tensile and compressively stressed regions are given as

$$N_v(\text{tension}) \cong \exp\left(-\frac{Q_f}{kT}\right) \exp\left(\frac{\sigma\Omega}{kT}\right) \tag{7.6a}$$

and

$$N_v(\text{compression}) \cong \exp\left(-\frac{Q_f}{kT}\right) \exp\left(-\frac{\sigma\Omega}{kT}\right) \tag{7.6b}$$

where Q_f is the vacancy-formation energy.[5] Provided the grain boundary is an excellent source or sink for vacancies if the grain of Fig. 7.5 is a polycrystal or, if it is a single crystal, the surface of it behaves likewise, the vacancy concentrations given by Eq. (7.6) are always maintained at the horizontal and lateral surfaces. The different concentrations there drive a net flux of vacancies from the tensile to the compressively stressed regions, and this is equivalent to a net mass flux in the opposite direction. As illustrated in Fig. 7.5b, this produces a change in grain shape. The grain elongates in one direction and contracts in the other; that is, creep deformation occurs.

The creep rate resulting from this process is estimated as follows. The vacancy flux, J_v, through the crystal is given by

$$J_v = -D_v\left(\frac{\delta N_v}{\delta x}\right) \tag{7.7}$$

where D_v is the vacancy diffusivity $[= D_{0V} \exp(-Q_m/kT)$, where Q_m is the vacancy motion energy] and $\delta N_v/\delta x$ is the vacancy concentration gradient. The term δx can be taken as a characteristic diffusion distance proportional to the grain size, d (cf. Fig. 7.5a), whereas δN_v is the difference between Eqs. (7.6b) and (7.6a). Multiplication of Eq. (7.7) by the diffusion area (proportional to d^2) gives the volumetric flow rate $\delta V/\delta t$; it represents the volume transferred per unit time from the lateral to the top and bottom sides of the crystal. According to the above reasoning,

$$\frac{\delta V}{\delta t} \cong D_{0V} d \exp\left[-\frac{Q_f + Q_m}{kT}\right]\left[\exp\left(\frac{\sigma\Omega}{kT}\right) - \exp\left(-\frac{\sigma\Omega}{kT}\right)\right] \tag{7.8}$$

---

[5]In Eqs. (7.6) the dimensions of Q_f are energy; e.g., in units of joules. The term Q_f/kT can also be written in the form Q_f/RT (R = the gas constant) if Q_f is expressed as the formation energy of a mole of vacancies. In this case, the units of Q_f are J/mole.

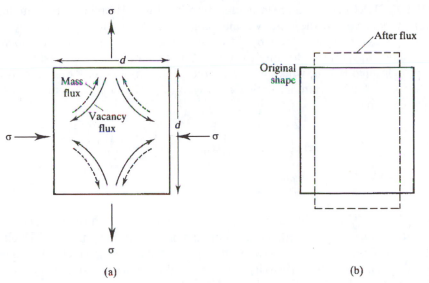

Figure 7.5
Nabarro-Herring creep results from a higher vacancy concentration in regions of a material experiencing a tensile stress compared to those subject to a compressive stress. This results in a vacancy flux from the former to the latter regions and a mass flux in the opposite direction (a). The resulting change in grain dimensions (b) is equivalent to a creep strain.

The change in length (δd) of the crystal along the tensile axis is related to δV by $\delta V \cong d^2\,\delta d$. The corresponding Nabarro-Herring creep rate ($\dot{\varepsilon}_{NH}$) is expressed as $(1/d)(\delta d/\delta t)$; thus

$$\dot{\varepsilon}_{NH} = \left(\frac{D_{0v}}{d^2}\right)\exp\left[-\frac{Q_f + Q_m}{kT}\right]\left[\exp\left(\frac{\sigma\Omega}{kT}\right) - \exp\left(-\frac{\sigma\Omega}{kT}\right)\right] \qquad (7.9)$$

The term $D_{0V}\exp[-(Q_f + Q_m)/kT]$ is identically equal to the lattice self-diffusion coefficient, D_L. Moreover, at the high temperatures and low stresses at which NH creep is important $\sigma\Omega \ll kT$, so that $\exp[\pm\sigma\Omega/kT] \cong 1 \pm \sigma\Omega/kT$. Using these relations, and letting a constant A_{NH} represent geometrical factors that we did not fully consider,[6] $\dot{\varepsilon}_{NH}$ can be written as

$$\dot{\varepsilon}_{NH} = A_{NH}\left(\frac{D_L}{d^2}\right)\left(\frac{\sigma\Omega}{kT}\right) \qquad (7.10)$$

As mentioned, NH creep is important at high temperatures and low stresses, i.e., in the temperature-stress regime where dislocation glide is not important. Nabarro-Herring is more important in creep of ceramics than in metals. This is so because dislocation mechanisms of creep can be considered competitive with NH creep, and dislocation motion is generally more difficult to effect in ceramics than in metals.

---

[6]More refined treatments show that A_{NH} is on the order of ten.

302

CHAPTER 7
High-Temperature
Deformation of
Crystalline
Materials

ii. COBLE CREEP. Coble creep is closely related to NH creep; for example, Coble creep is driven by the same vacancy concentration gradient. However, in Coble creep mass transport occurs by diffusion along grain boundaries in a polycrystal or along the surface of a single crystal. For polycrystals, the diffusion area is thus proportional to $\delta' d$, where δ' is an effective grain-boundary thickness for mass transport. Analysis similar to that employed for NH creep yields an expression for Coble creep:

$$\dot{\varepsilon}_C = A_C \exp\left(-\frac{Q_f}{kT}\right) D_{OGB}\left[\exp\left(-\frac{Q_m}{kT}\right)\right]\left(\frac{\delta'}{d^3}\right)\left(\frac{\sigma\Omega}{kT}\right)$$

$$= A_C\left(\frac{D_{GB}\delta'}{d^3}\right)\left(\frac{\sigma\Omega}{kT}\right) \tag{7.11}$$

In Eq. (7.11) Q_f represents, as it did previously, the vacancy formation energy, but Q_m is the activation energy for atomic motion along the grain boundary. The exponentials containing these terms have been incorporated into D_{GB} which represents an *effective* grain-boundary diffusivity[7] (or surface diffusivity if a single crystal is considered). As indicated by Eq. (7.11), Coble creep is more sensitive to grain size than NH creep. Thus, even though both forms of creep are favored by high temperature and low stress, Coble creep will be more important in very fine grained materials. In the general case, the diffusional creep rate should be considered a sum of $\dot{\varepsilon}_{NH}$ and $\dot{\varepsilon}_C$, since the mechanisms operate in tandem; i.e., they are *parallel* creep processes.

It is instructive to compare Eqs. (7.10) and (7.11) with Eq. (7.5). Some similarities are evident. For these extremes of creep mechanism, creep rates in all cases are thermally activated, and also related to a ratio $\sigma v/kT$, where v represents a volume. However, in physical terms, the activation energies and the volume term differ substantially between dislocation glide and diffusional creep. This notwithstanding, the phenomenological expressions for creep rate do bear similarities. We shall find this is also true for other creep mechanisms.

To prevent the formation of internal voids or cracks during polycrystalline diffusional creep, additional mass-transfer must occur at the grain boundaries. This results in *grain-boundary sliding* and the diffusional creep rate must be balanced exactly by the grain-boundary sliding rate if internal voids are not to form. The situation is illustrated in Fig. 7.6 for several grains within a polycrystal. As shown, "unaccommodated" creep leads to grain-boundary separation (Fig. 7.6b). The separation is prevented by concurrent displacement of the grains via their sliding over one another so as to "heal" the cracks (Fig. 7.6c) that would be formed by unaccommodated diffusional flow.

Diffusional flow and grain-boundary sliding, therefore, can be considered *sequential* processes in which mass is first transported by NH and/or Coble creep and a grain shape change and separation are effected. This is followed by "crack heal-

---

[7]The term D_{GB} in Eq. (7.11) is not the same as a grain-boundary diffusion coefficient. The reason is that D_{GB} of Eq. (7.11) includes an activation energy for vacancy formation within the crystal volume. A true grain-boundary diffusion coefficient has no such activation energy associated with it.

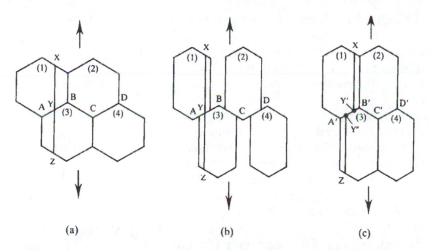

Figure 7.6
(a) Four grains in a two-dimensional hexagonal array before creep deformation. (b) Following diffusional creep, one dimension of the grains is increased and the other decreased and "voids" form between the grains. (c) The voids are removed by grain-boundary sliding. The extent of sliding is quantified by the distance Y'Y'', which is the offset along the boundary between grains 1 and 3 of the original vertical scribe line XYZ. (*Reprinted from A. G. Evans and T. G. Langdon, "Structural Ceramics," Prog. Matls. Sc., 21, 17, Copyright 1976, with permission from Elsevier Science.*)

ing" via grain-boundary sliding.[8] Since the grain-boundary sliding and diffusional flow processes occur sequentially, the net creep rate is the lesser of the separate creep rates. In general, the accommodating grain-boundary sliding creep rate is, well, accommodating. That is, it takes place rapidly relative to the diffusional flow creep. As a result, the expressions presented in this section generally describe well the creep rate. However, there are exceptions. For example, special boundaries—across which atoms match up more-or-less as they do across a coherent boundary in a two-phase material—exist; the sliding rates of these boundaries are rather low. Further, grain-boundary ledges (steps on the order of several atomic diameters in height) occur frequently and these structural features impede grain-boundary sliding. Except when grain-boundary sliding is restricted by second-phase particles (Sect. 7.3D), the diffusional flow processes causing sliding are the same as those causing Coble or NH creep. The result is that even when the grain-boundary sliding creep rate is inherently less than the diffusional flow creep rate, Eqs. (7.10) and (7.11) can still be used to describe the creep rate. However, in this circumstance the constants A_{NH} and A_C are reduced relative to their values corresponding to NH and Coble creep. Finally, we note that when accommodation does not occur, grain-boundary voids form. This is associated with the initiation of creep fracture (Chap. 11).

---

[8]Of course, this scenario is not followed. The sequential flow processes occur on an atom-by-atom basis rather than, as suggested by Fig. 7.6, on a much larger scale. Nonetheless, the representation of Fig. 7.6 is useful for illustrating the sequential nature of the diffusional flow–grain-boundary sliding processes.

C. Creep Mechanisms Involving Dislocation and Diffusional Flow

The linear dependence of creep rate on stress for diffusional creep is not observed under conditions of moderate applied stress. Instead the value of the stress exponent m' in Eq. (7.1) ranges from about 3 to 8 (with $m' = 4.5$ being observed as often as not).

Under these conditions, creep involves dislocation, as well as diffusional, flow. The process goes by several names; dislocation creep and power law creep (PLC) being the most common. The term power law creep arises because the creep rate (Eq. (7.1)) varies with stress to a power greater than unity.

A number of mechanisms have been proposed for PLC. We discuss two of these. The first, "solute drag" creep, appears to be well-understood and its applicability to certain metallic alloy systems is well-established. The second mechanism—dislocation "climb-glide" creep—is not so well understood in its details although the basic physics of this mechanism is not obscure.

i. SOLUTE DRAG CREEP. Solute drag creep is observed in certain metallic alloys which do not display the short-time creep behavior illustrated in Fig. 7.1. Instead, the creep rate of these materials *increases* during transient creep before assuming a steady-state value. The physics of solute drag creep are straightforward and relate to models of solid-solution strengthening discussed in Chap. 5. We saw there that the size misfit parameter between solute atoms and edge dislocations leads to restriction of dislocation motion. At low temperatures, the solute atoms are immobile; thus, the effect they have is to increase the flow stress required for dislocations to move by them.

At higher temperatures, solute atoms are mobile. And if the dislocation velocity is not too high (i.e., if the creep rate is not too high), the solute atoms move along with edge dislocations, acting as a "drag" on their motion. (Thus, the catchy term "solute drag.") How much of a drag they are depends on several factors. One is the solute atom diffusivity. Provided the solute atoms are able to keep up with the moving dislocation, high diffusivities lead to a lesser drag and vice versa. Second, greater size misfit parameters lead to a greater binding energy between the dislocation and the solute atoms and result in a greater drag. Third, the greater the solute atom concentration the greater the drag effect. An equation predicts how the velocity of the moving dislocation relates to these factors as well as to the applied stress, σ;

$$v \sim \frac{D_{sol}\sigma}{\varepsilon_b^2 c_0} \tag{7.12}$$

In Eq. (7.12), D_{sol} is the solute diffusivity, c_0 its concentration, and ε_b the misfit parameter introduced in Chap. 5.

The creep rate is then linked to dislocation velocity and mobile dislocation density by an equation analogous to Eq. (3.24), i.e.,

$$\dot{\varepsilon} = \rho b v \tag{7.13}$$

However, the dislocation density, ρ, is also a function of stress. It increases with σ and analysis indicates that $\rho \sim \sigma^2$ (cf. Eq. (5.5)). Thus, the solute drag creep rate, $\dot{\varepsilon}_{SD}$, can be expressed as

$$\dot{\varepsilon}_{SD} \sim \frac{D_{sol}\sigma^3}{\varepsilon_b^2 c_0} \qquad (7.14)$$

and m' for solute drag creep is predicted to be about 3, a value commonly observed in the alloys to which this treatment applies.

Of course, we haven't considered the "details" in the above treatment. A treatment more complete than ours yields

$$\dot{\varepsilon}_{SD} = \left(\frac{1}{64\varepsilon_b^2 c_0}\right)\left(\frac{D_{sol}}{b^2}\right)\left(\frac{kT}{G\Omega}\right)^2\left(\frac{\sigma\Omega}{kT}\right)\left(\frac{\sigma}{G}\right)^2 \qquad (7.15)$$

or

$$\dot{\varepsilon}_{SD} = A_{SD}\left(\frac{D_{sol}}{b^2}\right)\left(\frac{\sigma\Omega}{kT}\right)\left(\frac{\sigma}{G}\right)^2 \qquad (7.16)$$

In going from Eq. (7.15) to Eq. (7.16) we have included the constant, the material parameters (ε_b and c_0), and the dimensionless ratio $(kT/G\Omega)^2$ in the term A_{SD}. Thus, A_{SD} is not precisely a material constant since it is a function of temperature as well. However, the convenience of the representation justifies this. That is, Eq. (7.16)—as do Eqs. (7.10) and (7.11)—contains two dimensionless terms (the constant A and the ratio $\sigma\Omega/kT$) and another one having units of (sec)$^{-1}$. In NH creep, for example, this latter term is the ratio of a diffusion coefficient to the grain size squared. For solute drag creep, the Burgers vector substitutes for the grain size. Finally, Eq. (7.16) contains another dimensionless term $(\sigma/G)^2$. Such a term is absent in diffusional creep, and it is this additional stress-dependent term that makes solute drag creep more stress sensitive.

Solute drag creep takes place when the solute atoms and the dislocations move in concert. Dislocation velocity increases with the applied stress and, when the stress level becomes sufficiently high, the dislocations "break away" from the solute atoms. The breakaway phenomenon can also be observed in a tensile test conducted at an appropriate strain rate as illustrated in Fig. 7.7. As indicated there, the flow stress increases with plastic strain and, at a critical value of the stress, the dislocations break away from the solute atoms. Following breakaway, the stress decreases as shown in Fig. 7.7, and so does the dislocation velocity. This lower velocity permits the solute atoms to "catch up" to the previously departed dislocations and when they do the stress increases rapidly. When the stress again attains a value sufficient to induce breakaway, the process repeats itself and, as indicated in Fig. 7.7, *serrated flow*—indicative of the repetitive nature of the events just described—results. The phenomenon is called the Portevin–Le Chatelier effect, and it is observed in materials of the type considered in this section. The Portevin–Le Chatelier effect is found only over a limited range of strain rates. For example, if the strain rate is sufficiently high, the flow stress is always greater than the breakaway stress and serrated flow is not found.

ii. DISLOCATION CLIMB-GLIDE CREEP. Climb-glide creep is observed in materials that display the creep characteristics illustrated in Fig. 7.1. That is, it is found in materials for which the initial creep rate is greater than the steady-state creep rate. The microstructure evolves during the transition from Stage I to Stage II creep. With increasing strain, the high-temperature analog of a dislocation cell structure evolves during Stage I creep. At higher temperatures, though, the "cells"

306

CHAPTER 7
High-Temperature
Deformation of
Crystalline
Materials

resemble subgrains (Fig. 5.6) more than they do the cellular structures that form during low-temperature plastic flow. Nonetheless, the subgrain size scales inversely with the flow stress at high temperature just as does the cell size at low temperature (Eq. (3.21)). Recovery effects take place concurrently at high temperature. Eventually, the rate of these (e.g., dislocation annihilation) balances the rate of work hardening as reflected by the subgrain structure and the transition to Stage II creep takes place. That is, steady-state creep is characterized by an invariant microstructure in which recovery and work hardening occur at the same rate. As a corollary, the Stage II creep rate is constant.

An (over-)simplified description of what transpires during climb-glide creep is illustrated in Fig. 7.8. Here, a moving dislocation is held up by an obstacle on its slip plane. The applied stress is less than that needed to overcome the obstacle via dislocation glide alone. However, the dislocation can climb by diffusional processes to a parallel slip plane. The climb process permits the dislocation to then glide on the new plane until it encounters another obstacle and the process repeats itself. In a sense, the climb process is the high-temperature equivalent of cross-slip by which dislocations may circumvent slip-plane obstacles at low temperatures. Since dislocation motion involves both dislocation glide and climb, this type of creep is referred to as climb-glide creep.

Since glide and climb are sequential processes, the climb-glide creep rate is determined by the lesser of the glide and climb rates. In most circumstances, the climb velocity is less than the glide velocity; thus, the creep rate is determined by the climb rate.

While Fig. 7.8 is useful for illustrating the processes of glide and climb it does not address the issue of the steady-state microstructure of Stage II creep. Such a mi-

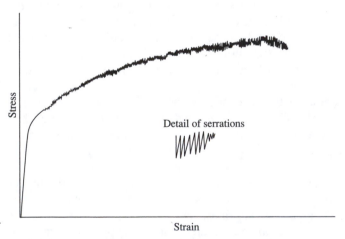

Figure 7.7
Schematic stress-strain curve of a material manifesting serrated flow. For the situation described in the text, local stress maxima correspond to the stress required to break dislocations away from the solute atoms following them. The local stress minima correspond to that stress required to move the dislocations absent the "drag" effect of the solute atoms. After a local minima is obtained, dislocation velocity is reduced and this permits the solute atoms to "catch up" with the moving dislocations; thus, the repetitive nature of the local stress maxima and minima. (*From Robert E. Reed-Hill, Physical Metallurgy Principles, 2nd ed., D. Van Nostrand, New York, 1973.*)

crostructure must reflect both dislocation annihilation and multiplication in a way that the dislocation density remains constant. Further, the microstructure must be consistent with the power law creep rate.

A large number of mechanisms/structures have been postulated to account for climb-glide creep. For the purposes of illustrating the reasoning behind them, we consider one such mechanism. It is illustrated in Fig. 7.9a. The steady-state microstructure—suggested by Hazzeldine and Weertman—is characterized by the following features. There are M dislocation sources per unit volume. These emit dislocations that glide a distance L in their slip plane. These interact with similar

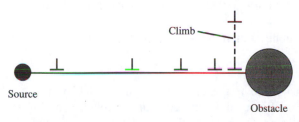

Figure 7.8
Schematic representation of the climb of a dislocation that has encountered an obstacle in its slip plane. Dislocations are generated by the "source," (e.g., a Frank-Read source) and glide until they encounter the obstacle. At high temperatures, the obstacle is surmounted by climb of the dislocations. (*From R. W. Evans and B. Wilshire, Introduction to Creep, The Institute of Metals, London, 1993.*)

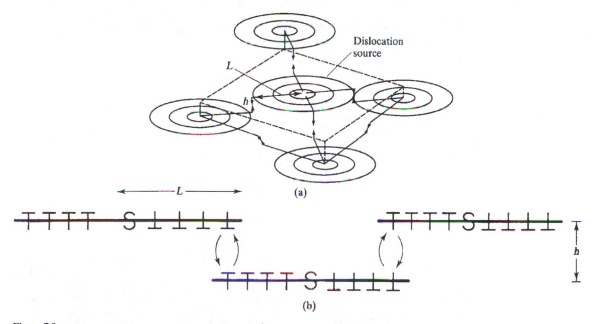

Figure 7.9
(a) M dislocation sources per unit volume emit dislocations over a radius L. Continued emission (and hence strain) requires that the outermost dislocations in each loop be annihilated by dislocation climb involving loops separated vertically by the distance h. (b) A two-dimensional view of the loop array illustrating how climb results in dislocation annihilation. (*Adapted from J. Weertman, Trans. ASM, 61, 681, 1968.*)

308

CHAPTER 7
High-Temperature
Deformation of
Crystalline
Materials

dislocations emitted from sources on parallel slip planes that are separated vertically by the distance h. Consistent with the constant microstructure, each source is assumed associated with a fixed number of loops. Thus, continued emission of dislocations from the sources (i.e., continued glide) is dependent on the annihilation rate of the outermost dislocation associated with the source. Annihilation is effected by climb over the distance h. A two-dimensional view of this climb is provided in Fig. 7.9b. Here dislocations of the type [⊣] climb and annihilate each other by addition of atoms (i.e., removal of vacancies) from the terminations of the respective dislocations. Conversely, dislocation pairs of the type [⊥] on the opposite side of the loop are removed by addition of vacancies (i.e., removal of atoms) between the respective terminations. Thus, the climb process involves mass transfer from one side of the loop to the other.

The associated strain rate can be written in the conventional form

$$\dot{\varepsilon} = \rho b v_g \qquad (7.17)$$

where ρ is the dislocation density and v_g the dislocation glide velocity which is greater than the climb velocity v_c. Since climb and glide are coupled, v_g and v_c are related through the geometrical ratio L/h as $v_g = (L/h)v_c$ where it is assumed that $L > h$. The dislocation density is obtained by multiplication of M by the average loop diameter ($\cong L$) and the number of loops per source which can be shown to be proportional to L/h; thus

$$\dot{\varepsilon}_{CG} \sim \frac{ML^3}{h^2} v_c \qquad (7.18)$$

Since the microstructure remains fixed, M is fixed. Thus, M multiplied by the volume per source ($\cong \pi L^2 h$) is constant, or $L \cong (Mh)^{-1/2}$. Hence

$$\dot{\varepsilon}_{CG} \sim \frac{v_c}{h^{3.5}M^{1/2}} \qquad (7.19)$$

Dislocation climb is driven by stress and accomplished by diffusion. Thus, it is reasonable to write $v_c \sim D_L\sigma$ where D_L is the lattice diffusivity and σ the applied stress. It is convenient to express σ in the usual normalized form, i.e., $\sigma\Omega/kT$. With these manipulations the climb-glide creep rate can finally be written as

$$\dot{\varepsilon}_{CG} = \frac{A_{CG}D_L}{h^{3.5}M^{1/2}}\left(\frac{\sigma\Omega}{kT}\right) \qquad (7.20)$$

where A_{CG} includes, among other terms, details of the loop geometry not considered here.

Since the substructure is finer at high stress levels, the separation distance, h, is less at high stresses. On the basis of previous discussion, it is reasonable to assume that h varies inversely with stress. If so, and if M is independent of stress (a questionable assumption), the exponent m' for this type of dislocation creep is 4.5. This value of m' is in accord with considerable experimental studies.

We note that the form of Eq. (7.20) is in accord with previous formulations. That is, Eq. (7.20) contains a diffusivity term and a $\sigma\Omega/kT$ term. The greater stress dependence of the climb-glide creep rate arises from the correlation between a microstructural dimension (and associated diffusion distance/area) and the applied stress. The model leading to Eq. (7.20) is but one of a number advanced to explain the stress dependence of power-law creep. They all have in common a correlation between a microstructural scale and the applied stress that results in the greater stress dependence of climb-glide creep compared to diffusional creep.

In addition to microstructural scale, dislocation creep is also found to depend on the material's stacking fault energy (SFE), increasing with this material parameter. In fact, numerous studies (see the paper by Argon in the references for details) have shown that the creep rate varies with the cube of the stacking fault energy. The effect can be included in the constant A_{CG} of Eq. (7.20) by incorporating into it a dimensionless term $(SFE/Gb)^3$.

As with creep taking place by diffusional flow, dislocation climb-glide also requires accommodating grain-boundary sliding. The accommodation can be accomplished by glide-climb processes in the boundary vicinity. Additionally, dislocations may be emitted or absorbed by the boundary, thereby providing the necessary accommodation. As with diffusional creep, if the boundary sliding rate is not in balance with the creep rate within the grains, cracks form.

D. Creep in Two-Phase Alloys

A dispersion of second-phase particles alters the creep rate of the matrix containing them. The means by which this takes place depends on where the particles are situated. If they are within grains, they can be considered an obstacle of the type illustrated in Fig. 7.8. Thus, if creep is controlled by dislocation climb-glide the effects of the particle can be incorporated into the empirical Eq. (7.1) even if the details of the climb-glide mechanisms remain incompletely known. There is one modification. The stress term in the equation is replaced by a term $(\sigma - \sigma_o)$, where σ_o is a temperature-dependent threshold stress below which climb does not take place. The threshold stress decreases with increasing temperature. From a physical standpoint this stress can be considered the high-temperature analog of the Orowan stress (Eq. (5.36)), but it is less than the Orowan stress because of the ability of edge dislocations to climb at high temperatures.

Intragranular particles do not much affect diffusional creep. They do influence the geometry of the diffusional flow causing NH creep; the effect is relatively minor.

When the particles are situated on grain boundaries, the situation is different. It is well known that a fine dispersion of nondeforming intergranular second-phase particles can markedly reduce the creep rate of some high-temperature materials. As indicated in Fig. 7.10, this reduced creep rate is undoubtedly related to the restraints that such particles place on grain-boundary sliding. It might be useful to compare Fig. 7.10 to Fig. 5.30, which illustrates the situation of a nondeforming particle within a grain during low-temperature deformation. In this situation, the displacement along the slip plane that would take place were the particle to deform is accommodated by geometrical dislocations. In Fig. 7.10, the grain boundary can be considered analogous to the slip plane. However, the accommodation at high temperature can be accomplished by diffusion. That is, since the particle does not deform, an amount of mass indicated by the shaded region of Fig. 7.10 must be "moved" by diffusional flow to permit the grain-boundary sliding displacement in-

310

CHAPTER 7
High-Temperature
Deformation of
Crystalline
Materials

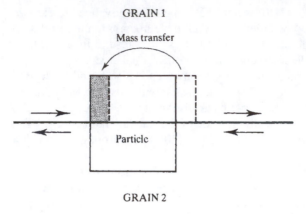

GRAIN 1

Mass transfer

Particle

GRAIN 2

Figure 7.10
Sliding displacement of two grains having a nondeformable particle situated at their boundary necessitates mass transfer of atoms of the particle. The solid lines outline the particle before displacement. The dotted lines show the position of the particle if it were to shear. Thus, to restore it to its original shape requires mass transfer. (*Adapted from R. Raj and M. F. Ashby, Metall. Trans., 2, 1113, 1971.*)

dicated. The mass transfer can take place by dislocation climb-glide processes in the boundary region. Or it can take place solely by diffusion. If so, the diffusion path might be along the boundary, within the particle, along the particle-matrix interface or within the matrix grain. Clearly, analysis of the restraint caused by nondeforming particles situated on grain boundaries is difficult. We will not touch it. However, we reiterate that the restraints associated with such particles have been put to good use in design of high-temperature creep-resistant materials.

E. Independent and Sequential Processes

As noted, different creep mechanisms may operate independently (i.e., in parallel) or in sequence (i.e., in series). For example, Nabarro-Herring creep and Coble creep operate independently, and the resultant diffusional creep rate is the sum of the respective creep rates, i.e.,

$$\dot{\varepsilon}_{\text{DIFF}} = \dot{\varepsilon}_{\text{NH}} + \dot{\varepsilon}_{\text{C}} \tag{7.21}$$

Since both NH and Coble creep depend linearly on stress, which of the two processes contributes most to the resultant creep rate depends on the respective coefficients (A_{NH} and A_{C}) and on the grain size. Coble creep depends more strongly on grain size ($\sim d^{-3}$) than does NH creep ($\sim d^{-2}$). Thus, Coble creep dominates in small-grain-sized materials and vice versa. The creep rate depends on grain size as shown schematically in Fig. 7.11, in which ln $\dot{\varepsilon}$ is plotted against ln d. While the total creep rate is the sum of the individual creep rates, it is usually found that—except in the vicinity of the critical grain size (d_c) at which the creep rates are equal—the net creep rate is approximately equal to the greater of the two rates. That is,

$$\dot{\varepsilon} \cong \dot{\varepsilon}_{\text{NH}} \quad (d > d_c) \tag{7.22a}$$

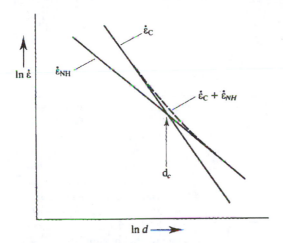

Figure 7.11
Variation of Nabarro-Herring and Coble creep rate with grain size. Coble creep dominates at grain sizes less than d_c; NH creep dominates when $d > d_c$. The total strain rate is the sum of $\dot{\varepsilon}_{NH}$ and $\dot{\varepsilon}_C$; however, one of these is usually much larger than the other.

and

$$\dot{\varepsilon} \cong \dot{\varepsilon}_C \quad (d < d_c) \tag{7.22b}$$

Thus, we conclude that for independent (parallel) creep processes, the overall creep rate is generally determined by the greatest of the individual creep rates.

The concept can be illustrated further by considering the variation of creep rate with stress at temperatures at which both diffusional flow and dislocation (power-law) creep must be considered. Since these, too, are independent processes the net creep rate is

$$\dot{\varepsilon} = \dot{\varepsilon}_{DIFF} + \dot{\varepsilon}_{DIS} = K_1\sigma + K_2\sigma^{m'} \tag{7.23}$$

In Eq. (7.23) the first term on the far right-hand side represents the linear strain rate-stress relationship of diffusional creep; the second term represents power-law creep. The total creep rate-stress relation is plotted schematically on logarithmic coordinates in Fig. 7.12. It can be seen that with increasing stress there is a transition from diffusional-dominated creep to dislocation-dominated creep at a critical stress σ_c. Such transitions have been observed experimentally in a number of materials (e.g., Fig. 7.13). Indeed, determination of the slope of $\ln \dot{\varepsilon}$ vs. $\ln \sigma$ is often used to define the operative creep mechanism; e.g., a slope of unity indicates diffusional flow is dominant.

A conclusion opposite to that above is obtained if the processes are sequential. For example, in discussion of diffusional flow we noted that grain-boundary sliding occurs sequentially to NH and/or Coble creep. Thus, in this circumstance the net creep rate is the lesser of the grain-boundary sliding and diffusional flow creep rates. Similar situations occur for dislocation climb-glide creep which necessitates the sequential steps of dislocation climb followed by glide (or vice versa). The glide velocity usually is greater than the climb velocity and, in fact, this was presumed

312

CHAPTER 7
High-Temperature
Deformation of
Crystalline
Materials

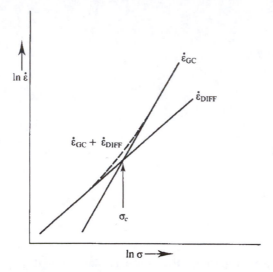

Figure 7.12
Schematic of strain rate–stress relationship for diffusional and dislocation creep. Dislocation creep (designated here as $\dot{\varepsilon}_{GC}$) dominates at high stress and has a greater stress dependence than diffusional creep. The creep rate is the sum of $\dot{\varepsilon}_{GC}$ and $\dot{\varepsilon}_{DIFF}$ but, except in the vicinity of the stress σ_c, one of the mechanisms dominates.

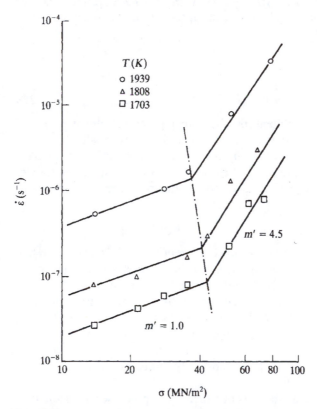

Figure 7.13
Steady-state creep rate vs. stress for UO_2 polycrystals with a grain size of 10 μm. At low stress levels, diffusional flow dominates; at higher stress levels dislocation creep does. (*From L. E. Poteat and C. S. Yust, Ceramic Microstructures, ed. R. M. Fulrath and J. A. Pask, Wiley, New York, 1968, p. 646.*)

in our discussion of dislocation climb-glide creep exemplified by Fig. 7.9 in which the creep rate was controlled by the dislocation climb rate. However, the climb and glide rates depend differently on stress. Then we find that there is a transition stress below which, for example, climb controls creep and above which, glide does. (Prob. 7.9)

To finally conclude! The net effective creep rate when parallel creep processes operate is approximately the *greater* of the creep rates of the respective processes. And when creep takes place by sequential processes, the net creep rate is the *lesser* of the individual creep rates.

F. Summary

Even though we have not discussed all possible creep mechanisms, enough is enough. The number of mechanisms treated, and the equations describing their creep rate, is daunting as it is. The understandable confusion can perhaps be lessened by noting that, in spite of the apparent diversity of the formulations of the several creep rates presented in this section, all of them can be expressed in a similar form, i.e.,

$$\dot{\varepsilon} = A\left(\frac{D}{\Omega^{2/3}}\right)\left(\frac{\sigma}{G}\right)^{m''}\left(\frac{\sigma\Omega}{kT}\right)\left(\frac{b}{d}\right)^{n'} \tag{7.24}$$

In Eq. (7.24), the last three parameters on the right-hand side are dimensionless as is the constant A. The term $(D/\Omega^{2/3})$ has units of s^{-1}; i.e., it has units of creep rate. As mentioned previously, the ratio $\sigma\Omega/kT$ is the ratio of a mechanical to a thermal energy. The parameter $(b/d)^{n'}$, where b is the Burgers vector (b is also about equal to $\Omega^{1/3}$) represents the grain-size dependence of the creep rate; e.g., $n' = 2$ for NH creep and $n' = 0$ for dislocation creep. The term σ/G is the ratio of the applied stress to the shear modulus; this ratio is important in determining the rate of power-law creep (note that $m'' = 0$ for diffusional creep). And the constant A represents geometrical aspects of diffusion (recall our discussion of Coble and NH creep, for example). The constant also implicitly contains other factors that affect the creep rate such as the stacking fault energy.

Table 7.1 summarizes all this. Listed there are values of the coefficients m'' ($= m' - 1$), n' and approximate values of the constant A for the creep mechanisms described. Comments in the table also indicate the approximate stress,

Mechanism	Favored by	A	m''	n'
Nabarro-Herring creep	High temperature, low stress, and fine grain sizes	7	0	2
Coble creep	Low stress, fine grain sizes and temperatures less than those for which NH creep dominates	50	0	3
Power-law creep	High stress	*	2–6*	0

* The terms A and m'' are stongly dependent on the mechanism controlling power-law creep at the substructural level. Values of A can range from several to several million.
Source: Data in table adapted from A. K. Mukherjee, *Treatise Matls. Sc. and Tech.*, ed. R. J. Arsenault, **6**, 163, 1975.

TABLE 7.1
Values of the parameters m'', n' and approximate values of the constant A in the expression for the steady-state creep rate $A(D/\Omega^{2/3})(\sigma/G)^{m''}(\sigma\Omega/kT)(b/d)^{n'}$

314

CHAPTER 7
High-Temperature
Deformation of
Crystalline
Materials

temperature and, if appropriate, grain-size regimes in which a particular mechanism may be expected to dominate creep.

Although Table 7.1 is useful for clarifying the conditions under which a particular creep mechanism is most important, a picture is worth a thousand words—or at least worth a number of equations. Such a graphical representation of the interrelation among stress, temperature, creep mechanism, and creep rate is provided by "deformation mechanism maps," the concepts and use of which are described in the following section.

7.4
DEFORMATION MECHANISM MAPS

A deformation mechanism map has axes of homologous temperature (T/T_m where T is the absolute temperature and T_m is the absolute melting temperature of the material considered) and stress. The stress axis is typically represented logarithmically, and stress is usually expressed in normalized terms; i.e., by the ratio of stress to the material's shear modulus. The homologous temperature scale ranges from 0 to 1, and the normalized stress scales from ca. 0.1 downward. (The value $\sigma = 0.1G$ represents an effective upper limit on the strength of a material.) Such a diagram is illustrated schematically in Fig. 7.14. Several deformation mechanism regions are included in the figure; these are the dominant deformation mechanisms at stress-temperature combinations for which they are listed. For example, at the stress σ_2,

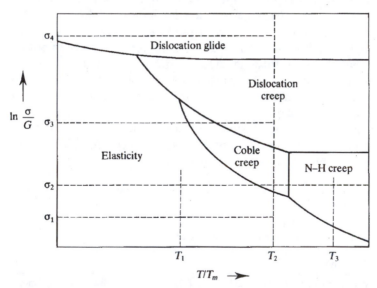

Figure 7.14
A schematic deformation mechanism map. The axes of the map are homologous temperature (T/T_m) and stress (normalized by the shear modulus). The stress-temperature combination determines the primary deformation mode. At the boundary lines, deformation is due equally to two mechanisms and, at the intersection of the lines, to three mechanisms.

elastic deformation, Coble creep, and NH creep are the dominant mechanisms at the temperatures T_1, T_2, and T_3, respectively. Likewise, at T_2, elastic deformation, Coble creep, dislocation creep[9], and dislocation glide are the dominant deformation modes at the successively higher stresses σ_1, σ_2, σ_3, and σ_4. The lines separating the regions represent stress-temperature combinations at which two mechanisms contribute equally to material deformation. Similarly, triple points—the intersection of three such lines—represent a particular stress-temperature combination at which three mechanisms contribute equally to the deformation.

Deformation mechanism maps constructed for specific materials are only as accurate as the equations—which relate strain, strain rate, stress, and temperature—on which they are based. The dominant deformation mechanism is determined from such equations as the one providing the greatest strain over the time scale of interest. Thus, at low stress and low temperatures, linear elasticity is dominant even though on a geological time scale diffusional creep could produce a greater strain than time-independent elasticity.

In some cases only plastic deformation is considered consequential, and this results in the elimination of the elastic region from the diagram. When this is done, the creep mechanisms occupy regions of the map previously occupied by elastic deformation.

As noted, construction of a deformation mechanism map depends on the availability of accurate constitutive equations. These have been developed for a reasonable number of engineering materials.[10] Several diagrams that have been constructed from such equations for specific materials are given in Figs. 7.15. The concept and schema of a deformation mechanism map are useful for distinguishing among dominant deformation mechanisms. For example, the transition from one dominant deformation mode to another as it depends on temperature and stress is visualized easily with such a map.

In addition to their value in delineating the dominant deformation mode, deformation mechanism maps can be used to estimate creep rates. Within a given region, for example, the equation relating strain rate to temperature and stress can be used to develop contours of equal strain (creep) rate, as illustrated schematically in Fig. 7.16a and for specific materials in Figs. 7.16b and 7.16c. Thus, both the dominant deformation mechanism and the resulting creep rate can be obtained from such an elaborated map.

The deformation mechanism maps discussed so far have not considered the effect of grain size; that is, a specific grain size has been assumed in their construction. With decreasing grain size, diffusional-flow mechanisms occupy a greater area of a deformation mechanism map (Fig. 7.17a). To account for varying grain size, an alternative formulation of a deformation mechanism map can be developed. Such a map, constructed at a constant temperature, has axes of σ/G and grain size

---

[9]More elaborate diagrams, such as those shown in the compilation of Frost and Ashby noted in the references, often divide the dislocation creep region into a low-temperature region and a high-temperature one. At the lower temperatures, dislocation climb, for example, occurs by diffusion along dislocations; at higher temperatures, by volume diffusion. Thus, the two dislocation creep regions are the dislocation creep analogues of the diffusional Coble and Nabarro-Herring creep regions.

[10]The Frost and Ashby reference is a good compilation of available deformation mechanism maps for specific materials.

316

CHAPTER 7
High-Temperature
Deformation of
Crystalline
Materials

as schematically shown in Fig. 7.17b. The vertical line (at $d = d_c$) in this figure represents the (temperature-dependent) grain size at which the transition from Coble to NH domination of creep takes place. The negative slopes ($-3/m'$ and $-2/m'$, respectively, on the logarithmic axes employed) of ln σ vs. ln d that separate the diffusional flow and dislocation creep regions reflect that diffusional mechanisms of creep are of less importance at large grain sizes.

The most complete representation of a deformation mechanism map is a three-dimensional construction having axes of stress, temperature, and grain size. In addition, to be of most use such a diagram would explicitly consider the variation of grain size resulting from grain growth during the life of the material. Such con-

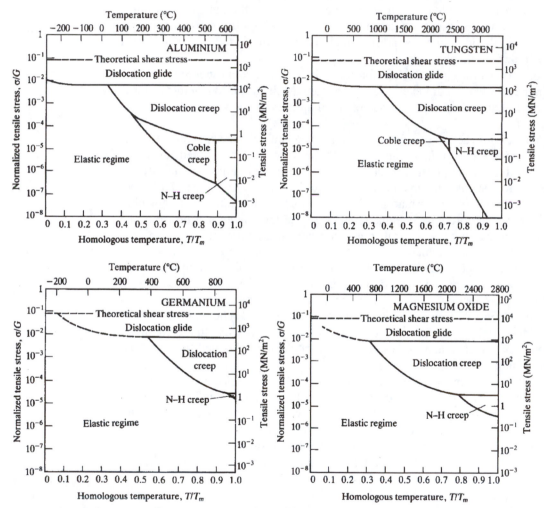

Figure 7.15

Deformation mechanism maps for several metals (Al and W) and nonmetals (Ge and MgO), all with a grain size of 32 μm. Nonmetals are somewhat more resistant to dislocation glide than metals. The covalent solid, Ge, is quite resistant to both dislocation and diffusional creep. (*From M. F. Ashby, "A First Report on Deformation Mechanism Maps," Acta Metall., 20, 887, Copyright 1972, with permission from Elsevier Science.*)

structions, while conceptually viable given the current state of computational display, are not yet possible for most materials since the ancillary data needed to develop them are not available. On the other hand, even in their present form, deformation mechanism maps are useful for design of high-temperature materials. With knowledge of the intended stress-temperature use, the dominant creep mechanism can be identified and alloy design can incorporate factors that reduce the associated creep rate. Further considerations of high-temperature creep-resistant materials are presented in the following section.

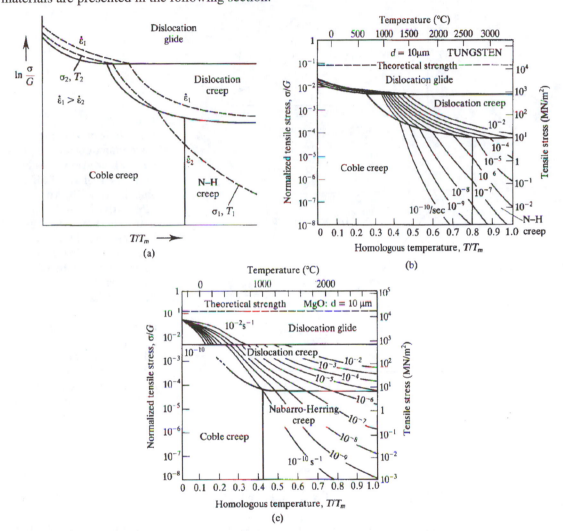

Figure 7.16
(a) Schematic deformation mechanism map with isostrain-rate contours imposed on it. For example, the strain rate $\dot{\varepsilon}_2$ is obtained at the stress-temperature combination (σ_1, T_1) where the creep mechanism is NH creep. The same strain rate is obtained by dislocation glide at the stress σ_2 and temperature T_2. (b) and (c) Deformation mechanism maps for W and MgO, both of 10-μm grain size, with imposed isostrain-rate contours. (*Part (b) from M. F. Ashby, "A First Report on Deformation Mechanism Maps," Acta Metall., 20, 887, Copyright 1972; part (c) from A. G. Evans and T. G. Langdon, "Structural Ceramics," Prog. Matls. Sc., 21, 171, Copyright 1976, with permission from Elsevier Science.*)

318

CHAPTER 7
High-Temperature
Deformation of
Crystalline
Materials

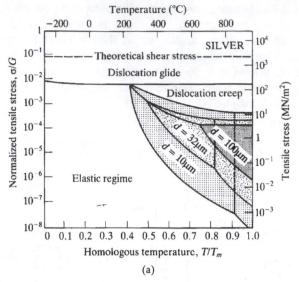

(a)

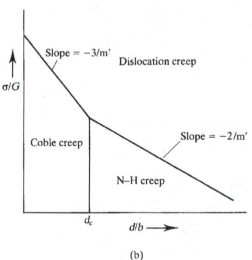

(b)

Figure 7.17
(a) A deformation mechanism map depends on grain size. Finer grain sizes expand the regions of diffusional creep. (b) An alternative formulation of a deformation mechanism map has axes of normalized stress and grain size at a specific temperature. In the figure, the dominant creep mechanisms are identified for various stress–grain-size combinations. (*Part (a) from M. F. Ashby, "A First Report on Deformation Mechanism Maps," Acta Metall., 20, 887, Copyright 1972, with permission from Elsevier Science.*)

7.5
MATERIALS ASPECTS OF CREEP DESIGN

Materials aspects of creep are the focus of this section. Means for improving material creep resistance are discussed in terms of materials properties and structure and are then illustrated by case studies. In the section that follows, predictive capabilities for creep life at long times and/or low stress levels are developed from data obtained under more severe stress/temperature combinations. Such extrapolations are required because the intended life of a material usually far exceeds that which can be measured during laboratory evaluation.

A. Creep Resistance as Related to Material Properties and Structure

As indicated in Eq. (7.24), creep rates vary with stress, material diffusivity and, in some cases, grain size. Creep resistance is improved if diffusion rates are reduced. Since within a given class of materials the diffusion activation energy scales with absolute melting temperature, low homologous operating temperatures clearly result in lower creep rates. Thus, design at a specified temperature calls for a material with a high melting point, even though this material may be less creep-resistant than a lower melting one at equivalent homologous temperatures.

There are, however, subtle differences in the relationship between the diffusivity and melting temperature for different material classes. Because of their slightly more open structures and increased atomic vibrational frequencies, many of the body-centered cubic transition metals have higher diffusivities at equivalent homologous temperatures than the face-centered cubic metals. This is reflected in correspondingly higher creep rates. Thus, bcc iron has a higher diffusion coefficient than does fcc iron at the allotropic transformation temperature, and the relative creep rates of these forms of iron scale almost exactly as their diffusivities. Hence, although the bcc transition metals possess an inherently greater resistance to dislocation glide than do fcc metals, they are not necessarily superior to them for high-temperature use. Ceramics are much better in this respect. Their relatively low diffusion coefficients make them attractive for high-temperature use and, if it were not for their brittleness, ceramics would be used far more often than they currently are.

When dislocation creep is the dominant creep mechanism, materials having higher shear moduli display better creep resistance (Eq. (7.24)). This is related to the driving force for, for example, climb-glide, and is analogous to similar conclusions reached when dislocation motion is the sole mechanism of plastic deformation (Chaps. 3–5). It must be remarked, though, that modulus variations among materials are much less than are differences among diffusion coefficients. Thus, "modulus hardening" plays a decidedly secondary role to "diffusivity hardening" in improving creep resistance.

Microstructure also influences creep behavior. When diffusional flow controls creep, creep rate is reduced by increases in grain size. On the other hand, generally no benefit is obtained by increases in grain size if dislocation creep dominates. High-temperature polycrystalline materials frequently contain second-phase intergranular particles. For example, many high-temperature alloys contain intergranular refractory metal carbides. The restriction these particles place on grain-boundary sliding results in improved creep resistance.

To summarize, creep resistance can be improved significantly by reductions in material diffusivity. Beyond this, creep behavior may be modified by other property and structural variations; an increased modulus improves resistance to dislocation creep, coarse grain size does likewise for diffusional creep, and intergranularly situated second-phase particles reduce grain-boundary sliding.

The above generalities are useful in the abstract. However, their implementation in development of creep-resistant materials is better illustrated by example. This is done in the paragraphs that follow.

B. Case Studies

i. SUPERALLOYS. When the steam engine was invented, there was little problem in finding materials suitable to use in it. After all, the operating temperature was a mere 100°C and a whole host of materials can operate satisfactorily under stress for extended periods at this temperature. But efficiency increases with the temperature at which an engine operates. Material operating temperatures in modern jet engines now routinely approach 1200°C.[11] This increase is linked intimately to improved materials. The heat-resistant materials that can be used in high-temperature engines are generically referred to as *superalloys*.

Superalloys are based on Fe, Ni, or Co, with the Ni-base superalloys being most important. Superalloys are made heat-resistant by a variety of means, as is discussed later. However, we first illustrate the use of deformation mechanism maps in alloy design with an example. The alloy MarM-200 is an Ni-base superalloy used at moderate temperatures, on the order of 600°C. Deformation mechanism maps for this material with two different grain sizes are shown in Figs. 7.18. The intended stress-temperature operating range for MarM-200 is indicated by the shaded box in the figures. Coble creep is the dominant mechanism for both grain

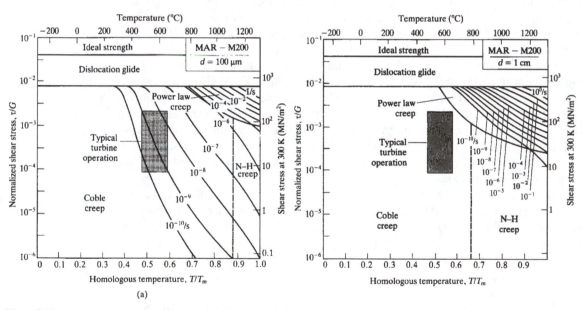

Figure 7.18

(a) A deformation mechanism map for the alloy MarM-200 with a grain size of 100 μm. The intended stress-temperature use for a turbine blade of this material is indicated. The dominant creep mechanism is Coble creep and the operational strain rates are in the range of 10^{-8} to 10^{-10} s^{-1}. (b) The deformation mechanism map for this same alloy with $d = 1$ cm. Coble creep still dominates creep, but the strain rates are much less. (*From M. F. Ashby, The Microstructure and Design of Alloys, Proc. Third Intl. Conf. on Strength of Metals and Alloys, Vol. 2, Cambridge, England, 1973, p. 8.*)

---

[11]Operating temperatures are even higher than this. This is because mechanical design incorporates cooling engine materials to a temperature less than the engine operating temperature.

sizes considered. Although the dominant creep mechanism is not altered with increases in material grain size,[12] the creep rate is reduced by orders of magnitude (see also Prob. 7.19).

The most demanding environments in turbine applications call for materials resistant to dislocation creep. The requirements are met with a series of Ni-base alloys based on the Ni-Al system. These are referred to as γ-γ' alloys; the γ refers to the fcc matrix of these materials, the γ' to the second-phase precipitate $Ni_3(Al,Ti)$ which is present in relatively large amounts. The precipitate phase provides particle strengthening. Moreover, the surface energy between the γ and γ' phases is quite low and, as a result, these alloys are strongly resistant to particle coarsening that would degrade properties. The particles are not enough to provide adequate creep resistance, though. Thus, solute elements such as Cr, Co, Fe, Mo, W, and Ta are added to the alloy to provide high-temperature solid-solution strengthening. When used in polycrystalline form, carbon is also added to γ-γ' alloys. This element reacts preferentially with carbide forming elements (e.g., Cr, W, Ti) to form carbide particles situated along grain boundaries where they restrict grain-boundary sliding.

A low creep rate is not the only desired property of turbine blades. They also must not fracture; that would not be good for the engine or the aircraft passengers who rely on the blade performing properly. In a turbine blade the principal stress axis is along the blade axis. In polycrystals containing carbide (or other) particles placed on grain boundaries, fracture initiates on boundaries, particularly those situated perpendicular to the stress axis.[13] The obvious solution is to get rid of these boundaries. This can be accomplished by directionally solidifying the blades along their axis. The process results in more-or-less cylindrical grains aligned along the blade axis. The grain boundaries are then parallel to the axis and thus not subject to a normal stress that causes them to fracture. More recently, blades have been produced in single crystal form thereby eliminating grain boundaries entirely. The single crystal materials not only manifest better creep resistance, they are also superior in terms of thermal fatigue resistance.[14]

ii. TUNGSTEN LIGHT-BULB FILAMENTS. Until 25 years ago or so, you could get tennis elbow changing incandescent light bulbs they had to be replaced so often. Why are they so much better now? The addition of potassium, an element that melts at a lowly 640°C, to the tungsten filament of the bulb is the reason. Why should a low-melting element such as potassium "strengthen" a material that, in a light bulb, operates at temperatures of 2500°C or higher? To answer, we must first consider how these filaments are made.

Tungsten light-bulb filaments are made through powder metallurgy processing. After pressing and sintering of W powders, they are subjected to extensive defor-

---

[12]Note, though, that an increase in grain size results in encroachment of the dislocation creep region into the diffusional-flow region.

[13]We are getting ahead of ourselves here. High-temperature fracture is discussed in more detail in Chap. 11.

[14]Individual phases in a polyphase alloy have different thermal expansion coefficients. Thus, when subjected to a cyclical varying temperature—as a turbine blade is subject to—the differing volume expansions of the phases causes a likewise cyclically varying internal stress. Thermal fatigue is a specific example of fatigue deformation, which is discussed in Chap. 12.

322

CHAPTER 7
High-Temperature
Deformation of
Crystalline
Materials

mation en route to becoming a filament. The grain structure of such an as-formed filament is schematically illustrated in Fig. 7.19a. It consists of interlocking grains elongated along the drawing (deformation) axis. When such a structure is exposed to high temperatures (the operating temperature of a W filament can exceed 75% of its absolute melting temperature), the W grains recrystallize. A "bamboo" structure, sketched in Fig. 7.19b, results from the recrystallization process. This structure consists of individual single crystals (diameter equal to the wire diameter) aligned along the filament axis. The lengths of the individual grains are several times the wire diameter.

Two essential requirements must be met in a good incandescent filament. As with the Ni-base superalloys, they relate to creep deformation resistance and creep fracture resistance. With respect to deformation, the filament must not sag (creep) during operation. This requirement is generally readily fulfilled (Prob. 7.14). Resistance to creep fracture is another matter. When a light bulb is switched on, the filament undergoes a transient nonuniform thermal expansion along its length. The gradient in expansion imparts a tensile force along the wire axis and, for recrystallized W filaments, the force is perpendicular to the grain boundaries of the bamboo structure. After a period of operation, the force is sufficient to cause filament fracture along the boundaries.[15] The situation is somewhat similar to that just described for the Ni-base alloys. The key to a long-life filament, then, is to prevent or delay recrystallization of the W grains present in the as-deformation processed filament.

How does potassium accomplish this? Potassium is inert in W; that is, it does not dissolve in it. As a consequence, the potassium (which is in the vapor state during bulb operation) is situated at grain and subgrain boundaries in the tungsten fila-

Drawing Direction

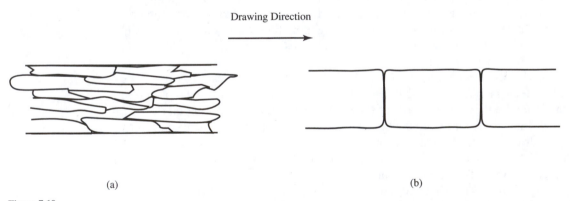

(a)   (b)

Figure 7.19
Schematic microstructures of (a) drawn W wire and (b) such a wire following high-temperature exposure. The drawing produces a fine-grained structure with the grains elongated along the drawing direction. On high-temperature exposure, the cold-worked structure recrystallizes, producing a "bamboo" structure. It consists of grains having diameter equal to the wire diameter and a grain length several times this diameter. The boundaries between grains display cusps. This feature, as well as that of the individual grain dimensions, gives rise to the term "bamboo" structure.

---

[15] Note that light bulbs almost invariably "burn out" when they are turned on (or less frequently, off). This is evidence that the transient thermally induced stress is responsible for their failure.

ment. There it restricts growth of recrystallized grains much as oxide dispersions do in the mechanically treated alloys described in Chap. 5 (cf. Fig. 5.41). A transmission electron micrograph of a potassium-doped W filament is shown in Fig. 7.20a. Potassium "bubbles" are seen both at grain boundaries and at subboundaries within the grains. However, because of the high bulb-operating temperature, the bubbles only delay recrystallization and do not prevent it. Even following recrystallization, though, the potassium serves beneficially. Figures 7.20b and c show two types of W boundaries in filaments exposed to the high operating temperature of a bulb. Figure 7.20b shows a "longitudinal" boundary in which the grain boundary is primarily aligned parallel to the filament axis. Figure 7.20(c) shows a "transverse" boundary akin to that illustrated in Fig. 7.19b. It is these boundaries that eventually

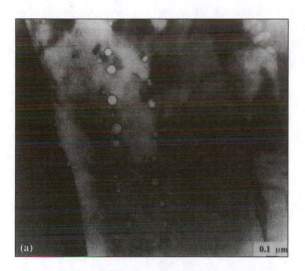

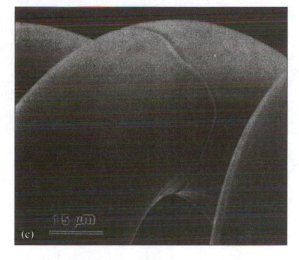

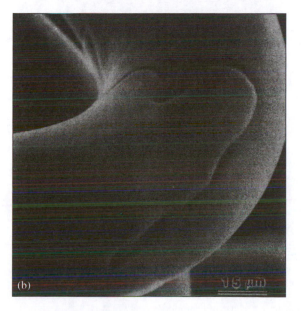

Figure 7.20
(a) Transmission electron micrograph of a drawn K-doped W light-bulb filament. The smaller potassium bubbles are situated on grain boundaries, and the larger ones on subgrain boundaries.
(b) "Longitudinal" grain boundary developed in a W filament following high-temperature exposure. These boundaries, in which a significant fraction of the grain-boundary area is aligned with the wire axis, do not lead to in-service filament fracture.
(c) A "transverse" grain boundary, also developed by high-temperature exposure, is prone to fail during service. Note the features of such a boundary resemble those of the bamboo structure schematized in Fig. 7.19b. ((a) from C. L. Briant, O. Horacsek, and K. Horacsek, Metall. Trans. A, 24A, 843, 1993; (b) and (c) from J. L. Walter, J. W. Pugh, and D. R. Sorenson, Metall. Trans. A, 23A, 2259, 1992.)

324

CHAPTER 7
High-Temperature
Deformation of
Crystalline
Materials

"snap" during operation. However, the time before they do is much greater than in filaments not containing potassium.[16]

7.6
ENGINEERING ESTIMATES OF CREEP BEHAVIOR

Many high-temperature components must operate efficiently and without failure for times on the order of decades. On the other hand, evaluation of a material's suitability for such service must take place on a much shorter time scale; creep tests, for example, are conducted for periods typically on the order of weeks to months. The necessary accommodation between service and testing is done by subjecting materials to more severe stress-temperature combinations during short-time testing than they will encounter in service, and then extrapolating test results to the long-time use environment. This approach has pitfalls for, as we have seen, different creep mechanisms dominate under different stress-temperature combinations, and accurate extrapolation is predicated on the same mechanism dominating under both short- and long-life conditions.[17]

The fundamentals underlying extrapolation techniques are contained in the empirical Eq. (7.1) relating strain rate to stress and temperature. On taking natural logarithms of both sides of this equation, it can be written as

$$\ln \dot{\varepsilon} = \frac{-Q_c}{RT} + g(\sigma) \qquad (7.25)$$

where $g(\sigma)$, a function of the applied stress, is the natural logarithm of all terms on the right-hand side of Eq. (7.1) excepting the term $\exp(-Q_c/RT)$. Stress-rupture tests, in which the time to (fracture) rupture, t_f, is measured as a function of the stress and temperature are commonly performed for short-time material evaluation. Provided t_f is inversely proportional to the steady-state creep rate, we can substitute $t_f = k/\dot{\varepsilon}$ in Eq. (7.25) to obtain

$$\ln t_f - \ln k + g(\sigma) = \frac{Q_c}{RT} \qquad (7.26a)$$

or

$$T[\ln t_f + g(\sigma) - \ln k] = \frac{Q_c}{R} \qquad (7.26b)$$

As Q_c/R is a material constant, the terms on the left-hand side of Eq. (7.26b), which are related to the Larson-Miller (LM) parameter, can be used to assess a material's creep resistance. The parameter is frequently expressed as

$$LM = T[\log t_f + C] \qquad (7.27)$$

---

[16]Thoria (ThO_2) had long been added to W filaments to achieve the same purpose that potassium accomplishes. However, thoria is mildly radioactive and as a result the use of thoriated-tungsten filaments was restricted to special applications (e.g., X-ray tube filaments).

[17]Precautions can be taken here. These may involve metallographic examination and determination of whether or not there are changes in the slope of the $\ln \dot{\varepsilon}$–$\ln \sigma$ curve observed in the short-time evaluations. Such changes reflect a change in controlling creep mechanism. But the best precaution, always a component of mechanical design with materials, is use of appropriate safety factors.

where log t_f represents the logarithm to the base 10 of t_f (when it is expressed in hours) and C is an empirically determined constant.

The LM parameter is used in the following manner. Since the LM parameter varies with stress, it is assumed that, at a given stress level, the temperature and fracture time are related to it; that is, different temperatures and fracture time combinations will have the same LM parameter at a specific stress. The usefulness of this procedure is illustrated in Figs. 7.21a and b. In Fig. 7.21a stress vs. fracture time data for an iron-base alloy are plotted for several different temperatures. These same data are plotted in Fig. 7.21b, where, however, the abscissa is now the LM parameter in which the constant C (20 in this case and usually close to this number for most engineering alloys) is adjusted so that the "fit" to the master curve is best. It is clear that the data cluster closely around this master curve. To determine an allowable stress at a temperature for which fracture must not occur in a specified period, the LM parameter is first calculated. The allowable stress level is then obtained directly from a curve of the type illustrated in Fig. 7.21b (see Example Prob. 7.1 and Prob. 7.21).

A number of other schemes have been developed for extrapolating laboratory results to service use. Although most of these are as fundamentally sound as the LM procedure, the latter remains widely used. In spite of this, extrapolation utilizing the LM parameter is nonetheless empirical. Further, as will be shown in Chap. 11 which discusses high-temperature fracture in more detail, the LM procedure is a nonconservative one. That it has been so successfully implemented is testimony to the caution exercised by those who select materials for high-temperature applications.

EXAMPLE PROBLEM 7.1.   Using Fig. 7.21b, estimate the stress allowed for safe operation of Alloy S590 at 1100°C if fracture is not to occur in 10 hours.

Solution.   The LM parameter is equal to $T(20 + \log t_f)$, where t_f is in hours and T in Kelvin. With $T = 1100°C = 1373$ K and $t_f = 10$ hr we have

$$LM = 1373(21) = 28.8 \times 10^3$$

With reference to Fig. 7.21b, this would correspond to an operating stress of 19 MN/m².

7.7
SUPERPLASTICITY

Superplasticity refers to a phenomenon in which certain materials subjected to high temperature (typically $T \geq 0.5\, T_m$) demonstrate remarkably high strains to failure at strain rates typically on the order of 10^{-3} s^{-1}. Tensile elongations to failure in superplastic materials can be remarkably high, well in excess of 5000 percent[18]; a striking example of such behavior is shown in Fig. 7.22.

---

[18]The *Guinness World Book of Records* has a category on superplastic tensile fracture strains. However, getting one's name in the recordbook seems to depend more on having constructed a long isothermal furnace than in having discovered the most superplastic material. For example, a 1-cm-long gage length extended by 5000 percent would be 51 cm long. It requires skill to construct a furnace with an isothermal zone that long!

326

CHAPTER 7
High-Temperature
Deformation of
Crystalline
Materials

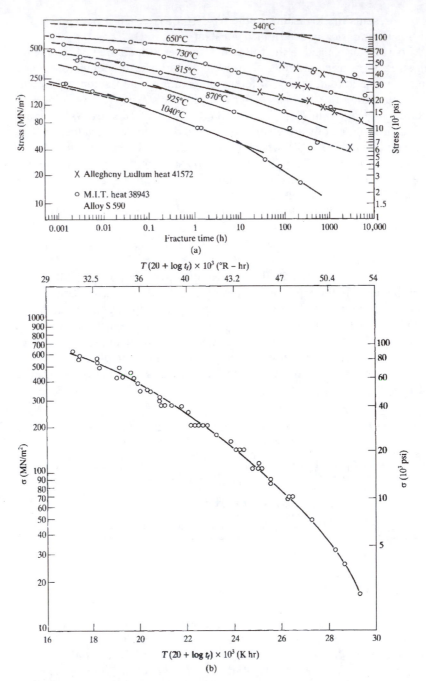

Figure 7.21
(a) Stress-fracture time plots for an iron-based alloy (S590) at different temperatures.
(b) Larson-Miller master plot of the data of (a); the abscissa is the parameter $T(C +$
$\log t_f)$. This diagram permits fracture times to be estimated for a variety of stress-
temperature combinations. (*Part (a) from N. J. Grant and A. G. Bucklin, Trans. ASM,
42, 720, 1950; part (b) reprinted from R. W. Hertzberg, Deformation and Fracture*
Mechanics of Engineering Materials, Copyright 1976, with permission from John Wiley
***& Sons, Inc.*)**

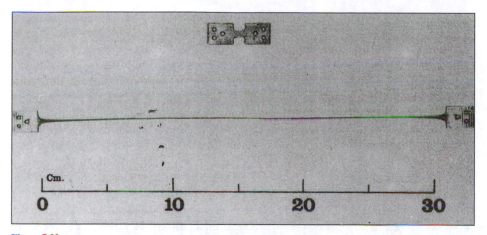

Figure 7.22
An example of exceptional superplasticity in a Pb-Sn eutectic alloy having a grain size of 6.9 μm. The top photograph illustrates the sample geometry before deformation. The other photograph is of the material extended in tension by 4850 percent at a temperature of 413 K and an initial strain rate of 1.33×10^{-4} s^{-1}. The test was discontinued prior to failure so the extension of 4850 percent represents a lower limit to the tensile failure strain. (*From M. M. I. Ahmed and T. G. Langdon, Metall. Trans. A, 8A, 1832, 1977.*)

Superplastic deformation is not only scientifically interesting, but also has important industrial ramifications. Superplastic deformation can be utilized in high-temperature forming of complex shapes for which forging operations, with their higher strain rates, are not suitable. In addition, ceramics and intermetallics also display superplastic behavior. This permits them to be formed into useful shapes by deformation processing. Given the inherent brittleness of these material classes, this is a definite plus in terms of processing them. In addition to its relevance to materials science and engineering, superplasticity and an understanding of the mechanisms causing it may also be useful in developing appreciation of the stress-assisted deformation of the earth's mantle, as it is thought that the stress, temperature, and strain rates associated with this deformation can, to a point, be compared directly to superplastic deformation of engineering materials.

Microstructure plays a role in superplastic deformation. A fine-grained equiaxed microstructure (grain size approximately < 10 μm) is required for superplastic behavior to be manifested. Moreover, the structure must be resistant to grain growth at the temperatures and time duration of superplastic deformation (excessive grain growth removes the requisite fine-grained structure). Hence, although superplasticity can be observed in single-phase materials, it is much more commonly associated with two-phase alloys which are considerably more resistant to grain growth. Intergranular second-phase particles, which pin grain boundaries (recall Ni-base superalloys and W light filaments), provide such an impediment, but grain growth is even more restricted in "microduplex" equiaxed alloys. These are two-phase materials with comparable volume fractions of the phases. Grain growth in them is markedly restricted as a result of frequent interphase grain contacts. Such microstructures can be produced by suitable processing of eutectic and eutectoid alloys; in fact, superplasticity was first observed in such materials.

328

CHAPTER 7
High-Temperature
Deformation of
Crystalline
Materials

Another microstructural feature of superplastic deformation is that *grain shape* is essentially preserved during the deformation. In nonsuperplastic deformation (excluding situations where the material recrystallizes, a feature seldom associated with superplasticity) individual grains change their shape commensurate with the overall strain. For example, in conventional deformation the average grain aspect ratio (= grain dimension along tensile direction divided by the grain dimension in the lateral direction) relates to the equivalent macroscopic dimensional ratio. This is not so during superplastic deformation, in which grains remain approximately equiaxed. For example, at superplastic strains on the order of several hundred percent, grain aspect ratios seldom exceed 1.5 and are typically less than this.

Microscopic examination of superplastically deformed materials shows that extensive grain-boundary sliding, accompanied by considerable grain rotation, occurs during superplasticity. Although evidence of extensive dislocation activity, commonly observed in power-law creep, is absent in superplasticity, this does not, as we shall see, preclude dislocation motion from contributing to the superplastic strain. These, and other, aspects of superplasticity are discussed in this section.

A. Strain-Rate Sensitivity and Superplastic Behavior

Superplastic behavior correlates with a high strain-rate sensitivity exponent, m, in the constitutive equation

$$\sigma_T = K'(\dot{\varepsilon}_T)^m \tag{1.14}$$

In general, K' can be considered a function of strain (cf. Eq. (1.15)), but in superplasticity this is not so; that is, the stress depends only on the plastic strain rate. The exponent m can be determined by conducting a series of tensile tests run at a constant true strain rate. Alternatively, m can be determined by a series of instantaneous changes in the strain rate during tensile testing (Prob. 7.23). For both types of test, m is found as the slope of the logarithmic plot of true stress versus true strain rate.

Since superplasticity is a high-temperature deformation phenomenon, m can also be determined by creep testing. The phenomenological Eq. (7.1) can be identified with Eq. (1.14) if we take $m' = (m)^{-1}$ and $K' = A^{-1} \exp(Q_c/RT)$. These two alternative ways of determining m are illustrated schematically in Figs. 7.23a and b. In the former, m is determined as the slope of a logarithmic plot of the measured stress versus the strain rate; in the latter m' is obtained as the slope of a similar plot of measured strain rate as a function of stress. Provided suitable experimental precautions are taken, the values of m determined by the different approaches are the same. Although discussion of creep has to this point been couched in terms of relating $\dot{\varepsilon}$ to σ, descriptions of superplastic behavior generally relate σ to $\dot{\varepsilon}$, and this will be the practice we follow here.

As the m value of a material increases, it displays a greater resistance to tensile neck development. This can be illustrated by substitution of F/A for σ and $\dot{\varepsilon} = (-1/A)(dA/dt)$ in Eq. (1.14). These substitutions, appropriate for a tensile test, lead to the relationship (Prob. 7.24)

$$\frac{dA}{dt} = -\left(\frac{F}{K'}\right)^{1/m} A^{(m-1)/m} \tag{7.28}$$

Values of m are bracketed between zero and unity. These correspond respectively to a nonstrain-rate-sensitive material (e.g., some metals at low temperatures) and to Newtonian viscous materials for which stress and strain rate are related linearly. For the latter, $m = 1$ and thus $dA/dt = -F/K'$; that is, the reduction in cross-sectional area per unit time is a constant along the gage length. This means that after a neck begins to form at some location along the tensile gage length, it deforms at the same, and not at a greater, rate than the material outside the incipient neck. The material eventually necks down to a point which is characteristic of tensile fracture of a viscous material. There is a general taper in the final gage length, and this is indicative of an extreme of a "diffuse" neck discussed in Chap. 4. For any value of $m < 1$, the necked region experiences a greater area reduction rate than the regions removed from the neck (Prob. 7.25). Nonetheless, a high value of m is always associated with a greater resistance to neck development and delayed tensile failure.

There is another way of viewing this "strain-rate hardening." As a neck forms, the strain rate in the necked region increases, and this leads to a greater resistance to flow if the material is strain-rate sensitive. When the material is viscous-like and $m = 1$, the increase in flow stress in the neck exactly compensates for the local reduction in area there, so that dA/dt is uniform along the gage length. As m becomes less than unity, the resistance to neck development is less. In this sense, strain-rate hardening at elevated temperatures is analogous to strain (work) hardening at lower temperatures. Both phenomena delay neck development as a result of the increased flow stress in the neck region. The extent to which development is retarded is related in strain hardening to the exponent n in the equation $\sigma \sim \varepsilon^n$ and in strain-rate hardening to the exponent m in $\sigma \sim \dot{\varepsilon}^m$.

Most superplastic materials do not display an m value of unity; rather, m is typically between 0.3 and 0.8 and an m value of 0.5 is commonly associated with superplasticity. Nevertheless, these values of m are sufficiently high to impart a strong

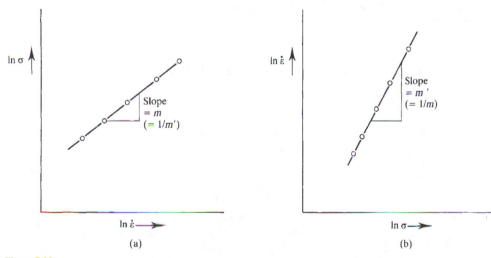

Figure 7.23
Alternative ways of determining a material's strain-rate sensitivity (m). In (a) the steady-state flow stress is determined as a function of strain rate and a logarithmic plot yields a slope of m. In (b), creep testing gives the steady-state creep rate as a function of stress. A logarithmic plot gives a slope of m' (cf. Eq. (7.1)), which is equal to m^{-1}.

330

CHAPTER 7
High-Temperature
Deformation of
Crystalline
Materials

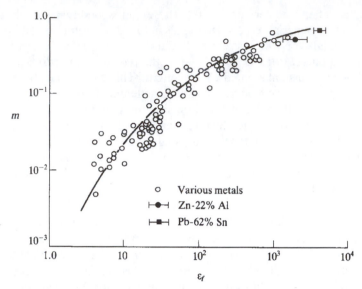

m

Figure 7.24
Correlation of tensile failure strain (ε_f in percent) with the strain-rate sensitivity of various materials. High values of m correlate with high tensile failure strains; i.e., with a high resistance to necking. (*From T. G. Langdon, Metall. Trans. A, 13A, 689, 1982.*)

resistance to neck development and enhanced tensile failure strains. The parameter m can, in fact, be correlated with tensile fracture strain in a wide variety of materials, as indicated in Fig. 7.24. Although there is considerable scatter of the individual datum about the line drawn in this logarithmic plot, it is clear that a high value of m is requisite for manifesting tensile failure strains characteristic of superplastic materials.

B. Experimental Observations of Superplasticity

A schematic of the variation of flow stress with strain rate for a superplastic material is provided in Fig. 7.25a; m, the instantaneous slope of the ln σ–ln $\dot{\varepsilon}$ curve of Fig. 7.25a, is shown as a function of strain rate in Fig. 7.25b. The behavior schematized in Figs. 7.25 is experimentally observed in superplastic materials (Fig. 7.26).

As indicated in Fig. 7.25, the stress-strain rate behavior of a superplastic material can be divided into three regions. Values of m are relatively low, and superplasticity is not manifested, in both the low-stress–low-strain rate Region I[19] and the high-stress–high-strain rate Region III. Rather, superplasticity is found only in Region II, a transition region in which the stress increases rapidly with strain rate. As shown in Fig. 7.25, Region II is displaced to higher strain rates as temperature is in-

---

[19]For some materials, Region I is not, or is only marginally, observed. This may be related to the very low strain rates of Region I; that is, experimental data at such low strain rates are not conveniently generated.

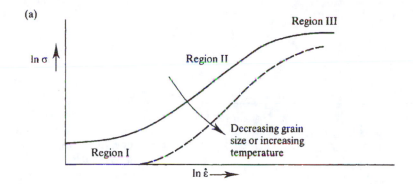

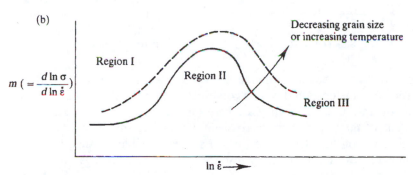

Figure 7.25
(a) The low-stress–strain rate behavior of a superplastic material. (b) In Regions I and III the strain-rate sensitivity is fairly low, whereas it is high in the superplastic Region II. As indicated in (a), increases in temperature or decreases in grain size shift the σ-$\dot{\varepsilon}$ curve downward and to the right. The same changes produce a somewhat higher value of m as shown in (b).

creased and/or grain size decreased. The maximum observed values of m also increase with similar changes in these parameters.

Although the deformation mechanisms operating in Region I and Region III are important insofar as they shed light on the mechanisms of superplasticity, it is the behavior in Region II that is of most interest. In addition to aspects of Region II behavior remarked on previously, it is observed that the activation energy for superplastic deformation is comparable to the material's grain-boundary-diffusion activation energy. Moreover, the strain rate at a constant stress varies as d^{-b} (conversely, the stress varies as d^{b}) where d is the grain size and b is in the range of 1–4, but typically about 2. Any successful model of superplastic behavior must rationalize such features as well as the stress-strain rate behavior shown in Figs. 7.25 and 7.26. Several such mechanisms are discussed in the following section.

C. Mechanisms of Superplasticity

Several mechanisms that fulfill the requirements noted above are discussed. All have in common that grain-boundary sliding in superplasticity is accommodated by

332

CHAPTER 7
High-Temperature
Deformation of
Crystalline
Materials

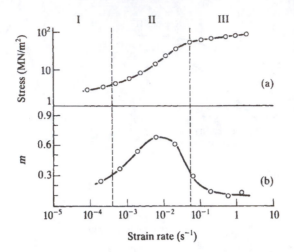

Figure 7.26
(a) The stress-strain rate and
(b) *m*-strain rate relations for
the superplastic Mg-Al eutectic
alloy (*d*=10.6 μm, *T*= 623 K).
Regions I–III can be seen.
(*Reprinted from D. Lee, Acta
Metall., 17, 1057, Copyright
1969, with permission from
Elsevier Science.*)

flow processes. As in creep deformation, the accommodation can be provided by either diffusional or dislocation flow or a combination thereof. Although both types of flow may operate in tandem, they are here discussed separately. It should be noted that none of the models we describe simulates all facets of superplastic deformation. In spite of this, the physical processes responsible for superplastic behavior are undoubtedly similar in many respects to these mechanisms.

    i.   **GRAIN-BOUNDARY SLIDING ACCOMMODATED BY DIFFU-SIONAL FLOW.** This accommodation mechanism is the superplastic analog of NH and Coble creep. As proposed originally by Ashby and Verrall, grain shape is preserved during deformation by a "grain-switching" mechanism that also produces a material strain. Grain-switching is illustrated in Fig. 7.27, which shows a two-dimensional configuration of four grains before (Fig. 7.27*a*), after (Fig. 7.27*c*), and at an intermediate stage of the grain-switching process. For the geometry illustrated, a true tensile strain of 0.55 is effected by the grain-switching event. The intermediate stage illustrates several significant features of the switching mechanism. First, there is an increase in grain-boundary area in the intermediate state compared to the initial and final states. This results in a "threshold" stress below which grain switching cannot occur. In effect, the applied stress must perform work associated with the formation of the increased grain-boundary area, and the stress to do this must be exceeded before it is capable of driving diffusional flow. Second, diffusional flow provides for the shape accommodation in the intermediate state (Fig. 7.28). In the Ashby-Verrall model, the flow can be either within grains (analogous to NH creep) or along grain boundaries (analogous to Coble creep). Provided the applied stress considerably exceeds the threshold stress, the strain rate for the grain-switching mechanism exceeds that of conventional diffusional creep. This is related to geometrical features of grain switching. First, the volume of material that must be transported to effect a given strain via grain switching is about 1/7 that required for diffusional creep. Additionally, the grain-switching diffusion distance is reduced by a factor of about 3 vis-à-vis the diffusional creep distance, and there are six such paths for grain switching as opposed to four for diffusional creep. Although these factors are mitigated to a degree by the fact that some of the grain boundaries are at angles of neither 0 nor 90° to the tensile axis (thus reducing the effective driving stress for diffusional flow), the net result is that the strain rate for grain switching is about an order of magnitude higher than for diffusional creep. Ashby and Verrall, considering both volumetric and grain-boundary mass transport, developed the following equation to describe the grain-switching creep rate;

$$\dot{\varepsilon}_{GS} \cong \frac{100\Omega}{kTd^2}\left(\sigma - 0.72\frac{\gamma}{d}\right)D_L\left(1 + \frac{3.3\delta'D_{GB}}{dD_L}\right) \tag{7.29}$$

The term $0.72\gamma/d$ represents the threshold stress for grain switching. If only boundary transport is important (i.e., if $3.3\delta'D_{GB} \gg dD_L$), Eq. (7.29) reduces to

$$\dot{\varepsilon}_{GS} \cong \frac{330\Omega}{kT}\frac{\delta'D_{GB}}{d^3}\left(\sigma - \frac{0.72\gamma}{d}\right) \tag{7.30}$$

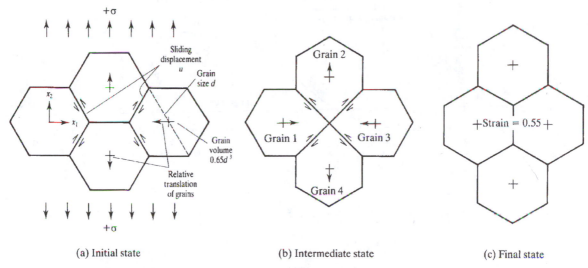

(a) Initial state

(b) Intermediate state

(c) Final state

Figure 7.27
Schematic of a grain-switching event. Relative grain-boundary sliding produces a strain (c) without a change in grain shape (compare (a) with (c)). However, the intermediate step (b) of the process is associated with an increased grain-boundary area. (*Reprinted from M. F. Ashby and R. A. Verrall, "Diffusion Accommodated Flow and Superplasticity," Acta Metall., 21, 149, Copyright 1973, with permission from Elsevier Science.*)

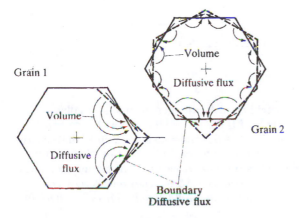

Figure 7.28
During the intermediate stage of grain switching, grains 1 and 2 (cf. Fig. 7.27) change their shape from that indicated by the solid lines to that of the dotted lines. The shape change is effected by diffusional flow, which can take place by volumetric or boundary diffusion. (*Reprinted from M. F. Ashby and R. A. Verrall, "Diffusion Accommodated Flow and Superplasticity," Acta Metall., 21, 149, Copyright 1973, with permission from Elsevier Science.*)

334

CHAPTER 7
High-Temperature
Deformation of
Crystalline
Materials

and grain switching can be considered competitive with ordinary Coble creep. Grain switching dominates Coble creep at stresses large in comparison to $0.72\gamma/d$ and vice versa (Example Prob. 7.2 and Prob. 7.29).

Ashby and Verrall considered grain-switching "creep," as described by Eq. (7.29), competitive also with dislocation creep with the latter dominating at high stress levels. Since the two creep processes operate in parallel, the net strain rate is

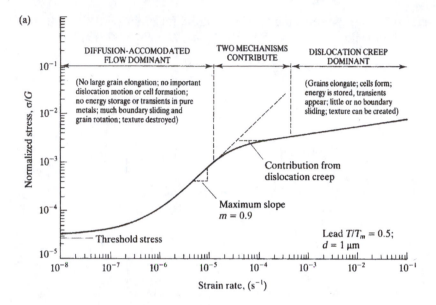

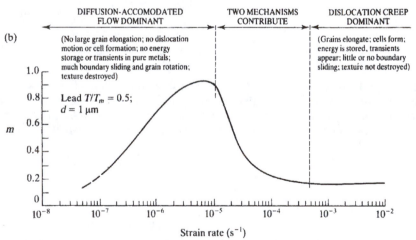

Figure 7.29
(a) The relationship between grain switching, dislocation creep, and superplasticity as proposed by Ashby and Verrall. Region I is considered dominated by grain switching and III by dislocation creep. The total strain rate is the sum of that due to both processes; this yields a high strain-rate sensitivity (b) in the transition region and associated superplastic behavior. Data used in the calculations are appropriate for Pb with $d = 1$ μm at $T = 0.5T_m$. (*Reprinted from M. F. Ashby and R. A. Verrall, "Diffusion Accommodated Flow and Superplasticity," Acta Metall., 21, 149, Copyright 1973, with permission from Elsevier Science.*)

the sum of the individual strain rates. This leads to stress-strain rate behavior similar to that shown in Fig. 7.25. In Region I, grain switching dominates and $\sigma \cong \sigma_0 +$
$A\dot{\varepsilon}$. Likewise, dislocation creep dominates Region III. The transition Region II is
characterized by a rapid increase in stress with strain rate and superplastic behavior.
The overall scheme as envisaged by Ashby and Verrall is summarized in Fig. 7.29.

Grain switching has been observed in emulsions (Fig. 7.30a) and in thin films
of the superplastic Zn-Al alloy (Fig. 7.30b). The mechanism is also conceptually
appealing, since it predicts the generally observed stress-strain rate relationship. It
also produces a tensile strain absent grain shape change and, as noted, this is consistent with experiment. Further, it does not, except as a transition event, invoke
dislocation mechanisms and, in view of the absence of observed concentrated dislocation activity in superplastic materials, this was for a time considered additional
support for the grain-switching mechanism. (Ashby and Verrall, however, realized
full well that dislocation and diffusional flow could occur concurrently.) However,
the grain-switching model is not without shortcomings. For example, it does not
generally predict accurately the stress level of Region I. Nor for that matter does it
consider in detail dislocation accommodation mechanisms (these are described in
the following section). Finally, there are some conceptual problems relating to the
grain shape in the intermediate stage of Fig. 7.27. Correcting them, however, does
not alter to any great extent the basic ideas of the grain-switching event as originally
put forth by Ashby and Verrall.

EXAMPLE PROBLEM 7.2. **a.** Schematically plot the grain-switching
creep rate (Eq. (7.30)) and the Coble creep rate (Eq. (7.11)) as functions of stress
and show that grain switching dominates creep at high stress levels and Coble creep at
low stress levels.
b. What is the value of the transition stress? (The answer will be expressed in terms of
γ, d, and the respective numerical coefficients in the creep-rate equations.)

Solution
a. Equations (7.11) and (7.30) are plotted below. Both creep rates vary linearly with the
stress. However, grain-switching creep does not commence until the threshold stress
($= 0.72\gamma/d$) is exceeded. But when this stress is exceeded, the grain-switching creep
rate increases more rapidly with stress than does the Coble creep rate. As a result, above
a certain stress (where the respective creep rates are the same) grain-switching creep is
dominant.

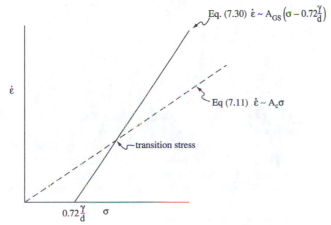

Eq. (7.30) $\dot{\varepsilon} \sim A_{GS}\left(\sigma - 0.72\frac{\gamma}{d}\right)$

Eq (7.11) $\dot{\varepsilon} \sim A_c\sigma$

transition stress

$0.72\frac{\gamma}{d}$ σ

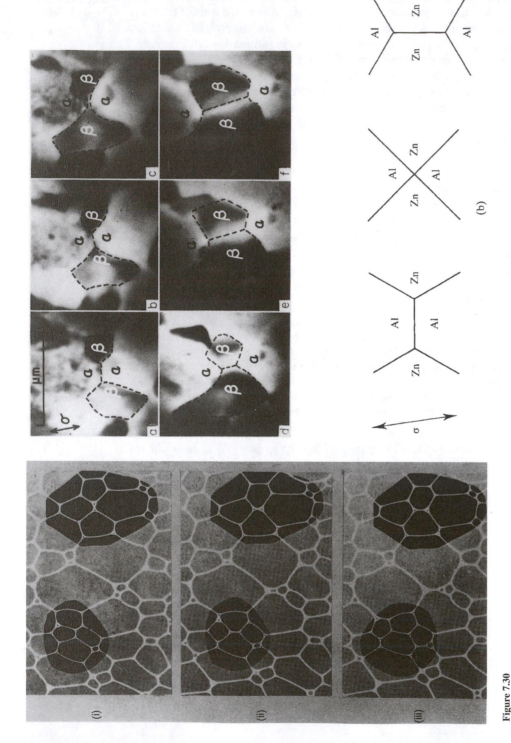

Figure 7.30
Grain switching in emulsions (a) and in the Zn-Al eutectoid alloy (b). The photographs (i–iii) of (a) represent a sequence of increasing "strain" in the emulsion, which is funneled between two glass slides. The photographs of (b) are electron micrographs of the eutectoid strained *in situ*. Thus, increasing time represents increasing strain. A schematic of the Zn-Al grain-switching event is shown in the lower portion of (b). (*Part (a) reprinted from M. F. Ashby and R. A. Verrall, "Diffusion Accomodated Flow and Superplasticity," Acta Metall., 21, 149, Copyright 1973, with permission from Elsevier Science; (b) from Naziri, R. Pearce, M. Henderson-Brown and K. F. Hale, Jl. Microscopy, 97, 229, 1973.*)

b. The transition stress is obtained by equating the respective creep rates. Both expressions contain a number of common terms. On factoring these out we find that the transition stress is a solution to the equation

$$A_C\sigma = A_{GS}\left(\sigma - \frac{0.72\gamma}{d}\right)$$

where the A's are the numerical coefficients in the creep rate equations. With $A_C = 50$ (Table 7.1) and $A_{GS} = 330$, we obtain for the transition stress

$$\sigma = \left[\frac{A_{GS}}{A_{GS} - A_C}\right]\left[\frac{0.72\gamma}{d}\right] = \frac{0.85\gamma}{d}$$

ii. GRAIN-BOUNDARY SLIDING ACCOMMODATED BY DISLOCATION FLOW.

Models of superplasticity invoking dislocation motion were developed even before microscopic observations confirming dislocation activity in superplastic materials were made. The observations were relatively late in coming, and for good reason. First, superplastic temperature-strain rate-stress combinations along with the associated fine grain size are such that the dislocation cell sizes in superplastically deformed materials should exceed the grain size. What this means is that dislocations tend to be emitted and captured at grain boundaries and are seldom observed in grain interiors in superplastically deformed materials. However, careful studies have demonstrated conclusively that some type of dislocation flow—in which dislocations strut rapidly across the grain and then redeposit themselves comfortably in the grain boundaries after having done their duty—is common to superplasticity. Dislocation accommodation mechanisms for superplasticity must also, though, provide a means by which grain shape remains relatively unchanged during deformation. Finally, since the grain size substitutes for a cell size in descriptions of superplastic behavior, dislocation models of superplasticity incorporate a grain (rather than cell) size in them.

Most (of the many) dislocation models for superplasticity arrive at a constitutive equation of the form

$$\dot{\varepsilon}_{DS} = CD\left(\frac{\sigma\Omega}{kT}\right)\left(\frac{\sigma}{G}\right)\left(\frac{b}{d}\right)^2 \tag{7.31}$$

where the constant C depends on the details of the model, and the diffusivity D is usually a grain-boundary diffusion coefficient. These descriptions are reasonable with respect to the magnitude of the stress required to produce a given strain rate, the m value of 0.5, and the grain-size dependence of strain rate. In common with the Ashby and Verrall model, dislocation models for superplasticity consider Region III to be controlled by conventional dislocation creep.[20] In contrast to the grain-switching model, however, Region II is viewed as due to an independent superplastic flow mechanism, and is not considered a "transition" stage.

According to some dislocation models of superplasticity, the transition from Region II to Region III happens when the subgrain size becomes less than the grain

---

[20]This is in agreement with activation energies (typically the same as the volume diffusion activation energy) and with the small values of m observed in Region III.

338

CHAPTER 7
High-Temperature
Deformation of
Crystalline
Materials

size. For many metals, the inverse relationship between subgrain size, d', and stress is approximated by[21]

$$\frac{d'}{b} = 10\frac{G}{\tau}$$ (7.32)

and Eq. (7.32), therefore, defines the shear stress at the transition from superplasticity to dislocation creep. That this description is reasonable is shown by Figs. 7.31a and b, which are deformation mechanism maps (in terms of d/b vs. τ/G) for superplastic Al-Zn and Pb-Sn alloys. The dotted lines in these diagrams represent Eq. (7.32), and it is seen that the equation agrees well with experimental results. Figures 7.31a and b also show that superplastic flow competes with diffusional creep, dominating it at higher stress levels. Although Region I is shown as separate areas in these figures, there is disagreement as to whether it represents a separate flow mechanism or is merely an obscure manifestation of Coble creep.

Although no model of superplasticity describes accurately the phenomenon in all of its quantitative and qualitative aspects, the models discussed are nonetheless useful. In effect, they can predict approximately the grain size, temperature, and strain-rate regimes where superplasticity is likely to be observed, and this has proven useful in engineering applications.

It is well to conclude this section with a description of some applications. For some time, superplasticity was thought to have little engineering relevance. The strain rates at which it is observed were considered too low to be of much use in forming engineering alloys. To some extent, this criticism still holds. For example, large volume productions (as might be found in the automotive industry) seem outside the capacity of superplastic forming. On the other hand, moderate volume productions are feasible. In shaping of complex parts, superplastic forming minimizes material waste compared to conventional (e.g., machining) operations. Thus, superplastic forming is widely used in production of aluminum- and titanium-base materials, particularly for aerospace applications. Further, the strain rates used for superplastic forming are now much higher than they were when the phenomenon was first investigated. An additional benefit of superplastic forming is the low forming pressures involved. This minimizes tool wear and contributes to the economy of forming. More recently, the discovery of superplasticity in ceramics also bodes well for forming of these materials into (nearly) flaw-free structures, thereby permitting them to be used in applications from which they have heretofore been excluded.

Superplastic forming has drawbacks, though. It is often found that internal voids form during superplastic forming, and the effect negatively impacts subsequent material performance. There are, however, ways to minimize void formation during superplastic forming. These are discussed in Chap. 11, which focuses on high-temperature fracture.

---

[21]Note that Eq. (7.32) is the high-temperature equivalent of Eq. (3.21) which relates dislocation cell size to stress at low temperature.

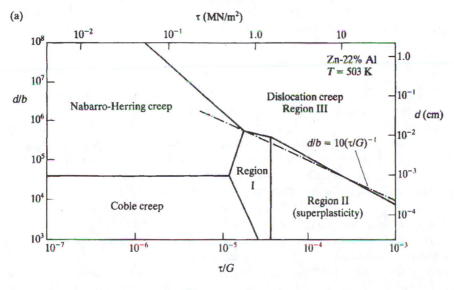

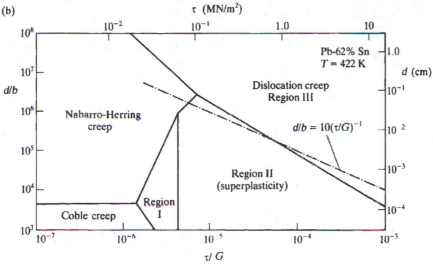

Figure 7.31

Deformation mechanism maps for (a) Zn-Al and (b) Pb-Sn in terms of grain size vs. shear stress. (Note that the axes are reversed from Fig. 7.17b.) According to the dislocation models of superplasticity, a transition from superplasticity to dislocation creep should occur when $d/b \cong 10G/\tau$. This is the approximate situation for these materials. (*Reprinted from F. A. Mohammed and T. G. Langdon, "Deformation Mechanism Maps for Superplatic Materials," Scripta Metall., 10, 759, Copyright 1976, with permission from Elsevier Science.*)

340

CHAPTER 7
High-Temperature
Deformation of
Crystalline
Materials

7.8
HOT WORKING OF METALS

A. Description of Hot Working

For several reasons, hot working of metals and their alloys is performed at temperatures greater than about $0.5T_m$. First, as a result of the lower flow stresses at higher temperatures, hot working offers an economical method for size reduction of large ingots into more readily handled sizes. Second, metals at high temperature are generally capable of undergoing much larger deformation strains without fracture than at lower temperatures. Finally, high-temperature deformation allows for ingot homogenization (chemical segregation on a microscopic, and frequently macroscopic, scale is inherent to the solidification process). Some processes involving hot and warm working also provide for control of microstructure and development of desirable properties. Ausforming of steels and thermomechanical treatment of microalloyed steels, discussed in Chap. 5, are cases in point.

Most hot working involves rolling, extrusion, or forging. Extrusion can be considered a variant of closed-die forging, and extrusion and hot rolling are illustrated schematically in Fig. 7.32. During rolling and, particularly, extrusion, the strain rate varies during the process and, moreover, the plastic strain during extrusion is not uniform throughout the material. Because of these processing features and also their relatively complicated stress states (not to mention difficulties in measuring stress, strain, and strain rate), rolling and extrusion are seldom used to characterize material hot-working deformation characteristics. Instead, the relations among stress, strain, and strain-rate are usually developed through compression and/or torsion testing. In these tests the pertinent parameters are well characterized and measured conveniently and accurately. Further, the stress state associated with torsion

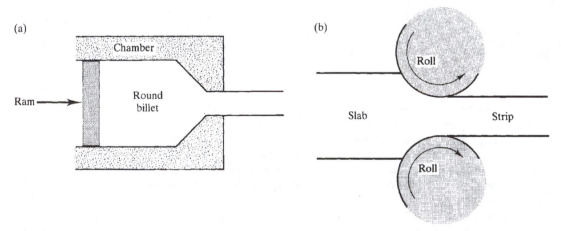

Figure 7.32
Schematic of two deformation processes used for hot-working metals and their alloys. (a) In extrusion, the reduction in area is accomplished by a ram forcing the material through the die. (b) In rolling, reduction is accomplished by drawing the material between rotating rolls that supply the necessary force for reducing the material thickness.

(cf. Chap. 1) is such that large plastic strains—commensurate with those of hot working—can be obtained.

Since it involves high temperature and, thus, recovery, hot working can in some respects be considered a creep phenomenon. There are, however, important differences between creep and hot working. First, in hot working the stress is a function of both strain rate and strain and not, as it is in creep, a function of strain rate alone. In assessing a material's hot-working behavior, a constant strain rate is usually imposed and the resulting deformation stress measured as a function of strain. Further, the strain rates common to hot working (typically 10^{-3} to 10^3 per second) are much higher than in superplastic deformation or creep. This results in higher flow stresses, and in turn this can alter microstructure significantly. For example, subgrain sizes developed in hot working are usually much finer than those in creeping materials. And, in some instances, the working stresses and strain rates are sufficiently high that recrystallization (which only infrequently accompanies creep) occurs simultaneously with, or subsequent to, deformation. Structural development during hot working is called *dynamic recovery* or, if recrystallization occurs, *dynamic recrystallization*. This distinguishes it from static recovery/recrystallization which can take place subsequent to hot working. Our discussion of hot working centers on the dynamic recovery/recrystallization processes.

B. Dynamic Recovery and Recrystallization

A schematic of the structural changes that can occur during a typical hot-working process is provided in Fig. 7.33. As indicated in Figs. 7.33a and b, dynamic recovery taking place during hot working can be followed by static recovery or static recrystallization.[22] Figure 7.33c illustrates dynamic recrystallization taking place during rolling. Immediately subsequent to the passage of the material through the rolls, grains recrystallized *during* deformation continue to grow at the expense of those that have not; this is termed *metadynamic recrystallization*.

Whether only dynamic recovery or dynamic recovery and recrystallization are observed during hot working depends on the stresses and strain rates involved. Generally, the higher these are, the greater the tendency for recrystallization. This generalization is not always followed. Some metals and alloys do not dynamically recrystallize; they only undergo dynamic recovery. Materials that do not dynamically recrystallize exhibit rapid recovery rates, and thus a dislocation density and configuration sufficient to nucleate and drive recrystallization is not developed in them. These materials include the high-stacking-fault-energy (SFE) face-centered cubic metals such as aluminum, the body-centered cubic transition metals such as iron, and most hexagonal close-packed metals like zirconium. In contrast, face-centered cubic metals with low SFE dynamically recrystallize under hot-working conditions. Thus, nickel and its alloys, face-centered cubic iron, copper, and brass dynamically recrystallize.

---

[22]Some static recovery inevitably precedes static recrystallization, since recrystallization requires an incubation time but recovery does not.

342

CHAPTER 7
High-Temperature
Deformation of
Crystalline
Materials

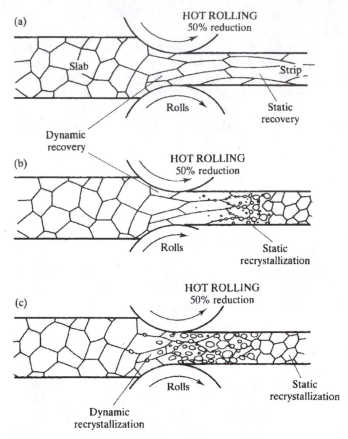

Figure 7.33
Schematic of possible microstructural changes during hot working. In (a) dynamic recovery occurs during forming, and is followed by static recovery subsequent to forming. If only recovery takes place, the grains change shape commensurate with the macroscopic shape change. In (b) the grains recrystallize after deformation (static recrystallization). In (c) dynamic recrystallization (i.e., recrystallization during deformation) takes place and is followed by static recrystallization. (*Adapted from H. J. McQueen and J. J. Jonas, Treat. Matls. Sc. Tech., ed. R. J. Arsenault, 6, 394, 1975.*)

    While static recovery/recrystallization following hot working influence properties of hot-worked materials, a discussion of the static processes belongs more in the field of phase transformations than in mechanical behavior. Thus, we discuss only the dynamic processes and how they relate to stress-strain–strain-rate behavior.

    i. DYNAMIC RECOVERY. Dynamic recovery can be viewed as creep occurring at the high strain rates of hot working. The true stress–true strain behavior of a material experiencing only dynamic recovery during hot working is illustrated in Fig. 7.34. It can be seen that the overall level of the flow curve increases with increases in strain rate and/or decreases in temperature. Following application of a

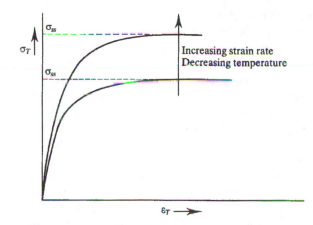

Figure 7.34
Schematic of the true stress–true strain behavior of a material undergoing dynamic recovery. The flow stress rises to a steady-state value—which increases with increases in strain rate and decreases in temperature—associated with a steady-state material substructure.

fixed strain rate, the stress increases and plastic flow commences at the (strain-rate dependent) high-temperature yield strength. If the material is initially in an annealed condition, the stress increases with further strain, eventually reaching a steady-state value (σ_{ss}, Fig. 7.34).[23] During the transient, a steady-state microstructure evolves in much the same way it does during steady-state creep. That the analogy with creep behavior is proper is further evidenced by σ_{ss} being related to strain rate and temperature by an equation similar to Eq. (7.1), i.e., $\sigma_{ss} \sim (\dot{\varepsilon})^m \exp(Q/RT)$. The steady-state microstructure closely resembles that of a material deformed during creep; i.e., grains elongate in the primary strain direction and contract in the others. Concomitantly, an equiaxed subgrain structure is formed with the subgrain size (usually finer during hot working than in creep) decreasing with increasing stress. Thus, our picture of dynamic recovery during hot working as being a manifestation of high-strain-rate creep is a good one.

  ii. DYNAMIC RECRYSTALLIZATION. A schematic of the true stress–true strain behavior of a material undergoing dynamic recrystallization during hot working is provided in Fig. 7.35a, and this behavior for a medium-carbon steel is shown in Fig. 7.35b. Material response can be divided broadly into two categories. At higher strain rates (or lower temperatures), the overall flow stress level is higher than at lower strain rates (or higher temperatures). At the higher strain rates, the flow stress increases following yielding and attains a maximum value (σ_{max}) after which the stress decreases, eventually attaining a steady-state value (σ_{ss}). At the lowest of the plastic strains, only dynamic recovery takes place. However, dynamic recrystallization begins prior to the stress reaching σ_{max}. At low strain rates and/or higher temperatures, the overall level of the flow curve is less. In addition, the stress-strain curve manifests an oscillatory nature, cycling above and below an approximate steady-state stress. The oscillatory behavior tends to dampen out at larger strains, as shown in Fig. 7.35b.

---

[23]If the material is initially cold worked, the yield strength may be higher than σ_{ss}. This is because the initial structure has a greater dislocation density than the steady-state one. However, σ_{ss} is independent of the material's initial state.

344

CHAPTER 7
High-Temperature
Deformation of
Crystalline
Materials

The contrasting behavior reflects the different substructures developed in the materials under the different conditions. At the high strain rate (low temperature) condition, recrystallized and cold-worked grains coexist throughout the material. Or, to put it another way, recrystallization occurs more or less continuously throughout the material volume and, prior to one recrystallization event being completed in one grain, one is initiated in another grain. If a series of high-resolution snapshots of the material were taken, successive snapshots would demonstrate that, in total, the structure is invariant even though the structure within each grain is continuously changing. The structure is also heterogeneous; some grains, just recrystallized, have a low dislocation density. Others, just about to recrystallize, possess a well-developed substructure with a high dislocation density. The heterogeneous distribution of the structure within different grains is a result of the high strain rate (and the correspondingly high stress). Within a specific grain, a critical substructure must be developed before recrystallizaton starts. Because of the high strain rate (and higher stress), the time needed to attain the critical structure is short for an in-

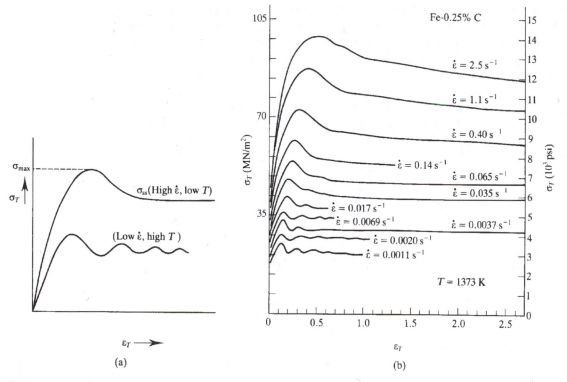

Fig. 7.35
(a) Schematic of the true stress–true strain behavior of a material undergoing dynamic recrystallization. The onset of recrystallization occurs before the stress σ_{max} is reached. An approximate steady-state stress (σ_{ss}), lower than σ_{max}, is eventually found; σ_{max} and σ_{ss} increase with increases in strain rate and decreases in temperature. At low strain rates and high temperatures, the high-strain behavior is such that the stress oscillates about a steady-state value. (b) True stress–true strain relationships for a Fe-0.25 wt.% C steel during hot working at 1373 K. (The crystal structure at this temperature is face-centered cubic.) The features displayed conform to those shown in Fig. 7.35a. *(Part (b) from C. Rossard, Metaux, 35, pp. 107, 140, 190, 1960.)*

dividual grain. After dynamic recrystallization starts, the aggregate develops the stochastic—yet steady-state—structure described earlier.

The situation is different at low strain rates. Under these conditions, it takes a longer time (and greater strain) to accumulate the critical dislocation density needed to trigger recrystallization. As a consequence, recrystallization "waves" pass through the material. For the condition of low strain rate, a structural snapshot taken at a specific time would show that all grains have approximately the same substructure. But this substructure would change with sequential snapshots. At a stress level corresponding to a local minimum, a low dislocation density characterizes each individual grain. This corresponds to the end of a "recrystallization wave." Successive photographs would reveal a progressively finer substructure—with increased dislocation density—until a new recrystallization wave is initiated (initiation takes place at a local stress maximum).

It is of interest to compare the oscillatory stress-strain curve of a material hot-worked at a low strain rate to the serrated yielding displayed in a tensile test of a material that creeps by dislocation solute drag (Fig. 7.7). While the causes for the local maxima and minima in stress are quite different, there are phenomenological similarities. In particular, stress maxima precede "softening" in both situations. For the solute drag case, softening is caused by the breaking away of dislocations from the solute atoms; for hot working it corresponds to the beginning of recrystallization. Further, in both situations, the stress minima precede a hardening event: the "catching up" to the dislocations by solute atoms in solute drag and the development of a deformation structure in hot-worked metals.

7.9
SUMMARY

High-temperature deformation of crystalline materials involves facets of diffusional flow and dislocation motion. These features of high-temperature deformation are important in discussing the creep, superplastic, and hot-working behavior of materials.

Creep is a high-temperature phenomenon in which a constant applied stress causes continuing material permanent deformation. As such, creep is frequently the design limiting factor for many high-temperature applications. Under a fixed stress, transient creep, often characterized by a decreasing creep rate, is first observed. After a certain time (strain), the strain rate attains a characteristic constant value called the steady-state creep rate. As expected, the microstructure of the material is invariant during steady-state creep. Empirical, as well as fundamentally based, constitutive equations have been developed to describe steady-state creep rates. All relate the creep rate to stress and a diffusivity. Additionally, diffusional creep rates are explicitly related to the material grain size.

The particular form of a constitutive creep equation depends on the rate-controlling mechanism. At low stresses and strain rates, diffusional flow is dominant, and dislocation aspects of creep are of secondary importance. Since for diffusional creep the characteristic diffusion distance and area depend on the grain size, the equations describing diffusional creep incorporate a grain-size term. The creep rate increases with decreasing grain size for volume-diffusion-controlled

346

CHAPTER 7
High-Temperature
Deformation of
Crystalline
Materials

(Nabarro-Herring) creep, and depends even more strongly on grain size when mass transport is along grain boundaries (Coble creep).

A number of models have been developed for dislocation creep, which dominates at higher stresses and results in higher strain rates. For dislocation creep, the diffusion distance is a dimension smaller than the grain size; e.g., the subgrain size. Grain size is hence not explicitly included in the equations describing dislocation creep. Since the characteristic microstructural dimension generally decreases with stress, dislocation creep depends more strongly on stress than does diffusional creep.

Provided the controlling creep mechanism is known, a material's creep resistance can be improved through engineering design. Creep rates are reduced by utilizing materials with high moduli (for dislocation creep) and low diffusivities. If diffusional flow is the dominant creep mechanism, creep resistance is improved by increasing the grain size. Since both diffusional and dislocation creep involve concurrent grain-boundary sliding, and since the sliding strain must be equivalent to that of the primary creep modes, creep resistance can also be improved by the presence of intergranular particles that restrict grain-boundary sliding. Particles situated within grains also often improve creep resistance. This is often manifested experimentally by a threshold stress below which dislocation creep does not take place in such materials.

Deformation maps, the axes of which are stress (normalized in terms of the shear modulus) and homologous temperature, are convenient diagrams from which the dominant deformation mechanism of a material can be identified at a given stress-temperature combination. The accuracy of deformation maps is only as good as the (usually) empirical equations on which they are based; the accuracy of strain-rate contours imposed on them is limited similarly. An alternative formulation of a deformation map has axes of normalized stress and grain size; such a diagram is appropriate for a fixed temperature. Once a dominant creep mechanism has been identified, it can be used in design considerations. Provided the creep mechanism is the same for long-time service as for short-time service, short-time tests are useful for determining the operating life of a material or defining stress-temperature combinations safe for long-time use of it. This is done using extrapolation procedures. The Larson-Miller parameter is one such widely used extrapolation procedure, even though it is not an inherently conservative procedure.

Superplasticity, in which extensive strains to failure can be obtained at strain rates and stresses intermediate to those for which diffusional and dislocation creep dominate, is observed in fine-grained ($d \lesssim 10$ μm) materials for which the grain structure is stable during deformation. Superplastic materials are characterized by a high strain-rate sensitivity; the exponent m in the equation $\sigma = K'(\dot{\varepsilon})^m$ is on the order of 0.3–0.8 for materials demonstrating superplasticity. An equiaxed grain structure is maintained during superplastic deformation, and there is evidence of extensive grain-boundary sliding and grain rotation during deformation. Several mechanisms involving diffusional and dislocation flow have been advanced to explain superplasticity. Most of these are reasonable insofar as describing the main features of superplastic behavior is concerned. Superplasticity is industrially useful. Complex shapes can be formed via superplastic forming with relatively little expenditure of power during the forming operation. However, the process suffers from the relatively low production rates associated with it.

This is in distinction to hot working of metals. In contrast to the low strain rates of ca. 10^{-3} s^{-1} characteristic of superplasticity, hot working can be carried out at strain rates as high as ca. 10^3 s^{-1}. Hot working in which only dynamic recovery is observed can be considered a dislocation-creep process occurring at the higher stresses and strain rates of hot working. If recrystallization occurs during hot working, the process cannot be viewed in this way, as a steady-state structure is not developed within each grain. Rather, substructure develops with increasing strain within individual grains until the stored energy is sufficient to nucleate recrystallization of the grains. At lower stresses (and strain rates), dynamic recrystallization is characterized by a periodic propagation of recrystallization waves through the material. These are associated with—and the cause of—the oscillatory shape of the stress-strain curve under these circumstances. With increasing strain rate, the oscillatory nature of the flow curve is replaced by one which manifests a constant steady-state stress.

REFERENCES

Argon, A. S.: "Mechanical Properties of Single-Phase Crystalline Media: Deformation in the Presence of Diffusion," *Physical Metallurgy—Vol. III,* ed. R. W. Cahn and P. Haasen, 4th ed., North-Holland, Amsterdam, 1957, 1996.

Edington, J. W., K. N. Melton, and C. P. Cutler: "Superplasticity," *Prog. Matls. Sc.,* **21,** 61, 1976.

Evans, A. G., and T. G. Langdon: "Structural Ceramics," *Prog. Matls. Sc.,* **21,** 171, 1976.

Frost, H. J., and M. F. Ashby, *Deformation Mechanism Maps*, Pergamon Press, Oxford, 1982.

McQueen, H. J., and J. J. Jonas: "Recovery and Recrystallization During High Temperature Deformation," *Treatise Matls. Sc. and Tech.*, ed. R. J. Arsenault, **6,** 394, 1975.

Mukherjee, A. K.: "High Temperature Creep," *Treatise Matls. Sc. and Tech.,* ed. R. J. Arsenault, **6,** 164, 1975.

Mukherjee, A. K.: "Superplasticity in Metals, Ceramics and Intermetallics," *Plastic Deformation and Fracture of Materials—Materials Science and Technology,* ed. H. Mughrabi, **6,** 407, 1993.

Mukherjee, A. K., and R. H. Mishra: "Creep Mechanisms in Dispersion Strengthened Materials," *The Johannes Weertman Symposium,* ed. R. J. Arsenault et al., TMS, Warrendale, PA, 119, 1996.

Pilling, J., and N. Ridley: *Superplasticity in Crystalline Solids,* The Institute of Metals, London, 1989.

Taplin, D. M. R., G. L. Dunlop, and T. G. Langdon: "Flow and Failure of Superplastic Materials," *Ann. Rev. Matls. Sc.,* **9,** 151, 1979.

PROBLEMS

7.1 Refer to Fig. 7.3. Determine the creep activation energy for (a) the bcc transition metals W, Nb, and αFe, (b) the fcc metals Pt, Ni, Au, Cu, and Al, (c) the hcp metals αTl, Mg, and Zn, (d) the ionic rock salts NaCl and AgBr, and (e) the polar covalent solids MgO and Al$_2$O$_3$. For these different material classes, divide the activation energy by RT_m, where T_m is the material's absolute melting temperature. Comment on any trends you may have uncovered.

348

CHAPTER 7
High-Temperature
Deformation of
Crystalline
Materials

7.2 Explain in physical terms why the vacancy formation energy is changed by $\pm\sigma\Omega$ for compression and tension, respectively. (Hints: Remember that Q_f is the work required to form a vacancy; i.e., it is the energy required to remove an atom from the interior of a crystal and place it on its surface. When a stress acts on a material, it does work $p\,\delta V$, where p is the mean pressure.)

7.3 Would you expect the activation energy for vacancy motion to be affected by compressive or tensile stresses? If so, would the effect be a large or small one? Explain your reasoning.

7.4 Consider Fig. 7.5b, which illustrates the change in shape undergone by a cubic crystal during creep. Show that when mass is transported from the edges of the crystal subject to compression to those experiencing tension, the crystal extends an amount $\delta d = \delta V/d^2$, where δV is the volume of matter transported and d is the original crystal edge length.

7.5 Compare the temperature dependence of Nabarro-Herring and Coble creep. Which is more temperature-sensitive? Which creep mechanism will dominate at high, and which at low, temperatures?

7.6 a In terms of relative diffusivities, etc., estimate the grain size below which Coble creep dominates diffusional creep and above which Nabarro-Herring creep is dominant.
b Determine the temperature above which NH creep dominates Coble creep. Express your answer in terms of the material grain size and the respective D_0's and diffusion activation energies.

7.7 a Power-law creep can be phenomenologically considered a competition between work hardening and recovery. Let the stress, σ, be represented as $\sigma = \sigma\,(\varepsilon,t)$. Express $d\sigma$ in terms of partial derivatives of the stress with respect to strain (ε) and time (t). The first partial derivative can be considered a work-hardening parameter (W) and the second a recovery parameter (R). Given that the stress is constant during a creep test, how is the steady-state creep rate related to R and W?
b Steady-state power-law creep can also correspond to a constant dislocation density, ρ. Let $\rho = \rho(\varepsilon,t)$ and express $d\rho$ in terms of partial differentials, one of which (W') represents work hardening and the other (R') recovery. Express the steady-state creep rate in terms of W' and R'.
c Are your answers in (a) and (b) two different ways of viewing the same phenomenon? That is, are the expressions complementary or contradictory?

7.8 Refer to Fig. 3.37. Assume that continued motion of the screw dislocation will not take place unless atom transport occurs to the vacancy sites created at the jog PP' (or, alternatively and equivalently, vacancy migration away from the jog takes place). The climb-creep rate scales with $(1/\Omega)(dV/dt)$ where Ω is the atomic volume and dV/dt is the volume transfer rate to the jog; dV/dt is proportional to $A_D D_L(\delta\mu/\delta x)$ where A_D is the diffusion area, D_L is the vacancy diffusion coefficient, $\delta\mu$ is the chemical potential driving force, and δx is the diffusion distance. Let $\delta\mu = \sigma\Omega$, and assume that δx is proportional to the dislocation spacing and A_D is proportional to the square of this spacing. Determine how the climb-creep rate depends on the dislocation density and stress. Assume a reasonable relationship between stress and dislocation density and show that the process produces a power-law creep rate expression.

7.9 Refer to Fig. 7.9 and the discussion pertaining to it. Note that the ratio of the dislocation glide to climb velocity is equal to the ratio L/h which relates to the geometry of the steady-state dislocation structure. Use the relation between h and σ stipulated in the text as well as that between L, h and M to derive a functional relationship relating the dislocation glide and climb velocities to the stress. Show that this predicts that climb will control the creep rate at high stresses and dislocation glide will control it at lower stresses.

7.10 Assume that cubic particles, of edge length a, are situated on grain boundaries; the particle center-to-center spacing is λ. With reference to Fig. 7.10, estimate the amount of mass that must be transferred to produce a relative grain-boundary offset equal to the interparticle spacing. What do your results tell you about the effect that particle size and particle grain-boundary volume fraction have on the grain-boundary sliding creep rate?

7.11 The following data were obtained for creep of a polycrystalline oxide having a grain size of 10 μm.

	Strain rate (s^{-1})		
Stress (MPa)	at $T = 1700$ K	at $T = 1810$ K	at $T = 1940$ K
10	2.0×10^{-8}	7.0×10^{-8}	4.2×10^{-7}
20	4.0×10^{-8}	1.4×10^{-7}	8.4×10^{-7}
30	6.0×10^{-8}	2.1×10^{-7}	1.3×10^{-6}
40	8.0×10^{-8}	2.8×10^{-7}	1.7×10^{-6}
50	2.3×10^{-7}	8.0×10^{-7}	4.8×10^{-6}
60	5.0×10^{-7}	1.8×10^{-6}	1.1×10^{-5}
70	9.9×10^{-7}	3.5×10^{-6}	2.1×10^{-5}
80	1.8×10^{-6}	6.3×10^{-6}	3.8×10^{-5}
90	3.1×10^{-6}	1.1×10^{-5}	6.5×10^{-5}
100	4.9×10^{-6}	1.7×10^{-5}	1.0×10^{-4}

a Plot (logarithmic coordinates) creep rate vs. stress for the three temperatures. Identify on your plot the stress regions that correspond to diffusional creep and those that correspond to power-law creep.
b Does the diffusional creep region correspond to Nabarro-Herring creep or to Coble creep? (Hint: Recall that the activation energies for NH creep and power-law creep are the same, and greater than that for Coble creep.)
c On the graph you developed for part (a), plot creep rate vs. stress at 1700 K for this material if the grain size were increased to 20 μm.

7.12 a A conventional deformation mechanism map has axes of (log) stress and temperature (at a fixed material grain size). Explain why the boundary line separating the Coble and Nabarro-Herring regions is a vertical line in such a representation.
b An alternative way of presenting a deformation mechanism map is to use axes of (log) stress and (log) d ($d =$ grain size) for a fixed temperature (cf. Fig. 7.17b). Explain why the lines representing the transition from the diffusional creep regions to the power-law creep region are straight lines in such a representation. Explain why the line separating the Coble and Nabarro-Herring regions is a vertical line.

350

CHAPTER 7
High-Temperature
Deformation of
Crystalline
Materials

7.13 Ashby used the following creep equations to calculate the deformation map for W in Fig. 7.16b. (Note: The form of these equations is slightly different than that in the text. Use the equations presented here in solving this problem.)

$$\dot{\varepsilon}_{NH} = 14\left(\frac{\sigma\Omega}{kT}\right)\left(\frac{D_L}{d^2}\right); \quad \dot{\varepsilon}_C = 44\left(\frac{\sigma\Omega}{kT}\right)\left(\frac{\delta D_C}{d^3}\right); \quad \dot{\varepsilon}_D = A\left(\frac{D_L Gb}{kT}\right)\left(\frac{\sigma}{G}\right)^{m'}$$

The subscripts NH, C, and D stand for Nabarro-Herring, Coble, and dislocation creep, respectively. D_L and D_C are the lattice and Coble creep diffusion coefficients, respectively; d is the grain diameter, G the shear modulus, b the Burgers vector, and the other terms have their usual meaning.

The following properties of tungsten were use in the creep equations.

$$\Omega = 1.59 \times 10^{-29} \text{ m}^3; \quad b = 2.74 \times 10^{-10} \text{ m}; \quad G = 15.5 \times 10^{10} \text{ N/m}^2;$$

$$D_{0L} = 5.6 \times 10^{-4} \text{ m}^2/\text{s}; \quad Q_L = 586,000 \text{ J/mol}; \quad D_{0C} = 10^{-3} \text{ m}^2/\text{s}$$

$$Q_C = 379,000 \text{ J/mol}; \quad A = 1.99 \times 10^{12}; \quad m' = 5.8; \quad \delta = 10^{-9} \text{ m}$$

a Redraw Fig. 7.16b, including in it an "Elasticity" region, based on the criterion that a strain rate of less than 10^{-10}/s is inconsequential. Note that in the diagram the line dividing Dislocation and NH creep is horizontal (i.e., is at a fixed stress; $\sigma/G = 6.5 \times 10^{-5}$) whereas the line dividing NH and Coble creep is vertical (i.e., at a fixed temperature; $T/T_m = 0.8$). The diagram you have drawn was constructed for a grain size of 10 μm. Now consider W with a grain size of 20 μm.
b What is the new value of σ/G separating the Dislocation and NH creep regions?
c What is the new value of T/T_m separating the NH and Coble creep regions?
d Note that for a grain size of 10 μm $\dot{\varepsilon} = 10^{-9}$ s for $T/T_m = 0.8$ and $\sigma/G = 10^{-8}$. For this same temperature and stress, what is the creep rate for W having a grain size of 20 μm?

7.14 Creep deformation, as well as fracture, is a design consideration for W light-bulb filaments. Here the "force" acting on the light bulb is its own weight. The maximum force arising from the gravitational effect is $F = ma$ where m is the filament mass and a the acceleration due to gravity. Suppose we are dealing with a 15-cm-long W filament having a diameter of 0.025 cm. We want the bulb to operate at 2500°C for at least 1000 hr with a total creep strain of less than 5%. Would you expect the W whose properties are shown in Fig. 7.16b to fulfill these creep requirements? (Hint: Work with standard SI units throughout (kg, m, s), and realize that $1 \text{ N} = 1 \text{ kg} \cdot \text{m/s}^2$.)

7.15 Refer to the deformation mechanism maps of Fig. 7.15. What can you say about the relative resistance to time-independent plastic flow of the face-centered cubic metal aluminum, the body-centered cubic metal tungsten, the diamond cubic covalent solid germanium, and the polar covalent solid MgO? Consider $T/T_m = 0.1$ and 0.4.

7.16 The diffusional flow regions in the deformation mechanism map of silver (Fig. 7.17a) are not labeled. Schematically redraw this figure and identify the controlling creep mechanisms in these regions.

7.17 The following table provides data needed to determine the strain rate of aluminum for Nabarro-Herring creep, Coble creep, and power-law creep.

Parameter	Value
Atomic volume (Ω)	1.2×10^{-29} m^3/atom
Lattice diffusion coefficient ($D_{0L} \exp[-Q_L/kT]$)	
D_{0L}	3.5×10^{-6} m^2/s
Q_L	2×10^{-19} J/atom
Grain size (d)	10^{-5} m
Coble creep diffusion coefficient ($D_{0C} \exp[-Q_C/kT]$)	
D_{0C}	10^{-5} m^2/s
Q_C	1.2×10^{-19} J/atom
A for power-law creep	2.5×10^9
m'' for power-law creep	4.8
Shear modulus (G)	2.5×10^{10} N/m^2

a Determine the dominant high-temperature deformation mechanism for all combinations of the stresses and temperatures listed below. (A short computer program will do the necessary calculations quickly and accurately.)

$$T \text{ (K) 600, 700, 800, 900, 1000}$$
$$\sigma \text{ (MN/m}^2) \text{ 1, 3, 10, 30, 100}$$

Based on your results, construct a deformation mechanism map for aluminum.
b You are asked to consider development of an aluminum-base material for service at 475°C. The requirements are that the material should creep only 1% in 1000 hr at this temperature. Estimate the stress required to cause pure aluminum of 10-μm grain size to creep at this rate. What is the dominant creep mechanism?
c Briefly describe an approach you would use for designing an aluminum alloy to meet the requirements stipulated in part (b).

7.18 The steady-state creep rate, $\dot{\varepsilon}_{II}$, is often considered to scale inversely with t_f. Under what conditions is this approximation most closely followed? As the relative fraction of a material's life spent in Stage III increases, how is the relationship between $\dot{\varepsilon}_{II}$ and t_f altered?

7.19 Consider the MarM-200 alloy whose deformation mechanism maps are provided in Figs. 7.18. At $T/T_m=0.5$, estimate the change in the Coble creep rate for this alloy if its grain size is increased from 100 μm to 1 cm. Are your results consistent with the creep-rate contours of Fig. 7.18b compared to those of Fig. 7.18a?

7.20 Consider the material MarM-200. Deformation mechanism maps for polycrystalline MarM-200 are provided in Figs. 7.18. Construct a deformation mechanism map for a single crystal of MarM-200. Pay particular attention to how the regions on the map are altered in area and displaced when the behavior of single crystals, rather than polycrystals, is represented.

7.21 a Considering Alloy S590, estimate the stress allowed for safe operation at a temperature of 500°C such that failure must not take place within the first 10^4 hr of operation.
b Estimate the failure time for Alloy S590 when $\sigma = 250$ MN/m^2 and $T = 870$°C.

7.22 a The steady-state creep rate relates to stress and temperature through Eq. (7.1). Apply this relationship to creep of a steel component in a power-plant boiler. The

352

CHAPTER 7
High-Temperature
Deformation of
Crystalline
Materials

component supports a tensile stress of 8000 psi at 1000°F and its creep strain must not exceed 10%. Using the data shown in the figure below, calculate the lifetime of the component. (Note the temperature in the diagram is given in degrees Rankine (R); (°R = °F + 460).

b The lines on the figure below look similar to those on a stress-time to rupture diagram. Discuss how they are related.

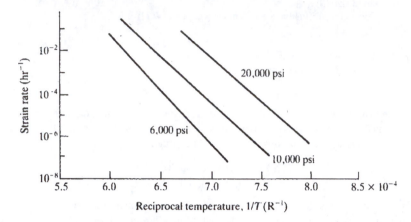

7.23 a Plot schematic stress-strain curves at several different strain rates for a material whose stress-strain rate behavior is described by $\sigma_T = K'(\dot{\varepsilon}_T)^m$.

b Show that m can be determined by conducting a tensile test at some initial low strain rate, and then periodically increasing the strain rate. What precautions must be taken when using such a procedure to determine m?

7.24 Derive Eq. (7.28) using $\sigma_T = K'(\dot{\varepsilon}_T)^m$, $\dot{\varepsilon}_T = -A^{-1}(dA/dt)$, and identifying σ_T as F/A (F is the force and A is the material cross-sectional area).

7.25 Plot the ratio of dA/dt for a "necked" region having a cross-sectional area A_n to dA/dt for the unnecked region (cross-sectional area A_u) as a function of A_n/A_u. Do this for several values of m between 0 and 1. Comment on your results.

7.26 Using geometry, show that the grain-switching mechanism (Fig. 7.27) produces a tensile strain of 0.55.

7.27 Show that to effect equivalent strain, the grain-switching mechanism requires only about ⅓ the amount of mass transport that Coble creep requires. (Hint: Refer to Fig. 7.28 for geometrical details, and establish a strain of 0.55 as your basis of comparison.)

7.28 Using the two-dimensional grain-switching geometry (Figs. 7.27 and 7.28), determine the increase in grain-boundary area per unit volume that accompanies the intermediate stage of the process. (Note: When multiplied by γ, your result is the two-dimensional analog of the term $0.72\gamma/d$—the threshold stress for grain-switching.)

7.29 a Refer to Example Prob. 7.2. Use the results from this problem to sketch a schematic deformation mechanism map which places a grain-switching creep region in the map.

b Again using the results of Example Prob. 7.2, calculate the value of the transition stress from Coble to grain-switching creep for grain sizes of 0.1 μm, 1 μm, 10 μm, and 32 μm. (Take $\gamma = 0.5$ J/m^2.)

c Repeat parts (a) and (b), but now consider grain switching due to volume diffusion being competitive with Nabarro-Herring creep.

d The deformation mechanism map for Al in Fig. 7.15 is appropriate for a grain size of 32 μm. Combine your results from parts (a) and (b) to place grain-switching regions in this map.

7.30 Explain why the transition from diffusional (NH and Coble) creep to superplasticity (Region II) takes place at higher stresses for Zn-Al alloys than it does for Pb-Sn alloys. (Deformation mechanism maps for these materials are provided in Figs. 7.31.)

CHAPTER 8

Deformation of Noncrystalline Materials

8.1
INTRODUCTION

The mechanical behavior of noncrystalline, or amorphous, materials is described in this chapter. The macroscopic mechanical behavior of amorphous and crystalline materials is similar. For example, increases in strain rate and decreases in temperature provide for a more "rigid" amorphous material, just as they do for a crystalline one. However, the micromechanics of flow differ substantially between the material classes. In crystalline materials, permanent shape changes can be related to clearly identified defects (e.g., vacancies, dislocations) within the crystal lattice. Although current discussions on flow in noncrystalline solids relate permanent deformation to analogous "defects" within the amorphous structure, these defects are nowhere near as readily identified as they are in crystals. Therefore, in this chapter flow in noncrystalline solids is discussed in terms of localized "slip" and/or viscous flow that occurs in response to an applied stress aided, in many situations, by thermal activation.

Regardless of whether the noncrystalline material is a metallic glass, a silica-based glass, or an organic glass (i.e., a polymer) the flow behavior at high temperatures is Newtonian viscous; i.e., the strain rate is related linearly to the applied stress. This reflects the fact that, at high temperatures, amorphous materials behave like viscous fluids, and localized "slip" processes are associated with permanent displacements of the basic molecular units of the material with respect to each other. For metallic glasses the basic units are simple, i.e., single atoms. For silicate glasses the units are the more complicated SiO_4 tetrahedra, whereas for polymeric materials the units are the long-chain molecules that constitute the polymer. High-temperature plastic flow of noncrystalline materials is distributed uniformly throughout the material volume. As temperature is decreased not only does the material become stiffer, but alterations in macroscopic and microscopic flow behavior take place. In silicate glasses, plastic flow is completely precluded at low temperature, and the glass behaves like a brittle, linearly elastic solid. Low-temperature flow can still occur in metallic and organic glasses, but it becomes macroscopically heterogeneous. For metallic glasses, the heterogeneous flow is associated with localized and concerted displacements of many atoms. In the case of long-chain organic polymers, the flow is associated with cooperative displacements of molecular chain segments rather than, as it is at higher temperature, with relative (and large) displacements of entire chains.

Many useful polymers, such as polyethylene and nylon, are composed of both amorphous *and* crystalline regions. Although deformation of semicrystalline polymers differs fundamentally from that of noncrystalline polymers at low strains, at higher strains the molecular rearrangements occurring during deformation are similar. Thus, we discuss in this chapter deformation of both semicrystalline and noncrystalline polymers.

Toughening and strengthening mechanisms in amorphous solids are also briefly discussed. These are particularly important for polymeric materials which are, especially as components of composites, increasingly being used as substitute structural materials for metals.

8.2
CRYSTALLINE VERSUS NONCRYSTALLINE STRUCTURES

A clear understanding of the different structural arrangements in crystalline and noncrystalline materials is necessary in order to appreciate differences in their flow behavior. Amorphous and crystalline arrays in a simple two-dimensional metal are illustrated in Fig. 8.1. The noncrystalline structure (Fig. 8.1a) displays no long-range atomic order, whereas the crystalline arrangement (Fig. 8.1b) does. The noncrystalline array is thermodynamically stable above the material's melting point (T_m) and, provided the material is cooled sufficiently slowly through T_m, a transition to a crystalline arrangement is expected for all but a few metals. The transition can be followed by measurement of a number of thermodynamic properties (e.g., enthalpy, entropy) or by measurement of certain physical properties. The specific volume (volume per mole) is one of these, and the variation of specific volume with temperature for a material that crystallizes at T_m is shown by curve (a) of Fig. 8.2.[1] Above T_m the amorphous arrangement can be thought of as a viscous liquid; below T_m crystalline plastic flow takes places by low- and high-temperature deformation mechanisms of the type described in Chaps. 3–5 and 7.

If a material is cooled sufficiently rapidly through and below T_m, the noncrystalline-crystalline transition can be suppressed. In this circumstance, the specific volume varies with temperature as shown by curve (b) of Fig. 8.2. At temperatures somewhat below T_m, the slope of the volume-temperature curve (the volumetric thermal expansion coefficient) is the same as that above T_m, and the material can be considered a supercooled liquid with a viscosity that increases with decreasing

---

[1]In this curve, the crystalline specific volume is, as it is for most materials, assumed less than the noncrystalline specific volume. However, certain substances, such as water and bismuth, expand on crystallization; i.e., the crystal specific volume (V_c) is greater than the amorphous one (V_a).

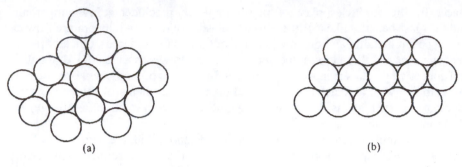

(a)   (b)

Figure 8.1
(a) Noncrystalline and (b) crystalline arrangements in a metal. The noncrystalline (amorphous) arrangement is thermodynamically stable above the melting point and the crystalline array below it. In almost all cases the more orderly crystalline arrangement is associated with a lower molar volume (i.e., a higher density).

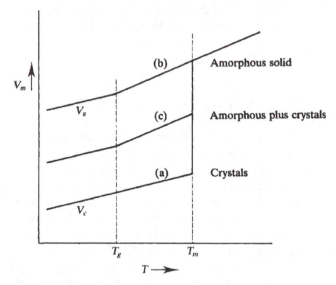

Figure 8.2
Molar volume-temperature curves on cooling from above T_m to below it. If the material crystallizes (curve (a)), a discontinuity in molar volume occurs at T_m. If it does not (curve (b)), the thermal expansion coefficient ($\sim dV_m/dT$) is unchanged at T_m and the liquid structure remains below T_m. At T_g—the glass transition temperature— there is a discontinuity in the thermal expansion coefficient. Below T_g, the material demonstrates mechanical characteristics of a solid rather than a supercooled liquid. Some materials partially crystallize at T_m and their molar volume-temperature behavior on cooling is illustrated by curve (c). In certain materials (e.g., long-chain polymers) crystallization is easily avoided on cooling; in others (e.g., metals) very rapid cooling is required to prevent crystallization.

temperature. However, at a certain temperature (the glass transition temperature, T_g) a decrease in the thermal expansion coefficient is noted, and the material assumes a "rigidity" comparable to that of the crystalline form. In the vicinity of, and below, T_g the mechanical characteristics of the glass are no longer the same as those of a supercooled liquid. Rather, the noncrystalline solid displays aspects of linear elasticity and viscoelasticity and, if the material is a metal or a long-chain polymer, aspects of permanent deformation as well. Finally, in some materials—particularly long-chain polymers—only partial crystallization takes place at T_m. The volume-temperature behavior for this situation is shown as curve (c) of Fig. 8.2, and the structure below T_m is a mixture of the crystalline and amorphous phases.

The effective viscosity of noncrystalline materials increases rapidly in the vicinity of T_g. This is a manifestation of the marked decrease in atomic or molecular mobility that takes place at this temperature. Whether a material assumes an amorphous structure on cooling from the liquid or crystallizes depends on several factors, including these mobilities and the material's structure. When the basic structural "building blocks" are simple, as they are for metals, crystallization usually occurs rapidly in the vicinity of T_m because the process involves only minor rearrangements of, for example, single atoms bonded together by the essentially nondirectional metallic bond. Thus, crystallization of pure metals is prevented only by rapid cooling from the liquid state through T_m and down to T_g. For example, it has been estimated that silver requires cooling rates on the order of 10^{12} K/s to prevent crystallization, and such cooling rates are not readily attained in laboratory environments let alone in commercial processes. For some metallic alloys, cooling rates "only" on the order of 10^2 to 10^6 K/s are required to produce as-solidified noncrystalline structures; these comprise the metallic glasses, the properties of which are discussed in Sec. 8.5.[2]

Silicate glasses and organic glasses can be rendered amorphous at cooling rates commonly encountered in industrial practice. In the case of silicates this is related to the strong, directional polar covalent bonding between silicon and oxygen within and between the basic tetrahedral building blocks. Two-dimensional schematics of the tetrahedral arrangement in a noncrystalline and a crystalline silicate are shown in Fig. 8.3. Transformation from the amorphous (Fig. 8.3a) to the crystalline (Fig. 8.3b) form requires breaking and reforming of these bonds, and this is not easily accomplished. Thus, even though the basic structural unit of a silicate glass is a small one—composed of but a few atoms—the strong, directional bonding between units effectively hinders crystallization.

Only the simplest of polymer chain molecules crystallizes easily. This is a direct result of the complex long-chain configuration of the basic chemical and structural units in polymers and, as shown in Fig. 8.4, crystallization necessitates molecular rearrangements over distances large in comparison to atomic dimensions. The bonding between the polymer chains is usually of the weak van der Waals type. Thus, in distinction to inorganic glasses, crystallization in polymers is prevented by the large chain size and not by strong interchain bonding. Even in what are termed "crystalline" polymers, the amorphous structure is never converted fully to the crystalline arrangement. Thus, the volume-temperature behavior for a typical

[2]Recently developed multi-component metal alloys can be solidified to a noncrystalline structure at cooling rates comparable to those employed in commercial practice. In addition, certain crystalline metallic alloys can be rendered amorphous by extensive mechanical deformation; e.g. by making "sandwiches" of the elemental components and rolling them to very large strains. A variant of this technique is found in mechanical alloying (MA) in which elemental mixtures may be converted to an amorphous structure by subjecting them to extensive mechanical deformation by high-energy milling. However, the MA process produces these amorphous substances in powder, rather than bulk, form and consolidating these powders without their crystallizing is a challenge.

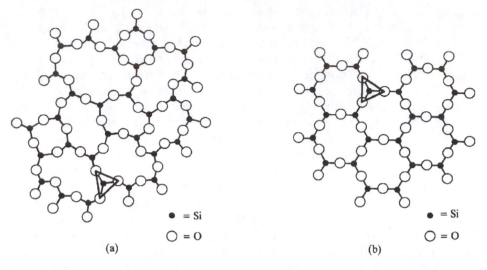

● = Si

○ = O

(a)

● = Si

○ = O

(b)

Figure 8.3
(a) Two-dimensional schematics of (a) noncrystalline and (b) crystalline arrangements in SiO$_2$, the prototype of a silicate glass. The intertriangular (tetrahedral in three dimensions) O–Si bond is a strong polar-covalent one. The transformation of the structure in (a) to that in (b) necessitates breaking and reforming of these bonds and, as this is difficult, crystallization of pure SiO$_2$ glass is difficult. (*From K. M. Ralls, T. H. Courtney, and J. Wulff, Introduction to Materials Science and Engineering, Wiley, New York, 1976.*)

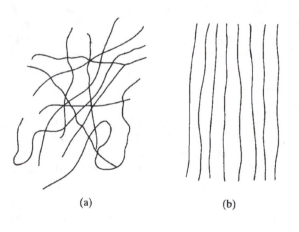

(a)

(b)

Figure 8.4
Schematics of (a) noncrystalline and (b) crystalline arrangements in a long-chain organic polymer. Although interchain bonding is weak, crystallization requires chain readjustments over distances large in comparison to interatomic dimensions. Thus, long-chain polymers, like silicate glasses, crystallize with difficulty, but for different reasons.

"crystalline" polymer on cooling from above T_m is represented by curve (c), rather than curve (a), of Fig. 8.2.

A measure of the ease of glass formation is the ratio T_g/T_m. High values of this ratio (> 2/3) favor glass formation, as T_g is associated with a reduction in atomic mobility; that is, crystallization occurs only in the temperature interval (T_m-T_g). When T_g/T_m is low (< 1/2), as it is for most metals and their alloys, the temperature range over which crystallization can occur is expanded and this, combined with the more rapid atomic movements in metals, makes the retention of the noncrystalline phase difficult.

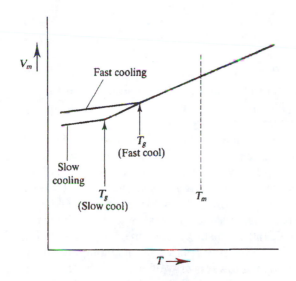

Figure 8.5
The glass transition temperature is a function of cooling rate; T_g increases with more rapid cooling. A material with a higher T_g also has a greater molar volume.

Heating an amorphous material at temperatures below T_m can lead to crystallization. The extent and rate of this crystallization is associated with the same factors influencing crystallization on cooling. For example, many metallic glasses crystallize at temperatures around T_g, in spite of the limited atomic mobility that characterizes this temperature. Inorganic glasses can be crystallized (devitrified) by heating at temperatures somewhat below T_m for periods on the order of hours, whereas most amorphous long-chain polymers will not crystallize at all.

The glass transition temperature is affected by cooling rate as well as by material structure. As shown in Fig. 8.5, T_g increases with increases in cooling rate and hence the specific volume of a noncrystalline solid below T_g depends on cooling rate. Extended heating of an amorphous material below T_g results in "relaxation" of the specific volume to a value characteristic of the "equilibrium" noncrystalline structure. The volume-temperature curves of Figs. 8.2 illustrate an additional concept relating to amorphous materials. The "free volume" of a noncrystalline material is defined as $(V_a - V_c)$, where V_a and V_c are the respective specific volumes of the amorphous and crystalline phases. Free volume is a useful concept in describing deformation of glasses, for it is a measure of the volume available for displacive flow of atoms or molecules. The greater the free volume, the greater the ease of such flow.

In this chapter we deal primarily with the mechanisms and phenomenology of flow in noncrystalline materials in the vicinity of T_g. At higher temperatures, permanent deformation is effected primarily by the simpler and more readily visualized mechanisms of viscous flow. We start our discussion of permanent deformation in noncrystalline solids with a description of such viscous behavior.

8.3
VISCOSITY

The plastic flow of noncrystalline solids above and slightly below T_m is viscous in nature and is described by the constitutive equation

$$\tau = \eta\dot{\gamma} \tag{8.1a}$$

or its tensile equivalent,

$$\sigma = 3\eta\dot{\varepsilon} \tag{8.1b}$$

The proportionality constant, η, relating shear stress and shear strain rate is the material's viscosity. If the material is *Newtonian* viscous, η is independent of the applied stress but depends on temperature.

The viscosity can be related to temperature and other factors via simple models. One model is based on kinetic principles and is similar to that used in Sect. 7.3A to describe dislocation glide creep at low temperatures. In fact, as we shall see, the equation that describes viscous flow by this approach is similar to Eq. (7.5) describing low-temperature dislocation glide creep. The viscosity "problem" is schematically illustrated in Fig. 8.6a, where viscous flow takes place by the stress-assisted motion of atoms (or molecules) that must surmount an energy barrier (Fig. 8.6b) to effect an atomic transit. The applied stress biases atomic motion in one direction. This is indicated in Fig. 8.6b by the lower energy of an atom after motion in the direction shown. The energy decrease is $F\delta$, where δ is the "jump distance" and F is the force acting on the atom. Substitution of τA for F, where A is an effective area on the "slip plane," shows that the energy decrease is $\tau A\delta = \tau V_{act}$, where V_{act} is the *activation volume*, having magnitude comparable to the atomic volume. The net rate (number per second) of atoms moving in the biased direction (J_{net}) is obtained as the difference in the rates of atoms jumping from left to right and right to left in Fig. 8.6b. Thus,

$$J_{net} = J_{L\rightarrow R} - J_{R\rightarrow L} = v\left[\exp\left(-\frac{\Delta U}{kT}\right) - \exp\left(-\frac{\Delta U + \tau V_{act}}{kT}\right)\right] \tag{8.2a}$$

$$= v\exp\left[-\frac{\Delta U}{kT}\right]\left[1 - \exp\left(-\frac{\tau V_{act}}{kT}\right)\right] \tag{8.2b}$$

where v is a fundamental atomic vibration frequency having units of s^{-1}. The shear strain rate is obtained by multiplication of Eq. (8.2) by δ and division of it by a characteristic vertical spacing (a) between atoms. Since δ and a are comparable, $\dot{\gamma}$ is given approximately by Eq. (8.2), and the absolute reaction rate viscosity, η_{rr}, is obtained from Eq. (8.1a) as

$$\eta_{rr} = \frac{\tau}{v}\left(\frac{\exp\left(\frac{\Delta U}{kT}\right)}{\left[1 - \exp\left(-\frac{\tau V_{act}}{kT}\right)\right]}\right) \tag{8.3}$$

For relatively low stresses and/or high temperatures (i.e., when $\tau V_{act}/kT \ll 1$), the exponential term in the denominator can be written as $[1 - (\tau V_{act}/kT)]$ and η_{rr} becomes

$$\eta_{rr} = \frac{kT}{v V_{act}}\exp\left(\frac{\Delta U}{kT}\right) \tag{8.4}$$

We see that the reaction-rate theory predicts a stress-independent viscosity at high temperatures for which the activation energy (ΔU) is, because of the atomic-jump nature of the flow process, related directly to the diffusion activation energy. At low temperatures and/or high stresses, the viscosity increases with stress; for example, when $\tau V_{act} \gg kT$, η increases linearly with stress (Prob. 8.3), but the activation energy is the same.

Correlation of experimentally measured viscosities with the reaction rate model is good at high temperatures and low stresses, where Newtonian flow is observed in all classes of amorphous materials. However, the agreement is best for fluid materials such as gases and liquids which have low viscosities. Moreover, as expected on discussions above, the viscosity activation energy under these conditions is the same as the diffusion activation energy. Viscosity values and activation energies correlate with material class. Liquid metals, for example, have low viscosities which are not very temperature sensitive because of the low activation energy associated with atomic motion in a liquid metal. Viscosities of liquid metals near their melting points are on the order of 0.005 N·s/m²;[3] these metals flow like water.[4] Nonmetals have higher viscosities because of their larger molecular sizes (for organic long-chain polymers) or stronger and more directional bonding (silicate glasses). For example, silicate-based glasses have viscosities on the order of 10 N·s/m² at temperatures at which they can be considered fluid and demonstrate a strong temperature dependence to their viscosity as well. The same applies to glucose, a highly viscous organic fluid, which has a viscosity close to 10^{13} N·s/m² at room temperature. However, the viscosity of glucose decreases by about eleven orders of magnitude when it is heated to the boiling point of water.

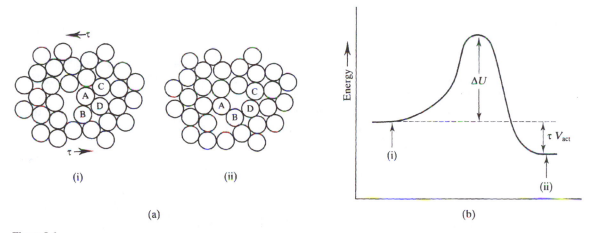

Figure 8.6
Viscous flow in an amorphous material. An applied shear stress (a(i)) induces flow of the upper portion of the material with respect to the lower. The flow is accomplished (a(ii)) by atomic or molecular jumps (e.g., atom A). As shown here, most such displacements occur in regions where the local atomic volume is highest. (b) Energy-reaction coordinate diagram for the process of (a). The atomic displacements require an activation energy (ΔU), but the displacements are stress-aided with an energy τV_{act} (V_{act} = activation volume). Kinetic principles (see text) determine the stress–strain rate relationship.

---

[3] A N·s/m² (or, equivalently Pa·s) is the appropriate viscosity SI unit. It is still common to express viscosity in cgs units, i.e., when stress is expressed in dynes/cm². In this case, viscosity is expressed in poises (P); 1 N·s/m² = 10 P.

[4] Liquid metals are so fluid they are unable to support a tensile stress; their viscosities are measured by determining their resistance to shear.

Newtonian flow, in which the strain rate varies linearly with stress, is predicted by the reaction-rate theory at high temperatures and/or low stresses. As noted, the reaction-rate theory best describes viscosities for fluid materials for which the viscosity and diffusion activation energies are the same. In "solid" viscous materials, the apparent viscosity activation energy is not that of diffusion. However, the flow often remains Newtonian. A model—the free volume model—is capable of accurately predicting viscosities of "solid" viscous materials down to temperatures in the vicinity of T_g.

In its simplest form, the free-volume model assumes Newtonian flow. However, the proportionality constant linking stress and strain rate depends on the probability of there being a relatively open volume in the proximity of an atom or molecule that wishes to move in response to an applied stress. The physics here does not differ much from that of the reaction-rate theory. However, in the reaction-rate theory thermal activation produces fluctuations in local volumes. The free-volume model is more structurally founded. Viscous flow in this model still devolves on there being a locally large open volume, and in this model the probability (P_v) of such a local volume being present is taken as proportional to $\exp(-V_f/V_c)$ where V_c is the crystalline atomic volume and V_f is taken as the difference between the amorphous and crystalline atomic volumes ($V_a - V_c$) (cf. Fig. 8.2). The shear strain rate is expressed as

$$\dot{\gamma} = P_v \tau \tag{8.5}$$

and the free-volume viscosity, η_{fv}, is then obtained as

$$\eta_{fv} = B \exp\left(\frac{V_c}{V_a - V_c}\right) \tag{8.6}$$

where B is an appropriate proportionality constant. The viscosity temperature variation comes from the temperature variation of V_c and V_a. That is, for $T > T_g$, $V_a - (V_c - V_a)_{T_g} [1 + \Delta\alpha(T - T_g)]$, where $\Delta\alpha$ is the difference in volumetric thermal expansion coefficients between the noncrystalline and crystalline phases. Using these relationships, the viscosity is obtained as (Prob. 8.4)

$$\eta_{fv} = B \exp\left(\frac{1}{(f_g)[1 + \Delta\alpha(T - T_g)]}\right) \tag{8.7}$$

where f_g is the fractional free volume at T_g.

The free-volume model well describes the magnitudes and temperature dependence of the viscosities for a number of organic and inorganic glasses over a broad temperature range (Figs. 8.7). The agreement noted in Figs. 8.7 is good over about six orders of magnitude variation in viscosity and breaks down only at low temperatures where the viscosity is so great that the material mechanically resembles a true solid more than a supercooled viscous glass.

As also shown in Figs. 8.7, the viscosity activation energy (the slope of the $\ln \eta$ vs. $(1/T)$ curve) is greater at low than at high temperatures. For silicate glasses, the apparent low-temperature activation energy is 2–3 times larger than the high-temperature one; for organic glasses, the ratio of the low- to high-temperature activation energy may be an order of magnitude. These observations lend further credence to the belief that different mechanisms of flow dominate as T_g is approached; several of these have been identified for polymeric glasses (Sect. 8.6).

The stress dependence of the viscosity is also of interest; even the reaction-rate theory for viscosity predicts that viscosity becomes stress dependent at high stress levels although at sufficiently low stresses Newtonian flow is observed in all classes of noncrystalline materials. The critical stress above which viscosity increases with stress is material class dependent. In polymers, this critical stress is so low (10^7 N/m^2) that non-Newtonian flow is the rule rather than the exception. In contrast, for oxide glasses, the critical stress is so high (typically 10^{10} N/m^2) that most viscous flow in them is Newtonian.

The viscosity of a material depends both directly and, as it is affected by free volume, indirectly on molecular structure. The following section provides examples of how the viscosities of silicate glasses can be altered markedly by changes in molecular architecture resulting from changes in chemical composition.

8.4
THE DEFORMATION BEHAVIOR OF INORGANIC GLASSES

The deformation response of glasses such as SiO$_2$ and ordinary window glass varies significantly with temperature. At high temperatures, Newtonian viscous flow is dominant. Conversely, at low temperatures, glass deforms almost exclusively in a linearly elastic manner. At intermediate temperatures, some viscoelastic component of strain is also observed. The various components of the strain can be written in terms of the response of a glass to an applied tensile stress. The total strain, ε, is

$$\varepsilon_c = \varepsilon_{el} + \varepsilon_v + \varepsilon_{ve} = a + bt + \varepsilon_{ve} \qquad (8.8)$$

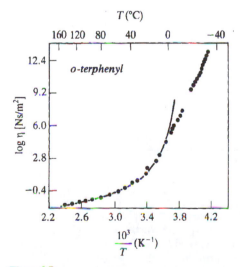
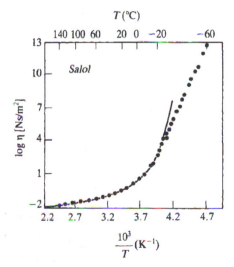

Figure 8.7
Viscosity-temperature relationships for two organics: (a) o-terphenyl; (b) salol. The solid lines are the free-volume-model predicted relationship (Eq. (8.7)). This model fits the data well at high temperatures and over a broad range of viscosities. (*From M. Cukierman, J. W. Lane, and D. R. Uhlmann, J. Chem. Phys., 59, 3639, 1973.*)

In Eq. (8.8), $\varepsilon_{el} = a$ (= σ/E for a tension test) is the elastic strain, ε_v (= bt where b is the creep rate and t the time over which the stress is applied) is the creep strain, and ε_{ve} is the viscoelastic strain. The time variation of these strains is illustrated in Fig. 8.8.[5] The linear elastic strain results from the stretching and distortion of chemical bonds within the glass network. The viscous flow term, ε_v, is due to permanent relative displacements of the SiO_2 tetrahedra, and the viscoelastic strain represents conformal distortions within the silicate network. The latter are time-dependent and fully recovered (albeit with a time lag) on removal of the applied stress (Fig. 8.8). As described below, additional viscoelastic response is found when the chemical composition of the glass deviates from the stoichiometric glass composition (SiO_2 for a silica-based glass).

The relative contributions of the various strains to the total strain are *strongly* temperature dependent. As mentioned, viscous flow dominates at high temperatures so that $a/bt \cong 0$. Conversely, $bt/a \cong 0$ at low temperatures where the strain is primarily linear elastic. The viscoelastic strain component is small at ordinary (e.g., room) temperatures for most glasses where $\varepsilon_{ve}/a \cong 0.001$–$0.03$. At the glass transition temperature, however, ε_{ve} and a are comparable. In contrast to noncrystalline polymers and metals, most silicate glasses do not exhibit permanent deformation at low temperatures. Instead, they fracture at stresses well below those required to induce permanent deformation.

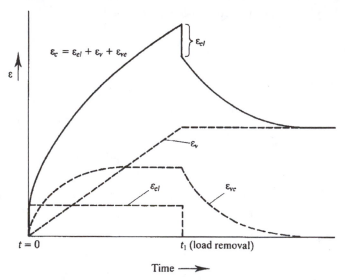

Figure 8.8
The creep strain of a silicate glass is composed of elastic, viscoelastic, and viscous strain components. Upon removal of the stress (at t_1), the elastic strain is instantaneously, and the viscoelastic component gradually, recovered. The relative contributions of the various strains are strongly temperature dependent. Viscous flow dominates at high temperature and elastic behavior at low, whereas viscoelastic and elastic deformation are comparable in the vicinity of T_g.

---

[5]Equation (8.8) can be derived on the basis of models described in Chap. 2 (Prob. 8.6).

The viscosities of inorganic glasses, as well as their tendencies for viscoelasticity at intermediate temperatures, relate to glass molecular structure. As shown in Fig. 8.9, viscosity is strongly temperature dependent and the viscosity can also be profoundly altered by changes in glass composition. Typical glass modifiers that produce such changes are the oxides of the alkali metals (e.g., Na_2O); these alter the glass structure (compare Fig. 8.10 with Fig. 8.3). The positive alkali ions do not enter into the silicate network, but instead occupy the rather open spaces within it. The alkali ions may be considered ionically bonded, with a bond strength less than that of the polar covalent SiO_2 bond, to the now negatively charged[6] silica network. The network disruption is responsible for the decreased viscosity of modified glasses as well as the reduced temperature at which they flow viscously. That is, as viscous flow of glass is caused by the displacement of silicate tetrahedra with respect to each other, and since in pure SiO_2 this necessitates breaking and reforming of bonds connecting tetrahedra, the modifiers catalyze this flow by chemical introduction of a certain fraction of "broken" tips.

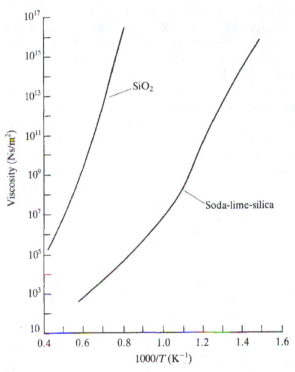

Figure 8.9
Viscosity-temperature relationships for a pure SiO_2 and a soda-lime-silica glass. (The latter contains soda (Na_2O) and lime (CaO).) Viscosity is strongly temperature and composition-dependent. (*From Introduction to Ceramics, W. D. Kingery, H. K. Bowen, and D. R. Uhlmann, 2nd ed., Wiley, New York, 1976.*)

[6]Since modifier ions are positive, charge neutrality dictates that the silicate network have a negative charge. This is accomplished by disrupting the Si–O bonds so that all tips of the SiO_4 tetrahedra are no longer shared. This increases the O/Si atomic ratio to a value greater than two, and gives rise to a negative network charge.

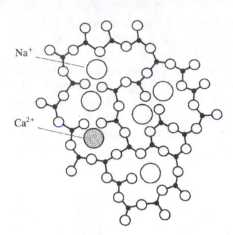

Figure 8.10

Soda (Na₂O) and lime (CaO) alter glass structure. The Na⁺ and Ca²⁺ ions do not enter the Si-O network; rather, they occupy the "holes" within it. Charge balance requires that this be accompanied by a "breaking up" of the network; i.e., not all of the Si are joined to other Si by a bonding O ion. This is a cause for the reduced viscosity of the soda-lime-silica glass. (*From K. M. Ralls, T. H. Courtney, and J. Wulff, Introduction to Materials Science and Engineering, Wiley, New York, 1976.*)

Modifying ions also affect the viscoelastic response of silicate glasses, usually increasing the viscoelastic strain. In many cases this is caused by the stress-assisted displacement, over interatomic distances, of the alkali ions. An applied stress alters the relative volumes of the open spaces within the network, and ions tend to jump into local volumes that are increased by the stress. Thus, this viscoelasticity closely parallels that discussed in Chap. 2, in which carbon atoms in body-centered cubic iron occupy preferential interstitial positions in response to an applied stress. In the case of glasses, the modifiers are analogous to the carbon atoms. Of course, removal of the stress also removes the stress-formed preferential sites, and the viscoelastic strain is recovered as the material "relaxes" by reversion of the ions to the sites they occupied prior to load application.

From a technological standpoint, the effect of molecular architecture on glass properties is better represented by Fig. 8.11. Here the viscosities of a number of commercially important glasses are plotted as a function of temperature.[7] Glass technology has, as all technologies do, its own nomenclature. In terms of Fig. 8.11, the *"melting point"* does not represent an equilibrium crystallization temperature but rather that temperature above which the glass is fluid enough to be considered a liquid. A glass must be cooled below the "melting point" to develop a rigidity sufficient to permit it to be formed into useful shapes, while at the same time remaining fluid enough so that this can be done easily and without glass fracture. Thus, working of glass takes place over the temperature range designated as the *"working range,"* which is the range encompassed by the *"working"* and the *"softening"* points. The *"annealing point"* is that temperature used to temper glass, i.e., to relieve internal residual stresses that may have been developed in it during thermal processing. And the *"strain point"* is that temperature at which the glass behaves mechanically more like an elastic solid than a supercooled liquid.

In addition to providing us with some technical jargon, Fig. 8.11 reinforces previous comments regarding the effect of molecular architecture on glass mechanical behavior. For example, we see that ordinary window glass (the soda-lime glass) can

---

[7]Compositions (in wt.%) of the glasses of Fig. 8.11 are: fused silica (> 99.5% SiO_2); 96% silica glass (known as Vycor—96% SiO_2, 4% B_2O_3); borosilicate glass (known as Pyrex—81% SiO_2, 3.5% Na_2O, 2.5% Al_2O_3, 13% B_2O_3); soda-lime glass (74% SiO_2, 16% Na_2O, 5% CaO, 4% MgO, 1% Al_2O_3).

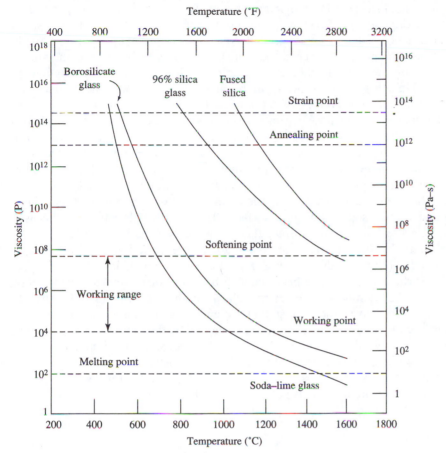

Figure 8.11
Viscosity (logarithmic scale) vs. temperature for fused silica and several other silicate glasses. Viscosity is reduced proportionally to the amount of modifier ions in the network. Fused silica (essentially pure SiO$_2$) does not soften until very high temperatures. The heavily modified soda-lime glass softens at a moderate temperature. The other glasses (compositions provided in the text) have modifying agent amounts intermediate to those in soda-lime and fused silica glass. The engineering significance of the various temperatures (melting point, working point, etc.) is described in the text. (*From E. B. Shand, Engineering Glass, Modern Materials, Vol. 6, Academic Press, New York, 1968, 262.*)

be worked at temperatures between about 700 and 1000°C. Conversely, fused silica (essentially pure SiO$_2$) must be heated to temperatures on the order of 1800°C before its viscosity is reduced to the point at which it can be molded. We add that the behavior of glasses at temperatures where they are elastic solids does not depend much on their molecular architecture. For example, room-temperature elastic moduli and fracture toughnesses of the various glasses of Fig. 8.11 are, for practical purposes, the same.

From a technological standpoint, glasses can be considered as either viscous or linearly elastic materials. Their viscous nature at high temperatures is used to form glass into useful shapes. The temperature range over which the viscosity increases rapidly is not great (Fig. 8.11) and, thus, at ordinary temperatures neither viscous nor viscoelastic behavior of glass need be considered important; glass is essentially a linear elastic material at lower temperatures. Although the bond strength within a glass is strong and, thus, its inherent strength is high (e.g., the high-strength glass fibers used in composites) ordinary monolithic glass is a brittle material that fractures at relatively low values of an applied tensile stress. Measured fracture strengths are well below those expected on considerations of bond strength, and this is a result of the presence of small cracks on the glass surface. These cracks form, or at least are made larger, by exposure of the glass to such seemingly innocuous environments as humid air, and preventing their occurrence is out of the question in most cases. Although fracture of glass is discussed at more length in Chap. 9, it is worth noting here that glass can be rendered more fracture-resistant by producing a compressive stress state on the glass surface. That is, since glass fracture is caused by a tensile stress acting on a surface flaw, it can be delayed if the surface is compressed initially. Such a stress state is imparted to tempered glass (Prob. 8.9) by appropriate cooling schedules from elevated temperatures. More recently, compressive surface stresses have been induced by ion implantation. Implantation is accomplished by bombarding the glass surface with high-energy ions. These enter into a thin surface layer, and the resulting decrease in local atomic volume results in an internal compressive stress similar to that caused by the presence of some impurity atoms in crystalline materials. Although expensive, ion implantation has proven useful in the development of fairly fracture resistant glasses.

As mentioned, silicate glasses do not exhibit any tendency for low-temperature permanent deformation. This is not so for metallic and organic glasses, and their behavior is discussed in the sections following.

8.5
DEFORMATION OF METALLIC GLASSES

As a result of the rapid crystallization kinetics of metals and their alloys, it had long been considered unfeasible to produce noncrystalline structures in metals by rapidly cooling them from the liquid. Some 40 years ago, however, small thin sections of certain metals were produced in the amorphous state by subjecting them to extremely rapid solidification rates. While initially considered laboratory curiosities, the intriguing electronic and mechanical properties of such materials have since rendered them of some commercial importance. Moreover, more recently developed alloys containing nonmetallic (e.g., phosphorus or boron) alloying elements are produced in noncrystalline form employing solidification rates orders of magnitudes less than those necessary to yield amorphous structures in "classical" metals. These alloys can be produced routinely (typically in ribbon form) at costs that, while high, are not exorbitant when measured against their resultant properties. And, as noted in Sect. 8.2, recently formulated alloy compositions permit retention of the amorphous structure using solidification rates common to many extant solidification processes. Finally, and as also noted in Sect. 8.2, extensive mechanical deformation can develop amorphous structures in certain metal combinations.

Glass composition	T.S. (MPa)	E (GPa)	v	K/E	T.S./E
$Pd_{80}Si_{20}$	1330	67	—	—	0.020
$Pd_{77}Cu_6Si_{17}$	1530	96	0.41	1.85	0.016
$Pd_{64}Ni_{16}P_{20}$	1560	93	0.41	1.85	0.017
$Pd_{16}Ni_{64}P_{20}$	1760	106	0.40	1.67	0.017
$Pt_{64}Ni_{16}P_{20}$	1860	96	0.42	2.08	0.019
$Fe_{80}B_{20}$	3530	166	—	—	0.021
$Fe_{80}P_{13}C_7$	3040	122	—	—	0.025
$Fe_{80}P_{16}C_3B_1$	2440	135	—	—	0.018
$Fe_{78}B_{10}Si_{12}$	3330	118	—	—	0.028
$Ni_{78}Si_{10}B_{12}$	2450	64	—	—	0.038
$Ni_{49}Fe_{29}P_{14}B_6Si_2$	2380	129	—	—	0.018
$Ni_{36}Fe_{32}Cr_{14}P_{12}B_6$	2720	141	—	—	0.019
$Co_{78}Si_{10}B_{12}$	3000	88	—	—	0.034

Source: Data from articles by H. S. Chen, K. A. Jackson, and L. A. Davis, *Metallic Glasses*, American Society for Metals, Metals Park, Ohio, 1978; H. Kimura and T. Masumoto, *Amorphous Metallic Alloys*, ed. F. E. Luborsky, Butterworths, London, 1983.

Table 8.1
Properties of selected glasses

Metallic glasses are characterized by some excellent mechanical properties and by interesting deformation characteristics. Some of their properties are listed in Table 8.1. Tensile moduli of noncrystalline metals are typically 30 percent lower than moduli exhibited in crystal form, and this reflects the slightly greater atomic volumes of amorphous metals. So, too, do their higher Poisson ratios, which indicate that noncrystalline metals are slightly less compressible than in their crystalline forms. The bulk moduli of amorphous metals are comparable to those in crystalline form; i.e., the somewhat greater glass Poisson's ratio compensates for the lower elastic modulus (recall—Chap. 2—that the bulk modulus, K, is related to E and v by $K = E/3(1 - 2v)$). Of especial interest are the high values of tensile strength that characterize many metallic glasses. These materials are inherently strong; a ratio of tensile strength to modulus of 0.02 is close to the theoretical strength of a material and corresponds, for example, to a steel having a tensile strength of 4.1 GPa (= 600,000 psi).

The deformation response of metallic glasses as affected by temperature, stress, and strain rate is illustrated in the schematic deformation mechanism maps of Fig. 8.12. Figure 8.12a includes regions in which elastic deformation is predominant, whereas these regions are not shown in Fig. 8.12b. (The two different representations are similar to those used in Chap. 7—e.g., Fig. 7.15 vs. Fig. 7.16.) We consider first Fig. 8.12a. We see that at low stresses, viscous flow is first supplanted by viscoelastic deformation and then by linear elastic deformation as temperature is decreased. Viscous flow, in which the strain rate scales linearly with the stress, is the dominant deformation mechanism at temperatures down to somewhat below T_g. Linear elastic deformation is the dominant deformation mode from 0 K up to a critical temperature above which viscoelastic deformation becomes more important. Although the elastic moduli of metallic glasses do not differ all that much from oxide glasses, their propensity for viscoelastic deformation does. The viscoelasticity of metallic glasses is caused by recoverable atomic adjustments occurring in response to an applied stress. These are similar in many respects to the viscoelastic

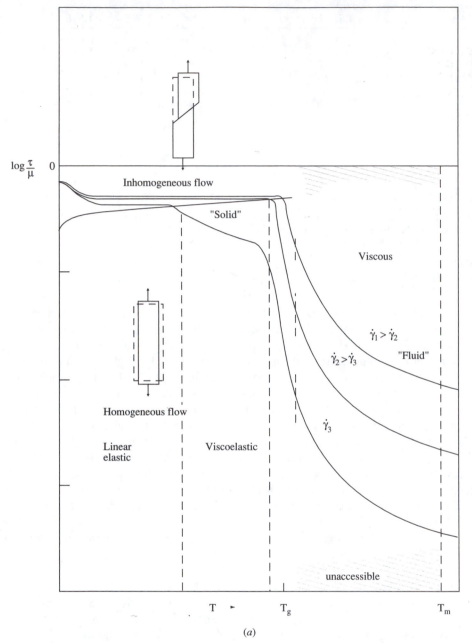

(a)

Figure 8.12

Schematic deformation mechanism maps for metallic glasses. (a) Includes linear elastic and viscoelastic regions; these are not shown in (b). At high temperatures (from the melting temperature downward to about T_g), the glass deforms in a Newtonian viscous manner. Further, permanent deformation is distributed uniformly throughout the material volume. At lower temperatures (less than about T_g) and low stresses,

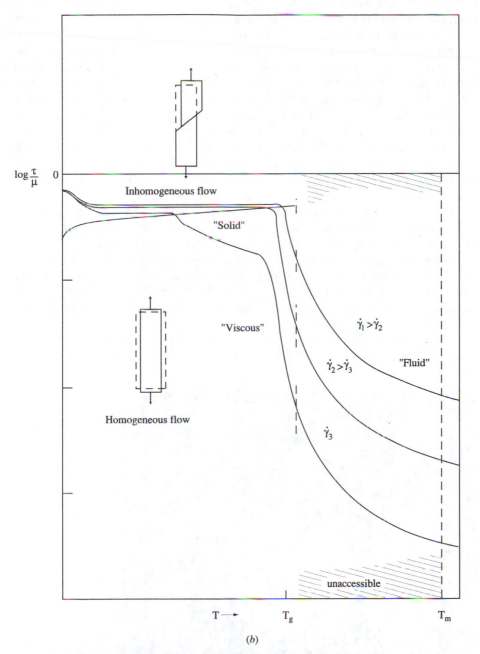

$\log \dfrac{\tau}{\mu}$ 0

Inhomogeneous flow

"Solid"

"Viscous"

$\dot{\gamma}_1 > \dot{\gamma}_2$

$\dot{\gamma}_2 > \dot{\gamma}_3$

"Fluid"

$\dot{\gamma}_3$

Homogeneous flow

unaccessible

T → T_g T_m

(b)

flow (whether elastic or permanent) is also homogeneous. However, at higher stresses
at low temperatures, permanent deformation is heterogeneous and takes place by the
formation of shear bands. Further, the strain-rate sensitivity of the flow stress is reduced at low temperatures. (*Modified from A. Spaepen, "A Microscopic Mechanism
for Steady State Inhomogeneous Flow in Metallic Glasses," Acta Metall., 25, 407,
Copyright 1977, with permission from Elsevier Science.*)

readjustments taking place at grain boundaries in crystalline materials. However, since the readjustments occur throughout the volume of the metallic glass, viscoelastic strains of metallic glasses are much greater than those arising from the very small fraction of atoms available to effect grain-boundary viscoelasticity in polycrystals. Likewise, the metallic glass viscoelastic strain is much greater than the corresponding strain in oxide glasses. At the highest temperatures in the viscoelastic region of Fig. 8.12a, for example, the ratio of viscoelastic- to linear-elastic strain can be as high as 200. This is additional evidence of the relative ease by which atoms rearrange themselves in metallic glasses in comparison to silicate tetrahedra rearrangement in oxide glasses, and the cause is the simpler fundamental atomic unit and the essentially nondirectional bonding in metallic glasses. This is also why viscosities of metallic glasses are much lower than those of oxide glasses.

Figure 8.12b does not include the elastic regions of Fig. 8.12a, but concentrates instead on plastic deformation modes. (Although at low temperatures and stresses, the strain rates arising from viscous flow might be inconsequential.) As noted in both Figs. 8.12a and b, the deformation at low stress levels, irrespective of the temperature, is "homogeneous," meaning that deformation takes place uniformly throughout the material volume. What truly differentiates Fig. 8.12 from prior discussions is the high stress–low temperature (i.e., $T < T_g$) region labeled "inhomogeneous flow."

What goes on there? In the inhomogeneous flow region, permanent deformation is confined to small shear bands (Fig. 8.13). These bands have a thickness on

Figure 8.13
The surface of a cold-rolled Pd-Cu-Si metallic glass. The surface markings are shear bands. *(H. S. Chen and K. A. Jackson, Metalic Glasses, ASM International, Materials Park, OH, 44073-0002, (formerly American Society for Metals, Metals Park, OH 44073), 1978 p. 86.)*

the order of 10–20 nm, and within such a band the shear strain can be quite large (on the order of unity). In a tension test, shear bands form at an angle to the tensile axis approximately equal to that corresponding to the maximum shear stress. Tensile flow behavior of a metallic glass is also characterized by a yield point. That is, the stress required to propagate a band is less than that needed to initiate it. Tensile ductilities of metallic glasses are very limited, however, because fracture takes place following propagation of the shear band across the sample thickness. Tensile strength is, therefore, coincident with the development of the first (and only) shear band. Multiple bands can be formed if the stress state is not conducive to geometrical instability as it is in a tensile test. Figure 8.13, in fact, illustrates multiple shear bands formed during rolling of a metallic glass. During such rolling, reduction in thicknesses of about 50% can be achieved before material fracture. However, this strain remains much less than that in the individual shear bands since the volume of material subject to plastic deformation remains but a small fraction of the total volume.

The essential distinction between homogeneous and inhomogeneous flow in metallic glasses is related to the amount and distribution of free volume within the glass. At higher temperatures, the greater free volume is fairly uniformly distributed so that many widely separated atoms are able to undergo atomic displacements, and this leads to homogeneous flow. At lower temperature, the free volume is less. A local dilatancy (expansion) accompanies the first atomic displacements at the high stresses associated with shear band formation. This induces a greater local free volume that catalyzes further localized flow, resulting in propagation of the shear band. That shear bands exhibit an increased chemical activity (e.g., a greater tendency to corrode) is considered supportive evidence of the greater atomic volume within the band. Further, the difference in volume between the material within a shear band and the undeformed adjacent material leads to a stress concentration at the interface between these regions. This also serves to propagate the band at a stress level less than that required to form a second band.

The tensile fracture behavior of metallic glasses is unique. Shear fracture takes place by separation along the active shear band. The fracture surface displays a "river" pattern, which consists of fine ridges; the pattern can be mimicked by shearing and pulling apart two glass slides separated by a thin layer of Vaseline. Because of this appearance, it has been suggested that the material about to fail in a shear band behaves in a liquid-like manner. The idea has appeal for the increased atomic volume in a shear band results in a lesser viscosity, one characteristic of a temperature greater than that of the test temperature.

Metallic glasses can be partially crystallized. This is accomplished by heating them to temperatures between T_g and T_c (T_c is the temperature at which rapid crystallization takes place on heating). Alternatively, intermediate cooling rates during crystallization can produce a mixed crystalline-amorphous structure (cf. Fig. 8.2). Crystallization is often associated with further "embrittlement" of the metallic glass, but this is not always so. Certain Al-base glasses exhibit both greater strengths and ductilities when they contain a fine dispersion (3–10 nm in diameter) of crystalline Al particles. The tensile strengths of these mixtures can be as high as 1500 MPa, a value much greater than that achievable in the highest-strength aluminum precipitation-hardened alloys. While the cause for the greater ductilities of these partially crystallized glasses has not been established, it is not unreasonable to believe that the dispersed crystalline particles—because they have a size comparable to the thickness of a shear band—interfere with shear band propagation. Such a mechanism of toughening would be similar to that found in certain glassy polymers, as is discussed in the following section. While localized deformation in glassy metals does have some parallels to the deformation of long-chain polymers, the more complicated geometry of long-chain polymers, as well as the much different chemical bonding in them, leads to a more complicated deformation response at the molecular level. These and other aspects of polymer deformation are described in the following section.

8.6
DEFORMATION OF POLYMERIC MATERIALS

In the decades following the end of the Second World War, polymeric materials have become widespread substitutes for metals and other structural materials. Polyethylene competes with metal and glass as a beverage container material and advanced fibrous-reinforced plastics are seeing increased use in aerospace where aluminum alloys once held sway. As a result of the relatively low melting temperatures of polymers, they cannot compete with metals and ceramics as high-temperature materials. Nonetheless, there is no reason to believe that polymers will not increasingly substitute for other materials at room- and slightly higher-temperatures. This is a natural consequence of the continuing development of high-strength organics and the economic advantages they have vis-à-vis metals and ceramics with respect to raw material and production costs.

The mechanical behavior of polymers is discussed in this section. Since the molecular architecture of polymers differs so dramatically from that of metals and ceramics, and is so important with respect to polymeric mechanical behavior, our discussion begins with a brief review of polymer chemistry and structure.

A. Polymer Chemistry and Structure

Polymers are large molecules consisting of very many repeating units of a basic molecule (mer). Most polymeric materials are organic; that is, they consist of carbon—bonded to hydrogen, oxygen, nitrogen, etc.—as the primary chemical constituent, although silicones, which are based on the oxygen-silicon bond, display properties similar to those of organic polymers.

As noted in Chap. 2, polymers can be classified on the basis of their mechanical characteristics which, in turn, are determined by their molecular architecture. Thus, elastomers are highly kinked long-chain polymers periodically cross-linked in such a way that permanent displacement of the chains (i.e., plastic deformation) does not take place in them. Thermosets are polymers in which the primary bonding lies not along a chain, as it does in a chain polymer, but assumes a skeletal or three-dimensional character. While thermosets are capable of elastic and viscoelastic deformation, their structure does not allow for permanent deformation at ordinary temperatures; thermosets decompose at a temperature below which permanent deformation could be accomplished in them. Thus, all that is necessary to say about the mechanical characteristics of elastomers and thermosets has been said in Chap. 2. You might think that would shorten our task. It does, but not by much. The mechanical characteristics (especially their permanent flow behavior) of the third polymer class—thermoplastics—are so intriguing that it takes more than a few paragraphs to describe them. To remind, thermoplastics consist of long molecular chains with covalent bonding along the chains. The chains are held together by the much weaker van der Waals bond or, in certain thermoplastics, hydrogen bonds.

There are two basic types of long-chain polymers, depending upon how they are made. Carbon-chain polymers are composed of the same basic "mer." Polyethylene (PE) is the simplest of these polymers and also the one most widely used. As the name implies, PE is synthesized from ethylene (C_2H_4), with synthesis being catalyzed by organic peroxides. The overall synthesis can be simply represented by

$$n \begin{bmatrix} & H & H \\ & | & | \\ -& C - C & - \\ & | & | \\ & H & H \end{bmatrix} \longrightarrow \begin{bmatrix} & H & H \\ & | & | \\ -& C - C & - \\ & | & | \\ & H & H \end{bmatrix}_n \tag{8.9}$$

The number of ethylene molecules (n) in Eq. (8.9) represents the degree of polymerization (D.P.) and is on the order of thousands for a typical long-chain polymer. The synthesis reaction is stochastic in nature, though, so that the D.P. varies substantially from one long-chain polymer to another in a typical polymer containing a large number of such chains (Prob. 8.12). Therefore, the D.P. generally refers to the average degree of polymerization in the material.

Polymer chemists are adept at controlling the D.P. of chain polymers. This permits them to produce polymers having substantially different characteristics even when the material is composed of the same chemical mer. Polyethylene serves as an example. Ethylene is a liquid at room temperature. As the D.P. of PE increases, the bonding energy per mer does also and this is manifested by changes in material properties and an increase in the melting temperature. For example, when its D.P. is in the range of 500–2500, PE is a greasy or soft-wax like material. Increasing the D.P. to values of 10,000 and more renders polyethylene an "engineering solid" at room temperature and this form of PE does not melt until ca. 425 K.

Polyethylene is the simplest of the vinyl series of polymers, which are formed from mers having chemical composition C_2H_3R. In the vinyl series, R can range from being a simple atom (e.g., H in PE or Cl in polyvinyl chloride) to a fairly complex molecular side-group (such as a benzene ring in polystyrene). Carbon-chain polymers can also be produced from mers having the base composition $C_2H_2R_1R_2$ (e.g., poly(vinylidiene chloride), $R_1=R_2=Cl$; polyisobutylene, $R_1=R_2=CH_3$; polymethylmethacrylate (PMMA), $R_1=CH_3$, $R_2=C_2O_2H_3$) and many other compositions as well. A partial list of some of the vinyl and other series of carbon-chain polymers is given in Table 8.2.

Long-chain polymers can also be made from chemically distinct constituents in which case they are called *heterochain* polymers. Nylon is perhaps the best known of these polymer types. The mer of Nylon 66 is constructed from the chemical species hexamethylene diamine [$NH_2(CH_2)_6NH_2$] and adipic acid [$COOH(CH_2)_4COOH$] that react according to

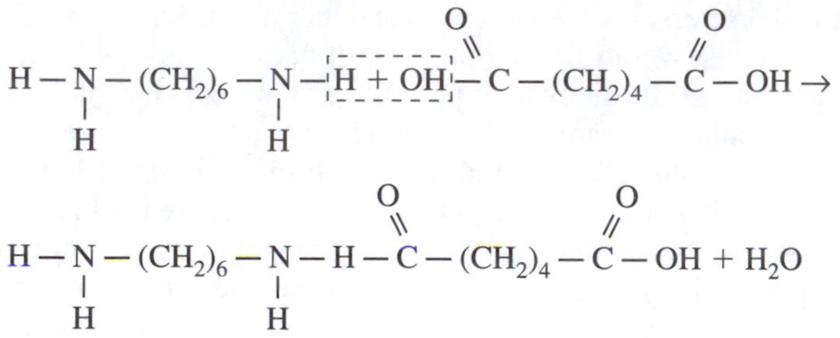

$$H-N-(CH_2)_6-N-\overset{}{H}+\overline{OH}-\overset{O}{\overset{\|}{C}}-(CH_2)_4-\overset{O}{\overset{\|}{C}}-OH \rightarrow$$
$$\overset{|}{H}\qquad\qquad\overset{|}{H}$$

$$(8.10)$$

$$H-N-(CH_2)_6-N-H-\overset{O}{\overset{\|}{C}}-(CH_2)_4-\overset{O}{\overset{\|}{C}}-OH+H_2O$$
$$\overset{|}{H}\qquad\qquad\overset{|}{H}$$

The long chain of Nylon 66 is obtained by repetition of the above reaction in which adipic acid and hexamethylene diamine alternately attach themselves to the growing chain. Several of the more common heterochain polymers, their structures, and typical properties are given in Table 8.3.

The response of long-chain polymers to an applied stress also depends on their degree of crystallinity; crystalline polymers[8] are generally stiffer and more resistant to flow than their noncrystalline analogs. Additionally, the mechanisms by which permanent deformation is accomplished are different in crystalline and noncrystalline polymers. The degree to which a polymer is prone to crystallization is related to the nature and arrangement of side groups along the chain. If these are simple, as in PE, crystallization occurs readily;[9] nylon also crystallizes fairly readily for the same reason. Polymers with bulky side-groups (e.g., polystyrene (PS) and PMMA, Table 8.2) are more difficult to crystallize. Likewise, when the side-groups are arranged in a regular manner (e.g., along the same side of the chain or alternating from side-to-side along it), crystallization is facilitated. Polymer "alloys" (copolymers[10]) are difficult to crystallize for the opposite reasons. And, as noted previously,

Table 8.2
A partial list of carbon-chain polymers and their uses

Polymer	Repeating unit in structure	Typical uses
Polyethylene	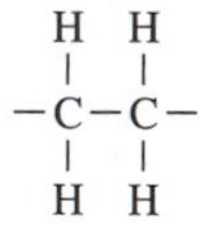 H H | | −C−C− | | H H	Films for packaging applications (branched PE); bottles and other containers (linear PE)

Polypropylene	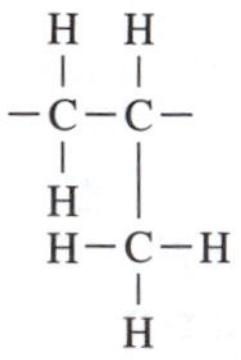 H H
	| |
	−C−C−
	| |
	H |
	    H−C−H
	      |
	      H
		Rope and filaments; sheet; pipes

(continued)

[8]Even the most "crystalline" polymers are not entirely crystalline, but contain amorphous regions. Thus, "semicrystalline" polymers is a more precise term. However, jargon persists and the term "crystalline" polymers is taken to imply a material which is a crystalline-noncrystalline mixture.
[9]The degree of crystallization within polyethylene can also be varied by the extent of "side-branching." *Linear* polyethylene, which consists almost entirely of long-chain molecules, crystallizes easily. *Branched* polyethylene, in which auxiliary branched chains are attached to, and abut from, the main chains, crystallizes to a lesser extent due to its more complicated structure.
[10]An example is the copolymer of styrene and butadiene discussed in Chap. 2. Another example is Saran, a blend of poly(vinyl chloride) and poly(vinylidene chloride).

Polystyrene	H H │ │ —C—C— │ │ H ⬡ (benzene ring)	Molded packaging containers; foams; plastic control components
Poly(methyl methacrylate) (PMMA; Lucite; Plexiglass)	H │ H—C—H H │ │ │ —C—C— │ │ H C=O │ O │ H—C—H │ H	Plastic windows; fixtures; sanitary ware
Polyacrylonitrile (Orlon)	H H │ │ —C—C— │ │ H—C— ‖‖‖ N	Acrylic fibers
Poly(vinyl acetate)	H H │ │ —C—C— │ │ H—O— │ C=O │ H—C—H │ H	Water-based emulsion paints; adhesives
Poly(vinyl alcohol)	H H │ │ —C—C— │ │ C OH	Textile fibers; wet-strength adhesives
Poly(vinyl chloride) (PVC)	H H │ │ —C—C— │ │ H Cl	Electrical wire and cable insulation; pipe; floor covering; film
Poly(vinylidene chloride) (Saran, if copolymerized with 20 percent PVC)	H Cl │ │ —C—C— │ │ H Cl	Transparent film; fiber
Polytetrafluoroethylene (Teflon, TFE)	F F │ │ —C—C— │ │ F F	Chemically resistant coatings; laboratory ware; coatings for nonstick food-processing equipment; bearings; gaskets

Source: Adapted from K. M. Ralls, T. H. Courtney, and J. Wulff, *Introduction to Materials Science and Engineering,* Wiley, New York, 1976.

we must realize that even "easily" crystallized polymers contain remnant amorphous material. Linear PE, the most readily crystallized polymer, still contains noncrystalline material in the amount of ca. 5 percent. Finally, many common long-chain polymers are completely amorphous. That is, the noncrystalline structure is the rule rather than the exception for long-chain polymers.

In polymers, the glass transition temperature is associated with a marked reduction in chain mobility. As a result of the weak interchain bonding, extensive permanent deformation in chain polymers is caused by interchain sliding. The sliding is promoted not only by stress but by temperature, and the reduction in chain mobility at T_g, therefore, corresponds to a transition in the mechanical characteristics of the polymer. Above T_g the polymer is a leathery or rubber-like substance with considerable flexibility. Below T_g it is much stiffer and less flexible. Although the glass transition is commonly construed as a ductile-to-brittle transition, this is not strictly so. At low stresses and strain rates, permanent deformation of glassy polymers can still be accomplished although it is effected by local, rather than massive, interchain displacements. This indicates that some atomic mobility along the chain and at its ends persists below T_g, although this mobility is much reduced compared to what it is at higher temperatures.

As noted, both the extensive interchain displacements at high temperature and the localized displacements at low temperatures are furthered by the temperature-assisted mobility of side-groups. This can perhaps be better visualized by consideration of the three-dimensional structure of PE in Fig. 8.14 which is a more accurate depiction than the two-dimensional structure implied by Eq. (8.10). The term used to describe the three-dimensional structure is "planar zigzag," indicating that the carbon atoms in the chain all lie in a plane, but the trace of their atomic positions there is a "zigzag." The hydrogen atoms are staggered about this line. Relative interchain displacement is hindered by atoms in close proximity to each other (steric hindrance). Temperature, though, provides localized energy which enables

Polymer	Repeating unit in structure	Typical uses
Poly(hexamethylene adipamide) (Nylon 66, a polyamide)		Gears; fibers; rollers; bearings; jacket over primary electrical insulation; wheels; light machinery components
Poly(ethylene terephthalate) (Dacron, Mylar, a polyester)		Textile fibers; films; recording tape
Acetal copolymer		Gears; instrument housing; plumbing valves; shower heads

Source: Adapted from K. M. Ralls, T. H. Courtney, and J. Wulff, *Introduction to Materials Science and Engineering,* Wiley, New York, 1976; and R. W. Hertzberg, *Deformation and Fracture Mechanics of Engineering Materials,* Wiley, New York, 1976.

Table 8.3
A partial list of heterochain polymers and some of their uses

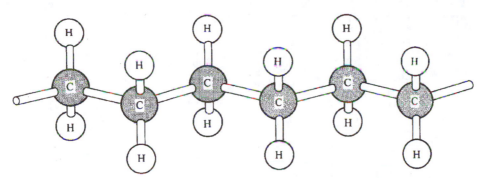

Figure 8.14
A three-dimensional representation of the polyethylene long-chain polymer also showing the arrangement of its hydrogen atoms. Thus, this "linear" PE is actually a "zigzag." The hydrogen atoms are staggered along the chain axis, with a bond angle characteristic of the tetrahedral bond angle of carbon.

carbon atoms—and the hydrogen atoms associated with them—to rotate about the chain axis thereby temporarily removing the geometrical hindrance and facilitating interchain sliding. The ease of carbon rotation is associated with the side-group size; rotation is fairly easy in PE, for example, which causes it to have both a low glass transition temperature and a fairly low yield strength. The carbon-oxygen bond in polyesters (Table 8.3) is also quite flexible insofar as carbon atom rotation is concerned, and polyesters also have relatively low glass transition temperatures. In effect, T_g is low when the energy required to induce chain mobility and carbon atom rotation is low.

B. Deformation of Noncrystalline Polymers

The tensile flow behavior of a glassy polymer as it depends on temperature is schematically illustrated in Fig. 8.15. Four different types of behavior are indicated there. Uniform viscous flow takes place at temperatures slightly above T_g and in accord with our discussion on viscosity of Sect. 8.3. At temperatures below about $0.8T_g$, "brittle" fracture takes place prior to macroscopic permanent deformation. However, the polymer extension before failure is, as a result of viscoelastic strain, much greater than the fracture elongation of a silicate glass. Viscoelastic strains can be on the order of several percent for a "brittle" glassy polymer provided the temperature is not too far below T_g. In this section, though, we primarily address the flow behavior designated in Fig. 8.15 as "extensive cold drawing" and as "onset of plasticity."

    i. DRAWING. A more detailed depiction of the flow of a polymer in the "cold drawing" region is shown in Fig. 8.16, which displays the evolving shape of a tensile bar as it varies with "drawing" strain. During initial loading, the material deforms elastically, displaying a limited linear elastic strain and a more extensive viscoelastic strain. Prior to attainment of the maximum in stress observed at relatively low strains, permanent deformation commences. This deformation is initially uniform throughout the material volume. However, at the stress maximum, massive

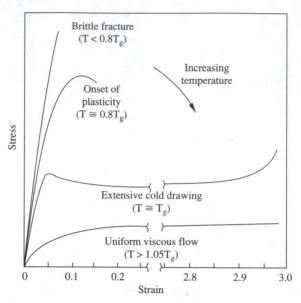

Figure 8.15
The temperature variation of the tensile flow behavior of a chain polymer. Flow characteristics change rapidly over a relatively narrow temperature range. For example, polymer behavior changes from that of a viscous liquid to that of a macroscopically brittle solid as the temperature is decreased from slightly above T_g to about $0.8T_g$. (*From L. J. Gibson and M. F. Ashby, Cellular Solids-Structure and Properties, Pergamon Press, Oxford, 1988.*)

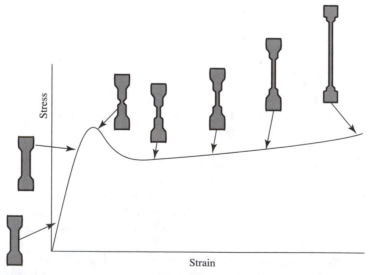

Figure 8.16
Schematic illustrating the development and propagation of a neck during tensile "cold-drawing" of a polymer. Plastic deformation precedes neck development, which takes place at the maximum stress. Following this, the neck propagates along the gage length, resulting in significant material extension. (*From J. M. Schultz, Polymer Materials Science, Prentice-Hall, Englewood Cliffs, N.J., 1974.*)

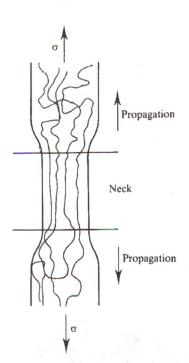

σ

Propagation

Neck

Propagation

σ

Figure 8.17
Schematic of process of molecular alignment during neck propagation of a glassy polymer. Extensive alignment takes place in the neck as this strong region propagates along the gage length and converts the less deformed and aligned material into the molecularly oriented structure.

flow—in which a neck forms and propagates along the gage length—initiates. In many cases, the associated stress is also the polymer tensile strength. Massive flow and neck initiation is associated with a decrease in the stress required to continue deformation. The formation of a neck in a drawn polymer is, as it is in a metal, a geometrical instability. However, unlike in a metal, once a neck forms in a glassy polymer, further deformation is not restricted to the neck region. Instead, the higher true stress in the necked area accelerates the process of molecular alignment along the draw (tensile) direction (Fig. 8.17). The molecular alignment results in strengthening in the neck region because a substantial portion of the stress there is borne by primary covalent bonds in the aligned molecules. Thus, post-tensile deformation is associated with neck propagation along the gage length and, at strains in excess of the necking strain, deformation in glassy polymers that initially deform homogeneously becomes macroscopically heterogeneous. The strain in the neck region is very large, whereas that in the material that has not yet necked is much smaller. As the neck propagates, the stress required to continue flow generally increases beyond that immediately following formation of the neck. This increase is associated with further deformation of the aligned molecular structure via primary bond stretching; this is somewhat similar to what is observed in the final stages of elastomeric tensile deformation (cf. Fig. 2.10).

As temperature is decreased to well below T_g the elongation to failure of the polymer is reduced considerably. However, glassy polymers at temperatures less than T_g still display localized plasticity as is discussed in the following paragraphs.

ii. LOCALIZED PLASTIC DEFORMATION. Two forms of localized low-temperature plasticity—*shear banding* and *crazing*—can occur in glassy polymers. Which of the two is the dominant plastic deformation mode depends on temperature and the state of the applied stress. Thus, shear banding and crazing can be considered competitive in much the same way that, for example, dislocation creep and diffusional creep can be considered competitive.

ii a. Shear Bands. Plastic flow in a glassy polymer that deforms by the formation and propagation of shear bands initiates at a stress level below the tensile strength. Although some homogeneous flow precedes the heterogeneous plastic deformation associated with shear bands, their formation is commonly taken as the yielding event in materials deforming in this manner. A schematic of shear bands in a glassy polymer is shown in Fig. 8.18a, and a photograph of them is provided in Fig. 8.18b. Shear bands are always aligned nearly parallel to the plane of maximum shear stress. For example, in a tension test, the normal to the shear band makes an angle close to 45° to the tensile axis.

As in metallic glasses, the boundary between a shear band and the adjacent material separates a region of intensely deformed material from one in which very little permanent deformation has taken place. The formation of a shear band in a polymer is, therefore, phenomenologically similar to that in a metallic glass in that a localized permanent displacement catalyzes flow in areas immediately adjacent, thus resulting in heterogeneous plastic deformation. However, in distinction to tensile deformation of a metallic glass, formation of a shear band in a polymer is not an

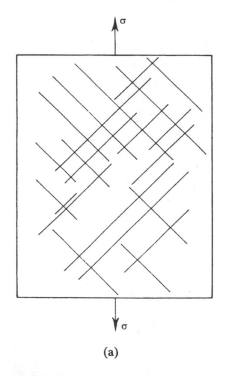

(a)

(b)

Figure 8.18
(a) Schematic of shear band pattern in an amorphous polymer. The bands are regions of localized plastic deformation. They form on planes very close to those on which the maximum shear stress operates. (b) Shear bands in plastically extended poly(ethylene terephthalate). (A. S. Argon, Polymeric Materials, ASM International, Materials Park, OH 44073-0002 (formerly American Society for Metals, Metals Park, OH 44073), 1975, p. 411.)

immediate precursor to fracture. That is, numerous shear bands form during deformation of a glassy polymer, and their number increases with strain.

The high-strain tensile behavior of glassy polymers deforming via shear band formation parallels that of polymers discussed in the preceding section. That is, polymers that initially form shear bands can be "drawn." Beyond the tensile point, a neck containing highly aligned molecules is formed and propagates along the sample length. The means by which shear bands are transformed into an aligned molecular structure within the neck are illustrated in Fig. 8.19. Immediately above the boundary region separating the necked volume from the unnecked one, the shear band density increases dramatically and the neck propagates into this heavily, albeit heterogeneously, deformed region, effecting molecular orientation.

To summarize! The difference between drawing of polymers that initially deform homogeneously and those that initially deform by shear banding is primarily that of the initial flow morphology. We add that the two manifestations of flow are competitive; shear banding dominates initial homogeneous flow as temperature is decreased. Further, over a limited temperature range, a polymer may deform by shear banding in compression whereas if it is deformed in tension a neck forms without prior shear band formation.

The stress at which shear banding initiates depends on temperature and stress state, as noted. This variation for a typical glassy polymer is illustrated in Fig. 8.20. As shown there, yield strengths are relatively high fractions of the shear modulus. Additionally, the compressive yield strength is greater than the tensile yield

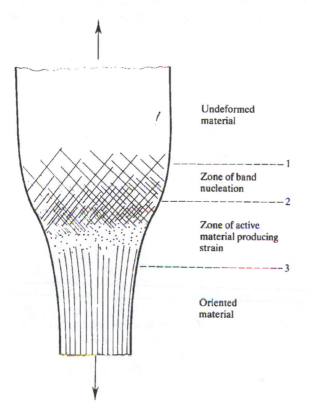

Undeformed
material

————————— 1

Zone of band
nucleation

————————— 2

Zone of active
material producing
strain

————————— 3

Oriented
material

Figure 8.19
Neck propagation in a glassy polymer deforming by shear banding. The region between that with coarse shear bands and the molecularly aligned neck is characterized by a high shear band density, which facilitates molecular alignment as the neck propagates. (*A. S. Argon, Polymeric Materials, ASM International, Materials Park, OH 44073-0002 (formerly American Society for Metals, Metals Park, OH 44073), 1975, p. 411.*)

strength. This suggests that the volume available for catalyzing localized deformation is increased by a dilational stress (i.e., one, such as a tensile stress, that increases the atomic volume), and decreased by a compressive stress. This is in contrast to a crystalline material, for which the critical resolved shear stress is the same in tension and compression and is unaffected by dilatancy. The effect of dilatancy on the shear yield condition for a glassy polymer is taken into account by modifying the von Mises yield criterion. For a material in which dilatancy plays no role in yielding, this criterion is

$$\frac{1}{\sqrt{6}}[(\sigma_1 - \sigma_2)^2 + (\sigma_2 - \sigma_3)^2 + (\sigma_1 - \sigma_3)^2]^{1/2} = S \geq \tau_y \qquad (8.11)$$

where τ_y is the yield stress in pure shear (e.g., as in a torsion test where $\sigma_1 = -\sigma_3$, $\sigma_2 = 0$). Equation (8.11) is modified for glassy polymers to

$$S + \alpha_p p \geq \tau_y \qquad (8.12)$$

where p is the mean pressure ($= (\sigma_1 + \sigma_2 + \sigma_3)/3$) and α_p is a constant. Agreement between the yield condition Eq. (8.12) and experiment has been obtained for a number of glassy polymers subjected to biaxial stress states (Fig. 8.21). The materials whose behavior is illustrated did not deform by an alternative heterogeneous plastic deformation mode called crazing. Crazing may occur in preference to shear banding at certain temperatures, and whether a material crazes or shear yields also depends on the character of the applied stress state.

ii. b. Crazing. Just as shear banding competes with initially homogeneous flow in polymers that draw, crazing competes with shear banding. The tendency for

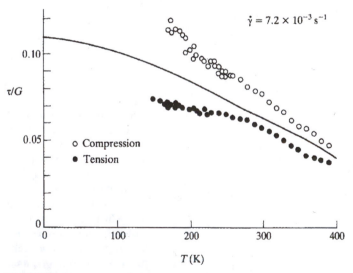

Figure 8.20
The plastic resistance to shear banding (in terms of the shear modulus, *G*) for a polycarbonate as a function of temperature for shear (solid line), compression, and tension testing. The flow resistance depends on stress state indicating that yielding is a function of pressure as well as the resolved shear stress. The high values of τ relative to *G* are an indication of the intrinsically high strengths of many polymers. (*From A. S. Argon, Phil. Mag.*, 28, 839, 1973.)

crazing increases as temperature is decreased, and crazing is also promoted by positive mean pressures.

What is crazing? Crazes give the visual appearance of cracks, as shown by the photograph of a tensile sample of polystyrene (PS) (Fig. 8.22) that has crazed. The normal to the long dimension of the craze is always parallel to the direction of maximum stress, as is apparent in the PS of Fig. 8.22. Despite their appearance, crazes are microscopic regions of highly localized deformation similar to those developed on a macroscopic scale at large tensile strains in drawn glassy polymers. This craze micromorphology is illustrated in Fig. 8.23. The craze contains fibrils (the dark regions of Fig. 8.23) of highly oriented molecules separated by porous regions. While the fibrils are quite strong, the overall strength of a craze is reduced by the presence of the pores. Due to them, the density of the craze is less (and the specific volume of it is greater) than that of the uncrazed material, and the resulting stress concentration at the boundary between these regions serves to propagate the craze along a direction *normal* to the principal stress axis. This is accomplished without a noticeable increase in craze thickness (which has dimensions typically on the order of micrometers) although this thickness is temperature dependent (it decreases with decreasing temperature). Although many crazes form in a tensile test, overall sample ductility is limited, as fracture intervenes prior to the development of crazes in sufficient number to provide for extensive macroscopic strain. Sample fracture generally initiates within the craze and is accompanied by necking and rupture of the

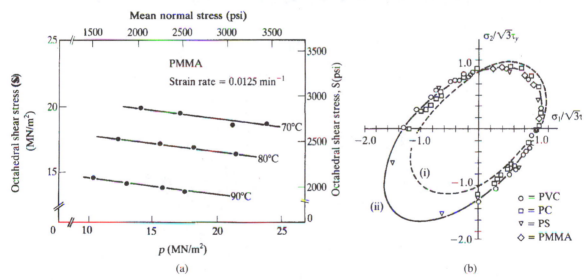

Figure 8.21
(a) The octahedral shear stress for yielding (S of Eq. (8.11)) of PMMA vs. the mean stress (p of Eq. (8.12)). The linear relationship validates Eq. (8.12) as the shear-banding yield criterion for glassy polymers. (b) The yield surface in biaxial loading (normalized in terms of the shear yield stress, τ_y) for a number of polymers deforming by shear banding. Curve (i) is the yield surface in the absence of a pressure effect on yielding. Pressure affects yielding of these polymers (curve (ii)); tensile stress components promote yielding and compressive components delay it. (*Part (a) from S. S. Sternstein and L. Ongchin, Polymer Preprints, 10, 1117, 1969; (b) From R. Raghava, R. M. Caddell, and G. S. Y. Yeh, J. Matls. Sc., 8, 225, 1973.*)

Figure 8.22
Crazes in polystyrene tested in tension.
The crazes are made visible by the in-
ternal reflection of light at the bound-
aries between the crazed and
noncrazed regions. (*From A. J. Kinloch*
and R. J. Young, Fracture Behavior of
Polymers, Applied Science, London,
***1983.*)**

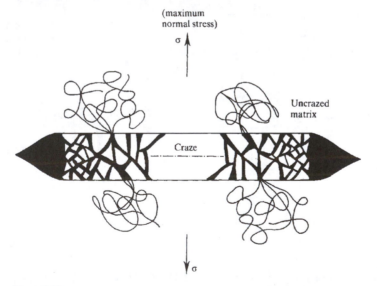

Figure 8.23
The molecular morphology in a craze. The molecules are heavily aligned in the form
of fibrils (the dark areas) in the craze. However, the craze also contains voids. These
are more predominant in the craze center (the first portion of the craze to form). The
voided areas eventually assume the characteristics of a crack as the craze spreads lat-
erally. (*From S. S. Sternstein, Polymeric Materials, American Society for Metals, Met-*
***als Park, Ohio, 1975, p. 369.*)**

fibrils within it. The reduced ductility accompanying craze formation is commonly associated with craze embrittlement, and this view is correct inasmuch as the toughness of a polymer that crazes is much less than that of one that flows by shear yielding. However, the localized plastic deformation accompanying crazing is extensive, and polymers deforming by crazing exhibit considerably higher toughnesses than do materials, such as silicate glasses at room temperature, that fracture prior to the advent of even localized plastic deformation.

As mentioned, crazing is favored by positive normal stresses and, for conditions of a multiaxial stress state, the craze normals are parallel to the direction of the maximum principal stress. Biaxial stress states for which $\sigma_1 > \sigma_2 > 0$ have been used to investigate the criteria for crazing in a number of glassy polymers; the results of such investigations for PMMA are shown in Fig. 8.24. We first note that the craze yield stress increases with decreasing temperature. Further, crazing is promoted by a positive dilatory stress. That is why the principal stress required to initiate craze yielding decreases as the magnitude of the secondary principal stress is increased from zero. These observations can be used to develop a craze yield criterion. First, we know that crazing does not occur unless the mean pressure is positive. Second, since crazing involves the opening up of regions experiencing the maximum tensile stress, it is reasonable that the yield criterion reflect this. We designate the critical "opening" strain as ε_c. Letting σ_1 be the maximum principal stress and ε_1 be the corresponding principal strain, a criterion for craze yielding, taking into account the effect of the mean pressure, can be written as

$$\varepsilon_1 = \varepsilon_c = A(T) + \frac{B(T)}{p} \tag{8.13}$$

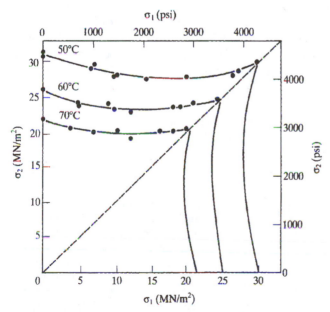

Figure 8.24
Biaxial stress combinations (tension-tension) that initiate crazing in PMMA. (*From S. S. Sternstein and L. Ongchin, Polymer Preprints, 10, 1117, 1969.*)

where $A(T)$ and $B(T)$ are temperature-dependent constants. For biaxial loading, $\sigma_3 = 0$, and from elasticity theory we have $\varepsilon_1 = (1/E)(\sigma_1 - \nu\sigma_2)$ (ν = Poisson's ratio) so that the biaxial craze yield condition becomes

$$\sigma_1 - \nu\sigma_2 = E\left[A(T) + \frac{B(T)}{p}\right] \qquad (8.14)$$

Expressions comparable to those of Eq. (8.14) can be written for triaxial loading conditions. However, the general yield criterion is expressed more conveniently by incorporating the terms that depend on the polymer-specific values of Poisson's ratio into different empirical and temperature-dependent constants. Thus, the general craze yield criterion is usually written as

$$\sigma_{max} - \sigma_{min} = C(T) + \frac{D(T)}{p} \qquad (8.15)$$

where σ_{max} is the algebraically largest principal stress and σ_{min} the algebraically smallest one. Of necessity, according to Eq. (8.15) $p > 0$ for crazing to be possible.

The competitive nature of crazing vs. shear yielding can be illustrated by drawing their respective yield loci under biaxial loading conditions (Fig. 8.25). In the first quadrant ($\sigma_1, \sigma_2 > 0$), crazing is most competitive with shear yielding, whereas crazing does not take place when $\sigma_1, \sigma_2 < 0$ (the third quadrant). In the equivalent second and fourth quadrants ($\sigma_1 > 0$, $\sigma_2 < 0$, or vice versa), crazing can compete with shear yielding provided that the algebraic sum of σ_1 and σ_2 is positive.

Experimental investigations in the second quadrant are especially useful for illustrating the competitive nature of crazing/shear yielding and also for testing the adequacy of the constitutive Eqs. (8.12) and (8.15). An expanded view of the second quadrant is shown in Fig. 8.26a. Biaxial stress states in region A produce no plastic deformation. Stress states lying within regions B and C should produce flow by crazing and shear yielding, respectively, whereas stress combinations in region D are conducive to simultaneous crazing and shear yielding. Experimental data for PMMA, which support this interpretation, are provided in Fig. 8.26b. Of especial

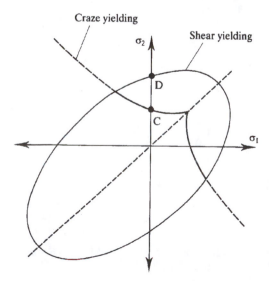

Figure 8.25
The biaxial loading yield loci for shear banding and crazing. Crazing dominates in the first ($\sigma_1, \sigma_2 > 0$) quadrant and in parts of the second and fourth ones ($\sigma_1 \geq 0$, $\sigma_2 \leq 0$ or vice versa). Only shear yielding is observed for biaxial compression ($\sigma_1, \sigma_2 < 0$). Points C and D are the uniaxial tensile stresses required to initiate crazing and shear yielding, respectively. (*From S. S. Sternstein and L. Ongchin, Polymer Preprints, 10, 1117, 1969.*)

interest are samples tested with stress combinations in region D that lead to both forms of plasticity. A schematic of the appearance of a tensile bar experiencing both types of flow is provided in Fig. 8.27.

The craze-shear yield transition is also affected by application of a third normal stress (σ_3). While both the shear- and craze-yielding criteria are affected by such a stress, the crazing criterion is more so. Thus when σ_3 is positive, the craze-yield

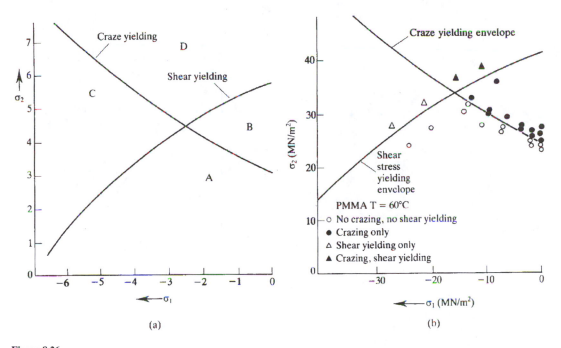

(a)   (b)

Figure 8.26
(a) An expanded view of the second quadrant ($\sigma_1 < 0$, $\sigma_2 > 0$) of Fig. 8.25. Stress combinations lying in region A produce no yielding; in region B, craze yielding; in region C, shear yielding; and in region D both craze and shear yielding. (b) Experimental verification of the scheme of (a) for PMMA. (*From S. S. Sternstein and F. A. Myers, J. Macromolec. Sc., B8, 539, 1973.*)

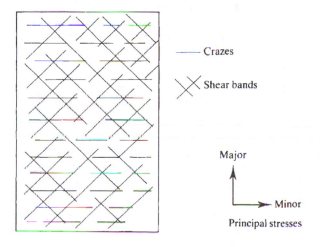

—— Crazes

✕ Shear bands

Major

Minor

Principal stresses

Figure 8.27
Schematic of craze–shear band morphology when both deformation modes are operative. The maximum principal stress is always perpendicular to the long dimension of the craze. Shear bands lie along planes nearly parallel to those characterized by the maximum shear stress.

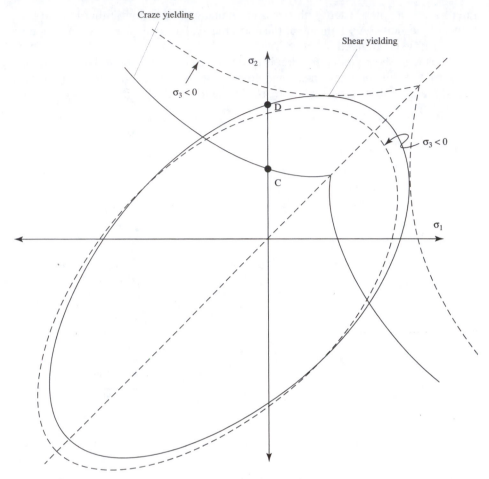

Figure 8.28
Schematic of the effect of a third hydrostatic ($\sigma_3 < 0$, dotted line) stress on the craze- and shear-yield loci. (The solid lines correspond to $\sigma_3 = 0$.) Both yield surfaces are affected by such a stress. However, the craze-yield surface is altered more. Thus, in distinction to Fig. 8.25, shear yielding precedes crazing on loading in tension when the magnitude of the third (algebraically negative) stress component becomes large enough. If the third stress component is a tensile one, the situation is reversed.

surface moves inward relative to the shear-yield surface, and crazing occurs over a greater range of σ_1 and σ_2 combinations. Conversely, if $\sigma_3 < 0$, the craze-yield surface moves outward with respect to the shear-yield surface and shear yielding becomes more favored. The effect is shown in Fig. 8.28.

Decreases in temperature also favor craze formation relative to shear yielding. Craze thickness, as noted, decreases with decreasing temperature and, in the absence of an increase in the number of crazes (which does not happen), macroscopic ductility and toughness likewise decrease. Concurrently, the ratio of linear elastic to plastic deformation increases, and a crazing glassy polymer becomes more and more like a brittle glass as temperature is lowered. However, glassy polymers are always capable of some degree of permanent deformation, at least at sufficiently low strain rates, even at temperatures approaching 0 K. Thus, they always demonstrate a toughness greater than do oxide glasses that are incapable of such deformation.

We conclude this section with a brief description of techniques used to toughen glassy polymers prone to craze deformation. Polystyrene is one such polymer. It is toughened by the introduction of colloidal-size elastomeric particles into it; the "composite" is called HIPS (high-impact polystyrene). The elastomer particles in HIPS interfere with propagation of crazes, thereby delaying fracture. Associated with the reduction in the length of the crazes is an increase in their number and a correspondingly greater tensile failure strain. The situation is similar to that surmised to apply to partially crystallized metallic glasses (Sect. 8.5). However, in distinction to what is found for some metallic glasses, the tensile strength of HIPS is less than that of ordinary PS. More about the causes for toughening of this nature is written in Chap. 10.

EXAMPLE PROBLEM 8.1. A hypothetical polymer yields by shear banding in room-temperature compression at a stress of 29 MPa. For room-temperature tension, shear banding initiates at a stress of 21 MPa. Develop an expression for the shear yielding criterion for this material applying to room-temperature biaxial loading ($\sigma_3 = 0$, $\sigma_{1,2} \geq, \leq 0$).

Solution. The shear yield condition (Eq. (8.15)) is

$$S + \alpha_p p = \tau_y$$

Let the tensile yield strength (21 MPa) be designated by σ_T and the compressive yield strength (29 MPa) by σ_c. In the following two equations the respective yield strengths are written in terms of their *magnitudes:*

$$\frac{1}{\sqrt{6}}[2\sigma_T^2]^{1/2} + \frac{1}{3}\alpha_p\sigma_T = \tau_y \quad \text{or} \quad \sigma_T\left[\frac{1}{\sqrt{3}} + \frac{1}{3}\alpha_p\right] = \tau_y \tag{1}$$

$$\frac{1}{\sqrt{6}}[2\sigma_c^2]^{1/2} - \frac{1}{3}\alpha_p\sigma_c = \tau_y \quad \text{or} \quad \sigma_c\left[\frac{1}{\sqrt{3}} - \frac{1}{3}\alpha_p\right] = \tau_y \tag{2}$$

We have two equations (those above) and two unknowns (τ_y and α_p). Equating Eqs. (1) and (2) and rearranging we find

$$\frac{1}{3}\alpha_p = \frac{1}{\sqrt{3}}\left[\frac{\sigma_c - \sigma_T}{\sigma_c + \sigma_T}\right] = \frac{1}{\sqrt{3}}\left[\frac{8}{50}\right] = 0.0924$$

To find τ_y use Eq. (1) (or Eq. (2), it doesn't matter which)

$$21 \text{ MPa}\left[\frac{1}{\sqrt{3}} + 0.0924\right] = 14.07 \text{ MPa} = \tau_y$$

Thus, for biaxial loading the shear yielding criterion for the polymer is expressed as

$$\frac{1}{\sqrt{6}}[(\sigma_1 - \sigma_2)^2 + \sigma_1^2 + \sigma_2^2]^{1/2} + 0.0924(\sigma_1 + \sigma_2) = 14.07 \text{ MPa}$$

when the respective stresses are expressed in MPa.

C. Deformation of Crystalline Polymers

Although this chapter is concerned primarily with deformation of noncrystalline materials, it is appropriate to also discuss the deformation response of semi-crystalline polymers. This is so because the mechanisms and phenomenology of crystalline polymer deformation in certain respects closely resemble those of amorphous polymers.

As mentioned, "crystalline" polymers always contain some remnant noncrystalline material. Since the crystalline regions possess a greater deformation resistance, the strength of amorphous-crystalline polymer mixtures increases as the degree of the crystallinity increases. This is illustrated in Table 8.4 for such mixtures of polyethylene. When the volume fraction of crystalline material is low, the deformation response may be likened to that of a composite material in which a relatively weak matrix (in this case the noncrystalline material) is stiffened by the presence of a harder phase (the crystalline material). When the polymer is primarily crystalline, plastic deformation also involves flow of the stronger crystalline material. Additionally, the amorphous material within such a polymer displays a greater flow resistance than does a wholly amorphous material of the same chemistry. The reason relates to the molecular configuration in semicrystalline polymers. In brief, in semicrystalline polymers packets of crystalline regions in which the polymer chains are arranged in a regular matter are connected by "tie molecules." That is, an individual polymer chain runs from one crystalline region (where it is arranged orderly) to another through the amorphous regions. The connecting polymer molecules are termed "tie molecules." The greater strength of the amorphous regions in a semicrystalline polymer, then, arises from the constraint placed on their flow by the stronger crystallites and because of the connecting tie molecules.

The tensile stress-strain behavior of two crystalline polymers is shown in Fig. 8.29. Initial elastic deformation (Region 1) is followed by yielding (Region 2). Yielding is associated with plastic deformation of the crystalline regions, which generally involves mechanisms of flow (e.g., twinning) other than the dislocation motion common to metals and ceramics. As with amorphous polymers that "draw," a neck forms in crystalline polymers following yielding, and this neck propagates along the sample gage length (Region 3). Prior to neck formation, the long axis of the polymer chains in the crystalline regions rotates towards the direction of the applied stress, and during necking the originally more or less randomly oriented structure is converted into a highly oriented one. The idea is schematized in Fig. 8.30.

Deformation following neck propagation along the whole of the gage length is accompanied by a significant increase in the flow stress (e.g., Region 4 for nylon, Fig. 8.29). This reflects the strengthening accompanying molecular orientation. However, the observed flow stress is still less than that required to sever atomic bonds. Instead, it is a measure of the stress required to displace segments of oriented crystals relative to

Percentage crystallinity	Density (10^3 kg/m^3)	T.S. (MN/m^2)
65	0.920	13.8
75	0.935	17.2
85	0.950	27.6
87	0.960	31.0
95	0.965	37.9

Source: From H. V. Boening, *Polyolefins: Structure and Properties,* Elsevier Press, Lausanne, 1966, p. 57.

Table 8.4
Tensile strength of polyethylene as affected by percentage crystallinity

each other. As a result of the small segment thickness (on the order of micrometers), the frictional stress retarding intersegment displacement is high. An expanded view of the structure between segments is shown in Fig. 8.31. Note the segments are still connected occasionally by tie molecules, and this also contributes to the difficulty of intersegment sliding.

As would be expected, considerable mechanical anisotropy is developed concurrently with formation of the highly oriented neck structure. This anisotropy, as reflected by the values of elastic modulus as they vary with orientation to the draw

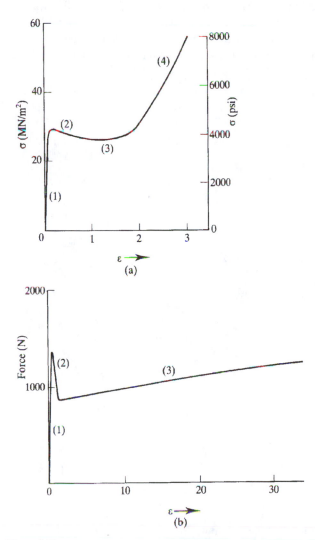

Figure 8.29
(a) Stress-strain curve for crystalline nylon and (b) load-strain curve for crystalline polyethylene. Both polymers display linear elastic behavior (Region 1), yielding (Region 2), and drawing (Region 3). Additionally, nylon manifests a stage (Region 4) corresponding to elongation of the highly aligned molecular structure. (*Part (a) from J. Rubin and R D. Andrews, Polymer Eng. Sc., 8, 302, 1963; (b) from A. Peterlin, J. Macromol. Sc., B7, 705, 1973.*)

direction, is shown for PE in Fig. 8.32. The modulus in the draw direction can be quite high at large drawing strains. However, this modulus does not reflect the inherently high modulus of the oriented crystalline regions which is expected to be on the order of 250 GPa. Amorphous PE is rubbery at room temperature (its modulus is on the order of 0.1 GPa), and the crystalline-amorphous mixture behaves rather much like a composite material subjected to an equal-stress condition. Thus, the modulus of the mixture lies much closer to that of the amorphous material than the crystalline material (Prob. 8.21).

It should also be remarked that the tensile strengths along the draw direction of heavily drawn crystalline polymers exceed similar strengths in comparably extended noncrystalline polymers. This is true even though comparable molecular orientation occurs in both types of material and shows that the remnant crystalline

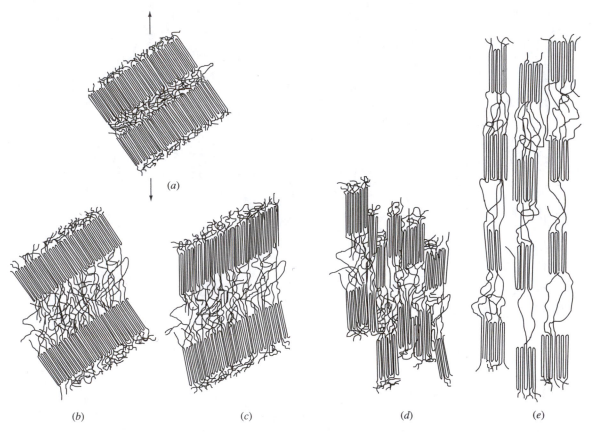

Figure 8.30
Schematic of stages of tensile deformation of a semicrystalline polymer. (a) The initial structure in which crystalline regions are connected by amorphous ones. Note that molecules in the amorphous region constitute the "ties." (b) The initial stages of deformation in which elongation of the amorphous tie molecules occurs. (c) With increasing strain the crystalline segments rotate, aligning themselves with the tensile axis. (d) The crystal segments undergo "fragmentation" by intersegmental displacement at a later deformation stage. (e) Orientation of the crystalline segments and the tie molecules in the last stages of deformation. (*From J. M. Schultz, Polymer Materials Science, Prentice-Hall, Englewood Cliffs, N.J., 1974, p. 500.*)

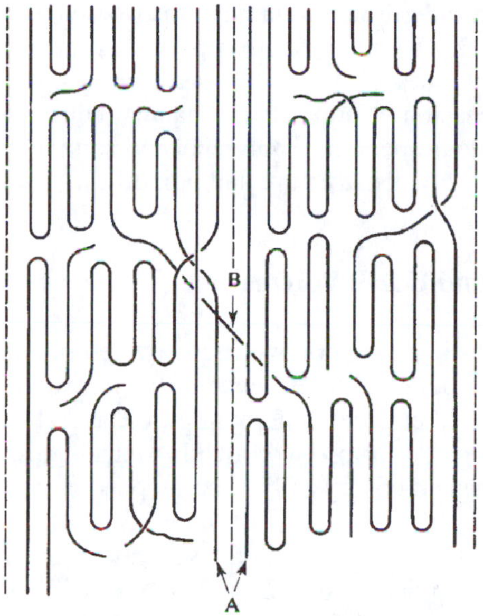

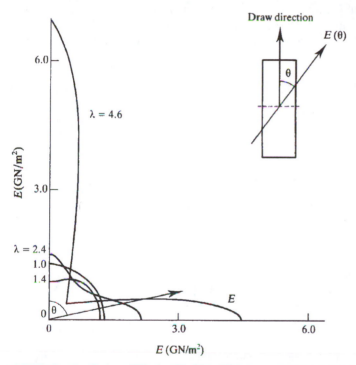

Figure 8.31

An expanded view of intersegmental sliding of the type illustrated in Fig. 8.30*d*. Fragmentation of the crystalline segments occurs by sliding along fibrillar "boundaries." These can be considered "subboundaries" within the original segments. (*From A. Peterlin, Polymeric Materials, American Society for Metals, Metals Park, Ohio, 1975, p. 175.*)

Figure 8.32

Angular dependence of elastic modulus of drawn crystalline polyethylene as a function of draw ratio λ (= 1 + ε, where ε is the drawing strain). During drawing, the material becomes more anisotropic and stiffer in the draw direction. At high draw ratios, the structure also displays a high modulus at 90° to the draw direction.
(*A. Peterlin, Polymeric Materials, ASM International, Materials Park, OH 44073-0002 (formerly American Society for Metals, Metals Park, OH 44073), 1975, p. 175.*)

segments in a drawn crystalline polymer provides it with strengths superior to those of an oriented glassy polymer. As a result, drawn crystalline polymers are the strongest of polymeric materials, possessing strength-to-density ratios comparable to those of many strong metals. Our discussion of polymer mechanical behavior concludes with a description of the ultimate properties of polymers and some further discussion of the relationship between polymer structure and properties.

D. Structure-Property Relationships and Use of Polymers

The essentials of polymeric elastic and plastic behavior have been covered in this chapter and in Chap. 2. In Chap. 2, the temperature dependence of the relaxation modulus of the several polymer categories (thermosets, thermoplastics, and elastomers) is encapsulated in Figs. 2.11. However, it is also worthwhile to distinguish between the relaxation moduli of crystalline and noncrystalline thermoplastics, and this is done here.

The temperature dependence of the relaxation modulus for several forms of polystyrene is shown in Fig. 8.33. The sharp decrease in E_r at T_g ($\cong 100°C$) for amorphous PS is caused by the increasing amount of viscoelastic strain at this temperature. At higher temperatures (ca. 175°C, which is still well below $T_m = 240°C$), viscous flow ensues and E_r decreases even more. The modulus of crystalline PS also decreases at T_g for the same reasons that it does for the noncrystalline form;

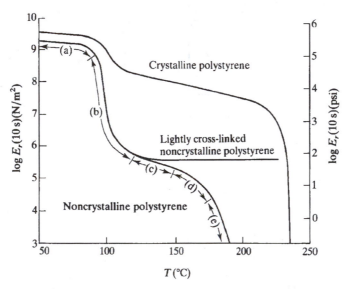

Figure 8.33
Relaxation modulus vs. temperature for structurally different polystyrenes. Noncrystalline PS manifests the five regions of behavior [(a) glassy, (b) leathery (near T_g = 100°C), (c) rubbery, (d) transition to liquid behavior, (e) viscous flow] described in Chap. 2. Lightly cross-linked PS also demonstrates regions (a) and (b) but is an elastomer over an extended range above 130°C. The (semi)crystalline polymer shows a much lesser decrease in E_r at T_g and remains fairly stiff until T_m is reached. (*After A. V. Tobolsky, Properties and Structures of Polymers, Wiley, New York, 1960.*)

however, the reduction in modulus of the crystalline form is fairly small. Viscous flow in crystalline PS occurs only when the temperature reaches T_m, at which the crystalline structure transforms into the amorphous liquid state. Lightly cross-linked PS displays elastomeric behavior at temperatures greater than ca. 130°C. As in room-temperature elastomers, the cross-links in this PS form preclude permanent interchain displacements. Although not evident in the logarithmic scale of Fig. 8.33, the modulus increases with temperature in the range of elastomeric behavior (as is common to all elastomers; cf. Chap. 2). It is clear that cross-linking of many amorphous polymers can render them elastomeric at temperatures somewhat above T_g provided that the polymer chains are highly coiled. Behavior of the type displayed in Fig. 8.33 can also be discussed in terms of creep curves, and this is left as an exercise (Prob. 8.23).

Highest polymeric strengths are found in fine filaments in which a highly oriented structure is developed by subjecting the polymer to extensive drawing strains. As mentioned, crystalline polymers so processed display strengths considerably greater than do like-processed amorphous polymers. Properties of some organic filaments are given in Table 8.5, where they are also compared there to several high-strength nonmetallic fibers (see also Table 6.1). Considering that polymers have low melting points and considerably lesser densities than inorganic materials, the properties displayed by strong organic filaments are impressive indeed. Such fibers find widespread use as "structural" materials in the textile (e.g., as wear-resistant carpeting and fabric) and other industries.

In monolithic form, most polymers display strengths much less than those of inorganic solids. This precludes the use of unreinforced polymers as structural materials. However, polymeric matrices are common to materials reinforced with

Material Class	Material	E (GN/m²)	T.S. (GN/m²)	ρ (10³ kg/m³)	E/ρ (MNm/kg)	T.S./ρ (MNm/kg)
Organic	Poly(p-phenylene terephthalamide)*	82	2.8	1.44	56.9	1.94
	Polyamide hydrazide	106	2.4	1.47	72.1	1.63
	Copolyhydrazide	57	2.7	1.47	38.1	1.84
	Poly(oxadiazole hydrazide)	54	2.5	1.36	39.7	1.84
	Nylon	4.9	1.05	1.1	4.5	0.95
	Polyester	14	1.1	1.1	12.7	1.00
	Kevlar	185	2.0	1.50	123	1.3
	PE (Ultra)**	170	3.0	1.00	170	3.0
Inorganic	Beryllium	315	1.3	1.8	175	0.72
	Pearlitic steel	210	4.2	7.9	26.6	0.53
	Tungsten	350	3.9	19.3	18.1	0.20
	S glass	72	6.0	2.5	28.8	2.40
	SiC whiskers	470	2–20	3.17	148	0.6–6.3

*Average of several forms.
**PE (Ultra) is the high-strength polyethylene filament referred to in the last paragraph of this chapter.
Source: Data for organic filaments obtained from W. B. Black, *Ann. Rev. Matls. Sc.,* **10,** 311, 1980, and A. H. Windle, *Physical Metallurgy, Vol. III,* 4th edition, ed. R. W. Cahn and P. Haasen, North-Holland, Amsterdam, 1996, p. 2663.

Table 8.5
Some properties of organic and inorganic filaments

high-strength glass, carbon, boron, and similar filaments and, as discussed in Chap. 6, these fiber-reinforced solids exhibit excellent room-temperature strengths. A recently developed polymer alloy class also displays surprising strengths. The "alloys" consist of "rigid-rod" molecules (this alloy constituent has a high glass transition temperature) dispersed in a softer polymeric matrix. In a simple view, the "rigid-rod" molecules can be considered molecular-sized high-strength fibers.

Other types of reinforcement are also used to enhance polymeric properties. High-impact polystyrene, discussed earlier, is a case in point. Fillers (inorganic particulates) are used to reinforce noncrystalline thermoplastics and thermosets. For example, automobile tires are reinforced with fine particles of carbon black (a form of graphite) because unreinforced rubber possesses insufficient stiffness for the heavy demands placed on tires. Additional tire reinforcement is provided by fibers, mostly steel fibers nowadays, but nylon and glass were used for this purpose not too long ago. The fibers serve as the primary load-bearing member in a tire, whereas the elastomeric matrix provides the required resiliency.

Polymers can also be used as components of macroscopic composites. Safety glass incorporates a thin thermoplastic layer between brittle glass layers. The polymer prevents crack propagation from one side of the layered structure to the other. Not only must polymers used in safety glass have optical properties compatible with the glass, but their glass transition temperatures must be sufficiently low that the crack blunting they provide is not lost at low temperatures.

8.7
SUMMARY

The elastic and permanent deformation behavior of amorphous solids have been discussed in this chapter. For all classes of noncrystalline solids (inorganic and organic), the primary deformation mode at low to moderate stress levels and high temperatures is viscous flow and the principal low-temperature deformation mode is linear elasticity. Viscoelastic deformation is an important component of the strain at low stress levels and intermediate temperatures in metallic and organic glasses.

Viscous flow is accomplished by permanent relative displacements of atoms or molecules and the flow, although taking place at discrete locations in the material, is macroscopically homogeneous. Viscosity values reflect the difficulty of the flow process. Thus, metallic glasses with small basic structural units (individual atoms) and nondirectional bonding have fairly low viscosities, whereas viscosities of inorganic glasses are high, reflecting the strong polar covalent Si-O bond that links silicate tetrahedra in such a glass. Viscosities of inorganic glasses can be altered profoundly by chemically induced structural changes. Soda (Na_2O) and lime (CaO), for example, reduce the viscosities of silica-based glasses by orders of magnitude because their presence disrupts the three-dimensional network of Si-O bonds.

With increasing stress at low temperatures, linear elastic deformation of silicate glasses is followed by brittle fracture. This is not so for metallic and organic glasses, which deform permanently at higher stress levels even when $T < T_g$. For the most part, this permanent deformation is heterogeneous in nature, although there are some organic glasses for which flow is homogeneous. Although permanent deformation at both low and high temperatures is caused by relative atomic/molecular displacements, the number of sites suitable for such action is reduced considerably at lower temperatures. This, in concert with the catalytic activity of the flow process, leads to the development of heterogeneous flow in the form of shear bands in metallic and organic glasses.

The mechanical response subsequent to shear-band initiation and propagation differs between metallic and organic glasses. For the former, "softening" accompanies band formation. That is, the stress required to propagate the band is less than that to initiate it and, in a tensile test, tensile strength is concurrent with (one) shear-band formation; fracture occurs after its propagation across the sample width. Only when stress states are such as to prevent tensile instability are multiple shear bands formed in metallic glasses. In contrast, shear-band formation in tension in organic glasses is not followed by tensile instability and fracture. Instead the shear-band density increases with increasing stress (strain) until the bands are converted into a neck that contains highly oriented chain molecules. Such a molecular reorientation is also observed in organic polymers that flow homogeneously prior to neck formation. In both cases, following neck formation the neck propagates up and down the gage length, and this "drawing" leads to a highly oriented and strong material.

Crazing is an alternative heterogeneous plastic deformation mode in glassy polymers. It competes with shear yielding, and crazing is favored by low temperatures and positive mean pressures. A craze is wedge-shaped (typical thickness on the order of micrometers) and consists of oriented molecular arrangements (fibrils) intermeshed with void volume. The voids increase the craze volume and decrease its strength, notwithstanding the inherently high fibril strength. In contrast to materials that shear yield, those that craze do not demonstrate massive flow. That is, materials that craze typically fail at several percent strain in a tensile test, and thus they are much less tough than materials that deform by shear banding.

Tensile deformation of crystalline polymers is phenomenologically similar to that of noncrystalline polymers that yield by shear banding, albeit the overall level of the flow stress of a crystalline material is higher. For crystalline polymers, plastic deformation, involving aspects of crystallographic flow, follows elastic deformation. This is then followed by massive flow involving neck formation and propagation of a highly oriented crystal segmental structure within the neck. When crystalline polymers are subjected to large drawing strains, they demonstrate high strengths; in terms of strength-to-density ratio, for example, they compare favorably to the highest-strength metallic materials.

In spite of advances in processing and understanding of the mechanisms of polymeric flow, owing to strength limitations polymers are usually not used by themselves as structural materials. Instead, they are combined with other materials that provide the necessary strength and/or toughness. Examples include fiber composites (Chap. 6), particulate-reinforced plastics (both types of composites are stronger than the polymer matrix), and "toughened" brittle plastics.

And, finally, a final note! As far as structural materials are concerned, organic polymers are still "new." If the advances of the last several decades are any indication, the future should witness an ever-increasing use of polymers as structural materials in their own right. As but one example, recent advances in the processing of bulk polyethylene (molecularly the simplest of all polymers) have generated strengths that only 20 years ago would have been considered characteristic of a highly drawn polymer.

REFERENCES

Alfrey, T., and E. F. Gurney: *Organic Polymers*, Prentice-Hall, Englewood Cliffs, N.J., 1967.

Argon, A. S.: "Plastic Deformation in Glassy Polymers," in *Polymeric Materials,* American Society for Metals, Metals Park, Ohio, 1975, p. 411.

Cahn, R. W., and A. L. Greer: "Metastable States of Alloys," in *Physical Metallurgy, Vol. II,* 4th edition, ed. R. W. Cahn and P. Haasen, North-Holland, Amsterdam, 1996, p. 1723.

Chen, H. S., and K. A. Jackson: "The Influence of Alloy Composition on Glass Formation and Properties," in *Metallic Glasses,* American Society for Metals, Metals Park, Ohio, 1978, p. 74.

Davis, L. A.: "Strength, Ductility and Toughness," in *Metallic Glasses,* American Society for Metals, Metals Park, Ohio, 1978, p. 190.

Kimura, H., and T. Masumoto: "Strength, Ductility and Toughness—a Study in Model Mechanics," in *Amorphous Metallic Alloys,* ed. F. E. Luborsky, Butterworths, London, England, 1983, p. 187.

Peterlin, A.: "Mechanisms of Deformation in Polymeric Solids," in *Polymeric Materials,* American Society for Metals, Metals Park, Ohio, 1975, p. 175.

Ralls, K. M., T. H. Courtney, and J. Wulff: *Introduction to Materials Science and Engineering,* Wiley, New York, 1976.

Schultz, J. M.: *Polymer Materials Science,* Prentice-Hall, Englewood Cliffs, NJ, 1974.

Spaepen, F., and A. I. Taub, "Flow and Fracture," in *Amorphous Metallic Alloys,* ed. F. E. Luborsky, Butterworths, London, England, 1983, p. 231.

Sternstein, S. S.: "Yielding in Glassy Polymers," in *Polymeric Materials,* American Society for Metals, Metals Park, Ohio, 1975, p. 369.

Windle, A. H.: "A Metallurgist's Guide to Polymers," in *Physical Metallurgy, Vol. III,* 4th edition, ed. R. W. Cahn and P. Haasen, North-Holland, Amsterdam, 1996, p. 2663.

Young, R. J.: *Introduction to Polymers,* Chapman & Hall, London, England, 1981.

PROBLEMS

8.1 a When plotted on a conventional isothermal transformation diagram, liquid-to-solid transformation rates are described by a "C" curve. Sketch the shape of such a "C" curve under the assumptions: (1) the transformation rate goes to zero at T_m and T_g, and (2) the maximum transformation rate occurs at $T = (T_m + T_g)/2$.

b A cooling rate on the order of 10^{12} K/s is required to prevent crystallization of silver quenched from the liquid. Assuming $T_g = T_m/2$, estimate the transformation time to convert liquid to solid if silver is cooled instantaneously to $(T_m + T_g)/2$ and then held at this temperature.

c What are the (sometimes insurmountable) difficulties in conducting isothermal transformation studies of the liquid-to-solid transformation in metals?

8.2 The activation volume for dislocation glide creep is ba_s (see Chaps. 3, 4, and 7 for definitions). How does the magnitude of this volume compare to V_{act}, the activation volume of the reaction-rate theory of viscosity?

8.3 Show that the reaction-rate viscosity (Eq. (8.3)) increases linearly with stress in the high-stress limit (i.e., when $\tau V_{act} >> kT$).

8.4 Using the relationships given in the text among V_f and the crystalline and amorphous atomic volumes, derive the expression for the free-volume model viscosity (Eq. (8.7)).

8.5 Calculate the apparent "high-" and "low-temperature" activation energies for viscous flow ($\eta \sim \exp(Q/RT)$) of the organics whose behavior is illustrated in Fig. 8.7. What are the ratios of the high- to the low-temperature activation energies? How do these ratios compare to the one mentioned in the text?

8.6 The strain-time behavior shown in Fig. 8.8 can be simulated on the basis of mechanical models discussed in Chap. 2. Is the behavior illustrated in Fig. 8.8 approximated best by (1) a standard linear solid, (2) a Voigt model, or (3) a Maxwell model? (This last is equivalent to a standard linear solid without the Voigt element; see Example Prob. 2.2 and Prob. 2.18.)

8.7 Addition of Na_2O to a silica glass lowers T_g and increases the viscoelastic strain. Some of this viscoelastic deformation is a result of the stress-assisted motion of the Na ions between "holes" in the glass network. Describe how temperature and frequency of the applied stress will affect this viscoelastic strain. What are the similarities and differences between this kind of viscoelasticity and the anelasticity of bcc iron alloys arising from the stress-assisted motion of interstitial carbon atoms?

8.8 Estimate the "low-" and "high-temperature" viscosity activation energies for SiO_2 and the soda-lime-silica glass whose viscosities are illustrated in Fig. 8.9. Compare the absolute values of the viscosities for the two glasses, and comment on any differences. Are the differences in low- and high-temperature activation energies consistent with what was said in the text regarding the temperature variation of such activation energies for silicate glasses?

8.9 The development of surface compressive stresses in glass can be deduced through the following "thought" experiment.
a Sketch the temperature variation from the surface to the interior of a glass at some intermediate stage of cooling from a high to a low temperature.
b For this situation, sketch the specific volume as it varies from the glass surface to its interior. Show that this leads to development of tensile surface stresses and compressive interior stresses.
c At this intermediate temperature, assume the outer surfaces are sufficiently cool that viscous flow cannot relieve any thermally generated stresses but that the interior is sufficiently hot that viscous flow does relieve such stresses there. Now consider cooling from this intermediate temperature to room temperature. Show that the later cooling stages lead to the development of surface compressive stresses and tensile interior ones. (Hint: What are the relative temperature-caused contractions between the surface and the interior during this last cooling stage?)
d Could you imagine being able to shatter glass by cooling it quickly from a high to a low temperature? If so, state the conditions required for this to happen.

8.10 As noted in the text, ion implantation can generate surface compressive stresses in silicate glasses. Using sketches (at the atomic level), explain how compressive states are attained by this process. If you heated such a glass to high temperature, would you expect viscosity differences between the glass surface and its interior? Explain.

8.11 What can you say about the relative activation energies for viscosity in metallic and silicate glasses? (Hint: The reaction-rate theory can be used as a basis for comparison. What sort of "energy" must be overcome in order for the atomic/molecular displacements resulting in viscous flow to take place?)

8.12 The molecular weight, x, of a polymer is related to the degree of polymerization, D.P., by $x = (D.P.)x_0$, where x_0 is the monomer molecular weight. For many polymers, the fraction of chains having a molecular weight between x and $x + dx$ is given by $p(x) dx$ where

$$p(x) = \left(\frac{2}{\pi}\right)^{1/2} \frac{1}{\sigma} \exp\left[\left(-\frac{(x - \bar{x})^2}{2\sigma^2}\right)\right]$$

where $\bar{x}$ is the average molecular weight and σ is the standard deviation of the distribution.

a Schematically plot $p(x)$ vs. x (the latter in units of x/σ).

b Using tables for the function $p(x)$ and its integral (these can be found in standard mathematical handbooks) estimate the fraction of molecules having molecular weights between $\bar{x}/2$ and $3\bar{x}/2$.

c What features of polymer fabrication can be used to vary the parameter σ?

8.13 Consider synthesis of Nylon 66. Extend the reaction illustrated by Eq. (8.10) one additional step to show how the chain propagates.

8.14 Lüders band formation often accompanies tensile yielding in mild steels (Chap. 1). The band propagates along the gage length at a lower yield stress after being initiated at an upper one. After the band has propagated the entire gage length, work hardening commences and the tensile strength is typically much greater than the upper and lower yield strengths. In what respects is this behavior similar to, and how does it differ from, the tensile behavior of a polymer that "draws"? Compare the magnitude of the Lüders strain to the tensile strain associated with neck propagation in the polymer. Also compare the ratio of tensile to initial yield strengths for the two materials.

8.15 Refer to Fig. 8.20, which shows the compressive and tensile yield strengths for shear banding in a polycarbonate. Are the results displayed in Fig. 8.20 qualitatively in accord with Eq. (8.12)? Can you determine a procedure to test if they are in quantitative agreement with this equation?

8.16 Show that the straight lines of Fig. 8.21a, describing shear yielding of PMMA, are consistent with the yield criterion, Eq. (8.12). From Fig. 8.21a, determine the constant α_p for the several test temperatures. Are these values of α_p "large" or "small"? (Hint: In terms of Eq. (8.12), what kind of an α_p value constitutes a large correction to the von Mises yield criterion?)

8.17 If the shear yield strength of polystyrene (PS) in compression is 40% higher than it is in tension, determine the coefficient α_p in Eq. (8.12) for PS.

8.18 Write Eq. (8.15) in a form appropriate to biaxial loading ($\sigma_1 > \sigma_2$, $\sigma_3 = 0$). Show that when the craze yield condition is plotted in biaxial stress space, curves of the kind illustrated in Figs. 8.24 and 8.25 are generated.

8.19 **a** The craze yielding criterion is given by Eq. (8.15). At room temperature a hypothetical polymer crazes in uniaxial tension at a stress of 22 MPa. For balanced biaxial tension, crazing initiates when $\sigma_1 = \sigma_2 = 20$ MPa. Develop an expression for the crazing yield criterion appropriate to room-temperature biaxial loading.

b Assume that this hypothetical polymer is the same one whose shear band tendencies were analyzed in Example Prob. 8.1. For this material, determine which deformation mode will be first observed under the following loading conditions: (i) simple (uniaxial) tension; (ii) simple compression; (iii) balanced biaxial tensile loading; (iv) balanced biaxial compressive loading; (v) biaxial loading with one stress negative and one positive, but both stresses having the same magnitude.

8.20 Chapter 6 discussed statistical aspects of fiber fracture. In this approach, distribution functions describe the extent of fiber fracture in a bundle or composite as it is affected by stress and the inherent distribution in fiber strengths.
a Could you similarly model polymeric shear banding and/or crazing?
b If so, briefly describe how you would go about modeling the microscopically heterogeneous tensile flow behavior of a polymer on this basis.

8.21 The modulus of heavily drawn polyethylene (PE) is on the order of 7 GN/m² (Fig. 8.32). Take the modulus of the PE crystalline regions as 250 GN/m² and that of the amorphous regions as 0.1 GN/m². Assuming that the drawn mixture is a composite that elastically deforms under an equal-stress condition (Eq. (6.8)), estimate the volume fraction of the drawn PE that is amorphous and that which is crystalline.

8.22 The temperature variation of the relaxation moduli for crystalline and noncrystalline polystyrene (PS) is shown in Fig. 8.33.
a Which of the two forms of PS manifests the greater time-dependent deformation at $T > T_g$? Give your reasoning.
b What additional information would enable you to determine how much of the time-dependent strain was viscoplastic and how much viscoelastic?

8.23 Polymer relaxation moduli are functions of time as well as temperature. For example, for a constant applied stress (σ) the modulus is defined as $E_r = \sigma/\varepsilon$, where ε is the (in general) time-dependent response strain. The time variation of the room-temperature relaxation modulus for crystalline and noncrystalline polypropylene (PP) is shown in the figure below.

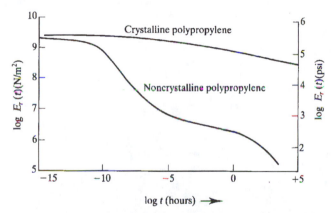

a On the basis of the time variation of the modulus of the two PP forms, what can you say about the tendency for room-temperature viscoelastic/viscoplastic strain in crystalline and noncrystalline PP?
b Assume that a constant stress is applied to crystalline and noncrystalline PP at room temperature. Schematically plot strain vs. time for the two PP forms. Is there a way to determine how much of the time-dependent strain is viscoplastic and how much of it is viscoelastic?

8.24 Sketch a cross-sectional view of a piece of safety glass. Indicate how fracture in the glass takes place when it is struck by a high-velocity object such as a small rock. Show how the plastic layer serves to prevent propagation of the crack through the whole of the safety glass.

CHAPTER 9

Fracture Mechanics

Frontispiece

Photograph of a Liberty ship that failed at pier in 1943. This failure represented but one (albeit one of the more catastrophic ones) of the 25% of the all-welded Liberty ships that fractured during World War II. About 5% of the ships experienced fractures so serious that the ships were lost or considered to be unsafe. *(From E. R. Parker, Brittle Behavior of Engineering Structures, Wiley, New York, 1957.)*

9.1
INTRODUCTION

This chapter initiates our discussion of material fracture. Fracture represents the ultimate deficiency in the engineering design and/or materials selection process. To be sure, the unanticipated permanent deformation of a structure is a serious matter, but the consequences of it pale in comparison to those of fracture. Fracture not only leads to economic loss but often, and most importantly, to loss of life. Material fracture can also lead to adverse environmental effects. Fracture of a toxic substance container, for example, may produce serious atmospheric and/or ground water contamination, leading to long-term harm to people and/or their surroundings. Thus, it is no surprise that design against fracture frequently represents the most serious challenge an engineer may face in his or her professional career.

A material may fracture in a variety of ways, depending on conditions such as temperature, the stress state and its variation with time, and the environment in which it functions. The several types of fracture can be characterized broadly according to the scheme of Table 9.1. Tensile fracture, the subject of this and the following two chapters, occurs by the stress-assisted separation of atomic bonds across the plane of fracture. Fracture occurs at a critical stress level, the value of which depends on factors we discuss in this and the following chapters. As with plastic deformation, tensile fracture can be discussed on the basis of whether or not it is aided by diffusion. When it is not, fracture is a "low-temperature" fracture, the focus of this chapter and Chap. 10. High-temperature fracture, which does involve diffusion, is discussed in Chap. 11.

Low-temperature fracture can take place under loading other than that of simple tension. However, since tensile forces act to open up, and hence propagate, a crack, they are our main (although not exclusive) focus. Moreover, the tensile fracture mode is by far the most common one.

When the stress applied to a material varies with time, the material is subject to cyclical or *fatigue fracture*. That is, it may fail at a stress level below (sometimes

Fracture type	Characteristics
Low-temperature tensile fracture	Separation of atomic bonds under static loading.
High-temperature tensile fracture (creep fracture)	Separation of atomic bonds under static loading; fracture is aided by diffusional flow.
Fatigue fracture	Fracture associated with cyclically applied strains and/or stresses; fracture may take place at stress levels well below those causing tensile fracture.
Embrittlement and static fatigue	Static fatigue is fracture taking place at a stress level less than that causing ordinary tensile fracture; fracture is aided by a hostile environment. Static fatigue is one manifestation of the adverse effect environment may have on material behavior. It and the other adverse effects are forms of material embrittlement.

Table 9.1
Classification of the various types of fracture

well below) that necessitated for tensile fracture or plastic yielding. The important phenomenon of fatigue is discussed in Chap. 12.

Static fatigue describes failure of a material subjected to a constant stress of magnitude below that required to cause tensile fracture. Static fatigue is associated with a hostile or corrosive environment. In effect, stress and the environment act in concert to facilitate atomic bond breaking and crack propagation. A variety of environments can cause static fatigue; each has a particular "claim" on the resulting material failure. Thus, hydrogen in steels and many other metallic alloys gives rise to premature fracture via *hydrogen embrittlement*. Liquid metals likewise cause a reduction in strength and ductility of many metals, and the phenomenon is called *liquid-metal embrittlement*. Aqueous solutions (e.g., common acids or electrolytes) give rise to static fatigue by *stress corrosion cracking*. The various kinds of environmentally assisted fracture are described in Chap. 13, under the generic title *embrittlement*.

It is not uncommon for fracture to take place by a combination of the above-mentioned processes. Thus, corrosion fatigue refers to fatigue accelerated by a hostile environment, and creep fatigue is due to diffusional-flow-assisted fatigue. Although failure interactions of these types can be quite important, we have but little space to devote to them.

Failure (and especially tensile failure) in metals and in thermoplastics can in a simple way be considered the outcome of a competition between the processes of plastic deformation (shear flow) and fracture (tensile separation). That is, if the stress required to initiate permanent deformation by shearing of atomic planes (in metals) or relative displacement of molecular chains (in long-chain polymers) is less than the stress necessary to permanently separate atoms/molecular chains by tensile distortion of their atomic bonds, flow occurs in preference to fracture and vice versa. As plastic deformation is generally a far more preferable "failure" event than fracture, and also because plastic flow preceding fracture markedly increases the work accompanying fracture, this is the preferred result. Hence, any factor (e.g., an increase in temperature or a decrease in strain rate) that tends to reduce a material's yield strength without a commensurate change of the fracture stress is considered beneficial as far as averting fracture is concerned, even though it is undesirable from the viewpoint of material strength. Conversely, increasing a material's resistance to plastic flow, either by microstructural alteration (e.g., by solid-solution hardening of a metal) or by external means (e.g., by imposition of a triaxial stress state), promotes fracture vis-à-vis plastic yielding. We thus see that one of the disconcerting effects of improving a material's fracture resistance is that it is often accompanied by a decrease in its plastic flow resistance. Hence, insofar as fracture behavior is concerned, modern materials development is essentially a quest to develop materials with high resistance to *both* fracture and plastic flow.

As noted, stress state affects a material's relative resistance to plastic flow and fracture. This is because the criterion for fracture is essentially determined by a dilational stress, whereas the onset of plastic flow is dictated by a critical shear stress (e.g., the von Mises or Tresca conditions; Chap. 1). Thus, application of a second tensile stress component raises the level of applied stress necessary to initiate flow, but it also increases the dilational stress promoting fracture. A quantitative measure of the effect is provided by the mean pressure, p (= $(\sigma_1 + \sigma_2 + \sigma_3)/3$), where σ_1, σ_2, and σ_3 are the principal normal (tensile or compressive) stresses. High positive values of p promote fracture. On the other hand, negative values of the mean pressure, connoting predominantly compressive stresses, favor plastic flow compared to fracture.

There are two related material parameters that define a material's ability to resist the crack propagation that results in material fracture. One is the *fracture toughness*, designated by the symbol K. The other is the *toughness*, given the symbol $\mathcal{G}$. Their relationship is discussed in this chapter and in the one following. However, we note here that this concept of fracture toughness and toughness as material properties every bit as much as is a yield strength is useful in engineering design. Methods for evaluating fracture toughness are discussed in this chapter; examples of its usefulness in engineering design are also provided.

The use of K and $\mathcal{G}$ in design is an outgrowth of the development of the engineering science of fracture mechanics. This development was spurred by incidents of the type illustrated in the frontispiece of this chapter, which is a photograph of a Liberty Ship that broke in half while docked. The Liberty Ships were constructed in response to the need for ocean transport vessels in World War II. Construction of conventional transport ships took considerable time. A primary reason was that the hull of the conventional transports consisted of a large number of individual sections riveted to the main ship support structure; placement of the individual sections was time consuming. The design for the Liberty Ships invoked an all-welded hull. That is, the individual hull sections were welded together rather than being mechanically joined. The time to construct a transport ship with a welded hull was much less.

There was one—one very major—problem with the welded-hull construction. Specifically, an advancing crack in such a hull could propagate through the whole of it; thus, the catastrophe illustrated in the frontispiece. Conventional transport ships did not suffer similarly. A crack running in them would generally be "arrested" at the juncture between individual hull sections. The Liberty Ship problem spurred intense research, first at the United States Naval Research Laboratory, where George Irwin and a number of his colleagues played a pivotal role in developing the engineering science of fracture mechanics. This engineering science is the thrust of most of this chapter. Fracture mechanics permits design against fracture, just as a tensile test permits design against permanent deformation.

The design philosophy of fracture mechanics is that the stress to which a material is subjected must be less than the stress required to fracture it. That sounds simple enough; application is more complicated. In particular, the fracture stress is that stress needed to propagate cracks, such cracks being assumed initially present in the material or having developed in it during service. Thus, this critical stress may be much less in "real" structures than in "flaw-free" ones (e.g., a tensile bar). Indeed, under the worst of circumstances, engineered structures may contain cracks of macroscopic dimensions. Thus, our discussion of fracture is couched in terms of its taking place in the presence, and as a result of, these flaws. Depending on flaw severity and material properties, a material that undergoes ductile tensile fracture may fail in a macroscopically brittle fashion with a low expenditure of fracture work even though the microscopic fracture processes accompanying crack extension are the same in both situations.

Other tests, although not as directly useful in design as those resulting from a fracture toughness measurement, have been found worthwhile in design against fracture. These tests frequently yield qualitative (comparative) measures of a material's fracture resistance. One test, the Charpy impact test, is conveniently performed and widely employed. An impact test imposes a stringent set of conditions—a preexisting notch and high rates of loading (straining)—which tend to promote fracture rather than plastic flow. Because of these features, its use is widespread in assessing a material's propensity for brittle fracture.

The microscopic mechanics of crack propagation are described briefly in this chapter and in more detail in Chap. 10. Microscopic fracture mechanisms are the same in flaw-free structures as in flawed ones. Thus, in metals brittle *cleavage fracture* (which involves minimal plastic deformation) and *ductile fracture* (in which extensive local plasticity precedes fracture), and sometimes combinations of these occur in regions immediately adjacent or slightly removed from the tip of a propagating crack. In high-yield strength materials, ductile fracture in the presence of a crack does not necessarily lead to a great resistance to crack propagation; that is, high-strength materials that manifest a measurable reduction-in-area in a tension test may fracture in a "brittle" manner if they contain flaws. A metal's toughness depends on the amount of plastic deformation accompanying crack extension. The physics of this "toughening" are discussed at length in Chap. 10.

The chapter concludes with a few remarks on ceramic fracture. Fracture in them is different than in metals on at least two accounts. First, ceramic compressive fracture strengths are much higher than their tensile fracture strengths. Second, because of processing conditions a wide variety of preexisting crack sizes may be present in a ceramic. These features result in a probabilistic component to design with ceramics.

Before discussing the relationship of fracture toughness to design, we address the question of just how resistant to fracture a material should be. That is, at what stress would a material be expected to fracture by the severing of atomic bonds? Since such bonds are intrinsically strong (van der Waals bonds are an exception), it is not surprising that this "ideal" fracture stress is quite high, and it is so for much the same reason the theoretical yield strength for plastic deformation is high. Likewise, the discrepancy between the ideal and measured fracture stress is reconciled by considerations analogous to those that explain the difference between observed and theoretical plastic yield strengths. That is, "defects" promote fracture at applied stress levels well below the ideal fracture strength. However, the nature of the "defects" responsible for fracture is quite different from those of the dislocations that catalyze plastic flow.

9.2
THE THEORETICAL STRENGTH OF A SOLID

Fracture in a perfect crystal would take place by the simultaneous rupture of atomic bonds across the fracture plane (Fig. 9.1a). It is for this reason that theoretical fracture strengths are very high. Estimates of this strength (σ_{th}) can be made in several ways. For example, knowledge of the bonding energy–atomic volume relation (cf. Chap. 2) leads directly to a relation between the hydrostatic stress and this volume. The integral of the stress-volume curve, from the equilibrium volume to one corresponding to large interatomic separations, can be equated to the material's cohesive energy. The maximum hydrostatic stress required is the theoretical hydrostatic (fracture) stress. This calculation allows the theoretical stress to be expressed in terms of the material's cohesive energy and bulk modulus. Indeed, the bulk modulus is one appropriate measure of a material's inherent resistance to fracture since it is a measure of the material's resistance to atomic bond stretching.

The theoretical fracture strength can also be estimated by considering uniaxial tensile deformation. In doing this, Young's modulus (E), rather than the bulk modulus, is the appropriate elastic constant. Figures 9.1b and c illustrate the stages of bond stretching preceding and accompanying tensile fracture. Increasing stress results first in linear, and then nonlinear but still reversible, bond stretching. At a critical strain, the atomic bonds across the fracture plane rupture simultaneously; the stress at which this occurs is the theoretical strength, σ_{th}. Prior to fracture, atomic bonds are stretched uniformly throughout the volume. However, fracture occurs preferentially on one plane (the fracture plane) as a result of an infinitesimally small defect present there. Following fracture, the elastic strain is recovered throughout the material volume. However, failure is accompanied by the creation of surface area across the fracture plane. The associated surface energy represents permanent irreversible work accompanying the fracture process.

The stress-displacement relationship during fracture is shown in Fig. 9.2. The relation can be approximated as a sine curve—at least up to the displacement $\lambda/4$ at which fracture takes place at the stress σ_{th}. If our "thought" experiment is assumed to take place under a controlled displacement, rather than stress, condition, the stress necessary to continue atomic bond stretching beyond $\lambda/4$ would decrease, as shown in Fig. 9.2. However, the theoretical fracture strength is the same irrespective of the "test" conditions. Estimates of σ_{th} can be made by first realizing that the initial (linear) slope of the stress-displacement curve is related to Young's modulus by

$$\left(\frac{d\sigma}{dx}\right)_{x=0} = \left(\frac{d\sigma}{d\varepsilon}\right)_{x=0}\left(\frac{d\varepsilon}{dx}\right)_{x=0} = E\left(\frac{d\varepsilon}{dx}\right)_{x=0} \qquad (9.1)$$

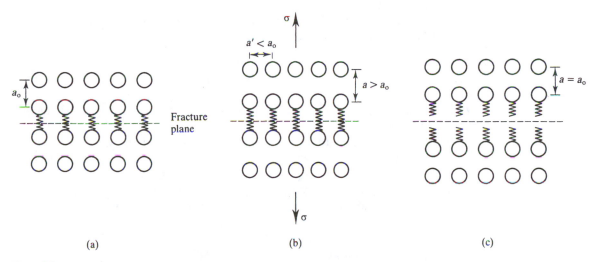

(a) (b) (c)

Figure 9.1

Atomistic model of theoretical tensile fracture. Upon application of a tensile stress (b), the equilibrium structure (a) with interatomic spacing (a_0) is altered. This spacing is increased in a direction parallel to the stress, and decreased in the transverse directions. On fracture (c), the load is released and the atoms revert to their equilibrium spacing, but two surfaces are formed by the fracture event.

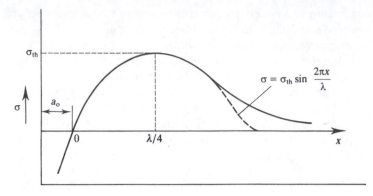

Figure 9.2
The stress-displacement profile (solid line) corresponding to the fracture process of Fig. 9.1. The theoretical fracture strength (σ_{th}) is the maximum in this curve, and is attained when the interatomic spacing is increased by $\lambda/4$. For purposes of estimating σ_{th}, the solid curve is approximated by the dashed sinusoidal curve.

The strain ε is related to the displacement x by $\varepsilon = x/a_0$ where a_0 is the interatomic spacing; thus, $(d\sigma/dx)_{x\to0} = E/a_0$. Using the sinusoidal relationship between σ and x, we also have

$$\left(\frac{d\sigma}{dx}\right) = \left(\frac{2\pi}{\lambda}\right)\sigma_{th}\cos\left(\frac{2\pi x}{\lambda}\right) = \left(\frac{2\pi\sigma_{th}}{\lambda}\right)_{x\to0} \tag{9.2}$$

Setting Eqs. (9.1) and (9.2) equal, the theoretical strength is obtained as

$$\sigma_{th} = \frac{\lambda E}{2\pi a_0} \tag{9.3}$$

It is reasonable to estimate λ as being on the order of a_0 (this corresponds to an elastic fracture strain at $x = \lambda/4$ of 25%) and, therefore,

$$\sigma_{th} \cong \frac{E}{2\pi} \cong \frac{E}{10} \tag{9.4}$$

where the approximation $2\pi \cong 10$ is justified on the basis of the approximate nature of our treatment.

The theoretical strength can also be estimated in a different way. The fracture work (per unit area of fracture plane)[1] is given by the integral of the stress-distance curve (Fig. 9.2). This can be equated to 2γ, where γ is the surface energy (per unit area) created by the fracture event (the factor of 2 arises because two surfaces form). Assuming that the stress is approximated by the sinusoidal representation, we have

$$\int_0^{\lambda/2} \sigma_{th}\sin\left(\frac{2\pi x}{\lambda}\right)dx = \frac{\lambda\sigma_{th}}{\pi} = 2\gamma \tag{9.5}$$

---

[1]The units of this work are N/m or, equivalently, J/m^2.

Substituting λ from Eq. (9.3) into Eq. (9.5), we obtain

$$(\sigma_{th})^2 = \frac{\gamma E}{a_0} \qquad (9.6a)$$

or

$$\sigma_{th} = \left(\frac{\gamma E}{a_0}\right)^{1/2} \qquad (9.6b)$$

Values of σ_{th} calculated from Eq. (9.6b) are presented in Table 9.2 for a number of materials. They are also compared there to $E/10$ ($\cong \sigma_{th}$ from Eq. (9.4)). Although the two approaches lead to somewhat different "numbers" (the discrepancy is not much worse than the ratio $10/2\pi$ used in obtaining Eq. (9.4)), the conclusion—that theoretical fracture strengths of solids are very high—is the same.

That fracture of real materials occurs at stress levels well below σ_{th} is a good indication that it does not take place as described above. The low observed fracture stresses also suggest that some type of defect must be responsible for fracture of most solids. In some cases these defects are preexisting interior or surface cracks that catalyze fracture, for example, because of the concentration of stress in their vicinity. This concentration provides for attainment of a localized stress at the crack tip that is much higher than the nominal one. Sometimes the "flaws" catalyzing fracture are introduced into the material by microscopic or macroscopic plastic deformation. That this can happen is evidence of the interplay between flow and fracture, a relation that makes the study of fracture so intriguing and simultaneously adds considerably to its complexity. We return to such factors in Chap. 10. But first we need to know how to assess the fracture resistance of a stressed material containing cracks.

9.3
CRACK-INITIATED FRACTURE

As noted, preexisting cracks (within or on a material's surface) can propagate, resulting in material fracture. This is a result of the stress concentration associated with a crack. We first discuss how crack propagation in fully brittle solids, e.g., silicate glass at room temperature, takes place as a consequence of such a stress concentration.

Material	$E/10$ (GN/m^2) $[= \sigma_{th}$, Eq. (9.4)]	$(\gamma E/a_0)^{1/2}$ (GN/m^2) $[= \sigma_{th}$, Eq. (9.6)]
Au	7.8	17.7
Cu	12.1	24.8
Ni	22.5	37.2
NaCl	3.7	6.3
MgO	31	40.8
TiC	$\cong 45$	28.5
Si	16	28.5

Table 9.2
Estimated theoretical fracture strengths for several materials

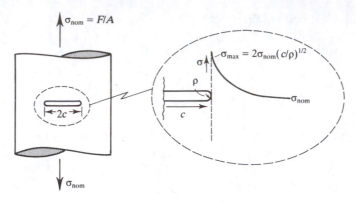

Figure 9.3
The tensile stress in the region close to an elliptical crack in an elastically stressed solid. The crack, of length 2c, is further characterized by its crack tip radius of curvature, ρ. The maximum stress occurs immediately in front of the crack tip; it is $\sigma_{max} = 2\sigma_{nom} (c/\rho)^{1/2}$, where $\sigma_{nom} = F/A$ is the nominal tensile stress. Fracture is possible when $\sigma_{max} = \sigma_{th}$, the theoretical strength.

The stress concentration at a crack tip depends on both the crack shape and size. For an elliptical-shaped interior crack of length $2c^2$ (Fig. 9.3), the maximum stress at the crack tip is

$$\sigma_{max} \cong 2\sigma\left(\frac{c}{\rho}\right)^{1/2} \tag{9.7}$$

where σ is the nominal applied tensile stress and ρ is the radius of curvature of the crack tip (assumed much less than c). We can reasonably assume that fracture occurs when this maximum crack stress tip equals the theoretical strength (Eq. (9.6b)). Equating the two stresses, we find crack propagation should take place at the nominal stress

$$\sigma_F = \left(\frac{\gamma E\rho}{4a_0 c}\right)^{1/2} \tag{9.8}$$

In discussion of fracture, Eq. (9.8) is usually written in its nearly equivalent form

$$\sigma_F = \left(\frac{2\gamma E\rho}{3\pi a_0 c}\right)^{1/2} \tag{9.9}$$

and in what follows Eq. (9.9), rather than Eq. (9.8), is used.

---

[2]The equations developed for an interior crack of length 2c apply also to a surface crack of length c. These equations strictly hold only for fracture occurring in thin sheets. Since the surface of a material does not support a stress normal to it, the stress state at the surface is defined by two principal stresses (a *plane stress* condition). In thick sections, the interior of the material containing a crack is subject to a triaxial stress state which leads to two principal strain components (*plane strain* conditions). Plane strain conditions facilitate brittle fracture more than do plane stress ones, as we describe later. When crack tip deformation accompanying fracture is only elastic deformation, the equations for σ_F differ but slightly for plane stress and plane strain conditions. This is not so when appreciable microscopic or macroscopic plastic deformation accompanies crack extension. The concepts of plane stress and plane strain are elaborated on in Sect. 9.4C. Here, we present results for both conditions, deferring further explanation of their differences to Sect. 9.4C.

Equation (9.9) represents a necessary, but not sufficient, condition for crack propagation. That is, although the stress required to sever atomic bonds at the crack tip must be attained for a crack to propagate, the propagation must also be accompanied by a reduction in system energy. The energetics accompanying crack propagation are illustrated in Fig. 9.4. Two energies are associated with a crack and its extension. One is the surface energy ($= 4c\gamma$ per unit sheet thickness for a crack of length $2c$). As the crack extends, this energy is increased; thus, this factor restricts crack advance. The other energy is a strain energy associated with the elastic strain near the crack tip that is recovered as the crack advances. The elastic strain energy per unit volume (cf. Chap. 2) is $\sigma^2/2E$ where E is the elastic modulus. The affected material volume (per unit sheet thickness) relaxed during crack advance is about $2\pi c^2$. As noted, this elastic energy is released as the crack advances; i.e., it represents a driving force for crack extension. The surface-energy retarding force and the elastic strain energy driving force depend differently on crack size. The resisting force scales linearly, and the driving force quadratically, with c. Thus, at a given stress level, cracks below a certain size will not propagate spontaneously because the surface energy term dominates; that is, the system energy increases with crack extension. On the other hand, at this same stress, cracks above a certain size advance spontaneously because the strain energy released during crack extension is greater than the surface energy created. These ideas are illustrated in Fig. 9.5, which plots the system energy as a function of c at several different stress levels. As shown, a crack having a half-length c^* will not propagate at a low stress (σ_1) because crack advance is accompanied by an increase in energy. However, at the high stress (σ_3) crack advance is accompanied by a decrease in system energy, and the crack spontaneously advances leading to material fracture. The critical stress at which the crack becomes unstable is designated as σ_2 in Fig. 9.5. It represents the stress at which an infinitesimal increase in either crack length or applied stress leads to spontaneous crack advance. The relationship between the critical crack propagation stress (i.e., σ_F) and crack length is associated with the maximum in the energy–crack length curve. That is, spontaneous fracture occurs when the magnitude of the crack-extension and crack-resistant forces (the respective derivatives of the energy–crack length curves) are equal. Letting U_{TOT} equal the total system energy $dU_{TOT}/dc = 0$ at $\sigma = \sigma_F$, and we obtain for the critical condition

$$\frac{d}{dc}\left[4c\gamma - \frac{\pi\sigma^2 c^2}{E}\right]_{\sigma=\sigma_F} = 0 \qquad (9.10)$$

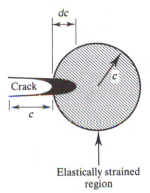

dc

Crack

c

Elastically strained region

Figure 9.4
The energetics accompanying crack extension. As the crack grows by dc, surface area is created. The associated energy represents irreversible work retarding crack extension. Concurrent with crack extension is recovery of elastic strain energy by relaxation of stretched atomic bonds above and below the fracture plane (cf. Fig. 9.1). The elastic energy is relaxed in a circular area having diameter about equal to $2c$. The relaxation in stored elastic energy promotes crack extension.

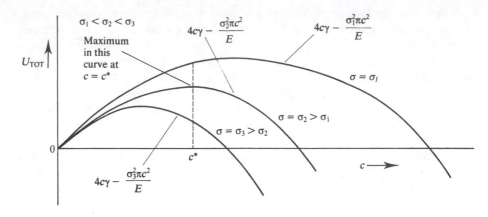

Figure 9.5
Total (surface plus stored elastic) system energy as a function of half crack length at three stress levels. With increasing stress, the general level of the curves decreases owing to the increased crack-extension driving force. A crack with $c = c^*$ does not spontaneously grow at the stress σ_1, for the system energy would increase if it did. Conversely, at σ_3 the crack is unstable and grows spontaneously owing to the decrease in system energy with increasing c. At the stress σ_2, the crack is in unstable equilibrium, i.e., $dU/dc = 0$ at $c = c^*$, and an infinitesimal increase in stress or crack length causes the crack to extend spontaneously.

and the relation between the crack propagation stress and crack size is obtained as

$$\sigma_F = \left(\frac{2\gamma E}{\pi c}\right)^{1/2} \tag{9.11}$$

In addition to satisfying Eq. (9.9), Eq. (9.11), the Griffith condition, must also be met if fracture is to occur.[3]

The fracture stress is the greater of Eqs. (9.9) and (9.11). Comparing them shows that Eq. (9.9) will define the fracture stress if $\rho > 3a_0$ whereas when $\rho < 3a_0$, Eq. (9.11) does so. However, for fracture taking place in fully brittle solids, there is little difference in the stresses predicted by the two equations. This relates to the shape of the crack tip in brittle solids; its radius of curvature is very small, on the order of the atomic spacing. Hence, for all intents and purposes the Griffith equation, Eq. (9.11), can be used to define the relationship between crack length and fracture stress. When plastic deformation in the crack tip vicinity accompanies crack propagation, this is no longer so. This deformation "blunts" the crack tip, increasing ρ, and raising σ_F. However, if this crack tip plasticity is not too extensive, a modified form of the Griffith equation can still be used to estimate σ_F.

The approaches leading to Eqs. (9.9) and (9.11) represent different, but complementary, ways of obtaining a fracture criterion. The approach leading to the Griffith equation is one of "fracture work" (or force). In essence, this fracture de-

---

[3]Equation (9.11) holds for thin sheets (plane stress conditions). For thick samples (plane strain) $\sigma_F = (2\gamma E/\pi c(1-\nu^2))^{1/2}$ where ν is Poisson's ratio. For fracture of brittle solids, the two fracture stresses are nearly the same. When fracture of other-than-brittle solids is concerned, the two fracture stresses can be quite different; we return to the point later.

scription states that fracture occurs when the energy release rate accompanying crack extension is greater than the inherent force resisting this extension, and vice versa. For fracture absent plastic deformation, the crack-resisting term is surface energy and the material "toughness," $\mathcal{G}_c$, is defined as 2γ. Using this definition of $\mathcal{G}_c$, Eq. (9.11) can be written as[4]

$$\sigma_F = \left(\frac{E\mathcal{G}_c}{\pi c}\right)^{1/2} \tag{9.12}$$

An advantage of using Eq. (9.11) in the form of Eq. (9.12) is that Eq. (9.12) can also be used to specify the fracture stress when plastic deformation accompanies crack extension. When this is the situation, though, $\mathcal{G}_c$ is no longer equal to 2γ.

The approach leading to the fracture criterion of Eq. (9.9) can be described as a "critical stress intensity" approach. Simply stated, in this description fracture occurs when the stress intensity at the crack tip reaches a critical value. Regardless of the exact crack shape and irrespective of whether or not plastic deformation accompanies crack propagation, the crack tip stress scales with the nominal stress and the square root of the crack length (Eq. (9.7)). Defining the product of these two terms as K^*, we have $K^* = \sigma(c)^{1/2}$. It is customary to take the proportionality constant relating stress intensity to σ and $(c)^{1/2}$ as $(\pi)^{1/2}$ rather than unity. The stress-intensity approach states that when $K \ (= (\pi)^{1/2}K^*)$ attains a critical value (K_c—a property called the fracture toughness[5]), fracture takes place. Thus,

$$\sigma_F = \frac{K_c}{(\pi c)^{1/2}} \tag{9.13}$$

The two approaches can be reconciled by comparing Eqs. (9.13) and (9.12). We see that they are the same for plane stress conditions if

$$K_c = (E\mathcal{G}_c)^{1/2} \quad \text{(plane stress)} \tag{9.14}$$

An almost identical expression applies to plane strain conditions;

$$K_c = \left[\frac{E\mathcal{G}_c}{(1 - \nu^2)}\right]^{1/2} \quad \text{(plane strain)} \tag{9.15}$$

On the basis of Eqs. (9.14) and (9.15), it would appear that the plane strain and plane stress fracture toughnesses are almost the same. However, in general they are not for the stress state profoundly affects the toughness ($\mathcal{G}_c$) of some materials (e.g., most metals). We further address this topic in Sect. 9.4C.

Both approaches to fracture are useful. For example, discussion of fracture in terms of $\mathcal{G}_c$ can be related directly to the irreversible work accompanying crack propagation. This is the approach we take in discussing material toughness (Chap. 10). Knowledge of $\mathcal{G}_c$ permits K_c to be calculated via Eq. (9.14) or (9.15). The fracture toughness approach is of more direct engineering relevance because K_c

---

[4]Again Eq. (9.12) applies to plane stress conditions. Incorporation of the factor $(1 - \nu^2)^{1/2}$ in the denominator of the right-hand side of Eq. (9.12) leads to an expression for σ_F for plane strain conditions.

[5]Terminology in the fracture community has evolved with time. Not too long ago, $\mathcal{G}_c$ was called the fracture toughness and K_c the critical stress-intensity. However, for reasons that remain obscure, it is now common to call $\mathcal{G}_c$ the *toughness* and K_c the *fracture toughness*. This shortens the vocabulary, but it is questionable whether the current terminology clarifies concepts.

can be determined by appropriate mechanical tests. As is discussed in Sect. 9.4E, values of fracture toughness have a direct bearing on estimating safe operating stresses below which fracture of a structure will not take place.

Fracture involving only elastic deformation at the tip of a crack has been discussed to this point. Preceding discussions, therefore, strictly apply only to those materials whose plastic yield strengths are greater than their theoretical fracture strengths. While this may be the situation for "completely brittle" solids such as glass, it is not so for most materials that fracture as a result of the stress intensification at the tips of cracks they may contain. That is, even though these materials may have yield strengths greater than the crack propagation stress (otherwise they would flow macroscopically prior to fracture), the high localized stress at the crack tip promotes plastic yielding in its vicinity. The extent of this yielding depends on the relative values of the yield and the theoretical fracture strengths. Further, the degree of crack-tip plastic deformation can be strongly dependent on temperature for those materials (e.g., the bcc transition metals) whose yield strengths vary significantly with temperature. These ideas are illustrated in Fig. 9.6. For truly brittle solids, fracture is associated only with elastic deformation processes (Fig. 9.6a). The stress at the crack tip is σ_{th}, the (elastic) strain here is about σ_{th}/E, and the crack tip radius of curvature is ca. $3a_0$. For somewhat less brittle materials (e.g., the bcc transition metals at low temperatures), the stress in front of the crack tip exceeds that required to effect plasticity there (Fig. 9.6b). The extent of this deformation is limited, the plastic strain being comparable to the elastic strain. Further, the crack tip stress is still about σ_{th}. The major distinction between linear elastic fracture and fracture involving microscopic plasticity of this type is that plastic deformation blunts the crack tip; i.e., the tip radius is increased above $3a_0$. When the plasticity accompanying crack propagation is sufficient to blunt the crack tip even further (as caused, for example, by increasing the temperature for a bcc transition metal) general yielding in front of the crack takes place (Fig. 9.6c). As indicated in this figure, the stress in front of the crack tip is equal to σ_y, the yield strength, out to a distance r_y from the crack tip, and the crack tip radius of curvature is blunted greatly compared to what it is for elastic fracture. The situation of Fig. 9.6c, therefore, corresponds to one in which the fracture work, $\mathscr{G}_c$, is much greater than for the situations of Figs. 9.6a and b.

Any plastic deformation accompanying crack extension represents additional irreversible work *expended*, rather than recovered, as is the case for elastic deformation. This plastic work hinders crack extension and *a fracture stress greater than that predicted by Eq. (9.11) is required*. How much greater depends on the extent of the plastic deformation, and this can be discussed further with reference to Fig. 9.6.[6] The plastic zone extends a distance ($\cong \rho$) above and below the fracture plane. For limited crack tip plasticity, i.e., when ρ is no greater than about $30a_0$, the Griffith equation in a modified form can be used to express the fracture stress. This can be done because the width of the plastic "ribbon" (Fig. 9.6b) is small and thus the fracture work scales linearly with crack length in the same manner as does the surface energy work. Thus, we replace γ in the Griffith equation by ($\gamma + \gamma_p$), with γ_p representing the plastic work done per unit surface area of crack advance. Likewise,

---

[6]We again caution that the current discussion applies only to materials such as metals and thermoplastics that yield plastically prior to their fracture.

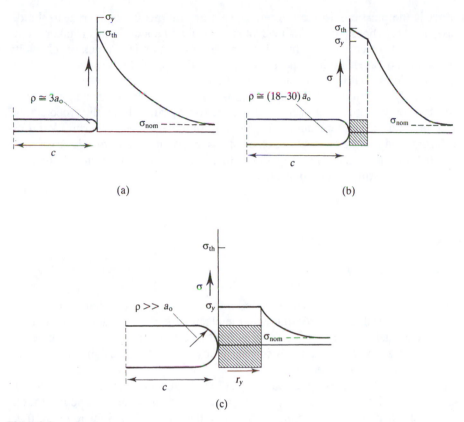

Figure 9.6
Crack propagation can be divided into whether it is entirely elastic in nature (a) or, as is common for metals and polymers, involves limited crack tip plasticity (b). For (a), σ_{th} is below σ_y (the yield strength), and $\mathcal{G}_c$ is defined by $\sigma_{th}\rho\varepsilon_f$ where $\varepsilon_f = \sigma_{th}/E$ and ρ compares to an atomic size. In (b), σ_y is slightly less than σ_{th} and a small region of plastic deformation exists in front of the crack tip. The fracture stress and strain are nearly the same as for elastic fracture, but ρ is somewhat higher owing to crack tip blunting. Thus, $\mathcal{G}_c(\cong \sigma_{th}\rho\varepsilon_f)$ is greater for the case of (b) than for (a). As σ_y becomes progressively less than σ_{th}, more extensive crack tip plasticity takes place. $\mathcal{G}_c$ is still defined by the product $\sigma\rho\varepsilon_f$, but σ is well below σ_{th} and ρ and ε_f are much above their values corresponding to (a) and (b).

the toughness, $\mathcal{G}_c$, of Eq. (9.12) now becomes $2(\gamma + \gamma_p)$. As mentioned, γ_p depends on the width of the plastic ribbon (and, thus, on ρ) and the extent of the strain within it. Thus, γ_p is a function of the yield strength and, if this strength is temperature sensitive, of temperature as well.

Metal toughness always relates to the irreversible work of crack growth. For truly brittle solids, the work is manifested in the creation of new surface. When limited plasticity at the crack tip takes place, this work is augmented by the plastic work as exemplified in the term γ_p. And when general plasticity accompanies crack extension, almost all of the fracture work is due to crack tip plastic deformation. Estimates on the toughness can be made by considering it proportional to the product of the yield stress (σ_y) and the plastic strain to fracture at the crack tip (ε_f). This product is the plastic deformation work per unit volume. To convert it to the dimensions of toughness (e.g., J/m^2) we must multiply by a length; this is taken as the crack tip radius of curvature. Thus, $\mathscr{G}_c \cong \sigma_y \varepsilon_f \rho$. For generalized crack tip plasticity $\sigma_y \ll \sigma_{th}$; however, both ρ and ε_f are increased appreciably compared to their values pertaining to brittle fracture propagation.

The "physics" underlying toughening is discussed later in this chapter and in Chap. 10. For the remainder of this chapter we focus primarily on the "mechanics" of fracture and the implementation of fracture toughness in engineering. We must also further address the concepts of plane stress and plane strain, for the stress state attending crack propagation profoundly influences the values of $\mathscr{G}_c$ and K_c for those materials manifesting crack tip plasticity.

9.4
FRACTURE MECHANICS

A. Design Philosophy

Materials "fail" by either their unintended plastic deformation or their fracture. Of the several materials classes, design with metals and their alloys (and some thermoplastics) considers both possible "failure" modes. Ceramics, thermosets, and composites of these materials fracture before they flow. Thus, their design involves only insuring that they do not fracture.

How is this design done? Equation (9.13) indicates that the crack propagation stress varies with flaw size as $\sigma_F \simeq K_c/\sqrt{c}$ where K_c is the material fracture toughness. Since, on a physical basis, K_c devolves on the work ($\mathscr{G}_c$) associated with fracture propagation, it (and $\mathscr{G}_c$) are material properties every bit as much as, for example, the yield strength is. Further, K_c, just like the yield strength, is microstructure sensitive. Provided K_c can be measured (means for doing this are discussed in Sect. 9.4D) the fracture stress can be approximated (Eq. (9.13)). Equation (9.13) applies to a through-crack in a sheet. Not all cracks that may be present in a material are so simply shaped. Some may be internal penny-shaped cracks; others may be elliptical-shaped flaws on the material surface. The shape of the crack influences its associated stress concentration. To take this into account the relationship between the flaw size, K_c and σ_F is written as

$$\sigma_F = \frac{K_c}{(\alpha \pi c)^{1/2}} \tag{9.16}$$

where the constant α ($= 1$ for a through-crack) accounts for the variation in flaw size geometry. The mechanics community has compiled a number of handbooks that provide values of α for almost every conceivable crack geometry and loading condition.

Design against fracture also presupposes that cracks are present in a structure. The challenge is to know how large these may be, and here is where a measure of engineering judgment is required. For example, the largest crack that may be present can be taken as that crack size that would go undetected by nondestructive testing before the structure is placed in service. Or the size may be estimated as the maximum-sized fatigue crack (Chap. 12) that could develop during the intended life of the component. In either or any case, the selected operating stress must be less (adjusted by an appropriate safety factor) than σ_F calculated on the above basis.

Metals are differently treated. As noted, their design entails that they neither plastically flow nor fracture. A simple outline of the design procedure for them is illustrated in Fig. 9.7. As can be seen in the figure, both flow and fracture are considered in this inherently iterative procedure. However, one consideration usually dominates. In high-strength metals (those for which $\sigma_y \gtrsim E/150$), σ_F is usually less than σ_y so it is design against fracture that is paramount. Conversely, for low-strength metals ($\sigma_y \lesssim E/300$) design against flow is more important. When the material yield strength lies between $E/150$ and $E/300$, both fracture and flow are of concern and the iterative design suggested in Fig. 9.7 is fully employed.

So we can design against crack propagation if we have a good idea of the size and shape of preexisting surface or interior cracks and we know the material fracture toughness. We describe measurement of this property in Sect. 9.4D. But before we do this, more details about crack propagation must be provided. In addition, we must direct more attention to plane stress and plane strain conditions for while Eqs. (9.14) and (9.15) that predict fracture propagation stresses under plane stress and plane strain conditions are similar in form, the values of K_c (and $\mathcal{G}_c$) used in them are different. That is, fracture toughnesses depend on stress state. As a corollary, K_c and $\mathcal{G}_c$ also depend on sample or component size.

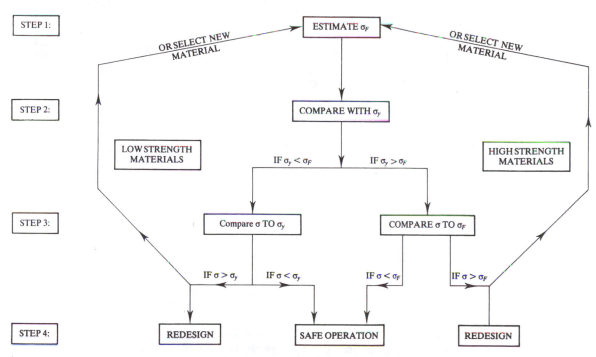

Figure 9.7

Schematic of the logic employed in design against failure. The material fracture stress, σ_F, is compared to its yield strength, σ_y (Steps 1 and 2). Whichever is less is then compared to the design operating stress, σ (Step 3). If σ is less than both σ_y and σ_F, safe operation is usually assured; if not, redesign or selection of a new material is called for (Step 4). For the latter circumstance, the procedure is repeated. Note that if $\sigma_y < \sigma_F$, design is based on preventing plastic flow; if $\sigma_F < \sigma_y$, design is based on preventing fracture.

B. Crack Propagation Modes

Fracture requires crack advance under the influence of external and/or internal stresses. The resulting stress states can be quite complex. Analysis is further complicated by the myriad of crack shapes and their orientations with respect to the stresses causing them to grow. Thus, thorough analysis of fracture is complicated. However, it has been shown that there are only three basic modes of crack propagation, the general case being some combination of the basic modes. Hence, discussion of fracture is simplified by considering only the primary fracture modes; they are illustrated in Fig. 9.8.

Mode I fracture (Fig. 9.8a) formed the basis of previous discussions. In this fracture mode a tensile stress acting normal to the crack surface serves to propagate

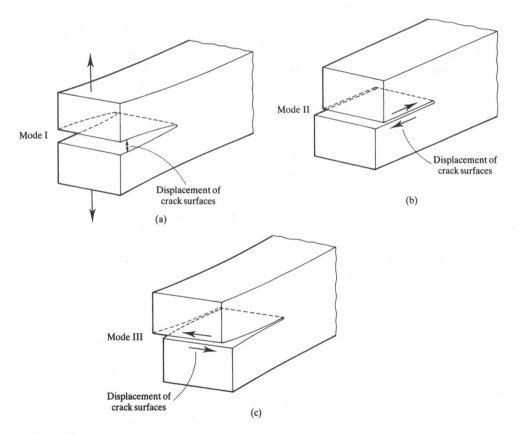

Figure 9.8
The three primary macroscopic fracture modes. (a) Mode I fracture (the tensile opening mode) is the most common tensile fracture. It is caused by a force tending to open up the crack and its tip. The crack propagates in a direction normal to the applied stress. (b) In Mode II fracture (a shear fracture), separation of the bar is achieved by propagation of the crack in a direction parallel to the sense of the applied shear stress. (c) In Mode III fracture (antiplane strain fracture), a shear stress causes displacement of the crack surfaces. This results in the crack propagating in the direction normal to the shear stress, and produces a tearing type of fracture.

the crack in a direction normal to the stress. In Mode II fracture (Fig. 9.8*b*), a shear stress is applied normal to the leading edge of the crack, which propagates in a direction parallel to the sense of this stress. Antiplane strain, or Mode III, fracture (Fig. 9.8*c*) takes place when a shear component of stress is applied parallel to the leading edge of the crack. The displacement of the crack surfaces parallels that of the applied stress, but the crack propagates normal to this direction.

The three primary fracture modes will, as a result of the differing work accompanying crack extension, have different values of fracture toughness and toughness. These are denoted as K_I, K_{II}, and K_{III} (or $\mathcal{G}_I$, $\mathcal{G}_{II}$, and $\mathcal{G}_{III}$), respectively. The general case of fracture involves some combination of the three modes (a simple example is provided in Fig. 9.9), and the fracture toughness/toughness then is an appropriate average of the respective K and $\mathcal{G}$ values. By far the most important fracture mode in engineering is the tensile opening mode; because of this Mode I fracture is the focus of most of our discussion. However, Fig. 9.9 does indicate that when a crack propagates in a direction not normal to that of an applied tensile stress the propagation involves a component of Mode II fracture. This becomes important in defining the fracture toughness of materials whose fracture paths reflect such crack path diversions. The fracture surfaces of certain fiber composites, for example, display such characteristics.

C. Plane Stress and Plane Strain

A thin sheet containing a surface crack subjected to a tensile force facilitating Mode I fracture is illustrated in Fig. 9.10.[7] The tensile stress (σ_{x2}) variation in the x_1 direction has been illustrated previously in Fig. 9.3 for the situation pertaining to crack tip elastic deformation. The stress variation has also been approximately

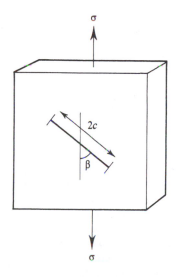

Figure 9.9
When a crack is inclined at an angle other than 90° to the tensile axis, both Mode I and Mode II fractures contribute to crack advance. For the case shown, Mode I fracture dominates for 60° < β < 120°; Mode II fracture dominates for other values of the angle β.

---

[7]In Fig. 9.10 the coordinate axes x_1, x_2, x_3 are used in preference to the conventional x, y, z notation. Doing this reduces confusion between the tensile stress in the x_2 direction (σ_{x2}) and the material's tensile yield strength, σ_y.

shown in Figs. 9.6*b* and *c* for the case when plastic deformation accompanies crack propagation. The stress σ_{x2}, however, is not the only one acting on the material close to the crack. A constraint effect, rather like that occurring in thin brazed joints, exists in this region and this restricts deformation in the x_1 direction. As a result, an additional tensile stress component (σ_{x1}) develops. The variation of σ_{x2} and σ_{x1} with distance in front of the crack is shown schematically in Fig. 9.10 for the elastic case; σ_{x1} is zero at the crack tip (it is a free surface and cannot support a tensile stress), but increases in front of it. Beyond a certain distance, σ_{x1} scales both with σ_{x2} and Poisson's ratio (i.e., $\sigma_{x1} \cong \nu\sigma_{x2}$) since the term $\nu\sigma_{x2}$ is proportional to the elastic deformation that would take place absent the constraint. No tensile stress exists in the x_3 direction; that is, σ_{x3} is zero because the surface does not support a tensile stress and the sheet is assumed "thin" enough that this situation prevails throughout the sheet thickness. Hence, the deformation leading to Mode I fracture in thin sheets is referred to as plane stress (i.e., $\sigma_{x1},\sigma_{x2}, \neq 0$, $\sigma_{x3} = 0$) deformation.

The stresses are altered when σ_{x2} attains a value sufficient to induce plastic flow in front of the crack. Nonetheless, the work accompanying crack extension with crack tip plasticity under plane stress can be discussed using concepts previously introduced. Crack tip plastic deformation results in an appreciable thinning of the material in the x_3 direction (Fig. 9.11), as a thin sheet is not constrained from contraction in this direction. The plastic deformation zone in front of the crack tip extends a distance comparable to r_y (cf. Fig. 9.6*c*) and also a distance above and below the crack plane by this same amount; i.e., the crack tip is blunted to the extent that $\rho \cong r_y \cong t$ (Fig. 9.11). Since $\mathcal{G}_c \cong \sigma_y\varepsilon_f\rho$ we have for the fracture of thin sheets

$$\mathcal{G}_c \cong \sigma_y\varepsilon_{of}t \tag{9.17}$$

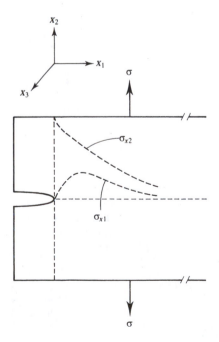

Figure 9.10
Tensile stress variation in front of a crack in a thin sheet when only *elastic* deformation takes place in front of the crack. The primary tensile stress component (σ_{x2}) varies with distance from the crack tip as $r^{1/2}$. The constraint provided by the crack gives rise to a secondary tensile stress, σ_{x1}. It is zero at the crack surface, but rises at distances removed from it. At larger distances, σ_{x1} scales with σ_{x2}. Thus, a state of biaxial stress develops in front of a crack in a thin sheet.

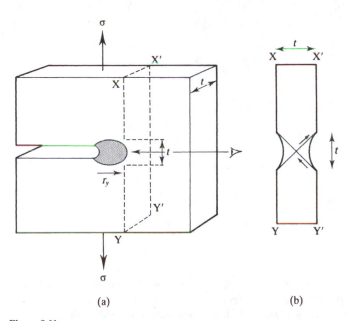

Figure 9.11
**Crack propagation accompanied by plastic deformation in a thin sheet. (a) The plas-
tic zone extends a distance r_y in front of the crack; r_y is comparable to the sheet thickness t. Deformation in the plastic zone in front of the crack is accompanied by a
thinning down of the sheet in the thickness direction. For example, by observing the
projection XX′YY′ (b), we see that the deformation results in a strain in this direction and a consequent thinning of the sheet.**

where ε_{of} is the plane stress fracture strain. Hence, toughness of a thin sheet scales
linearly with sheet thickness; this is shown in a plot of $\mathcal{G}_c$ versus sheet thickness (at
small t) in Fig. 9.12.

   Two other aspects of plane stress fracture deserve attention. First, the toughness of very thin sheets can be quite low. For example, this is the reason that metallic foils are so easily torn. (This type of low-energy fracture is accomplished at
stress levels below σ_y.) Second, it is clear that *plane stress toughness* is *not* a material property. To be sure it is related to some such (e.g., to σ_y), but it is also a function of sample size and for this reason $\mathcal{G}_c$ (or $K_c = (E\mathcal{G}_c)^{1/2}$ for plane stress fracture)
is not generally used in engineering design. Instead, toughnesses and fracture
toughnesses obtained from tests of thick sheets are. We shall see that this corresponds to plane strain conditions.

   Before discussing plane strain fracture, it is worthwhile considering the appearance of a plane stress fracture surface. The thinning down of the sheet preceding final shear fracture results in a slant-type fracture (Fig. 9.11b) rather like the
shear lips observed on the rim of many fractured tensile bars. Indeed, the appearance of slant fracture under Mode I loading is a clear indication of plane stress fracture. As the sheet thickness is increased, however, a region of macroscopically (but
not microscopically) flat fracture is observed first in the center of the sheet and then,
as the thickness of the sample is further increased, in regions removed from the center. The relative percentage of flat fracture area also increases with increases in

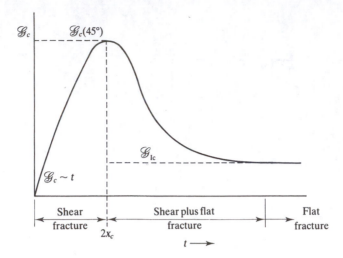

Figure 9.12
**Material toughness as a function of sheet or plate thickness (*t*). For thin sheets, frac-
ture is accomplished by plastic thinning in front of the crack, followed by shear fracture. For this case, the volume of plastically deformed material scales with *t*, and
since this is a measure of the plastic work accompanying fracture, $\mathcal{G}_c \sim t$ for thin
sheets. A maximum in $\mathcal{G}_c$ [$\mathcal{G}_c(45°)$] is obtained at a sheet thickness $2x_c$. At this thickness, the center of the sheet first begins to experience plane strain conditions. The latter is characterized by a smaller plastic-zone size than for plane stress and a triaxial,
rather than biaxial, stress state. Both effects result in a diminution in toughness,
which becomes more pronounced as the sheet thickness increases, i.e., as the fraction
of the sheet subject to plane strain conditions increases. Toughness approaches a limiting value, $\mathcal{G}_{Ic}$, at large plate thicknesses. Thus, we see that absent plane strain conditions, toughness depends on both the material and its geometry. In contrast, $\mathcal{G}_{Ic}$ is a
material property.**

sheet thickness. This flat area corresponds to fracture taking place under plane strain
conditions.

   To illustrate plain strain fracture we first consider fracture as it takes place in a
moderately thick sheet (Fig. 9.13*a*). Regardless of sample thickness, the sample surfaces are stress-free and this results in a slant or shear fracture there (Fig. 9.13*b*).
However, the stress state in the interior of a moderately thick sheet differs from that
on its surface. This leads to a different type of fracture and a different toughness. In
particular, at the midplane of a thick sheet, and for some distances to either side of
it, a deformation constraint (rather like that causing the stress σ_{x1} of Fig. 9.10) exists. This gives rise to a tensile component of stress in the x_3 direction which, for
elastic deformation, scales with $\nu(\sigma_{x1} + \sigma_{x2})$ (this product being proportional to the
strain in the absence of such a constraint). The variation of σ_{x3} across the sample
thickness is shown in Fig. 9.14; σ_{x3} is zero at the sample edges and reaches its maximum value (= $\nu(\sigma_{x1} + \sigma_{x2})$) at some distance removed from it. The distance x_c at
which this occurs represents a transition from plane stress to plane strain conditions.
Based on previous discussions, x_c scales with r_y. Notice, too, that if the sheet thickness, t, is $< 2x_c$, conditions are those of plane stress; conversely if $t >> 2x_c$, they are
mostly plane strain.

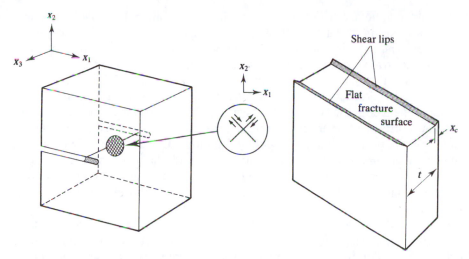

Figure 9.13
**Fracture in a plate, the center of which is subject to plane strain. (a) The sheet sur-
faces, unable to support a stress, deform under plane stress and a strain in the x_3
(thickness) direction (the shaded area) is produced. Thus, shear fracture is observed
at the sheet surface. However, the center portion of the sheet (the cross-hatched circle) is subject to plane strain conditions. As shown by the insert (a projection of the
deformation in the x_1-x_2 plane), plastic deformation in front of the crack tip takes
place only in this plane; no contraction in the x_3 direction is possible at the sheet center because of the constraint provided by material to either side in the x_3 direction.
(b) The fracture appearance of a cracked plate of finite thickness. The shear lips on
the surfaces are a result of plane stress fracture. The central flat fracture surface is
characteristic of plane strain fracture in this region. As the sheet thickness increases,
the percentage of plane strain fracture does likewise. (The distance x_c, over which
shear lips are found, is essentially independent of sheet thickness.)**

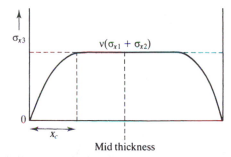

Figure 9.14
**The variation of σ_{x3} across plate thickness for a finite-sized plate subject to elastic
crack tip deformation. σ_{x3} is zero at the stress-free surface and rises to its maximum
value [$= \nu(\sigma_{x1} + \sigma_{x2})$] at a distance ca. x_c from the plate surface. Thus, plane stress
conditions obtain over the plate fraction $2x_c/t$, and plane strain over a fraction
$[1 - (2x_c/t)]$. When plastic deformation takes place in front of the crack tip, the
factor 1/2 replaces ν in the expression for σ_{x3}.**

When the zone in front of the crack tip in plane strain conditions experiences plastic deformation, the details of the stress distribution become more complex. Nonetheless, the most important results can be presented. First, the triaxial stress state of plane strain reduces the plastic zone size in comparison to the plane stress zone size,[8] and the plane strain fracture ductility is also reduced. These effects are accompanied by a reduction in material toughness. Second, the triaxial stress state is most pronounced at the boundary between the plastic and elastic zones. Since this stress state corresponds to the highest mean pressure (which is conducive to tensile fracture), crack propagation may occur by fracture at the elastic-plastic boundary, with subsequent linkup of these secondary cracks with the main crack tip. This effect may also alter material toughness. As a result of both factors, toughnesses obtained from plane strain Mode I fracture can be well below those pertaining to plane stress conditions.

How then does fracture take place in a "thick" sheet and what values of toughness/fracture toughness may be expected in such a situation? Figure 9.15 illustrates fracture progression in a finite-thickness sheet sample. Fracture advance in the sheet center proceeds more rapidly than at the surface because of the plain strain conditions prevailing in the sheet center and the lesser plane strain toughness. The flat interior fracture surface is bounded on the sheet edges by slant fracture. The shear lips form there because of the plane stress conditions at the surface; formation of the lips is further facilitated by the greater crack advance in the sheet interior. Again, the analogy with fracture in a tension test is apt. Shear lips form during tensile fracture both because the surface of the bar is incapable of supporting a stress

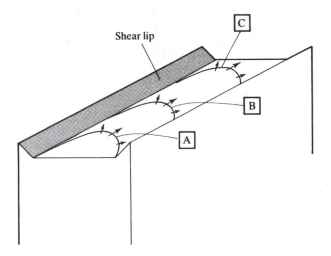

Figure 9.15
Fracture progression in a finite-sized sheet. The central portion of the sheet fractures under plane strain, and the surface under plane stress, conditions. Crack advance is more rapid in the plane strain region as a result of the smaller plastic zone size there and the triaxial stress state that exists. Thus, the crack assumes the shapes shown in this figure; the designations A, B, and C represent crack position and shape as the crack progressively advances through the sheet.

[8]Best estimates are that the zone size is reduced by ca. 50 to 67 percent.

and because a thin surface rim is formed around the previously fractured test bar interior.[9] As is evident from these discussions, the greater the plate/sheet thickness the larger the fraction of macroscopically flat fracture surface; the shear lips are restricted to a thickness approximately equal to the characteristic distance, x_c, of Fig. 9.12.

The variation of $\mathcal{G}_c$ with plate thickness has been shown in Fig. 9.12. At a certain critical thickness ($\cong 2x_c \cong 2r_y$), $\mathcal{G}_c$ has a maximum value called $\mathcal{G}_c(45°)$ in recognition of the slant fracture surface appearance of plane stress fracture. For plate thicknesses greater than ca. $2x_c$, the central region of the sheet undergoes plane strain fracture and the percentage of this flat fracture increases on testing progressively thicker sheets. Indeed, the thickness of material undergoing plane stress fracture remains at approximately $2x_c$. Accompanying the gradual change in fracture mode is a reduction in toughness. With further increases in plate thickness, the percentage of flat fracture surface approaches 100 percent and the toughness approaches a minimum value called $\mathcal{G}_{Ic}$, the plane strain toughness. This value of toughness (or the plane strain fracture toughness, $K_{Ic} = [E\mathcal{G}_{Ic}/(1-v^2)]^{1/2}$) is preferred for design because it represents a conservative approach to design.

How does one know whether a fracture toughness test determines K_{Ic} and $\mathcal{G}_{Ic}$ or values of K and $\mathcal{G}$ greater than these? Visual inspection of the fracture surface provides a qualitative answer. That is, if the fraction of flat (plane strain) fracture is large, K_c is close to K_{Ic}. This assessment is equivalent to estimating the plastic zone size, for when this size is much less than the plate thickness, plane strain conditions predominate, and when it is comparable to the thickness, plane stress conditions do. This idea can be used to quantitatively stipulate conditions for plane strain testing. The stress in front of the crack tip (or more precisely, in front of the elastic-plastic zone boundary) varies with distance, r, from it as

$$\sigma_{x2} = \frac{K}{(2\pi r)^{1/2}} \tag{9.18}$$

where K is the stress intensity factor. The parameter r_y is estimated by setting $\sigma_{x2} = \sigma_y$ and $K = K_c$. Thus, r_y (plane stress conditions) is obtained as

$$r_y = \frac{K_c^2}{2\pi\sigma_y^2} \tag{9.19}$$

This estimate on r_y for plane stress is somewhat low; the actual plastic zone size R (greater than that for plane strain conditions) is approximately $2r_y$ and thus, R can be compared directly with the plate thickness. If R is small in comparison to t, plane strain conditions may be assumed and the K_c and/or $\mathcal{G}_c$ determined is K_{Ic} and/or $\mathcal{G}_{Ic}$. We see that the required sample thickness depends on the material's fracture toughness.[10] Tougher materials necessitate thicker plates to measure a plane strain fracture toughness because the plastic zone size is larger in them. Thus, initial estimates of the necessary sample thickness require educated a priori "guesses" of the material's fracture toughness. Fortunately, accumulated experience is such that good

---

[9]Further description of ductile tensile fracture is provided in Sect. 10.11.
[10]Empirical correlations have yielded the guideline, $t \geq 2.5 \, (K_{Ic}/\sigma_y)^2$, as one that must be met for a valid determination of K_{Ic}.

estimates of this type can be made and this reduces the effort resulting from the inherently iterative nature of the procedure. Moreover, standard-sized and -shaped samples have been developed for such testing, and accurate expressions for the stress intensity as related to crack size and shape and sample geometry are available. Additionally, corrections for the effect that the plastic-zone size has on the effective crack length are incorporated into some of the test procedures. Some simple fracture toughness testing schemes are described in the following section. After that we (finally!) provide several examples of the use of fracture mechanics in "real" engineering situations.

D. Test Methods

A number of standard procedures for determining K_{Ic} have evolved since this approach to fracture control was developed. Several standard samples and the formulae used to derive fracture toughness values on fracturing them are provided in Fig. 9.16. Of the schemes illustrated in Fig. 9.16, the compact tension and the three-point bend test are most widely employed. Detailed discussion of fracture toughness testing lies outside our scope. However, references listed at the chapter end—particularly that of Anderson—provide information for those interested in knowing more about the topic. Here only the essentials of the test procedures are provided. Subsequent to fabrication of the test piece containing a premachined notch of length c', the notch is sharpened so as to produce a crack tip that will emulate those likely found in service. The honing of the tip can be accomplished by cyclical prestressing of the sample so as to induce a fatigue crack; alternatively, if the sample is susceptible to thermal shock, a sharp crack tip can be formed by subjecting it to a rapid temperature change. In any case, the crack length is increased (from c' to c) by such a process and, as shown in Fig. 9.16, this is the value of crack length used in subsequent analysis; thus, accurate measurement of c is required. The sample is then loaded in a manner to open up and extend the crack, and a load (F) versus sample deflection (δ) curve is obtained. Prior to crack propagation, the F-δ curve is essentially linear, and knowledge of the slope of this curve (or its inverse, the compliance) is required for analysis. Compliances can be obtained empirically or through the use of formulae derived from elastic mechanical analysis; the latter is generally suitable, since the sample usually fractures at a load well below that causing generalized plastic flow.[11] These analyses give rise to rather cumbersome expressions (as can be seen in Fig. 9.16) relating K_{Ic} to sample geometry. Nonetheless, knowledge of the elastic compliance, coupled with measurement of the critical load (F_c) inducing unstable crack growth, allows fracture toughness to be determined.

The above procedures apply primarily to metals and their alloys for which material class fracture mechanics was first developed and applied. In principle, the same tests can be used for polymers and ceramics. However, principle does not al-

---

[11]For this reason, fracture mechanics tests that rely on elastic analysis are most accurate when the fracture stress is well below the yield strength (typically, $\sigma_F \leq 0.6\sigma_y$). Tests for determination of fracture toughness when σ_F approaches (or exceeds) σ_y have also been developed. These show great promise for defining fracture toughnesses under conditions of fairly extensive plastic flow, an area well outside the realm of the linear-elastic fracture mechanics approach described here.

ways translate to practice. In the case of polymers, attention must be paid to viscoelastic strain that can occur during crack propagation; this complicates the resultant analysis. In addition, test procedures differ in detail between polymers and metals. Ceramics, as noted in Chap. 1, can have their fracture toughnesses estimated by hardness testing, provided this test is accompanied by microcracks that emanate from the hardness impression. In addition, the notches used in fracture toughness evaluation of ceramics sometimes have a different shape than those shown in Fig. 9.16. This alteration in geometry makes crack tip honing somewhat more feasible in

Specimen type		$f(c/a)$	Comments
Center notched tension $$K_{Ic} = \sigma_F \sqrt{\pi c}\, f(c/a)$$ $$\sigma_F = F_c/ta$$	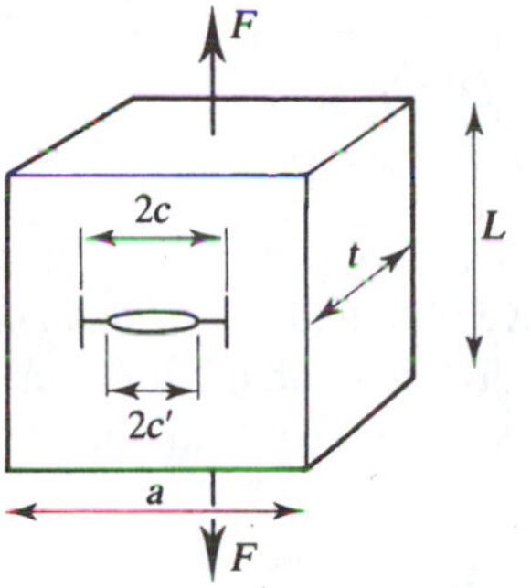	$$\left(\frac{a}{\pi c} \tan \frac{\pi c}{a}\right)^{1/2}$$	Specifications $L = 4a$, $2c = a/3$ For K_{Ic} $10 > a/t > 5$
Compact tension $$K_{Ic} = \sigma_F \sqrt{\pi c}\, f(c/a)$$ $$\sigma_F = F_c/ta$$	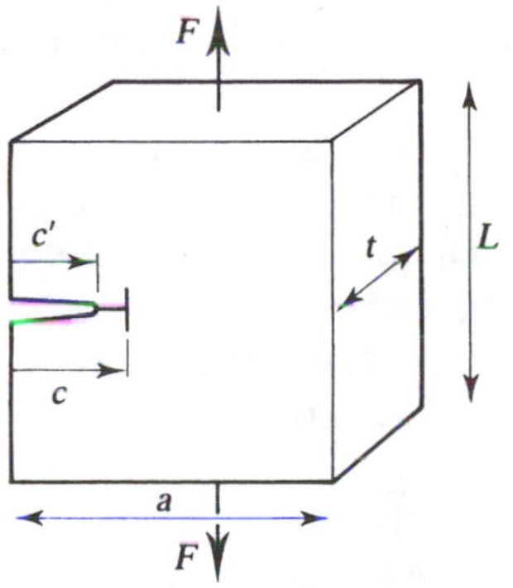	$$\frac{1}{\sqrt{\pi}}\left[29.6 - 185.5\left(\frac{c}{a}\right) + 655.7\left(\frac{c}{a}\right)^2 \right.$$ $$\left. - 1017\left(\frac{c}{a}\right)^3 + 638.9\left(\frac{c}{a}\right)^4\right]$$	Specifications $c = a/2$ For K_{Ic} $a/t > 2$
Three-point bend $$K_{Ic} = \sigma_F \sqrt{\pi c}\, f\left(\frac{c}{a}\right)$$ $$\sigma_F = F_c/ta$$	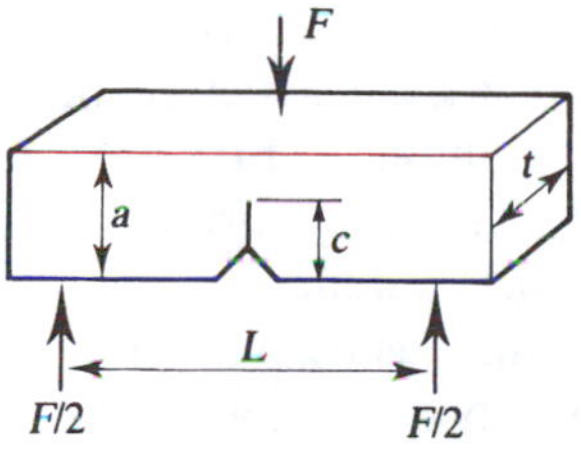	$$\frac{L}{\sqrt{\pi}\,a}\left[2.9 - 4.6\left(\frac{c}{a}\right) + 21.8\left(\frac{c}{a}\right)^2 \right.$$ $$\left. - 37.6\left(\frac{c}{a}\right)^3 + 387\left(\frac{c}{a}\right)^4\right]$$	Specifications $L = 8a$, $c = a/5$ For K_{Ic} $8 > a/t > 2$

Figure 9.16
Several sample geometries suitable for measuring plane strain fracture toughness. K_{Ic} is defined by $\sigma_F(\pi c)^{1/2} f(c/a)$, where σ_F is the fracture stress and $f(c/a)$ is a function of the test sample geometry. The crack length (c) used in calculating K_{Ic} is that of the initial machined notch (c') plus the increment provided by sharpening of the crack tip prior to testing. Specifications for producing plane strain conditions are noted. However, the sheet thickness should always be compared to the plastic zone size ($r_y \cong K_c^2/2\pi\sigma_y^2$) calculated from the measured fracture toughness to verify that plane strain conditions have been met.

brittle materials such as ceramics. Honing by fatigue or thermal shock often causes catastrophic, rather than controlled, crack propagation in them. Details are provided in the reference by Anderson.

Now that we have gotten the preliminaries out of the way, we can apply fracture mechanics to "real" situations. This is done in the next section.

E. Case Studies and Examples

Either the property of toughness ($\mathcal{G}_c$) or the fracture toughness (K_c) can be used in fracture analysis; the two properties are related through formulations previously provided. Current design utilizes the fracture toughness approach. The fracture toughness is related to the crack propagation stress and crack size by Eq. (9.16);

$$K_c = \sigma_F (\alpha \pi c)^{1/2} \qquad (9.16)$$

where α is a geometrical constant, typically of order of magnitude unity but varying with crack geometry. Values of α have obtained via mechanics analysis for a large number of crack geometries; they are available in the literature.

The plane strain fracture toughness is usually employed in design and analysis. Thus, we use

$$K_{Ic} = \sigma_F (\alpha \pi c)^{1/2} \qquad (9.20)$$

with the material property K_{Ic} being obtained from tests of the type described in the previous section.

In analysis of, and particularly in design against, fracture, some assumptions about crack geometry that permit estimation of α and of the potential maximum crack size are necessary. This information, obtained from empirical engineering "know how," allows σ_F to be calculated via Eq. (9.20). If σ_F so calculated lies above (usually well above) the design operating stress, the part is considered safe for use. If not, the design stress must be reduced and/or a material with a higher fracture toughness substituted.

Fracture analysis proceeds in much the same way as design. Postmortem inspection of a fractured part may lead to detection of preexisting flaws. Knowledge of their shape and size allows estimation of the stress required for their propagation. If this stress is less than the operating stress it can generally be assumed that the detected flaw(s) was (were) responsible for failure. If the calculated σ_F exceeds the operating stress, some other cause for the fracture must be found.

These procedures are illustrated in the examples that follow. Before doing this, however, we note that some of the formulations used to relate σ_F, K_{Ic}, and c are not of the simple form of Eq. (9.20). This is because corrections that incorporate the plastic-zone size into the crack length are used in them. We quickly summarize the bases for these corrections. The stress intensity factor for an internal through-crack in a plate is given as

$$K^2 = \sigma^2 \pi c \qquad (9.21)$$

The plastic zone increases the effective crack length. This is because the maximum triaxial stress is found at the boundary between the plastic zone and the elastic material in front of it. For plane strain conditions we assume the zone size to be half of what it is for the plane stress case; i.e., the effective crack length is $(c + r_y)$, where $r_y = K^2/2\pi\sigma_y^2$. Substituting $(c + r_y)$ for c in Eq. (9.21) and rearranging we have[12]

$$K^2 = \frac{\sigma^2 \pi c}{1 - \frac{1}{2}\left(\frac{\sigma^2}{\sigma_y^2}\right)} \quad (9.22)$$

Different crack geometries alter the effective plastic-zone size differently. Thus, a variation of Eq. (9.22) is used in analysis. In many cases, the relationship among K_{Ic}, σ_F, and c is given by

$$K_{Ic} = \frac{\sigma_F(\pi\alpha c)^{1/2}}{\left[B - A\left(\frac{\sigma^2}{\sigma_y^2}\right)\right]^{1/2}} \quad (9.23)$$

with α, A, and B being dependent on crack geometry. As mentioned, α is typically on the order of unity; A usually varies from 0 to about 0.5. Values of B and A, as well as those of α, can be obtained from the fracture mechanics literature.

CASE STUDY 1: DESIGN OF A STEEL PRESSURE VESSEL. Pressure vessels are often constructed of high-strength materials, and operate at stresses that are significant fractions of their yield strengths. A particular steel—Grade A—is to be used as a pressure vessel. This steel can be thermomechanically processed to various strength levels; as shown in Table 9.3, the higher the yield strength, the lower the fracture toughness of Grade A steel.

Experience suggests that elliptical cracks having the geometry shown in Fig. 9.17 may be present on the surface of the pressure vessel when it is put into service. Nondestructive evaluation and inspection suggest that cracks may go undetected if $c < 0.05$ in (=0.13 cm) or, equivalently, if $a < 0.10$ in (= 0.25 cm). The 220 ksi yield strength version of Grade A steel is suggested for use at a design operating stress of 180 ksi. Will this be a fracture-safe operating condition, given the possibility of the presence of cracks described above?

For an elliptical surface flaw, the relationship among K_{Ic}, σ_F, and c is

$$K_{Ic}^2 = \sigma_F^2 \frac{(1.20\pi c)}{f(c/a) - 0.212\left(\frac{\sigma}{\sigma_y}\right)^2} \quad (9.24)$$

σ_y (ksi, MN/m²)	K_{Ic} (ksi(in)$^{1/2}$, (MN/m²)(m)$^{1/2}$
190, 1310	82, 90
200, 1380	62, 68
210, 1450	49, 54
220, 1520	41, 45

Table 9.3
Properties of Grade A steel

where $f(c/a)$ (= B of Eq. (9.23)) depends on the ratio of the major and minor axes of an elliptical crack. For our case $f(c/a) = 2.439$. Thus, if fracture is *not* to occur at the operating stress of 180 ksi, we must have

---

[12]Equation (9.22) cannot be used if the applied stress is too high a fraction of the yield strength; an upper limit is $\sigma \geq 0.7\sigma_y$.

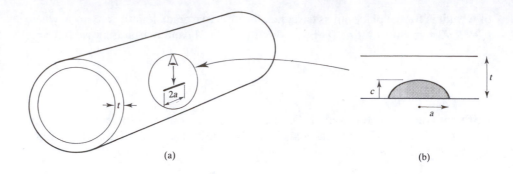

(a)   (b)

Figure 9.17
**(a) An elliptical-shaped crack on the surface of a pressure vessel. The crack is of
length 2a and projects into the wall the distance c (Fig. 9.17b). Flaws of this shape are
common in structural members. Stress-intensity factors are available in the literature for these flaws; see Case Studies 1 and 2 of Sect. 9.4E.**

$$K_{Ic} \geq \frac{180(\pi \cdot 0.05 \cdot 1.20)^{1/2}}{[2.439 - 0.212(180/220)^2]^{1/2}} \text{ ksi(in)}^{1/2} \geq 51.5 \text{ ksi(in)}^{1/2} \qquad (9.25)$$

This value is greater than the K_{Ic} of the 220-ksi strength steel. The solution, left for
an exercise (Prob. 9.9), is to either reduce the operating stress or to use a lower
yield-strength steel having a higher fracture toughness.

CASE STUDY 2: FAILURE ANALYSIS OF A ROCKET MOTOR CASE.
A rocket motor case having a yield strength of 215 ksi (=1485 MN/m²) and a K_{Ic} of
53 ksi(in)$^{1/2}$ (= 58 (MN/m²)m$^{1/2}$) failed at 150 ksi. Examination subsequent to failure revealed an elliptical surface crack of depth 0.039 in (= 0.099 cm) and surface
length 1.72 in (= 4.37 cm). Could this flaw have been responsible for failure?

The value of $f(c/a)$ for this flaw is 1.38. Rearrangement of Eq. (9.23) allows the
anticipated fracture stress to be calculated from

$$\sigma_F = \frac{\left[1.38 - 0.212\left(\frac{\sigma}{\sigma_y}\right)^2\right]^{1/2}}{(1.20\pi c)^{1/2}} K_{Ic} \qquad (9.28)$$

The fracture stress is best estimated iteratively. We substitute into Eq. (9.28) the
values of K_{Ic} and c and *assume* σ on the right-hand side of the equation is 150 ksi.
Doing this, σ_F is calculated as 156 ksi, which is reasonably close to the observed
fracture stress of 150 ksi. In the second iteration, 156 ksi is substituted for the stress
σ on the right-hand side of Eq. (9.26); σ_F is now found to be 156 ksi. Further itera-
tions (Prob. 9.10) do not substantially alter this predicted value. We conclude there
is a strong likelihood that this flaw was responsible for the failure.

CASE STUDY 3: SAFE USE OF A FLAWED PART. A cylindrical support
structure is used in the launching of a space shuttle. Radiographic inspection on the
day before liftoff detects a half-inch-long interior crack in the structure. The design

operating stress is 10 ksi (= 69 MN/m^2) and the properties of the support material are σ_y = 100 ksi (= 690 MN/m^2) and K_{Ic} = 80 ksi(in)$^{1/2}$ (= 88 (MPa·m$^{1/2}$). Must the launch be scuttled or will the support structure maintain its integrity during liftoff?

We assume the interior crack has the worst possible geometry, i.e., that it extends along the whole length of the structure. For such a crack, the fracture mechanics references tell us that $\alpha = 0.5$ and $A \cong 0$. Thus,

$$K_{Ic}^2 = \frac{\pi c \sigma_F^2}{2} \tag{9.27}$$

The fracture stress calculated from Eq. (9.27) is 90 ksi. Hence the launch need not be canceled on the basis of this defect. Indeed (Prob. 9.11), the structure could contain a far more severe flaw before it would fail. This is a result of the conservative design (i.e., $\sigma/\sigma_y = 0.10$) and the high toughness of the support structure, for which the ratio of K_{Ic}/σ_y is quite high (compare this ratio for the support structure to the similar ones for the materials of the other case studies).

The illustrations given here are simple ones. The situations arising in practice are more complex, but nonetheless involve analyses much like those used here. Moreover, our examples clearly indicate the utility of fracture mechanics in engineering design, particularly when high-strength materials are employed.

Although fracture mechanics is the most quantitative tool in design against fracture, other, more qualitative, tests are used for assessing a material's fracture resistance. Detailed discussion of all of them is beyond our task. However, in Sect. 9.6, we describe the widely used impact test. Before we do that, though, it is worthwhile to consider the variation in fracture toughness that exists among and within the several material classes. This helps to set the stage for our discussion of "toughening" mechanisms in Chap. 10; "toughening" mechanisms are to fracture resistance what strengthening mechanisms are to plastic flow resistance.

9.5
FRACTURE TOUGHNESS AND MATERIAL CLASS

Fracture toughnesses, just as elastic moduli and strengths, vary among the material classes. This variation for fracture toughness is shown in Fig. 9.18. Rather than presenting this information in bar chart form, as was done for modulus and strength, it is presented here as a logarithmic graph of fracture toughness vs. strength. The reasons for this will become apparent.

Variations in the fracture toughness of materials span more than four orders of magnitude, comparable to the like variations in strength. If we discount polymer foams (they typically are only about 10% dense) the variations in toughness and strength are less, but still about three orders of magnitude. Metals are by far the most fracture tough of materials. High toughness steels have K_{Ic} values in excess of 100 MPa·m$^{1/2}$. Engineered ceramics (e.g., MgO, ZrO$_2$, etc.) have fracture toughnesses comparable to those of engineering polymers,[13] and these toughnesses exceed those

---

[13]Polymers are a bit shortchanged in the representation of Fig. 9.18. Their moduli are much less than those of ceramics. Thus, if the materials are compared on the basis of their *toughnesses* ($\mathcal{G}_c \cong K_c^2/E$), as a class, polymers are tougher than ceramics.

of common (porous) ceramics such as brick, concrete and the like (such materials nevertheless see widespread use in compressive loading). Composite (e.g., GFRP—glass fiber reinforced plastics; CFRP—carbon fiber reinforced plastics) fracture toughnesses are surprisingly high. In distinction to the composite properties of modulus and strength—which are intermediate to these properties of their constituents—

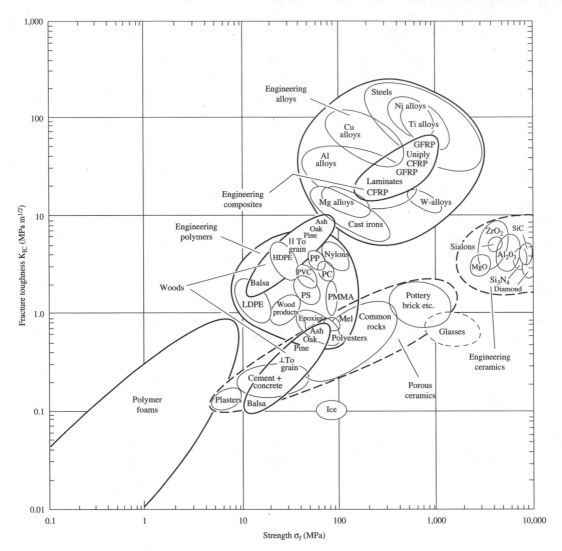

Figure 9.18
Fracture toughness vs. strength for various materials and material classes. The strengths displayed for metals and polymers are yield strengths. For ceramics and glasses, they are compressive strengths, and for composites they are tensile strengths. Metals are the toughest of the material classes, although some composites give them a run for their money. Materials having low strengths and high toughnesses (upper left of diagram) yield before they fracture; use of them is based on design against flow. Materials with high strengths and low toughness (lower right of diagram) fracture before yielding; design with them involves fracture considerations. (*Adapted from M. F. Ashby, Materials Selection in Mechanical Design, Pergamon Press, Oxford, 1992.*)

composite fracture toughnesses exceed (considerably) those of their constituents. Composite toughnesses range from about 10 to 50 MPa·m$^{1/2}$. Yet an epoxy—the matrix of many of these composites—has a K_{Ic} value of about 1 MPa·m$^{1/2}$, and so do glasses which are often used as their reinforcements. The reasons for the surprising fracture toughnesses of composites are discussed in Sect. 10.4.

It is interesting to explore the relationship between fracture toughness and strength. First, the metals. The strengths used in constructing Fig. 9.18 are their yield strengths. Within the large "circle" of Fig. 9.18 (the one designated "engineering alloys") there are several different "balloons" representing the various alloy systems; steels, Ni alloys, etc. These balloons usually run from upper left to lower right; that is, they run from low strength–high toughness combinations to high strength–low toughness combinations. This behavior reflects that, within a given alloy class, toughness and strength do not go hand in hand; for example, high-strength steels are not as fracture tough as low-strength steels are. Indeed, metallurgists are prone to think of a high-strength material (in a given alloy class) as being one of low fracture toughness and vice versa. Metallurgists would likely get agreement from polymer engineers on the matter. For example, the HDPE (high-density polyethylene) balloon in the engineering polymers circle of Fig. 9.18 also slopes downward. Similar behavior is found for other engineering polymers; PS (polystyrene), PVC (polyvinyl chloride), etc. But this correlation between fracture toughness and strength does not hold for the other material classes.

Consider now polymer foams, woods, and porous ceramics. They have one common characteristic; they are porous and the pores in them behave much like preexisting cracks. In Fig. 9.18 the "balloons" for these material classes extend from lower left to upper right.[14] That is, strength goes hand in hand with fracture toughness for porous solids. This behavior of porous ceramics is relatively easy to understand. Their strengths shown in Fig. 9.18 are compressive fracture strengths. A ceramic tensile strength is a small fraction of its compressive strength; thus, the shapes (but not their location in the figure) of the ceramic balloons would remain the same were tensile fracture strengths used in constructing Fig. 9.18. Since, the crack propagation stress varies directly with fracture toughness ($\sigma_F \sim K_{Ic}/\sqrt{c}$) we see that strength and fracture toughness for ceramics are expected to be linked. The variations in strength of a specific porous ceramic, indicated by the finite size of their balloon, are a manifestation of differing degrees of preexisting porosity and/or internal flaws in them.

Understanding the shape of the balloons of the porous organics (the foams and the wood products of Fig. 9.18) is more difficult. These materials are cellular solids. That is, they are engineered (for good reasons) to contain considerable porosity. (Nature is the engineer for wood.) For such materials, both fracture toughness and strength scale with relative density (i.e., the density relative to that of a fully dense solid of the same material). Typically, K_{Ic} scales with relative density to the ³⁄₂ power, and the strength with the square of this density. Thus, strength and fracture tough-

---

[14]It is not apparent on the logarithmic coordinates of Fig. 9.18 that the "balloons" for the engineering ceramics also have this shape, but they do. Engineering ceramics are more carefully processed and display lesser variations in strength and toughness than do porous ceramics. Thus, the "balloons" for the engineered ceramics occupy less area in a figure like Fig. 9.18 and this makes discerning their actual shape more difficult.

ness in cellular solids are linked through their relative densities such that in the polymer foam balloon of Fig. 9.18, high toughness–high strength combinations correspond to denser foams and low toughness–low strength combinations to the most porous foams.

So there are differences in fracture behavior of the material classes just as there are differences in their other mechanical properties. Improving the toughness of the various classes consequently involves different approaches. These are discussed in Chap. 10.

9.6
THE CHARPY IMPACT TEST

Although fracture mechanics is the preferred design tool against fracture, it is a relatively new procedure. Prior to its widespread implementation, other empirical tests had been developed. We describe one—the impact test—in this section. This test is widely employed in the ferrous metals and in the plastics industries. The reason is that both of these material classes are prone to a change in their fracture behavior with changes in temperature. At high temperatures, fracture is accomplished via modes of plastic deformation (ductile fracture). At low temperatures, their fracture is brittle (e.g., recall the effect of the glass transition temperature on the mechanical characteristics of long-chain polymers). We discuss these changes in more detail in Chap. 10. However, the details need not deter us from discussing the impact test. There is one caveat, though. In bcc ferrous materials, low-energy (brittle) fracture is accomplished by *cleavage*. A cleavage fracture is one in which atomic bonds are severed across preferred crystallographic (cleavage) planes, and the fracture surface is shiny and faceted as a result. At high temperatures, higher-energy ductile fracture is manifested by a rough surface appearance, which is called *fibrous*.

An impact test assesses a material's tendency to brittle fracture. Several such tests have been devised; in the United States the Charpy impact test (Fig. 9.19) is common. In the Charpy test, a hammer is mounted on a very nearly frictionless pendulum. The hammer is released from a specified height, h, and strikes the test sample at the bottom of its arc. When it does, the sample is subjected to a high strain rate, which favors fracture rather than flow. Moreover, the notch in the sample on the side of the bar subject to tensile loading induces a triaxial state of stress; this also tends to promote fracture. Thus, an impact test is associated with a high strain rate and a strong degree of triaxial loading. As such it is a rather severe test of a material's toughness. The sample test temperature can also be varied, thus allowing the determination of the temperature variation of the energy absorbed in impact fracture.

Subsequent to striking—and breaking—the sample, the hammer rises to a height h' less than h. The difference in potential energy ($= mg(h - h')$, Fig. 9.19) is the energy expended in fracturing the sample. This energy involves plastic work prior to fracture as well as the work associated with crack nucleation and propagation. Since the latter is often a fairly small fraction of the total energy expended, we see that the impact test does not bear the direct relationship to design against crack propagation that a fracture toughness test does.

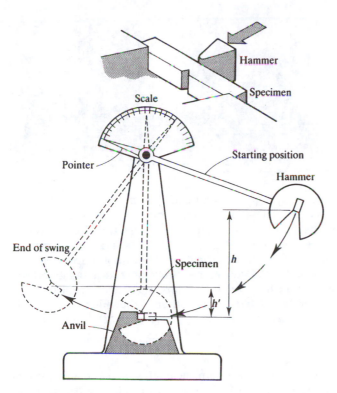

Figure 9.19
**Schematic of a Charpy V-notch impact test. A hammer attached to the end of a fric-
tionless pendulum is released from the height h. At the bottom of its arc it strikes a
notched sample. The high strain rate concurrent with impact, and the constraint on
plastic deformation provided by the notch, are conducive to sample fracture. Subsequent to fracturing the sample, the pendulum rises to a height h'; the potential energy difference $mg(h - h')$ represents the energy expended in sample fracture. This
energy (measured by the pointer on the scale) represents plastic deformation work
prior to fracture as well as that associated with fracture initiation and propagation.
(*From H. W. Hayden, W. G. Moffatt, and J. Wulff, The Structure and Properties of Materials, Vol. III, Wiley, New York, 1965.*)**

Visual inspection of the impacted sample's fracture surface can provide useful
information. For a metal, the surface may be fibrous (indicating ductile fracture) or
shiny and "crystalline" (giving evidence of cleavage). Or, for those materials that
undergo a change in fracture mode with temperature, the surface may be part fibrous and part cleavage. The cleavage portion is found in the central section of the
sample and is surrounded on its periphery by a region of fibrous failure. The percentage of fibrous fracture increases with increasing test temperature (Fig. 9.20).

That the impact test is a severe one in the sense of assessing fracture resistance
is evidenced in Fig. 9.21. Here the temperature variation of fracture energy and/or
ductility is shown for three tests (tension, torsion, and impact) conducted on iron, a
metal that displays a ductile-to-brittle transition as temperature is decreased. In a
torsion test, the maximum shear stress inducing plastic flow (τ_{max}) is *equal* to the
maximum tensile stress; the tensile stress is a component of the mean pressure that

Figure 9.20
The fracture appearances of Charpy impact samples, broken at different test temperatures, of a steel that undergoes a ductile-to-brittle transition. (A) At low temperatures, the fracture surface is flat and shiny, indicative of cleavage fracture. (B) At intermediate temperatures, the interior of the sample still manifests a shiny "crystalline" appearance, but the periphery is dull, indicative of fibrous or ductile fracture. (C) At higher temperatures still, the surface is entirely dull indicating that the fracture of the entire sample was ductile. Ductile fracture is also accompanied by a lateral contraction at the root of the notch (e.g., the bottom of the samples in (B) and (C)). (*From R. E. Reed-Hill, Physical Metallurgy Principles, 2nd ed., D. Van Nostrand, New York, 1973.*)

facilitates fracture. In a tension test, the maximum shear stress is *half* that of the tensile stress, and for the impact test a strong triaxial stress state is imposed so that τ_{max} is much less than the mean pressure. These considerations are reflected in the results shown in Fig. 9.21. The impact test yields the highest transition temperature by far and because of this conservative feature of it, the test is a useful tool in design against low-energy brittle fracture.

The temperature variation of the Charpy impact energy is shown schematically in Fig. 9.22 for several types of metals. For low-strength metals that do not exhibit a change in fracture mode with temperature (e.g., the fcc and some hcp metals and their alloys), the impact energy is high and relatively temperature insensitive. These features reflect the high toughness of these materials and the relative insensitivity of this property to temperature changes. For these reasons, the impact test is not widely used for assessment of fracture resistance of low- to moderate-strength materials that do not exhibit a change in fracture mode with temperature.

The temperature variation of the impact energy for a low-strength bcc transition metal or alloy (e.g., most steels) shows a typical ductile-to-brittle transition (Fig. 9.22). This is a macroscopic manifestation of the different microscopic deformation and fracture modes that take place at different temperatures. To a certain extent, the temperature variation of the impact energy for these metals parallels that of the toughness, and because of this the impact test is a good qualitative tool in assessing fracture resistance of the bcc transition metals. As mentioned, these metals exhibit a change in fracture mode from cleavage to fibrous with increasing temperature, and the temperature at which this occurs is related to (but not equal to) the transition temperature as defined by changes in the impact energy.

Impact energies are low for high-strength materials (Fig. 9.22). For high-strength materials other than steels (and other bcc transition metals), the impact

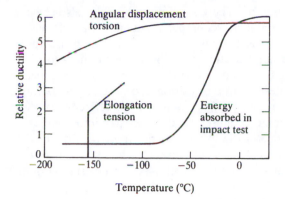

Figure 9.21
The temperature dependence of the relative "ductility" of iron as measured by angular displacement in a torsion test, elongation in a tensile test, and energy absorbed in an impact test. The ductile-to-brittle transition temperature depends strongly on the stress state (i.e., the test conditions). This temperature is lowest for torsion, where the shear stress is large vis-à-vis the maximum tensile stress. It is highest for the impact test, where a strong triaxial stress state is developed in conjunction with a high strain rate. (*After K. Heindlhofer, Trans. TMS-AIME,* 116, 232, 1935.)

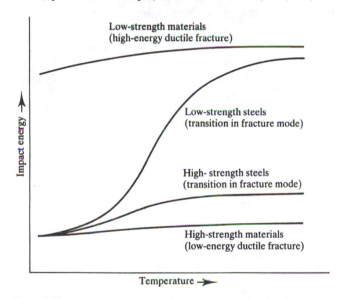

Figure 9.22
Schematic of the temperature variation of the impact energy for several classes of metals. For low-strength metals that do not exhibit low-temperature brittle fracture, the impact energy is high and temperature insensitive. Low-strength steels display high values of impact energy at high temperatures (at which they fail in a microscopically ductile manner), but at low temperatures (where their yield strengths are high and they fail by cleavage) their impact energies are low. This transition in fracture mode is mirrored by a like transition in impact energy. High-strength steels also manifest a transition in microscopic fracture mode. However, since the work required to fracture such steels is always relatively low, their impact energies are low even at temperatures at which they fail in a ductile manner.

energy is fairly insensitive to temperature. These low impact energies also attest to the ease with which fracture initiates and cracks propagate in high-strength materials. This can lead to in-service macroscopically "brittle," low-energy fractures. High-strength steels display a greater variation of impact energy than do high-strength metals that do not have the body-centered cubic structure. This is because steels undergo a microscopic ductile-to-brittle transition. However, the maximum (or shelf) impact energies of high-strength steels are still low—a reflection of their low-energy ductile fracture at higher temperatures. The change in microscopic fracture mode is noted by the appearance of the impact energy fracture surface (Fig. 9.23); a relatively sharp transition from cleavage to fibrous fracture takes place with increasing temperature and this is in contrast to the broad, diffuse variation in the corresponding impact energy.

Before considering the impact test further, it is worth reviewing design philosophy as it relates to the various material classes whose impact energies are illustrated in Fig. 9.22. Low- to moderate-strength metals (the nonferrous ones at all temperatures and the low-strength steels at high temperatures) display high impact energies; they also manifest high fracture toughnesses. These are materials for which $\sigma_F > \sigma_y$, and design against "failure" involves designing against plastic flow. Conversely, the low-strength steels at low temperatures, and the high-strength materials at all temperatures, have low impact energies and fracture toughnesses. For these materials, $\sigma_F < \sigma_y$ typically, and design is preoccupied by considerations of fracture.

The art of impact testing has been around for some time. Because of this, empirical design guidelines have been developed from it. These are most useful for the low- to moderate-strength steels and several concepts relating to the effect of microstructure on impact energy are illustrated in Fig. 9.24. In addition, the ductile-to-brittle transition temperature obtained from an impact test depends on the

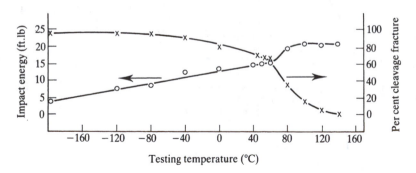

Figure 9.23
Charpy impact energy and percentage cleavage fracture surface vs. temperature for a high-yield strength ($\sigma_y = 216$ ksi $= 1490$ MN/m^2) 4340 alloy steel. The maximum (shelf) impact energy (obtained for $T > 80°C$) is low, and indicative of low-energy ductile fracture. At low temperatures, the impact energy is even less; it increases slowly with temperature over the temperature range in which cleavage fracture is found. The impact energy increases somewhat more rapidly over the temperature range (ca. 60–90°C) in which the failure mode changes. However, the change in impact energy does not as clearly delineate the ductile-to-brittle transition as does the change in fracture-surface appearance. (*From F. R. Larson and R. L. Carr, Metal Progress, 74, 1964.*)

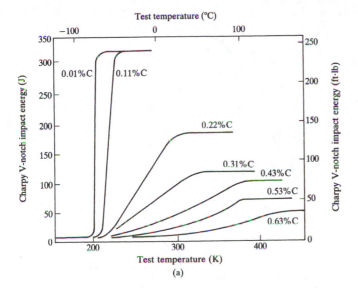

(a)

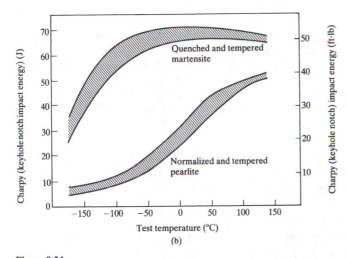

(b)

Figure 9.24

The impact energy and its temperature variation are structure sensitive. (a) Impact energy as a function of temperature for plain carbon steels. The transition temperature increases with the carbon content (and, therefore, with the pearlite content). Shelf energies are also higher for low-carbon steels. (b) The Charpy keyhole impact energy as a function of temperature for a 4340 alloy steel in two different microstructural conditions. The yield strengths of the pearlite and tempered martensitic structures were made the same by appropriate heat treatment. But the pearlitic steel, which contains Fe$_3$C in lamellar form, has a much higher transition temperature than the quenched and tempered steel in which Fe$_3$C is dispersed as particles within a ferrite matrix. (In the keyhole test, the "notch" is shaped like a keyhole rather than like a V in the normal impact test. The lesser constraint against plastic deformation provided by a keyhole results in a generally higher impact energy and lower transition temperature than found in a V-notch test.) *(Part (a) after J. A. Rineholt and W. J. Harris, Jr., Trans. ASM, 43, 1175, 1951; (b) after Society of Automotive Engineers, SP65, Low Temperature Properties of Ferrous Materials.)*

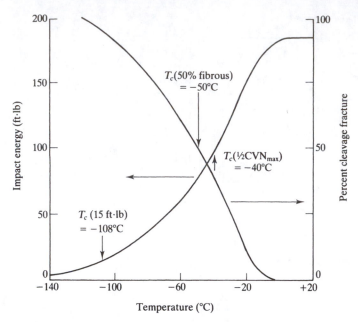

Figure 9.25
Impact energy and percentage cleavage fracture as a function of temperature for a 3.5% Ni, 0.1% C steel. The transition temperature varies, depending on the criterion used to define it. The 15 ft-lb transition temperature is $-108°C$, whereas T_c defined by the half shelf-energy criterion is $-40°C$. This temperature is close to that obtained using a 50% fibrous–50% cleavage fracture criterion ($-50°C$). (*Adapted from A. S. Tetelman and A. J. McEvily, Jr., Fracture of Structural Materials, Wiley, New York, 1967.*)

criterion used to define it (Fig. 9.25). For example, a transition temperature (T_c) can be defined on the basis of: (1) a certain percentage of fibrous fracture; (2) an impact energy halfway between the shelf and the "brittle" energy (the impact energy at $T \ll T_c$); (3) a specified level of energy (e.g., 15 ft-lb); or on the basis of several other criteria. There is no a priori reason to prefer any one of these. However, one engineering guideline has proven particularly useful in designing against brittle fracture in low-strength steels. If the impact energy is greater than 15 ft-lb at the service temperature, the steel is not likely to undergo brittle low-energy fracture.

As noted, the change in impact energy accompanying the cleavage-to-fibrous transition in high-strength steels is slight (and shelf energies are low) reflecting their low-energy "ductile fracture." Shelf energies of high-strength steels, and the impact energies of other high-strength metals, correlate with their fracture toughnesses. As such, these impact energies are good qualitative guides in assessing the fracture resistance of high-strength materials.

Impact energies are only a qualitative guide to a material's fracture resistance. Yet the impact test is conveniently performed. And since much experience has come from results of the test, it serves as a guide to fracture in ways that do not appear justified fundamentally. Indeed, this test remains, as it has for many years, useful in the selection and use of materials chosen on the basis of their fracture resistance.

FRACTURE OF BRITTLE NONMETALLICS

In principle the approach to tensile fracture to this point in this chapter applies to all material classes. However, ceramics and inorganic glasses display fracture behavior that makes it worthwhile to spend some additional time on them. As has been mentioned, the inherently high strengths of ceramics render them possible high-performance structural materials. This is especially true at high temperatures, where their strength retention is impressive and they are relatively inert to environmental interactions. Design with ceramics differs from that with metals, on at least two important accounts. As a result of the brittleness of ceramics, design is predicated only on preventing their fracture. That is, the design operating stress is made less than σ_F, since ceramic fracture strengths are (almost always) less than their plastic yield strengths. In addition, most ceramics contain preexisting defects within them or on their surfaces. This is a result of the way ceramics are manufactured. For example, they are often produced by powder processing, with attendant residual porosity that acts much like internal flaws. In addition, microcracks often form during the handling and machining of ceramics and inorganic glasses. And two-phase ceramics are prone to microcracking caused by thermal stresses. Thermal cracking can also be found in single-phase ceramics that have been subjected to thermal treatment in which thermal gradients between the material interior and its surface are present during processing. Finally, small surface cracks, particularly in inorganic glasses, can be sharpened via exposure to ordinary environments. As a result of all of these factors, the Mode I brittle fracture strength of a ceramic is subject to a degree of variability not found in metallic materials. This introduces a probabilistic component to design with ceramics.

A. Ceramic "Strengths"

Ceramic toughnesses are low. Even at the tip of a crack, where the stress intensity is high, microscopic plasticity is limited in ceramics. The result is a low fracture work, and a typical ceramic has a fracture toughness much less than that of even a "brittle" metal for which significant crack tip plasticity may accompany crack propagation. A "rule of thumb" is that K_{Ic} values of ceramics are about 5% those of the metals. Thus, the K_{Ic} for a "typical" ceramic is on the order of 2 MN/m$^{3/2}$. The tensile fracture stress is coupled directly to K_{Ic}; i.e., for a crack of internal length $2c$ or a surface crack of length c, $\sigma_F \cong K_{Ic}/(\pi c)^{1/2}$.

The discussion above pertains to tensile loading of ceramics. However, most structural ceramics are loaded in compression; bricks and concrete provide "concrete" examples. This use of ceramics in compression is so much the rule that, in fact, we have honored it in our display of ceramic "strengths" in Fig. 9.18; the strengths shown there are compressive failure strengths. These are much higher than their tensile fracture strengths, and this is related to the ways in which cracks propagate in ceramics in the different loading modes (Figs. 9.26a and b). Figure 9.26a indicates that in a tensile test the largest suitably oriented crack propagates unstably when the critical stress intensity for it is attained. In contrast, differently oriented cracks are precursors to compressive failure. Moreover, the cracks that propagate in

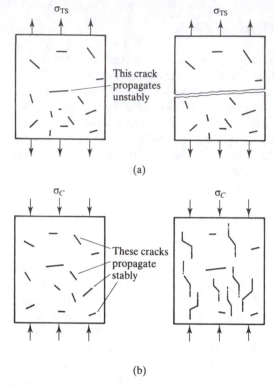

Figure 9.26
Fracture due to preexisting flaws in ceramics takes place differently in tension (a) and compression (b). In a tensile test the largest, most favorably oriented crack propagates unstably when its propagation stress (σ_F) is reached. In compression, cracks oriented at an angle to the compression axis propagate stably and in a direction more or less parallel to the loading axis. The multitude of microcracks formed in compression leads to failure by sample "crushing." As a result the compressive fracture stress is much greater than the tensile one. (*From M. F. Ashby and D. R. H. Jones, Engineering Materials 2: An Introduction to Microstructures, Processing and Design, Pergamon Press, Oxford, 1986.*)

compression do so in a stable manner and in a direction *parallel* to the compression axis (Fig. 9.26*b*). The result is that a large number of microcracks form prior to final compressive failure, which takes place by "crushing" of the material.[15] The different crack propagation mode in compression accounts for the higher ceramic compressive strengths. Another good "rule of thumb" is that ceramic compressive fracture strengths are about ten times their tensile fracture strengths.

Ceramic fracture in situations other than uniaxial loading can be described by a Coulomb law;

$$\sigma_{max} - B\sigma_{min} = (T.S.) \tag{9.28}$$

---

[15]A test with chalk will illustrate. Bend a piece of chalk. Its outer surface is in tension, and the chalk breaks into two pieces when the most deleterious flaw in it propagates. Now step on a piece of chalk. If you are heavy enough, the chalk "crumbles" into a number of pieces.

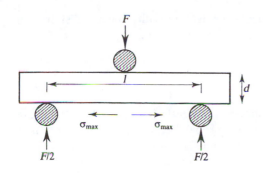

Figure 9.27
A three-point bend test can be used to measure the modulus of rupture of a ceramic. The maximum tensile stress is found on the bar surface. The modulus of rupture is equal to this stress at fracture and is obtained as $\sigma_r = 6M/bd^2$, where M is the moment causing fracture, d the sample thickness, and b the dimension of the sample (its width) into the plane of the drawing.

Here σ_{max} is the maximum, and σ_{min} the minimum, principal normal stress, B is an empirical constant, and (T.S.) is the measured fracture stress in tension. Thus, for example, if the ratio of the compressive to the tensile fracture stress is 10, then the value of the coefficient B is 0.1. The constitutive fracture Eq. (9.28) bears resemblance to the Tresca plastic yield condition, with two primary differences. First, (and obviously) the Tresca condition pertains to plastic flow. Second, the shape of the fracture surface is distorted compared to the Tresca yield surface. An example of how "distorted" it is can be obtained by solving Prob. 9.15.

Compression tests are relatively easy to conduct on ceramics and other brittle solids. Tension tests are not. Because of the material brittleness, tensile samples are prone to break in the sample grips. The results of an alternative test, a "modulus-of-rupture" test (Fig. 9.27), provide a reasonable approximation to the tensile fracture strength of a ceramic. In such a test, the force required to break a beam by bending it is measured. The maximum tensile stress in the beam surface at failure is called the modulus of rupture (MOR) (or, sometimes, the bend strength, σ_r). Since the material is linearly elastic through fracture, elastic beam theory can be used to estimate MOR. For an elastic beam, σ_r is related to the maximum moment in the beam, M, by

$$\sigma_r = \frac{6M}{bd^2} \qquad (9.29)$$

where d is the beam thickness and b its width. It might be thought that the MOR would equal the tensile fracture stress, but this is not so; σ_r is typically 50% greater than the tensile fracture stress. The reasons for this are discussed next.

When a MOR test is executed on a ceramic, an *average* fracture stress of a large number of samples is reported. This procedure is required because there is a significant variation in MOR from sample to sample as a result of the statistical distribution of flaws and flaw sizes on the surface of or within the volume of the material. This distribution in flaw sizes has other consequences. In particular, the average fracture strength of a large sample is less than that for a small sample for the simple reason that there is a greater probability of finding a large-sized flaw in the material with the greater volume. For the same reason, the average MOR is higher than the average tensile fracture stress because a bent beam is subject to the maximum tensile stress only over a thin surface layer whereas a tensile bar is subject to a uniform stress throughout. In effect, for a given total sample volume, the stressed volume of a tensile bar is greater than that of a MOR bar. Ceramists, and others who deal with brittle materials, have developed protocols for dealing with aspects of statistical failure. Some of these are described in the following section.

B. The Statistics of Brittle Fracture

A number of approaches consider the statistics of brittle fracture. One commonly used is due to Weibull. He defined a survival probability, $P_s(V_0)$, as the fraction of samples of volume V_0 that survive loading to a tensile stress σ, and proposed that

$$P_s(V_0) = \exp\left[-\left(\frac{\sigma}{\sigma_0}\right)^m\right] \tag{9.30}$$

where σ_0 and m are material constants.[16] The Weibull distribution ($P_s(V_0)$) is plotted in Fig. 9.28. As indicated there, σ_0 is that stress at which the survival probability is $1/e = 0.37$; that is, σ_0 is the tensile stress that 37% of samples can be subjected to without failing. The Weibull coefficient m is a measure of the material strength variability. As shown in Fig. 9.28, low values of m indicate a high variability in strength, and vice versa. Engineering (i.e., high quality) ceramics typically have values of m of about 10. They are less prone to fracture strength variations than are "routine" ceramics such as brick and concrete; for these materials m is about 5. Metals also exhibit slight variations in strength, and this variation can be described

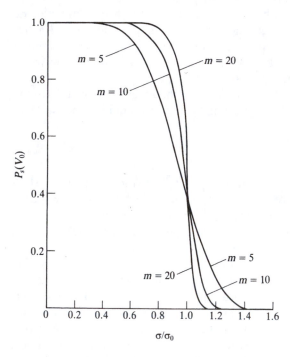

Figure 9.28
The survival probability ($P_s(V_0)$) of a brittle solid as a function of stress (normalized in terms of σ_0) and for several values of the Weibull coefficient (m). Large values of m indicate a small variability in strength and vice versa.

---

[16]These constants depend both on the material and the manner in which it is processed. Thus, for example, different grades of silicon carbide might have different values of σ_0 and m reflecting that the grades contain differing amounts of all of residual porosity, flaw sizes, and flaw-size distributions.

by a Weibull distribution as well. However, since strength variations in metals are small, their Weibull coefficients are high; e.g., $m \cong 100$ for a typical steel.

The parameters σ_0 and m are determined experimentally. A batch of samples of volume V_0 is tested at the stress σ_1 and the number that survive this stress is determined. Then a batch is tested at the stress σ_2, the survival probability determined there, and so on. The survival probability can then be plotted as a function of stress. Such a plot permits direct determination of σ_0 (it is the stress at which the survival probability is 0.37). The coefficient m can be determined by curve-fitting the graph of $P_s(V_0)$ vs. σ. However, it can be determined with more precision by an alternative technique. On taking double logarithms of both sides of Eq. (9.30) and rearranging, we have

$$\ln\left[\ln\left(\frac{1}{P_s(V_0)}\right)\right] = m \ln\frac{\sigma}{\sigma_0} \qquad (9.31)$$

Thus, a plot of $\ln[\ln(1/P_s(V_0)]$ vs. $\ln(\sigma/\sigma_0)$ yields a straight line of slope m.

The survival probability as it depends on sample volume can also be found with the Weibull distribution. The principles are similar to those used in discussion of fiber breaking strengths in Chap. 6. Consider a volume V that is related to a reference volume V_0 by $V = nV_0$. The survival probability for the reference volume has been determined; it is given by $P_s(V_0)$. The probability of the volume V surviving at a specified stress, σ, is obtained by multiplying $P_s(V_0)$ by itself n times; i.e.,

$$P_s(V) = [P_s(V_0)]^n = [P_s(V_0)]^{V/V_0} = \exp\left[-\frac{V}{V_0}\left(\frac{\sigma}{\sigma_0}\right)^m\right] \qquad (9.32)$$

Equations (9.30) and (9.32) are design equations for ceramics. The parameters σ_0 and m are determined as just described and if the material volume to be used in the application is different from V_0, its survival probability is calculated from Eq. (9.32). An "acceptable" survival probability is then arrived at. This depends on the application. Manufacturers of chalk, for example, need not worry much about its fracture (although they might be worried about its declining use) because it is not a catastrophe when a piece of chalk breaks. After all, there is (usually) another piece handy. Manufacturers and users of, say, a ceramic cutting tool might not be so sanguine. Replacing a broken cutting tool causes down time so a high tool survival probability is desired. However, this is accomplished only by more stringent processing and/or quality control (these reduce the incidence and size of preexisting flaws) which adds to material cost. So in use of a cutting tool some compromise is made between tool survival and tool cost.

While the Weibull distribution and its philosophy of accepting failure on a statistical basis is useful for many applications, it will not suffice when ceramics are used in critical situations, such as those when the loss of human life might result from material failure. To use ceramics in these applications requires significant increases in their fracture toughness and/or improvement in their processing so as to lead to a reduction in the number and sizes of the flaws they contain. Advances in ceramic processing are taking place rapidly. This has resulted in parts having relatively controlled initial flaw sizes and distributions, as well as a reduction in flaw content.

9.8
SUMMARY

In this chapter, the philosophy of design against fracture was described. We began by discussing the theoretical fracture strength of a material. This is the stress required to simultaneously rupture atomic bonds across a fracture plane. The theoretical fracture strength ($\cong E/10$, where E is Young's modulus) is much higher than observed fracture strengths. There is a reason. It is that preexisting cracks on the surface or the interior of the material concentrate stress in their vicinity. For a completely brittle solid, the stress concentration can result in the theoretical strength being attained at the tip of the crack. This is a necessary, but not sufficient, condition for the crack to advance, resulting in material fracture.

An additional condition for crack advance is that it be accompanied by a reduction in the system energy. For a brittle solid, the fracture surfaces formed by crack advance have an associated surface energy. This is one component of the system energy and it serves to retard crack extension. However, crack growth is also accompanied by a reduction in a second component of the system energy, the stored elastic energy. Its reduction during crack extension promotes crack growth. The two energies vary differently with crack size with the result being that the stress required to continue crack growth decreases with increasing crack size. In a nutshell, the elastic strain energy reduction is consumed in the creation of new surface area as the crack advances.

This description can be extrapolated to the situation where crack tip plasticity occurs, as it often does with metals and polymers. The crack tip stress concentration causes these materials to plastically flow at the crack tip (their yield strengths are less than their fracture strengths). The crack tip plastic deformation accompanying crack advance provides a component to the fracture work over and above (often well above) that due to creation of new surface. This plastic work, though, has the same units (J/m^2) as the surface energy. The total fracture work per unit area of crack surface is termed the material toughness ($\mathcal{G}_c$). Closely related to $\mathcal{G}_c$ is another material property, the fracture toughness (K_c). For a brittle material, the fracture toughness is that stress intensification necessary to attain the critical crack propagation stress (essentially the theoretical strength) to cause crack growth. Thus, K_c is proportional to the nominal crack propagation stress (σ_F) multiplied by the square root of the crack size (K_c has units of $MPa \cdot m^{1/2}$). A fracture toughness for other than brittle materials can also be defined on the same operational basis; i.e., a critical crack tip stress intensification required for crack advance. In this instance, though, the parameter K cannot be linked directly to a theoretical fracture strength. As might be expected, K_c and $\mathcal{G}$ are linked; they approximately relate through $K_c \cong (E\mathcal{G}_c)^{1/2}$.

Fracture toughnesses vary among material classes. Metals have the highest K_c values, and this reflects both their high $\mathcal{G}_c$ values (resulting from crack tip plasticity) and generally high moduli. Above their glass transition temperatures, thermoplastic polymers are relatively tough, although their K_c values are not that high because polymers have low moduli. The fracture toughnesses of a given polymeric or metal alloy class decrease as the strength level of the class increases. This is because $\mathcal{G}_c$ decreases with increasing strength in these materials. Brittle solids, such as ceramics (and composites based on them), behave differently. Increasing strength and fracture toughness go hand in hand in ceramics even though ceramic fracture toughnesses remain low regardless of their strength level. The fracture toughnesses of fiber-reinforced composites are impressive; they are typically much greater than the fracture toughnesses of their constituents.

The means for determining, and design use of, a material's fracture toughness have been addressed. Design philosophy depends on the value of a material's crack propagation stress versus its yield strength. If the latter is less than the crack propagation stress of a material containing as large a flaw as can be reasonably expected present, then the design operating stress is made less than σ_y; that is, design is based on preventing plastic flow. Materials falling into this category are generally low-strength ($\sigma_y \lesssim E/300$), high ductility metals that also have an appreciable fracture toughness. For high-strength metals ($\sigma_y \gtrsim E/150$), the fracture stress for structures containing flaws is usually less than σ_y. This necessitates design against crack propagation. For this circumstance the engineering science of fracture mechanics is most useful, and design involves insuring that the operating stress is less than σ_F for the flawed structure. Owing to the strong temperature dependence of their yield strengths, some steels constitute a special case. At high temperatures, where $\sigma_y < \sigma_F$, design is based on the plastic flow criterion. At low temperatures, σ_y increases to the extent that it becomes greater than σ_F. Use of such steels then entails design against fracture. Since the temperature defining the brittle-ductile transition is flaw-size dependent, considerable engineering art and science must be employed for the safe use of such steels.

The discussion of the previous paragraph has been couched in terms of metals, but it also applies to long-chain polymers since these materials can "fail" either by plastic flow or fracture. For ceramics, thermosets, and composites of them, only fracture needs to be considered in design. This is because the stresses at which they flow plastically at low temperature always exceed the stresses to propagate cracks in them.

Fracture can be caused by different types of stress state that lead to different crack propagation modes. The most important mode is that resulting from application of a tensile stress inducing crack opening in a direction parallel to the stress and effecting crack propagation in a direction normal to it. Indeed, this Mode I fracture has been the focus of this chapter. The details by which Mode I fracture takes place depend on the geometry of the material; e.g., the thickness of a sheet or plate sample. Plane stress fracture happens in thin sheets containing cracks. It is accompanied by appreciable thinning of the material in front of the crack as it advances, and this leads to development of shear lips on the fracture surface, a characteristic of plane stress fracture. The center of a thicker sheet is subject to plane strain conditions. That is, material removed from the plate surface is constrained from deformation in the thickness direction, and this material experiences tensile strains only in the two orthogonal directions. The absence of a strain in the thickness direction results in a stress in this direction, and the crack itself produces a similar effect in the other direction orthogonal to the applied stress. Thus, a triaxial state of stress is developed within the interior of a cracked thick plate, whereas a biaxial stress state is found in a thin sheet subject to plane stress conditions. In a finite-sized plate, the surface fractures under plane stress conditions and develops shear lips. The interior fracture surface is flat, a characteristic of plane strain fracture. As the thickness of the sheet or plate increases, the percentage of (plane strain) flat fracture surface does likewise.

The biaxial stress state of plane stress and, especially, the triaxial stress state of plane strain, promote fracture rather than flow. The greater effect of plane strain is manifested (generally) in a lower toughness. For plane stress, the toughness increases linearly with sheet thickness since the volume of plastically deformed material in front of the crack scales with this thickness. A maximum in toughness ($\mathcal{G}_c(45°)$) is reached at a thickness at which plane strain fracture just begins to develop in the sheet center. Further increases in sheet thickness result in a monotonically decreasing toughness and a limiting value, $\mathcal{G}_{Ic}$—the plane strain toughness—is approached. $\mathcal{G}_{Ic}$ (or K_{Ic}, the plane strain fracture toughness) is a material property just as the yield strength is. Use of K_{Ic} in design represents conservative design.

Plane strain fracture toughnesses can be determined by suitable tests. The fracture toughnesses obtained can then be used in design. For example, if a good estimate of the size of the largest expected in-service crack can be made, the crack propagation stress can be calculated, and the design operating stress is made less than σ_F. If σ_F is found to be less than the desired operating stress, the latter must be reduced or a material substitution made. Several examples of this procedure were provided in Sect. 9.4E.

Other, more qualitative, tests are used for assessing a material's fracture resistance. A Charpy impact test is one such test and, because of its convenience, is used widely. Moreover, correlation of service performance with data from this test exists, and this expands the test's utility. Results from impact tests are most significant for steels having low strength at high temperature, and vice versa. Absolute values of impact energy for these steels—obtained by measuring the energy required to fracture a notched bar under high-strain rate conditions—provide guidelines for avoiding brittle fracture in service. The impact energy value is less useful for high-strength materials—certainly not as useful as values of K_{Ic}, which can be directly utilized in design. Impact energies of high-strength materials are not useful for design primarily because these energies are low, even when the microscopic fracture process is a ductile one. The high-temperature impact energies of these materials do, however, correlate approximately with their fracture toughnesses. In a sense this is surprising, for the latter measures resistance only to crack propagation whereas an impact energy includes a (usually major) contribution due to the plastic deformation work preceding crack initiation and propagation.

Ceramics—and other brittle inorganics—received some special attention. One reason for this is that their fracture strengths in tension and compression differ appreciably. The difference arises because cracks propagate differently in the two loading situations. In tension, catastrophic fracture is triggered by the unstable propagation of the most deleterious flaw within or on the surface of the ceramic. In compression, a large number of cracks grow slowly and final fracture corresponds to material crushing by linkup of these cracks. Measured fracture strengths of ceramics are subject to considerable statistical variation as a consequence of the like variability in the preexisting flaw content and flaw size distribution. Means for dealing with this variability in design have been presented.

REFERENCES

Anderson, T. L.: *Fracture Mechanics, Fundamentals and Applications*, CRC Press, Boca Raton, Fla., 1991.

Ashby, M. F., and D. R. H. Jones: *Engineering Materials 2: An Introduction to Microstructures, Processing and Design*, Pergamon Press, Oxford, 1986.

Ewalds, H. L., and R. J. H. Wanhill: *Fracture Mechanics*, Edward Arnold, London, 1984.

Reed-Hill, R. E.: *Physical Metallurgy Principles,* 2d ed., D. van Nostrand, New York, 1973.

Sih, G. C. M.: *Handbook of Stress Intensity Factors*, Lehigh University, Bethlehem, Pa., 1973.

Srawley, J. E., and W. F. Brown: *Fracture Toughness Testing, ASTM STP* 381, 133, 1965.

Tetelman, A. S., and A. J. McEvily, Jr.: *Fracture of Structural Materials*, Wiley, New York, 1967.

PROBLEMS

9.1 The potential energy between two atoms in a solid can be expressed as a function of the interatomic separation distance (r) in terms of a Morse potential function,

$$U = U_0\{\exp[-2a(r - r_0)] - 2 \exp[-a(r - r_0)]\}$$

where U_0 is the binding energy and a is an empirical constant.

a Calculate the modulus of the material in terms of a, U_0, and r_0.

b Derive an expression for the theoretical fracture strength of the material in terms of the modulus and the parameters (a, r_0, U_0) as needed.

c Briefly discuss the *physical* significance of this expression for the ideal fracture strength. (That is, what physical parameters make a material theoretically strong or weak?)

9.2 Consider a potential energy–atomic volume curve of the kind considered in discussion of elastic behavior (Chap. 2).

a Estimate the theoretical fracture stress under conditions of hydrostatic tensile loading. (Note: The bulk modulus, K, rather than the tensile modulus, enters into your calculations. Your answer can be expressed in terms of E by using the relationship between K and E for isotropic materials and by assuming that Poisson's ratio $= \frac{1}{3}$.)

b Based on your analysis, what kind of relationship is expected to hold between the cohesive energy of a crystal (U_0) and its surface energy? Is this a reasonable correlation? Explain.

9.3 Compare the fracture stress as given by Eqs. (9.9) and (9.11). Under what conditions are the fracture criteria the same for the two formulations? For most kinds of brittle fracture, which of the two equations is generally most appropriate?

9.4 Using the relationship between σ_F and $\mathcal{G}_c$ (Eq. (9.12)), show that $\sigma_F < \sigma_y$ for a thin sheet containing small cracks. At what sheet thickness will σ_F exceed σ_y?

9.5 A mixed mode (i.e., plane stress and plane strain) fracture takes place for sheet thicknesses greater than $2x_c$ (cf. Fig. 9.12). Approximate $\mathcal{G}_c$ on the basis of a "volume-fraction rule" for toughness, and derive an expression for $\mathcal{G}_c$ in terms of sheet thickness (t), $\mathcal{G}_c(45°)$, and $\mathcal{G}_{Ic}$. Schematically plot $\mathcal{G}_c$ vs. t. Does your sketch resemble Fig. 9.12?

9.6 Consider a compact tension test sample (Fig. 9.16) for which crack propagation takes place at an applied load of 10^5 N. Sample dimensions are $a = 10$ cm, $c = 5$ cm, and $t = 5$ cm. Calculate K_c for this material. If the material's yield strength is 500 MN/m^2, is K_c equal to K_{Ic}?

9.7 If a through-crack of length 2.5 cm is present in a material, fracture occurs at a stress of 700 MN/m². What is this material's fracture toughness?

9.8 For a titanium alloy having a surface elliptical flaw shaped like that of Case Study 2 of Sect. 9.4E, calculate the deepest crack length that can be tolerated and still avoid crack propagation at an operating stress equal to 70% of the alloy's yield strength. Material properties: $K_{Ic} = 115$ MN/m$^{3/2}$, $\sigma_y = 900$ MN/m².

9.9 Refer to Case Study 1 of Sect. 9.4E for this problem.
  a Let the operating stress be 180 ksi. Consider the several versions of Grade A steel, and select the one best suited for use as the pressure vessel. Give the reason(s) for your choice.
  b If one were to use the 220 ksi Grade A steel, what would be the maximum safe operating stress?

9.10 Continue the iterative procedure in Case Study 2 of Sect. 9.4E, and obtain increasingly accurate estimates of σ_F.

9.11 Calculate the largest size crack the rocket motor case of Case Study 3, Sect. 9.4E, can withstand without catastrophic failure.

9.12 The fracture strength of tungsten at 225 K is 280 MN/m² when 5-cm-long cracks are present on its surface. The yield strength of tungsten at this temperature is 700 MN/m². If $\mathscr{G}_{Ic}$ increases linearly with temperature, by 0.3 MN/m per degree K, what is the maximum safe operating stress in a welded tungsten structure at 200 K if the minimum detectable flaw size is 2.5 cm and a residual tensile stress of 70 MN/m² exists within 1.25 cm from the weld? Assume the residual stress is parallel to the applied tensile stress. Data for W: $E = 410$ GN/m²; $\nu = 0.3$.

9.13 As shown in the figure below, a hot-rolled 2-cm-thick 4340 steel plate containing a bolt hole develops cracks at this hole after heat treatment. The steel has a yield strength of 1000 MPa, a tensile strength of 1300 MPa, and a plane strain fracture toughness of 90 MPa·m$^{1/2}$. For a flaw of the type shown, $f(c/a) = 1.4$. The design operating stress for this steel is 650 MPa. Will the plate fail?

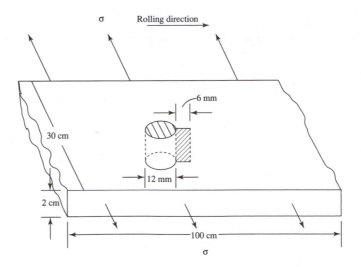

9.14 A certain ceramic has a fracture toughness, K_{Ic}, of 5 MN/m$^{3/2}$. A batch of samples of this material contains surface flaws, existing along the circumference and of depth 10 μm. Estimate (a) the tensile strength and (b) the compressive strength of the samples.

9.15 a Construct a fracture surface for biaxial loading ($\sigma_1, \sigma_2 \neq 0; \sigma_3 = 0$) of a ceramic. (Normalize your stress axes in terms of the tensile fracture stress, T.S.)
b Draw a second fracture surface which corresponds to a triaxial stress state for which $\sigma_3 = 0.2$ T.S.

9.16 A series of cylindrically shaped ($l = 25$ mm, $d = 5$ mm) ceramic samples are tested in tension. The Weibull distribution coefficient is 5, and 50% of the samples fracture at a stress of 100 MN/m^2 or less. Suppose we construct a cylindrical component of this material having a length $l = 50$ mm and diameter $d = 10$ mm. It is required to have a survival probability for this structure of 95% when it is subjected to a tensile stress σ'. What is the allowable value of σ'?

9.17 When a brittle material of volume V is subjected to a uniform tensile stress, the survival probability, $P_s(V)$, is given by

$$P_s(V) = \exp\left[-\frac{V}{V_0}\left(\frac{\sigma}{\sigma_0}\right)^m\right]$$

where V_0, σ_0, and m are constants.

If, instead of being constant, the tensile stress varies with position in the sample, the survival probability becomes

$$P_s(V) = \exp\left(-\frac{1}{V_0\sigma_0^m}\int \sigma^m \, dV\right)$$

MOR tests were carried out on a beam of silicon nitride (dimensions: length l, thickness d, width b). Fifty percent of the beams broke at a tensile stress (σ_r) less than or equal to 500 MN/m^2. An identical sample is to be loaded in tension (gage length $= l$). Calculate the tensile stress at which 50% of the samples should be broken. Hints and formula: For silicon nitride, $m = 10$. For the MOR test you should integrate $\sigma^m \, dV$ only over the lower half of the beam (why?). The tensile stress in a MOR test at the position (x, y) (see accompanying sketch) is given by

$$\sigma = \sigma_r \frac{y}{(d/2)}\left(\frac{(l/2 - x)}{l/2}\right)$$

Toughening Mechanisms and the Physics of Fracture

10.1
INTRODUCTION

This chapter is concerned with how materials fracture and ways to improve their fracture resistance. The latter topic, *toughening*, occupies the first several sections of this chapter. In them we describe the irreversible work accompanying crack propagation which is, of course, tied directly to the material toughness, $\mathcal{G}_c$. Since the toughness is related to the fracture toughness, knowledge of $\mathcal{G}_c$ translates directly into design against crack propagation.

The different material classes are toughened in different ways. That is, the work accompanying crack propagation differs among the classes. In metals and long-chain polymers, this work is associated with crack tip plasticity. Crack tip plasticity is limited in ceramics. Thus, other means are used to toughen them. Fiber composites of brittle materials have toughnesses that are generally quite high, typically much greater than the toughnesses of their constituents. This is a result of the work accompanying pullout of fibers that bridge the fractured matrix surfaces during crack propagation. Details relating to these mechanisms are described later. For here, though, it is important to note that the commonalty to our description of material toughness is that it is the irreversible work accompanying crack propagation. Yes, this work is manifested in different forms in different materials. But if we can identify the nature of this work we can estimate material toughness.

Material toughness is useful for design because it is coupled to the fracture toughness, which is the design parameter for materials containing cracks or flaws. Crack propagation features in a flawed material have parallels in the tensile fracture behavior of these same materials not containing flaws. Thus, we describe the types of low-temperature tensile fracture and how they are related to the bonding and crystal structure of the various material classes. This is followed by a more extended description of the details by which low-temperature fracture takes place. This treatment is worthwhile for while it is true that the material toughness/fracture

454

toughness are the design properties, these properties relate to the manner in which cracks propagate (and sometimes nucleate) in a material.

10.2
TOUGHENING IN METALS

In this section, and the several that follow it, we discuss the primary means by which fracture toughness can be increased. The material classes—metals, ceramic, composites, and polymers—are considered separately. On one account this is reasonable because the mechanisms of toughening are different in the material classes. On the other hand, it is artificial in that all materials are toughened by increasing the work, $\mathcal{G}_c$, accompanying crack propagation. Our description of toughening is couched in terms of $\mathcal{G}_c$; conversion to the fracture toughness, K_c, is accomplished through Eqs. (9.14) and (9.15). We start with metals. The engineering science of fracture mechanics was first developed in response to their unintended fracture, so it is appropriate to discuss them first. Further, the physics of toughening in metals is perhaps the easiest to visualize.

Figure 10.1 is a variation on previous figures. It illustrates the geometry of, and the work associated with, crack advance in a metal displaying crack tip plasticity. The geometry is characterized by the radius of the crack tip, ρ. As the crack advances a thin layer of plastically deformed material is left in its wake; the region is the remnant of the ligament in front of the crack which was strained plastically prior to its fracturing. This region extends a distance $\pm\rho/2$ above the fracture plane. The plastic fracture work per unit volume of the ligament is the product of the flow stress and a failure strain. The volume of material affected is the crack plane area (A) multiplied by ρ. Dividing the total fracture work by the fracture plane area gives the toughness, $\mathcal{G}_c$. The appropriate values of flow stress and fracture strain depend on the stress state. For plane strain conditions, for example, the flow stress is increased above that observed for tensile loading (the triaxial stress state at the crack tip sees to that) and the plane strain fracture strain is below that found in tensile loading (again the triaxial stress state is the culprit). However, the plane strain fracture stress and strain both scale with the corresponding stress and strain found

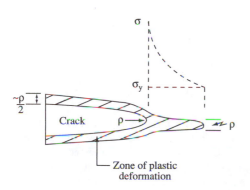

Figure 10.1
Schematic of the plastic work accompanying crack propagation. The gage length of the plastic zone in front of the crack tip is on the order of the crack tip radius of curvature, ρ. This length is extended to a strain, ε_{fr}, before it fails. The plastic work per unit area of fracture surface, therefore, is proportional to the product of the material yield strength, σ_y, the strain, ε_{fr}, and the gage length, ρ. A proportionality constant takes into account the complications arising from the stress state at the crack tip.

456

CHAPTER 10
Toughening
Mechanisms
and the Physics
of Fracture

in a tension test. We incorporate the effects of stress state into a constant, B, and write the plane strain toughness, $\mathcal{G}_{Ic}$, as

$$\mathcal{G}_{Ic} = B\sigma_{yf}\varepsilon_{fr}\rho \tag{10.1}$$

where σ_{yf} is the tensile flow stress at fracture and ε_{fr} is the true tensile fracture strain.

Increases in material flow strength increase $\mathcal{G}_{Ic}$. So do increases in ductility as exemplified by the term ε_{fr} in Eq. (10.1). However, the crack tip radius of curvature is also greater for low-strength materials because of the greater crack tip blunting in low-strength materials. Thus, as a general rule we can state that, within a given metal class (e.g., quenched and tempered martensitic steels), materials having a low yield strength (high ductility) have a greater toughness than do higher-strength metals of the same alloy class.

Figure 10.2 and Table 10.1, which illustrate the relationship between fracture toughness and yield strength for several metal alloy classes, further make this point. For each of these alloy classes, K_{Ic} decreases with increases in σ_y. The resulting philosophy inherent to development of fracture-resistant metals is thus as illustrated

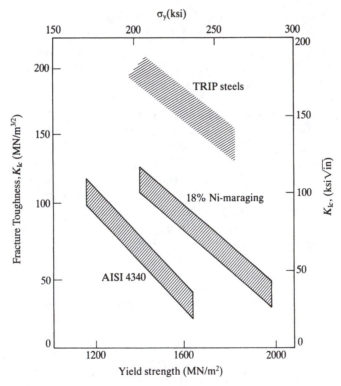

Figure 10.2
Fracture toughness–yield strength relationships for three ferrous alloy classes. The 4340 steel is a quenched-and-tempered martensitic steel. The maraging steel is a low-carbon martensite strengthened by nickel in solid solution and by fine precipitates. Maraging steels demonstrate fracture toughnesses superior to quenched and tempered steels. The characteristics of TRIP steels have been discussed in Sect. 5.9. The energy required to induce the austenite-martensite transformation in them contributes to their high fracture toughnesses. (*After V. F. Zackay, E. R. Parker, J. W. Morris, Jr., and G. Thomas, Matls. Sc. Eng., 16, 201, 1974.*)

in Fig. 10.3. That is, materials development focuses on raising the level of the K_{Ic} curve while maintaining the yield strength level. This can be accomplished by either development of new alloys or by improvements in materials processing.

Figure 10.2 provides some inkling of how alloy development can improve fracture toughness. The materials of Fig. 10.2 are ferrous alloys. The alloy steel AISI 4340 is a quenched and tempered martensitic steel. As described in Sect. 5.9, such a steel consists of a bcc matrix containing interstitial carbon in excess of the equilibrium solubility. The matrix also contains dispersed carbide particles. Both microstructural features affect the yield strength of quenched and tempered 4340. The "maraging" steel is strengthened somewhat similarly. However, the maraging steel contains very little carbon. The martensite in maraging steels is strengthened by substitutional Ni atoms which produce a matrix inherently tougher than the 4340 matrix whose interstitial carbon negatively impacts toughness. Precipitation hardening increases further the strengths of maraging steels while maintaining their high fracture toughness. The TRIP steels have also been briefly described in Sect. 5.9. These steels have even more impressive fracture toughnesses than maraging steels. The additional toughness in TRIP steels is due to a strain-induced martensitic transformation developed in the plastic zone during crack propagation. The fcc matrix of TRIP steels is, in effect, metastable and the fcc→ martensite (a low carbon and tough martensite in the case of TRIP steels) transformation is catalyzed by plastic deformation. Thus, in addition to the toughness term of Eq. (10.1), the additional work associated with the martensite transformation contributes to the toughness of

Alloy class	$\sim\sigma_y$(MN/m^2)	$\sim K_{Ic}$(MN/m$^{3/2}$)
2000 series Al alloys	300–350	37
	350–400	32
	400–450	26
	450–500	26
	500–550	22
7000 series Al alloys	400–450	34
	450–500	28
	500–550	28
	550–600	25
4340 alloy steel	1300–1400	105
	1400–1500	88
	1500–1600	59
	1600–1700	55
Maraging steels	1400–1700	83
	1700–2000	80
	2000–2300	40
Ti-6Al-4V alloys	800–900	105
	900–1000	105
	1000–1100	55

Note: Values listed are approximate guidelines on the relationship between K_{Ic} and σ_y for the alloy systems illustrated, and should not be used for design purposes. Table compiled from data provided by R. W. Hertzberg, *Deformation and Fracture Mechanics of Engineering Materials,* Wiley, New York, 1976.

Table 10.1
Approximate relationship between yield strength and K_{Ic} for several metallic alloy systems

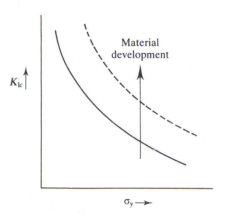

Figure 10.3
Philosophy employed to develop high-fracture-toughness materials. New materials are sought so that, at an equivalent σ_y, their fracture toughnesses are increased or vice versa. This means that the K_{Ic}-σ_y curve should be displaced vertically (or, equivalently, to the right) in the figure.

458

CHAPTER 10
Toughening
Mechanisms
and the Physics
of Fracture

TRIP steels. As we discuss later, certain ceramics are toughened by a somewhat similar transformation.

So why not just use TRIP or maraging steels in applications calling for high-toughness ferrous materials? Alas, engineering is always a compromise. Economic considerations may preclude the use of the more expensive, high-fracture-toughness alloys. Moreover, in some applications (e.g., aerospace), weight considerations are at a premium and steels, even the highest toughness ones, cannot be used because of their high densities. Instead, lower-density materials such as Al and Ti alloys are used, even though they are not as fracture tough. Titanium alloys are also often used in chemically corrosive environments for which ferrous materials are not suitable. These, and like considerations, make the materials selection component of design frustrating but also challenging.

In addition to developing new materials, materials engineers can increase (sometimes appreciably) the fracture toughness of a given alloy class by improvements in the techniques used to process them. For example, many high-strength steels are now melted in vacuum, and this minimizes the presence of nonmetallic inclusions in them; the fracture of such inclusions is a critical step in the low-energy ductile fracture of many steels as is discussed later. Additionally, steels and other alloy classes can be made less fracture prone by controlling their chemistry, particularly with respect to "tramp" elements. For example, phosphorus is particularly detrimental to the fracture resistance of steels. Phosphorus segregates to grain boundaries in them, promoting brittle intergranular fracture. Reducing the phosphorus content of steels, therefore, increases their fracture toughness. It is distressing that while the tramp elements and inclusions in steels do not much affect their yields strengths, they rather dramatically and negatively impact their fracture toughnesses. Thus, for example, the scatter band of fracture toughness–yield strength of AISI 4340 in Fig. 10.2 is likely due to variations in processing. The highest K_{Ic}'s correspond to "clean" steels; the lowest to "dirty" ones.

How do these microstructural factors affect the various parameters in Eq. (10.1)? To understand how they do, we revisit crack propagation. We use Fig. 10.4 to illustrate. It illustrates the crack propagation process in materials—such as steels—that undergo a ductile-to-brittle transition. At the lowest temperatures, where the steel is most brittle, the plastic zone in front of the crack tip is minimal (Fig. 10.4c). Crack propagation takes place by repeated cleavage crack initiation at the crack tip. Although the fracture here is brittle in nature we note that it is preceded by some plastic deformation since the yield strength of the steel is less than its theoretical fracture strength; thus, material toughness remains represented by Eq. (10.1). At these low temperatures, the yield strength is high, but the fracture strain and the crack tip radius of curvature are low, resulting in low toughness.

As temperature is increased, material yield strength decreases. Thus, a plastic zone forms in front of the crack. Since the conditions for fracture are most favorable at the boundary between the elastic and plastic zones, cleavage cracks initiate at this boundary and crack propagation involves linkup between the boundary and the main crack tip. The fracture of the grains in the linkup zone is usually a mixture of cleavage and fibrous (ductile) fracture (recall our discussion of the impact test in Sect. 9.6), with the percentage of fibrous fracture and the size of the plastic zone increasing with temperature. At the highest temperatures of brittle fracture (Fig. 10.4a), linkup is entirely by fibrous fracture. As temperature is increased, the crack tip radius of curvature also becomes increasingly blunted. This factor, along with

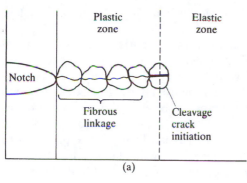

Plastic zone | Elastic zone

Notch

Fibrous linkage | Cleavage crack initiation

(a)

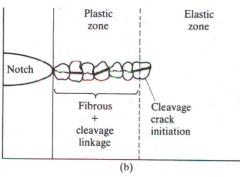

Plastic zone | Elastic zone

Notch

Fibrous + cleavage linkage | Cleavage crack initiation

(b)

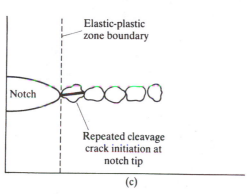

Elastic-plastic zone boundary

Notch

Repeated cleavage crack initiation at notch tip

(c)

Figure 10.4
Brittle fracture initiation in front of a crack. A cleavage crack is nucleated at, or close to, the elastic-plastic zone boundary. Linkup with the main crack depends on temperature and plastic-zone size. (a) At the highest temperatures of brittle fracture, the plastic-zone size is large and linkup takes place primarily by fibrous fracture of the grains within this zone. (b) As the temperature decreases, the zone size shrinks and cleavage fracture of grains within the zone becomes dominant. (c) At even lower temperatures, the plastic-zone size is comparable to the grain size. Hence, fracture is initiated at the crack tip and crack propagation no longer involves linkup. Instead, fracture occurs by repeated cleavage of grains at the crack tip. With decreasing temperature, both the plastic-zone size and the fraction of grains failing in a fibrous manner decrease. Both effects result in a decrease in fracture toughness with decreasing temperature.

the higher fraction of ductile fracture in the linkup zone, causes material toughness to increase with temperature.

Crack propagation in a material failing in a ductile manner is illustrated in Fig. 10.5. Ductile fracture progression involves a linkup process but it is of a different nature than that for brittle fracture. Crack progression is often tied to second-phase particles (inclusions) that fracture at the elastic-plastic zone boundary.[1] These

---

[1]Other microstructural features that cause plastic strain incompatibility (e.g., twin-twin and slip-band intersections, etc.) can also nucleate microcracks at the elastic-plastic zone boundary. Such events are common to inclusion-free single-phase metals. In our discussion, however, we focus on second-phase fracture or decohesion as the microcrack nucleation event. This covers the important case of most high-strength metals. Further, the substance of the fracture process is the same irrespective of the origin of the microcrack.

460

CHAPTER 10
Toughening
Mechanisms
and the Physics
of Fracture

particles are rather like the grains that cleave at this boundary in materials displaying brittle crack propagation.

The stress that causes inclusion fracture/decohesion arises from the external stress (taking into account the stress alteration due to the crack presence, e.g., Fig. 9.10) and from that coming from geometrical dislocations generated by the requirement of material contiguity across the inclusion-matrix interface. Both stresses are expected to be most conducive to particle fracture at the elastic-plastic zone boundary. Whether crack linkup takes place directly (Fig. 10.5a) between the crack tip and the first particle fractured there depends on several factors. The plastic-zone size is one; the smaller it is, the greater the chance of fracture taking place by direct linkup. Likewise, if the particle density is low (or spacing high), it is also expected that crack linkup occurs in the direct manner. When these conditions do not apply, other internal microcracks form within the plastic zone and linkup occurs by a stepwise progression through the zone, as illustrated in Fig. 10.5b.

As noted, ductile fracture crack propagation bears a resemblance to the mixed cleavage–fibrous fracture mode discussed. In ductile fracture, the fractured or decohered particles are analogous to the cleaved grains, and the matrix between the particles to the fibrously fractured material, in the brittle fracture case. Toughness is also determined by similar considerations for the two fracture modes. For example, ductile fracture may be of the low-energy type if there is a large density of decohered/fractured particles and a small linkup distance between them. (The latter is manifested in Eq. (10.1) by a reduced crack tip radius, which scales with the linkup distance.)

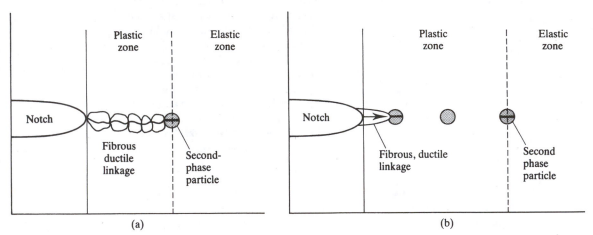

(a)   (b)

Figure 10.5
Ductile fracture crack propagation. (a) When the plastic-zone size is small, fracture initiates by second-phase particle decohesion/fracture at the elastic-plastic zone boundary, and linkup across the zone occurs by fibrous fracture of matrix grains. The energy associated with this type of fracture depends on the size of the plastic zone and whether or not other second-phase particles are located along the linkup path. When they are, the fracture work will be less than for when they are not (the situation shown here). (b) With decreasing yield strength, the plastic-zone size increases and this is associated with an increased toughness. This effect is mitigated to a degree if other fracture initiation sites (e.g., other second-phase particles) are present in the plastic zone. For the case shown here, linkup occurs by a stepwise progression from the crack to the closest fracture-initiation site within this zone.

Estimating a material's toughness becomes increasingly difficult as its ductility and linkup path increase and its yield strength decreases. Indeed, the detailed mechanics of high-energy ductile fracture remain incompletely known. Fortunately, this is not a particularly serious drawback in design. High-energy ductile fractures are associated with low-yield strength materials and, as such, the use of these materials is dictated by design against yielding and not fracture. In distinction, microscopically ductile—but low-energy—fracture is well characterized and fracture toughnesses pertaining to these conditions are readily measured. For these materials, $\sigma_F < \sigma_y$, and design is predicated on insuring that the operating stress lies below σ_F.

10.3
TOUGHENING IN CERAMICS

The toughness of ceramics, the most brittle solids, devolves on crack propagation work different from that responsible for the (generally) high toughnesses of metals. Metal toughness derives from the plastic deformation work accompanying crack propagation. However, ceramics display essentially nil of this type of deformation. Thus, different—sometimes radically different—means are employed to improve ceramic fracture resistance as is discussed in this section. The approach we take, though, is basically the same as that used to describe metal toughnesses. That is, we determine the irreversible work associated with crack propagation. Before going into this in detail, we need to introduce the concept of thermal residual stresses. These can be present in all materials. However, it is only in ceramics that these stresses may profoundly impact fracture behavior.

A. Thermal Stresses

Consider Fig. 10.6, which depicts a solid consisting of the two phases β and γ. The solid is heated to a temperature T_1 and then cooled to the temperature T_2. All solids expand on heating and contract on cooling. The extent to which they do is gaged by their *linear coefficient of thermal expansion*, α_l. This coefficient is a material-specific property and is defined by

$$\alpha_l = \left(\frac{1}{l}\right)\left(\frac{dl}{dT}\right) \tag{10.2}$$

In Eq. (10.2), l is the length of a sample at a standard temperature. Thus, the change in length (the thermal expansion or contraction) on heating/cooling the material from one temperature to another is found by integrating Eq. (10.2). The fractional change in length can be considered a thermal strain, ε_{th}. This strain can be expressed as

$$\varepsilon_{th} = \int \frac{dl}{l} = \alpha_l \Delta T \tag{10.3}$$

where $\Delta T (= (T_2 - T_1))$ is the difference in the two temperatures.

462

CHAPTER 10
Toughening
Mechanisms
and the Physics
of Fracture

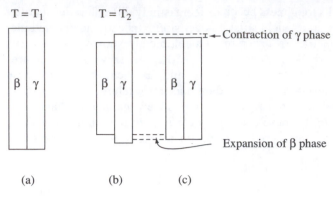

Figure 10.6
(a) Plates of two phases (β and γ) are bonded together and have the same length at the temperature T_1. (b) If the phases were not bonded and were cooled from T_1 to a temperature T_2, they would have different lengths because their thermal expansion coefficients differ. In the example shown here the β phase would have a lesser length because it expands on heating (and contracts on cooling) more than the γ phase; i.e., the β phase has a greater thermal expansion coefficient (c). The bonding between the phases causes them to have the same length at all temperatures between T_1 and T_2. Thus, at T_2, the β phase is stretched and the γ phase is compressed so as to maintain the same lengths; thus, the β phase experiences a tensile thermal stress and the γ phase experiences a compressive thermal stress.

The two phases of Fig. 10.6 have different thermal expansion coefficients. Figure 10.6*b* shows how they would contract were they not bonded together. We have taken the β phase to have the greater coefficient of thermal expansion. Thus, on cooling the β phase would "separate" itself from the γ phase were it not bonded to it as indicated in Fig. 10.6*b*. The bonding prevents this separation by producing a tensile traction across the interface which stretches the β phase and contracts the γ one (Fig. 10.6*c*). That is, the γ phase is in compression and the β phase is in tension. There is a stress generated by this accommodation. Its magnitude can be estimated. The contraction strain of the unconstrained β phase is $\alpha_\beta(T_2 - T_1) = \alpha_\beta \Delta T$. That of the unconstrained γ phase is $\alpha_\gamma \Delta T$. Thus, the strain accommodated is the *difference* in the individual thermal strains; i.e., $\varepsilon_{th} = (\Delta\alpha)\Delta T$ where $\Delta\alpha$ is the difference in thermal expansion coefficients. The magnitude of the thermal stress is given by multiplication of the differential strain by an appropriate average modulus (E).[2] Thus, the magnitude of the thermal residual stress, σ_R, is approximated as

$$\sigma_R \cong E(\Delta\alpha)\Delta T \qquad (10.4)$$

For one of the phases (the one that contracts the most), $\sigma_R > 0$. For the other, the one that contracts less, $\sigma_R < 0$.

Thermal residual stresses are important in ceramics, for these materials are almost always subject to thermal treatment during their manufacture. Thus, in a two-phase ceramic one of the phases is subject to a tensile, and the other to a

---

[2]This relation assumes that the thermal strain is accommodated by elastic deformation. This is the situation for ceramics. In metals, any "thermal" mismatch can also be accommodated by plastic strain. Values of the appropriate modulus to use in Eq. (10.4) are illustrated by Prob. 10.1.

compressive, residual stress. If the magnitude of these stresses is great enough, fracture across interphase boundaries takes place resulting in the formation of microcracks.[3] (These microcracks are often the preexisting flaws whose propagation gives rise to fracture.) Thus, processing of two-phase ceramics requires care that this not happen.

Even single-phase ceramics may experience thermal residual stresses. In general, the linear coefficient of thermal expansion varies with crystallographic direction. For cubic structures, however, the expansion coefficient is isotropic and, thus, residual thermal stresses are not found in single-phase cubic ceramics. However, many ceramics are not cubic and manifest anisotropic thermal expansion coefficients. In single-phase materials of these ceramics, thermal residual stresses may therefore be present. For example, if a grain wishes to contract more in one direction than another during cooling and a neighboring, differently oriented, grain did the same, a thermal stress would be developed across their boundaries. However, the magnitude of thermal residual stresses in single-phase ceramics is less than in two-phase ceramics because the anisotropy-caused difference in thermal expansion coefficients is generally much less than the difference in thermal expansion coefficients among materials.

B. Toughening Due to Crack Deflection and Geometry

Crack propagation in polycrystalline ceramics is often intergranular in nature. A photograph illustrating this in alumina is provided in Fig. 10.7. The toughness associated with intergranular crack propagation differs from that of transgranular crack propagation. For the latter situation, and in a fully brittle material, the toughness is (cf. Sect. 9.3) 2γ, where γ is the material surface energy. At first glance, it might appear that the work associated with intergranular crack propagation is less than this because the two fracture surfaces formed via intergranular fracture replace a previously existing grain boundary. That is, the fracture work per unit area is $(2\gamma - \gamma_{gb})$ where γ_{gb} is the grain-boundary energy. However, intergranular crack propagation is associated with a fracture area greater than the nominal fracture plane area. This mitigates to a degree the reduced surface energy term. Further, crack deflection from the nominal fracture plane is associated with Mode II crack propagation (recall Fig. 9.9). This increases the toughness further. The net result is that the toughness associated with intergranular fracture typically exceeds that of transgranular fracture by a factor of about 2–4. However, this is not much of an increase. The related increase in K_{Ic}, for example, would be a factor of 2 at most. In addition, the effect is usually implicitly incorporated into the measured fracture toughness which is taken as the "base" fracture toughness of the material.

Although crack deflection in single-phase ceramics does not much affect their toughness values, the situation can be different in two-phase ceramics. In them, the residual stress can result in additional frictional work associated with final separation along the macroscopic fracture plane. This *crack bridging* can result in significant increases in toughness in certain cases (Sect. 10.3E).

---

[3]The process by which microcracks form, release of elastic strain energy and formation of surface energy, is similar to that accompanying brittle fracture (Sect. 9.3).

464

CHAPTER 10
Toughening
Mechanisms
and the Physics
of Fracture

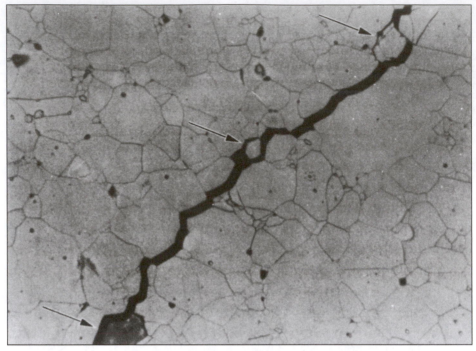

Figure 10.7
A crack in polycrystalline alumina with a mean grain size of 25 μm. The fracture is along grain boundaries and, as a result, the actual fracture surface area is greater than the nominal one which does not take into account such crack path diversions. The diversions add to the fracture work. (*From P. L. Swanson, C. J. Fairbanks, B. R. Lawn, Y.-W. Mai, and B. J. Hockey, J. Amer. Cer. Soc., 70, 279, 1987.*)

C. Microcrack Toughening

The stress concentration in front of a crack (acting sometimes in concert with thermal residual stresses) can result in additional work associated with crack propagation. This is illustrated in Fig. 10.8, which indicates that this stress can cause additional microcracking (presumed here along grain boundaries) above and below the fracture plane. The amount of this additional work is, in principle, easy to estimate. One need but determine the additional fracture area associated with the microcracks (A_{mc}) and multiply this by the associated toughness ($\cong (2\gamma - \gamma_{gb})$ for intergranular fracture). Since the toughness associated with the additional microcracks is presumed the same as for the main crack ($= \mathcal{G}_{co}$), the total toughness is given by

$$\mathcal{G}_c = \mathcal{G}_{co}\left\{1 + \frac{A_{mc}}{A_f}\right\} \tag{10.5}$$

where A_f is the nominal fracture plane area.

Equation (10.5) can be expressed differently. The distance above and below the fracture plane over which microcracking exists is r_c (Fig. 10.8b). Assuming that the residual stress is paramount in nucleating microcracks, the work expended in form-

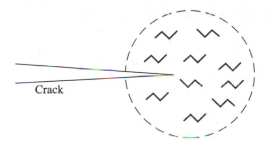

(a)

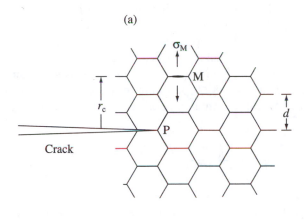

(b)

Figure 10.8
(a) Schematic of secondary microcracks that form in front of a crack in a brittle material. The work associated with forming the secondary microcracks contributes to material toughness. (b) Expanded view illustrating coordinates used to describe microcrack toughening. The grain size is *d*, and secondary microcracks form at sources (M) up to a distance r_c above and below the primary crack plane (P). The stress, σ_M, that initiates the secondary microcracking arises from the applied stress plus any residual thermal stresses. (*From Brian Lawn, Fracture of Brittle Solids, 2nd ed., Cambridge University Press, Cambridge, 1993.*)

ing them is an elastic work (which scales with the residual stress, σ_R,[4] multiplied by the thermal strain $(\Delta\alpha\Delta T)$ where $\Delta\alpha$ is the effective difference in thermal expansion coefficient between neighboring grains) multiplied by the distance over which microcracks form. Thus,

$$\Delta\mathcal{G}_c \cong 2r_c\sigma_R(\Delta\alpha)(\Delta T)V_f \tag{10.6}$$

where V_f in Eq. (10.6) represents the fraction of grains within the affected volume that are associated with microcracks; an upper bound on $\Delta\mathcal{G}_c$ is obtained with $V_f = 1$.

Both Eqs. (10.5) and (10.6) express the essential physics of microcrack toughening. However, we can go further though doing so makes life more complicated. We wish to understand how the additional toughness depends on material properties and structure. We will use single-phase solids to illustrate even though the added toughness in them coming from microcracking is generally much less than it is in two-phase ceramics. The analysis presented here, though, can be easily extrapolated to two-phase ceramics (Prob. 10.3).

---

[4]The actual stress at the crack tip is, because of the presence of the applied stress, greater than σ_R. The approximation used here focuses on the physics of toughening, even though precision is lost by our approach.

466

CHAPTER 10
Toughening
Mechanisms
and the Physics
of Fracture

For microcracks to retard crack propagation, they must form *during* crack propagation and not before it. If they are a priori present, they add nothing to (in fact, they can detract from) material toughness. The condition to develop preexisting microcracks is estimated from Eq. (9.16) relating stress, stress intensity, and flaw size, the latter here taken as the grain size. Doing this, we see that there is a critical grain size (d_c) *above which* microcracks form *spontaneously*. It is given by

$$d_c = \left(\frac{K_{co}}{\sigma_R}\right)^2 \tag{10.7}$$

where K_{co} is the fracture toughness of the matrix and σ_R the thermal residual stress.[5] Thus, for a material to be capable of microcrack toughening it must have a grain size *less* than that stipulated by Eq. (10.7). We note that when residual stresses are absent (as in cubic single-phase ceramics), no spontaneous microcracking is expected (i.e., d_c goes to infinity when the residual stress goes to zero).

As noted, the extent of microcrack toughening depends on r_c. If r_c is less than the grain size, then microcracking makes no contribution to toughness; the "additional" microcracks are subsumed within the main crack. Analysis shows that r_c scales with grain size (d) as

$$\frac{r_c}{d} \cong \left\{\frac{0.23}{\left[1 - \left(\frac{d}{d_c}\right)^{1/2}\right]}\right\}^2 \tag{10.8}$$

Equation (10.8) is plotted in Fig. 10.9. Note that when $d/d_c \to 1$, r_c goes to infinity. This is not as desirable as you might think; it corresponds to spontaneous microcracking. Thus, to achieve microcrack toughening, the grain size must be less than d_c but not too much less for when $r_c/d < 1$ the width of the microcracked zone is essentially zero. Thus, there is a relatively narrow "window" of grain size for which microcracking can toughen a ceramic. Provided the grain size permits such toughening, we see that the extent of the microcracked zone increases with grain size. So what is desired is a grain size below a critical one (to prevent spontaneous microcracking), but not much below this one (so as to maximize the microcrack zone size). That the window of grain size is small means that the ratio of d/d_c is on the order of (but somewhat less than) unity (cf. Fig. 10.9). If we assume that r_c scales with d_c, and making this substitution in Eq. (10.6), we find that the expression for the microcracking-effected toughness can be written as

$$\Delta\mathscr{G}_c = \frac{\beta\mathscr{G}_{co}(\Delta\alpha)(\Delta T)E}{\sigma_R} \tag{10.9}$$

where we have taken V_f of Eq. (10.6) as equal to one. The coefficient, β, turns out to be on the order of 0.1. Note that there is an implicit grain-size effect in Eq. (10.9). In particular, the stress required to effect microcracking (taken as σ_R in Eqs. (10.6) and (10.9)), varies inversely with grain size. Thus, provided we are in the grain-size domain where microcrack toughening exists, the toughening is greater for materials with larger grain sizes, a result arrived at just a few sentences ago.

---

[5]We are cavalier with proportionality constants in our discussions. The book by Lawn listed at the end of this chapter deals with matters more precisely.

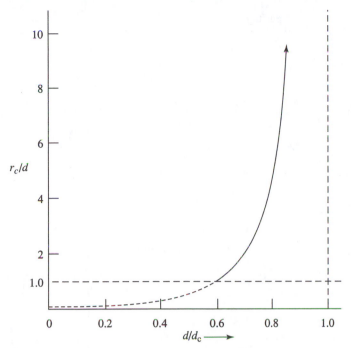

Figure 10.9
The ratio of the distance r_c over which microcracks form vs. the material grain size,
d; r_c is normalized in terms of the grain size, and the grain size is expressed in terms
of the grain size, d_c, that results in spontaneous microcracking on cooling. When this
normalized grain size is greater than one, microcracks so form and do not contribute
to toughness. When the radius r_c is less than the grain size, secondary microcracks
are subsumed within the main crack. Thus there is a narrow window of grain size
(illustrated by the solid line) over which microcrack toughening is found.

Microcrack toughening in single-phase materials is modest. However, this is
not so for two-phase ceramics with their greater residual thermal stresses. In two-
phase ceramics, the fraction of active grains can be made quite large and the incidence of microcracking in the "process zone" is high (i.e., $V_f \rightarrow 1$).

D. Transformation Toughening

During the past 20 years the ceramic analog to the martensitic transformation found
in TRIP steels was discovered in partially stabilized zirconias (PSZ). The difference
between the two phenomena is that the martensite transformation in TRIP steels is
induced by plastic strain; that in PSZ is induced by stress and is analogous to an
elastic transformation. This *transformation toughening* has resulted in fracture
toughnesses approaching 20 MPa · m$^{1/2}$ in certain PSZ. While this is not a high
fracture toughness by the standards of tough metals, it is high enough to permit serious consideration of these materials in critical applications.

We first discuss the basis for transformation toughening. A portion of the
ZrO_2-CaO phase diagram is shown in Fig. 10.10. Pure zirconia has two allotropic
forms; a high-temperature tetragonal form and a low-temperature monoclinic form.

468

CHAPTER 10
Toughening
Mechanisms
and the Physics
of Fracture

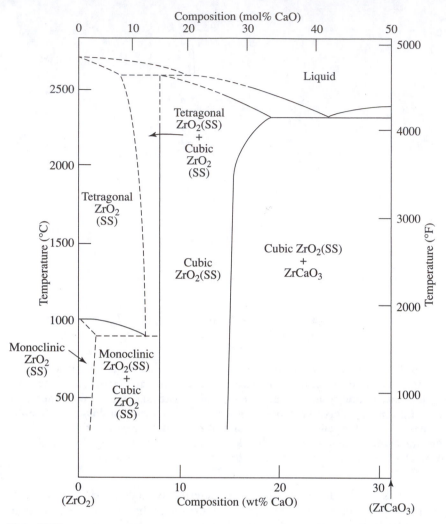

Figure 10.10
The ZrO$_2$-rich portion of the zirconia-calcia phase diagram. Adding CaO produces a stable cubic form of zirconia. Partially stabilized zirconias (PSZ) have compositions corresponding to two-phase equilibrium between tetragonal and cubic zirconia (at high temperatures) and monoclinic and cubic zirconia (at low temperatures). In a typical PSZ the initial structure consists of particles of unstable tetragonal zirconia imbedded in cubic zirconia. The stress intensification associated with a crack triggers a martensitic transformation from the tetragonal to the monoclinic form. The work expended in the transformation adds to the toughness of PSZ. (*From P. Duwez, F. Odell, and F. H. Brown, Jr., J. Am. Cer. Soc., 35, 109, 1952.*)

In pure zirconia, the transformation from the high- to the low-temperature form on cooling occurs martensitically. There are relatively large dilational (ca. 4%) and shear strains (ca. 7%) associated with the transformation. In pure zirconia, these can lead to thermal cracking and loss of structural integrity if the material is cooled rapidly from high temperature.

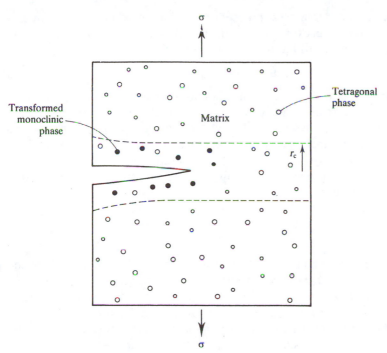

Figure 10.11

Schematic of a stress-induced martensite transformation in a partially stabilized zir-conia (PSZ). The stress intensification at the crack tip causes some of the tetragonal particles to transform to the monoclinic phase. The work of transformation adds to the PSZ toughness. The additional toughness scales with the volume of the trans-formed zone ($\sim r_c$) and the volume fraction of the particles within it that transform.

Adding CaO (MgO serves the same purpose) to zirconia produces a cubic phase, termed "cubic zirconia,"[6] which is clearly a misnomer in light of the fact that it is a zirconia-lime ceramic "alloy." Cubic "zirconia" is stable with CaO contents in excess of about 8 wt.% (about 15 mol.%) CaO. For compositions in the range of 7 wt.% CaO, the tetragonal and cubic forms coexist in equilibrium at high temper-ature. At lower temperatures, the cubic and monoclinic structures are the equilib-rium phases. Partially stabilized zirconias have compositions corresponding to these two-phase equilibrium structures.

In PSZ the tetragonal-monoclinic transformation takes place sluggishly on cooling to room temperature. In fact, most of the tetragonal phase does not trans-form. Further, the unstress-assisted martensite start temperature for the tetragonal-monoclinic transformation lies below room temperature in certain PSZ compositions. In these materials, the transformation may be triggered by the stress field associated with a crack. This is illustrated in Fig. 10.11. As the crack ad-vances, tetragonal particles transform to the monoclinic form in a zone lying $\pm r_c$ above and below the fracture plane. The work expended in effecting this transfor-mation adds to the material toughness.

---

[6]This is the same "zirconia" sold as a gemstone.

470

CHAPTER 10
Toughening
Mechanisms
and the Physics
of Fracture

Analysis of the added toughness proceeds along lines used in analyzing microcrack toughening. In fact, an expression for the associated transformation toughness very much parallels Eq. (10.6) for microcrack toughening; that is,

$$\Delta \mathcal{G}_c \cong 2r_c \sigma_M \varepsilon_M V_f \qquad (10.10)$$

In Eq. (10.10) σ_M is the stress initiating the martensite transformation, ε_M is the corresponding martensite transformation strain, and V_f is the fraction of the triclinic phase in the distance $\pm r_c$ above and below the fracture plane that transforms martensitically. The distance r_c varies with the matrix fracture toughness and the stress σ_M as $r_c \cong (K_c/\sigma_M)^2$ (recall Eq. (10.7)). Substituting $K_{co} = (E\mathcal{G}_{co})^{1/2}$, we obtain an equation for transformation toughening having the same form of Eq. (10.9) for microcrack toughening;

$$\Delta \mathcal{G}_c \cong \frac{\mathcal{G}_{co} \varepsilon_M E V_f}{\sigma_M} \qquad (10.11)$$

The similarities between microcrack and transformation toughening do not end here. There is a particle-size effect in transformation toughening, analogous to a grain-size effect in microcrack toughening. The stress required to drive the martensite transformation decreases as the tetragonal particle size increases. This is because the restraint placed by the cubic matrix on the transformation decreases with increasing particle size (as does the stress σ_M). When the particle size is too large, the transformation occurs spontaneously (recall $r_c \sim \sigma_M^{-2}$) during cooling. (This spontaneous transformation is analogous to the situation of spontaneous microcracking.) Note that if the particle size is too small, r_c is decreased (because σ_M is high) and the added toughness is also low. Thus, there is an optimum tetragonal particle size for maximizing transformation toughening; this is analogous to there being an optimum grain size for maximizing the toughness arising from microcracking.

Transformation toughening is, at present, only associated with PSZ, and the mechanism can be used to toughen zirconia and its "alloys." Alternatively, the unstable tetragonal phase can be added to other ceramics (e.g., alumina) although to do this properly requires some fancy materials processing. Because of the rather large incremental toughnesses associated with the effect in zirconia, much current research seeks to explore other material systems capable of exhibiting similar behavior.

Finally, we note that for the ceramic toughening mechanisms discussed to this point, the added toughness scales with the inherent matrix toughness (i.e., with $\mathcal{G}_{co}$). In this sense, it is of benefit to have an inherently tough ceramic matrix to begin with. Other toughening mechanisms, to which we now direct our attention, do not manifest this feature.

E. Crack Bridging

Fractured ceramics often exhibit some structural integrity. That is, "failed" samples often remain "intact," and can be completely separated only by application of a post-fracture stress. It is as if there were contacting ligaments between the halves of the "fractured" sample. In fact, this is the situation. To understand this phenomenon, reconsider Fig. 10.7. Note that the irregular crack path in the polycrystalline

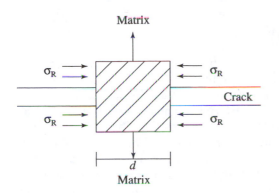

Matrix

σ_R σ_R

Crack

σ_R σ_R

d

Matrix

Figure 10.12
**Pull-out of a cubic grain
(shaded) bridging a crack plane.
Because of residual thermal
stresses, a frictional (clamping)
stress exists across some grain
boundaries (for a noncubic single-phase ceramic) or interphase
boundaries (for a two-phase ceramic). The "frictional" work
of grain pull-out adds to the material toughness.**

Al_2O_3 implies that some grains to each side of the main crack protrude into the other side. Complete fracture requires dislodgment of these bridging grains. Were there no force resisting this separation, there would be no additional work associated with this final severance. However, when thermal residual stresses are present, this grain "pull-out" work can be appreciable. The residual stresses are manifested by a clamping force between grains that is overcome in accomplishing their final separation. The work done in this separation represents additional fracture work.

The added fracture work can be calculated on the basis of it arising from a frictional clamping force. We consider a cubic grain of edge-length d that is "pulled out" of a mating surface across the fracture plane (Fig. 10.12). The pull-out work is given by the integral of the frictional force over the pull-out length, l. The frictional stress is $\mu\sigma_R$ where μ is the coefficient of friction and σ_R the "gripping" pressure, here reasonably taken as the residual compressive stress. The area over which this stress operates is (for a cubic grain) $4dl$ where l is the instantaneous distance over which the grain is clamped. This length varies with the extent of pull-out. It initially has the value l_o, but it decreases with the pull-out distance, x, as $l = l_o - x$. Thus, the pull-out work is given by

$$\int_0^{l_o} 4\mu\sigma_R d(l_o - x)\, dx = 2\mu\sigma_R dl_o^2 \tag{10.12}$$

The maximum pull-out distance would be $d/2$; the minimum would be zero. We take the average distance as $d/4$. Thus, the pull-out work is about $\mu\sigma_R d^3/8$.

Dividing Eq. (10.12) by d^2 (the area of the pulled-out grain) yields the fracture work per unit area. This can be equated to $\Delta\mathcal{G}_c$. However, only a fraction, V_f, of the grains are subject to crack bridging. (For monophase alumina, this fraction is on the order of 1/3.) Thus, for grain crack bridging,

$$\Delta\mathcal{G}_c \cong \frac{1}{8}\mu\sigma_R V_f d \tag{10.13}$$

Bridging of this type provides reasonable toughening (Prob. 10.4). However, toughening by crack bridging is most effective in composites, as is discussed in the next section.

472

CHAPTER 10
Toughening
Mechanisms
and the Physics
of Fracture

10.4
TOUGHENING IN COMPOSITES

Polymers have not been forgotten; they are discussed in the following section. However, the connection between grain crack bridging and toughening of certain ceramic (and polymer) composites is so direct that we treat this matter first.

Recall why composites are used in the first place, and it is understandable why we will not discuss toughnesses of metal matrix composites. The dispersions they contain are stronger and typically stiffer than their matrix. That is, the dispersions are used to strengthen the metal and this is accompanied by a *reduction*, and not an increase, in the material toughness. In fact, to a good approximation crack propagation in MMCs can be considered similar to crack propagation in "dirty" metals and their alloys.

The additions in ceramic matrix composites (CMCs) are used to toughen, and not to strengthen, them. Ceramics are strong enough to begin with; what they need is an improved fracture resistance. The additives to CMCs can be ductile (e.g., metal particles or fibers) or they can be almost as brittle as the matrix; graphite-graphite composites are an example here. In both situations, though, the composite toughness is greater than that of the ceramic matrix.

Graphite fibers in CFRPs, and glass fibers in GFRPs, both strengthen *and* toughen their polymer matrices. The added strength is due to the strong fibers, and the additional toughness comes from interfacial pull-out work similar to that just discussed for "bridging" in ceramics. This type of toughening is also dominant in CMCs containing brittle fibers.

A. Crack Bridging with Brittle Fibers

When reinforcing fibers fail at a greater fracture strain than the matrix containing them, the fibers bridge the wake of a propagating crack (Fig. 10.13). Bridging is generally accompanied by fiber debonding along a portion, l_d, of its length. The toughening due to this bridging can be discussed in two ways. In one, we can imagine the load carried by the fibers in the crack wake to produce a crack-closing force. This force reduces the stress intensity in front of the crack. Further, the crack deflection accompanying debonding provides a component of Mode II fracture, and this affects the critical stress intensity in the same way as does the crack deflection associated with intergranular fracture (Sect. 10.3B). However, we focus here on the second way of viewing the toughness provided by bridging; that is, in terms of the additional toughness, $\Delta \mathcal{G}_c$, resulting from it. We first consider continuous fibers (more precisely, fibers having a length, l, much greater than the fiber critical length l_c; cf. Sect. 6.5).

The work associated with fiber bridging is schematically illustrated for the general case in Fig. 10.14a. The fiber elongates a distance, u, before it fails. Following fiber failure, it is pulled out of the matrix. Both processes (plastic deformation and fiber pull-out) increase composite toughness. Any plastic deformation of the fiber contributes to the fracture work. (If the fiber does not plastically deform, the elastic energy stored prior to its fracture is recovered and is expended in creating the fiber fracture surfaces.) The pull-out work also contributes to toughness. For a

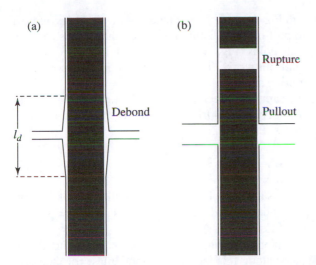

Figure 10.13
**Fiber bridges behind the main crack tip in a brittle-matrix composite. (a) Some
debonding of the fiber along a length l_d takes place. The debonding adds a component of shear fracture, which increases toughness. (b) Eventually the fiber fractures,
and it is subsequently pulled out of the matrix. There are two components of additional work coming from fiber bridging. One is the pull-out work; it is most important for brittle fibers. Another is the plastic deformation that precedes fiber fracture;
it is the main contribution to the added toughness when the fiber is ductile. (***Adapted
from B. Lawn, Fracture of Brittle Solids, 2nd ed., Cambridge University Press, Cambridge, England, 1993.***)**

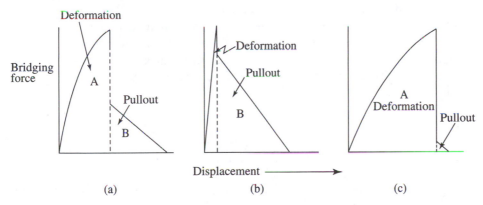

Figure 10.14
**Bridging force–displacement curve for a fiber in a brittle matrix. (a) The general sit-
uation indicates the two contributions bridging makes to toughness. The work has
two components; the area marked (A) represents plastic deformation work on the
fiber before its fracture, while that denoted (B) is fiber pull-out work subsequent to
fiber fracture. (b) The force-displacement curve when the fiber is brittle; almost all
of the work is due to pull-out. (c) When the fiber is ductile, the force-displacement
curve is as shown here. In this case, fiber plastic deformation is the main contributor
to the toughness increase.**

474

CHAPTER 10
Toughening
Mechanisms
and the Physics
of Fracture

brittle fiber, Fig. 10.14a reduces to Fig. 10.14b. We see that for brittle fibers, almost all of the fracture work is due to pull-out. In distinction, when the fiber is capable of plastic deformation, the opposite holds (Fig. 10.14c); i.e., the plastic deformation work is much greater than the pull-out work. Thus, for all practical purposes, we need only consider these two limiting situations.

For a brittle fiber, we use the same approach as that of Sect. 10.3E. The only difference is that we consider pull-out of a cylindrical fiber rather than a cubic grain. The pull-out work for such a fiber is (Prob. 10.5)

$$W = \frac{1}{2} \tau \pi d l^2 \qquad (10.14)$$

where τ is the interface friction stress and l the length of fiber pulled out of the matrix. (When the interfacial stress is due to residual stresses, we substitute $\tau = \mu \sigma_R$ and obtain a result similar to that of Eq. (10.12).)

A fiber can fracture anywhere along the length over which it is debonded (the stress borne by the fiber is the same along this length). A good guess for the average pull-out length is about $l_d/4$, so the fracture work of Eq. (10.14) is $\tau \pi d l_d^2/32$. Dividing by the fiber cross-sectional area ($= \pi d^2/4$) gives the fiber pull-out work per unit area of fiber. Recognizing that the fibers constitute only a fraction, V_f, of the total fracture area lets us express $\Delta \mathcal{G}_c$ as

$$\Delta \mathcal{G}_c = \frac{1}{8} \frac{\tau l_d^2 V_f}{d} \qquad (10.15)$$

Inspection of Eq. (10.15) suggests that a large debond length is desirable. That is true, but only within limits. A large debond length indicates a relatively weak fiber-matrix bond and, for ceramic-ceramic and polymer-matrix composites, the interfacial stress τ scales with this bond stress. That is, an increased debond length correlates with a decreased value of τ in Eq. (10.15).

We can go further, although we speculate when we do. It might be expected that the debond length should scale with the fiber critical length, l_c, because both lengths depend on the bond strength between the matrix and the fiber. Letting $l_d = \beta l_c$, and substituting $l_c/d = \sigma_f/2\tau$ (Eq. (6.26)) where σ_f is the fiber strength, we obtain

$$\Delta \mathcal{G}_c = \frac{1}{32} \frac{\beta^2 \sigma_f^2 d V_f}{\tau} \qquad (10.16)$$

According to Eq. (10.16), maximum toughness is obtained by increasing fiber volume fraction (this is expected), reducing the interfacial stress, and maximizing fiber strength. Thus, increasing fiber strength, at least on this argument, benefits both composite strength and toughness.

When fiber lengths are less than the critical length, bridging still contributes to the fracture work. In these circumstances, the average pull-out length is liable to be on the order of $l/4$ where l is the fiber length. Substituting l for l_d in Eq. (10.15) gives the toughness increase for this situation.

Finally, we are dealing with a composite. The composite toughness is expressed as a "volume fraction rule;" that is

$$\mathcal{G}_c = V_m \mathcal{G}_m + V_f \mathcal{G}_f + \Delta \mathcal{G}_c \qquad (10.17)$$

where $\mathcal{G}_{m,f}$ is the matrix/fiber toughness and $\Delta\mathcal{G}_c$ the added toughness. Conversion to fracture toughness is made with knowledge of the composite modulus, using the relationship $K_c \cong (E\mathcal{G}_c)^{1/2}$.

EXAMPLE PROBLEM 10.1. We estimate the toughness of aligned glass-fiber composites in this problem. We consider a composite containing 30 vol.% of 10-μm-diameter glass fibers in an epoxy matrix. The following properties apply: $E_{glass} = 150$ GN/m^2; $E_{epoxy} = 5$ GN/m^2; $\sigma_{f\text{-glass}} = 5000$ MN/m^2; $\sigma_{f\text{-epoxy}} = 50$ MN/m^2; τ (interfacial shear stress) $= 25$ MN/m^2; $K_{Ic\text{-glass}} = 0.8$ MN/m$^{3/2}$; $K_{Ic\text{-epoxy}} = 1$ MN/m$^{3/2}$; $K_{Ic\text{-composite}} = 70$ MN/m$^{3/2}$.

a. Let the debond length equal the fiber critical length for this composite. Calculate $\Delta\mathcal{G}_c$ arising from crack bridging in this material.

b. The composite toughness is given by $\Delta\mathcal{G}_c$ plus the inherent matrix and fiber toughness. Show that these latter terms are much less than $\Delta\mathcal{G}_c$, so that we can reasonably assume the composite toughness is represented by $\Delta\mathcal{G}_c$.

Solution **a.** Equation (10.16), with a value of $\beta = 1$, is appropriate for this problem. Thus, on substituting in the appropriate property values, we obtain

$$\Delta\mathcal{G}_c = \frac{1}{32}\frac{\sigma_f^2 d V_f}{\tau} = \frac{(25 \times 10^{18})(10^{-5})(0.3)}{(32)(2.5 \times 10^7)} = 9.4 \times 10^4 \ \frac{J}{m^2}$$

This is a rather substantial toughness.

b. The inherent toughness of the matrix/fibers is determined from their fracture toughnesses via $\mathcal{G}_c \cong K_c^2/E$. Substituting in the properties given, we find $(\mathcal{G}_c)_{glass} = 4$ J/m^2 and $(\mathcal{G}_c)_{epoxy} = 200$ J/m^2. The composite base toughness follows a volume fraction rule so that $(\mathcal{G}_c)_{composite\text{-base}} = (0.7)(200) + (0.3)(4) = 141$ J/m^2. The composite "base" toughness (and those of its constituents) are much less than the toughness arising from fiber bridging.

Problem 10.6 expands on the ideas and concepts of this sample problem.

B. Crack Bridging with a Ductile Phase

Figure 10.14c illustrates the situation pertaining to a ductile fiber imbedded in a brittle matrix. In this circumstance, fiber plastic deformation contributes by far the most to the increase in material toughness. We estimate this toughness, using methods similar in principle (although the details differ) to those used previously.

We consider a fiber debonded over a length l_d. The plastic work (per unit volume) through fiber fracture is proportional to the fiber flow stress (σ_y) multiplied by its failure strain (ε_f). Multiplication by the fiber volume that deforms (the transverse cross-sectional area times l_d) and division by the fiber cross-sectional area gives the plastic work per unit fiber area. Further multiplication by V_f, the fiber volume fraction, gives the toughness resulting from fiber plastic deformation;

$$\Delta\mathcal{G}_c \cong \sigma_y \varepsilon_f l_d V_f \tag{10.18}$$

In Eq. (10.18) we take the flow stress and failure strain as those obtained from a tension test of the fiber. When we do this, though, modification of Eq. (10.18) is required. The situation illustrated in Fig. 10.14c corresponds to a constrained (multiaxial stress) version of a tension test. When the debond length is very short (i.e., when the length of the fiber that plastically deforms is about equal to the

476

CHAPTER 10
Toughening
Mechanisms
and the Physics
of Fracture

crack-opening displacement) the value of flow stress is increased greatly compared to that in a tension test. The situation is a microscopic version of a brazed joint (Sect. 5.8). On this basis, a short debond length is beneficial to composite toughness. Increasing the debond length, though, improves toughness through the l_d term of Eq. (10.18). With large debond lengths, the flow stress is more closely approximated by the tensile flow stress because of the reduced constraint accompanying debonding. Thus, two effects compete in much the same way as do the effects of fiber debond length and interfacial shear stress in the situation of fiber bridging. The result is that optimum toughening is associated with a *limited* amount of debonding. The effects described—ones we cannot quantify—are incorporated into Eq. (10.18) in our usual arbitrary fashion. That is, when we don't know the answer but have identified the important variables, we just insert a proportionality constant. Thus,

$$\Delta \mathcal{G}_c = C\sigma_y \varepsilon_f l_d V_f \tag{10.19}$$

where the constant C is estimated as being between about 1.5 and 6.

It costs more to place fibers into a matrix than it does to put particles into one. As a result, development of ductile-particle reinforced ceramic composites has also been of recent interest. Dispersions of aluminum and nickel particles, for example, can toughen alumina and magnesia. In particle-toughened composites, it is imperative that the particles do not fully debond in the crack tip wake. If they do (or, at least if they do prior to their undergoing significant plastic flow), the contribution to fracture work is no more than that due to crack deflection/bridging. Analysis of particle toughening, indicates that a modification of Eq. (10.19) can be used to express the increased toughness. The particle radius (r) substitutes for l_d and the constant C is changed to account for the different geometrical constraints for particles compared to fibers. As with the situation for fibers, some debonding of the particle and matrix where they meet at the fracture plane maximizes toughness (Prob. 10.8).

Before we leave ceramic toughening, one more point needs to be made. Refining microstructural scale in metals usually improves fracture resistance (Sect. 10.8). The toughness of a "virgin" ceramic is often also improved by reductions in grain size. Yet, in discussion of toughening we noted that in some cases (e.g., fiber bridging), the toughness increases with increases in microstructural scale. And in our discussion of microcrack and transformation toughening, there was an optimum microstructural scale associated with maximum toughness. These complicating factors make design for improved toughness quite challenging in ceramics. The factors can also be tough to keep straight. So, just as in our discussion of strengthening mechanisms, we summarize the basic physics in a few short paragraphs. But before we do this, we must address polymers. They, too, deserve attention.

10.5
TOUGHENING IN POLYMERS

Toughening in fiber-reinforced PMCs has been covered in our discussion of toughening of brittle matrices with brittle fibers. (That is what an epoxy containing glass fibers is.) However, the toughnesses of polymers whose glass transition temperatures lie above their use temperatures can be increased in other ways. The tensile behavior of high-impact polystyrene (HIPS) was mentioned in Chap. 8. The crack

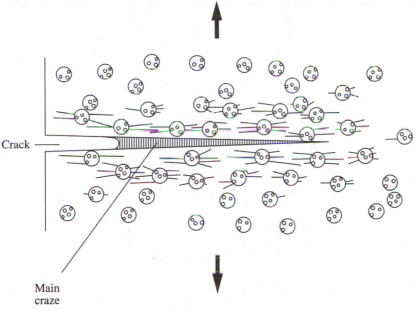

Crack

Main
craze

Figure 10.15
Secondary crazes form around rubber particles in a glassy polymer during crack
propagation. The secondary crazing occurs in a finite distance above and below the
fracture plane. This is the region, analogous to that in microcrack and transforma-
tion toughening in ceramics, in which additional fracture work takes place. (*From*
A. H. Windle, Physical Metallurgy, Vol. III, 4th ed., R. W. Cahn and P. Haasen, eds.,
***North-Holland, Amsterdam, 1996, 2663.*)**

propagation resistance, as well as tensile failure strain, is also increased by the elastomeric dispersion in HIPS. Figure 10.15 depicts crack propagation in HIPS. Note
that additional crazes form in a volume lying above and below the fracture plane.
The associated toughening parallels that of microcrack and transformation toughening in ceramics and is analyzed similarly. That is, the additional fracture work associated with craze formation (treated as microscopic plastic deformation work)
represents the added toughness. An expression for the toughness of HIPS can be
written in a form similar to Eqs. (10.19), (10.10), and (10.6) (Prob. 10.9).

Toughening by elastomeric dispersions is also common to polymers other than
PS; polyvinyl chloride, for example. The only difference in the crack propagation
behavior of this type of PVC and HIPS is that local plasticity is manifested in the
form of shear bands in PVC. Even epoxies, generally considered thermosets and incapable of plastic deformation, can be toughened by elastomeric dispersions. Epoxies are capable of limited plasticity, at least at the microscopic level, and the
dispersion catalyzes this deformation in the vicinity of a crack thereby providing an
added component of fracture work.

Glass particles imbedded in epoxies improve toughness, although nowhere near
to the extent that glass fibers do. This type of particle toughening is due to crack deflection. Analysis of it proceeds in much the same way as does analysis of grain
crack bridging in ceramics.

478

CHAPTER 10
Toughening
Mechanisms
and the Physics
of Fracture

Mechanism	Work per unit volume	Volume affected per unit area of fracture plane	$\mathcal{G}_c \sim$
Crack tip plasticity	$\sigma_y \varepsilon_f$	ρ	$\sigma_y \varepsilon_f \rho$
Microcracking	$\sigma_R \Delta\alpha\Delta T$	$2r_c V_f$	$\sigma_R \Delta\alpha\Delta T 2r_c V_f$
Transformation toughening	$\sigma_M \varepsilon_M$	$2r_c V_f$	$\sigma_M \varepsilon_M 2r_c V_f$
Bridging with brittle fibers	$\tau V_f l_d / 2d$	$l_d/4$	$\tau V_f l_d^2 / 8d$
Bridging with ductile fibers	$V_f \sigma_y \varepsilon_f$	l_d	$V_f \sigma_y \varepsilon_f l_d$

Key to Symbols: σ_y = plastic flow stress; ε_f = failure strain; ρ = crack tip radius of curvature.
σ_R = thermal residual stress; $\Delta\alpha\Delta T$ = thermal residual strain.
V_f = volume fraction of microcracked grains (for microcrack toughening).
  = volume fraction of transformed particles (for transformation toughening).
  = volume fraction of fibers (for crack bridging).
r_c = distance above and below crack plane where microcracking (for microcrack toughening)
  or martensite transformation (for transformation toughening) takes place.
τ = friction stress resisting fiber pull-out.
l_d = fiber debond length.
Notes: Distance r_c depends on grain size (for microcrack toughening) or particle size (for transformation toughening). Bridging with ductile particles produces a toughness equation having the same form as that for ductile fibers, but with particle radius substituting for fiber diameter and with a different proportionality constant.

Table 10.2
Features of some common toughening mechanisms

Polymers are also toughened by addition of "plasticizers." Plasticizers are low-molecular-weight monomeric liquids that separate the long chain clusters in the polymer. As such, this type of solid can be viewed as a "composite" of a viscous phase (the plasticizer) imbedded in a glass-like (the polymer) matrix. Plasticizers not only improve polymer toughness, they also decrease the glass transition temperature of the polymer. However, the addition of plasticizers is also accompanied by a reduction in material strength.

We have not spent much time discussing toughening of polymers. This is not because they are less important than other materials, but because the concepts of toughening them are similar to those discussed previously for other material classes. Because of this commonalty, it is worthwhile to synopsize "toughening" before going on to other matters. Such a summary is also needed because—as with our treatment of strengthening mechanisms—quite a bit of ground has been covered. So Table 10.2 is provided. It may give some intellectual relief. The table lists many of the toughening mechanisms we have discussed, provides expressions for the toughness, and adds some explanatory comments. But note one feature common to all of the expressions for the toughness. It is a work per unit volume multiplied by a distance above and below the fracture plane over which this work is manifested. The primary difference among the several toughening mechanisms lies in the nature of this irreversible work.

10.6
TYPES OF LOW-TEMPERATURE TENSILE FRACTURE

In this and the following section, we describe the various forms of low-temperature fracture observed in a tension test. In such a test, the sample may or may not contain preexisting cracks. If it does, the sample is a variant of a fracture toughness test and the fracture strength is defined by the relation among this strength, flaw size, and fracture toughness (Eq. (9.13)). However, the fracture toughness to be used in the equation depends on the sample size. For example, for a ductile metal the sample dimensions may be such that plane stress conditions apply. If so, the plane stress fracture toughness is the one to be employed.

When the sample does not contain preexisting flaws (or if they are too small to be the cause of fracture), the sample nonetheless eventually fractures. The various tensile fracture modes are discussed in this section. In the next, the relative tendencies for the various fracture modes in several material classes are discussed. Discussion of the modes of tensile failure in this manner also facilitates a deeper appreciation of the toughnesses that characterize these materials.

Low-temperature tensile fracture may be preceded by varying degrees of plastic deformation. The extremes of this variation are illustrated in Fig. 10.16. Figure 10.16a illustrates fracture taking place prior to the onset of macroscopic plasticity; here fracture occurs by tensile separation of atomic bonds across the fracture plane. A fracture taking place prior to plastic yielding is termed brittle. The fracture mechanism is called cleavage (if fracture occurs within grains) or brittle intergranular fracture (BIF, if it progresses along grain boundaries).

Figures 10.16b and c illustrate fracture caused by plastic deformation mechanisms. The fracture shown in Fig. 10.16b can happen in a tensile test of a single crystal oriented for single slip (e.g., the metal zinc with the hexagonal crystal structure). Crystal separation is achieved when the relative displacement over the active slip planes equals the bar diameter. *Rupture fracture* corresponds to a 100% reduction in area (R.A.). It can be observed in single crystals (Fig. 10.16c) or polycrystals (Fig. 10.16d). As suggested by Fig. 10.16c, rupture in single crystals involves multiple slip whereas in polycrystals it is associated with necking (Fig. 10.16d). Such a 100% R.A. can be observed in some pure polycrystalline metals at low temperatures. However, rupture is by far more common at high temperatures where it is often associated with dynamic recrystallization; further discussion of this type of rupture is given in Chap. 11.

Depending on the applied stress, the stress state, the temperature, the material's grain size, the strain rate, etc., a material may fracture in either a ductile or a brittle manner. Since plastic deformation precedes ductile fracture (associated with a sample neck in a tensile test), rupture is the extreme case of ductile fracture. In many materials there is a gradual transition from "brittle" to "ductile" fracture as temperature is increased, as has been mentioned. Thus, even "brittle" fracture may be accompanied by some material plastic deformation and, therefore, we (somewhat arbitrarily) subdivide brittle fracture into categories depending on the extent of plastic deformation that precedes fracture (Fig. 10.17). The first type of fracture, which we call Mode I,[7] occurs in inherently brittle materials and may take place by crack propagation within grains (Cleavage I, Fig. 10.17a) or along grain boundaries (Fig. 10.17b, BIF I). This fracture mode has been treated in the fracture mechanics

---

[7]Alas, terminology can wreak confusion; apologies are tendered. In discussion of fracture mechanics, we also defined Mode I–Mode III types of fractures. These are the terms used by mechanicians to refer to the effect of stress state on crack propagation. In the present treatment, the terms Mode I, etc., come from material, and not mechanical, considerations and they relate to the microscopic mechanisms of crack nucleation and propagation.

480

CHAPTER 10
Toughening
Mechanisms
and the Physics
of Fracture

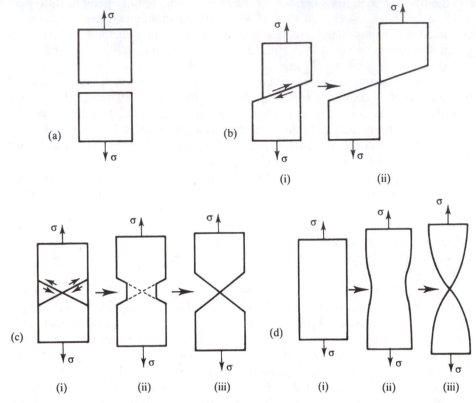

Figure 10.16
Extremes in the variation of low-temperature tensile fracture. (a) Brittle fracture by
tensile-induced separation of atomic bonds leads to a macroscopically flat fracture
plane normal to the tensile axis. (b) Fracture occurring by plastic displacement
across a slip plane. Such a fracture might happen in a single crystal oriented for sin-
gle slip. The fracture plane here may also be macroscopically flat even though frac-
ture is induced by shear, and not tensile, forces. (c) Rupture fracture (100% R.A.) in
a single crystal as a result of multiple slip. The progression in crystal shape change is
shown in the series of figures c(i)–(iii). (d) Rupture in a polycrystal occurs via neck-
ing, which continues until a 100% R.A. is effected. This type of fracture takes place
in some pure metals at low temperature, but is more common at high temperatures
where it is often associated with dynamic recrystallization.

discussions of Chap. 9. Specifically, Mode I fracture is caused by the propagation
of preexisting cracks and the fracture stress for it is given by Eq. (9.13). Mode I
fracture in a tension test is initiated by flaws, provided they are of a size typically
having dimensions exceeding the material grain size.

Equation (9.13) suggests that the fracture stress increases indefinitely as the
crack size becomes progressively smaller. Would that this were so, but it is not.
When the crack size becomes less than about the grain size, low-temperature Mode
I fracture gives way to Mode II fracture. Microscopic yielding precedes Mode II
fracture, which can propagate transgranularly (Cleavage II, Fig. 10.17c) or intergranularly (BIF II, Fig. 10.17d). The higher stress levels associated with Mode II

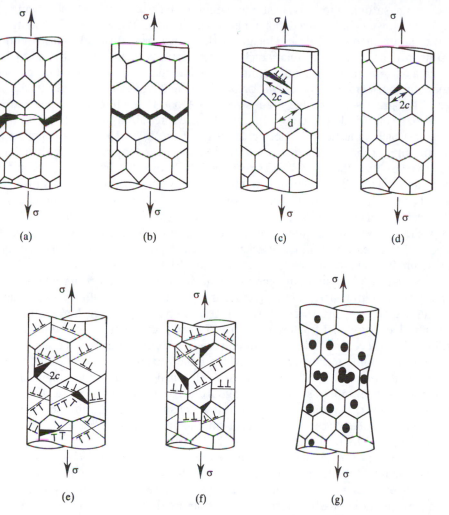

Figure 10.17

Further subdivision of fracture modes. The most brittle of the fracture modes (Mode I, (a) and (b)) occurs without plastic deformation (except for that which may occur in the material adjacent to the crack tip). Mode I fracture may propagate transgranularly (by cleavage, (a)) or intergranularly ((b) brittle intergranular fracture, BIF). Microscopic plastic deformation precedes Mode II brittle fracture. This deformation nucleates cracks that may propagate by cleavage (c) or by BIF (d). Generalized plastic deformation precedes Mode III brittle fracture (e, f). A limited reduction in area is found in a tensile test, but fracture propagates by cleavage (e) or BIF (f) prior to sample necking. Ductile tensile fracture (g) is preceded by necking. Microscopic voids form in the material, and they subsequently grow and coalesce by plastic deformation mechanisms. Voids nucleate frequently at inclusions, but may also form in regions of heterogeneous slip. Void linkup is restricted to the necked volume of a tensile sample.

482

CHAPTER 10
Toughening
Mechanisms
and the Physics
of Fracture

fracture are sufficient to activate slip or twinning on certain preferred slip/twin systems. This results in localized, but not general, yielding. While general yielding removes strain incompatibilities, localized flow does not. Instead, microyielding results in the formation of microcracks. These form typically at the boundary between a deforming and a nondeforming grain, and the crack so formed propagates rapidly across a grain. In other words, localized yielding nucleates cracks on the order of the grain size, and the grain size is to Mode II fracture as the preexisting crack size is to Mode I fracture. Indeed, the stress required to *propagate* (but not *nucleate*) such a plasticity-induced crack is given by Eq. (9.13), with the grain size substituting for the crack size. In some cases, the stress for propagating the microcrack is less than the microyield stress (which defines the fracture stress in this case), and fracture is concurrent with microyielding; the crack propagates instantaneously from the grain in which it is formed through adjacent ones. In other cases, the crack propagation stress is greater than the yield strength, and then the crack propagation stress defines the fracture stress. That is, the stress must be increased beyond that necessary to initiate cracks before crack propagation takes place. Mode II fracture is discussed in greater length in Sect. 10.8.

The plastic yield strength decreases with increasing temperature and does so rapidly for certain materials (e.g., the bcc transition metals). With increasing temperature, therefore, macroscopic (general) yielding takes place before fracture. When it does, Mode III fracture (Figs. 10.17*e* and *f*) can occur. Even though general yielding precedes Mode III fracture, it is still a "brittle" fracture in that it occurs by the same atomic mechanism as do Mode I and Mode II fractures. Since plastic strains on the order of 1–10% precede Mode III tensile fracture, it is somewhat of a misnomer to refer to it as simply Cleavage III or BIF III fracture. Yet, as noted, the classification is consistent with the fracture mechanism. Crack initiation in Mode III fracture can take place by the same, or similar, processes that cause Mode II fracture initiation. Mode III fracture is described further in Sect. 10.9.

Mode III "brittle" fracture is preempted at higher temperatures by ductile fracture. Although Mode III and ductile fracture both follow macroscopic yielding, the processes of crack growth differ between them. For Mode III, crack propagation occurs by cleavage or BIF. During ductile fracture (Fig. 10.17*g*), however, microcracks (or, more commonly, voids) nucleate, grow, and link up by mechanisms involving plastic flow. Ductile fracture frequently initiates at second-phase particles or inclusions (recall Sect. 10.2). The plastic strain incompatibility between the particles and the matrix is the cause of particle fracture or decohesion. The material between the particles eventually necks down, rather as if it were a microscopic tensile bar, and final fracture is effected by void linkup due to matrix rupture or plastic shearing. Ductile fracture is further described in Sect. 10.10.

The various fracture modes involve differing amounts of fracture work. Mode I fracture is a low-toughness fracture. The fracture work is that associated with the propagation of preexisting cracks. Plastic deformation accompanying Mode I crack growth is restricted to material in close proximity to the fracture plane. Mode II fracture, if it occurs at the same temperature as Mode I fracture, has the same toughness as Mode I fracture. However, Mode II fracture takes place at a higher stress level because the cracks causing it are not preexisting ones. When general yielding precedes fracture (Mode III and ductile fracture), the toughness is greater because of the higher temperatures associated with these fracture modes. That is, the amount of material undergoing crack tip plastic deformation is greater at higher temperatures (see also Sect. 10.2) and this correlates with a higher toughness. Before discussing in more detail the various fracture modes, it is worthwhile to describe the differing tendencies for them among the material classes.

483

SECTION 10.7
The Relation
Among Bonding,
Crystal Structure,
and Fracture

10.7
THE RELATION AMONG BONDING, CRYSTAL STRUCTURE, AND FRACTURE

The different chemical bonding and, to a lesser extent, crystal structure are the reasons the various material classes exhibit differing tendencies for brittle and ductile fracture. We treat this topic in this section. Specifically, the fracture characteristics of metals (which are relatively ductile) are compared to those of ionic or covalent solids (which typically are far more prone to brittle fracture). We begin by describing the fracture of the most ductile material class—the face-centered cubic metals—and then progress in order of increasing tendency toward brittle behavior.

A. Fracture of Face-Centered Cubic Metals

Face-centered cubic metals are inherently ductile. Only two of them (iridium and rhodium) display any tendency at all for brittle fracture. The ductility of these metals is related to their intrinsically low (and relatively temperature insensitive) yield strengths and to the large number of slip systems for the fcc crystal structure. Because of their low yield strengths, fcc metals almost invariably plastically flow at stresses less than those required to fracture them, and this is true at low, as well as at high, temperatures. The availability of slip systems also allows for the relaxation of plastic strain incompatibilities, which often initiate the microcracks that are precursor to brittle fracture of the other metals and the nonmetals.

Fracture mechanism maps conveniently illustrate how a material's fracture mode varies with temperature and applied stress. These maps are to fracture as deformation mechanism maps (Chap. 7) are to elastic, plastic, and creep deformation. A fracture mechanism map for nickel, a typical fcc metal, is shown in Fig. 10.18. The coordinates are the applied tensile stress (normalized by the elastic modulus, E) and homologous temperature (T/T_m). The uppermost stress typically displayed on a fracture mechanism map is the theoretical fracture strength; i.e., $E/10$ (cf. Eq. (9.4)). Three different fracture regions are shown in Fig. 10.18. *Dynamic fracture* occurs at the highest of stress levels. These are sufficient to cause propagation of stress waves in the material. In effect, the material cannot be subjected to a uniform stress at these high stresses and strain rates. Dynamic fracture lies outside of our scope and we do not discuss it further. *High-temperature fracture* occurs at stress and temperature combinations indicated by the shaded region of Fig. 10.18. High-temperature fracture takes place by rupture or by one of several creep mechanisms; these are discussed in Chap. 11. The only type of low-temperature fracture displayed by nickel is *ductile transgranular fracture*, and this is also the situation for the other fcc metals excepting Ir and Rh. The stress range encompassed by this fracture mode extends from the tensile strength (ca. 10^{-3} to $10^{-2}\,E$ for Ni and most fcc metals) upward to the stress at which dynamic fracture intervenes. In Ni, ductile transgranular fracture is observed from 0 K up to $T/T_m \cong 0.7$; above this temperature, high-temperature fracture is observed.

484

CHAPTER 10
Toughening
Mechanisms
and the Physics
of Fracture

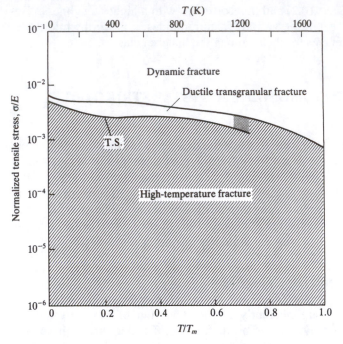

Figure 10.18
A fracture mechanism map for the face-centered cubic metal Ni. Dynamic fracture occurs at high stress levels and strain rates. The only low-temperature fracture mode observed in Ni is ductile transgranular fracture. It occurs at stress levels above the tensile strength but below those resulting in dynamic fracture. High-temperature fracture, involving aspects of diffusional flow (see Chap. 11), dominates in Ni at low stress levels and high temperatures. The uncertainty in the transition between fracture modes in this and the other fracture mechanism maps presented is indicated by shaded boundaries, and reflects sample and stress-state variability as well as uncertainty in the data used to construct the map. (*Adapted from M. F. Ashby, C. Gandhi, and D. M. R. Taplin, "Fracture-Mechanism Maps and Their Construction for FCC Metals and Alloys," Acta Metall., 27, 699, Copyright 1979, with permission from Elsevier Science.*)

Figure 10.18 is not as simple to interpret as may appear at first glance. Thus, several comments regarding it, and fracture mechanism maps in general, are in order. First, fracture mechanism maps for fcc metals parallel their deformation mechanism maps. For example, stress-temperature combinations leading to creep are also those that lead to creep fracture. And the combinations that result in ductile fracture are associated with macroscopic (essentially time independent) plastic flow. Second, the various regions in a fracture map define the fracture that *would* be observed if the material *were* to fail at a stress-temperature combination in the region. For example, a fracture mechanism map says nothing about the time to cause creep fracture. Thus, as shown in Fig. 10.18, "high-temperature" fracture of Ni occurs down to 0 K at low stress levels. This is because these stresses are too low to result in time-independent plastic flow, but they are high enough to provide a finite (albeit extremely low) creep rate at any temperature greater than 0 K. In practice, these low-temperature creep rates are so minimal that creep fracture would not be observed during the ordinary lifetime of a material. Thus, the fracture mechanism map of Ni states only that creep fracture would eventually occur if Ni were exposed to the specified stress for a sufficiently long time (eons, in some cases). Moreover, if a region of Mode I fracture is found in a fracture mechanism map, this only means that if fracture takes place, it is by this mechanism. Since Mode I fracture results from the propagation of preexisting cracks, if these are absent or are not of sufficient severity then Mode I fracture will *not* occur at the specified stress-temperature combination. We should also emphasize that a fracture mechanism map of a particular material is appropriate only for a specified material composition, grain size, and microstructure. That is, since these factors influence a material's yield strength and toughness, fracture mechanism maps are (as are deformation mechanism maps) structure sensitive. Fracture maps are sensitive, too, to the applied stress state. The ones shown in this chapter are, strictly speaking, appropriate only for uniaxial tensile loading. Thus, superposition of a second positive tensile stress component, which facilitates fracture but not plasticity, would expand the brittle fracture regions in a map at the expense of the ductile fracture ones. The above factors must be kept in mind when utilizing a fracture mechanism map. Nonetheless, on the coordinates used in Fig. 10.18, and in the other fracture maps that follow, microstructural and stress state conditions do not generally alter transitions between fracture modes to a degree greater than that indicated by the diffuse boundaries that are depicted on the maps. The "uncertainty" reflected by these boundaries is a measure of ordinary microstructural variations within a material (e.g., grain size and nominal composition) as well as of the uncertainties in the data or the equations used to empirically or theoretically construct a fracture mechanism map.

485

SECTION 10.7
The Relation
Among Bonding,
Crystal Structure,
and Fracture

B. Fracture of Body-Centered Cubic Transition Metals

The bcc transition metals have a proclivity for brittle fracture. This is a result of the strong temperature dependence of their yield strengths, which results in a low-temperature flow stress higher than the fracture stress. Thus, as shown in Fig. 10.19, a fracture mechanism map for molybdenum, a "typical" bcc transition metal, the three brittle fracture modes are observed at appropriate stress-temperature combinations and ductile fracture only occurs for temperatures greater than about 0.3 T_m. Although there are differing tendencies for brittle fracture among the bcc transition metals (e.g., tantalum is the most, and chromium the least, ductile of them), there is a remarkable similarity to their fracture mechanism maps. As with the fracture maps for the fcc metals, those for bcc metals parallel their deformation maps. Thus, high-temperature fracture and ductile fracture occur at stress-temperature combinations at which creep and time-independent plastic deformation are the respective dominant deformation modes. Likewise, the region of Mode I fracture in such a map corresponds to linear elastic deformation.[8] It is of interest, too, that the general level of stress in the fracture map of a bcc metal is higher than that in the fracture map for a typical fcc metal. This reflects the inherently greater resistance to flow and/or fracture of the bcc metals.

---

[8]As with fcc metals, creep deformation, which eventually culminates in fracture, can occur at any temperature greater than 0 K in the bcc transition metals. The temperature separating Mode I (and Mode III) fracture from high-temperature fracture in Fig. 10.19 is established at a creep rate so low that creep failure cannot be expected to happen in any realistic time frame.

486

CHAPTER 10
Toughening
Mechanisms
and the Physics
of Fracture

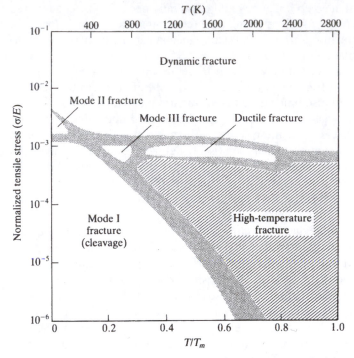

Figure 10.19
**A fracture mechanism map for the body-centered-cubic metal molybdenum.
Dynamic, ductile, and high-temperature fracture occur in Mo as they do in Ni.
However, at low temperatures, brittle fracture is observed in Mo. At the lowest
temperatures and stresses, Mode I fracture (by cleavage) takes place. At somewhat
higher stresses, Mode II fracture dominates and, at the highest brittle fracture
temperatures, Mode III fracture does.** (*Adapted from C. Gandhi, and M. F. Ashby,
"Fracture-Mechanism Maps for Materials Which Cleave," Acta Metall., 27, 1565,
Copyright 1979, with permission from Elsevier Science.*)

Mode I fracture dominates at low temperatures and low stresses in molybdenum
and the other bcc transition metals. Its occurrence, of course, requires the presence
of preexisting cracks. The upper-bound stress for Mode I fracture is the *greater* of
the stresses required to *nucleate* internal microcracks by inhomogeneous plastic flow
or to *propagate* such cracks once formed. At lower temperatures, the microyielding
stress is the critical one and fracture is concurrent with initiation of microscopic plasticity (Mode II fracture). At higher temperatures, the crack propagation stress may
exceed the microyield stress and, if so, the propagation stress defines the Mode
I–Mode III fracture transition. Any initiation-to-propagation transition is accompanied by a slight increase in the observed fracture strength with increasing temperature. This is not evident in the fracture map for molybdenum or in those for other bcc
metals, but can be seen in some of the fracture maps for ionic materials and refractory oxides (we discuss them later). As with the fcc metals, the highest stresses in the
fracture maps for bcc metals are associated with dynamic fracture.

Mode II fracture for molybdenum is supplanted at higher temperatures by
Mode III fracture. This is due to the lesser yield strength of Mo at higher tempera-

487

SECTION 10.7
The Relation
Among Bonding,
Crystal Structure,
and Fracture

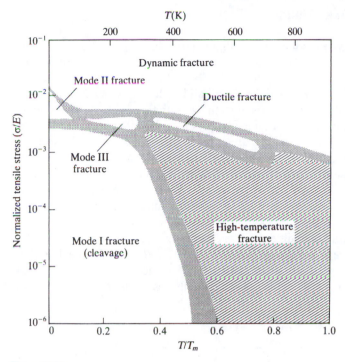

Figure 10.20

A fracture mechanism map for the hexagonal close-packed metal magnesium. The various low-temperature fracture modes displayed by Mg are the same as those for Mo. At low temperatures, however, the general strength level for Mg and the other hcp metals is somewhat greater than that for the bcc and fcc metals. (*Adapted from C. Gandhi, and M. F. Ashby, "Fracture-Mechanism Maps for Materials Which Cleave," Acta Metall., 27, 1565, Copyright 1979, with permission from Elsevier Science.*)

ture, and this feature allows for generalized plastic flow prior to fracture. For Mode III fracture, work hardening follows yielding. This increases the flow stress to a level at which fracture takes place in a brittle manner. At still higher temperatures this is not the situation. Fracture then is of a ductile variety and, in a tensile test, is accompanied by a significant %R.A.

C. Fracture of the Hexagonal Close-Packed Metals

Figure 10.20 is the fracture mechanism map for magnesium, a typical hcp metal. At low temperatures, hcp metals exhibit the three brittle fracture modes. In magnesium, this generally happens at stresses slightly higher than those associated with fracture of the bcc metals. Thus, low-temperature fracture of hcp metals is somewhat akin to that of ceramics. The hcp metal fracture maps are also similar to those for ceramics in another sense. Specifically, low-temperature Mode II fracture in hcp polycrystals often initiates as a result of localized slip due to the limited number of slip systems in this crystal structure. More slip systems are activated in hcp metals at higher temperatures, and for this reason the high-temperature regions of the hcp fracture mechanism maps closely resemble those of the fcc metals.

488

CHAPTER 10
Toughening
Mechanisms
and the Physics
of Fracture

D. Fracture of the Alkali Halides

Ionic bonding predominates in the alkali halides (e.g., NaCl). This bonding is typically no stronger than metallic bonding, and this is reflected in the general level of the fracture stress in the fracture map for NaCl (Fig. 10.21). However, the regions of brittle fracture in the fracture maps for the alkali halides, of which NaCl is a prototype, are much expanded in comparison to those in the maps of "brittle" metals. Indeed, fracture maps for polycrystalline alkali halides do not contain ductile fracture regions, even though these halides are capable of limited plastic deformation (e.g., the regions of Mode III, and high-temperature fracture in Fig. 10.21). As with the hcp metals, the cause of Mode II fracture in the alkali halides is their low number of low-temperature slip systems. Thus, localized slip initiates on the "soft" slip systems and the resulting plastic strain incompatibility nucleates microcracks, usually at grain boundaries. These cracks then propagate quickly across a grain. In NaCl, and several other alkali halides, the fracture propagation stress is higher than the soft system yield stress and the propagation stress, therefore, defines the lower bound for Mode II fracture. This can be seen in NaCl's fracture mechanism map, in which the "soft" system [{110}<1$\bar{1}$0>] yield strength lies below the stress defining

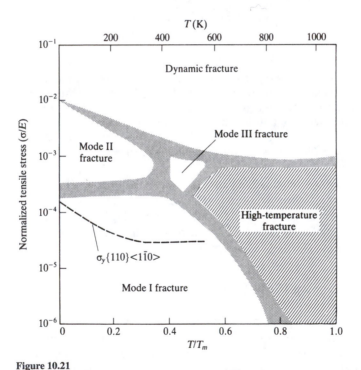

Figure 10.21
A fracture mechanism map for NaCl, an alkali halide. All three brittle fracture modes are displayed by NaCl. The transition from Mode I to Mode II fracture is dictated by the crack propagation stress, which is greater than the soft system [{110}<1$\bar{1}$0>] yield strength. In common with many nonmetallics, NaCl does not exhibit low-temperature ductile fracture (although limited macroscopic plasticity is found during Mode III fracture). (*Adapted from C. Gandhi, and M. F. Ashby, "Fracture-Mechanism Maps for Materials Which Cleave," Acta Metall., 27, 1565, Copyright 1979, with permission from Elsevier Science.*)

489

SECTION 10.7
The Relation
Among Bonding,
Crystal Structure,
and Fracture

the Mode I–Mode II transition. That the transition stress increases slightly with increasing temperature reflects the similar temperature variation of the crack propagation stress. At even higher temperatures, additional slip systems are activated in the alkali halides. This allows for their Mode III fracture. However, ductility to the extent required to produce ductile fracture is not observed in NaCl.

E. Fracture of the Refractory Oxides

Refractory oxides such as MgO and Al_2O_3 are potential high-temperature structural materials because their ionic-covalent (polar covalent) bonding provides an inherently great resistance to flow and fracture. Further, since they are oxides they are chemically inert in high-temperature oxidizing environments. This is not the situation for structural metals, which often must be protected from high-temperature oxidation by expensive coating processes.

As expected, a reduced tendency for ductile behavior is concomitant with the higher strengths of the refractory oxides. Thus, for Al_2O_3, whose fracture mechanism map is given in Fig. 10.22, the lowest temperature at which tensile ductility is observed is ca. $0.50T_m$, and even here its manifestation is limited to Mode III

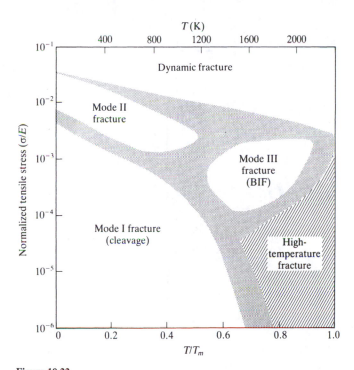

Figure 10.22

A fracture mechanism map for Al_2O_3, a polar covalent oxide. Alumina does not exhibit ductile fracture, although limited plasticity precedes Mode III fracture by BIF. The general stress level for fracture in Al_2O_3 and other covalent oxides is greater than that in the metals and alkali halides; this reflects the inherently stronger bonding of these oxides. (*Adapted from C. Gandhi, and M. F. Ashby, "Fracture-Mechanism Maps for Materials Which Cleave," Acta Metall., 27, 1565, Copyright 1979, with permission from Elsevier Science.*)

490

CHAPTER 10
Toughening
Mechanisms
and the Physics
of Fracture

fracture. Thus, the refractory oxides are stronger, but more brittle, than alkali halides and metals. Current work focuses on "toughening" these materials by microstructural manipulation (cf. Sect. 10.3). Indeed, the potential of these oxides is evidenced by the fact that even "poorly processed" oxides, e.g., those containing considerable porosity, are currently used as moderate-load-bearing members in devices such as high-temperature ovens.

F. Fracture of Covalent Solids

What holds for the refractory oxides holds even more for covalently bonded solids such as silicon carbide (SiC) and silicon nitride (Si_3N_4). Their covalent bonding imparts to them considerable resistance to flow and fracture and, paradoxically, at the same time renders them prone to brittle fracture. These characteristics are shown in the fracture mechanism map of Si_3N_4 (Fig. 10.23). Covalently bonded solids are even more attractive materials for high-temperature structural use than are the refractory oxides. Although not shown in the fracture mechanism map of Fig. 10.23, the creep rates of covalent solids are also low in comparison to those of metallics. This reflects the fact that diffusional rates in covalent solids remain low up to fairly high homologous temperatures.

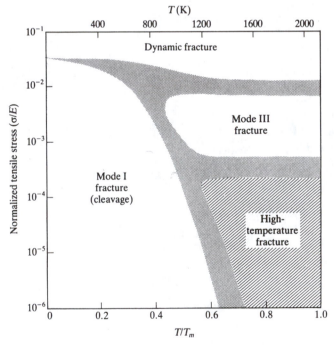

Figure 10.23
A fracture mechanism map for the covalent solid, Si_3N_4. Silicon nitride and similarly bonded materials exhibit the highest strengths of all materials, yet are also the most brittle; e.g., Mode III fracture dominates at high stress levels even when the homologous temperature is unity. (*Adapted from C. Gandhi, and M. F. Ashby, "Fracture-Mechanism Maps for Materials Which Cleave," Acta Metall., 27, 1565, Copyright 1979, with permission from Elsevier Science.*)

G. Synopsis

491

SECTION 10.7
The Relation
Among Bonding,
Crystal Structure,
and Fracture

Fracture mechanism maps depict the fracture mode of a material as it is affected by stress and temperature. Such maps are material specific and should also take into account microstructural features and stress state. These affect the tendencies for the various fracture modes and the relative areas occupied by them in a fracture map. For example, the ones presented here are appropriate to uniaxial tensile loading; application of a second positive principal normal stress would expand the regions of the map occupied by the brittle fracture modes.

Some regions of a fracture mechanism map are permissive, not definitive. For example, in the Mode I fracture region a material fails only if it contains preexisting cracks. Likewise, the creep rates concurrent with high-temperature fracture may be so low that the inevitable fracture may take so long that it need not be considered for all practical purposes. Other areas of the map are definitive. Thus, a material will fail by Mode II, Mode III, or ductile fracture if it is exposed to the stress levels appropriate to these fracture modes as indicated in the fracture mechanism map.

The stress required to cause fracture, and the mode of fracture, depends on crystal structure and chemical bonding. Metals are the most ductile materials. In fact, almost all fcc metals do not fail by brittle fracture in tensile loading. Body-centered cubic transition metals and hexagonal close-packed metals do exhibit brittle fracture at low temperatures, but they become ductile at relatively low homologous temperatures.

At low temperatures, polycrystalline nonmetals fracture in a brittle manner and, consequently, the regions of brittle fracture occupy a relatively large fraction of their fracture mechanism maps. In nonmetals, strength levels increase with the degree of covalent bonding. Thus, primarily ionically bonded alkali halides are not as strong as polar-covalent oxides, and covalent solids are stronger still. These tendencies are reflected in the temperature at which failure involving limited general plasticity (Mode III fracture) or creep ensues. These generalities have been noted in the materials whose specific fracture mechanism maps have been presented. They are also summarized in Figs. 10.24, which display the general form of fracture mechanism maps as they vary with chemical bonding and crystal structure. Although the details in a map differ from material to material, the maps for a given class of material are remarkably similar. Thus, the maps of Fig. 10.24 are useful guides in delineating the fracture modes of the material classes as they vary with stress and temperature.

As noted, fracture is triggered when the fracture stress is less than the stress required to initiate or continue plastic flow. In this sense, fracture represents a competition between shear (plastic flow) and tensile (fracture) deformation. An intrinsic measure of the plastic-flow resistance is the yield strength at 0 K. This can be estimated by extrapolating low-temperature yield strength or hardness values to 0 K. Young's modulus, E, is a measure of the resistance to bond stretching. So is the bulk modulus, K, and for reasons previously mentioned it is a somewhat more direct measure of bond strength than E. Hence, the ratio of the yield strength at 0 K (σ_{y0}) to K should reflect a material's tendency to "brittleness." High values of this ratio indicate a tendency to brittle behavior; low values to ductile behavior. Figure 10.25 plots the homologous temperature at which macroscopic plasticity (either by high-temperature creep or low-temperature ductile tensile behavior) is observed

492

CHAPTER 10
Toughening
Mechanisms
and the Physics
of Fracture

versus σ_{y0}/K for the several material classes. This figure concisely summarizes a material's fracture tendencies as they relate to bonding and crystal structure.

10.8
MODE II BRITTLE FRACTURE

Mode I brittle fracture has been treated in our discussion of fracture mechanics (Chap. 9). In effect a tensile bar containing preexisting cracks can be considered similar to a structure containing cracks, and the crack propagation stress is given by Eq. (9.13).[9] However, the physics of the other forms of fracture, as observed in a tensile test containing none or fairly innocuous flaws, differs. Thus, these other fracture modes are discussed in this and subsequent sections.

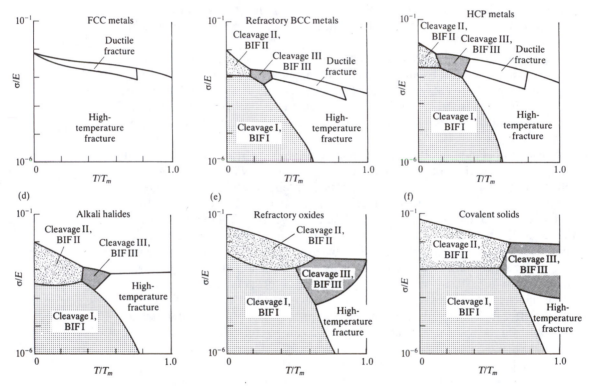

Figure 10.24
Schematic fracture mechanism maps for the various classes of crystalline solids. The general stress levels and temperatures at which the fracture transitions take place are similar within a given material class. The specific features for each material class are discussed in the text and in the captions to Figs. 10.18–10.23. (*Adapted from C. Gandhi, and M. F. Ashby, "Fracture-Mechanism Maps for Materials Which Cleave," Acta Metall., 27, 1565, Copyright 1979, with permission from Elsevier Science.*)

---

[9]As noted previously, there is only one complication; it has to do with the tensile sample size. The crack propagation stress depends on whether the sample size is characteristic of plane strain or plane stress conditions (or a mixture of the two). Thus, the value of K_c appropriate to Eq. (9.13) must take this into account.

Mode II tensile fracture is preceded by microscopic plastic deformation. This plasticity is distributed through the material volume and is not confined to the crack tip vicinity as it is in Mode I fracture. On the other hand, this plasticity is *microscopically heterogeneous*. Indeed, it is the strain incompatibility associated with localized "packets" of plasticity that leads to crack nucleation. Thus, the sequence leading to Mode II fracture is

$$\text{microscopic plasticity} \rightarrow \text{crack nucleation} \rightarrow \text{crack propagation}$$

and the observed fracture stress is the *greatest* of the stresses required to effect the three individual processes. There is another difference between Mode II and Mode I fracture. In Mode I fracture, the fracture stress is the crack propagation stress (e.g., σ_F of Eq. (9.13)) because cracks preexist when this fracture mode is observed. Because the various stresses σ_y (microscopic yield strength), σ_{nuc} (crack nucleation stress), and σ_F (crack propagation stress) depend differently on temperature, we often find a change in the criterion for Mode II fracture with temperature alterations.

The Mode II fracture *propagation* stress varies with grain size to the $-\frac{1}{2}$ power. This is because the crack size nucleated by plastic deformation scales with the grain size and so the crack length in Eq. (9.13) is replaced by the grain diameter for Mode II fracture. However, the type of plasticity preceding microcrack nucleation in Mode II fracture differs among materials. In nonmetals (e.g., the alkali halides and the polar covalent solids) and in some hcp metals, the number of slip systems required for polycrystalline plasticity is not available at low temperatures. Thus, the

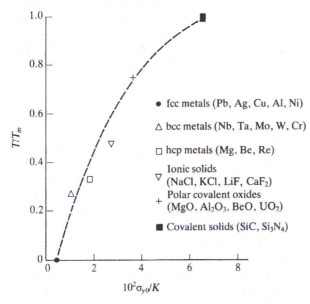

Figure 10.25
Homologous temperature at which plastic flow is observed vs. the ratio of the yield strength at 0 K (σ_{y0}) to bulk modulus (K). The material classes display different values of this ratio, and this is reflected in differing tendencies to brittleness; i.e., different homologous temperatures must be attained before plastic deformation takes place in the different classes. (*Constructed from data compiled by C. Gandhi and M. F. Ashby, Acta Metall., 27, 1565, 1979.*)

494

CHAPTER 10
Toughening
Mechanisms
and the Physics
of Fracture

heterogeneous plastic flow resulting from slip on the "soft" slip systems (i.e., the ones with the lowest value of τ_{CRSS}) is what nucleates cracks. On the other hand, the bcc transition metals have a reasonable number of slip systems at all temperatures. Thus, the stress initiating heterogeneous flow in them is also the *macroscopic* yield strength. However, because this low-temperature plastic deformation is heterogeneous (it often takes place by twinning), crack nucleation results from it. Thus, although Mode II fracture propagation is similar in all materials, there are differences in crack nucleation behavior for this mode between the bcc transition metals and other materials. For this reason, we discuss separately Mode II fracture in bcc metals and their alloys.

A. Mode II Fracture Initiated by Slip Incompatibility

As described in Chaps. 3 and 4, most nonmetals and some hcp metals (especially at low temperatures) do not have a sufficient number of slip systems to provide for generalized polycrystalline plasticity. However, this factor does not rule out microscopic plasticity. The latter initiates at a tensile stress producing the critical resolved shear stress for slip on the "soft" slip system(s). General yielding cannot occur until the tensile stress is sufficient to activate flow on the "hard" systems. At low temperatures this latter stress is much higher than the "soft" system flow stress.

Following initiation of microscopic flow, a crack must be nucleated for fracture to occur. A number of crack nucleation mechanisms have been suggested; two are discussed here. Crack nucleation can be caused by dislocation pileups or intersections. In one pileup model (Fig. 10.26a), unaccommodated displacements across the slip plane where the dislocations have piled up against an obstacle (e.g., a grain boundary) produce a wedge-shaped cavity (a microcrack). In Fig. 10.26b, glide dislocations coalesce when they intersect and produce a crack. In principle, cracks so nucleated can continue to grow as long as dislocations continue to run into obstacles or to intersect each other. The stress necessary for this dislocation "pumping" is somewhat greater than the initial flow stress. Provided microcracks do grow in this way, they can either extend across the entire grain (where the grain boundary provides an obstacle to their further growth), or grow to a critical size at which they spread spontaneously across the grain.

An estimate of the crack nucleation stress is obtained by reconsidering the stress concentration at a grain boundary resulting from a dislocation pileup. From Chap. 5, the shear stress acting at a distance r within an adjacent grain is approximately $(\tau - \tau_0)(d/4r)^{1/2}$, where the slip distance is taken as the grain size d, τ_0 is the intrinsic flow strength, and τ is the stress necessary to continue plastic deformation in the presence of the pileup. We assume a crack nucleates at the boundary (thus $r \cong a_0$, the interatomic spacing) when the effective tensile stress ($\cong 2\tau$) there equals the theoretical fracture strength. Substituting $\tau = \sigma/2$ and $\tau_0 = \sigma_0/2$, the crack nucleation condition is

$$(\sigma - \sigma_0)(d/4a_0)^{1/2} = \sigma_{th} \qquad (10.20a)$$

Substituting for σ_{th} from Eq. (9.6b)

$$\sigma_{nuc} = \sigma_0 + (4E\gamma/\pi d)^{1/2} \qquad (10.20b)$$

In Eq. (10.20b), σ_0 is the tensile stress required for "soft" system slip. Clearly, $\sigma_{nuc} > \sigma_0$, but at the high stress levels at which Mode II fracture occurs, the second term of Eq. (10.20*b*) is, at most, comparable to σ_0 and generally less than it (Prob. 10.14). Thus, we approximate the crack nucleation stress as being about equal to the soft system yield strength.

On this approach, whether or not plastic-flow initiation is concurrent with fracture depends on the value of the "soft" system yield strength compared to σ_F. The respective stresses depend differently on temperature as indicated in Fig. 10.27. The yield strength decreases with temperature, but σ_F (because of the temperature variation of the toughness) increases with it. Thus, below the temperature T_1 in Fig. 10.27 fracture is concurrent with flow ($\sigma_y > \sigma_F$). Conversely, if $T > T_1$ flow occurs and cracks are nucleated thereby. But the stress must be increased to σ_F for these cracks to propagate. Thus, in a tension test the measured fracture stress is σ_y when $T \leq T_1$ and it is σ_F when $T \geq T_1$.

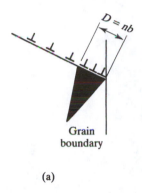

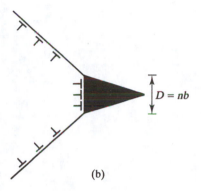

(a)   (b)

Figure 10.26
Dislocation mechanisms for crack initiation. In (a) a dislocation pileup at a grain boundary produces a net offset (equivalent to a microcrack) across atomic planes. (b) Dislocation intersections and resulting reactions can produce a similar effect within a grain. In both cases, the typical length of a microcrack formed is ca. *nb*, where *n* is the number of dislocations constituting the microcrack and *b* is their Burgers vector. Microcracks formed in these manners can continue to grow provided the stress is sufficient to continue to push dislocations into, for example, a pileup at the grain boundary.

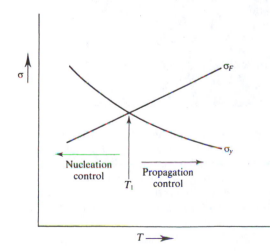

Figure 10.27
Mode II fracture of other than the bcc transition metals can be subdivided into two classes, depending on whether it is controlled by yielding or crack propagation. At temperatures below T_1, microcracks nucleate at σ_y. Since this stress is greater than σ_F, microcracks so formed propagate spontaneously. Above T_1, cracks nucleate at σ_y, but they do not propagate until the stress is increased to σ_F. Thus, the stress required for fracture is the greater of σ_F or σ_y (= σ_y for $T < T_1$; = σ_F for $T > T_1$).

496

CHAPTER 10
Toughening
Mechanisms
and the Physics
of Fracture

Not all materials exhibit such a change in controlling fracture mechanism. For example, the soft system flow strength is always less than the crack propagation stress in the alkali halides; the propagation stress, therefore, defines the fracture stress. Conversely, fracture in the polar covalent oxides (e.g., MgO and Al_2O_3) almost always is concurrent with microplasticity initiation; this reflects their inherently low fracture toughnesses and the resulting low crack-propagation stresses.

Crack nucleation by the mechanisms just described has been observed. Examples illustrating nucleation by slip-band pileups and intersections are provided in Figs. 10.28. Likewise, nucleated microcracks that have not propagated have been

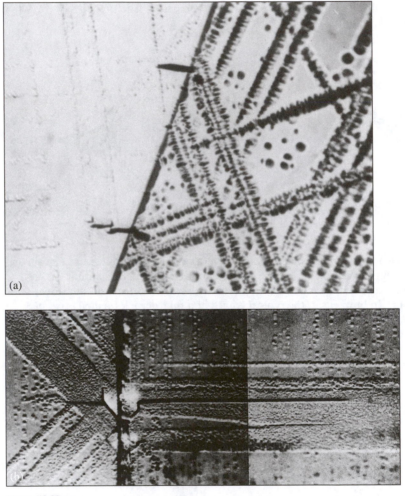

Figure 10.28
(a) Microcracks formed in MgO at tips of slip bands impinging on a grain boundary. Magnification, 1170 ×. (*From T. L. Johnston, R. S. Stokes, and C. H. Li, Phil. Mag., 7, 23, 1962.*) (b) Cracks in MgO that formed at the intersection of slip bands. This photograph shows two adjacent sides of a crystal that meet at a common edge (the dark vertical line). The crack is associated with the intersection of the crystal faces. (*From E. R. Parker, Fracture, The Technology Press and Wiley, New York, 1959, p. 181.*)

observed in materials for which the crack-nucleation stress is less than the crack-propagation stress (Fig. 10.29). Thus, our description of Mode II fracture is in agreement with experiment.

B. Mode II Fracture in bcc Transition Metals

As mentioned, the main distinction between Mode II fracture in bcc transition metals and in other materials is that the bcc structure is "strain compatible." That is, slip takes place on a sufficient number of systems to provide for polycrystal plasticity. Thus, the yield strength for Mode II fracture in the bcc transition metals is the macroscopic yield strength.

If the number of slip systems for the bcc structure can provide for general plasticity, why do the bcc transition metals exhibit Mode II fracture? The reason is that low-temperature plasticity in them is heterogeneously manifested. Slip-band pile-ups or intersections (at the higher temperatures of Mode II fracture) or similar twin configurations (twinning usually precedes slip at lower temperatures) produce local strain incompatibilities that are not accommodated. Instead, microcracks form by means similar to those described in the previous section.

Mode II fracture of the bcc transition metals, therefore, is always concurrent with yielding. That is, if the crack-propagation stress were greater than σ_y, generalized plastic flow would occur before fracture. But this is Mode III, not

Figure 10.29
Cleavage cracks in polycrystalline iron that have been arrested at grain boundaries. Further crack propagation requires a higher stress than that needed to grow these cracks to a size comparable to the grain diameter. Magnification, 145 ×. (*From G. T. Hahn, B. L. Averbach, W. S. Owen, and M. Cohen, Fracture, The Technology Press and Wiley, New York, 1959, p. 91.*)

498

CHAPTER 10
Toughening
Mechanisms
and the Physics
of Fracture

Mode II, fracture. This is a situation different from that for the nonmetals considered in the previous section. The difference can be elaborated on using Fig. 10.30. In Fig. 10.30a, the temperature dependence of the soft system yield strength (σ_s), the hard system yield strength (σ_h, which corresponds to general yielding), and the crack-propagation stress (σ_F) are illustrated for materials of the type discussed in Sect. 10.8A. Below the temperature T_1, σ_s defines the Mode II fracture stress; above it, σ_F does. General plasticity takes place only at temperatures in excess of T_2 when σ_F becomes greater than σ_h; i.e., T_2 represents the transition from Mode II to Mode III fracture. Analogous curves for the bcc transition metals are provided in Fig. 10.30b. For these metals σ_h and σ_s are identical and define the Mode II fracture stress. For the bcc transition metals, the temperature T_2 is the "nil-ductility" temperature (i.e., below T_2 no measurable R.A. is observed in a tensile test) and also the Mode II to Mode III transition temperature; i.e., for $T > T_2$, crack propagation controls fracture of the bcc transition metals.

Materials failing by Mode II fracture do not exhibit macroscopic plastic deformation prior to fracture; thus, they can be considered "brittle" solids. However, at temperatures at which failure is controlled by flow ($\sigma_y > \sigma_F$) they are stronger than materials failing by Mode I fracture. They also become tougher above the Mode II–Mode III transition temperature, for not only does general plastic flow precede fracture then, but $\mathcal{G}_c$ is also increased.

Before discussing Mode III fracture, we consider the effect of grain size on the Mode II fracture stress. For materials other than the bcc transition metals and for

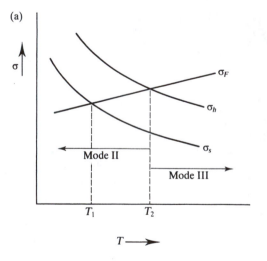

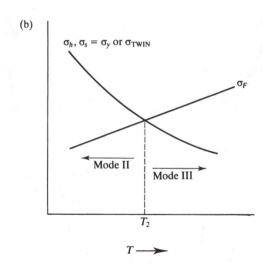

Figure 10.30
The Mode II–Mode III fracture transition in (a) hcp metals and nonmetallics and (b) the bcc transition metals. In (a) (cf. Fig. 10.27) Mode II fracture is crack nucleation controlled below T_1 and crack propagation controlled above it. Mode III fracture ensues when the stress for general yielding (σ_h) becomes less than the crack-propagation stress; this takes place at the temperature T_2. The fracture stress is thus σ_s for $T < T_1$ and σ_F for $T > T_1$. (b) Owing to their large number of slip systems, there is no distinction between "hard" and "soft" system yield strengths for bcc metals. Thus, the transition from Mode II to Mode III fracture takes place at temperature T_2 above which $\sigma_F > \sigma_y$ (or the twinning stress, if appropriate). The fracture stress is thus σ_y for $T < T_2$ and σ_F for $T > T_2$.

which yielding (defined by σ_s) determines the fracture stress, there is *no effect* of grain size on fracture behavior (other than that associated with the second term of the crack nucleation stress of Eq. (10.20b) which we have neglected). However, when crack propagation determines the fracture stress, fine-grain-sized materials are more fracture resistant than coarse-grained ones; that is, σ_F scales with $d^{-1/2}$. Because of this, grain-size refinement is used widely to "toughen" such materials. It is clear, however, that this is of benefit only when exercised in the temperature range where the crack propagation stress represents the fracture criterion.

Grain-size refinement always improves the fracture resistance of the bcc transition metals. This is so even though a fine grain size results in an increased yield strength, a feature ordinarily associated with a greater tendency to brittle behavior. We show why this is by considering the equations describing the grain-size dependence of σ_y and σ_F:

$$\sigma_y = \sigma_0 + k_y d^{-1/2} \tag{10.21}$$

and

$$\sigma_F = \left(\frac{E\,\mathcal{G}_c}{\pi}\right)^{1/2} d^{-1/2} \tag{10.22}$$

The coefficient of $d^{-1/2}$ in Eq. (10.22) is greater than it is in Eq. (10.21) (Prob. 10.18). Thus, plots of σ_F and σ_y vs. $d^{-1/2}$ are as shown in Fig. 10.31. At large grain sizes (small $d^{-1/2}$), $\sigma_F < \sigma_y$ and fracture is concurrent with yielding (Mode II fracture). At fine grain sizes, $\sigma_y < \sigma_F$ and general plasticity precedes fracture (Mode III

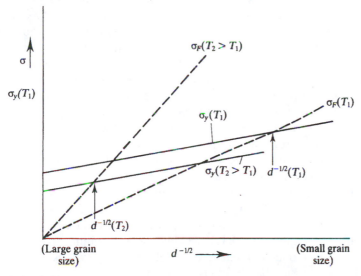

Figure 10.31
The Mode II–Mode III fracture transition is affected by grain size for the bcc transition metals. At large grain sizes (small $d^{-1/2}$), $\sigma_y > \sigma_F$ and fracture is concurrent with yielding and microcrack formation (Mode II fracture). At small grain sizes, $\sigma_F > \sigma_y$ and thus Mode III fracture prevails. Increases in temperature (e.g., the curves marked T_2) raise σ_F ($\mathcal{G}_c$ increases) and reduce σ_y; thus, the grain-size transition takes place at larger grain sizes at higher temperatures.

500

CHAPTER 10
Toughening
Mechanisms
and the Physics
of Fracture

fracture). Thus, at a given temperature there is a grain-size transition from Mode II to Mode III fracture. At higher temperatures, σ_y is decreased (due to decreases in both σ_0 and k_y) and σ_F is increased ($\mathcal{G}_c$ increases with temperature). Thus, the nil ductility transition occurs at larger grain sizes with increasing temperature. Hence, we arrive at the remarkable conclusion that grain-size refinement of the bcc transition metals is unique; it not only increases the yield strength but also the fracture resistance. As mentioned, most other strengthening mechanisms (e.g., solid-solution hardening, work hardening, etc.) increase σ_y without much affecting σ_F and this leads to an increase in the nil-ductility transition temperature (Prob. 10.24).

10.9
MODE III BRITTLE FRACTURE

The transition from Mode II (or Mode I) to Mode III fracture occurs when the yield strength for macroscopic flow equals the crack-propagation stress. At temperatures above that at which this occurs, $\sigma_F > \sigma_y$ and generalized plasticity precedes fracture. During Mode III tensile fracture the extent of plastic deformation remains limited (ca. several to 10 percent strain in a tensile test), and final fracture takes place in a brittle manner (by cleavage or BIF) and through linkup of microcracks that have been nucleated via plastic deformation.

From previous discussions, the stress (a crack-propagation stress) for Mode III fracture varies with material grain size. However, the microcrack whose propagation results in final fracture need not necessarily have a size equal to that of a typical grain. At the lowest temperatures of Mode III fracture, such microcracks typically do, but at higher temperatures several such microcracks may link up ("slow" crack growth) prior to attaining a size sufficient to induce catastrophic ("fast" crack growth) failure. Moreover, in "dirty" materials, inclusions or dispersoids are often the source of the critical microcracks. However, the critical crack length in both "dirty" and "clean" materials undoubtedly scales with microstructural scale, as typically represented by the grain size. Thus, the above complexities can be considered incorporated into the toughness term of Eq. (10.22).

Mode III fracture happens when work hardening raises the material's flow stress to the level of the crack-propagation stress where previously formed microcracks link up. Since the propagation stress (which controls Mode III fracture) increases with temperature, the tensile fracture stress does likewise (Fig. 10.32). Indeed, this temperature variation of σ_F is responsible for the "bump" in the tensile fracture stress of all materials manifesting a transition from Mode II to Mode III fracture. When the fracture propagation stress eventually attains a value equal to the material's (ductile) tensile strength (Fig. 10.32), necking precedes failure. This temperature represents the onset of ductile fracture, for which fracture proceeds by plastic deformation mechanisms rather than brittle fracture ones.

Mode III fracture differs from both Mode I (or Mode II) fracture and ductile fracture. During Mode II fracture, the first suitably oriented microcrack initiated by plastic deformation is the precursor to fracture. Microcracking is much more prevalent during Mode III fracture. That is, as the yield strength progressively falls below σ_F, more and more microcracks form prior to final material fracture. As the temperature continues to increase, however, the yield strength becomes so low that

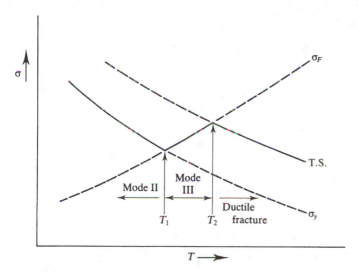

Figure 10.32
**Yield strength, tensile strength, and crack-propagation stress variation with tempera-
ture. At low temperatures, $\sigma_y > \sigma_F$ and microcracks formed on yielding propagate
spontaneously; σ_y defines the fracture stress for this Mode II fracture. At intermediate temperatures, $\sigma_F > \sigma_y$, but σ_F is less than the tensile strength. Following yielding
and when the stress has been increased to σ_F, Mode III fracture occurs. At the highest temperatures, $\sigma_F > $ T.S. Thus, necking and ductile fracture take place. The overall fracture stress ($= \sigma_y, T < T_1; = \sigma_F, T_1 < T < T_2; = $ T.S., $T > T_2$) is shown by the
solid line.**

crack nucleation is harder to effect, and the incidence of microcracking decreases
(Fig. 10.33a). Finally, ductile fracture ensues at the temperature at which $\sigma_F = $ T.S.
Profuse microcracking is then restricted to the necked region of a tensile bar; that is,
for ductile fracture, microcracks are not generally found in abundance throughout
the material volume.

Changes in material ductility also accompany the progression towards ductile
fracture. Below the nil-ductility transition temperature, the %R.A. for tensile fracture is zero (Fig. 10.33b). During Mode III fracture, the %R.A. is small but finite;
the limited plastic deformation preceding brittle fracture is unaccompanied by sample necking. As the temperature for ductile fracture is approached, and then exceeded, the %R.A. increases dramatically (Fig. 10.33b).

Mode III toughnesses exceed those for Mode I or Mode II fracture. This is reflected in higher σ_F values. In contrast to Mode I or II fracture, crack tip plasticity
is fairly extensive during Mode III fracture. This is illustrated schematically in Fig.
10.34 in which, because of the low values of σ_y compared to the theoretical fracture
stress, the plastic zone extends a considerable distance in front of the crack tip.
Estimates of this distance r_y yield (for plane stress conditions, see Sect. 9.4C) $r_y \cong$
K^2/σ_y, where K is the stress-intensity factor. It is clear that with decreasing σ_y the
plastic-zone size increases. The plastic zone actually extends a distance R from the
crack tip that is greater than r_y (typically $R \cong 2r_y$, Fig. 10.34). This is because the
displacements within the plastic zone must be accommodated by elastic displacements outside it, and this is more easily accomplished if the plastic-zone size is

502

CHAPTER 10
Toughening
Mechanisms
and the Physics
of Fracture

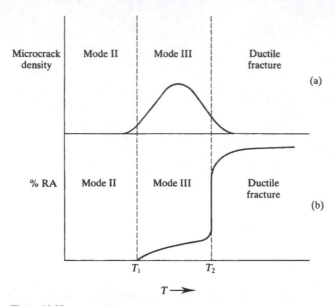

Figure 10.33
(a) Microcrack density and (b) %R.A. in a tensile test as they vary with temperature during the brittle-to-ductile fracture transition. For Mode II fracture, the first microcrack formed results in fracture. Thus, microcrack density and %R.A. are minimal. For Mode III fracture, microcracks form as a result of plastic deformation at stress levels less than σ_F. Thus, the microcrack density is relatively high and a small %R.A., corresponding to the uniform tensile strain, precedes brittle fracture. At the highest temperatures, necking precedes ductile fracture. Some microcracking may take place, but its incidence decreases as temperature is increased (i.e., as σ_y decreases).

increased. Irrespective of these details, the important result is that the extent of the plastic-zone region, both in front of the crack tip and above and below the fracture plane, is increased during Mode III fracture. Thus, crack tip blunting is greater as is the crack tip radius of curvature, ρ. Since $\mathcal{G}_c$ is proportional to the product of σ_y, ρ, and ε_f (ε_f = fracture strain), we see that the product $\rho\varepsilon_f$ must increase with temperature more rapidly than σ_y decreases with it in order to account for the increase in $\mathcal{G}_c$ (and σ_F) with temperature.

Ductile fracture toughnesses are high. This is a result of the larger strains (and hence plastic work) associated with this type of fracture. Indeed, as discussed later, $\mathcal{G}_c$ may be a factor of several hundred higher for ductile fracture compared to what it is for Mode III fracture.

A brief discussion of the appearance of brittle fracture surfaces is worthwhile before concluding our discussion of this form of fracture. On a macroscopic level, brittle fracture surfaces are "flat," but this is not so when the surface is examined at higher magnifications. Such a micrograph (fractograph) is shown in Fig. 10.35a for a cleavage fracture. The faceted or crystalline appearance of the surface indicates that fracture occurs preferentially on certain crystallographic planes—cleavage planes—across which the tensile stress required to sever atomic bonds is a minimum. The cleavage plane depends on a material's crystal structure and bonding,

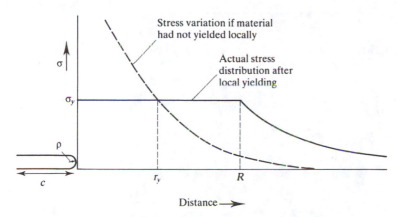

Figure 10.34
Tensile stress variation in front of a crack for a material for which $\sigma_y \ll \sigma_{th}$. The broken line illustrates the stress-distance relation appropriate to an elastic solid. The stress exceeds σ_y at distances $< r_y$ from the crack tip, and this causes crack tip plasticity and then the actual stress distribution is shown by the solid line. The plastic zone extends a distance R ($\cong 2r_y$ for plane stress conditions) in front of the crack tip. The stress distribution is modified in this way to allow for accommodation in the elastic region of the displacements in the plastic one.

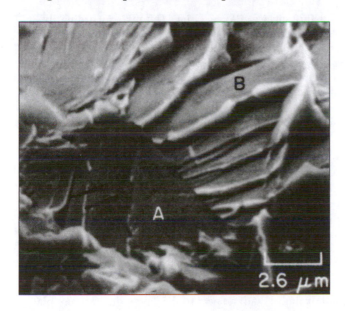

Figure 10.35
(a) Cleavage fracture in a plain carbon steel at 77 K. The fracture surfaces of individual grains (e.g., A and B) correspond to preferred crystallographic fracture (cleavage) planes. This results in a microscopically irregular (but macroscopically flat) fracture surface. Within an individual grain (e.g., B), the fracture may propagate along several parallel cleavage planes.

but the appearance of a cleavage-fracture surface is independent of these factors. The topographical features of the fracture surface show that in a polycrystal the cleavage crack propagates across crystallographically equivalent planes in grains oriented differently with respect to the tensile axis. Note that this feature leads to a "true" fracture surface greater than the nominal one and, as a result, a corresponding greater fracture work. In addition, and as can be seen in grain B of Fig. 10.35a, the fracture plane is periodically displaced within a specific grain. This "hip hopping" also results in an increased toughness. The distinctive appearance of a

504

CHAPTER 10
Toughening
Mechanisms
and the Physics
of Fracture

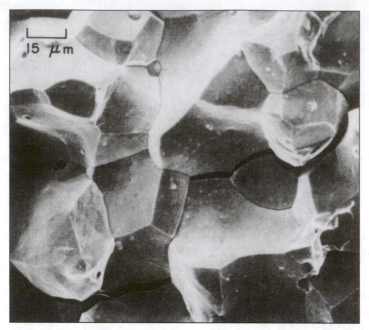

Figure 10.35 (continued)
(b) The brittle intergranular fracture surface of Armco iron. This material ordinarily fails transgranularly; however, it has been oxygen "embrittled" leading to intergranular fracture. The fracture is microscopically irregular; its topography mirrors that of the individual grains within the polycrystal. (*Metals Handbook, 8th ed., Vol. 9, Fractography and Atlas of Fractographs, ASM International, Materials Park, OH 44073-0002 (formerly American Society for Metals, Metals Park, OH, 44073-0002), 1974.*)

cleavage-fracture surface, when viewed at the proper magnification, can be a useful tool in fracture diagnosis.

Brittle intergranular fracture surfaces (Fig. 10.35*b*) are also easily recognized. The macroscopically flat fracture surface is again irregular on a microscopic level, and its "pattern" reflects the topography of the polycrystalline grain boundaries across which fracture has occurred. As Fig. 10.35*b* indicates, fractography can also be a useful tool in identifying BIF as a failure mode.

Inorganic glass brittle-fracture surfaces do not resemble fracture surfaces in crystalline materials. On a microscopic level, the fracture surface of a glass is smoother than in a brittle polycrystal, but it is not "atomistically" smooth. Small ridges (river markings) on the surface indicate that the propagating crack periodically displaces itself by small distances with respect to the "average" fracture plane. This is caused by slight structural irregularities that make it energetically favorable for the crack to occasionally change its fracture plane. The variation in the fracture plane produces steps which are the noncrystalline analog of the similar steps observed on cleavage planes in crystals. As with the cleavage steps, the ones in glass contribute to an increased toughness.

EXAMPLE PROBLEM 10.2. The plastic-flow strength of metal "B" is given by $\sigma = \sigma_0(A + \varepsilon)^n$, where $\sigma_0 = 700$ MN/m^2 and $n = 0.5$; σ_0 and n are temperature independent, but A is not. For temperatures greater than 100 K, the yield strength is proportional to $T^{-1/2}$, where T is absolute temperature. Moreover, at 400 K the yield

at a true tensile strain of 24%. If the crack-propagation stress is temperature independent, what is the cleavage fracture strain of metal "B" at 300 K?

Solution. The problem solution is straightforward, but it involves some number
crunching. The idea here is that the material plastically flows and then work hardens.
When the work hardening has increased the flow strength to a level equal to the cleavage fracture stress, fracture takes place. We proceed in steps. The units of stress in the
following calculations are all in MN/m².

We first must use the provided information to determine the (temperature depen-
dent) constant, A, in the plastic-flow equation. The yield strength is defined for $\varepsilon = 0$.
Thus, we have

$$\sigma_y = 700A^{0.5}$$

We are told that the yield strength varies as $T^{-1/2}$, so we can write $\sigma_y = 700A_0^{0.5}/T^{1/2}$
where A_0 is a temperature-independent constant. We are also told that at 400 K, $\sigma_y = $
350 MN/m². So

$$350 = 700A_0^{0.5}/(400)^{0.5}; \qquad 700A_0^{0.5} = 7000 \text{ MN/m}^2; \qquad A_0 = 100$$

and

$$\sigma = 700\left(\frac{100}{T} + \varepsilon\right)^{0.5}$$

Cleavage fracture takes place at a strain of 0.24 at 400 K. This happens when the flow
stress (written above) equals the cleavage fracture stress. Letting this latter stress be sig-
nified by σ_F, we have

$$\sigma_F = 700(0.25 + 0.24)^{0.5} = 490 \text{ MN/m}^2$$

So now we can determine the cleavage fracture strain at 300 K. This is done by equating the plastic flow and the cleavage stresses. We see that the critical strain is a solution
to

$$700\left(\frac{100}{300} + \varepsilon\right)^{0.5} = 490; \qquad \left(\frac{1}{3} + \varepsilon\right) = 0.49 \qquad \varepsilon = 0.16$$

After all this arithmetic, it is gratifying to find that the fracture strain is less at 300 K
than at 400 K. That is as it should be!

EXAMPLE PROBLEM 10.3. Suppose the tensile yield strength of a bcc metal
varies with absolute temperature (T) as $\sigma_y = (2000 - 4T)$ MN/m², and the metal's crack
propagation stress varies with T as $\sigma_F = (600 + 2T)$ MN/m². Calculate the Mode II to
Mode III fracture-transition temperature that would be found in tensile testing of this
metal.

Solution. This is a variation of the previous problem. But the arithmetic here is much
easier. We set the yield strength equal to the crack propagation stress and solve for the
temperature. Below the temperature that we find, fracture is concurrent with yielding
(Mode II fracture); above it the metal macroscopically flows before it fractures (Mode
III fracture). Thus, (again working with stress in MN/m²)

$$2000 - 4T = 600 + 2T; \qquad T = \frac{1400}{6} = 233 \text{ K}$$

Several other problems at the end of this chapter deal with fracture transitions of the
type considered in these two examples.

506

CHAPTER 10
Toughening
Mechanisms
and the Physics
of Fracture

10.10
DUCTILE FRACTURE

At higher temperatures (but not so high that creep is important), tensile fracture of most metals and some other materials is ductile in nature. In ductile fracture, crack nucleation *and* crack linkup involve plastic flow. Thus, ductile fracture differs fundamentally from the brittle fracture modes discussed previously.

The extreme manifestation of ductile fracture is rupture (Fig. 10.16*d*). It occurs at high temperatures in many metals and some alkali halides, and can also take place at ordinary temperatures in high-purity metals. The important role that plastic deformation plays in rupture is made clear by inspection of a tensile bar failing in this way. The 100% R.A. is effected by plastic deformation continuing beyond necking; this deformation eventually separates the bar into two pieces.

Most ductile tensile fracture is not accompanied by a 100% R.A. The lesser R.A. is, in many cases, a result of the presence of second-phase particles (or inclusions) having relatively low fracture strains. The particles' presence leads first to nucleation of microcracks or microvoids at them. The essential mechanics can be summarized as follows. The strain incompatibility between the particles and the matrix leads to generation of geometrically necessary dislocations in the matrix.[10] The internal tensile stress due to these dislocations (σ_{gd}) adds to that of the applied stress (σ_{Te}) so that the total stress (σ_T) acting to produce particle fracture/decohesion (Fig. 10.36) is

$$\sigma_T = \sigma_{Te} + \sigma_{gd} \tag{10.23}$$

Both σ_{Te} and, especially, σ_{gd} increase with plastic strain. Thus, at some critical strain, σ_T, as given by Eq. (10.23), reaches a value sufficient to cause particle fracture or decohesion; that is, microcracks or microvoids nucleate.

Crack nucleation for Mode III fracture may also take place in the way just described. However, crack linkup in Mode III is primarily by brittle fracture mechanisms whereas plastic deformation controls linkup during ductile fracture. The reason for the different behavior is straightforward. Prior to necking, microcracks/voids may form throughout the material volume. However, for ductile fracture, the tensile strength is less than the crack-propagation stress and thus necking takes place prior to material failure. Following necking, the true tensile stress in the necked region is greater than without it, so further void linkup is restricted to this region (Fig. 10.37*a*).[11] Final fracture takes place after void linkup has caused a

---

[10]If the second-phase particles deform only elastically, all such dislocations are located in the matrix. If, on the other hand, the inclusion is capable of plastic deformation (but is still less ductile than the matrix), the geometrical dislocations are partitioned between the phases. Further, the geometrical dislocation density is reduced because of the lesser strain incompatibility between the phases. This delays fracture nucleation and is one reason why the ductilities of two-phase alloys containing deformable inclusions are generally greater than those of similar alloys containing second-phase particles that are incapable of plastic deformation.

[11]The extent of microcracking that precedes necking depends on a number of factors, including the volume fraction and size of the second-phase particles and the yield strength and work-hardening characteristics of the matrix. When the particle volume fraction and the matrix work hardening are low, the presence of microcracks is more likely to be restricted to the necked region and vice versa. In any case, the incidence of generalized microcracking is *much* lower in ductile fracture than in Mode III fracture.

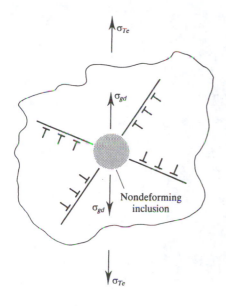

Figure 10.36
Plastic-deformation-induced inclusion fracture can lead to nucleation of voids that are precursors to ductile fracture. Geometrically necessary dislocations form in the region around the elastic inclusion. They give rise to an additional tensile stress component (σ_{gd}), which adds to the external stress (σ_{Te}). Eventually, these stresses cause inclusion fracture or decohesion from the matrix.

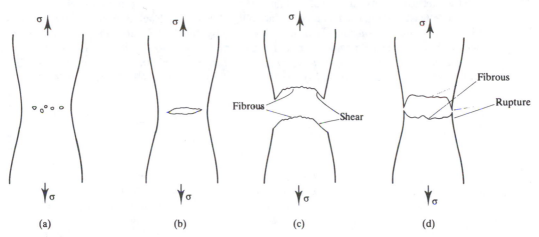

Figure 10.37
Stages of ductile tensile fracture. (a) Voids, often associated with inclusions, nucleate and grow throughout the material volume but do so preferentially in the neck volume as a result of the triaxial stress state there. (b) The voids link up by plastic flow and form a penny-shaped internal crack. (c) Final fracture often occurs by shear failure of the outer rim of the sample; the "shear lip" gives rise to the oft-observed "cup and cone" appearance of a fractured tensile sample. (d) Occasionally the outer rim may neck down, rather than shear, to failure. This produces a "double cup" fracture appearance.

penny-shaped crack to form in the sample interior (Fig. 10.37*b*). The final separation is accomplished by a shear, or tear, fracture of the rim surrounding this crack. Such shear lips are ubiquitous to ductile fracture. In a tensile test they are responsible for the characteristic "cup and cone" appearance (or variants thereof) of the fracture surface (Figs. 10.37*c* and *d*).

This description of ductile fracture is substantiated by Figs. 10.38 and 10.39. Figure 10.38 is an electron micrograph of a ductile fracture surface. It has a

508

CHAPTER 10
Toughening
Mechanisms
and the Physics
of Fracture

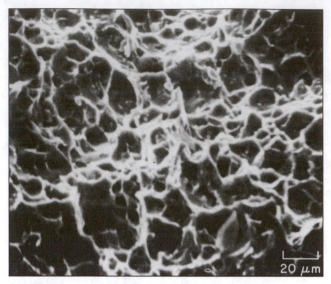

Figure 10.38
The ductile fracture surface of a plain carbon steel. The rough, dimpled surface is characteristic of ductile fracture surfaces in polycrystals. Each dimple corresponds to a fracture-nucleation site; these have linked up by plastic deformation processes. Inclusions may be spotted at the centers of some dimples. (*Metals Handbook, 8th ed., Vol. 9, Fractography and Atlas of Fractographs, ASM International, Materials Park, OH 44073-0002 (formerly American Society for Metals, Metals Park, OH 44073-0002), 1974.*)

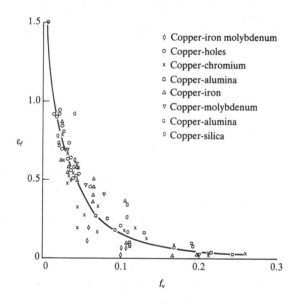

◊ Copper-iron molybdenum
○ Copper-holes
× Copper-chromium
□ Copper-alumina
△ Copper-iron
▽ Copper-molybdenum
◁ Copper-alumina
▷ Copper-silica

Figure 10.39
Tensile ductility (true strain at fracture in the neck) as a function of inclusion (or hole) volume fraction (f_v) in various copper alloys containing mostly nondeforming second-phase particles. The marked reduction in ductility with increases in f_v is evidence of the important role that inclusions play in ductile fracture. (*Adapted from B. Edelson and W. Baldwin, Trans. ASM, 55, 230, 1962.*)

characteristic "dimpled" look to it, in marked contrast to the faceted appearance of a brittle-fracture surface. The inclusions responsible for void initiation are located at the dimple centers. The material between these centers is the matrix whose separation has been accomplished by void linkup by plastic flow. Figure 10.39 illustrates the important role that inclusion content has on tensile ductility. The marked

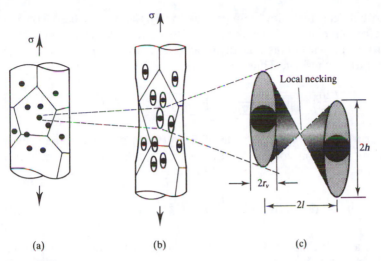

Figure 10.40
A model for ductile fracture initiated by voids or fractured/decohered inclusions. (a) Once formed, voids concentrate stress in their vicinity, resulting in void growth (b). (c) An expanded version of (b) shows that linkup can occur (by microrupture of the shaded regions between the particles in the case shown) when the void length (2h) is on the order of the void separation (2l − 2r_v). (Reprinted from M. F. Ashby, C. Gandhi, and D. M. R. Taplin, "Fracture-Mechanism Maps and Their Construction for FCC Metals and Alloys," Acta Metall., 27, 699, Copyright 1979, with permission from Elsevier Science.)

decrease in ductility with increases in second-phase volume fraction is a result of more extensive void nucleation and the greater ease with which void linkup is effected when the inclusion spacing is reduced.

The linkup mechanism can be discussed in more detail. One possible linkup mode is illustrated in Fig. 10.40. After void formation (the current discussion is in terms of particle decohesion, but the description also applies to failure initiated by particle fracture), the voids begin to grow/elongate in the direction of the applied stress. This *void growth* is accomplished by plastic deformation. On geometrical considerations, the strain rate ($\dot{\varepsilon}_v$) and hence strain in the void region is increased in comparison to that in the matrix material ($\dot{\varepsilon}_s$) removed from the voids. Detailed considerations show that $\dot{\varepsilon}_v = C\dot{\varepsilon}_s$, where the constant C is between 1 and 2. The increased strain in the intervoid region eventually produces a "microtensile" bar there and the final fracture stage—void linkup—can be accomplished by rupture of the matrix material (Fig. 10.40c). The effective true strain in the matrix between the voids is

$$\varepsilon_t = \ln\left(\frac{2h}{2r_v}\right) \tag{10.24}$$

where r_v is the particle size and h the axial extent of void elongation along the tensile axis. We assume that microrupture occurs when ε_t, as given by Eq. (10.24), reaches a critical value ε_g (the critical strain for void linkup). The sample true tensile strain at fracture, ε_{sf}, is linked to ε_g by the constant C, i.e.,

$$\varepsilon_{sf} = \frac{\varepsilon_g}{C} \tag{10.25}$$

510

CHAPTER 10
Toughening
Mechanisms
and the Physics
of Fracture

The relationship between inclusion volume fraction (f_v) and ε_g is based on the premise that microrupture occurs when the ratio of the length ($2h$) to spacing ($2l - 2r_v$) of the microtensile bar attains a critical value; i.e., when $2h = \alpha(2l - 2r_v)$ where α is an empirical constant. Thus,

$$\varepsilon_g = \ln\left(\frac{2h}{2r_v}\right) = \ln\left(\alpha\frac{(2l - 2r_v)}{2r_v}\right) = \ln[\alpha(f_v^{-1/2} - 1)] \tag{10.26}$$

where the last term on the right-hand side of Eq. (10.26) uses the relationship, $f_v = (r_v/l)^2$, among inclusion volume fraction, spacing, and size. In terms of a tensile failure strain

$$\varepsilon_{sf} = C^{-1}\varepsilon_g = C^{-1}\ln\left(\alpha\frac{(2l - 2r_v)}{2r_v}\right) = C^{-1}\ln[\alpha(f_v^{-1/2} - 1)] \tag{10.27}$$

Equations (10.26) and (10.27) illustrate the important effect of inclusion volume fraction on tensile ductility.

It is noteworthy that the reduction in ductility accompanying the presence of nondeforming second-phase particles does not reduce the *inherent* ductility of their matrix. That is, the matrix *microscopic* reduction in area remains substantial, notwithstanding that the macroscopic ductility may be considerably reduced. However, the toughness of the material is lowered when microrupture, rather than macroscopic rupture, occurs. This is primarily a result of geometrical effects. The decohered or fractured inclusions serve as defects that localize rupture along gage lengths on the order of $2h$, a microscopic dimension. In macroscopic rupture, the plastic flow leading to a 100% R.A. takes place over a macroscopic length (on the order of the sample diameter in a tensile test). Thus, the volume of material undergoing the extensive plastic deformation of rupture is reduced considerably for microrupture, and this is reflected in a lowered toughness.

Void linkup by microrupture is one of several possibilities and is, in fact, the linkup mechanism leading to maximum toughness. Any other such mechanism, e.g., void linkup by shear bands (Fig. 10.41), reduces the localized failure strain between voids, resulting in a lower toughness. The associated toughness reduction is due in part to the smaller volume of heavily deformed material accompanying this kind of linkup compared to that in microrupture.

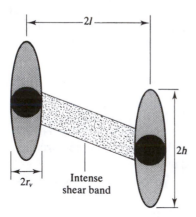

Figure 10.41
An alternative mode of void linkup in ductile fracture. In this case, linkage is effected by shear deformation between voids which takes place when the voids have elongated some critical amount. The plastic work accompanying linkup by shear is typically less than that for rupture (a lesser volume of intervoid material deforms during shear linkup).

$2l$

$2h$

$2r_v$

Intense
shear band

Toughnesses of ductile materials are considerably greater than those of brittle ones. However, this does not mean ductile materials are immune to service fracture of a nominally brittle nature. That is, material and loading geometries may be such that ductile cracks propagate at nominal stress levels below those required for yielding, and when this happens plasticity is restricted to the vicinity of the fracture plane. Indeed, such fracture is described by the fracture mechanics approach of Chap. 9.

A table and a summarizing figure might help in digesting the information presented in the last several sections. Table 10.3, which lists characteristics of the several forms of tensile fracture, represents such an attempt. Figure 10.42 also might help. The figure illustrates the features of a tensile test conducted on materials manifesting the various forms of tensile fracture. In addition, the correlation between fracture stress and $\mathcal{G}_c$ presented there helps in relating the tensile behavior of a material to its toughness. In this regard, attempts to correlate toughness (particularly for metals) with tensile properties (e.g., yield strength, ductility, work-hardening rate) have been made, but with only limited empirical success. A fracture toughness test must be conducted to obtain a meaningful material property to be used in design against fracture. This is a pity for a tensile test is (somewhat) easier to carry out. This situation is somewhat different for brittle ceramics. A hardness test, provided it is accompanied by secondary cracking, can be used to estimate (within about 25%, Sect. 1.4A) fracture toughnesses of ceramics.

Fracture mode	Fracture nucleation	Initial crack propagation	Final crack propagation	Fracture preceded by macroscopic yielding
Mode I brittle fracture	By presence of preexisting cracks	None	By cleavage or intergranular fracture with limited crack tip plasticity	No
Mode II brittle fracture	By microscopic yielding at yield strength	To something on the order of the size of a grain	By cleavage or intergranular fracture with limited crack tip plasticity	No
Mode III brittle fracture	By microscopic yielding at yield strength	To something on the order of the size of a grain; since $\sigma_y < \sigma_F$, multiple micro-cracking is found	By cleavage or intergranular fracture between microcracks with limited crack tip plasticity	Yes, limited
Ductile fracture	By inhomogeneous plastic flow, frequently at sites of second-phase particles	Void growth (by plastic deformation processes) at the locations of fracture initiation	By void linkup via ductile microrupture or shearing	Yes, extensive

Table 10.3

Characteristics of the various low-temperature fracture modes

512

CHAPTER 10
Toughening
Mechanisms
and the Physics
of Fracture

10.11
SUMMARY

Two main topics—toughening and tensile fracture behavior—have been treated in this chapter. The toughness of a material is defined by the irreversible work (per unit area of fracture surface) accompanying crack propagation. This work takes on different forms for different materials. In metals, it is the plastic work accompanying crack propagation. In the most brittle of solids, e.g., an inorganic glass at room temperature, it is twice the surface energy associated with the formation of the fracture surfaces. The latter toughness represents the minimum possible value of a material toughness.

Room-temperature toughnesses of metals vary depending on their yield strengths and ductilities. Metal toughnesses can also vary substantially with temperature if they are prone to a ductile-to-brittle transition. And, within a given alloy class, toughness is altered by microstructural considerations relating to alloy "cleanliness." "Dirty" alloys—those containing nonmetallic inclusions—have lower

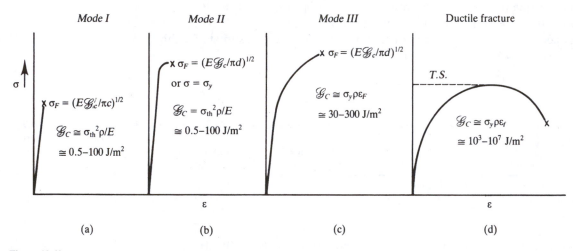

Figure 10.42
A synopsis of tensile stress-strain curves, fracture stresses, and toughnesses for different tensile fracture modes. **(a) Mode I fracture is preceded only by linear elastic deformation. Fracture occurs at the crack-propagation stress for preexisting flaws.** $\mathscr{G}_c \cong \sigma\rho\varepsilon_f$, and for this type of fracture, $\sigma = \sigma_{th}$, $\rho = (3\text{–}30a_0)$, and $\varepsilon_f = \sigma_{th}/E$. **Typical** $\mathscr{G}_c$ **values range from 0.5–100 J/m², depending on the values of surface energy and modulus, and the extent of crack tip plasticity accompanying crack extension. (b) The stress-strain curve for Mode II fracture is also essentially linear elastic even though microplasticity precedes fracture. The fracture stress is the greater of the yield strength or the crack-propagation stress. The work (i.e., toughness) accompanying Mode II crack extension is essentially the same as that for Mode I. (c) Mode III fracture is preceded by macroscopic yielding. However, fracture at the stress** σ_F **occurs prior to necking and at a small value of plastic strain (3–10 percent).** $\mathscr{G}_c$ **values for Mode III fracture are typically higher than for Mode II fracture as a result of increased crack tip blunting (which raises** ρ**) and greater values of** ε_f**. (d) The ductile fracture stress is the tensile strength. Values of** $\mathscr{G}_c$ **for ductile fracture are much greater than for brittle fracture. The several orders of magnitude variation in** $\mathscr{G}_c$ **(from 10^3 to 10^7 J/m²) for ductile fracture result from the several modes of ductile fracture. Rupture, for example, manifests the highest** $\mathscr{G}_c$ **values (10^6–10^7 J/m²), void growth by coalescence and microrupture the next highest ($\mathscr{G}_c \cong 10^4$–10^6 J/m²), and shear-induced ductile fracture the lowest (10^3–10^5 J/m²).**

toughnesses than do clean alloys of the same nominal composition and yield strength.

Ceramic toughnesses are due to causes different from those that control metallic toughnesses. Because of their high yield strengths, crack tip plasticity is very limited in ceramics. Thus, different means must be found to increase the work associated with crack propagation in them. Some of the most important are additional surface energy (associated with secondary microcracking), the work accompanying stress-induced phase transformations, and crack bridging. The latter leads to interfacial pull-out work (if the phase extracted is brittle) or additional plastic deformation work (if the phase is ductile). Bridging also plays an important role in toughening of ceramic-ceramic composites and brittle fiber reinforced plastics. The toughnesses of these composites considerably exceed those of their constituents because of the effects of bridging.

Polymers also can be toughened. Additions of plasticizers, which behave rather like a viscous second phase, improve plastic toughnesses although this is achieved at the expense of strength. Elastomeric dispersions also favorably impact toughness. They do so by providing for additional (localized) plastic deformation during crack propagation.

Tensile fracture behavior correlates with toughness. Some tensile bars contain preexisting cracks. When they do their fracture stress can be calculated via the fracture mechanics analysis of Chap. 9, provided allowance is made for any sample-size effect (i.e., whether plane stress or plane strain conditions pertain to the tensile test). This mode of tensile fracture is termed Mode I fracture.

When preexisting flaws are not present in a tensile bar (or when they are so small they are ineffective in causing crack propagation), material fracture still takes place in a tensile test. This can be either a ductile or brittle fracture. The criterion used to differentiate between the fracture modes is that a measurable sample reduction in area and evidence of necking are found in ductile fracture. Only the R.A. associated with uniform deformation prior to necking is observed for a material that fails in a brittle way; i.e., a brittle material does not neck.

Ductile fracture occurs at the highest temperatures of "low-temperature" fracture. Ductile fracture involves, sequentially, void (crack) nucleation, growth, and linkup. Crack nucleation is frequently accomplished by the fracture or decohesion of second-phase particles. Decohesion is caused by a tensile stress arising from both the applied stress and the internal stress generated as a result of the plastic strain incompatibility between the matrix and the second-phase particle. Void growth, like nucleation, involves plastic deformation. When voids have grown or elongated to a critical extent, they link up by mechanisms of plastic deformation. Since all of the sequential processes leading to ductile fracture are often linked to second-phase particles, it is not surprising that tensile ductility is strongly affected by the particle size, its spacing, and—particularly—the volume fraction of the particles.

Brittle fracture is accompanied by a nil or minimal R.A. in a tensile test, and never by necking. However, it may be preceded by microscopic, or even macroscopic, plastic deformation. In this regard, the criterion for brittle fracture is that crack linkup and/or growth takes place by tensile separation of atomic bonds. At temperatures somewhat below those at which ductile fracture is observed, Mode III brittle fracture is found in materials undergoing a ductile-to-brittle transition. In a

514

CHAPTER 10
Toughening
Mechanisms
and the Physics
of Fracture

tensile test, Mode III fracture is associated with generalized plastic flow followed by fracture. No necking takes place, but a small R.A., corresponding to the uniform plastic strain, is observed. Fracture initiation in Mode III is caused by plastic-flow mechanisms. They may be similar to those for crack nucleation in ductile fracture or may correspond to the formation of a microcrack having a size comparable to the material grain size.

At even lower temperatures (for materials that are brittle there), Mode II fracture is the dominant fracture mode, at least at fairly high applied stress levels. Mode II fracture is associated with heterogeneous microscopic plasticity. This leads to nucleation of microcracks of size comparable to the grain size. In the bcc transition metals, slip bands or twins are the source of the deformation nucleating microcracks. For the bcc transition metals, the stress at which this takes place is the macroscopic yield strength, σ_y (although the extent of macroscopic deformation preceding crack nucleation is minimal), and since σ_y is greater than the crack-propagation stress for these metals, fracture immediately follows yielding. That is, their Mode II fracture strengths are their yield strengths.

In the hcp metals and in many nonmetallics, the heterogeneous deformation leading to Mode II crack initiation relates to their limited number of low-temperature slip systems. This feature does not allow for strain compatibility among grains in a polycrystal. Microyielding ensues at the "soft slip system" yield strength, and microcracks nucleate thereafter. If the soft system yield strength is greater than the crack-propagation stress (this is common at low temperatures), fracture follows microyielding and the fracture stress is the soft system yield strength. If it is not, the stress level must be increased to the crack-propagation stress before fracture takes place. This is more common in the high-temperature region of Mode II fracture. In this case, significant microcracking occurs prior to final failure.

The various classes of crystalline materials have different tendencies for the several low-temperature fracture modes. With but a few exceptions, the face-centered cubic metals are inherently ductile at all temperatures; their tensile fracture strengths are their tensile strengths. The other important engineering metals—the hcp metals and the bcc transition metals—do demonstrate a tendency to low-temperature brittle behavior.

The alkali halides, prototypes for ionic solids, are not particularly strong. In fact, their "soft" system yield strengths are on the order of those of soft metals. Yet, due to their low number of slip systems, polycrystals of the alkali halides seldom undergo ductile fracture at low temperature. Thus, even at the highest temperatures of low-temperature fracture, polycrystalline alkali halides display Mode III fracture.

Polar covalent solids display higher strengths than metals and alkali halides. Covalent solids are stronger still. Concurrent with this trend is a marked incapacity of these materials for ductile fracture. Instead, they fail in a brittle manner even at fairly high homologous temperatures.

The tendencies just remarked on can be illustrated with fracture mechanism maps. The axes of such a map are homologous temperature and stress (usually normalized in terms of Young's modulus). The regions within a map define stress-temperature combinations at which various fracture modes dominate. In the region of Mode I fracture, this does not mean that fracture *must* occur, but only that if it does it is by the Mode I mechanism. As with deformation mechanism maps, fracture maps for specific materials are only as accurate as the information used in their construction. Moreover, the regions of the map shrink or expand depending on microstructural considerations (grain size, solid-solution strengthening, etc.) and on the stress state. Biaxial and triaxial positive stress states expand the brittle fracture regions at the expense of the ductile fracture one. Likewise, solid-solution strengthening, and most other mechanisms which increase yield strength, favor brittle fracture. The principal exception to this rule is grain size. Grain-size refinement increases both a material's yield strength and its fracture propagation stress. Thus, grain-size refinement is one structural variable that can sometimes be used to simultaneously increase both a material's plastic flow and fracture resistance.

REFERENCES

Ashby, M. R., C. Gandhi, and D. M. R. Taplin: "Fracture Mechanism Maps and Their Construction for FCC Metals and Alloys," *Acta. Metall.*, 27, 699, 1979.

Gandhi, C., and M. F. Ashby: "Fracture Mechanism Maps for Materials Which Cleave; FCC, BCC and HCP Metals and Ceramics," *Acta Metall.*, 27, 1565, 1979.

Lawn, B.: *Fracture of Brittle Solids*, 2nd ed., Cambridge University Press, Cambridge, England, 1993.

Tetelman, A. S., and A. J. McEvily, Jr.: *Fracture of Structural Materials*, Wiley, New York, 1967.

PROBLEMS

10.1 Refer to Fig. 10.6. We now specify the thicknesses of the β and γ phase plates as l_β and l_γ, respectively. After cooling to the temperature T_2, there is no net *force* on the system. Thus, the tensile force on the β phase equals the compressive force on the γ phase. Equating the magnitude of the forces we have $\sigma_\beta l_\beta = \sigma_\gamma l_\gamma$, where the σ's are the magnitude of the stresses borne by the β and γ phases, respectively. Further, the volume fractions of the phases relate to their thicknesses as $V_\beta = l_\beta/(l_\gamma + l_\beta)$ with a similar expression for V_γ. Derive an expression for the stresses carried by the two phases in terms of their differences in thermal expansion coefficients, the temperature difference (ΔT), and the phase volume fractions and elastic moduli. On this basis, identify the modulus of Eq. (10.4). Compare this modulus to those of composite materials presented in Chap. 6. Comment on your results.

10.2 Metallic thermostats are often bimetal plates. The two metals are bonded along the plate length, and the metals have different thermal expansion coefficients. One plate end is clamped; the other is free. Explain how thermal stresses arising from changes in temperature permit a thermostat to function as a switch in an electrical circuit.

10.3 In this problem we estimate the magnitude of microcrack toughening in a two-phase ceramic, containing 50 vol.% of both phases. Both phases have the same modulus (400 GPa) and fracture toughness (3 MN/m$^{3/2}$). The thermal expansion coefficients of the phases differ by an amount $\Delta\alpha = 5 \times 10^{-6}$/K where K is temperature in Kelvin. Assume that residual stresses form during cooling over a range of 1000°C and that these stresses are responsible for secondary microcracking.
a Calculate the grain size below which microcrack toughening might be observed in this system.
b What ceramic grain size (d) will provide a value of the ratio of microcrack zone size, r_c, to d of 3? Is this a realistic grain size?

516

CHAPTER 10
Toughening
Mechanisms
and the Physics
of Fracture

c For the situation of (b), estimate (via Eq. (10.6) and assuming $V_f = 1$), the toughness due to secondary microcracking. Is this a "large" or a "small" increase relative to the matrix toughness?

10.4 Use Eq. (10.13) to estimate the toughness that might be expected from grain bridging in a single-phase ceramic with a grain size of 10 μm. Take μ = 0.1, and assume a residual thermal stress of 1000 MPa. How does this toughness compare to those illustrated for fracture of metals in Fig. 10.42?

10.5 Use the procedure employed to develop Eq. (10.13) (which pertains to grain bridging effects) to derive Eq. (10.15), which describes similar toughening due to fibers dispersed in a ceramic matrix.

10.6 Refer to Example Prob. 10.1. The property data used in this problem are the same as those of the example problem.
a Assume that the debond length (l_d, Eq. (10.15)) varies between the limits of $l_c/4$ (which would apply to a composite containing discontinuous fibers of length equal to the critical length) to $2l_c$. Calculate $\Delta\mathscr{G}_c$ for these two situations. The results from this calculation, plus those from the example problem, should more or less bracket the expected toughness increase.
b The composite fracture toughness is $K_{Ic} \cong (\Delta\mathscr{G}_c E_c)^{1/2}$ where E_c is the composite modulus (calculated from the volume fraction rule, Chap. 6). Calculate the composite fracture toughnesses for the bracketed debond lengths considered. How do your "numbers" compare to the actual value of composite fracture toughness?

10.7 Consider Eq. (10.17), which represents the toughness of a brittle fiber–brittle matrix composite arising from fiber pull-out.
a Separately plot the terms in this equation as a function of the fiber volume fraction, V_f. Show that if the inherent fiber toughness is less than the inherent matrix toughness there is a critical value of the parameter $\tau l_d^2/d$ that must be exceeded in order for the composite toughness to exceed that of the matrix. Express this critical value in terms of the fiber and matrix toughnesses.
b Let the debond length range between $l_c/4$ and $2l_c$. Calculate this critical fiber volume fraction for a composite with $d = 20$ μm and with constituent properties given in Example Prob. 10.1.
c Repeat part (a) for a ductile-particle-toughened ceramic. Use Eq. (10.19) and take the constant $C = 1$. In this case determine the critical particle volume fraction in terms of its size, yield strength, fracture strain, and the inherent matrix toughness.
d For the particle-toughened situation, consider a silicon nitride matrix ($K_{Ic} = 3$ MN/m$^{3/2}$ and $E = 380$ GN/m^2). Let this material be toughened by a dispersion of 15 vol.% of metal particles of a metal having a yield strength of 350 MN/m^2 and a fracture strain of 0.8. What particle size must be exceeded in order that the particles toughen the silicon nitride? Is this a particle size that can be reasonably achieved?

10.8 Consider toughening of a brittle ceramic by dispersed ductile particles. The particles only appreciably improve toughness if they plastically deform in the zone behind the main crack tip. This requires that the particles partially decohere along a fraction of the interface between them and the matrix; this permits an appreciable particle "gage length" to be developed. While decohesion also contributes to fracture work, we consider only how the extent of decohesion affects the plastic work performed on the particle. Assume that decohesion precedes particle deformation. A pertinent sketch is shown below. Particles are assumed spherical in shape, and the fractional surface area of the sphere that is decohered is expressed in terms of the angle ϕ; this fractional area is cos ϕ.

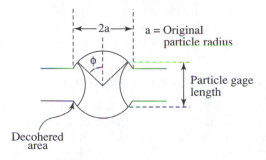

a Express the effective "gage length" of the particle in terms of its radius, a, and ϕ.

b The plastic work of fracture is $\int F\, dx$, where F is the force corresponding to the particle-flow stress and the integral is taken over the particle extension from the onset of plastic deformation through the critical displacement, x_c, at which the particle fractures. Determine the plastic work per particle in terms of the particle yield strength (σ_y), a, ϕ, and the particle fracture strain (ε_f). (Assume the flow stress does not vary during deformation. To keep the arithmetic simple, work with engineering stress and strain.)

c Determine the added toughness due to plastic particles of volume fraction, V_f.

d Determine an optimum "decohesion angle" that maximizes the toughness arising from particle plastic deformation. To what fractional decohered area does this angle correspond?

10.9 a Write an equation to describe "craze" toughening in a glassy polymer. The equation should contain the essential physics of toughening, and it should incorporate a distance (r_c) above and below the fracture plane in which additional crazes form, the work associated with elastomeric particles that induce crazes, and the fraction of such particles in the zone r_c associated with secondary crazes.

b Would you expect the size of the elastomer particles to affect the toughness? Explain.

10.10 Describe the change in low-temperature tensile response of a material prone to Mode I fracture as samples containing finer and finer preexisting cracks are tested.

10.11 Schematically sketch how the regions of a fracture map are altered when the loading is changed from simple to biaxial tension. Consider how such a stress-state alteration changes both the fracture and the plastic-flow criterion.

10.12 Mode I fracture may or may not occur in the regions of a fracture map where it is designated as the fracture mode. Is this also true for Mode II fracture? Give your reasoning.

10.13 Suppose grain size were refined in a face-centered cubic metal and in a body-centered cubic metal. How would the fracture mechanism maps of these materials be changed by grain-size refinement?

10.14 Consider a bcc metal having a grain size of 20 μm. Assume $\gamma = 1.0$ J/m². Refer to Chaps. 2, 4, and 5 to obtain reasonable values for σ_0 and E for some bcc metals to use in Eq. (10.20b). Compare the two terms on the right-hand side of this equation. Which term is more important in defining the fracture-nucleation stress? What must the grain size be in order for the second term on the right-hand side of this equation to equal the first one?

518

CHAPTER 10
Toughening
Mechanisms
and the Physics
of Fracture

10.15 If 350 MN/m² and 280 MN/m² are the shear stresses required to nucleate cracks in bcc molybdenum having grain sizes of 0.3 mm and 2 mm, respectively, what shear stress would be required to nucleate a crack in molybdenum having a grain size of 0.02 mm?

10.16 Describe how the transition from Mode I to Mode II fracture depends on grain size.

10.17 Consider Fig. 10.30, which compares the Mode II–Mode III fracture transition in hcp metals (and nonmetallics) with this transition in bcc metals.
a A colleague says that the temperature T_1 (at which $\sigma_s = \sigma_F$) of Fig. 10.30a actually represents the Mode II to Mode III fracture transition. Explain why this view is inconsistent with the definition of Mode III fracture. Also explain why, from certain conceptual points of view, your colleague has a point.
b Another colleague insists that the onset of Mode III fracture corresponds to the temperature at which the fracture stress begins to increase with increasing temperature. Explain why this assertion is incorrect.

10.18 Compare the coefficients of $d^{-1/2}$ in Eqs. (10.21) and (10.22). Refer to Chap. 5 for typical values of k_y, and make reasonable estimates for E and $\mathcal{G}_c$ of typical bcc transition metals. Show that your results are consistent with the relative slopes of σ_y and σ_F vs. $d^{-1/2}$ in Fig. 10.31.

10.19 Discuss the efficiency of grain-size refinement in altering the Mode II fracture stress in (a) NaCl, (b) Mo, (c) Al_2O_3, and (d) Si_3N_4. (Recall what factors control the Mode II fracture stress.)

10.20 In some materials the stress defining the Mode I–Mode II transition in a fracture map decreases with increasing temperature at low temperatures, and then increases with temperature at higher temperature. Explain the cause for this.

10.21 Consider NaCl, whose fracture mechanism map is given in Fig. 10.21.
a Suppose NaCl had the requisite number of low-temperature slip systems to permit polycrystalline plasticity. How would the fracture map of NaCl be altered if this were the situation?
b Suppose the crack propagation stress were greater than this hypothetical macroscopic yield strength for NaCl. Would the fracture map of NaCl be altered? If so, how?

10.22 The yield strength of a bcc metal varies with temperature as $\sigma_y = (1500 - 3.5T)$ MN/m², where T is in Kelvin. The cleavage strength is independent of temperature and is equal to 1200 MN/m². The transition from Mode II to Mode III fracture takes place at 86 K for this metal. If the metal yield strength is increased 150 MN/m² by solid-solution strengthening, what will be the new Mode II–Mode III transition temperature? (Assume that the solid-solution strengthening is temperature independent.)

10.23 Crystallographic orientation plays a role in determining the brittle fracture tendencies of single crystals. Cleavage of single crystals can occur when a critical *normal* stress, σ_c, is attained on the preferred cleavage planes.
a Show that cleavage will occur at a tensile stress, $\sigma = \sigma_c/\cos^2 \phi$, where ϕ is the angle between the cleavage plane normal and the tensile axis. (Hint: Recall the derivation of Schmid's law, Chap. 4. How is this case different?)
b A colleague finds considerable scatter in the "ductile-to-brittle" tensile transition temperature of variously oriented single crystals of a material prone to cleavage. Explain the cause of this.

c Will single crystals of a material generally be more or less prone to brittle fracture than polycrystals of the same material? Give your reasoning.

10.24 Show, with sketches, that any effect that increases a material's plastic-flow resistance without changing its crack-propagation stress will result in an increase in the Mode II–Mode III fracture transition temperature.

10.25 a Discuss the temperature variation of $\mathcal{G}_c$ for Mode III fracture as temperature affects (i) the number of cleaved grains in front of a crack tip and (ii) the plastic-zone size.
b How does the number of cleaved grains in front of a crack tip for a material failing via Mode III fracture correlate with the number of microcracks observed in a tensile sample of the same material?

10.26 Compare the processes of crack nucleation and "growth" for Mode III brittle fracture and for ductile fracture. Consider mechanisms at the atomic and microscopic levels that give rise to crack nucleation and to crack linkup and/or crack propagation.

10.27 A figure from the Gandhi and Ashby reference is provided below. The solid bold line at low temperature (up to $T \cong 0.6T_m$) represents the tensile fracture strength for a material containing preexisting cracks having a size less than the grain size. Assume the figure pertains to a ceramic with a grain size of 10 μm.

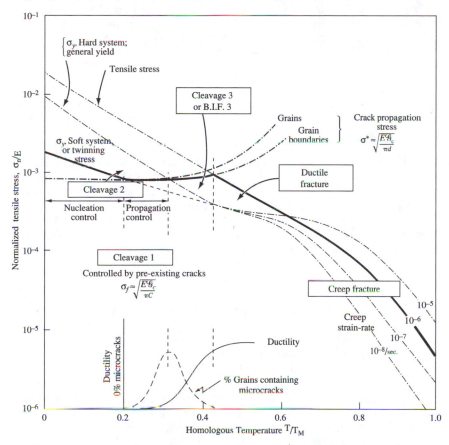

a Reproduce the Gandhi-Ashby figure. On it, graph the tensile fracture strength for a material containing internal cracks of length $2c = 40$ μm. (Note: $2c$ substitutes for

520

CHAPTER 10
Toughening
Mechanisms
and the Physics
of Fracture

d in the expression for the crack-propagation stress.) Be as quantitative as possible in constructing your graph.

b How would the graph of microcrack density vs. temperature (see insert at the bottom of the Gandhi-Ashby figure) change when the situation of part (a) applies rather than the situation for which the graph was initially constructed?

c For a ceramic, the soft system yield strength is generally not a function of grain size. However, the hard system yield strength obeys a Hall-Petch relationship. Schematically sketch (for a material not containing preexisting cracks), the tensile fracture stress if the grain size of the material pertaining to the Gandhi-Ashby figure is reduced by a factor of two.

d How would the graph of microcrack density vs. temperature change for the situation of part (c) compared to that of part (a) and the original situation?

10.28 The following properties apply to W: k_y (Hall-Petch coefficient)$= 0.8$ MN/m$^{3/2}$; $E = 4.11 \times 10^{11}$ N/m^2. The table below lists additional properties of this metal.

T/T_m	$\mathscr{G}_c$ (J/m^2)	σ_y (MN/m^2)	T.S. (MN/m^2)
0	130	4110	4200
0.1	130	2010	2880
0.2	130	450	1300
0.3	165	410	700
0.35	400	370	480
0.4	770	330	450
0.45	1600	290	430

Values of σ_y in the table are for W with a grain size of 100 μm. For W with this grain size and for the temperature range, $T/T_m < 0.15$, $\sigma_y = (4.11 - 21(T/T_m))$ GN/m^2. For this problem assume that the difference between tungsten's T.S. and yield strength is independent of grain size.

Determine the tensile fracture stress of W as a function of temperature between 0 K and $T/T_m = 0.5$ for the following combinations of grain size (d) and preexisting crack size (c): (i) $d = 100$ μm, $c = 200$ μm; (ii) $d = 100$ μm, $c = 5$ μm; (iii) $d = 10$ μm, $c = 200$ μm; (iv) $d = 10$ μm, $c = 5$ μm. Graph tensile fracture stress vs. temperature for these four grain size–crack size combinations.

10.29 Most discussion in Chaps. 9 and 10 has focused on fracture under uniaxial loading. In this problem, we consider other stress states. A simple criterion for brittle fracture of a metal is that the mean pressure reaches a critical value. That is, if σ_F is the tensile fracture stress, the critical value of mean pressure is $\sigma_F/3$.

a Schematically plot this brittle fracture criterion for biaxial tension (i.e., in the first quadrant of σ_1-σ_2 space). (Normalize your stress axes by the brittle fracture stress in tension.)

b Using the Tresca yield condition, show that it is possible for a material to plastically deform prior to fracturing in a brittle manner when it is loaded in uniaxial tension, but that this same material may fracture before plastic flow when biaxially loaded.

c Assume that the tensile strength (for ductile fracture) scales with the plastic yield strength. Construct a graph similar to that for part (a), but which also displays the tensile strength under biaxial loading. Show that it is possible for a material to undergo ductile fracture during uniaxial loading, but that the fracture mode may give way first to Mode III fracture and then to Mode II fracture as the degree of biaxial loading increases.

10.30 Schematically plot the density of inclusion (not grain) microcracking prior to tensile fracture as a function of temperature. Consider the temperature range spanning Mode II, Mode III, and ductile fracture.

10.31 Compare Fig. 10.5a, which illustrates ductile fracture crack propagation, with what is expected for crack propagation at the highest temperatures of Mode III brittle fracture. Is it reasonable to consider the transition from Mode III to ductile fracture a continuous, rather than a discontinuous, one? Give your reasoning.

10.32 Consider Fig. 10.39, which depicts the relationship between true fracture strain and second-phase particle volume fraction in copper-base materials. Rearrange Eq. (10.27) and replot Fig. 10.39 in a manner that allows the validity of this equation to be tested.

CHAPTER 11

High-Temperature Fracture

11.1
INTRODUCTION

In this chapter we discuss various aspects of high-temperature fracture. Just as low-temperature fracture may take place by different mechanisms, materials can also fail in different ways at high temperature. The type of high-temperature fracture observed depends on the material, the temperature, the applied stress level, and the stress state.

Some high-temperature fracture involves crystallographic flow, in common with most low-temperature fracture. However, high-temperature fracture also involves, to one extent or another, diffusional flow and this is what distinguishes it from low-temperature fracture. Moreover, there are parallels between high-temperature deformation and fracture just as there are similarities between low-temperature flow and fracture.

The most technologically important of the high-temperature fracture processes is creep fracture, which takes place at low to moderate stress levels. This is the stress (and temperature) regime in which materials are often called on to function as high-temperature load-bearing members. Examples, some cited previously, abound; turbine blades operating at high temperatures, high-temperature pressure vessels in energy generating facilities, etc. Because of its importance, particular attention is given to this type of high-temperature failure. We consider the nucleation, growth, and coalescence of internal voids and cracks that lead to failure. This provides us with a basis for further consideration of design against creep fracture. In the section following, we initiate our discussions by describing the various modes of high-temperature fracture and the stress/temperature regimes in which they are observed.

11.2
HIGH-TEMPERATURE FRACTURE MODES

The three primary modes of high-temperature fracture—rupture, transgranular creep fracture (TCF), and intergranular creep fracture (ICF)—are illustrated schematically in Fig. 11.1. This figure is appropriate for fracture taking place under conditions of uniaxial tensile loading, as in a tensile test.

Rupture, characterized by a reduction in area of, or close to, 100 percent (Fig. 11.1a), takes place at high stress levels and high temperatures. Such stress levels are those commonly associated with the high strain rates characteristic of hot working. Hence, dynamic recovery and/or recrystallization is often associated with rupture fracture. Indeed, rupture is frequently, although not always,[1] concurrent with dynamic recrystallization.

For a material to be capable of undergoing a 100 percent reduction in area, the nucleation of internal cavities must be suppressed or, if cavities do form, they must

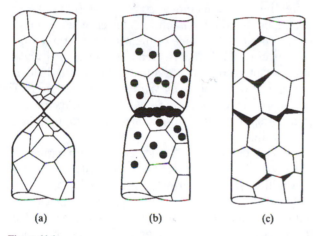

(a)   (b)   (c)

Figure 11.1

The three modes of high-temperature tensile fracture. (a) Rupture, which occurs at high temperatures and high stress levels in some materials, is characterized by (or close to) a 100% reduction in area. Rupture is often associated with dynamic recrystallization as reflected in this schematic. (b) Transgranular creep fracture (TCF) takes place at temperatures less than those causing rupture. TCF results in a macroscopically ductile fracture, characterized by a finite reduction in area. Intragranular voids (the dark circles) form, and their growth and coalescence lead to fracture. (c) Intergranular creep fracture (ICF) takes place at stresses lower than those leading to rupture or TCF (if these happen), and is observed in all classes of crystalline solids. Intercrystalline voids or (as shown here) cracks are precursors to ICF. Following their nucleation they grow, and fracture follows their eventual coalescence. This mode of fracture is macroscopically brittle, with nil or little reduction in area.

---

[1]Aluminum does not dynamically recrystallize, for example, yet it does rupture. In fact, very high purity Al ruptures at temperatures well below those at which diffusional flow is of consequence. However, ordinary "high-purity" aluminum ruptures only at higher temperatures at which diffusional flow is concurrent with deformation.

not grow to an extent that permits their coalescence and subsequent growth of an internal crack. For reasons not fully understood, dynamic recrystallization apparently prevents the nucleation of internal voids. This is the reason why the high strain rates of hot working are associated with extensive malleability in many metals.

Transgranular creep fracture (TCF, Fig. 11.1*b*) is the high-temperature analog of low-temperature ductile fracture. Internal cracks or voids first nucleate—usually around inclusions, but they can also form in other regions of heterogeneous microscopic strain. As in low-temperature ductile fracture, TCF is preceded by the growth and eventual coalescence of these holes. The fracture, which is associated with a considerable reduction in area, is a macroscopically ductile one, and typically takes place at stress levels somewhat below those required for ordinary ductile fracture and in approximately the same temperature range. Hence, TCF is associated with high strain rate—or equivalently, high stress—creep. In essence, the stress is not high enough so that crystallographic flow can cause completely the process of ductile fracture. Instead, fracture is furthered by diffusional flow and/or power-law (dislocation) creep, both processes facilitating void growth and coalescence.

At the stresses and temperatures of TCF, the internal stresses due to microscopic strain inhomogeneities are diminished in comparison to those found during the deformation leading to low-temperature ductile fracture. This is because at high temperature recovery processes can provide for some of the necessary strain compatibility at, say, an inclusion-matrix boundary. That is, creep deformation provides for some of the high-temperature strain accommodation that geometrical dislocations provide during crystallographic flow.[2] Hence, *nucleation* of internal voids is delayed during TCF in comparison to low-temperature flow. Moreover, the greater strain-rate sensitivity of creep deformation acts to restrict neck development and this, too, should lead to greater ductilities of TCF.

However, other effects reduce TCF ductilities. The decreased strain hardening at high temperatures acts oppositely to that of the strain-rate sensitivity in terms of delaying neck development. Additionally, void growth is accelerated and augmented by diffusional flow and/or dislocation creep. The net result is that transgranular creep fracture ductilities are roughly comparable to those observed in low-temperature ductile fracture. Hence, the appearance of a TCF fracture surface (Fig. 11.2) does not differ appreciably—either macroscopically or microscopically—from a low-temperature ductile fracture surface.

Intergranular creep fracture (ICF, Fig. 11.1*c*), which takes place at stress levels and strain rates less than those causing rupture or TCF, is macroscopically brittle (Fig. 11.3*a*). During ICF, "voids"—often associated with inclusions or second-phase particles—form along grain boundaries. These voids can form on all such boundaries; i.e., they can be present irrespective of the inclination of the boundary with respect to the tensile axis. However, only those voids situated on boundaries normal to the tensile axis grow and coalesce so as to cause fracture (Fig. 11.1*c*). The microscopic appearance of an ICF (Fig. 11.3*b*) clearly illustrates this feature. In this figure, one type of void—wedge-like in shape (a "w" void)—is associated with fracture. However, spherical voids ("r" voids, Fig. 11.4) are also found during ICF. Generally, "w" voids are associated with high-stress ICF and "r" voids with low-stress ICF.

---

[2]Alternatively, the dislocation stress field generated to satisfy strain compatibility can be considered relieved by recovery processes.

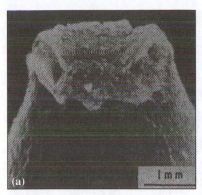

Figure 11.2
Transgranular creep fracture in nickel. (a) This macroscopic view shows the ductile fracture features of TCF. (b) The microscopic appearance of the fracture surface does not differ substantially from that of a low-temperature ductile fracture surface (e.g., Fig. 10.38); the dimpled surface is characteristic of fracture caused by void nucleation, growth, and coalescence. (*Reprinted from M. F. Ashby, C. Gandhi, and D. M. R. Taplin, "Fracture-Mechanism Maps and Their Construction for FOC Metals and Alloys," Acta Metall., 27, 699, Copyright 1979, with permission from Elsevier Science.*)

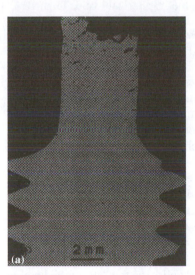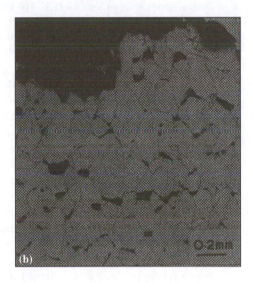

Figure 11.3
Intergranular creep fracture in nickel. (a) The fracture is a macroscopically brittle one, characterized by little or no tensile reduction in area. (b) A microscopic view shows a profusion of wedge-shaped cracks situated on grain boundaries. The nucleation, growth, and coalescence of these leads to ICF. Hence, ICF is characterized by some localized permanent deformation in the vicinity of grain boundaries. (*Reprinted from M. F. Ashby, C. Gandhi, and D. M. R. Taplin, "Fracture-Mechanism Maps and Their Construction for FCC Metals and Alloys," Acta Metall., 27, 699, Copyright 1979, with permission from Elsevier Science.*)

The basic mechanisms leading to ICF—void nucleation, growth, and coalescence—do not differ phenomenologically from those culminating in macroscopically ductile fracture. However, during ICF, appreciable permanent deformation is restricted to regions adjacent to the grain boundaries on which the growing

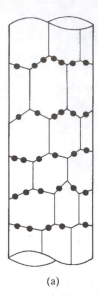

(a)

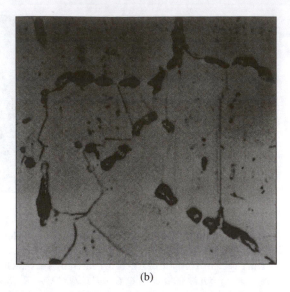

(b)

Figure 11.4
(a) Schematic of spherical ("r") voids situated on grain boundaries. Their growth and coalescence can also lead to ICF. (b) Photograph of "spherical" voids formed during creep of copper at a temperature of 753 K and a stress of 326 MN/m². The tensile axis is vertical. Magnification, 1650 ×. (*From A. J. Kennedy, Processes of Creep and Fatigue in Metals, Wiley, New York, 1963.*)

voids are situated. As a result ICF is characterized by moderate local permanent deformation, but by minimal macroscopic deformation. And, of course, ICF differs fundamentally from low-temperature ductile fracture by the manner in which void growth, affected by diffusional flow, takes place. Intergranular creep fracture occurs at stress levels and temperatures at which many structural parts operate. Because of this, more extended discussion of ICF is provided in Sect. 11.4.

The tendencies of the several material classes for the various creep fracture modes vary in the same way as do their deformation tendencies. Rupture, for example, common in the metals, is not observed in covalent (and brittle) solids and is often not found in polar covalent solids such as Al_2O_3. Moreover, TCF does not occur in covalent crystalline solids. That is, these materials *always* exhibit macroscopically brittle ICF at high temperatures, albeit microscopic permanent deformation may precede this form of fracture. Likewise, as the "strength" level of the solid is increased, the transition to high-temperature fracture ensues at progressively higher temperatures. Discussion of these tendencies is facilitated by reference to high-temperature fracture mechanism maps, as is done in Sect. 11.3. Before considering such maps, some comments with regard to the effect that stress state has on creep fracture are in order.

In this chapter, we concern ourselves primarily with creep fracture taking place under uniaxial tensile loading, i.e., as in a tension or a creep test. This is reasonable for an introductory discussion, and has the further advantage of minimizing mathematical complexity while still emphasizing the basic physics of creep fracture. However, it is important to consider briefly the effect that stress state has on high-temperature fracture.

Internal voids formed at high temperature have a natural tendency to shrink, rather than grow. This is because void growth is accompanied by an increase in the internal solid-pore surface area, and an applied stress must do work to overcome the associated surface energy. This is also true at low temperatures, of course. However, void shrinkage does not occur at low temperature, because it takes place by diffusional processes that are absent there. Several examples illustrate the point. Porous solids densify when exposed to elevated temperatures. That is, during this *sintering* process, voids shrink and disappear rather than nucleate and grow. Moreover, the void shrinkage rate can be accelerated by certain stress states. For example, hydrostatic compression (utilized commercially in HIP—hot isostatic pressing) greatly enhances densification (i.e., pore shrinkage) rates. Hence, it is clear that at high temperatures materials exhibit a natural tendency to pore closure, and this tendency is enhanced by certain stress states. Thus, whether or not pores grow and coalesce and result in material creep fracture depends on the stress state and its magnitude.

For voids to grow, an external stress of greater "strength" than the surface energy stress, which acts to shrink voids, must be applied. This retarding stress scales with the surface energy, γ, and the void surface-to-volume ratio. Thus, for a spherical void, the retarding force is proportional to γ/r where r is the void radius. More detailed considerations for a general void shape show that the retarding stress is equal to $(\gamma/r_1 + \gamma/r_2)$ where r_1 and r_2 are the principle radii of curvature along the void surface. For a spherical void, $r_1 = r_2 = r$ (the sphere radius) and, thus, the retarding stress is $2\gamma/r$ (see also Example Prob. 11.1). The dimensions of this parameter are energy per unit volume or stress (e.g., J/m^3 or, equivalently, N/m^2). For a void to grow, rather than shrink, a pore-opening stress algebraically greater than this so-called "sintering limit" must be applied. In most of our discussions we assume that the applied stress is well in excess of the sintering limit. But it should be kept in mind that there exists a stress level below which voids "heal," rather than grow.

In this chapter, which deals primarily with tensile loading, the stress needed to overcome the sintering limit is taken as the applied tensile stress. When other than uniaxial loading is the situation, some appropriate "effective" stress must be so used. For "brittle" solids, the maximum normal tensile stress is used; i.e., this stress must exceed the sintering limit for void growth to take place. For ductile materials, such as metals, the mean hydrostatic stress or pressure $[= (\sigma_1 + \sigma_2 + \sigma_3)/3$, where σ_1, σ_2, and σ_3 are the principal stress components] is the effective stress. We see that, for tensile loading, both criteria, apart from a proportionality constant which we neglect, are comparable. This is not so for other loading conditions. Several of the references at the end of this chapter provide additional guidelines in the matter.

EXAMPLE PROBLEM 11.1. Consider an isolated spherical void in a material. Let the void grow by transfer of mass from its periphery to the material surface. Show that the accompanying change in surface energy per unit volume of void is $2\gamma/r$, where γ is the surface energy and r is the void radius.

Solution. The accompanying sketch illustrates growth of an interior spherical void from a radius r to $r + dr$. Growth is accomplished by deposition of mass from the annular rim of thickness dr to the material surface, which is flat. The void volume change, dV, is given as $d(4\pi r^3/3) = 4\pi r^2\, dr$. The associated increase in surface area, dA, of the void is $d(4\pi r^2) = 8\pi r\, dr$. Thus, for the spherical void, $\gamma\, dA/dV = 2\gamma/r$. Note that the system energy increases by void growth. If the void were to shrink (by transfer of mass from the material surface to the void), the system energy would decrease.

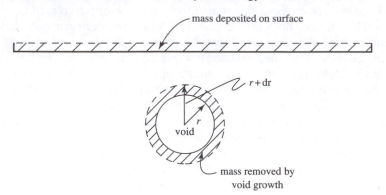

11.3
HIGH-TEMPERATURE FRACTURE-MECHANISM MAPS

High-temperature fracture-mechanism maps for Ni, Mo, Mg, NaCl, Al_2O_3, and Si_3N_4 are given in Fig. 11.5. These are the same materials for which low-temperature fracture maps were provided in Chap. 10 (Figs. 10.18–10.23). As mentioned there, these materials are representative of the various classes of crystalline solids, as distinguished by crystal structure and chemical bonding. As in the corresponding figures of Chap. 10, the axes of Fig. 11.5 are homologous temperature (T/T_m where T is absolute temperature and T_m is the material's absolute melting temperature) and normalized stress (σ/E, where σ is the applied tensile stress and E is the elastic modulus). In Figs. 10.18–10.23, however, the various low-temperature fracture modes were distinguished, whereas the high-temperature fracture modes were not specified. The opposite holds for the fracture maps of Fig. 11.5.

Discussion of the low-temperature fracture modes was extensive in Chap. 10. Because many of the structural and chemical characteristics that lead to the different tendencies for fracture among the material classes are common to both high- and low-temperature fracture, a similarly labored discussion is not necessary here. However, it is worthwhile to briefly consider how structure and bonding correlate with manifestation of different high-temperature fracture modes.

The high-temperature fracture-mechanism map for nickel (Fig. 11.5a) is typical for many pure fcc metals. Rupture, facilitated by dynamic recrystallization, takes place at temperatures as low as ca. $0.7T_m$ in Ni and in many other pure fcc metals. When these are alloyed, however, rupture occurs only at higher temperatures and, in some cases, not at all. Such considerations are addressed later in this section and also in Sect. 11.5. Transgranular creep fracture in Ni occurs at stress levels well below the tensile strength (T.S.); the lower limit of stress at which TCF takes place is typically 0.2 T.S., below which stress nickel fails by ICF. The lowest temperature at which TCF is manifested in Ni is ca. $0.2T_m$; this, too, is typical of many fcc metals. As shown in Fig. 11.5a, ICF can occur in Ni at all temperatures above 0 K.

This is a direct result of the absence of low-temperature brittle fracture in Ni (and in almost all fcc metals). While ICF may occur in Ni at low stresses and temperatures, a fracture-mechanism map says nothing about the *time to fracture*. At the lower temperatures and stresses of ICF, the time to fracture is much longer than any reasonably expected service lifetime. Indeed, such times are on the scale of geological time and, in fact, fracture-mechanism maps at low-stress–low-temperature combinations have applications in this context. Thus, we see that high-temperature fracture-mechanism maps must be augmented by additional data (e.g., fracture time, fracture strain, steady-state creep rate, etc.) in order for them to be of greatest utility. These other factors are discussed later.

The ICF region in the fracture map of Ni is subdivided into two areas according to the type of internal cracks that are found. Wedge-shaped cracks precede fracture at high stress levels, whereas cavities ("r" voids) do so at lower stresses. We conclude our discussion of the fracture-mechanism map for Ni by noting that all

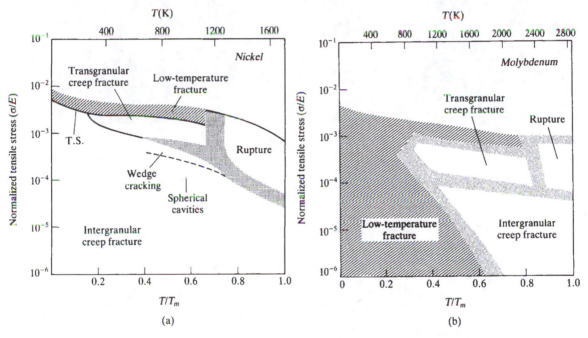

(a)   (b)

Figure 11.5
High-temperature fracture-mechanism maps for (a) nickel, (b) molybdenum, (c) magnesium, (d) sodium chloride (NaCl), (e) alumina (Al_2O_3), and (f) silicon nitride (Si_3N_4). The maps illustrate the high-temperature fracture modes of these materials, which serve as prototypes for the material classes on the basis of bonding and crystal-structure considerations. Since fcc nickel does not fail by any of the low-temperature brittle fracture modes, ICF *can* take place in it at very low temperatures. Nickel, and the other metals, display all three of the high-temperature fracture modes, as does sodium chloride. However, NaCl's region of TCF is limited in comparison to those of metals. Al_2O_3 and Si_3N_4 are strong brittle solids. As such, neither of them fail by rupture and TCF takes place in Al_2O_3 only at very high temperatures and does not take place at all in Si_3N_4; ICF is the only high-temperature tensile fracture mode for Si_3N_4. ((a) Adapted from M. F. Ashby, C. Gandhi, and D. M. R. Taplin, "Fracture-Mechanism Maps and Their Construction for FCC Metals and Alloys," Acta Metall., 27, 699; (b)–(f) Adapted from C. Gandhi and M. F. Ashby, "Fracture Mechanism Maps for Materials Which Cleave," Acta Metall., 27, 1565, Copyright 1979, with permission from Elsevier Science.)

boundaries separating the regions of fracture are of finite width; this is true for both high- and low-temperature fracture. The reasons are those mentioned in Chap. 10; viz. uncertainties in the data or equations used to construct the map and also to make allowance for reasonable variations in stress state.

Molybdenum, whose high-temperature fracture-mechanism map is given in Fig. 11.5*b*, is a prototype of the bcc transition metals. Rupture can take place in Mo

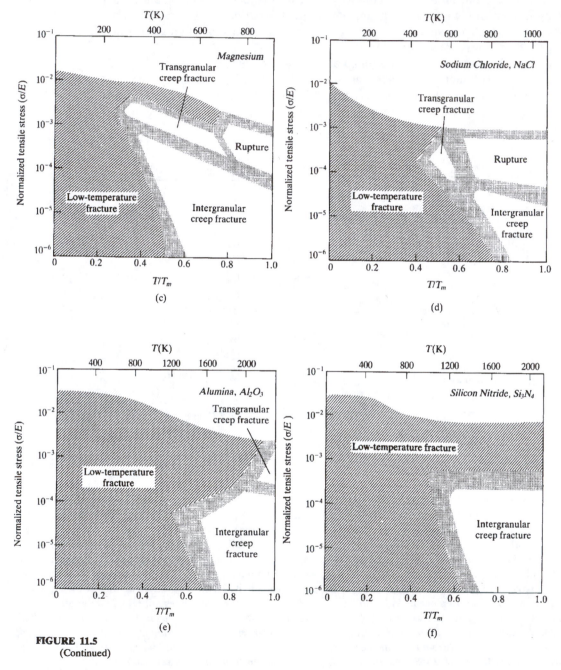

FIGURE 11.5
(Continued)

Figure 11.5 (continued)

at temperatures greater than ca. $0.8T_m$, a temperature slightly above the minimum rupture temperature of typical fcc metals. Molybdenum can also fail by either TCF or ICF. The lowest temperature at which TCF takes place in Mo is about $0.3T_m$; low-temperature brittle fracture is dominant in Mo (and in most of the other bcc transition metals) at temperatures less than this. Likewise, low-temperature cleavage, rather than ICF, is the dominant fracture mode at lower stresses. To construct the boundary between ICF and Mode I brittle fracture requires specifying a minimum creep rate below which ICF can reasonably be expected not to occur in the intended lifetime of a part. (Recall that creep *can* take place at all temperatures, although there is a creep rate, and temperature, below which it can be considered inconsequential.) In constructing the diagram for Mo, and the others that follow, this critical creep rate is taken as 10^{-10} s^{-1}, a rate low enough to satisfy most engineering requirements (Prob. 11.3).

Magnesium may be considered a typical hcp metal. Its high-temperature fracture map is provided in Fig. 11.5c. Magnesium exhibits high-temperature fracture tendencies similar to those of the bcc transition metals, and Mg is similar to all of the metals in that it manifests all three (rupture, ICF, and TCF) high-temperature fracture modes.

Sodium chloride is a typical ionically bonded alkali halide; its high-temperature fracture-mechanism map is given in Fig. 11.5d. While the alkali halides are not structural engineering solids, their fracture behavior is interesting on several accounts. Sodium chloride (and also KCl) dynamically recrystallizes at fairly low temperatures ($\cong 0.6T_m$), and this results in rupture fracture. Dynamic recrystallization and rupture fracture is common, in fact, to the alkali halides, even though these materials do not manifest low-temperature ductile fracture.[3] Moreover, since polycrystalline alkali halides are brittle at low temperatures, creep fracture (TCF or ICF) only takes place at temperatures above about $0.4T_m$. It is the advent of additional slip systems at higher temperatures in the alkali halides that provides for their ductility concomitant with TCF and rupture.

Alumina (Al_2O_3) behaves similarly to many polar covalent solids; its high-temperature fracture map is given in Fig. 11.5e. Polycrystalline Al_2O_3 does not rupture at any stress-temperature combination.[4] In fact, Al_2O_3 is so "brittle" that TCF takes place in it only at temperatures exceeding about $0.9T_m$. The small region of TCF shown in Fig. 11.5e represents the only stress-temperature combinations at which polycrystalline Al_2O_3 exhibits ductility under uniaxial loading. All of the other combinations result in a minimal reduction in area at fracture.

As expected on the basis of its chemical bonding, covalently bonded silicon nitride, whose high-temperature fracture map is given in Fig. 11.5f, is brittle for all stress-temperature combinations. Neither rupture nor TCF is observed in Si_3N_4, and microscopic ductility, manifested in the deformation preceding ICF, is found only at low stress levels and fairly high (> 0.5) homologous temperatures. As mentioned, Si_3N_4 and like materials are potential high-temperature structural materials because they are inherently strong. However, as both the high- and low-temperature fracture maps of Si_3N_4 show, care must be taken in their use. Materials displaying such high

---

[3]Rupture does not occur in all of the halides at the low temperatures at which it does in NaCl; e.g., temperatures greater than ca. $0.8T_m$ are required for rupture of LiF.

[4]This is common to most polar covalent solids but, as with all generalities, there are exceptions. MgO, for example, is expected to fail by rupture at high temperatures.

strengths are also inherently brittle, and their engineering use requires careful and sophisticated design.

As mentioned, a fracture-mechanism map is not definitive. That is, times to fracture, creep rates, etc., are not explicitly incorporated into them—at least, not in the form of Figs. 11.5. They can be made so, however, by plotting lines of constant time to failure on them. This is done in the fracture map of Monel (Fig. 11.6), a solid-solution alloy of nickel containing 30 wt.% copper. As a result of the solution strengthening provided by Cu, the boundaries in the map of Monel are displaced to higher stress levels than they are in the fracture map of Ni (cf. Fig. 11.5a). Moreover, since Cu, like other alloying elements in Ni, retards recrystallization, rupture fracture in Monel takes place at temperatures higher than it does in Ni (above about $0.8T_m$ for Monel). The lines of constant fracture times shown in the map reinforce comments made previously with respect to ICF; for example, at stress levels below ca. $10^{-4} E$ and temperatures below about $0.4T_m$, fracture times for Monel are on the order of eons. Hence, for all practical purposes, Monel is "fail-safe" under these conditions.

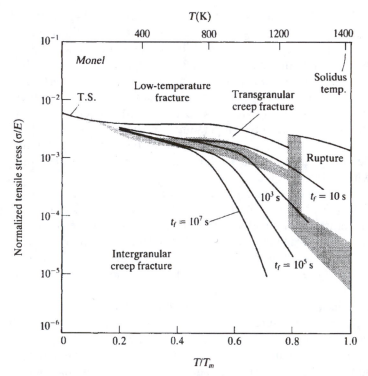

Figure 11.6
The high-temperature fracture-mechanism map for Monel, a solid-solution nickel alloy containing 30 wt.% copper. The stress levels in the map for Monel are higher than in the map for nickel, and the rupture region is displaced to higher temperature for Monel. The map also displays curves of constant fracture time. For example, at $T = 0.62\,T_m$ and an applied tensile stress of $10^{-4}\,E$, Monel would fracture in 10^7 seconds. (Adapted from M. F. Ashby, C. Gandhi, and D. M. R. Taplin, "Fracture-Mechanism Maps and Their Construction for FCC Metals and Alloys," Acta Metall., 27, 699, Copyright 1979, with permission from Elsevier Science.)

High-temperature fracture-mechanism maps may be presented in a different way. For example, the data given in Fig. 11.6 for Monel can be replotted using axes of normalized stress and fracture time. Such a map (Fig. 11.7) is a variant of an ordinary stress-time-to-fracture diagram of the type introduced in Chap. 7. In Fig. 11.7, lines of stress-time to fracture are plotted for various homologous temperatures. A map like this has an advantage over a conventional stress-time-to-fracture map in that the fracture modes are clearly indicated in a diagram like Fig. 11.7. Moreover, the changes in slope on the log-log coordinates employed clearly correlate with changes in fracture mode.

A final comment with regard to rupture is in order. We know that this fracture mode is associated with high temperatures, as is shown directly in the map of Fig. 11.6 and indirectly in Fig. 11.7. However, the latter figure also shows that rupture-fracture times are short; this results from the typically high strain rates that accompany rupture fracture.

The processes resulting in TCF[5] have been described adequately in previous discussions on ductile fracture (Chap. 10). However, ICF warrants further consideration because the processes of void growth, coalescence, etc., that lead to it are

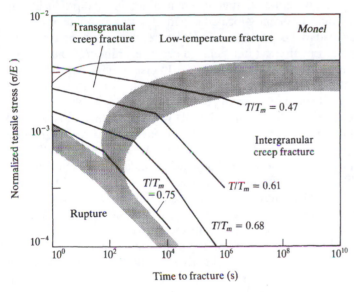

Figure 11.7
A different way of plotting the data contained in Fig. 11.6, the fracture map of Monel. In this figure the data are displayed in the form of stress-time-to-fracture curves at different temperatures. Thus, this figure is similar to Fig. 7.21a. However, the regions representing the various fracture modes are delineated on this map, and the stress-time-to-fracture lines are marked by changes in slope as the boundaries between the regions are traversed. (*Adapted from M. F. Ashby, C. Gandhi, and D. M. R. Taplin, "Fracture-Mechanism Maps and Their Construction for FCC Metals and Alloys," Acta Metall., 27, 699, Copyright 1979, with permission from Elsevier Science.*)

---

[5]The major difference between low-temperature ductile fracture and TCF is that the deformation resulting in void nucleation and growth are time dependent for TCF whereas they are time independent for low-temperature fracture.

facilitated by diffusion, and these considerations have not been addressed previously. Additionally, the stress-temperature combinations resulting in ICF are often of technological importance. Hence, the mechanisms that lead to ICF are discussed next. Although focus is on this fracture mode, some parts of our discussion also pertain to TCF.

11.4
INTERGRANULAR CREEP FRACTURE

A. Overview

As mentioned, creep fracture involves the same events—crack (void) nucleation, growth, and coalescence—that low-temperature fracture does. We initiate our discussion of these processes with an overview; the individual processes are later discussed at greater length.

Just as it is at low temperature, void/crack nucleation at high temperature is often associated with microscopic strain inhomogeneities. Additionally, certain manufacturing processes (e.g., powder fabrication of ceramics and metals) produce parts containing preexisting voids, frequently located at particle or grain boundaries, and the growth of these can lead to creep fracture. High-temperature void nucleation effected by microscopic deformation is often delayed because strain gradients can be relieved by recovery processes at high temperatures. In discussions that follow, we sometimes assume that the time to nucleate voids is small in comparison to creep fracture times. This is equivalent to assuming that void growth controls creep fracture, and in many cases this is a good assumption. However, in others it is not, and this should be kept in mind. The point is illustrated by creep deformation/fracture studies on high-purity silver (Fig. 11.8). In these, silver was processed so that internal voids (diameter $\cong 1$ μm) were situated on grain boundaries in one condition, but not in the other. The decrease in fracture times and strains when voids are initially present compared to when they are not is substantial. Thus, in the example shown, were the times for void nucleation to be disregarded, the predicted fracture times would be the same for both materials and this would be much in error.

Void growth occurs by deformation involving diffusional flow and/or dislocation creep. Moreover, the driving force for void growth is the same as that for creep deformation. Void growth also contributes an additional increment to creep strain; that is, the tertiary creep strain, over and above that due to steady-state creep, is often a result of the increase in material volume due to the growth of internal voids. The tertiary strain depends on the void shape; if the voids are shaped rather much like penny-shaped cracks or wedges that grow in a direction normal to the applied tensile force, rather than as spherical cavities that expand equally in all directions, the contribution of void growth to creep strain is minimal. Void growth leading to transgranular creep fracture is not all that different from the like growth resulting in ICF. The primary difference is that for TCF void growth results only from power-law creep. When cavities are on the boundaries, as they are for ICF, surface and grain-boundary diffusion can also cause void growth.

As voids grow, they occupy an ever-increasing fraction of the material's cross-sectional area. This leads to an enhancement of stress on the nonvoided areas of the

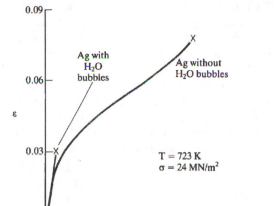

Figure 11.8
Creep curves of ordinary polycrystalline silver and of silver containing grain-boundary voids (i.e., H_2O bubbles). The creep curves of the two materials are identical at low strains ($\varepsilon \leq 0.02$). However, the voided material fractures prematurely ($\varepsilon_f \cong .03$, $t_f \cong 1$ hour). The results show that the initial presence of voids of this size ($\cong 1\mu m$) significantly reduces the fracture time and strain. (*Reprinted from S. H. Goods and W. D. Nix, "The Kinetics of Cavity Growth and Creep Fracture in Silver Containing Implanted Grain Boundary Cavities," Acta Metall., 26, 739, Copyright 1978, with permision from Elsevier Science.*)

boundary (if ICF is to occur) or on the like areas within grains (if TCF is to follow). As a result, the creep rate is accelerated by void growth; in effect in the later stages of void growth, voids catalyze further growth, and the final stages of creep fracture can be considered "catastrophic." When the voids have grown to an extent that they occupy a critical area fraction normal to the applied stress, they quickly coalesce by shear or microrupture. Hence, void coalescence at high temperature is similar to what it is at low temperature. For example, extensive microplasticity accompanies void coalescence during creep fracture, and this is so irrespective of the macroscopic ductility, which may be limited as it is in ICF. The time spent, and strain accumulated, during final void coalescence is small compared to creep fracture times and total creep failure strains, and hence we will not consider further the coalescence process. Instead, our concerns are primarily with void nucleation and growth. A consensus within the high-temperature materials community regarding the mechanisms responsible for void growth has emerged within the last 25 years or so. This is not so for the more complicated void nucleation problem. As a result, nucleation, which is discussed in the section following, cannot be treated with the same rigor as can void growth.

B. Void Nucleation

A number of mechanisms has been proposed for creep void/crack nucleation. Undoubtedly most of these may occur under certain stress-temperature combinations.

However, there is as yet no universally accepted mechanism for creep void nucle-ation and, given the diverse creep deformation processes that operate at various stress-temperature combinations, there is no reason to believe there should be. In this section we present several physically plausible models for crack nucleation in creeping solids.

It has been argued that wedge-shaped cracks can be formed by unaccommo-dated grain-boundary sliding (gbs);[6] schematics of proposed mechanisms are given in Fig. 11.9. The shapes of the cracks speculated to form bear a distinct resem-blance to "w" cracks (e.g., Fig. 11.3), and this lends credence to the mechanisms il-lustrated in Fig. 11.9. However, crack shapes found following fracture may bear no resemblance to crack shapes following their nucleation; we shall see, for example, that when surface diffusion controls crack growth, wedge-shaped cracks result naturally, regardless of the initial crack shape. Moreover, gbs should be minimal on boundaries whose normals are parallel to the applied tensile stress, but it is primarily on these boundaries that "w" cracks are most often found (Fig. 11.3). However, we point out that crack growth *rates* are greatest on such boundaries, and this may account for the apparent discrepancy. All things considered, the nucleation of wedge-shaped cracks by unaccommodated gbs is a plausible hypothesis. Indeed, the occurrence of wedge-like cracks is always associated with gbs. Finally, gbs-induced wedge cracking is consistent with the concept of microscopic strain in-homogeneities causing crack nucleation; these will be greatest, for example, at grain-boundary triple points, where it is believed "w" type cracks have the greatest propensity to form.

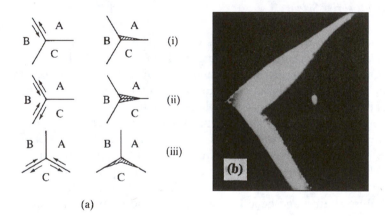

(a)

Figure 11.9
Schematic illustrating how unaccommodated grain-boundary sliding can nucleate wedge-shaped cracks. Depending on the type of sliding, differently oriented and shaped cracks can form, as shown in (i)–(iii). A wedge-shaped crack, similar to that shown in Fig. 11.9a(iii), formed in a crept aluminum alloy. Magnification, 90 ×. (*H. C. Chang and N. J. Grant, reprinted with permission from Trans. TMS-AIME, 206, 544, 1956.*)

---

[6]Of course, gbs per se does not lead to crack formation. For example, gbs accompanies and accommo-dates diffusional flow (Chap. 7). However, if the strain caused by diffusional flow does not match that due to gbs, the misaccommodation can result in formation of cracks.

Spherical shaped ("r" type) cavities also form during creep, and their growth and coalescence can lead to fracture. For ICF, these cracks are commonly associated with grain-boundary precipitates or second-phase particles. Although these particles limit grain-boundary sliding, they are also the sites of crack nucleation. If the idea of unaccommodated gbs being responsible for crack nucleation is correct, we can say that "w" cracks form by unaccommodated sliding taking place over a distance comparable to that of a grain edge length. In contrast, the like, but lesser, misaccommodation that takes place on a scale comparable to the interparticle separation distance is responsible for the nucleation of cavities at boundary particles.

Other kinds of strain incompatibility can also lead to void nucleation. For example, voids can be formed at grain boundaries as a result of heterogeneous matrix slip due to slip bands. The intersection of these bands with grain boundaries may result in void formation as a result of subsequent unaccommodated gbs (Fig. 11.10). Grain-boundary structural ledges (i.e., atomistically sharp discontinuities along the boundary) can facilitate nucleation of this type.

Crack nucleation at grain-boundary particles leads to cracks with initial shapes as shown in Fig. 11.11a. In addition, these cracks can form by a "nucleation" process driven by the external stress. The crack nucleation is handled in much the same manner as is nucleation taking place in phase transformations; the primary difference is that the external stress substitutes for the thermodynamic driving force of a phase transformation. The resulting mathematical treatment bears a resemblance to that used in the analysis of Griffith cracks. It appears that this type of nucleation does not take place in the absence of some strain gradients at the particle-matrix-boundary junction or at the precipitate-matrix interface. For the latter situation, a crack of the type illustrated in Fig. 11.11b is formed. This type of crack can also be formed in grain interiors, where it is the precursor to TCF.

As mentioned, "w" shaped voids form preferentially at high stress levels. The transition from "r" to "w" voids also depends on microstructural features, such as the grain size. This is illustrated for a high-alloy stainless steel in Fig. 11.12. We see that "r" voids form preferentially at small grain sizes. A factor complicating subsequent discussion relates to the initial (nucleated) void volume. As is described

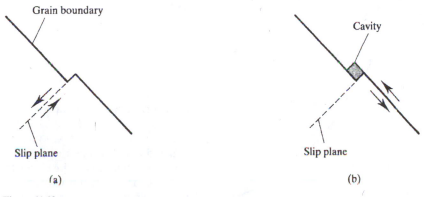

(a)

(b)

Figure 11.10
Heterogeneous plastic flow via slip bands or (in the situation shown) planes can lead to cavity formation. In (a) slip produces an offset (ledge) at the grain boundary. (b) Subsequent grain-boundary sliding results in formation of a cavity.

below, this is often an assumed quantity in analyses of creep fracture. Moreover, the fraction of initial voided area depends on the types of voids formed. Generally, "w" voids are characterized by a greater initial void volume fraction than are "r" voids. This is a result of the rapid initial spreading of an as-nucleated "w" void by time-independent plasticity promoted by the stress concentration at the tip of the wedge. It is only after this initial spreading that time-dependent plasticity controls the growth of this kind of cavity.

As the above discussion shows, it is clear that our understanding of crack nucleation during creep is incomplete. In developments that follow, we make simplifications regarding the initiation of cracks/voids in terms of describing how creep fracture times can be estimated. In the approach taken, it is often estimated that voids nucleate at some nucleation time (or, equivalently, some creep nucleation strain), and that the nucleation event produces some initial area fraction of voids

(a)   (b)

Figure 11.11
(a) Initial shape of a crack formed at the junction between an inclusion and a grain boundary. Crack nucleation is caused by unaccommodated grain-boundary sliding. (b) A different type of crack formed by unaccommodated flow at an inclusion-matrix interface. This type of crack can also form at inclusions within grains. It is important to note that these are *as-nucleated* crack shapes; the equilibrium shape of a growing crack can differ substantially from the nucleated shape. (*Adapted from R. Raj and M. F. Ashby, "Intergranular Fracture at Elevated Temperatures," Acta Metall., 23, 653, Copyright 1957, with permission from Elsevier Science.*)

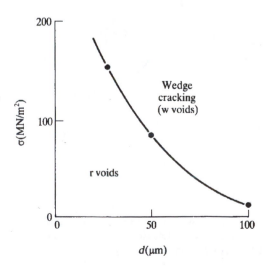

Figure 11.12
The relative tendencies for wedge (w) and cavity (r) voids. Wedge cracks are found at high stress levels and large grain sizes; spherical cavities form under the opposite situations. Data are for a 20% Cr–24% Ni high-alloy stainless steel. (*Data taken from J. S. Waddington and K. J. Lofthouse, J. Nucl. Mater., 22, 205, 1967.*)

along the boundaries or within the grains. The assumption makes analysis easier, and it is often consistent with experiment. However, it is also often found that cavities nucleate throughout the creep process. While a continuous nucleation sequence can be incorporated into the description of creep fracture, it makes the mathematics of it much more complicated than they already are. So we shy from it.

C. Void Growth

i. AN OVERVIEW. The schematic microstructures of Fig. 11.1 depicting creep fracture are expanded on in Figs. 11.13 and 11.14. Although voids/cavities may form on all grain boundaries, Fig. 11.13*a* considers only those whose growth leads to ICF; these voids are located on boundaries normal to the tensile axis. Later, in extending the discussion to TCF, spherical voids are assumed present within grain interiors (Fig. 11.13*b*). Although both grain-boundary and intragranular voids are associated commonly with second-phase particles, for clarity the latter are not shown in the schematics of Figs. 11.13 and 11.14.

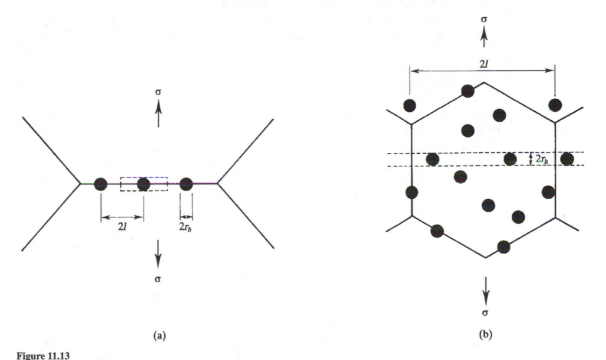

(a)   (b)

Figure 11.13
Idealized geometry of voids whose growth results in creep fracture. (a) Spherical voids of radius r_h are separated by the center-to-center distance 2*l* along grain boundaries. Spherical voids are found when void growth is controlled by boundary diffusion; the shape is altered when surface diffusion controls growth. The dotted "pillbox," used in calculating void growth rates, is a cylindrical "unit cell" of the void array. (b) Void geometry appropriate to power-law-creep-controlled void growth leading to TCF. The "unit cell" of this void array is similar to that shown in (a); the main difference is that voids are separated by a distance along the tensile axis comparable to the grain diameter for ICF, whereas the separation distance is about $2r_h$ when TCF is the fracture mode. (*Adapted from A. C. F. Cocks and M. F. Ashby, Prog. Matls. Sc., 27, 189, Copyright 1982, with permission from Elsevier Science.*)

Simplified assumptions on void shape and arrangement are made in analyses of void growth. For example, spherical grain-boundary cavities are assumed to all have the same radius r_h and to be separated by a distance $2l$ along the boundary (Fig. 11.13a). Similar geometrical assumptions apply to spherical intragranular cavities when TCF is considered (Fig. 11.13b).

The void-growth processes illustrated in Fig. 11.14 parallel those responsible for creep deformation. Figure 11.14a illustrates the void growth leading to ICF when it is controlled by the rate of diffusion of atoms along the boundaries between the grains. The void surfaces are stress-free, whereas the boundaries are subject to a tensile stress. This results in a driving force for mass transfer from the regions near the void periphery to those along the grain boundaries. Therefore, the void-growth mechanism is a mass transfer process, bearing a strong resemblance to diffusional creep (cf. Chap. 7); indeed, the "driving force" for mass transport is the same in both situations. The transport takes place by a series process. Matter is transferred first by surface diffusion[7] along the void periphery to the boundary-void intersection, and then by grain-boundary diffusion along the boundary. This void-growth process contributes to tertiary creep strain on two accounts. First, void growth itself leads to extension of the sample along the tensile axis and, second, the material plated out on the grain boundaries produces a tensile extension. Intergranular creep fracture occurs when the void boundary area fraction ($f_h \cong r_h^2/l^2$) reaches some critical value, at which point rapid final failure occurs. A reasonable estimate for this critical value of f_h is about 0.25.

The void geometry of Fig. 11.14a applies when surface diffusion is rapid in comparison to grain-boundary diffusion. This enables the void to maintain its spherical shape as it grows (as it would like to do since a spherical shape has the lowest possible surface area to volume ratio). When the opposite situation applies, that is, when the grain-boundary diffusion is more rapid than surface diffusion, the void becomes progressively more elliptical-shaped and, in the limit when surface mass transport is minimal compared to grain-boundary transport, the void assumes a "wedge"-like shape. Given that this void shape is no longer one of minimum surface energy, it is clear that surface-diffusion-limited growth dominates only at stress levels higher than those for which grain-boundary diffusion dominates. That is, the additional surface energy associated with a penny-shaped crack requires more work and this is reflected in a greater external stress to accomplish this.

The crack shape for surface-diffusion control is approximated as that of a penny-shaped crack (Fig. 11.14b). Crack growth is accomplished by an increase in crack radius r_h, which grows at a much faster rate than its height $2r_0$. Surface-diffusion-limited void growth contributes less to tertiary creep strain than does void growth limited by boundary diffusion. This results from the void growing comparatively rapidly in the direction transverse to the tensile axis and only slowly along the tensile direction during surface-diffusion-limited growth. Tertiary creep that does take place results from the plating-out of material along the grain boundaries; the plating effectively "jacks" the grains apart. In addition, a given amount of mass transport increases the void grain-boundary fraction more for surface-diffusion-limited than for boundary-diffusion-limited growth. Thus, the critical fraction of

---

[7]Although volume diffusion can contribute to mass transfer during void growth (as it does, for example, in diffusional creep), the rate of such is much less than that due to surface or boundary diffusion. Thus, the contribution of volume diffusion to void growth can be neglected.

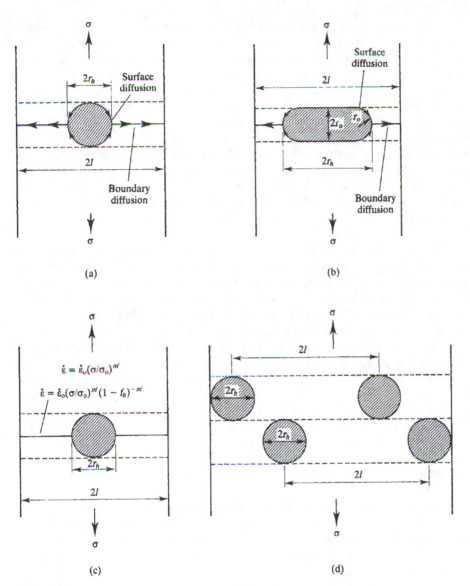

Figure 11.14
Schematics of the processes leading to void growth for ICF ((a) to (c)) and TCF (d).
In (a) grain-boundary diffusion controls void growth; the voids remain spherical as
they grow. This is because mass transport by surface diffusion along the void periph-
ery is rapid in comparison to the sequential boundary-diffusion transport. The dot-
ted lines represent the unit cell noted in Fig. 11.13. (b) When surface diffusion is slow
vis-à-vis boundary diffusion, the cavity no longer remains spherical; it becomes elon-
gated in a direction transverse to the tensile axis. The shape is approximated as a
penny-shaped crack of height $2r_0$ and diameter $2r_h$. The unit cell "pillbox" is outlined
by the dotted lines. (c) Power-law-creep-controlled void growth. The growth is
caused by dilation of the void due to a higher strain rate in the voided area; it is
greater than that in the grain interior by the factor $(1 - f_h)^{-m'}$, where f_h is the void
area fraction on the boundary. (d) Power-law creep also controls growth of (spheri-
cal) voids within grains which results in TCF. The geometry is similar to that in (c);
the difference lies in the vertical separation distance of voided areas.

voided grain-boundary area is obtained at a lesser tertiary strain *and* with a smaller amount of mass transport when surface diffusion controls the void growth.

Void growth can also take place by power-law creep; the situation is shown for ICF in Fig. 11.14c. The voids on the boundary cause a higher local stress on the material in their vicinity compared to that experienced by material well removed from the boundary. A force balance between the voided and unvoided regions shows that the stress on the former is a factor $(1 - f_h)^{-1}$ higher than on the latter. For power-law creep, $\dot{\varepsilon}$ varies as $\sigma^{m'}$; thus the strain rate in the voided region (of approximate size $2r_h$) is a factor $(1 - f_h)^{-m'}$ higher than that in the nonvoided grain interiors. This causes void growth and a tertiary creep strain.

Void growth leading to TCF takes place only by power-law creep. It can be described by a variant (Fig. 11.14d) of the geometry used to describe boundary void growth via power-law creep; i.e., the situation of Fig. 11.14d differs from that of Fig. 11.14c only in geometry.

Which of the void-growth mechanisms leads to ICF depends on a number of factors. Among them are stress, temperature (this controls the diffusion rates), intervoid spacing, and the *instantaneous* void area fraction. Boundary diffusion typically controls growth at low stresses. At higher stresses, surface diffusion or power-law creep controls it; the latter dominates at higher temperatures (it involves volume diffusion) and the former at lower temperatures.

As mentioned, the controlling mechanism depends also on the area fraction of voids, which increases monotonically during void growth. Generally, we find that when voids, and their boundary area fraction, are small, growth is controlled by one of the diffusional mechanisms. As the voids increase in size, their area fraction does also, and power-law creep supersedes the diffusional mechanisms. Transitions like this are complicated further by considerations of "coupling." That is, a transition in controlling mechanism encompasses a range of f_h, as well as temperature and stress, values. Over a relatively wide range of these variables, void growth rates are comparable for two mechanisms, and this necessitates consideration of processes acting in tandem.

In the following section we address void growth in more detail. But before we do this, a summary of where we are going and some of the associated pitfalls is in order. We wish to know how long it takes for the material to fracture. We would also like to know the creep-fracture strain. *If* we know the time to initiate voids and their initial spacing and location (and we don't), then we can trace void growth provided we have sufficient ancillary information (e.g., surface energy, diffusion coefficients, power-law creep equations). But usually we do not have this information, either. But *if* we did, we could estimate how long a time is required to attain the critical void fraction at which fracture occurs. Moreover, since void growth is linked to the tertiary strain and strain rate, we can estimate these parameters as well. So, in principle, the void growth problem is "solved." But in practice it is not, because of the just-mentioned deficiencies in our knowledge. Because of this, we describe the void growth/fracture scenario in only a semi-quantitative way.[8] Perhaps at some later date we will have enough information to do full justice to the creep-fracture problem.

---

[8]The Cocks and Ashby reference provides a more-complete treatment. In fact, the information provided here is a distillation of some of their more important findings.

ii. VOID GROWTH. In this section, we describe the approaches that result in expressions for void volume fraction and tertiary creep strain resulting from void growth. As noted, this growth can take place by boundary-diffusion control (spherical voids), surface-diffusion control (wedge-shaped voids), or power-law creep. We consider only one mechanism—boundary-diffusion control—in detail, although results for the other mechanisms are given. The Cocks and Ashby reference can be consulted for details relative to these other mechanisms.

The flux of matter from the void surface/grain boundary junction to a boundary position intermediate between two voids is given by[9]

$$J_B \cong \frac{D_B}{\Omega kT} \frac{\Delta\mu}{\Delta x}$$ (11.1)

In Eq. (11.1), D_B is the boundary-diffusion coefficient (m²/s), Ω is the atomic volume (m³/atom), kT has its usual meaning (dimensions of energy, e.g., joules (J)), and $\Delta\mu$ is the change in chemical potential ($\cong \sigma\Omega$, dimensions of Nm/atom) which is dissipated over the distance Δx. We note that J_B has dimensions of [number (of atoms)]/(m²s). Taking the diffusion distance as half the intervoid spacing, l, we have

$$J_B \cong \frac{D_B\sigma}{kTl}$$ (11.2)

Multiplying J_B by the area available for diffusion (proportional to $r_h\delta_B$, where δ_B is the grain-boundary thickness) gives the number of atoms per second leaving the void volume. And multiplying this term in turn by the atomic volume gives the rate of increase in void volume due to atoms departing it. Calling this rate of change in volume dv/dt, we have

$$\frac{dv}{dt} \cong \frac{D_B\delta_B r_n\sigma\Omega}{kTl} = \frac{D_B\delta_B\sigma\Omega}{kT(l/r_h)}$$ (11.3)

The above expression is not exact; it does not consider geometrical details, for example. However, the physics of void growth is essentially described by Eq. (11.3). Further, knowing dv/dt permits us to obtain expressions for the instantaneous values of the void radius and voided grain-boundary area (f_h, the "damage"; the void radius and damage are related through $f_h = r_h^2/l^2$). Beyond that we can calculate the tertiary strain rate and strain resulting from void growth. This strain arises from two effects. One is the strain directly associated with void growth. The other is a result of the plating of material on the grain boundaries; this causes the grains to elongate, i.e., to produce a creep strain. Performing the necessary (and labored) "arithmetic," the following expressions are obtained for the damage and tertiary strain rate for boundary-diffusion-controlled void growth;

$$\frac{df_n}{dt} = \left[\frac{D_B\delta_B\sigma\Omega}{kTl^3}\right]\left[\frac{1}{f_h^{1/2}\ln(1/f_h)}\right]$$ (11.4)

[9]Since surface diffusion is rapid vis-à-vis boundary diffusion, the chemical-potential driving force is dissipated along the grain boundary and not the void periphery.

$$\dot{\varepsilon}_t = \left[\frac{1}{\ln(1/f_h)}\right]\left[\frac{2D_B\delta_B\sigma\Omega}{kTl^2d}\right] \tag{11.5}$$

Equation (11.5) stipulates only the tertiary creep rate ($\dot{\varepsilon}_t$). The total creep rate is obtained by adding to it the steady-state creep rate. Note that the expression for $\dot{\varepsilon}_t$ contains the grain size, d. This is because the tertiary extension must be divided by an appropriate gage length. Since the growing voids are separated by the grain diameter along the tensile axis, the grain size is the appropriate gage length.

A similar analysis applies to surface-diffusion-controlled void growth, with but two alterations. One is geometrical. The analysis assumes that the void grows laterally only, neglecting any extension of the void dimension, r_0, along the gage length. In addition, the analysis considers the effect of the surface energy, γ_s, which acts to restrict the formation of wedge-shaped voids. Because of the latter consideration, it is found that the damage and tertiary creep rates depend more strongly on stress for surface-diffusion-controlled void growth than for boundary-diffusion-controlled void growth. Results of the analysis for the growth of "wedge-like" voids lead to the following expressions for the damage and tertiary strain rates;

$$\frac{df_h}{dt} \cong 0.7\left[\frac{f_h^{1/2}}{(1-f_h)^3}\right]\left[\frac{D_S\delta_S\sigma^3\Omega}{kTl\gamma_s^2}\right] \tag{11.6}$$

$$\dot{\varepsilon}_t \cong 9\left[\frac{f_h^{1/2}}{(1-f_h)^3}\right]\left[\frac{D_S\delta_S\sigma^2\Omega}{kTld\gamma_s}\right] \tag{11.7}$$

In Eqs. (11.6) and (11.7), D_S is the surface-diffusion coefficient and δ_S is the effective thickness over which surface diffusion occurs. The other terms have their previous meanings.

Voids can grow by power-law creep, too. As indicated in Fig. 11.14c, for power-law creep the strain rate in the voided region is increased by the factor $(1-f_h)^{-m'}$ in comparison to that in material within the grains. The higher strain rate in the boundary regions causes the voids to dilate and grow, and this leads to an increase in damage and to tertiary creep. Analysis shows that the damage and strain rates are given by

$$\frac{1}{\dot{\varepsilon}_0}\frac{df_h}{dt} \cong 0.6[(1-f_h)^{-m'} - (1-f_h)]\left(\frac{\sigma}{\sigma_0}\right)^{m'} \tag{11.8}$$

$$\frac{\dot{\varepsilon}_t}{\dot{\varepsilon}_0} \cong \left[\frac{1.2f_h^{1/2}l}{d}\right][(1-f_h)^{-m'} - 1]\left[\frac{\sigma}{\sigma_0}\right]^{m'} \tag{11.9}$$

In Eqs. (11.8) and (11.9) new terms, $\dot{\varepsilon}_0$ and σ_0 are introduced. These are linked to the steady-state power-law creep rate ($\dot{\varepsilon}_{ss}$) through $\dot{\varepsilon}_{ss} = \dot{\varepsilon}_0(\sigma/\sigma_0)^{m'}$, and can be couched in terms of the analysis of Chap. 7 (e.g., Table 7.1). However, for the present purposes it is easier to consider the terms $\dot{\varepsilon}_0$ and σ_0 as a normalized strain rate and a normalized stress such that when the applied stress is equal to σ_0, the material's steady-state creep rate is $\dot{\varepsilon}_0$.

Transgranular creep fracture—which is controlled by power-law creep—can be handled in the same way used here to describe power-law creep ICF. Indeed, an expression the same as that of Eq. (11.8) holds for the damage rate for TCF. However, because voids are present throughout the material during TCF—and not restricted to grain boundaries—the TCF tertiary creep strain is greater. The proper correction is made by multiplying Eq. (11.9) by the ratio d/l to obtain the expression for the TCF tertiary creep strain rate.

What is one to make of, or to do with, the above equations? First, some of their physics. We see that the damage and tertiary creep rates depend on material parameters (e.g., diffusion coefficient, surface energy, etc.), the instantaneous value of the damage, and the applied stress. The dominant void-growth mechanism, therefore, depends on these factors. And, since the damage rate depends differently on stress depending on the controlling mechanism, we expect different dominant void-growth mechanisms at different stress levels. Boundary-diffusion-controlled void growth should dominate at low stress levels, for it depends only linearly on the stress. At somewhat higher stresses, surface-diffusion control—for which the damage rate varies with the cube of the stress—should supersede boundary-diffusion control. This is in agreement with experimental observations that "w" cavities are found at higher stress levels than are "r" cavities. Finally, at even higher stress levels, power-law creep should control because the damage rate for it depends on the stress to the m' power (m' is usually greater than three for power-law creep). Had we knowledge of the pertinent parameters in the damage rate equations and the pertinent scale to the voids (particularly their spacing, $2l$), Eqs. (11.4), (11.6), and (11.8) would permit us to estimate fracture times. How to go about conducting such an estimate is discussed in the next section.

EXAMPLE PROBLEM 11.2. Compare grain-boundary- and surface-diffusion-controlled void growth in the following way. For equal volumetric void-growth rates, determine the ratio of the damage accumulation rate (df_h/dt) for the two mechanisms.

Solution. The damage is the voided area ratio in the unit cell pillboxes of Figs. 11.13 and 11.14. The radius of the pillbox is l; the radius of the void is r_h. Thus, the damage is given by $f_h = r_h^2/l^2$ in both cases and the damage rate is

$$\frac{df_h}{dt} = \frac{2r_h}{l^2}\frac{dr_h}{dt}$$

The volume rate of change of a spherical void (volume $= 4\pi r_h^3/3$) is $dv/dt = 4\pi r_h^2(dr_h/dt)$. For the penny-shaped void (volume $= 2\pi r_h^2 r_0$), $dv/dt = 4\pi r_h r_0(dr_h/dt)$. Expressing the damage rate in terms of the volume mass transfer rate, we have

$$\frac{df_h}{dt}\text{(boundary-diffusion control)} = \left(\frac{1}{2\pi r_h l^2}\right)\frac{dv}{dt}$$

$$\frac{df_h}{dt}\text{(surface-diffusion control)} = \left(\frac{1}{2\pi r_0 l^2}\right)\frac{dv}{dt}$$

Thus, for equal values of dv/dt, the damage rate for surface-diffusion control is greater by the factor r_h/r_0. This means for equivalent mass transfer, damage accumulation is greater when the voids are penny-shaped.

---

iii. DOMINANT VOID-GROWTH MODES. In this section, we restrict our discussion to ICF; extrapolation of the approach to TCF can be easily done. In analyzing creep fracture, we must first determine—for diffusion-controlled void growth—whether voids grow by boundary-diffusion control (spherical voids) or surface-diffusion control (penny-shaped voids). This is done by comparing the respective damage rates (Eqs. (11.4) and (11.6)) and determining which is less. (Recall that the diffusion processes are series processes.) The assessment depends on the instantaneous value of the damage and further presumes the proper ancillary data are available. Following this, we determine whether the damage rate for power-law creep is greater than that for the dominant diffusion-control mechanism; this depends on the stress level as well as the instantaneous damage value. We then reasonably take the damage rate as that of the mechanism having the higher damage rate.[10]

Determination of the dominant void-growth mechanism is conveniently done graphically; examples are shown in Figs. 11.15a and b. Figure 11.15a plots normalized damage rate (i.e., the damage rate divided by $\dot{\varepsilon}_0$) vs. normalized stress (the stress divided by σ_0). Here we have assumed that boundary diffusion controls diffusional void growth so that the voids are spherical. The graph is also constructed for specific values of the microstructural parameters (d and l) and thermophysical parameters (e.g., the diffusion coefficient). The graph also pertains to a specified value of f_h; 0.01 in this instance. Note that in Fig. 11.15a there is a transition in the dominant void-growth mechanism with increasing stress. Boundary diffusion controls the damage rate at low stress levels and power-law creep does so at high stress levels. This transition is the creep-fracture analog to the like transition in creep-deformation mode when the dominant creep mechanism changes from diffusional- to power-law creep (Fig. 7.13) with increasing stress. The total damage rate is that resulting from both processes and is represented by the solid line in Fig. 11.15a. Over a reasonable stress range, the total damage rate can be taken as that resulting from the dominant mechanism.

Just as increases in stress cause a transition in the controlling mode of damage accumulation, a similar transition is effected by increases in f_h. This is illustrated in Fig. 11.15b, in which boundary-diffusion and power-law creep damage rates are plotted vs. f_h at a constant stress (material properties are the same as in Fig. 11.15a and in Fig. 11.15b we have taken $\sigma = \sigma_0$). For $f_h \lesssim 5 \times 10^{-2}$, damage accumulation is controlled by boundary diffusion; when f_h is greater than this value, power-law creep controls damage accumulation. The solid line in Fig. 11.15b represents the total damage rate. As before, it is approximately the sum of the individual rates.

The intersection of the two curves of Fig. 11.15b represents a transition in void-growth mechanism with damage accumulation. That is, voids initially grow by boundary diffusion until a transition value of f_h (designated f_t; $\cong 5 \times 10^{-2}$ in Fig. 11.15b) is realized. The later stages of void growth are controlled by power-law creep. A similar discussion of the transition from surface-diffusion-controlled to power-law-creep-controlled damage accumulation can be carried out. It is left as an exercise (Prob. 11.13).

We can also have transitions from boundary-diffusion-controlled void growth to surface-diffusion-controlled growth. However, since these processes operate in

---

[10]This assumption neglects "coupling" of mechanisms, and is not always a good assumption. See Cocks and Ashby for amplification. However, our assumption makes discussion of the already complicated physics of void growth a little less hairy.

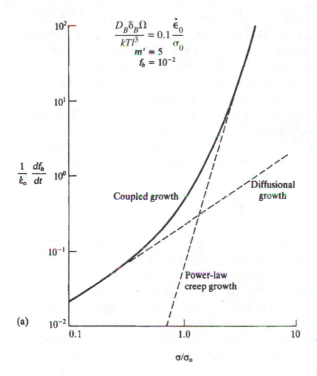

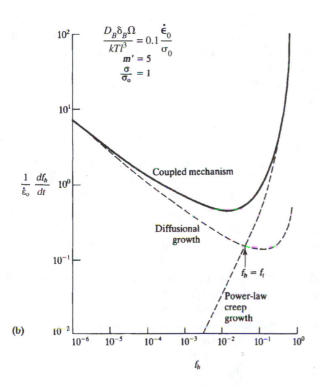

Figure 11.15
(a) **Damage rate as a function of stress for void growth controlled by grain-boundary diffusion and by power-law creep. The rates of the individual mechanism (Eqs. (11.4) and (11.8)) are shown by the dotted lines. The solid line represents a more detailed development that considers coupling of the mechanisms and represents the total damage rate; it can be** *approximated* **as the sum of the individual rates. For low stresses ($\sigma \lesssim 0.4\sigma_0$), damage rate is controlled by diffusional void growth; at high stresses ($\sigma \gtrsim 3.0\sigma_0$), power-law creep void growth determines the damage rate. The stress at which the dominant void-growth mode changes is slightly greater than σ_0. (Note: The curves are drawn for a specific material, as reflected in the value of** m' **and for specific values of diffusion coefficient, etc., and for a fixed damage.) (b) Damage rate as a function of damage for void growth controlled by grain-boundary diffusion and by power-law creep. Again, the dotted lines represent Eqs. (11.4) and (11.8). The solid line represents the damage rate due to both mechanisms. Boundary diffusion controls damage rate when the damage is low and power-law creep controls it when** f_h **is high. The transition in dominant void-growth mode takes place at** $f_h = f_t$**. The curves are drawn for the same material properties as in (a) and for a stress** $\sigma = \sigma_0$**.** (*Reprinted from A. C. F. Cocks and M. F. Ashby, Prog. Matls. Sc., 27, 189, Copyright 1982, with permission from Elsevier Science.*)

series, rather than parallel, it is the lesser of their rates that determines damage accumulation. While the damage rates for the two mechanisms depend on the instantaneous damage in different ways, the important kinetic variable that differentiates the mechanisms is the diffusion coefficient. To a first approximation, for example, we would expect that if $D_B > D_s$, penny-shaped cracks would form, surface diffusion would control void growth, and the damage rate would be given by Eq. (11.6). Conversely, if $D_B < D_s$, void growth is limited by grain-boundary diffusion, the cavities are spherical, and the damage rate is expressed by Eq. (11.4). Since D_B and D_s vary differently with temperature, the transition from boundary-diffusion to surface-diffusion control depends on temperature. Likewise, the transitions from one or the other of the diffusion-controlled growth mechanisms to power-law creep control is also temperature sensitive. The complexities associated with temperature- and stress-induced transitions in void-growth mechanisms can be reduced by representing them in the form of a *void-growth map.*

Void-growth maps have the same axes (σ/E and T/T_m) as fracture-mechanism maps; indeed, they can be considered a refinement of fracture maps. Void-growth maps constructed for the metal Cu are shown in Figs. 11.16a and b. The lower dotted lines in these diagrams represent the sintering limit; applied stress levels less than this lead to void shrinkage and not growth. The solid line separating the two diffusion-controlled void-growth regions is the *approximate* locus for which $D_s = D_B$. The transitions from surface-diffusion control (at low temperatures) and boundary-diffusion control (at higher temperatures) to power-law creep control are obtained from plots similar to those of Fig. 11.15a. The stress at which the transitions occur is temperature sensitive, as can be seen in this diagram. The upper boundary line at high stress levels in Figs. 11.16 corresponds to void-growth controlled by time-independent plasticity. Note, too, that void-growth maps are constructed for a specific value of f_h; thus, each map is a "snapshot" in time and Figs. 11.16a and b are different because they have been constructed for different damage values. Figure 11.16b applies at a longer creep time (the damage value for it is higher) than does Fig. 11.16a. In addition, the maps are constructed for a specific value of intervoid spacing (= 24 μm for the maps of Figs. 11.16).

As shown in Figs. 11.16, contours of constant damage rate can be drawn on void-growth maps and this increases their utility. As in deformation- and fracture-mechanism maps, the boundaries between various regions in a void-growth map are diffuse. For void-growth maps this reflects both the uncertainty in the data used to construct them and the "coupling" effects associated with void growth.

Although void-growth maps are conceptually useful for determining the controlling damage accumulation mechanism, they have limitations. For example, they are "permissive" maps. As noted, they are constructed for a specific f_h value, and tell us only what mechanism would control damage accumulation *if* cavities of the specified amount were present. As an illustration, copper dynamically recrystallizes and undergoes rupture fracture over much of the temperature-stress regime occupied by "power-law creep growth" in Figs. 11.16. In this case, while the maps tell us the rate and mechanism of void growth if voids are present, as a result of dynamic recrystallization they are not. We also see that the boundaries of the map "sweep downward" as f_h increases (compare Fig. 11.16b ($f_h = 10^{-1}$) to Fig. 11.16a ($f_h = 10^{-2}$)). This is a manifestation of the kinds of transition shown in Fig. 11.15b; i.e., power-law creep dominates void growth at higher values of f_h. The sintering limit also decreases as

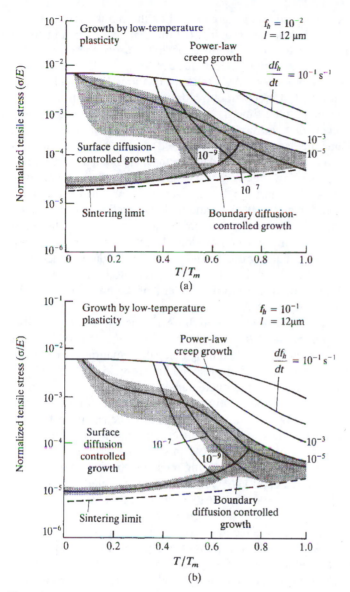

Figure 11.16
Void-growth mechanism maps appropriate to Cu with $l = 12$ μm, and for a damage
of (a) 10^{-2} and (b) 10^{-1}. The boundary lines separating the regions are loci of equiva-
lent damage rate due to two mechanisms. Within the power-law creep region, dam-
age rate due to this process is greater than that due to the controlling diffusion
process. Likewise, when (either of the) diffusional damage rates is greater than that
due to power-law creep, a diffusion-controlled growth region is shown in the dia-
gram. The shaded regions indicate stress-temperature combinations for which cou-
pled growth must be considered. The lower broken line represents the sintering limit;
at stresses less than this, voids shrink rather than grow. Contours of constant dam-
age rate are also shown in the diagrams. (*From A. C. F. Cocks and M. F. Ashby,* Prog.
Matls. Sc., *27, 189, Copyright 1982, with permission from Elsevier Science.*)

void volume (more precisely, void size) increases. For the maps of Fig. 11.16, which are constructed for a constant void spacing, the limit decreases by a factor of $(10)^{1/2}$ as f_h increases from 10^{-2} to 10^{-1}. This is expected because the limit scales with r_h^{-1}, and since $f_h \sim r_h^2$, the sintering limit varies as $(f_h)^{-1/2}$.

iv. TIMES AND STRAINS TO FRACTURE. Equations (11.4)–(11.9) permit (in principle) determination of creep-fracture strains and fracture times. It is assumed that voids nucleate at some time, t_n, and with some initial void volume fraction, f_{ho}. The dominant damage accumulation mechanism for these initial conditions is identified, and the damage rate is integrated until a value of damage is attained at which a transition in the controlling damage mechanism takes place (e.g., from diffusion control to power-law creep control, Fig. 11.15b). Then the damage rate is further integrated from the transition damage value, f_t, until the critical damage value (reasonably approximated as 0.25) is attained. Doing this permits a fracture time to be calculated. It is the sum of the void nucleation time and the times spent during the two stages of damage accumulation. Likewise, the tertiary creep strain is the sum of the strains occurring during the two different damage accumulation modes. The creep-fracture strain is obtained by adding to it the steady-state creep strain accumulated over the fracture time. Details of the procedures can be found in the Cocks and Ashby reference.

The procedure appears straightforward enough. And it is—in principle! Now we repeat why its practical implementation is difficult. First, we do not know the void nucleation time. If voids nucleate early during creep deformation,[11] we can approximate the creep-fracture time as that time associated with void growth until the critical damage value is reached. If cavities nucleate late during creep deformation, the fracture time just arrived at is a conservative estimate of the material life time. We must also estimate a value of the as-nucleated damage. (One can probably reasonably bracket initial damage values. Then the uncertainty in the fracture time is that time needed for the damage to increase between its initial bracketed values.) More serious, though, is that the initial void spacing must also be stipulated to realistically employ the damage-rate equations. Our present state of understanding of void nucleation does not permit us to do this. (Although here, too, we can likely realistically bracket this spacing.) Finally, the ancillary data to effectively use the damage-rate and tertiary-creep-strain rate equations are, for the most part, missing. Surface energies can be guessed at within a factor of two or so, but grain-boundary and surface diffusivities are seldom known to precision. So we are left with the distasteful situation where a physically appealing model cannot be employed in engineering design because we have too many unknowns to contend with. Let's hope that this will not always be the situation.

Before leaving fracture-mechanism and void-growth maps, some final comments are in order. We have treated high- and low-temperature fracture as if they were entirely separable, but they are not. Let us reconsider Fig. 11.5c, the fracture map of Mg. This map implies that low-temperature Mode I fracture is supplanted by ICF at a critical (stress-dependent) temperature. There is a temperature range, however, over which the transition from Mode I low-temperature fracture to ICF in

---

[11]The approach to creep fracture described here permits incorporation of a time-varying void nucleation rate. However, to be of practical use this rate must be expressed mathematically; this is seldom realistic.

Mg is gradual. During this transition, preexisting cracks in Mg still propagate as a result of the stress intensification at their tips. But, because creep deformation takes place at the crack tip, the propensity for rapid crack advance is lessened relative to what it is at low temperature. On the other hand, the stress intensification associated with the crack can enhance the growth of voids somewhat removed from the crack. Ice serves as a good example here. We think of ordinary low-temperature (i.e., temperatures slightly below 0°C) fracture of ice as brittle. However, we are familiar with fracture of ice at high stresses. At lower stress levels, crack advance in ice involves aspects of both low-temperature fracture mechanics and diffusion. In fact, a new field of fracture mechanics describes this type of "combined" fracture, not only in ice but in other materials that behave similarly. Discussion of this lies outside our scope, but essential features of the description can be found in the references listed at the end of this chapter.

11.5
DESIGN AND MATERIALS CONSIDERATIONS

A. Design Considerations

The Larson-Miller parameter, discussed in Chap. 7, permits engineering estimates of material creep lives. This is so even though the procedure is not a conservative one; engineering experience, "know how," and utilization of appropriate safety factors alleviate the deficiency. The LM procedure, however, does not address the important problem of materials subject to varying stress-temperature combinations. Most structural components are so exposed; thus, estimating fracture times for such situations is an important design consideration.

One such "cumulative damage" rule has been proposed by Robinson. It estimates the time to fracture under varying stress-temperature conditions in the following way. If the time spent at one stress σ_i and temperature T_i is t_i and the fracture time for this stress-temperature combination is t_{fi}, then Robinson's rule states that fracture takes place when

$$\Sigma \left(\frac{t_i}{t_{fi}} \right) = 1 \qquad (11.10)$$

This rule can be extended to the situation where stress and temperature vary continuously. The rule is then written in integral form, in which case it reads

$$\int_0^{t_f} \frac{dt}{t_f} = 1 \qquad (11.11)$$

It can be shown that Robinson's rule is valid when void growth is controlled by one (and only one) mechanism. Of course, damage does not accumulate by any one mechanism. Even for most stress-temperature combinations that remain fixed during service, voids grow initially by diffusional processes and later by power-law creep. In general, it is found that Robinson's rule is most closely followed when power-law creep dominates almost the whole of the damage-accumulation process. It is clear that accurate estimations of fracture times cannot generally be made on the basis of appealing approximations such as Robinson's rule. Nonetheless, simplifications like this, when tempered by empirical knowledge and judicious use of safety factors, have their place in high-temperature design.

B. Materials Considerations

We conclude our discussion of creep fracture by considering microstructural features that affect it. Creep fracture depends importantly on void growth which, as we have seen, takes place by mechanisms common to creep deformation. Thus, structural alterations beneficial to enhancing creep-deformation resistance also likely improve a material's creep-fracture resistance.[12] In this regard, design of materials for high-temperature fracture resistance differs fundamentally from similar design at low temperatures. In the low-temperature case, flow and fracture are, in many cases, competitive. That is, an increase in a material's resistance to permanent deformation is usually accompanied by a diminution in its fracture resistance and vice versa. This is not so at high temperatures.

These points can be clarified with the aid of fracture-mechanism maps. Such a map for Monel, a solid-solution nickel alloy, is given in Fig. 11.6. (Monel can be considered a typical solid-solution nickel alloy in this regard.) We see that the stress levels in the map of Monel are higher than they are in the corresponding map of Ni (Fig. 11.5a), and this is indicative of a greater resistance to creep fracture. The effect of a fine, stable dispersoid on fracture of Ni is illustrated by the map for thoriated nickel (Fig. 11.17). This alloy contains a fine (ca. 40-nm diameter) dispersion of ThO_2 particles, the oxide comprising about 2 vol.% of the alloy. Thoria does not dissolve in nickel, even at temperatures close to nickel's melting point, and the ThO_2 particles do not appreciably coarsen at high temperatures. The result is that not only are the stress levels of the map for thoria-dispersed nickel higher than they are in the map of Ni (compare Fig. 11.17 with Fig. 11.5a), but the thoria dispersion impedes matrix recovery/recrystallization even at temperatures close to nickel's melting point. Thus, the map for this alloy does not contain a dynamic-recrystallization-induced-rupture fracture region.

For these reasons, thoria (or other stable oxide) dispersion-hardened nickel alloys are potential high-temperature materials. However, the strengths exhibited by such alloys at ordinary and moderately high temperatures are not all that high, and they do not yet compete with the nickel-base "superalloys" as high-temperature structural materials. Nimonic 80-A serves as a prototype for such alloys; its high-temperature fracture-mechanism map is given in Fig. 11.18. The Ni-base superalloys are hardened in several ways. They typically consist of a solid-solution Ni matrix containing a dispersion of ordered precipitates based on the composition Ni_3Al. The combination of solid-solution and particle strengthening provides a resistance to power-law creep at high temperatures and to dislocation flow at low temperatures. Polycrystalline Ni-base superalloys also contain particles—usually metallic carbides—situated at grain boundaries, and these resist grain-boundary sliding. These features are all reflected in the high stress levels of the fracture map of Nimonic 80-A, and in the relatively long fracture times exhibited by this alloy.

---

[12]An important caveat relates to void nucleation. For example, void nucleation can be facilitated by grain-boundary-situated particles that restrict grain-boundary sliding, which is a component of creep deformation.

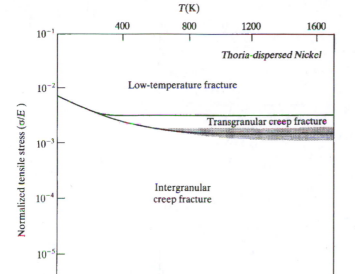

Figure 11.17
The high-temperature fracture mechanism map for thoriated nickel. The fine disper-
sion of ThO_2 particles prevents recovery/recrystallization; hence, a high-temperature
rupture field is not shown in the diagram. The thoria also "hardens" the nickel; this
is reflected in the higher stress levels in this map compared to those in the equivalent
map for Ni (Fig. 11.5a). (*Reprinted from M. F. Ashby, C. Gandhi, and D. M. R. Taplin,*
"Fracture-Mechanism Maps and Their Construction for FCC Metals and Alloys," Acta
***Metall., 27, 699, Copyright 1979, with permission from Elsevier Science.*)**

We note that a rupture fracture field is shown in the map for Nimonic 80-A. This is
a result of the Ni_3Al phase dissolving in the matrix at temperatures on the order of
$0.8T_m$, and the carbides doing likewise at somewhat higher temperatures. Thus, at
temperatures near its melting point, Nimonic 80-A is a solid solution susceptible to
dynamic recrystallization and rupture.

The Ni-base superalloys are the highest performing of the metallic superalloys.
That is, they are employed at higher stresses and temperatures than are Co- and Fe-
based superalloys. In turn, Co-based superalloys are used at somewhat higher
stress-temperature combinations than the Fe-base superalloys. All of the superalloy
classes, though, employ somewhat similar means to strengthen them. The Fe alloys, for example, often are extensively solid-solution strengthened by additions of
relatively large amounts of Cr and Ni. In these materials, the Cr serves also to
improve oxidation resistance because a relatively adherent and impervious Cr_2O_3
oxide surface layer forms during high-temperature oxidation. Some alloys of the
higher-melting temperature bcc transition metals—Nb, Ta, Mo, and W, the so-called
refractory metals—are used in specialized high-temperature applications where
weight is not a consideration. The refractory metals, in common with many of
the superalloys, require some protection from the environment since the refractory

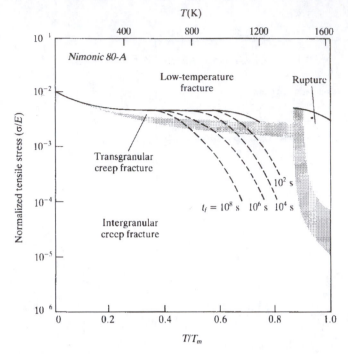

Figure 11.18
The high-temperature fracture-mechanism map for Nimonic 80-A, a nickel-base
superalloy. The several strengthening mechanisms utilized in this alloy make it a
more useful structural material than thoria-dispersed nickel up to fairly high tem-
peratures. (Compare the stress levels in this map to those in Fig. 11.17.) At high tem-
peratures both the intragranular Ni_3Al phase and the intergranular carbides dissolve
in the matrix of Nimonic 80-A. This permits dynamic recrystallization and rupture.
Contours of constant fracture time are shown in the diagram. (*Adapted from M. F.*
Ashby, C. Gandhi, and D. M. R. Taplin, "Fracture-Mechanism Maps and Their Con-
struction for FCC Metals and Alloys," Acta Metall., 27, 699, Copyright 1979, with per-
***mission from Elsevier Science.*)**

metals are inherently reactive. They are thus coated with oxidation- or other
gaseous-resistant coatings. Structural ceramics are much more resistant to the deleterious effects caused by reactions with the environment; indeed for some material-
environment combinations (e.g., Al_2O_3 in oxygen) they are inherently nonreactive.
This is another reason why their potential for high-temperature service is being ac-
tively investigated.

11.6
FAILURE IN SUPERPLASTIC MATERIALS

Superplastic materials display extensive tensile ductility. This capability is manifested over a limited range of temperature and strain rate, as described in Chap. 7.
In this section, we consider tensile failure of superplastic solids.

Photographs of failed tensile samples of several Pb-Sn superplastic alloys
are shown in Fig. 11.19. In some of the materials (e.g., pure Pb-Sn and the alloy

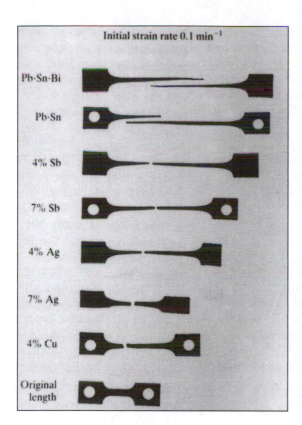

Figure 11.19
Photographs of fractured samples of several Pb-Sn-based superplastic alloys. The binary alloy and the ternary alloy containing Bi do not cavitate, and tensile failure is associated with a 100% reduction in area. Other ternary elements promote internal cavitation. When this happens, the material's tensile reduction in area is reduced. (*From N. Ridley and J. Pilling, Superplasticity, ed. B Baudelet and M. Suery, Centre Nationale de la Recherche Scientifique, Paris, 8.1, 1985.*)

containing Bi), the material necks down to, or close to, a point; i.e., the reduction in area at fracture is 100% or close to it. In other materials (e.g., the alloys containing Cu and Ag), ductility is much less, even though these materials display the high values of strain-rate sensitivity associated with superplastic behavior. As can be seen in Fig. 11.19, fracture surfaces of these less ductile materials are macroscopically flat; indeed, at this level they have a "brittle" appearance. However, when observed at higher magnifications they evidence features of ductile fracture.

The extremes of fracture behavior illustrated in Fig. 11.19 reflect two inherently different failure modes in superplastic solids. Materials manifesting 100 percent reductions in area fail by continuing plastic deformation until a diffuse "neck" shrinks to a point. This type of failure is similar to rupture. During it, no—or only minimal—internal cavitation takes place. In contrast, superplastic materials that display lesser tensile reductions in area manifest considerable internal cavity formation. It is the linkup or coalescence of these cavities that results in a reduced ductility. Void linkup in superplastic materials bears resemblance to the same process in creeping or low-temperature plastically deforming solids, but also differs in some respects from it. The point is amplified later.

We first consider failure in superplastic materials that do not cavitate. The issues to be considered have several aspects. First, nucleation of cavities must be suppressed or, if they are present initially or form during deformation, they must not grow appreciably. Factors inhibiting void nucleation during creep also are effective in preventing cavitation during superplastic deformation. Thus, heterogeneous flow must be suppressed, and the potential for strain incompatibilities minimized. The latter is often accomplished by working with "clean" materials, i.e., ones in which processing renders the material free of inclusions to begin with. The binary Pb-Sn alloy illustrated above, for example, can be made to cavitate by adding hard, non-deforming second-phase particles to it. The interface between the particles and the grain boundaries on which they are often situated serves as a void nucleation site.

We must also consider what factor(s) permit some superplastic solids to neck to a point, whereas others do not (cf. Fig. 11.19). The problem is related to neck development during quasi-uniform (i.e., superplastic) tensile flow. The topic is difficult. A complete solution entails a description of the true strain as it varies along the gage length and as it is controlled by material properties, including strain-rate sensitivity and, if present, strain hardening. A number of treatments dealing with the matter have been put forth. One, due to Ghosh, is physically appealing and unburdened by mathematical complexity. It forms the basis for our discussion of failure of noncavitating superplastic solids.

Although superplastic materials that do not cavitate are typically "clean" materials, this does not mean they are "defect-free" or entirely homogeneous. Materials fabrication always yields heterogeneous structures, albeit proper control of processing can minimize the extent and degree of heterogeneity. Thus, in Ghosh's model it is assumed the sample length of a tensile bar can be divided into two regions: (1) an unnecked or homogeneously deforming region and (2) an inhomogeneously deforming one where a neck initiates and then develops. The situation is illustrated in Fig. 11.20; the neck is assumed to form in a region where the "defect"

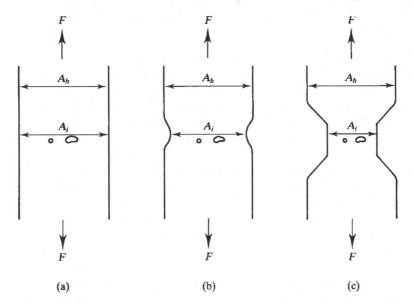

Figure 11.20
Ghosh's model of quasi-uniform flow. A neck forms in a region of high defect concentration. The effective load-bearing area (A_i) in the neck is less than that (A_h) for the unnecked region. The sample is assumed to consist of two regions: a homogeneously deforming one (area = A_h) and an inhomogeneously deforming one (area = A_i). The ratio of the initial areas for these regions (A_{0i}/A_{0h}) is defined as f_0. (*From B. Harriprashad, T. H. Courtney, and J. K. Lee, Metall. Trans. A, 19A, 517, 1988.*)

concentration (inclusions, small voids, etc.) is highest. A parameter f_0 is defined as the ratio of the initial effective load-bearing cross-sectional area of the neck region, A_{0i}, to that of the unnecked region, A_{0h}. That is, f_0 describes the variation in the defect concentration in the volume associated with the neck relative to that in the remainder of the material. In fact, its complement $(1 - f_0)$ represents the ratio of the defect density in the two regions; e.g., if $f_0 = 0.99$ [so that $(1 - f_0) = 0.01$] there is a 1 percent difference in defect density between the respective regions.

If we assume that strain hardening can be neglected, the material flow behavior can be represented by $\sigma = K'(\dot{\varepsilon})^m$, where m is the strain-rate sensitivity. A force balance between the homogeneously (h) and inhomogeneously (i) deforming regions yields

$$\sigma_h A_h = \sigma_i A_i \tag{11.12}$$

where σ_h and σ_i are the respective stresses borne by the homogeneously and inhomogeneously deforming regions and A_h and A_i are their respective areas. With strain defined as $\varepsilon = \ln(A/A_0)$ (A_0 = initial cross-sectional area), and using the above constitutive equation, Eq. (11.12) can be expressed as

$$(\dot{\varepsilon}_h)^m A_{0h} \exp(-\varepsilon_h) = (\dot{\varepsilon}_i)^m A_{0i} \exp(-\varepsilon_i) \tag{11.13}$$

By definition $A_{0i}/A_{0h} = f_0$. On taking the mth root of Eq. (11.12), it becomes

$$\left[\exp\left(-\frac{\varepsilon_h}{m}\right)\right]\left(\frac{d\dot{\varepsilon}_h}{dt}\right) = (f_0)^{1/m}\left[\exp\left(-\frac{\dot{\varepsilon}_i}{m}\right)\right]\left(\frac{d\dot{\varepsilon}_i}{dt}\right) \tag{11.14}$$

Integration of Eq. (11.14) with respect to strain and rearrangement of the resulting expression yields an expression showing how the strain away from the imperfection (ε_i) is affected by ε_i, m, and f_0;

$$\varepsilon_h = -m \ln\{(f_0)^{1/m}[\exp(-\varepsilon_i/m) - 1] + 1\} \tag{11.15}$$

For samples that neck to a point, $\varepsilon_i \rightarrow \infty$ and

$$\varepsilon_h = -m \ln[1 - (f_0)^{1/m}] \tag{11.16}$$

Figure 11.21 shows how the strain away from the imperfection is strongly dependent on the factor f_0 and on m. Applications using this analysis often substitute total sample elongation for ε_h in Eq. (11.16), and correlations between total elongation and m (and f_0) have been made on this basis.[13] As indicated in Fig. 11.21, these correlations are moderately successful except at low values of m (which corresponds to lower temperatures). Modification of the analysis to account for finite strain hardening at low temperatures removes much of the discrepancy.

There is one important deficiency in the model just described. It does not mimic well the sample geometry found during this kind of deformation; i.e., sample cross-sectional area varies monotonically along sample length in most situations (Fig. 11.19). In this sense all of the material can be considered to be deforming inhomogeneously, and none of it homogeneously. Thus, in many respects it is inappropri-

[13]Figure 11.21 should be compared with Fig. 7.24 which correlates tensile ductility with strain-rate sensitivity. It is clear that Fig. 7.24 is an experimental verification of the model described by Fig. 11.21.

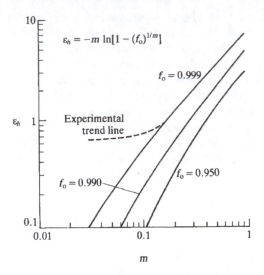

$$\varepsilon_h = -m \ln[1 - (f_o)^{1/m}]$$

$f_o = 0.999$

Experimental
trend line

$f_o = 0.990$

$f_o = 0.950$

ε_h

m

Figure 11.21
**A graph of uniform strain (ε_h) vs. m
(Eq. (11.16)) for several values of
the parameter f_0. Tensile failure
strains are often plotted vs. m.
When they are, curves similar to
the experimental trend line are often found (cf. Fig. 7.24). At low m
values, experimental fracture
strains are higher than predicted,
in part because of work hardening
which is not incorporated into
Eq. (11.16). (*From B. Harriprashad,
T. H. Courtney, and J. K. Lee,
Metall. Trans. A, 19A, 517, 1988.*)**

ate to identify ε_h with total elongation as is often done. A better approach would involve determination of the distribution of strain within the inhomogeneous region,
and then obtaining an integrated value of it for correlation with m and f_0. (This is a
complex undertaking.) In spite of these reservations, it is clear that total fracture
strain depends critically on the strain-rate sensitivity and initial defect density.
Moreover, for materials that do not cavitate during quasi-uniform flow, the dependence of fracture strain on these factors is sensibly described by the Ghosh model.

Superplastic materials that cavitate do not neck to a point, even though they
may manifest extensive fracture strains. As described above, cavitation correlates
with microstructure and slip behavior in a way similar to the correlation existing for
ordinary creep cavitation. Cavities typically form at strains well below the fracture
strain and continue to grow as strain is increased (Fig. 11.22). Final cavity volume
fractions can be substantial (10 percent or even higher), and cavity sizes are typically on the order of micrometers or greater (i.e., on the scale of the grain size).
What is remarkable about superplastic cavity growth is that extensive quasi-uniform
flow continues in the presence of cavities. That is, their formation and growth do
not result in rapid neck development. This is related to the mechanical characteristics of superplastic solids.

It is now accepted that cavity growth in superplastic materials occurs by dislocation creep, rather than diffusional flow. This is in accord with our description of
creep cavity growth (recall that power-law creep dominates cavity growth when
cavities are large and their volume fraction is high). However, cavity coalescence
differs in superplasticity from that in ordinary creeping solids. Because of the pronounced resistance to tensile instability of superplastic materials, ligaments between
cavities often neck down to a point (or close to one). Thus, the cavity coalescence
can reasonably be described as cavity growth until the cavities overlap. That is, the
critical void linkup event is not caused by shear across intervening ligaments.
When a sufficient number of cavities have joined in this manner, final fracture takes
place. Thus, the delay in final fracture is related to the resistance of the ligaments
to neck development; on a microscopic level, the ligaments experience a tensile

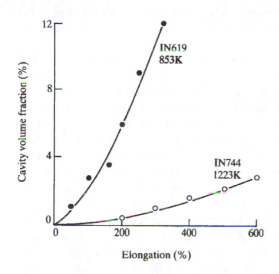

Figure 11.22
Cavity volume fraction vs. sample elongation for alloy IN744 ($T = 1223$ K) and a nickel-silver alloy (alloy IN619; $T = 853$ K). Cavity volume fraction increases appreciably with strain. Even so, tensile instability is delayed in these superplastic materials. (*Data for IN744 from C. I. Smith, B. Norgate, and N. Ridley, Met. Tech., 1, 191, 1974; for the nickel-silver alloy from D. W. Livesey and N. Ridley, Metall. Trans. A, 9A, 519, 1978.*)

reduction in area the same as (or close to) that of a noncavitating solid. However, the presence of the cavities prevents this ductility from being manifested macroscopically.

Cavity nucleation in superplastic deformation is commercially important on at least two accounts. First, it limits the formability of superplastic materials. Second, products made from parts that have undergone cavitation contain porosity and this is deleterious to their performance. Quality control is used in superplastic forming to minimize the presence of inclusions or other second-phase particles that facilitate cavitation. Additionally, forming can be carried out in the presence of a hydrostatic stress that retards or prevents cavity growth. In some cases the presence of such a stress leads to the disappearance of cavities (e.g., by pressure-assisted sintering). In closing, we note that the suggestion of using superplasticity in forming operations was initially greeted with skepticism. The process was, it was thought, destined to remain a laboratory curiosity. This is not what has happened. Manufacture of a number of different thin-walled and other complex-shaped parts is routinely done via superplastic forming, and the process results in considerable savings in material and machining costs.

11.7
SUMMARY

Material fracture at high temperatures is manifested in one of three ways: (1) rupture, (2) transgranular creep fracture (TCF), or (3) intergranular creep fracture (ICF). Although crystallographic flow is common to all of these (the extent of it is stress- and temperature-dependent), high-temperature fracture is distinguished from low-temperature fracture by the contribution that diffusion makes to it.

At high temperatures and stress levels, rupture occurs in metals, alkali halides, and some other ionic solids. Rupture is commonly associated with dynamic recrystallization. That is, the high stresses associated with rupture fracture result in high strain rates, e.g., as in hot working. The extensive malleability of metals found during their hot working correlates with tensile reductions in area equal or close to 100 percent.

Transgranular creep fracture occurs at lower temperatures than rupture (assuming the latter takes place). Transgranular creep fracture is observed in all crystalline solids excepting covalent solids such as SiC and Si_3N_4; these are the most brittle of crystalline solids and they exhibit neither TCF nor rupture. The processes resulting in TCF involve void nucleation within grains (usually at second-phase particles but occasionally at other locations of inhomogeneous plastic flow). This is followed by void growth, the rate of which is determined by power-law creep processes. When the void fraction attains some critical value, the voids coalesce rapidly by microscopic flow. We see that the processes culminating in TCF—void nucleation, growth, and coalescence—are similar to those leading to low-temperature ductile fracture. Moreover, measurable reductions in area also accompany TCF. Thus, the distinction between low-temperature ductile fracture and TCF is made on mechanistic grounds. In particular, void nucleation and growth resulting in TCF are caused to some degree by diffusional processes, and these are absent for low-temperature ductile fracture.

At certain stress-temperature combinations, all crystalline solids fail by ICF. Intergranular creep fracture takes place at stresses less than those causing rupture and/or TCF (provided these other failure mechanisms are observed). Intergranular creep fracture differs from TCF on several accounts. First, the voids whose presence leads to ICF are situated on grain boundaries rather than in grain interiors. This leads to a considerable reduction in tertiary creep strains for ICF compared to those associated with TCF. Indeed, since void growth is restricted to the grain-boundary regions, and as the voids there occupy but a small fraction of the material volume, ICF tertiary creep strains are minimal. While the void coalescence leading to ICF takes place by the same shear/rupture processes that accompany TCF void coalescence, void-growth mechanisms are distinctly different for the two fracture modes. In particular, the void growth leading to ICF involves diffusional flow as well as power-law creep. Diffusional void growth is accomplished by removal of mass from the void periphery (by surface diffusion) followed by its deposition (by grain-boundary diffusion) along the boundaries between the voids. Tertiary creep strain is due to both the removal of mass from the void (which enlarges it) and by the plating-out of this mass along the boundary (which jacks the grains apart and dilates the void). The shape of the growing void is determined by the *slower* of the surface- or boundary-diffusion mass transfer rates. For example, if the boundary transport rate is less than the corresponding surface rate, the voids maintain a spherical shape as they grow. If the opposite situation holds, the voids are like penny-shaped cracks.

During the processes culminating in ICF, void growth is controlled by diffusional flow when the void area fraction on the boundary is small and by power-law creep when it is "large." This results because void-growth rates for power-law creep increase rapidly with void boundary area fraction, and because the void-growth rate is determined primarily by the *greater* of the diffusion/power-law creep rates.

This physical description of creep fracture can be complemented graphically. For example, the tendencies of the various material classes for the several high-temperature fracture modes can be represented in high-temperature fracture mechanism maps (Figs. 11.5). In a sense, these maps are refinements of high-temperature deformation-mechanism maps, for high-temperature fracture also involves permanent deformation. Hence, high-temperature fracture maps parallel deformation-mechanism maps. The covalent solids manifest brittle fracture at all stress-temperature combinations (i.e., they exhibit neither rupture nor TCF). Any ductility in them is limited to microscopic flow near grain boundaries at temperature-stress combinations leading to ICF.

Void-growth mechanism maps (Fig. 11.16) identify the dominant damage accumulation mechanism at a given stress-temperature combination. However, since void-growth rates depend also on the instantaneous value of the damage, the maps are *permissive*, and not definitive; that is, they show the dominant growth mechanism for a stipulated value of f_h. In spite of this deficiency, these maps are useful for understanding how dominant damage-accumulation mechanisms are altered by changes in stress and temperature.

Creep fracture can be described quantitatively on the basis of void geometry (e.g., intervoid spacing) and the dominant damage mechanism. An important deficiency of this approach is a lack of a quantitative criterion for defining void-nucleation times. At high, as at low, temperatures, voids often nucleate in the vicinity of second-phase particles; either within grains (where their growth and coalescence lead to TCF) or along grain boundaries (where they are a precursor to ICF). Since diffusional flow and/or recovery is common to creep fracture, the stress fields of dislocations generated near matrix–second-phase interfaces are reduced at high temperatures compared to what they are at low temperatures. Nonetheless, strain incompatibility eventually leads to void nucleation at high temperatures, manifested typically by decohesion at particle-matrix interfaces (for TCF) or at grain boundary–particle intersections (for ICF). Quantification of void nucleation is also made difficult by the several mechanisms that can cause it. For example, grain-boundary sliding is ubiquitous to void nucleation leading to ICF, but it is unrelated to nucleation leading to TCF.

Creep fracture can be modeled if void nucleation times and other related features are known. The latter include the initial void size, spacing, and volume fraction. Further, ancillary data (creep equations, diffusion coefficients, etc.) must also be on hand if quantitative predictions are to be made. Unfortunately, neither the microstructural variables nor the ancillary property data are known to a precision to enable this promising approach to creep fracture to be implemented in a practical way. Perhaps in the future we will be able to do better.

We have also seen that high-temperature creep fracture, as it is related to void growth, is caused by deformation processes. Thus, resistance to creep fracture is accomplished by the same means that resistance to creep deformation is. In this respect, design against creep fracture differs fundamentally from design against low-temperature fracture. At low temperatures, an increase in a material's yield strength (if it is a metal) is usually accompanied by a decrease in its fracture toughness and vice versa. This is not so at high temperatures; improvement in a material's resistance to creep deformation is usually concomitant with an increase in creep fracture times and a reduction in the tertiary creep strain rate. The results of structural alterations which improve creep deformation resistance are observed readily in fracture-mechanism maps; they raise the stress levels in the map and increase the fracture times.

Some superplastic materials do not cavitate during tensile deformation. Their reductions in area equal 100 percent and failure is accomplished by continuing plastic flow of material within a diffuse neck. A model developed by Ghosh is adequate for phenomenologically describing this kind of tensile failure in terms of material strain-rate sensitivity and microstructural variations within the material. Some superplastic solids cavitate. For them, tensile reductions in area are reduced, although tensile strains may still be extensive. The large strains are a consequence of the reluctance of the material between cavities to undergo microscopic tensile instability.

We close our discussion of high-temperature fracture by recalling some of the limitations to the discussion. In this chapter we have only considered in detail creep fracture resulting from tensile loading. Stress state, however, plays an important role in determining void-growth and damage rates. This can be easily remembered by realizing that voids are inherently unstable. That is, in the absence of an appropriate external stress, voids shrink and disappear rather than grow. A tensile force facilitates void growth, but a compressive one retards it. Indeed, hydrostatic compressive stresses will not result in creep fracture, but rather in void "healing."

REFERENCES

Ashby, M. F., C. Gandhi, and D. M. R. Taplin: "Fracture Mechanism Maps and their Construction for FCC Metals and Alloys," *Acta Metall.*, 27, 669, 1979.

Beeré, W. B.: *Cavities and Cracks in Creep and Fatigue*, ed. J. H. Gittus, Elsevier Applied Science, London, 1975.

Cocks, A. C. F. and M. F. Ashby: "On Creep Fracture by Void Growth," *Prog. Matls. Sc.*, 27, 189, 1982.

Evans, H. E.: *Mechanisms of Creep Fracture*, Elsevier Applied Science, London, 1984.

Evans, R. W., and B. Wilshire: *Introduction to Creep*, The Institute of Materials, London, 1993.

Gandhi, C., and M. F. Ashby: "Fracture Mechanism Maps for Materials which Cleave: FCC, BCC and HCP Metals and Ceramics," *Acta Metall.*, 27, 1565, 1979.

Ghosh, A. K.: "Tensile Instability and Necking in Materials with Strain Hardening and Strain Rate Hardening," *Acta Metall.*, 25, 1413, 1977.

Gittus, J. H.: *Creep, Viscoelasticity and Creep Fracture*, Elsevier Applied Science, London, 1975.

Pilling, J., and N. Ridley: *Superplasticity in Crystalline Solids*, The Institute of Metals, London, 1989.

Stowell, M. J.: "Cavitation in Superplastic Failure," *Metal Sc.*, 17, 1, 1983.

PROBLEMS

11.1 Consider a surface crack in a glass rod at a temperature at which viscous flow can take place. Assume that the stress in front of the crack tip scales with the square-root of the crack length. (This elastic-mechanics approximation is really not valid for a viscous medium.) Illustrate how the crack will become quickly blunted via viscous flow. (Recall the relationship between stress and strain rate for viscous materials.)

11.2 a Sintering of materials (i.e., densification by reduction of void volume through diffusional flow) can take place in the absence of an applied stress. What is the "stress" driving sintering?

b Sintering occurs more rapidly when a compressive stress is applied to a body as in, for example, hot pressing. If the applied stress is σ, estimate the ratio of the sintering rate with stress to that without it. Assume sintering takes place only by diffusional processes, and not by plastic flow. The diffusion coefficients can be taken as independent of the stress. (Hint: What is the "stress" driving sintering in the two situations?)

11.3 A structural member is subjected to a stress-temperature combination such that the steady-state creep rate is 10^{-10}/s. The material must not creep more than 0.1% over its intended lifetime. If this were the only design criterion, estimate how long the structure could remain in service.

11.4 Explain why NaCl is capable of dynamic recrystallization while it is not capable of macroscopic low-temperature plastic deformation.

11.5 By judiciously extrapolating data from Figs. 11.6 and 11.7, estimate the fracture time for Monel at $T = 0.2T_m$ and for an applied tensile stress equal to $10^{-3} E$.

11.6 The final stages of both creep fracture and low-temperature ductile fracture involve void coalescence. How does the local strain rate affect the ease of void coalescence at low temperatures? How does it affect this ease at high temperatures?

11.7 Void nucleation (e.g., as spherical cavities) will not take place without a driving force (such as stress) because the system energy increases with the increase in surface area accompanying void nucleation.
a Why is void nucleation possible if a tensile component of stress is present?
b Show that for spherical voids that may nucleate, only those having radius $r > 2\gamma/\sigma$ are stable (that is, these voids will grow rather than shrink).

11.8 Consider a penny-shaped crack on a grain boundary. If the material were not subject to an external stress, how would the crack shape change with time?

11.9 Consider a spherical void situated on a grain boundary. Show how the void volume and shape change with increasing time when (1) the grain-boundary mass transport rate is much greater than the surface-diffusion mass transport rate, and (2) the opposite conditions apply.

11.10 a The transition from grain-boundary- to surface-diffusion-controlled intergranular void growth corresponds to a change in void shape from that of a sphere to a penny-shaped crack. The transition is *not* controlled *solely* by the respective values of the surface and grain-boundary diffusion coefficients. Explain why this is so. (Hint: What is the equilibrium void shape?)
b A simple calculation can provide some idea as to how stress enters into the grain-boundary–surface-diffusion control transition. The surface energy "stress" is equal to $\gamma \, dA/dV$, where dA/dV is the differential of void surface area with respect to void volume (see Example Prob. 11.1). For voids of the same volume, compare the ratio of dA/dV for a sphere to that of a penny-shaped crack. (Take the minor dimension, r_0, of the penny-shaped crack as fixed.) The difference in these "stresses" is a measure of how much additional external stress is required to grow penny-shaped voids compared to that needed to cause spherical voids to grow.

11.11 a In general, which of the boundary-diffusion-controlled or surface-diffusion-controlled void-growth processes will lead to longer fracture times? Why?

b In general, which of the mechanisms will lead to greater tertiary creep strains? Why?

11.12 For diffusion-controlled intergranular void growth, mass transfer results in "plating" of mass on the grain boundaries. This jacks them apart, contributing to tertiary creep strain. Will such "plating" accompany power-law-creep-controlled intergranular void growth?

11.13 Plot schematic graphs similar to Figs. 11.15, but for the situation in which power-law creep void growth and surface-diffusion-controlled growth are the competing processes. Comment on differences expected in terms of transitions in mechanisms (with damage accumulation and with stress) for this situation compared to that of Fig. 11.15.

11.14 **a** The volume mass transfer rate for diffusion-controlled void growth is the product of a flux (e.g., Eq. (11.1)) by a diffusion area. Compare the ratio of the volume mass transfer rate for surface-diffusion-controlled intergranular void growth and grain-boundary-diffusion-controlled intergranular void growth. (Note that for surface-diffusion-controlled growth, the chemical potential is dissipated along the curved edge of the penny-shaped crack. The surface-diffusion-controlled case can be handled in a manner similar to that employed for the boundary-diffusion case, cf. Eqs. (11.1)–(11.3).)
b The volume mass transfer rates can also be deduced from the respective equations for the damage rates, because the damage relates directly to the void volume. Use the damage rate equations (Eqs. (11.4) and (11.6)) for the two mechanisms to estimate the respective volume mass transfer rates. Compare the results obtained in this way to your answer of part (a). Comment on any differences between the expressions.
c Refer to the result obtained in part (b) (the ratio of the respective volume transfer rates). In terms of parameters in this ratio, under what conditions will surface diffusion control void growth and under what conditions will grain-boundary diffusion do likewise? (Note that this analysis compensates to an extent for the simplification suggested in the text that the relative diffusion coefficients dictate the transition from one controlling mechanism to another. Likewise, it indicates effects due to the nonequilibrium shape of a penny-shaped crack [see Prob. 11.10].)

11.15 Here we estimate creep-fracture times on the assumption that voids are initially present in the material (f_i = initial damage value) and that creep fracture takes place when the damage attains a value of 0.25.
a First consider void growth controlled initially by surface diffusion and later by power-law creep. Obtain an expression for the fracture time if surface diffusion were to control the *whole* of the void-growth process. Do likewise for the power-law creep case. For the actual case (i.e., when there is a transition in controlling mechanism with increasing damage) determine the relative times spent in each of the controlling regions and divide by the times just calculated. Take the sum of these fractions and compare it with unity. This is sort of a test for a modified Robinson's rule. Is the modified Robinson's rule a good one? Does it lead to a conservative estimate of fracture life?
b Repeat part (a) for when boundary-diffusion control is the controlling mechanism at low damage values.

11.16 What are the similarities and the differences between void-growth maps and high-temperature (creep) fracture-mechanism maps? What kind of information is provided by each kind of map?

11.17 Show that Robinson's rule is conservative if, as a result of variations in the applied stress and temperature, the damage rate decreases with time. Using physical reasoning, explain why this is so.

11.18 Fit the curve drawn through the data points of Fig. 7.24 to Eq. (11.16). What is the value of f_0 at which a reasonable "fit" can be made? Is this f_0 value a reasonable one? Explain.

CHAPTER 12

Fatigue of Engineering Materials

Frontispiece
The fatigue-fracture surface of a steel I-beam that was part of a drag line excavator. The top and bottom flange sections have dimensions of 40 by 40 by 4 cm. The connecting web is 2.5 cm thick. Each flange was reinforced by a plate welded to it (the welds are the protrusions on the top and bottom flange surfaces). A fatigue crack originated near the bottom weld and then grew slowly during service. This crack penetrated the entire bottom flange, all of the web, and a significant portion of the top flange before final fracture of the I-beam took place. The fatigue (slow crack growth) area of the fracture surface is light; the darker area corresponds to final fast fracture. (You would have thought someone would have noticed this disaster about to happen before it did, wouldn't you?) (*ASM Handbook, Vol. 12, Fractography, ASM International, Materials Park, OH 44073-0002, 1987.*)

12.1
INTRODUCTION

Fatigue fractures are the most commonly identified kinds of failure of structural metals. They are also observed frequently in polymeric solids. The term fatigue is an appropriate one, for it refers to the time-delayed fracture of materials subjected to cyclical stresses below those causing plastic yielding and/or tensile failure. In order for a fatigue failure to occur, at least some portion of a time-varying stress must be tensile in nature. That is, a stress state for which the maximum principal stress is always algebraically negative (i.e., it is a compressive stress) does not lead to fatigue fracture.

Almost all structures are prone to fatigue. In some, such as those common to aircraft, design against fatigue is every bit as much a consideration as is design against yielding or conventional fracture. This is a natural consequence of the inherently stochastic loading conditions to which such structures are subjected. For the same reason, fatigue is an important consideration in applications involving rotating parts, such as automotive drive axles and turbine blades in energy generating devices. However, fatigue fractures are also observed in components we do not ordinarily think of as being cyclically loaded. For example, as illustrated in the frontispiece of this chapter, I-beams can fracture by a fatigue process. So can bridges which for many years were thought immune to fatigue fracture because the primary load supported by a bridge structure is that of its own, invariant weight. However, recent bridge failures, some after many decades of service, have in some cases been attributed to fatigue (often in concert with corrosion) caused by a cyclical stress arising from variations in bridge traffic density. Our understanding of the processes involved in fatigue has increased considerably recently, and this has been translated into more efficient design against this type of failure. However, dealing with fatigue remains complicated by the random nature of the many time-varying stress applications in engineering service and relating these applications to fatigue studies conducted in a laboratory setting.

As with most fracture, fatigue fracture involves crack nucleation, growth, and "coalescence." Crack nucleation in fatigue, as in most ductile fracture processes, is related to inhomogeneous plastic flow occurring (usually) at a microscopic level, and such flow can take place even when a structure is only elastically stressed in a macroscopic sense. Fatigue cracks nucleated thusly grow during, and as a result of, the time-varying or cyclical stress. During this "slow growth" process, the crack advance rate relates to the frequency of the stress or strain cycle. Slow crack growth is interrupted by final fast fracture, resulting from the inability of the cross-sectional area, diminished by slow crack growth, to sustain the maximum tensile load (resulting in an overload failure), or because the fatigue crack has grown to a size sufficient to permit crack propagation of the type described in our discussion of fracture mechanics in Chap. 9. We see this scenario of the fatigue process implies that a structure will withstand a certain number of stress or strain cycles prior to failure. Later, we describe how this number of cycles is related to variables such as the stress range, the maximum stress, and material properties.

As mentioned, correlation of laboratory results to a material's service fatigue resistance is complicated by the more or less random stress-time variation seen by engineered structures. Nonetheless, several mechanical tests have been developed that qualitatively, and in some cases quantitatively, correlate well with engineering requirements. Perhaps the most common of them subjects a material to a specified variation in the applied stress or strain, and measures the number of such cycles required to cause fracture. This type of engineering-oriented testing scheme is useful in design. Other tests involving aspects of fatigue behavior are common, and provide fundamental information on the fatigue process. As one example, the rate of slow crack growth can be measured as a function of the crack length and range of applied cyclic stress. The results from such a procedure are useful for estimating the number of cycles a material spends during the slow-crack-growth stage of fatigue. Other tests determine how a material's flow stress varies with the cyclical strain range. Results from such studies lead to the construction of a cyclic stress-strain curve that usually differs from the analogous tensile-flow curve. Characterization of a material in this manner allows differentiation between material substructure developed during cyclic and uniaxial loading.

More demanding use of materials at high temperatures has led engineers to also consider the interaction between creep and fatigue. An example is provided by turbine blades subjected to cyclical stresses at elevated temperatures, which use, therefore, necessitates consideration of both fatigue and creep. Engineering design here is still developing. Nonetheless, because of its importance, we briefly consider the creep-fatigue problem in this chapter.

Most discussion of fatigue centers on the phenomenon as it occurs in metals and their alloys. This is reasonable, as most fatigue failures are observed in this material class. Ceramics, for example, are generally not considered prone to fatigue failure, although this view is a bit oversimplified. However, it is correct in the sense that if a fatigue crack were initiated in a ceramic at ordinary temperatures, it would result in a relatively short fatigue life because of the low ceramic fracture toughness.[1]

Polymers are prone to fatigue failure. From a phenomenological viewpoint, fatigue of metals and polymers is similar in that an applied cyclical stress or strain can be correlated with the number of cycles required to cause fracture. But there are important differences in the processes attendant on fatigue in the two material classes. For one, a relatively large viscoelastic strain component exists in many polymers, and this leads to hysteretic heating during polymer fatigue. In and of itself, this effect can lead to material "damage." There are also significant microstructural differences between metals and polymers, and this leads to differing fatigue-deformation mechanisms for the two material classes. Thus, we discuss separately polymeric and metallic fatigue.

12.2
CHARACTERISTICS OF FATIGUE FRACTURE

Fracture surfaces of metals that have failed by fatigue frequently have distinctive features, and these are often used to identify fatigue as the probable failure cause.

---

[1]Slow crack growth is observed in some ceramics at fairly low values of an applied stress (or strain) range and when the crack size is small. Nonetheless, catastrophic crack propagation takes place in ceramics when the stress range is increased or the crack has grown to a size modest in comparison to observed fatigue-crack lengths in metals or polymers.

Macrographs of fatigue-fracture surfaces are given in Fig. 12.1 and in the chapter frontispiece. At this level, the surface morphology has two distinct characteristics. One, noted as clamshell or beach markings, represents that portion of the fracture surface over which crack progression was gradual (i.e., slow crack growth). This is a characteristic of fatigue failures.

The examples of Fig. 12.1 give telltale evidence of intermittent crack growth, and textbooks commonly present such clear-cut pictures to illustrate fatigue fracture. In reality, the fracture areas on a fatigue-fracture surface cannot always be so easily delineated. For example, corrosion products may be present on the slow-growth area, and may obscure it. Similarly, the surface may, in the absence of a corrosive medium, be a shiny one on which the characteristic clamshell markings are obliterated as a result of intermittent rubbing of the mating portions of the fracture surface. On the other hand, either of these artifacts also indicates that a crack has propagated intermittently.

The rough, fibrous area of the remaining portion of the fracture surface corresponds to the final fracture process. The relative areas of the slow- and fast-fracture regions are a rough measure of the magnitude of the maximum cyclical stress and/or the material's fracture toughness or tensile strength. For example, at a given fracture toughness (if final crack advance is dictated by fracture-mechanics considerations) or tensile strength (if fast fracture is caused by overload), the fast-fracture area increases with the maximum applied stress. Likewise, for a fixed maximum stress, the relative area corresponding to slow crack growth increases with a material's fracture toughness or tensile strength.

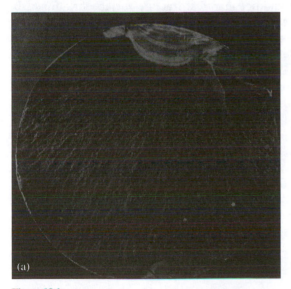 

Figure 12.1
Macrographs of fatigue-fracture surfaces in steel rods. (a) Failure originated at the top edge of this rod. The smooth area with "clamshell" markings corresponds to the slow fatigue-crack growth; the dull, fibrous section is the fast-fracture region. (b) Fracture of this rod originated at the location marked by the arrow (a discontinuity in a cut thread). The slow-crack-growth region here occupies almost the whole of the fracture surface; the fast fracture area (lower right corner) comprises only about 10% of the fracture surface. Magnification, ca. 0.65 ×. (*Metals Handbook, 8th ed., Vol. 9, Fractography and Atlas of Fractographs, ASM International, Materials Park, OH 44073-0002 (formerly American Society for Metals, Metals Park, OH 44073-0002), 1974.*)

The distances between the "ring markings," macroscopically visible on the slow-fracture surface, are not a measure of the crack advance per stress/strain cycle. Since fatigue failures typically occur only after many thousands (sometimes millions) of such cycles, it is clear that the slowly growing crack advances only microscopic distances per cycle. Provided a fatigue-fracture surface is relatively clean—i.e., free from corrosion or oxidation debris and not marred by abrasion—microscopic examination often reveals characteristic *fatigue striations* (Fig. 12.2).[2] As is shown later, the spacing between these is a measure of the slow crack advance per stress/strain cycle. When fatigue is caused by application of a fixed stress cycle, as is frequently the situation in a laboratory test, striation spacings are relatively constant, as in the examples of Figs. 12.2. During fatigue of engineering structures, however, striation spacing varies, reflecting the more or less random application of the cyclical stress causing slow crack growth. The crack-propagation rate is related closely to the stress range (i.e., to the algebraic difference of the maximum and minimum stress in each cycle); the greater this is, the greater the crack advance rate (compare Figs. 12.2*a* and *b*). We also caution that the presence of clamshell markings and/or striations is per se evidence only of intermittent crack growth, and does not necessarily mean that the failure is caused by fatigue of the kind discussed here. Other phenomena (e.g., corrosion fatigue, stress corrosion cracking, and other forms of "static" fatigue discussed in Chap. 13) may also give rise to intermittent crack growth, and can produce fracture-surface characteristics similar to those associated with fatigue. Hence, while it is reasonable in many cases to identify fatigue as a cause of failure on the basis of the presence of clamshell markings/striations, this should not always be done.

Materials can fail by fatigue and not manifest fracture-surface striations. In some cases, noted above, they are obliterated by other processes. In others, they never form. This is the case when the crack advance per cycle is relatively large (as a result of a large stress range and/or high maximum stress). In these circumstances, crack advance is effected by microscopic crack linking of the kind described in our discussions on low-temperature fracture. As a result, the slow-growth fracture area resembles a ductile fracture one.

Fatigue-fracture initiation has not been discussed to this point. There is some reason for this since, at least from a quantitative standpoint, the nucleation event is the least understood of the various aspects of fatigue. A large number of studies devolving on microstructural observations have been conducted, however, and these allow us to qualitatively describe fatigue initiation as it takes place in metals.

Fatigue cracks in metals originate almost exclusively at internal or external surfaces, the latter being more common. In all materials there are regions of local inhomogeneity that result in local "softening" or surface flaws that cause local stress concentration.[3] Either or both of these factors can result in localized plastic flow, which, under the action of a cyclical stress (strain) can produce surface features that

---

[2]Fatigue striations are most easily observed on fracture surfaces developed during laboratory testing. They are much less often found on fatigue-fracture surfaces of parts used in service. This is a result of the factors mentioned above and, in addition, the complexities that arise from the stochastic loading of engineered structures which leads to large differences in the crack advance per cycle.

[3]Indeed, even in materials that are made as uniform as possible the localized plastic deformation that is the cause of fatigue crack initiation is heterogeneously distributed, often in the form of "persistent slip bands" (PSBs). These PSBs are often associated with surface crack initiation, as is shown in Fig. 12.4*a*.

bear a resemblance to a crack or a flaw. A marking of this kind is illustrated schematically in Fig. 12.3a, where surface "extrusions" and "intrusions" arise from the inhomogeneity of microscopic plastic flow. Continued cyclical stressing enhances these surface features and, at some point, a surface crack can be considered nucleated. Note that this nucleation stage, Stage I of fatigue fracture, is dictated by plastic flow, rather than fracture, considerations. As a consequence, the initial fatigue-crack plane normal is not parallel to the principal tensile axis, and the nucleated crack first propagates at an angle other than 90° to this axis. Following these nucleation and initial crack-propagation events, slow crack growth (Stage II of fatigue) occurs when the Stage I crack has grown to some critical size determined by material properties and the applied stress level and state (Fig. 12.3b). As shall be shown, Stage II (slow) crack-growth rates can be correlated with the stress and stress range to an extent that Stage I crack-growth rates cannot. The latter are related to the a priori unpredictable (although sometimes correlatable) microscopic flow morphologies developed during cyclical stressing.

The schematic of Fig. 12.3a can be elaborated on with reference to photographic evidence of these surface manifestations of microscopic plasticity. Figure 12.4a is a photograph of a surface of a single crystal of pure Cu, a material prone to local fatigue deformation in the form of persistent slip bands (PSBs). In this figure

(a)   1 mm   (b)   2 μm

Figure 12.2
Characteristic fatigue striations in (a) copper subjected to a high, and (b) an aluminum alloy subjected to a much lower, stress amplitude. The different striation spacings on the two surfaces (this spacing for Cu is much greater; note the different magnifications of the figures) reflect the different stress amplitudes. *(Part (a) from A. S. Tetelman and A. J. McEvily, Jr., Fracture of Structural Materials, Wiley, New York, 1967; (b) Metals Handbook, 8th ed., Vol. 7, Atlas of Microstructures, ASM International, Materials Park, OH 44073-0002 (formerly American Society for Metals, Metals Park, OH 44073-0002), 1972.)*

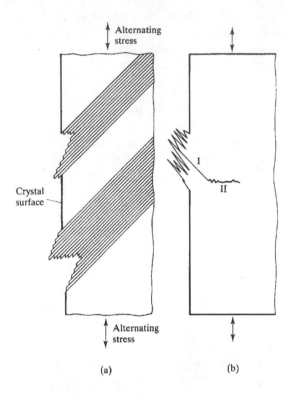

(a)   (b)

Figure 12.3
**(a) Surface intrusions and extru-
sions on a crystal subjected to an
alternating stress. This surface
morphology, not observed during
monotonic loading, arises from
heterogeneous cyclical plastic
deformation. At some point the
feature assumes a crack-like nature, and a Stage I fatigue crack
is considered nucleated. (b) The
Stage I crack propagates in a
direction dictated by plastic-flow
considerations (this direction is
not normal to the stress axis). At
some point the crack alters direction and the propagation direction becomes normal to the
stress axis, and Stage II slow
crack growth commences. (*Part
(a) after R. Reed-Hill, Physical
Metallurgy Principles, 2d ed.,
D. van Nostrand, New York, 1973;
(b) after A. S. Tetelman and A. J.
McEvily, Jr., Fracture of Structural Materials, Wiley, New York,
1967.*)**

the surface protrusions are the PSBs. Closer examination of PSBs indicate that the
protrusions contain sections of material that are "extruded" during cyclic deformation as well as material intrusions. The site of fatigue-crack initiation is on the surface—at the interface between the PSBs and the surrounding material not plastically
deformed in this way. Even when localized deformation is not as heterogeneous as
it is with PSBs, surface intrusions and extrusions are found during cyclic deformation. This is shown by the photograph of a fatigued Cu-Al single crystal in Fig.
12.4b. The "hill and valley" surface morphology illustrated also is intimately linked
with the initiation of Stage I fatigue cracks.

While fatigue-crack initiation can be caused by plastic strain inhomogeneities
inherent to material cyclical strain behavior, initiation can also come about through
other causes of plastic strain inhomogeneity. For example, fatigue cracks have been
observed to nucleate at grain boundaries in polycrystals where localized flow is unaccommodated. In principle, the crack initiation here is rather like that due to high-
temperature crack initiation via unaccommodated grain-boundary sliding (although,
of course, the flow mechanisms resulting in crack initiation are much different for
the two situations). Second-phase particles are known to be sources for crack initiation during low-temperature ductile fracture because of the stress concentration resulting from the different flow behavior of the particles and the matrix. And, as
shown in Fig. 12.5, fatigue-crack initiation can also be facilitated by the presence of
inclusions within the material. Similarly, surface and interior flaws can accelerate
fatigue-crack initiation and shorten fatigue life. This is particularly so when the

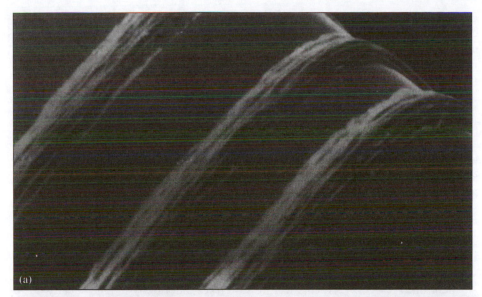

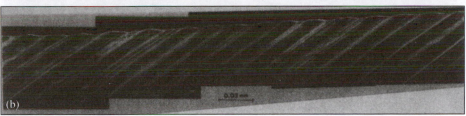

Figure 12.4
(a) Protrusions on the surface of a cyclically strained Cu single crystal. The protrusions are stacks of individual extrusions and intrusions. Cracks initiate at the interface between the protrusions and the surrounding material. (b) The surface of a cyclically strained Cu-Al single crystal. This material plastically deforms differently from the Cu single crystal of (a). This accounts for the different surface features (termed "hill and valley") which, nonetheless, result from cyclical plastic deformation. Surface cracks nucleate at the confluences of the hills and valleys. (*(a) From B. T. Ma and C. Laird, Small Fatigue Cracks, eds. R. O. Ritchie and J. Lankford, The Metallurgical Soc., Warrendale, Pa., 1986; (b) from S. I. Hong and C. Laird, Mat. Sci. and Eng., A128, 155, 1990.*)

time required to develop a reasonably sized fatigue crack is a significant portion of the total material life. Thus, a material with a smooth, polished surface usually exhibits a longer fatigue life than one with a rough surface. Likewise, regions of macroscopic stress concentrations (such as notches) diminish fatigue life.

To summarize, fatigue fracture takes place in three stages. In the first, crack nucleation results from heterogeneous microscopic plastic flow.[4] The Stage I crack so formed propagates in a direction dictated by plasticity, rather than fracture, consid-

---

[4] The current discussion assumes the absence of preexisting cracks within or on the surface of the material. When such cracks are present and when their size is sufficient, there is no longer a need for crack nucleation; the propagation of these preexisting cracks leads to fatigue fracture.

erations. When it reaches a certain size, a Stage II crack develops from it, and this crack propagates at a rate defined principally by the applied stress range and does so in a direction normal to the principal tensile axis. (Further aspects of Stage II crack-growth rates are discussed in Sect. 12.4.) Final fatigue fracture (Stage III) takes place when the slow crack growth has proceeded to the extent that the unfractured cross-sectional area is unable to sustain the maximum applied tensile load. Thus, final fracture is a tensile overload fracture or one caused by the maximum stress being such that the material's fracture toughness is exceeded.

We will return to consideration of fatigue-crack growth, particularly with reference to the slow-crack-growth process. Before doing this, however, we must consider additional aspects of fatigue, especially with regard to how fatigue life is affected by the stress and strain ranges. This is done in the following section.

12.3
EVALUATION OF FATIGUE RESISTANCE

As mentioned in Sect. 12.1, engineering structures that fail by fatigue usually do so as a result of stochastic loading. The underwing skin of an aircraft serves as an example. The tensile stress borne by it during smooth flight can be calculated, and the results of the calculations can be tested experimentally. However, owing to maneuvering and also as a result of turbulence, the actual stress-time history of the wing skin is likely to vary as schematized in Fig. 12.6a. Furthermore, the stress history will be unique to a particular aircraft; others will experience different, but still essentially random, variations in stress. Despite these considerations, systematic

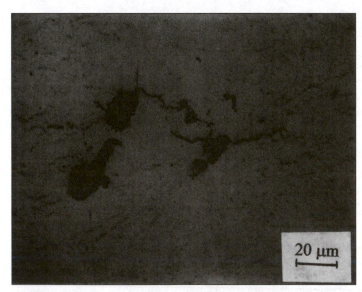

Figure 12.5
A fatigue crack that nucleated at an inclusion (the dark areas of the micrograph) in an alloy steel. *(L. F. Coffin, Jr., M. F. Henry, and L. A. Johnson, original source. Reproduced with permission, from the Ann. Rev. of Materials Sc., 2, 123, © 1972 by Annual Reviews Inc.).*

analysis of a material's fatigue resistance usually involves application of a well-characterized stress/strain cycle, even though this type of testing scheme does not simulate service very well. Perhaps the oldest of such evaluations is the rotating-beam fatigue test (Fig. 12.7a). In this test a load is applied so as to produce a stress $\pm\sigma$ on the tensile and compressively stressed beam surfaces. As the beam rotates, the respective surfaces are interchanged so that each portion of the beam surface experiences an alternating stress of the kind shown in Fig. 12.6b.

Stress parameters needed to characterize the rotating-beam (and other types of) fatigue test are related to the maximum (σ_{max}) and minimum (σ_{min}) stresses the part is subjected to each cycle. These include the mean stress σ_{mean} [$= (1/2)(\sigma_{max} + \sigma_{min})$], the stress range $\Delta\sigma$ (= $\sigma_{max} - \sigma_{min}$), and the stress amplitude σ_a [$= (1/2)(\sigma_{max} - \sigma_{min}) = \Delta\sigma/2$]. Frequently, the stress ratio $R = \sigma_{min}/\sigma_{max}$ is also used; clearly R is redundant provided σ_{mean} and σ_a are known (Prob. 12.1). For the rotating-beam fatigue test, $\sigma_{mean} = 0$, $\Delta\sigma = 2\sigma_{max}$, $\sigma_a = \sigma_{max}$ and $R = -1$.

Only a few of the types of structures prone to fatigue failure experience cycles that are simulated by alternating compressive and tensile stresses of equal magnitude. For these situations, tests other than the rotating-beam test are better suited for assessing fatigue resistance. A cyclical tensile test is one such test (Fig. 12.7b). In it, a specified stress amplitude is cyclically imposed on a finite mean stress; a typical stress-time history for such a test is shown in Fig. 12.6c.

The frequency of the applied stress can have an effect on a material's fatigue resistance. Indeed, this is so for polymeric materials even at relatively low frequencies (Sect. 12.8). However, for metals and their alloys, fatigue behavior depends on cyclical frequency only at high frequencies. If the frequency remains below about 200 Hz, fatigue behavior of metals can be considered frequency independent. Since frequencies of 200 Hz or less encompass the vast majority of engineering

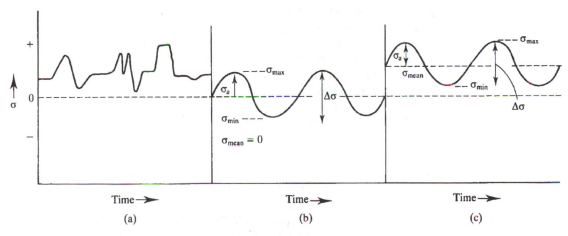

Figure 12.6
Characteristic stress-time variations in (a) a structure subject to random loading superimposed on a positive mean stress, (b) a rotating-beam fatigue test in which the material experiences alternating compressive and tensile stresses of equal magnitude, and (c) a cyclical tension test in which a time-varying stress is superimposed on a constant mean stress. The stress range ($\Delta\sigma = \sigma_{max} - \sigma_{min}$), the stress amplitude ($\sigma_a = \Delta\sigma/2$), and the mean stress (= $(\sigma_{max} + \sigma_{min})/2$) are shown in (b) and (c).

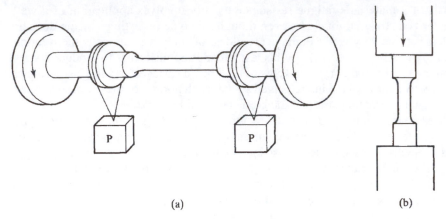

(a)   (b)

Figure 12.7
(a) A rotating-beam fatigue test. A constant load (P) results in a constant moment along the sample gage section. Sample rotation results in a stress-time variation on the sample surface like that shown in Fig. 12.6*b*; i.e., alternating compression/tension with $R = -1$, $\sigma_a = \Delta\sigma/2 = \sigma_{max}$. (b) A cyclical tension test for investigating fatigue behavior for $R \neq -1$. Here a cyclical stress is imposed on a constant mean stress; the stress varies with time as shown in Fig. 12.6*c*. (Note: A cyclical tension-compression test ($R < 0$) can also be performed.) (*Adapted from R. W. Herzberg, Deformation and Fracture Mechanics of Engineering Materials, Copyright 1976. With permission of John Wiley and Sons, Inc.*)

applications, only this frequency range—for which a metal's fatigue response is frequency-independent—is considered in this chapter.

Because of the important role of plasticity in metal fatigue, it is fundamentally more sound to assess a material's fatigue response under conditions of a specified cyclical strain, rather than stress. Tests of this kind are described later, and with modern testing machines they are as convenient to conduct as are stress-controlled fatigue tests. Nonetheless, the latter are traditional, and the results from them are widely used in design against fatigue fracture. What is usually done is to perform a series of tests in which the stress amplitude is varied (at a fixed mean stress), and to measure the number of stress cycles (N_f) necessary to cause fatigue fracture. The results of such a series of tests on two engineering metals are shown in Fig. 12.8; the coordinates here are σ_a (linear scale) and N_f (logarithmic scale). As expected, N_f increases as σ_a decreases. For the steel whose behavior is shown in Fig. 12.8, there apparently exists a stress amplitude below which fracture will not take place regardless of the number of stress cycles. This is called the fatigue (or endurance)[5] limit of the steel; a good "rule of thumb" is that the fatigue limit is about 40% of the tensile strength of the steel when the limit is obtained for $\sigma_{mean} = 0$ ($R = -1$). The aluminum alloy whose fatigue response is illustrated in Fig. 12.8 apparently does not manifest a fatigue limit. That is, this material will always fracture by the fatigue process provided it is subjected to a sufficiently large number of stress reversals,

---

[5] The terms fatigue limit and endurance limit are synonymous and used interchangeably. We do that here (despite the resulting (minor) confusion) to keep pace with engineering practice.

Figure 12.8
Stress amplitude–number of cycles to failure relationships for a plain carbon (1045) steel and an aluminum alloy. The steel manifests an endurance (fatigue) limit; the aluminum alloy does not. The data were obtained for a mean stress of zero. (*Adapted from H. W. Hayden, W. G. Moffatt, and J. Wulff, The Structure and Properties of Materials, Vol. III, Wiley, New York, 1965.*)

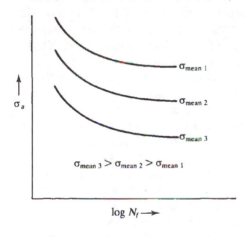

Figure 12.9
Stress amplitude (σ_a, linear) vs. N_f (log) showing the effect of mean stress (σ_{mean}) on the $\sigma_a - N_f$ relationship. As σ_{mean} increases, N_f decreases at a constant σ_a. Moreover, the endurance limit (if one exists) also decreases with increasing σ_{mean}.

although, to be sure, the number of cycles to cause this may be very large indeed. For materials showing a fatigue limit, this is often used as a fatigue design parameter, just as the yield strength is used in design against plastic deformation. For materials not displaying such a limit, an alternative design parameter or an "effective endurance limit" is used in design. Frequently, this is taken as the stress amplitude required to cause failure at a very large number of cycles (e.g., $N_f = 10^7$). In following discussions, the term endurance limit (or fatigue limit) is conveniently used to define this stress amplitude for materials that do not display a true endurance limit.

The endurance limit (as just defined) is altered by the magnitude of the mean stress. It is expected that as σ_{mean} increases, the endurance limit decreases, and in fact this is the case as is shown in Fig. 12.9. Several empirical engineering estimates have been developed that relate the endurance limit to the mean stress. A simple one, which has the advantage of generally being a conservative approximation, is due to Goodman. The Goodman equation is written as

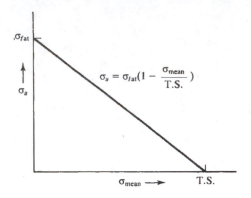

Figure 12.10
**The Goodman relationship (Eq. (12.1))
is graphically represented as a straight
line in which the allowable stress amplitude decreases linearly with the mean
stress. To construct the diagram only
two material datum are needed: the endurance limit (or permissible stress amplitude) at a mean stress of zero, and
the tensile strength.**

$$\sigma_a = \sigma_{fat}\left(1 - \frac{\sigma_{mean}}{T.S.}\right) \tag{12.1}$$

where σ_{fat} is the endurance limit when $\sigma_{mean} = 0$, and T.S. is the material's tensile strength. A graphical representation of Eq. (12.1) is a straight line in a coordinate system using axes of σ_a and σ_{mean} (Fig. 12.10). Equations similar to, but different in form from, Eq. (12.1) have also been developed. Some are considered in Prob. 12.3.

There are deficiencies, over and above those involved with extrapolation of data obtained under uniform cyclic loading to the random cycling encountered in service, in implementing the above approaches in fatigue design.[6] For example, some material use corresponds to a cyclical strain, rather than stress, condition. Of course, cyclical stress and cyclical strain tests are related. For example, during high-cycle fatigue (when the number of cycles to failure is large, typically greater than 10^3), the macroscopic stress level is such that the structure as a whole undergoes only elastic deformation. In this case, the elastic strain range ($\Delta\varepsilon_{el}$) is coupled to the stress range by $\Delta\varepsilon_{el} = \Delta\sigma/E$, where E is Young's modulus. Conversely, in the low-cycle fatigue range (where usually $N_f < 10^3$), the material is typically subject to macroscopic, as well as microscopic, plastic strain. For relatively low values of N_f, the plastic strain range is much greater than the elastic one, so that $\Delta\varepsilon \cong \Delta\varepsilon_{pl}$; $\Delta\varepsilon_{pl}$ is related to $\Delta\sigma$ via the material's cyclic hardening response (Sect. 12.6). In the general case, the total strain range, $\Delta\varepsilon$, is the sum of $\Delta\varepsilon_{el}$ and $\Delta\varepsilon_{pl}$ with, as noted, the latter dominating in low-cycle, and the former in the high-cycle regions. The transition from the low- to high-cycle fatigue regions for metals takes place at a N_f value on the order of 10^3 for which situation, $\Delta\varepsilon_{el} \cong \Delta\varepsilon_{pl}$.

The results from cyclically strain-controlled fatigue tests are plotted schematically in Fig. 12.11 and for several metals in Fig. 12.12. In comparison to Fig. 12.8, the strain, rather than the stress, amplitude is the ordinate in this figure. The other noticeable difference is that $\Delta\varepsilon/2$ is plotted logarithmically, whereas σ_a is plotted linearly when the results from stress-controlled tests are graphically displayed. A number of empirical correlations relating strain amplitude and N_f have been advanced. For example, for high-cycle fatigue ($\Delta\varepsilon \cong \Delta\varepsilon_{el}$), the elastic strain range is correlated to N_f through the equation

---

[6]The design "issue" is returned to in Sect. 12.5, after we have discussed other aspects of fatigue behavior required to establish a design protocol.

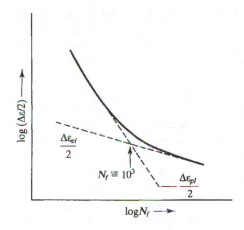

Figure 12.11

In a strain-controlled fatigue test the cyclic strain amplitude ($\Delta\varepsilon/2$) can be related to the number of cycles to failure, N_f. During high-cycle fatigue ($N_f \gtrsim 10^3$) most of the macroscopic strain is elastic, and the slope of log ($\Delta\varepsilon/2$) vs. log N_f is less negative than it is during low-cycle fatigue ($N_f \lesssim 10^3$). For the latter, most of the applied strain is permanent. Note that $\Delta\varepsilon/2 = (\Delta\varepsilon_{el} + \Delta\varepsilon_{pl})/2$; $\Delta\varepsilon_{el} > \Delta\varepsilon_{pl}$ during high-cycle fatigue and $\Delta\varepsilon_{pl} > \Delta\varepsilon_{el}$ for low-cycle fatigue.

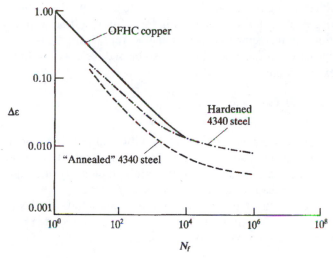

Figure 12.12
Relationship between strain range ($\Delta\varepsilon$) and N_f for several engineering metals. The low-cycle behavior of the differently heat-treated alloy steels does not differ all that much. However, the hardened 4340 steel manifests longer fatigue lifetimes during high-cycle fatigue. The ductile copper alloy, which work hardens rapidly, is a superior fatigue-resistant material for low-cycle applications. However, it would perform poorly compared to either of the steels in high-cycle applications. (*Data from R. C. Boettner, C. Laird, and A. J. McEvily, Jr., Trans TMS-AIME, 233, 379, 1965, and S. S. Manson, Exp. Mech., 5, 193, 1965.*)

$$\frac{1}{2}\Delta\varepsilon_{el} = \frac{\sigma_f'}{E}(2N_f)^{-b} \qquad (12.2)$$

where E is the modulus, and σ_f' and b are determined by curve fitting. The parameter σ_f' has been found to scale with the material's tensile strength (Prob. 12.4). The implication, therefore, is that a strong material is best in high-cycle fatigue applications (i.e., for a given $\Delta\varepsilon$, N_f increases with σ_f'). This is consistent with the view that a significant portion of a material's high-cycle fatigue life is spent in nucleating cracks. Engineering design recognizes this aspect of high-cycle fatigue. Since crack nucleation is caused by local plastic deformation, usually on the material's surface, increases in surface strength can delay crack nucleation. Thus, surface shot peening (cold working) is used to improve high-cycle fatigue behavior and so are specialized chemical and thermal surface treatments (e.g., carburizing, nitriding, and surface martensite formation in steels) which increase the surface yield strength.

The exponent b can be related to the material's cyclic work-hardening coefficient (n'); n' is determined through the empirical relation $(\Delta\sigma/2) = K'(\Delta\varepsilon/2)^{n'}$, appropriate to this kind of loading (Sect. 12.6). The coefficient b varies with n' as $b \cong n'/(1 + 5n')$. Since a larger value of b produces a more negative slope in the $\Delta\varepsilon/2 - N_f$ plot, it is clear that a material having a high cyclical work-hardening coefficient is not as useful for high-cycle fatigue as is one with a low value of n'. On the other hand, n' is usually of minor importance compared to σ_f' in defining high-cycle fatigue life. As noted, proper use of the relationship between b and n' requires knowledge of the cyclical hardening response. In the absence of such information, the uniaxial work-hardening exponent, n, can be used in lieu of n' in a first estimate of b; however, the limitations on this approximation must be kept in mind.

Empirical relationships between $\Delta\varepsilon_{pl}$ and N_f have also been developed for low-cycle fatigue. One such expression is

$$\frac{1}{2}\Delta\varepsilon_{pl} = \varepsilon_f'(2N_f)^{-c} \tag{12.3}$$

where c is typically 0.5–0.7 and correlates approximately with n' as $c \cong 1/(1+5n')$. Additionally, ε_f' is close to the tensile ductility. Thus, for design against low-cycle fatigue it is desirable to have a material that manifests both high ductility and high work hardening, for each of these features leads to increased N_f values.

Since Eqs. (12.2) and (12.3) adequately represent the strain amplitude–cycles to failure relationship at the "extremes" of high-cycle and low-cycle fatigue, it is reasonable to assume that the general relationship between $\Delta\varepsilon$ (= $\Delta\varepsilon_{el} + \Delta\varepsilon_{pl}$) and N_f is the sum of them. That is,

$$\frac{1}{2}\Delta\varepsilon = \frac{1}{2}\Delta\varepsilon_{el} + \frac{1}{2}\Delta\varepsilon_{pl} = \frac{\sigma_f'}{E}(2N_f)^{-b} + \varepsilon_f'(2N_f)^{-c} \tag{12.4}$$

This relationship is schematically plotted in Fig. 12.13, where we see, as expected, that when N_f is low and $\Delta\varepsilon \cong \Delta\varepsilon_{pl}$, Eq. (12.4) reduces to Eq. (12.3), whereas when N_f is large and $\Delta\varepsilon \cong \Delta\varepsilon_{el}$, Eq. (12.4) becomes Eq. (12.2). Application of the approach exemplified by Eq. (12.4) to engineering metals has been relatively successful; examples are provided in Fig. 12.14.

We conclude with a reminder that different material properties are desirable for low- versus high-cycle fatigue applications. A high ductility is wanted for low-cycle fatigue, because fatigue cracks are nucleated early on in the material's life. Since slow crack advance is inhibited by a material that work hardens adequately and manifests good malleability, these properties are needed for low-cycle fatigue applications. In contrast, crack nucleation should be delayed for materials intended to perform safely over a large number of stress or strain reversals. This is accomplished by using high-strength materials. These remarks are summarized in the schematic of Fig. 12.15 which sketches $\Delta\varepsilon/2$-N_f curves for both "strong" and

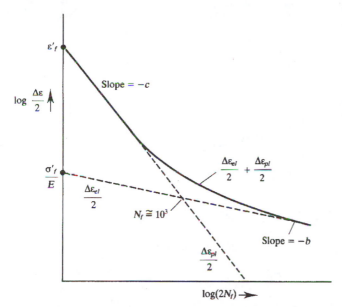

Figure 12.13
A schematic of the strain amplitude–number of stress (strain) reversals as given by Eq. (12.4). At low N_f values, $\Delta\varepsilon \cong \Delta\varepsilon_{pl}$, and the slope of this line on logarithmic coordinates is $-c$ with an intercept (at log $(2N_f) = 0$) of $\varepsilon_f{}'$. At high cycles the logarithmic slope is $-b$ ($b < c$), and the extrapolation of this portion of the line (where $\Delta\varepsilon \cong \Delta\varepsilon_{el}$) produces the intercept $\sigma_f{}'/E$.

"ductile" materials. From the above discussion, strong materials are better for fatigue applications for high, and ductile ones for low, values of N_f. The "crossover" point is typically in the range of $N_f \cong 10^3$. Optimum materials performance over the whole range of strain amplitudes[7] is obtained by use of the "tough" material whose behavior is also indicated in Fig. 12.15. Such a material can be made by reducing the strength level and enhancing the ductility of the "strong" material or by increasing the strength (usually at the expense of the ductility) of the "ductile" one. It is clear that this "ideal" fatigue-resistant material manifests both high strength and ductility. On this basis, criteria for selection of a material resistant to fatigue do not differ from criteria used for selecting a material resistant to tensile fracture.

It has been noted that slow crack growth often occupies a large fraction of a material's low-cycle fatigue life. Further, many structures can be presumed to contain preexisting cracks (recall our discussion of fracture mechanics, Chap. 9). Because of these factors, techniques that monitor Stage II crack growth rates have been developed. Results provide data suitable for estimating fatigue lifetimes in certain situations. The tests are also useful for shedding light on the mechanisms responsible for, and the material properties' relationship to, Stage II crack growth. Thus, the slow-crack-growth problem is discussed in the following section. Following this, we are in a position to discuss engineering design against fatigue.

---

[7]Of course, this is seldom required in a specific engineering application.

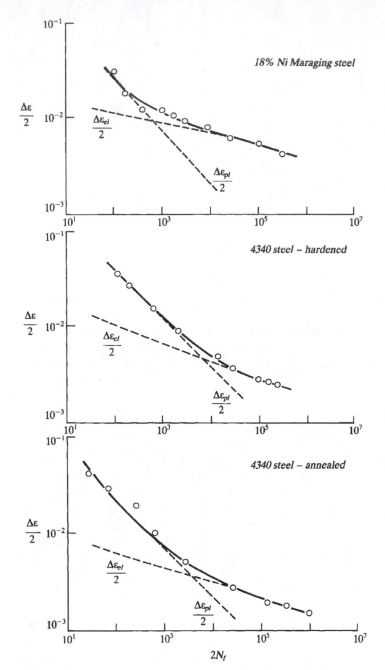

Figure 12.14
Strain amplitude ($\Delta\varepsilon/2$)-stress reversal ($2N_f$) relationships for several ferrous alloys.
The solid line represents Eq. (12.4); the circles are experimental data. The dotted
lines represent elastic and plastic strain amplitudes. These strain amplitudes are
comparable when $N_f \cong 10^3$. (*Data from S. S. Manson, Exp. Mech., 5, 193, 1965, and
R. W. Landgraf, ASTM STP 467, Philadelphia, Pa., 1970, p. 3.*)

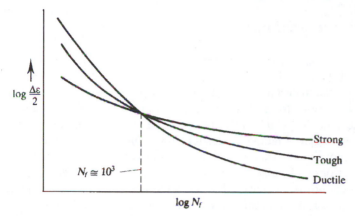

Figure 12.15

Schematic of strain amplitude–cycles to failure relationships as generally observed for "strong" (but not ductile) alloys, "ductile" (but not strong) alloys, and "tough" materials (those having both reasonable strength and ductility). Strong alloys are best for high-cycle fatigue applications, since it is more difficult to nucleate fatigue cracks in them (the crack-nucleation process takes up much of high-cycle fatigue life). Ductile alloys are better for low-cycle fatigue because it is more difficult to propagate cracks in them (Stage II crack propagation occupies most of the low-cycle fatigue life). A "tough" alloy is a good compromise when both high- and low-cycle fatigue resistance is necessary (although this is seldom a design requirement). (*Adapted from R. W. Landgraf, ASTM STP 467, Philadelphia, Pa., 1970, p. 3.*)

EXAMPLE PROBLEM 12.1. Refer to Fig. 12.14. For the several steels whose behavior is illustrated, estimate the number of cycles to failure at total strain ranges of 5×10^{-2}, 10^{-2}, 5×10^{-3}, and 10^{-3}. Comment on your results in terms of material properties affecting high- vs. low-cycle fatigue behavior.

Solution. The following information was obtained from Fig. 12.14. (Some data were obtained by extrapolation at high and low values of $2N_f$.)

	$2N_f$		
$\Delta\varepsilon/2$	**Maraging steel**	**4340 hardened steel**	**4340 annealed steel**
5×10^{-2}	37	75	21
10^{-2}	170	180	560
5×10^{-3}	1.3×10^5	1.3×10^4	3.2×10^3
10^{-3}	3.2×10^9	8.7×10^7	1.0×10^7

The Maraging steel is a much superior material for high-cycle fatigue applications; being the strongest of the three materials, it resists fatigue-crack nucleation. For low-cycle fatigue conditions, the Maraging steel is slightly inferior to the hardened 4340 steel. It is somewhat surprising that the annealed 4340 steel is not best for low-cycle applications (although it is the best material at a total strain amplitude equal to 10^{-2}). At high *total* strain amplitudes, the *plastic* strain range is greater for the annealed steel than for the other materials. This likely accounts for its inferior performance at the highest total strain range.

12.4
FATIGUE-CRACK GROWTH RATES

Measurements of Stage II crack growth rates are useful in design, and they also add to our understanding of the fatigue process. For example, knowledge of the Stage II crack growth rate and the material's fracture toughness permits estimation of the number of Stage II cycles prior to catastrophic final fracture. Thus, for a material subject to low-cycle fatigue, for which Stage II occupies a major portion of the material's fatigue life, the number of fatigue cycles it can withstand prior to failure can be approximated. Moreover, as discussed in Chap. 9, many structures contain pre-existing surface or interior cracks that can be precursors to fatigue (as well as tensile) failure. The presence of such cracks eliminates the necessity of nucleating a fatigue crack, and for materials containing preexisting cracks, knowledge of their flaw size and geometry permits estimation of fatigue life.

Measurement of Stage II crack growth rates is now readily performed. A pre-cracked sample of the kind used for fracture mechanics tests (cf. Chap. 9) suffices also for fatigue-crack growth rate measurements.[8] The sample is typically subjected to a fixed stress (or in some cases, strain) amplitude at a specified stress ratio, and the crack length is monitored as a function of the number of cycles. Crack length can be measured in a variety of ways, including direct measurement with a microscope or by measurement of the electrical resistance across the fractured portion of the sample, accompanied by a suitable calibration. The results obtained from such testing are illustrated schematically in Fig. 12.16. Here, c, the length of a surface fatigue crack, is plotted versus the number of stress cycles. As shown by the figure, the rate of crack advance (the slope of the c-N curve) increases continuously with the number of cycles; more precisely, the rate of crack advance increases with crack size. This result suggests that Stage II fatigue-crack advance is associated with concepts related to stress intensity, just as is the propensity for tensile fracture. Figure 12.16 also shows that, for a fixed stress range, crack advance is promoted by higher values of the stress ratio, R. This points out the important effect that the mean stress and the maximum tensile stress have on fatigue-crack propagation.

As noted, the slopes of the curves of Fig. 12.16 are a measure of the fatigue-crack growth rate, dc/dN. In some situations dc/dN is to fatigue design as void-growth rates are to design against creep fracture. Studies have concluded that the Stage II crack growth is driven principally by the same kinds of forces responsible for tensile fracture. For the latter, the "driving force" is the stress intensity, which scales with the product of stress and the square-root of crack length. When extended to fatigue fracture, the same approach is taken with the exception that, in recognition of the necessity of a cyclical stress for fatigue, the stress range, $\Delta\sigma$, substitutes for σ in the stress-intensity factor. A large number of studies has shown that, for a given material and stress ratio, Stage II crack growth rates are a unique function of ΔK ($\sim \Delta\sigma(c)^{1/2}$), and that over an appreciable range of this variable, dc/dN is related to it by

$$\frac{dc}{dN} \sim A(\Delta K)^m \tag{12.5}$$

---

[8]The size of the preexisting crack must well exceed a critical microstructural scale (e.g., the grain size) in order for the analysis that follows to be valid.

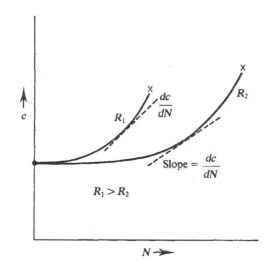

Figure 12.16
Schematic of crack length–number of cycles relationship. Crack growth rate increases as crack length increases. This rate is measured by the instantaneous slope of the curves. These schematics are constructed for a fixed stress amplitude, but the stress ratio R ($= \sigma_{min}/\sigma_{max}$) differs between the two curves. Crack growth rates increase and fracture (denoted by X) takes place at smaller crack sizes with increasing R values.

where A is a constant that depends on the material and the stress ratio, and m is an empirical constant ($2 \leq m \leq 7$, usually).

Although dc/dN may be uniquely related to ΔK, it must be emphasized that Eq. (12.5) applies only over a portion of the dc/dN-ΔK curve. This is illustrated in Fig. 12.17, which shows both schematic plots of dc/dN-ΔK and measured values of this relationship for some engineering metals. From these figures, we see that at low ΔK values, crack growth rates are very low (Region I, Fig. 12.17a). Indeed, there is a threshold value of ΔK ($= \Delta K_{th}$) below which fatigue cracks essentially do not propagate. This value, ΔK_{th}, represents inherently safe design against fatigue as is discussed in Sect. 12.5. For many metals, however, ΔK_{th} is much less than their fracture toughness so that employing these metals at such low ΔK values places a great restriction on their utilization. Thus, design in such cases accepts the presence of fatigue cracks, assumes some of them will grow, but also ensures that they do not grow to catastrophic length during the intended lifetime of the part. Ceramics, and many polymers as well, have values of ΔK_{th} which are close to their K_{Ic} values. When this is so, a different design approach is taken. Further discussion of these matters is deferred to the next section.

As ΔK increases, K_{max} ($\sim \sigma_{max}(c)^{1/2}$) approaches the material's critical fracture toughness, and Stage II fatigue cracks propagate at an ever-increasing rate as K_{Ic} is approached (Region III, Fig. 12.17a). During this "near-final" stage of fatigue-crack propagation, tensile failure occurs concomitantly with fatigue-crack advance.

As mentioned, increases in the stress ratio (R) increase the crack growth rate; the effect is most pronounced within Regions I and III of Fig. 12.17a, i.e., at low and high values of ΔK. During Region II, the R value has a lesser effect on crack growth rate. In fact, for R values less than zero (i.e., when there is a compressive stress applied during some portion of the stress cycle), the stress ratio has a minimal influence on crack propagation rates. The situations are compared in Figs. 12.18a and b.

Changes in fracture morphology frequently parallel changes in crack growth rates. As mentioned, at high ΔK values, crack advance is often accomplished in a tensile-like mode, and fatigue striations are not prominent. Because of the change in crack-advance mechanism at high ΔK values, thought has been given to, and evidence advanced for, relating crack advance rates to the plastic strain range, $\Delta\varepsilon_{pl}$, rather than to the stress range in this crack growth stage. This has met with considerable success, and is consistent with the idea that low-cycle fatigue is inherently related to the plastic strain range. (The high values of dc/dN associated with high ΔK values are often characteristic of low-cycle fatigue.)

At low ΔK values, characteristic fatigue striations are also frequently not evident on a fracture surface or, if they are present, they are often obscure, in part because of a fine striation spacing. During Region II, fatigue striations are often apparent. Indeed, measurement of fatigue striation spacings has been used to verify the accuracy of dc/dN values obtained by other methods. A number of mechanisms have been advanced to account for the characteristic appearance of fatigue striae. One that is physically plausible and conceptually appealing is shown in Fig. 12.19.

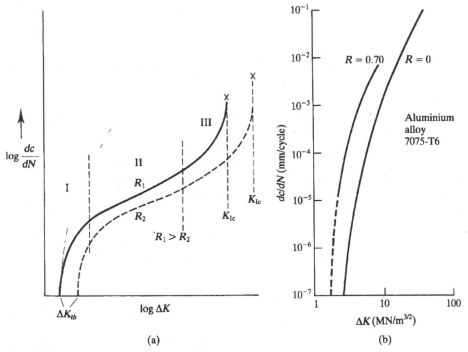

(a) (b)

Figure 12.17
(a) Schematic of crack-growth rate as a function of the cyclical stress intensity factor ΔK ($= \alpha\Delta\sigma c^{1/2}$) for different R values. At very low ΔK values, dc/dN is quite low; so much so that ΔK_{th}, a stress-intensity factor below which Stage II cracks will not propagate, can be identified. As R is increased, ΔK_{th} decreases. In Region II the crack growth rate–ΔK relationship is described by the power-law Eq. (12.5), which yields a straight line on the logarithmic coordinates of this figure. As in Region I, dc/dN increases with R, but in Region II dc/dN is less sensitive to R. Crack growth rates are also sensitive to R in Region III, where high ΔK values promote rapid crack growth rates. Fracture (marked by X) takes place when $K_{max} = K_{Ic}$; for a given stress amplitude, K_{Ic} decreases as R increases. (Alternatively, final fracture may be a tensile fracture.) (b) dc/dN vs. ΔK for a 7075-T6 aluminum alloy. Only the first two of the three regions schematized in (a) are evident here. (Data from C. M. Hudson, NASA TN D5390, 1969.)

In it, a crack with characteristic striation spacing, x, has the configuration shown in Fig. 12.19a at the stage of the stress cycle when $\sigma = 0$. During tensile loading (Fig. 12.19b), the crack opens up and slip is concentrated in bands at the crack tip. With increased loading, the crack becomes blunted and advances (Fig. 12.19c). On reversing the loading (Fig. 12.19d), the slip at the crack tip is also reversed and the crack begins to close and the new crack surface just created is folded into the crack tip. This produces the initial "double notch" morphology (Fig. 12.19e), and the process then repeats itself (Fig. 12.19f). The cycle accomplishes crack advance by the distance x, the striation spacing. It should be clear that on this model, x will increase with the stress range (or more precisely, ΔK). It is clear, too, that stress reversal is necessary to alter the crack-tip shape; that is, to remove the blunting and to resharpen the crack so as to accomplish crack advance.

Knowing the dc/dN-ΔK relationship allows determination of the number of cycles spent in Stage II crack growth. This requires knowledge of the critical crack length, c_f, required to cause final fracture as well as that of the preexisting (or as nucleated) crack length, c_0. In the general case, the analysis is done numerically, since dc/dN cannot be expressed analytically over all values of ΔK pertaining to Stage II growth. The general expression for the number of cycles in Stage II (N_{II}) is thus

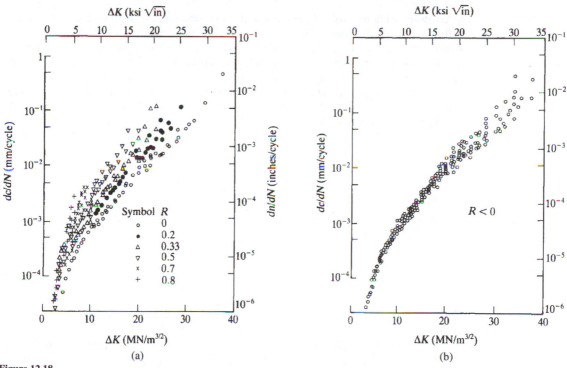

Figure 12.18
Fatigue-crack growth rate–ΔK relationships in 7075-T6 aluminum alloys as affected by the stress ratio (R) for (a) $R \geq 0$ and (b) $R < 0$. For $R \geq 0$, dc/dN increases with R at a fixed ΔK. Conversely, when $R < 0$, dc/dN depends primarily on ΔK and only secondarily on R. Indeed, for $R < 0$ stress ratio only has an obvious effect on dc/dN in Region III where crack advance often takes place by ductile rupture, in which circumstance σ_{max} (or K_{max}) predominately influences fatigue-crack growth rates. (Data from C. M. Hudson, NASA TN D5390, 1969.)

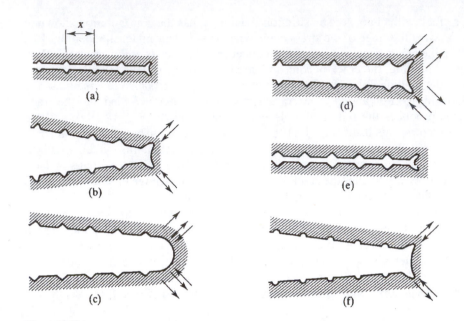

Figure 12.19
**Schematic mechanism of fatigue-striation formation and its relationship to crack ad-
vance. (a) A fatigue crack with characteristic striation spacing, x. (b) As the tensile
stress is increased, the crack opens and plastic slip bands develop at the crack tip.
(c) With further loading the crack tip blunts. (d) On stress reversal, crack-tip slip is
reversed, and the plastically deformed material folds into the crack tip. (e) When
the stress returns to zero, the crack has advanced the characteristic distance, x. The
process then repeats (f). (Note: the stress axis is vertical.)** (*Adapted from C. M. Laird,
ASTM STP 415, Philadelphia, Pa., 130, 1966.*)

$$N_{II} = \int_{c_0}^{c_f} \frac{dc}{dc/dN} \tag{12.6}$$

Estimates for N_{II} can be made by using dc/dN, as expressed in the form of Eq.
(12.5). When this is done, Eq. (12.6) becomes

$$N_{II} = \int_{c_0}^{c_f} \frac{dc}{A\alpha^m(\Delta\sigma(c)^{1/2})^m} \tag{12.7}$$

In Eq. (12.7), $\Delta K (= \alpha\Delta\sigma(c)^{1/2})$ varies as the crack grows and also varies because
the parameter α often depends on crack depth vis-à-vis sample dimension (cf. Chap.
9). However, there are a few geometries for which α does not vary, and for others
a suitable average value of α can be defined. Calling this $\bar{\alpha}$, integration of Eq.
(12.7) yields (for $m \neq 2$),

$$N_{II} = \frac{A^{-1}(\bar{\alpha}\Delta\sigma)^{-m}}{(m/2) - 1}\left[(c_0)^{1-(m/2)} - (c_f)^{1-(m/2)}\right] \tag{12.8a}$$

When $c_f \gg c_0$ (as is usually the situation), Eq. (12.8a) can be approximated as

$$N_{\text{II}} \cong \frac{A^{-1}(\overline{\alpha}\Delta\sigma)^{-m}}{(m/2) - 1}(c_0)^{1 - (m/2)} \qquad (12.8b)$$

or $N_{\text{II}}(\Delta\sigma)^m$ is equal to a material specific constant. Equation (12.8b) is useful in several ways. It can, for example, be used as a (nonconservative) estimate for fatigue life assuming that preexisting cracks of known size are present (Prob. 12.10). Likewise, when Stage II occupies the greatest portion of a material's fatigue life, it can be used to correlate N_f with $\Delta\sigma$. This is the case, for example, for high-cycle fatigue. Equation (12.2), used to correlate the stress range with N_f, is of the form $N_f(\Delta\sigma)^m = $ constant (Prob. 12.11), and this empirically reflects the fundamental basis of Eqs. (12.8).

It is somewhat surprising, but nevertheless true, that microstructure plays a relatively minor role in determining Stage II fatigue-crack growth rates of metals. Attempts to relate these rates with structure-dependent properties have met with only limited success. On the other hand, correlation of crack growth rates with the elastic modulus, a structure-insensitive property, is better. This is shown in Figs. 12.20a and b. In the former, crack growth rates vs. ΔK for several different metal classes are shown. In Fig. 12.20b, these same rates are plotted vs. $\Delta K/E$; it is clear that a significant relationship between modulus and crack growth rate holds.

An exception to the generality concerning microstructure and crack growth rates occurs at high ΔK values. In this case $K_{\max}$ is so close to K_{Ic} that microstructural features promoting high fracture toughness also reduce crack growth rates. This is consistent with earlier conclusions that ductile, tough materials are best for low-cycle (i.e., high ΔK) fatigue. Also, we note that while microstructure plays a minor role in high-cycle fatigue-crack propagation, this *does not* mean it has an insignificant effect on fatigue properties *in toto*. For high-cycle fatigue, strong materials are desired because crack initiation is delayed in them. Once formed, these cracks grow at a rate primarily dependent on ΔK and approximately independent of microstructure.

Perhaps one reason for the lack of correlation of structure and tensile flow behavior with fatigue-crack propagation rates is that materials respond differently to cyclically imposed stresses and/or strains from the way they do to tensile ones. In fact, the cyclical flow behavior of materials, say heat treated to quite different yield strength levels, usually differs much less than their tensile flow behavior. The cyclical flow curve is important with respect to understanding of fatigue mechanisms. So we address the matter in Sect. 12.6. However, even though this topic has not been yet treated (nor has the fatigue behavior of polymers), the discussion of this section and that of the preceding one has us in a position to treat in more detail engineering design against fatigue fracture and material properties that affect fatigue-fracture resistance. This is done next.

12.5
DESIGN AGAINST FATIGUE

In this section we discuss design against fatigue on the basis of (1) whether or not preexisting cracks are present in the material and (2) the resistance to fatigue manifested by the various material classes. As means of introducing the topic, let us

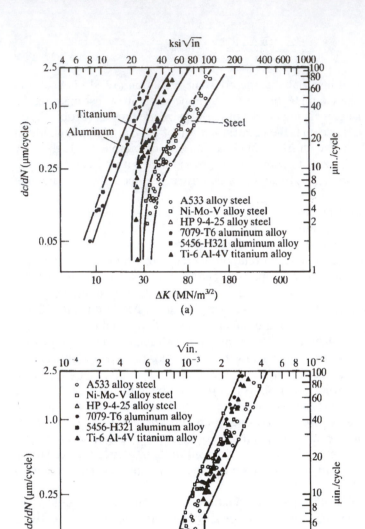

Figure 12.20

(a) *dc/dN* vs. Δ*K* for some titanium, aluminum, and steel alloys. (b) The data of (a) replotted as *dc/dN* vs. Δ*K/E*. Normalizing Δ*K* by the modulus produces a curve (with some scatter) in which crack growth rates in the several materials cannot be as clearly differentiated as they are in (a). This suggests that, in contrast to most mechanical properties, fatigue-crack growth rates are not structure sensitive. (*From R. C. Bates and W. C. Clark, Jr., Trans. ASM, 62, 380, 1969.*)

recapitulate some of the important related concepts that have been treated in the previous two sections. Section 12.3 dealt with fatigue lives for tensile laboratory samples. Such samples are carefully prepared. They often have their surfaces polished, and care is also taken that internal flaws or cracks are not present in the samples. Thus, these materials are the fatigue equivalent of tensile bars similarly processed and the process of fatigue fracture in them is the fatigue equivalent of tensile fracture of flaw-free materials. Fatigue lives under these conditions are given by Eqs. (12.2)–(12.4) and, provided the stress amplitude remains below the endurance limit,[9] the material can be considered immune to fatigue fracture.

What if materials contain preexisting cracks? Their propagation can lead to fatigue fracture provided their value of ΔK_{th} is exceeded. This requires a threshold stress range ($\Delta \sigma$) on the order of $\Delta K_{th}/\alpha(c_0)^{1/2}$ for cracks of initial length c_0. If the threshold stress range is not exceeded, materials are not prone to propagation of preexisting cracks; i.e., they will not fail by fatigue. Obviously, the critical stress range associated with this "fail safe" design depends on the initial crack size. Further, this fatigue design mode is the fatigue equivalent of design against tensile fracture using a fracture mechanics approach.

But the parallels do not end here. We can think of the "smooth" (i.e., unflawed) bar as succumbing to fatigue by nucleation of cracks via heterogeneous deformation (cf. Sect. 12.2) in much the same manner as unflawed tensile bars fracture as a result of cracks similarly nucleated. And, as with the transition in tensile fracture mode (from fracture due to propagation of preexisting cracks to that caused by propagation of cracks generated by plastic deformation), something similar goes on in fatigue. This can be illustrated by Fig. 12.21a. Here, the endurance limit (recall this is a stress amplitude) and $\Delta \sigma_{th}$ are plotted as a function of preexisting crack size. The endurance limit, of course, does not depend on this size; therefore, it is represented by a horizontal line in Fig. 12.21a. But $\Delta \sigma_{th}$ increases with reductions in initial crack size ($\Delta \sigma_{th}$ varies with $c_0^{-1/2}$). The intersection of the two stress amplitudes corresponds to a transition in fatigue failure mode from "extrinsic" fracture (i.e., due to the propagation of preexisting cracks) to "intrinsic" fracture (due to propagation of cracks nucleated by plastic deformation). That is, fracture dictated by the endurance limit is crack-initiation-limited fracture; that controlled by $\Delta \sigma_{th}$ is crack-growth-limited failure.

Application of stress amplitudes that lie within the envelope of the two curves of Fig. 12.21a represents intrinsically safe design against fatigue fracture. But sometimes it is not feasible to design in a "fail safe" manner. Such a situation might arise if a material is needed to serve only for a number of cycles less than those (e.g., $N_f \cong 10^7$) associated with the endurance limit. Or perhaps ΔK_{th} is sufficiently low so that fail safe design at the corresponding $\Delta \sigma_{th}$ value would seriously limit material use. When materials are used at stress amplitudes lying outside the "fail safe" envelope, it is still possible to employ them safely. For example, as indicated in Fig. 12.21a, material lifetimes can be estimated with Eq. (12.6) or Eq. (12.4), depending on whether we are dealing with extrinsic or intrinsic fatigue failure. In

---

[9]To keep our treatment simple, fully reversed loading, i.e., $R = -1$, is treated here. The Fleck et al. reference at the chapter end provides some guidelines for considering situations for other R values. Example Prob. 12.3 also deals with the issue. And Prob. 12.13 should enable the student to figure this matter out by herself.

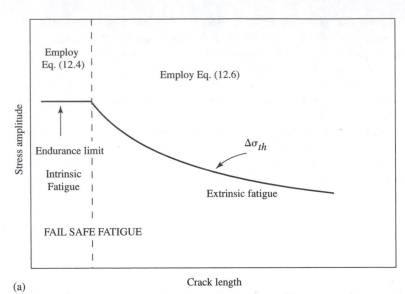

Figure 12.21
(a) Allowable stress amplitude vs. crack size for "fail safe" fatigue. At "large" crack sizes, the allowable stress amplitude is determined by ΔK_{th}. This stress amplitude increases with decreasing crack size. However, when the crack size becomes sufficiently small, "intrinsic" fatigue cracks nucleate and their propagation leads to fatigue fracture. The stress amplitude below which this will not occur is the endurance limit. Employing stress amplitudes lying within the envelope of the two curves constitutes fail safe fatigue design. Materials can be used in non-fail-safe conditions. When they are, equations can be employed to estimate fatigue lifetimes. As shown here, which equation to use depends on whether fracture occurs by propagation of preexisting cracks (extrinsic fatigue) or by the nucleation and propagation of fatigue cracks (intrinsic fatigue). **(continued)**

either case, design incorporates appropriate safety factors so that the material fatigue life well exceeds the intended lifetime. In addition, periodic inspection (as is common to the aircraft industry) adds a further safety component.

Figures 12.21b and c elaborate on intrinsic/extrinsic fatigue fracture design. Figure 12.21b illustrates a situation where the endurance limit (σ_e) is increased for a constant value of ΔK_{th}. It is clear that increasing σ_e increases the range of crack sizes over which design is dictated by crack propagation considerations. Figure 12.21c schematizes the situation for which σ_e is fixed, but ΔK_{th} is increased. For this situation, the range of crack sizes over which crack initiation determines fatigue life is increased. Thus, increasing ΔK_{th} results in greater consideration being paid to crack initiation; increasing σ_e results in more consideration being given to crack propagation.

The above considerations are summarized in the form of a material property chart in Fig. 12.22. Here, ΔK_{th} is plotted (logarithmically) vs. σ_e (also logarithmically). As indicated on the diagram, high values of σ_e and low values of ΔK_{th} correspond to crack-growth-limited (extrinsic) fatigue. Conversely, high values of ΔK_{th} and low values of σ_e correspond to intrinsic fatigue fracture. Material classes cluster in balloons in this figure just as they did in previous material property charts (e.g., Fig. 9.18). On first glance there does not seem to be much difference between

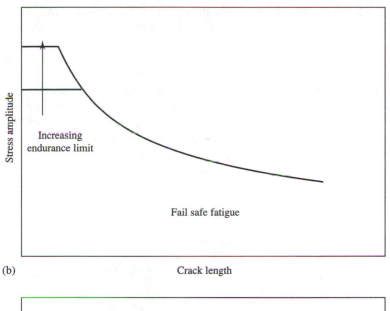

(b)   Crack length

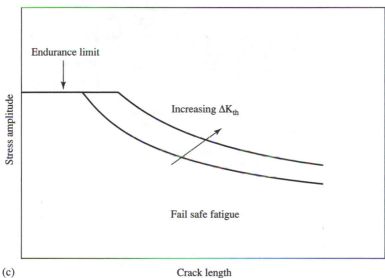

(c)   Crack length

Figure 12.21 (concluded)
(b) Increasing the endurance limit while maintaining ΔK_{th} constant, increases the range of allowable stress amplitudes for fail safe intrinsic fatigue. It also reduces the range of preexisting crack sizes over which this form of fracture dominates fatigue. (c) Increasing ΔK_{th}, while keeping the endurance limit fixed, permits a greater allowable stress amplitude for fail safe extrinsic fatigue. However, it reduces the range of crack size over which this form of fatigue causes fracture.

the ratios of ΔK_{th} to σ_e among the material classes. Closer inspection of Fig. 12.22 reveals that this is not so. Engineering ceramics have high values of σ_e and low values of ΔK_{th}. Thus, their fatigue behavior is dictated by crack-propagation considerations. Moreover, as was mentioned before, most ceramic parts contain residual porosity, microcracks, etc. Thus, fatigue fracture of ceramics is seldom of the intrinsic variety. Metals and polymers behave differently than ceramics. Tough materials of either class (e.g., nylons for polymers, and Cu alloys and steels for metals) often fail in fatigue by propagation of cracks nucleated in them. On the other hand, less tough materials (e.g., PMMA for polymers, and Mg alloys for metals) are more prone to fatigue fracture by propagation of preexisting cracks.

The apparent correlation between "toughness" and the relative tendency to intrinsic or extrinsic fatigue fracture bears further investigation. The material property chart of Fig. 12.23 plots (again logarithmically) ΔK_{th} vs. fracture toughness, K_{Ic}. First, the ceramics. Their ΔK_{th} values are only slightly less than their K_{Ic} values (see the line marked $\Delta K_{th}/K_{Ic} = 1$ in the figure). What this implies is that for

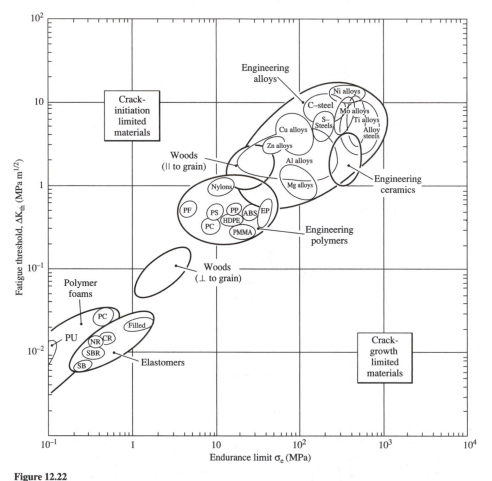

Figure 12.22
A material property chart displaying the fatigue threshold stress intensity (ΔK_{th}, obtained at $R = 0$) vs. endurance limit (σ_e, appropriate for $R = -1$). Although these two properties correlate for the several material classes, there are some subtleties. Ceramics, for example, have relatively high values of the ratio $\sigma_e/\Delta K_{th}$. Thus, they are more prone to crack-growth-limited fatigue fracture (extrinsic fatigue, cf. Fig. 12.21). Conversely, materials having high values of ΔK_{th} vis-à-vis σ_e (e.g., some of the tough metals) are more prone to intrinsic fatigue, which involves nucleation of the fatigue cracks that result in fracture (also see Fig. 12.21). (Adapted from N. A. Fleck, K. J. Kang, and M. F. Ashby, "The Cylic Properties of Engineering Materials," Acta Metall. et Mater., 42, 365, Copyright 1994, with permission from Elsevier Science.)

this material class, we may just as well design against tensile fracture due to the propagation of preexisting cracks and not even bother with fatigue. Or to put it another way, if a preexisting crack is not sufficiently large to propagate when the maximum value of applied tensile stress in a fatigue cycle is attained, such a crack will grow to a critical size in relatively few cycles. So design against static fracture is the rule for ceramics, and this accounts for the statement made in the chapter introduction that "ceramics are not considered prone to fatigue." The tougher of the metals (e.g., Cu and some steels) have low values of ΔK_{th} compared to their fracture toughness values. Thus, as mentioned previously, these materials are employed at alternating stress amplitudes that would eventually lead to their fatigue fracture. To only

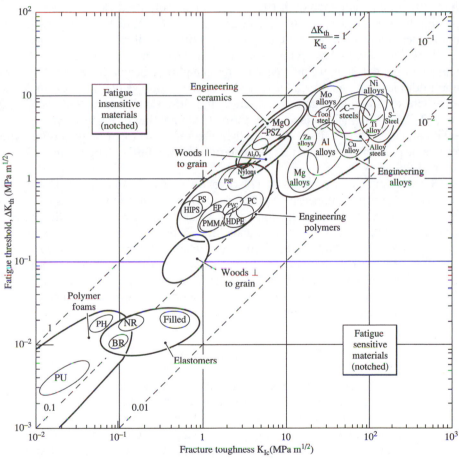

Figure 12.23
A material property chart illustrating the relationship between the fatigue threshold stress intensity (ΔK_{th}, values shown are for $R = 0$) and fracture toughness (K_{Ic}) for several material classes. Materials having ΔK_{th} values close to K_{Ic} are fatigue "insensitive," meaning that preexisting cracks they contain propagate readily. Design with such materials (e.g., ceramics), therefore, is based on a fracture toughness criterion. Most of the metals have high values of K_{Ic} relative to ΔK_{th}. Thus, metals containing preexisting cracks are prone to have these cracks propagate under cyclic loading; most metals, therefore, are considered fatigue "sensitive." (*Adapted from N. A. Fleck, K. J. Kang, and M. F. Ashby, "The Cylic Properties of Engineering Materials," Acta Metall. et Mater., 42, 365, Copyright 1994, with permission from Elsevier Science.*)

employ stress amplitudes that result in fail safe conditions would seriously limit the use of these metals. Polymers are intermediate to metals and ceramics in terms of their $\Delta K_{th}/K_{Ic}$ ratios. Generally, tough polymers (including those for which room temperature is above their glass transition temperature) have relatively high values of fracture toughness compared to their values of ΔK_{th}. In this sense they are like tough metals. The less tough polymers, including those for which room temperature lies below the glass transition temperature (e.g., polystyrene), behave more like ceramics in terms of our current discussion.[10]

Let us summarize the most salient points of this discussion. To design against fatigue we first determine whether fatigue fracture will result from the propagation of preexisting cracks or will be "intrinsic" in nature. This is done by comparing the value of the endurance limit (σ_e) to that stress amplitude defined by the relationship between it, the preexisting crack size, and ΔK_{th}. Whichever of these stress amplitudes is *lower* defines the types of cracks (intrinsic if $\sigma_e < \Delta K_{th}\sqrt{c}$; preexisting if $\Delta K_{th}\sqrt{c} < \sigma_e$) whose propagation leads to fracture. If we wish entirely fail safe design, we operate at a stress amplitude below the lesser of the two amplitudes just mentioned. Fail safe design is often not feasible. If that is the case, we use the constitutive equations defining fatigue life (Eq. (12.4) for "intrinsic," and Eq. (12.6) for "extrinsic," fracture) to estimate the number of loading cycles a material can last without failure. And we then use safety factors (coupled often with service inspection) to see to it that the material does not fail by fatigue during its intended lifetime.

Sounds simple enough, doesn't it? Not really. Recall the stochastic nature of the loading cycle during engineering applications. How do we handle this aspect of fatigue? Moreover, this random loading implies that R values vary during the life of a structure. Variations in R affect endurance limits more than they do fatigue thresholds (e.g., Figs. 12.10 and 12.18), particularly when R values less than zero are the situation. Numerical simulations can help. We can use a fatigue life equation (Palmgren-Miner's rule) which is similar to Robinson's rule for creep life (Eq. (11.10)) under varying stress-temperature combinations. Palmgren-Miner's rule is expressed in discrete form (it also has an integral form) as

$$\sum \frac{N_f}{N_{fi}} = 1 \qquad (12.9)$$

In Eq. (12.9), N_i is the number of cycles spent at the stress amplitude σ_{ai} for which the number of cycles to failure is N_{fi}. Rules such as this help. So does the conservative design approach embodied in this section. However, we must also sadly admit that engineering implementation is not perfect. Fatigue fractures still occur far more often than they should. Some fatigue fractures are merely a nuisance. Others can cause loss of life. In both situations, attorneys (and their technical consultants) are kept occupied by the deficiencies in the knowledge base of engineers and/or in our understandable inability to foresee all the possible "bad things" that can happen when a structural part is placed into service.

The constitutive fatigue equations (Eqs. (12.2)–(12.4)) that apply to crack-nucleation-limited fatigue fracture contain parameters that relate to a material's

---

[10]As with all generalizations, there are exceptions. Perhaps the reader can find a few of these in Fig. 12.23.

cyclical stress-strain behavior. As noted, this behavior differs from monotonic loading behavior, and we address cyclic stress-strain behavior in the next section. But first, two example problems.

597

SECTION 12.5
Design Against
Fatigue

EXAMPLE PROBLEM 12.2. Use the Palmgren-Miner rule Eq. (12.9) to estimate whether or not an initially crack-free 2014-T6 Al alloy (the alloy's fatigue behavior is shown in Fig. 12.8) will experience fatigue fracture under the following situations (all with a mean stress of zero): (i) stress amplitude of 200 MPa for 10^5 cycles followed by a stress amplitude of 100 MPa for 10^7 cycles; (ii) stress amplitude of 200 MPa for 10^7 cycles followed, successively, by a stress amplitude of 100 MPa for 10^5 cycles and a stress amplitude of 300 MPa for 10^4 cycles.

Solution. Figure 12.8 was used to estimate the following lifetimes at the several stress amplitudes; $\sigma_a = 100$ MPa, $N_f = 4 \times 10^9$; $\sigma_a = 200$ MPa, $N_f = 10^6$; $\sigma_a = 300$ MPa, $N_f = 2.5 \times 10^4$.

For the first situation we get $\Sigma(N_i/N_{fi}) = (10^5/10^6) + (10^7/4 \times 10^9) = 0.1025$. This is, therefore, a safe operating schedule, according to the Palmgren-Miner law.

For the second situation, we get $\Sigma(N_i/N_{fi}) = (10^7/10^6) + (10^5/4 \times 10^9) + (10^4/2.5 \times 10^4) = 10.4$. This is clearly not a safe operating schedule. The conclusion is also intuitive; e.g., consider the number of cycles intended to be spent at 200 MPa.

EXAMPLE PROBLEM 12.3. Refer to Fig. 12.21a. Assume that this curve was constructed for an R value of -1. Schematically sketch how the curve would change if it were constructed for an R value of zero. (Note: ΔK_{th} is not much affected by R in the range where R is negative, see Fig. 12.18b.) Assume the Goodman relationship holds for the relationship between the allowable stress amplitude (σ_a) and the mean stress. From your results how would the respective regions of intrinsic and extrinsic fatigue change by this alteration in R value?

Solution. The stress amplitude (σ_a) required to cause extrinsic fatigue will not change as R increases from -1 to 0. This is so because $\sigma_a = \Delta\sigma/2$ and $\Delta\sigma$ is related to ΔK_{th} via $\Delta K_{th} \sim (\Delta\sigma)\sqrt{c}$ where c is the crack length. However, the allowable stress amplitude for intrinsic fatigue will change (Eq. (12.1)). To use this equation, we first obtain a relation between σ_{mean}, the allowable stress amplitude σ_a, and R ($= \sigma_{min}/\sigma_{max}$). We find

$$\sigma_{mean} = \frac{1}{2}(\sigma_{max} + \sigma_{min}) = \frac{1}{2}\sigma_{max}(1 + R) \text{ and } \sigma_a = \frac{1}{2}(\sigma_{max} - \sigma_{min}) = \frac{1}{2}\sigma_{max}(1 - R)$$

Using the above equations we see that $\sigma_{mean} = \sigma_a[(1+R)/(1-R)]$. Thus, for $R = 0$ we have $\sigma_{mean} = \sigma_a$. On using the Goodman relation, we then find for $R = 0$,

$$\sigma_a = \sigma_{fat}\left(1 - \frac{\sigma_{mean}}{T.S.}\right) = \sigma_{fat}\left(1 - \frac{\sigma_a}{T.S.}\right)$$

Rearranging the above equation, we obtain (also for $R = 0$)

$$\frac{\sigma_a}{\sigma_{fat}} = \frac{1}{\left(1 + \dfrac{\sigma_{fat}}{T.S.}\right)}$$

Thus, the allowable stress amplitude for intrinsic fatigue is reduced when R is increased. (Recall that when $R = -1$, $\sigma_a = \sigma_{fat}$.) This means a schematic sketch of our results will very much resemble that of Fig. 12.21b with one major difference. We are diminishing, and not increasing as in Fig. 12.21b, the value of the horizontal line associated with fail safe design against intrinsic fatigue. This implies that the region of intrinsic fatigue fracture controls up to greater crack sizes.

We can go further. Doing so may provide hints for solving Prob. 12.13. We assume, based on previous discussions that $\sigma_{fat}/T.S. = 0.4$. Thus, the permissible stress amplitude is now $(1/1.4)\sigma_{fat} = 0.71\sigma_{fat}$. How much does this extend the crack range over which intrinsic fatigue controls fatigue fracture? The original crack size (c_0) at which the transition occurs is defined by $\sigma_{fat}\sqrt{c_0} = A\Delta K_{th}$ where A is an appropriate constant. The crack size (c_1) for the transition when $R = 0$ is defined by $0.71\sigma_{fat}\sqrt{c_1} = A\Delta K_{th}$. Thus, $(c_1/c_0)^{1/2} = 1/0.71$ or $c_1/c_0 = 1.98$.

12.6
CYCLIC STRESS-STRAIN BEHAVIOR

Materials behave "differently" when subjected to cyclical straining from the way they do when subjected to monotonic loading, as in a tension test. A knowledge of this cyclic behavior adds to our understanding of fatigue. Moreover, data obtained from such studies are useful for design against fatigue, as noted just above.

The behavior of a material under cyclical stressing/straining can be investigated by subjecting it to either a specified cyclical stress or strain amplitude. The latter is more common, and we focus on the results from this kind of test. If a fixed strain amplitude (consisting of some plastic strain component) is imposed on a metal, the stress range is not fixed but varies with the number of cycles or strain reversals. The metal may either soften (the stress amplitude decreases with time) or harden (this amplitude increases with time). The behavior of a cyclically hardening material is illustrated in Fig. 12.24a. Here the stress amplitude required to maintain the specified strain range increases with the number of cycles. When a material cyclically

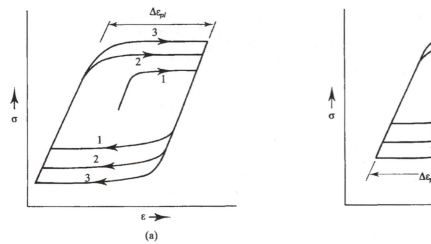
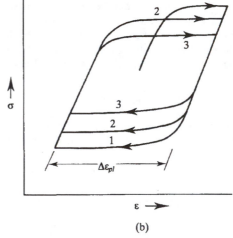

(a)   (b)

Figure 12.24
Cyclical stress-strain behavior for (a) a material that cyclically hardens and (b) one that softens. The tests are conducted for a fixed plastic strain range ($\Delta\varepsilon_{pl}$). For a material that cyclically hardens, the stress amplitude necessary to maintain a stipulated strain amplitude increases. (The numbers on the curves indicate successive cycles.) Conversely, this stress amplitude decreases for a material that softens cyclically. The phenomena are accompanied by respectively increasing and decreasing areas of the stress-strain loops generated during cyclical straining.

softens (Fig. 12.24*b*), the required stress amplitude decreases with the number of cycles.

The extent and rate of cyclical hardening/softening is ascertained by plotting the stress amplitude against the number of cycles (or time). This is illustrated for both cyclical hardening and softening materials in Fig. 12.25. As indicated there, after a sufficient number of cycles (typically fewer than 100), the stress amplitude attains a constant value. This "steady state" stress amplitude constitutes one datum for a material's cyclical stress-strain curve.

The phenomena of cyclic hardening and softening illustrate the inadequacy of using tensile properties in fatigue design. Although use of tensile properties for such design is still sometimes done, this results primarily from a lack of data obtained under cyclical loading conditions. Hence, considerable effort has been expended in the last several decades on obtaining cyclical stress-strain properties.

A number of attempts has been made to determine what kinds of materials cyclically soften and what kinds harden. Studies on particular alloy systems, e.g., copper in the annealed and work-hardened states, show that the "soft" (annealed) alloys cyclically harden and the "hard" (cold-worked) ones soften. This is a common result, and it is often stated as a general rule that "hard, strong" alloys soften cyclically whereas "soft, ductile" ones harden. As with most generalities, there are exceptions to this one.

The extent and rate at which hardening/softening occurs during cyclical loading depends on a number of factors. An important one has to do with the nature of plastic slip in the material. Materials manifesting "wavy" slip (i.e., slip not confined to single slip planes) display cyclical hardening/softening characteristics of the kind shown in Fig. 12.26*a*. That is, hard materials soften cyclically, and soft materials harden likewise, to a degree that the steady-state stress amplitude is independent of the material's starting condition. The cyclical behavior of a material behaving in this way is called "history independent." As might be expected on the basis of the cyclical stress-strain behavior of such materials, the dislocation structure they develop is also independent of the material's initial strength and dislocation structure. Further, as is discussed later, the dislocation structure developed in

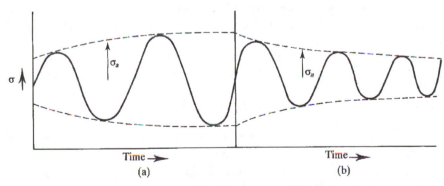

Figure 12.25

Stress vs. time (solid lines) for (a) a cyclically hardening, and (b) a cyclically softening, material. The dotted lines (the envelopes of the curves) represent the time variation of the stress amplitude. Typically, the stress amplitude attains a steady-state value after relatively few cycles.

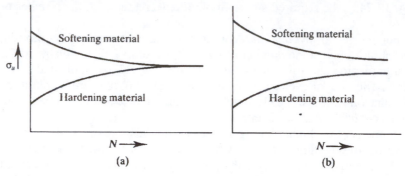

Figure 12.26
(a) Stress amplitude vs. number of cycles for "history-independent" materials. The steady-state stress amplitude for an initially hard (e.g., cold-worked) alloy that cyclically softens is the same as that of an initially soft alloy (e.g., an annealed metal) that hardens cyclically. (b) Stress amplitude–number of cycles behavior for a material whose cyclical response is history dependent. The initially hard material (which softens) has a steady-state stress amplitude greater than that of the initially soft material (which hardens).

cyclically strained materials differs (sometimes substantially) from that which forms in the same material under monotonic loading.

Materials that display planar slip are "history-dependent" under cyclical straining. Their behavior is shown in Fig. 12.26b. In common with "history-independent" materials, "hard" materials of this type usually cyclically soften and "soft" ones harden. In contrast to history-independent materials, however, the "hard" materials always remain "harder" than the originally "soft" ones. Clearly, planar slip does not permit the attainment of an "equilibrium" dislocation structure under cyclic loading.

A cyclical stress-strain curve is the loci of points representing the steady-state stress amplitude as a function of plastic strain range (actually $\Delta\varepsilon_{pl}/2$). Such a curve is developed in the following way. A number of samples of the material investigated is subjected to varying plastic strain ranges until a steady-state stress amplitude is attained for each strain range. A curve drawn through the loops (Fig. 12.27) defines the cyclical stress-strain curve.

The above procedure is cumbersome and time-consuming, involving numerous tests on a given material. An alternative, and far less time-consuming, means of determining a cyclical stress-strain curve has been developed, and it is valid for "history-independent" materials. It involves cyclically straining a material at some small initial value of $\Delta\varepsilon_{pl}$ until a steady-state stress amplitude (and associated stable hysteresis loop) is found. Following this, the plastic strain range is increased and the procedure repeated. This is done for as many plastic strain ranges as are necessary to define the cyclical stress-strain curve. It should be clear why this procedure is valid only for "history-independent" materials. This restriction must be kept in mind when cyclical stress-strain curves are derived via this routine.

It is instructive to compare cyclical stress-strain curves to tensile flow curves. Generally speaking, the cyclical flow curve lies below the corresponding tensile curve for materials that are initially "hard." Conversely, the cyclical curve lies

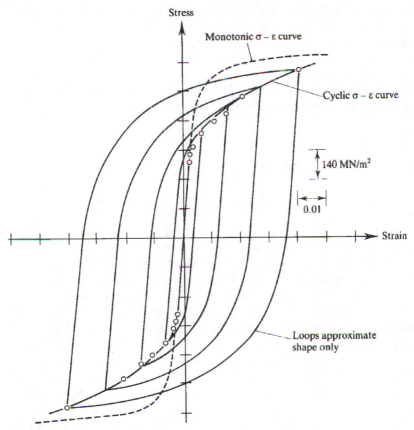

Figure 12.27
The monotonic and cyclic stress-strain curves for a 4340 steel. The hysteresis loops shown are the steady-state ones, obtained for cyclically straining at different plastic strain ranges. A curve drawn through the loop tips (i.e., through the steady-state stress amplitudes) defines the cyclical stress-strain curve. This steel cyclically softens; that is, the cyclical stress-strain curve lies below the monotonic one. (*From R. W. Landgraf, ASTM STP467, Philadelphia, Pa., 1970, p. 3.*)

above the tensile flow curve for materials that are initially soft. This is further indication that the dislocation structure in cyclically strained materials is different from that in materials subjected to monotonic loading. In brief, in cyclically strained materials, the fatigue equivalent of a tensile low-energy dislocation structure (cf. Chap. 3) develops. However, the manifestation of the low-energy structure is different in cyclically strained materials. The cyclical dislocation structure permits (at steady state) reversible plastic flow on strain reversals. Persistent slip bands, for example, are one manifestation of the low-energy structure in cyclically strained solids. As noted in Sect. 12.2, these bands, if present, are also related to crack initiation in flaw-free solids.

Empirical relationships between (the steady-state) stress and strain amplitude during cyclic straining have been developed. These parallel similar expressions for monotonic loading. For the latter, the tensile flow curve is often expressed as

$\sigma_T = K\varepsilon_T^n$, where σ_T and ε_T are the true stress and true plastic strain, K is the strength coefficient, and n is the strain-hardening coefficient. For the cyclical stress-strain curve, the stress amplitude ($\sigma_a = \Delta\sigma/2$) correlates to $\Delta\varepsilon_{pl}/2$ by $\sigma_a = K'(\Delta\varepsilon_{pl}/2)^{n'}$, where K' and n' (n' often has a value between 0.1 and 0.2 for metals) are cyclical coefficients analogous to the tensile ones. Data of this type have been gradually accumulated over the last several decades and they can be correlated to material fatigue life as has been discussed in Sects. 12.3 and 12.5.

12.7
CREEP-FATIGUE INTERACTIONS

Material use frequently entails simultaneous exposure to more than one "hostile" environment. Examples include cyclic straining in a corrosive environment and, the focus of this section, fatigue at temperatures where diffusional processes can operate. Our understanding of creep, fatigue, and environmentally induced fracture is considerable when these are the *sole* factors responsible for material failure. However, when they interact things become messy. As a result, empirical correlations, which may or may not have a fundamental basis, are used often in design in such situations. In this section, fatigue at high temperature or, equivalently, creep under cyclical loading, is considered. This is an important problem, for a number of high-temperature structures are subjected to fatigue environments. High-temperature aircraft turbine blades, one of our favorite examples, is such a structure. At aircraft cruising speeds, the environment the blade experiences is primarily that of creep. However, during engine start-up and when engine power is varied, fatigue plays a role in deterioration of blade performance. There are many other similar examples. In spite of the technological importance of these combined effects, our knowledge base of them is limited. As a result, we treat the problem in a simplified way.

Whether we should view the problem as creep enhanced by the fatigue environment or vice versa depends on several factors. When the cyclical stress (strain) amplitude is small in comparison to the mean stress, it is proper to view the phenomenon as one of creep perturbed by fatigue. This also holds when the applied frequency is low and/or the temperature is high. Under circumstances opposite to these, failure can be considered a fatigue failure accelerated by diffusional processes. This division is consistent with fracture-surface observations. In the creep-dominated regime, fracture typically takes place intergranularly, just as it does when only a uniform tensile stress is the cause of high-temperature fracture. Likewise, fatigue striations and other hallmarks of fatigue fracture, including transgranular fracture, are manifested at lower temperatures and/or higher cyclical frequencies.

From Chap. 11, we know that intercrystalline high-temperature creep fracture comes from grain-boundary cavitation followed by cavity growth and coalescence. These processes are enhanced by a tensile stress, but a compressive stress causes cavity shrinkage (crack "healing"). Thus, it is somewhat surprising that a cyclical stress, which may include a compressive component, can accelerate creep fracture. A number of analytical and conceptual models have addressed the issue. A simple one that illustrates how cyclical stresses *may* accelerate void growth is shown in Fig. 12.28. While the model is speculative to an extent, it is useful nevertheless for

(a)  Grain boundary

(b)  Sliding

(c) 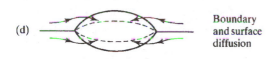 Surface diffusion

(d)  Boundary and surface diffusion

(e) Reversed sliding

Figure 12.28
Schematic demonstrating how a cyclical strain may accelerate void growth in a creeping solid. (a) A cavity situated on a grain boundary. (b) During boundary sliding the respective halves of the cavity are displaced. (c) A diffusive flux that attempts to maintain the equilibrium dihedral angle at the boundary-cavity junction results. (d) This causes cavity growth. (e) The process is repeated on stress reversal. (*From C. Wigmore and G. C. Smith, Met. Sc. Jl., 5, 58, 1981.*)

illustrating the possibility of fatigue-enhanced cavity growth. In the description, plastic deformation is associated with grain-boundary sliding (Fig. 12.28b). This disturbs the equilibrium dihedral angle at the grain boundary–cavity interface, and a surface and grain-boundary diffusive flux attempts to restore this angle (Fig. 12.28c). In the process, cavity volume increases (Fig. 12.28d). The process then repeats, irrespective of the algebraic sign of the stress. A mechanism of this kind can play a role in the fracture process only at low cyclical frequencies and/or high temperatures where diffusion is sufficiently rapid vis-à-vis the cyclical stress frequency to allow for this kind of cavity growth.

A number of empirical creep-fatigue correlations has been developed in the absence of a good fundamental understanding of the creep-fatigue interaction. One combines the Palmgren-Miner rule for fatigue life (Sect. 12.5) with Robinson's rule for creep life under varying stress-temperature combinations. We have noted that Robinson's rule has limitations for predicting creep lives, and the Palmgren-Miner rule has like restrictions. In spite of these limitations, a fracture criterion under combined creep-fatigue conditions can be specified as

$$\sum \frac{N_i}{N_{fi}} + \sum \frac{t_i}{t_{fi}} = 1 \qquad (12.10)$$

In Eq. (12.10), N_i is the number of cycles spent at the stress amplitude σ_{ai} for which the number of cycles to failure is N_{fi}, and t_i is the time spent at the stress-temperature combination resulting in a creep fracture life of t_{fi}. The limitations of Eq. (12.10) can be revealed by simple tests. These can be done by first subjecting the material to a creep environment, and then to a fatigue one; the sequence can then be reversed. According to Eq. (12.10), material life should be independent of how this sequence is conducted. Yet it is not and, thus, the results complicate predictions of material lifetime under combined creep/fatigue environments. Other experiments indicate that the combined law of Eq. (12.10) is conservative for materials that cyclically harden; that is, the sum of the terms on the left-hand side (LHS) of Eq. (12.10) exceed unity at material fracture. Conversely, Eq. (12.10) represents "unsafe" design for materials that cyclically soften (the sum of the terms on the LHS of Eq. (12.10) are less than one at material fracture).

Environment also affects the interaction between creep and fatigue, as is illustrated schematically in Fig. 12.29. Figure 12.29a shows how temperature generally influences the relationship between plastic strain range and the number of cycles to failure, and Fig. 12.29b shows how the relationship is affected by oxygen, ubiquitous in high-temperature air. As expected, generally (but not always) exposure to high temperatures reduces N_f at a given $\Delta\varepsilon_{pl}$. We see, too, that air has a pernicious effect on high-temperature fatigue life, and the degree to which it does depends on the frequency, ν, of the cyclical straining.[11] As ν is decreased, sample lives are shorter at a given plastic strain range.

A convenient way of conceptualizing creep, fatigue, and tensile fracture, and the interactions among these failure modes, is provided by the equivalent of a "fatigue-

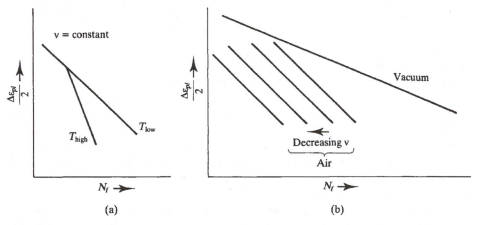

(a)   (b)

Figure 12.29
(a) Effect of temperature on the plastic strain amplitude–cycles to failure relationship. Except at very high strain amplitudes (and correspondingly low N_f values), temperature generally reduces a material's fatigue life. The graph is constructed for a fixed cyclical loading frequency. (b) Schematic illustrating the effect of cyclical frequency and environment on the fatigue life of stainless steel, whose fatigue-creep behavior is typical of many engineering alloys. Air environments decrease fatigue life. This effect is greatest at low frequencies; elevated-temperature fatigue of stainless steel in air also involves "embrittling" effects. The curve is constructed for a fixed temperature. (*Part (b) after L. F. Coffin, Adv. in Rsch. on the Strength and Fracture of Materials, ed. D. M. R. Taplin, Vol. 1, Pergamon Press, New York, 1977, p. 263.*)

---

[11]Note that this is in distinction to low-temperature fatigue for which frequency does not affect fatigue lives of metals and their alloys (Sect. 12.3).

fracture map." Such a map, applicable to a copper alloy (Alloy 175) is given in Fig. 12.30. The axes of the map are temperature (or homologous temperature) and the plastic strain range. As with conventional fracture maps, the several regions in a fatigue fracture map correspond to plastic strain range–temperature combinations that lead to a particular fracture mode. For example, at low temperatures and low cyclical plastic strain ranges (i.e., $\Delta\varepsilon_{pl} << \Delta\varepsilon_{el}$), a cyclical environment results in transgranular high-cycle fatigue fracture for Alloy 175. At these same temperatures, but at somewhat higher plastic strain ranges ($\Delta\varepsilon_{pl} > \Delta\varepsilon_{el}$), we have a low-cycle transgranular fatigue fracture. And at even higher values of $\Delta\varepsilon_{pl}$, fracture takes place by development of a tensile instability. At higher temperatures (cavitation and notch-sensitive cavitation failure), fracture surfaces (for values of $\Delta\varepsilon_{pl}$ greater than about 10^{-2}) assume a morphology characteristic of a creep-fracture surface. Further, as is discussed in the following paragraph, creep lifetimes at a specified plastic strain range are reduced. The region marked "intergranular oxidation fracture" in Fig. 12.30 is a new one in a fracture map. It represents the oxygen-enhanced high-temperature creep-fatigue fracture behavior observed commonly in air, and is an additional manifestation of the types of effects schematized in Fig. 12.29.

Lines of constant N_f values can be superimposed on fatigue-fracture maps, as is done in Fig. 12.30. The contour for $N_f = 1/4$ corresponds to tensile fracture on the

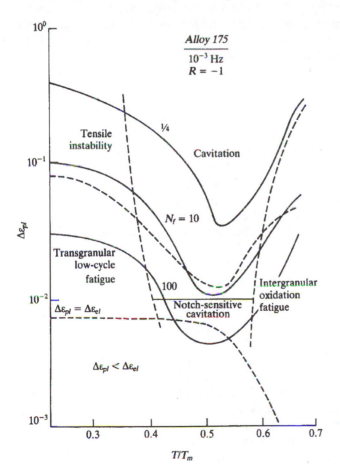

Figure 12.30
Fatigue-fracture map for copper alloy 175. At low homologous temperatures, low-cycle fatigue is supplanted by tensile fracture as the plastic strain range increases. Two types of high-temperature fracture are observed for this material; cavitation at $0.4 \le T/T_m \le 0.6$ and oxidation-instigated failure at higher temperatures. The regions on the map indicate the type of failure expected at different strain range–temperature combinations. Contours of constant N_f values are also placed on this map. Cavitation in copper alloy 175 is accompanied by a notable diminution in fracture strain. The dotted line at the lowest values of $\Delta\varepsilon_{pl}$ represent the transition at which the elastic and plastic strain ranges are equal. (D. M. R. Taplin and A. L. Collins, Reproduced with permission from the Ann. Rev. Matls. Sc., 8, 235, © 1978 by Annual Reviews Inc.)

first stroke of an applied strain cycle. Thus, the transitions in fracture mode obtained by following the line marked $N_f = 1/4$ in Fig. 12.30 are the same ones seen in a conventional fracture-mechanism map (cf. Chaps. 10 and 11). The line also reflects fracture ductility. For example, for Alloy 175, cavitation creep fracture is accompanied by a notable diminution in fracture strain as the transition from low-temperature transgranular fracture to high-temperature intergranular fracture takes place. This same loss of ductility is also seen in low-cycle fatigue. For example, a plastic strain range of ca. 3×10^{-2} is allowed for $N_f = 100$ at the low temperature of $T/T_m = 0.22$ ($T = 300$ K), whereas at $T/T_m \cong 0.5$ ($T = 675$ K), where fracture takes place intergranularly, the allowed plastic strain range is only about 7×10^{-3}. In closing our discussion of this kind of map, we note that such a map is constructed for a *specific* frequency and a *specific* value of stress ratio; for example, the map of Fig. 12.30 corresponds to $\nu = 10^{-3}$ Hz and $R = -1$ (completely reversed tension/compression).

It is clear that temperature plays an important role in metal fatigue at temperatures at which diffusion becomes significant. Temperature plays an even greater role in the room-temperature fatigue of polymers since room temperature is usually a relatively high homologous temperature for this material class. Moreover, the viscoelastic component of deformation in polymers provides for significant hysteretic stress-strain behavior even without permanent deformation (cf. Chap. 2). This dissipates heat which can serve to increase material temperature. Since polymers are becoming increasingly used as substitutes for metals, a discussion of polymeric fatigue is called for. This is attempted in the next section.

12.8
POLYMERIC FATIGUE

As has been emphasized before, polymeric use, primarily as substitutes for metals, has increased substantially since the end of the Second World War. Originally "cheap" substitutes in noncritical applications, polymers have been increasingly used as structural materials in their own right in addition to serving as matrices of composites. As for metals, polymer applications often involve subjecting the material to a cyclical stress or strain. Thus, a discussion of polymeric fatigue is in order. We have separated discussion of metal and polymeric fatigue, for several reasons. One involves the relationship between fatigue and temperature in polymers. For example, room temperature for many polymers is a substantial fraction of their melting temperature; this is not the case for structural metals. Moreover, hysteretic elastic effects are common to many polymers as a result of their (frequently substantial) viscoelasticity, and this is not so for metals, which usually only manifest linear elastic behavior. As described in Chap. 2, the area within an elastic hysteresis loop represents energy dissipated per strain cycle, and this can lead to temperature increases during cyclic deformation of polymers, with resulting complications. The topic is addressed in Sect. 12.8D. Even when polymers are subjected to only cyclical linear elastic strains on a macroscopic level, the deformation taking place at the tip of a propagating fatigue crack in them is partially permanent and this can lead to localized crack-tip heating. Another feature complicating discussion of polymeric fatigue is the variety of their permanent deformation modes. For example, during polymer tensile deformation, homogeneous plastic flow is distinguished from microscopic inhomogeneous flow, and within the latter shear banding and crazing are differentiated. Since permanent deformation plays a key role in fatigue, these features must also be considered in discussing the response of polymers to cyclical deformation.

Our discussion of polymer fatigue closely parallels that of metals, although topics are not treated in exactly the same order. Various subheadings (e.g., Sect. 12.8A) are used to approximately distinguish between topics similar to those covered in metallic fatigue. Discussion of polymer fatigue behavior is initiated with a description of the cyclical stress (strain)–cycles to failure relationship found in polymers.

A. Stress (Strain) Amplitude and Polymeric Fatigue Life

The relationship between stress (or strain) amplitude and fatigue life is ascertained for polymers in the same manner as it is for metals.[12] Moreover, certain phenomenological similarities in the fatigue behavior of these two material classes are evident when their fatigue resistance is characterized in this way. This is illustrated in Fig. 12.31, where the σ_a-N_f relationship for polymers is shown schematically. We have (somewhat arbitrarily) divided this figure into three regions. Region III, the high-cycle fatigue region, is characterized by a fairly small negative slope to the σ_a-N_f curve. In fact, some polymerics can be considered to possess endurance limits just like those exhibited by some metals. As with high-cycle metallic fatigue, the nucleation of fatigue cracks occupies the predominant fraction of a polymer's high-cycle fatigue life. At the low stress and strain amplitudes characteristic of Region III, fatigue cracks nucleate by heterogeneous, microscopic flow processes. The kind of flow associated with crack nucleation depends on the material and, perhaps more importantly, the test temperature relative to the material's glass transition or melting temperature. However, at temperatures at which fatigue (rather than creep) controls the polymer's cyclic behavior, crazing is very frequently the flow mechanism associated with high-cycle fatigue-crack nucleation.

The slope of the stress amplitude–cycles to failure curve is considerably more negative in Region II than in Region III; this is akin also to observations in metals. Crack propagation rates (Sect. 12.8B) become increasingly important in determining material lifetime as strain or stress amplitude is increased, and this, too, is common with fatigue of metals. At the higher stress amplitudes of Region II, crack propagation frequently is associated with crazing. However, as with tensile flow, shear banding is competitive with crazing during polymer fatigue as temperature is increased.

Whether a distinct Region I is found (cf. Fig. 12.31) in the σ_a-N_f curve depends on the material's mechanical characteristics. If the maximum stress during the first cycle of a fatigue test exceeds the material's craze yield strength, crazes so formed

---

[12]Care must be taken not to increase the polymer temperature by hysteretic heating during testing. Otherwise, the *true* fatigue response is not measured. Thus, fatigue testing of polymers requires more precautions than does similar testing of metals. In this section, data presented are for isothermal conditions, except when temperature effects are explicitly discussed (Sect. 12.8D).

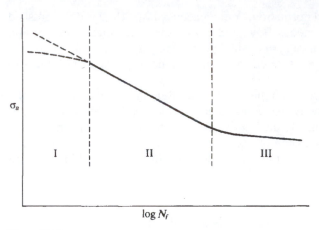

Figure 12.31
Schematic stress amplitude–log cycles to failure curve for polymers. The relationship between these variables is similar in polymers and metals. For example, high-cycle fatigue (Region III) is characterized by a small negative slope in a σ_a–log N_f plot, and endurance limits are also found in some polymers. At intermediate stress amplitudes (Region II), N_f is a more sensitive function of σ_a, just as it is for metals. For low-cycle fatigue two kinds of behavior are found for polymers. When crazing does not take place, the σ_a-N_f curve is an extrapolation of the σ_a-N_f relationship found in Region II. When crazing occurs, N_f values are less than predicted by this extrapolation. Thus, a Region I may or may not be present in the diagram, depending on the polymer deformation behavior.

serve, as it were, as crack nuclei. When this happens the stress amplitude in Region I lies below that obtained by extrapolation of the Region II σ_a-N_f curve. When crazing does not intervene in the first load cycle, then no distinct Region I is found in the σ_a-N_f curve, and the low-cycle fatigue portion of the curve is an extrapolation of the Region II σ_a-N_f relation. Thus, a low craze yield stress, and a material prone to crazing, are requisites for Region I behavior. Examples of the behavior just described are shown for engineering polymers in Fig. 12.32.

Although on a macroscopic basis there are similarities in the fatigue response of polymers and metals, there are considerable differences in the microscopic mechanisms of fatigue between the material classes. The variety of flow mechanisms possible in polymers leads to some intriguing fatigue characteristics. Understanding these, in turn, provides guidelines for the development of fatigue-resistant engineering plastics. Thus, we now direct our attention to fatigue-crack growth in polymers.

B. Fatigue-Crack Growth

Fatigue-crack growth rates can be measured for polymers in the same way as they are for metals. Further, polymer crack growth rates, dc/dN, can be related to stress-intensity ranges, ΔK. Because we are concerned with temperatures below which diffusion is relatively inconsequential, we deal with polymers wherein crack-tip plasticity is microscopically heterogeneous (i.e., via either shear banding or

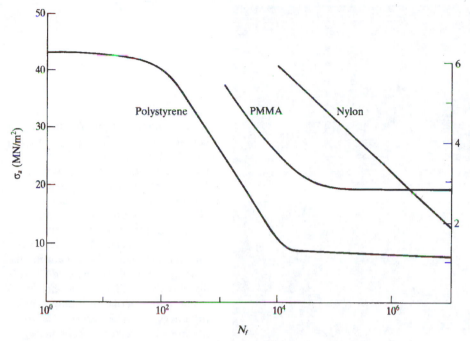

Figure 12.32
Stress amplitude–cycles to failure relationships for nylon, polymethyl methacrylate
(PMMA), and polystyrene (PS). Polystyrene, which crazes, exhibits all three regions
of the schematic Fig. 12.31. Polymethyl methacrylate, which also crazes, likewise dis-
plays three distinct regions, but its low-cycle fatigue data are not shown here. (Note:
PMMA also exhibits a fatigue limit.) (*Data for nylon and PMMA from M. N. Riddell,*
G. P. Koo, and J. L. O'Toole, Polymer Eng. Sc., 6, 363, 1966. Data for PS from
P. Beardmore and S. Rabinowitz, Treat. Matls. Sc. and Tech., 6, 267, 1975.)

crazing). Fatigue in which craze formation is the dominant crack-tip plasticity
mode has been studied most extensively, and describing the behavior of polymers
deforming in this manner occupies a good portion of our discussion. On the other
hand, at certain stress-temperature combinations, polymer fatigue also involves
shear-band formation; we also discuss some of these effects.

The microscopic fatigue-fracture surface appearance of polymers varies, de-
pending on the value of ΔK. At moderate ΔK levels (well above ΔK_{th} but below
K_{Ic}), fatigue striations are often evident on such surfaces (Fig. 12.33a). As for metals, striation spacings in polymers correlate with crack-advance rates measured in
other ways, and the striation spacing represents the crack advance per cycle.

A different characteristic "spacing" is found when polymers are cycled at low
ΔK values (Fig. 12.33b). This spacing is much larger than the average crack advance per cycle obtained by direct measurement of dc/dN. The conclusion is that
the crack advances discontinuously (the process is termed discontinuous crack
growth, DCG). During DCG the crack tip does not advance for a period of several
to many cycles, during which period fatigue "damage" accumulates at the crack tip.
When this damage attains a critical value (a function of the material, ΔK, and the
test conditions, including cyclical frequency and temperature), the crack advances a

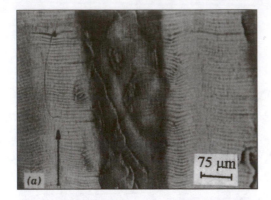

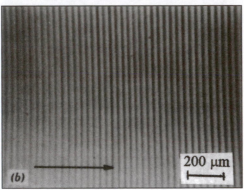

Figure 12.33
Characteristic fatigue-fracture
surfaces in amorphous polymers.
(a) Striations, similar to those
found in metals, have a spacing
that correlates with the crack
advance per cycle. (b) A feature
similar to striations, but having a
spacing much greater than the crack
advance per cycle. The spacing be-
tween markings is the discontinuous
crack advance distance; *dc/dN* **is this**
spacing divided by the number of
cycles between discontinuous crack
advances. (*R. W. Hertzberg, Defor-*
mation and Fracture Mechanics of
Engineering Materials, Copyright
1976. Reprinted by permission of
John Wiley and Sons, Inc.)

distance equal to that between the markings seen in Fig. 12.33*b*. The average crack advance per cycle, therefore, is this distance divided by the number of cycles between crack-advance events.

At the very lowest of ΔK values at which DCG occurs the process involves microscopic crack-tip crazing. DCG is illustrated schematically in Fig. 12.34. Damage accumulates at the craze in front of the crack tip; this craze is somewhat analogous to the plastic zone in front of a crack in a metal. As a result of fatigue damage, the crack-tip opening displacement (COD) increases during the period when the crack is static, and when the COD reaches a critical value a DCG event takes place. The hypothesized progression of the crack through the craze is illustrated in Fig. 12.34*b*. Cracking is presumed to mainly occur along the craze-matrix interface and along fibril boundaries in the craze. Fracture of the craze midrib only happens near the end of the DCG event. The microscopic fracture surface, therefore, is not a flat one, but is irregular over a distance comparable to the craze thickness (however, this is typically a small dimension; see Chap. 8). As shown in Fig. 12.34*b,* the proposed model assumes that the crack is halted prior to its complete progression through the craze. However, this may not be the case and, in any event, it is not a necessary condition for describing DCG in the way just done. Following a DCG event, the process repeats; i.e., the stress at the crack tip causes craze formation there, the craze lengthens, and the COD increases until another critical state is attained. Speculation has been made as to the cause of the "damage" that eventuates in DCG. For example, it has been argued that microscopic hysteresis in the

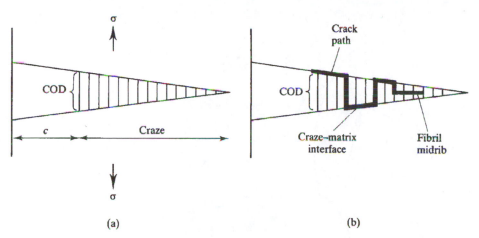

Figure 12.34
Relation of discontinuous crack advance to crazing. (a) A craze is formed in front of
the crack tip as a result of the stress concentration there. During cyclical straining,
and as a result of accumulated fatigue damage, the craze lengthens and the crack
opening displacement, COD, increases. (b) After some critical COD is attained, the
crack advances by propagating along the craze-matrix interface and the fibril
midribs. According to the model shown here, the midribs are severed only near the
end of the DCG event, and the crack does not propagate through the whole of the
craze. (*Model in (b) developed by L. Konczol as reported by M. T. Takemori. Repro-*
duced with permission from the Ann. Rev. of Matls. Sc., 14, 171, © 1984 by Annual Re-
view Inc.)

craze adjacent to the crack tip leads to local heating there that softens the material,
thus allowing for a subsequent increase in the COD and extension of the craze.

In certain materials, those having a tendency to shear-band deformation, the latter is also often involved with DCG. In these circumstances DCG takes place at
stress amplitudes greater than those at which it is accomplished solely through crazing. A schematic of an εDCG event, as it is called, is provided in Fig. 12.35a, and
a photograph of it is given in Fig. 12.35b. We see that shear bands form at angles inclined to the direction of crack propagation, and that both shear banding and crazing are intimate to εDCG.[13] Fatigue-crack growth rates during εDCG are generally
less than expected on the basis of those found for DCG involving only crazing, and
they are also less than expected by extrapolation of dc/dN-ΔK curves obtained at
higher ΔK values for which εDCG does not occur. This effect is evidently related
to the plastic work expended during shear-band formation, which makes crack propagation more difficult. We note that this correlation represents one difference between fatigue-crack growth in metals and polymers. As we have seen, for metals the
fracture toughness has little effect on fatigue-crack growth rate. This is not so for
polymers; additional examples of this are provided later.

---

[13]However, at even higher stress amplitudes, fatigue cracks can propagate absent crazing, i.e., propagation is associated only with shear bands. In fact, the crack in the photograph of Fig. 12.35b is in the
process of developing shear-band cracks.

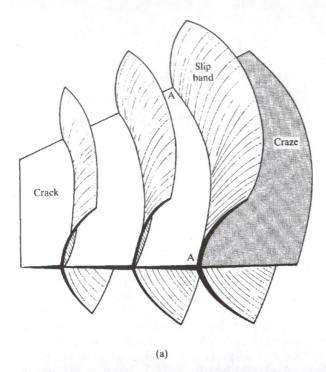

(a)

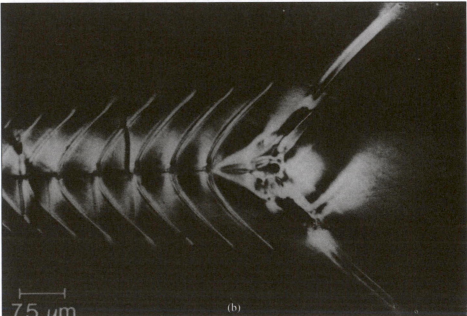

75 μm

(b)

Figure 12.35
(a) Schematic and (b) actual appearance of εDCG morphology. As for conventional DCG, crack-tip crazing is associated with εDCG. However, slip bands also form at the crack tip in εDCG; this makes crack advance more difficult. (*M. T. Takemori, Reproduced with permission from the Ann. Rev. of Matls. Sc., 14, 171, © 1984 by Annual Reviews Inc.*)

As noted, materials manifesting shear-band-related DCG exhibit several differ-ent fatigue-crack growth modes. At the lowest values of ΔK, DCG is only associ-ated with crazing. At progressively higher stress amplitudes, first εDCG happens and then fatigue-crack growth is accomplished only in the presence of shear-band-induced cracks (shear cracking). For a material behaving in this way, a simplified fatigue-fracture map can be constructed, and the various stress amplitudes associ-ated with the several crack modes can also be delineated in it. This is done in Fig. 12.36, where we see that increases in stress amplitude are associated with the tran-sitions just described. Higher temperatures diminish the stress range of craze-effected DCG, favoring instead shear-band-related fatigue-crack propagation. This is similar to craze–shear banding transitions in tensile deformation, where higher temperatures also favor shear banding. Correlations between the stress amplitudes at which the transitions shown in Fig. 12.36 occur also exist. As a rule of thumb, conventional DCG, involving craze formation only, takes place at stress amplitudes up to about one-third of the material's tensile strength. Shear cracking occurs when this stress amplitude exceeds two-thirds of this strength value, and εDCG happens in the intermediate stress-amplitude range. It should be clear that εDCG is restricted to materials for which shear-band formation is the dominant *tensile* de-formation mode. When crazing is dominant in tension, fatigue-crack propagation—whether it is continuous or discontinuous—is accomplished only by mechanisms involving crazing, regardless of stress amplitude.

We conclude discussion of polymer fatigue-crack propagation by consider-ing again how its rate relates to other mechanical properties, particularly fracture

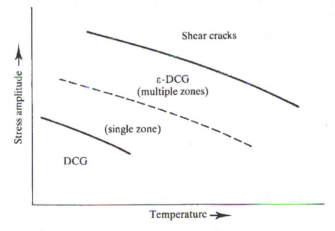

Figure 12.36
A schematic fatigue-crack map for polymers displaying both DCG and εDCG. The map delineates stress-temperature combinations resulting in different types of fatigue cracks. At low stresses and temperatures, conventional DCG takes place. At high stresses and temperatures, fatigue cracks are associated with shear bands, and crack advance is not discontinuous. At intermediate stresses and temperatures, εDCG takes place. The εDCG region is subdivided into stress-temperature combinations for which shear bands are widely formed and those where they are not. (*M. T. Take-mori, Reproduced with permission from the Ann. Rev. of Matls. Sc., 14, 171, © 1984 by Annual Reviews Inc.*)

toughness. Fatigue-crack growth rates as a function of ΔK and test frequency in polystyrene (PS), a material that crazes, are shown in Fig. 12.37. We see that increases in cyclical frequency are accompanied by a substantial decrease in fatigue-crack propagation rates.[14] It is believed the effect arises from the strain-rate sensitivity of the craze stress. Higher strain rates result in a higher flow stress in the crack-tip vicinity, and a lesser amount of damage per cycle as a consequence.

Additional evidence for the role that "toughness" plays in reducing polymer fatigue-crack growth rates is provided in Fig. 12.38. Crack growth rates as a function of ΔK for several polymers are shown in Fig. 12.38a. Details of the behavior-structural correlation are provided in the caption to this figure. Here it is sufficient to note that low fatigue-crack propagation rates correlate with high fracture toughness. We also add that correlation of crack growth rates with modulus, although appropriate for metals (cf. Fig. 12.13) is not so for polymers. For example, normalizing the crack growth rates of the materials whose behavior is shown in Fig. 12.38a by plotting these rates vs. $\Delta K/E$ nowhere near accounts for the great differences in the polymer fatigue-crack growth rates.

Polymer structure also affects fracture toughness and, therefore, fatigue-crack growth rates. The effect of one structural alteration in PS on these growth rates is illustrated in Fig. 12.38b. We see that HIPS (polystyrene containing finely dispersed elastomeric spheres) is more resistant to fatigue-crack propagation than is PS, and ABS (which contains an even finer dispersion of butadiene spheres) is even better than HIPS in this respect. As noted, PS crazes during both tensile and cyclic deformation, and PS also has a low fracture toughness. The HIPS and ABS polymers are more fracture tough than ordinary PS, and they are also more fatigue "tough" for much the same reason. During cyclic loading, multiple craze-initiated cracks form in HIPS and ABS. However, these cracks are arrested by the elastomeric dispersoids during cyclic deformation just as they are during tensile loading. During both kinds of deformation, this process makes crack advance more

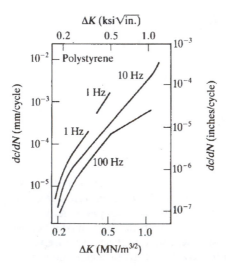

Figure 12.37
Fatigue-crack growth rates in polystyrene for several cyclical loading frequencies. Increasing frequency reduces crack-growth rates, presumably as a result of the strain-rate sensitivity of polystyrene's flow stress. Thus, it is more difficult to effect permanent deformation associated with crack advance at the higher strain rates associated with the increased frequency. (*From R. W. Hertzberg, J. A. Manson, and M. D. Skibo, Polymer Eng. Sc., 15, 252, 1975.*)

---

[14]This is clearly not a temperature effect. If it were, higher frequency would lead to greater hysteretic heating (see Sect. 12.8D), resulting in higher crack growth rates.

difficult. From this and other examples, it is clear that fracture toughness and fatigue-crack propagation resistance are complementary properties for polymers.

C. Polymeric Cyclical Stress-Strain Behavior

The cyclical stress-strain behavior of polymers differs from their tensile stress-strain behavior. Of course, this also holds for metals, but the cyclical response of polymers differs profoundly from that of metals in a number of respects. For one thing, polymers do not cyclically harden; they either soften or are cyclically stable (a cyclically stable material is one that neither softens nor hardens; thus the stress amplitude does not vary with the number of cycles). However, cyclical stability is a rarity in polymerics. It is observed only for small strain ranges, ones in which

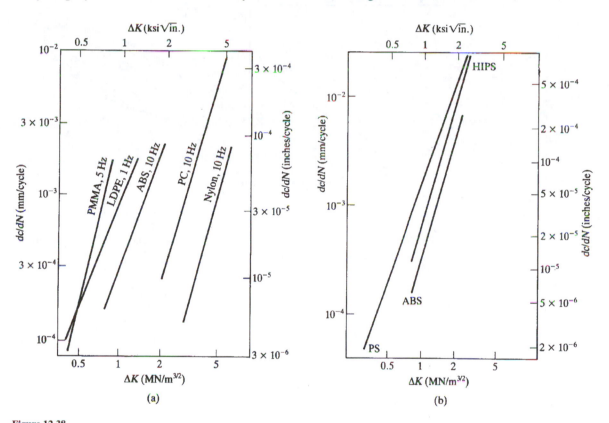

Figure 12.38
(a) Crack-growth rates vs. ΔK for several polymers: poly(methylmethacrylate) (PMMA), low-density polyethylene (LDPE), acrylonitrile-butadiene-styrene (ABS), polycarbonate (PC), and nylon. Crack-growth rates correlate with polymer toughness. For example, they are highest in PMMA, the most "brittle," and lowest in nylon, one of the toughest, of the polymers whose behavior is shown. (*From R. W. Hertzberg, H. Nordberg, and J. A. Manson, J. Matls. Sc., 5, 521, 1970.*) (b) Fatigue-crack growth rates vs. ΔK for several kinds of polystyrene (PS). Crack growth rates are greatest for ordinary PS, but are reduced considerably when it is "toughened" by elastomeric dispersions, as in high-impact polystyrene (HIPS) and ABS. (*R. W. Hertzberg, Deformation and Fracture Mechanics of Engineering Materials, Copyright 1976. Reprinted by permission of John Wiley and Sons, Inc.*)

deformation is primarily elastic—and linearly elastic at that. The degree to which cyclic softening takes place in a polymer depends on its mechanical characteristics. The greater the polymer ductility, the more likely it is to significantly cyclically soften. Thus, nylon and polypropylene cyclically soften to a great extent, whereas PMMA softens only moderately.

It is worthwhile to consider some details of the cyclic stress-strain behavior of polymers for these relate importantly to fatigue behavior. The cyclical stress-strain behavior of a polycarbonate (PC) is shown in Figs. 12.39a and b. The unusual, but characteristic, propeller shape of the hysteresis loop differs substantially from the symmetric hysteresis loop manifested by a metal subject to the same

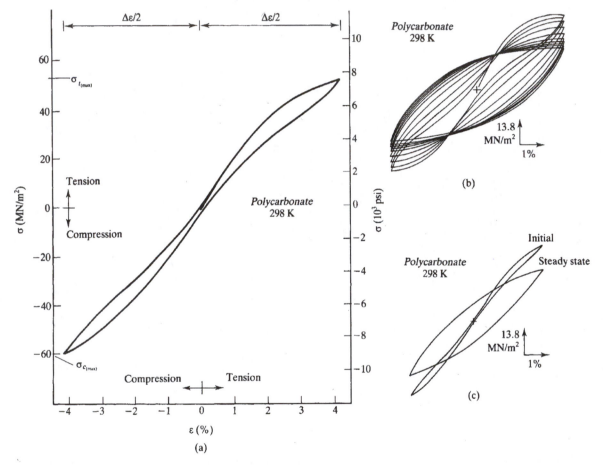

(a)

(b)

(c)

Figure 12.39
The cyclical stress-strain behavior of polycarbonate (PC) at 298 K. (a) The initial hysteresis loop shape is typical of that displayed by many ductile polymers. Most of the cyclical deformation is viscoelastic; permanent deformation is marked by the small offset on the strain axis at zero stress. (b) Cyclical softening in PC as shown by the progressive decrease in stress amplitude with the number of cycles. There is a significant change in the loop shape concurrent with cyclical softening. (c) Comparison of the initial and steady-state hysteresis loops in PC. (*From P. Beardmore and S. Rabinowitz, Treat. Matls. Sc. Tech., 6, 267, 1975.*)

kind of test. The "strange" shape of the PC loop reflects the large viscoelastic strain PC experiences. The plastic strain component, evidenced by the strain offsets at the zero stress condition, is fairly small. It is clear, too, that the loop is not balanced insofar as tensile and compressive loading is concerned. For equal compressive and tensile strain ranges, the maximum compressive stress is greater than the corresponding tensile one. This is expected on the basis of polymer tensile characteristics (Chap. 8).

The steady-state hysteresis loop developed by the carbonate is shown in Fig. 12.39c, and compared there to the initial loop. Cyclic straining eventually leads to the development of a symmetrical hysteresis loop, and this is commonly found to be the case for amorphous homopolymers. In other plastics, for example, semicrystalline polymers, the steady-state hysteresis loop is asymmetrical.

Other features of the cyclical stress-strain behavior are of interest, and also provide information on fatigue-fracture mechanisms. At low temperatures (e.g., 77 K), PC crazes under tensile loading. The cyclical stress-strain behavior for PC at 77 K is shown in Fig. 12.40. We see evidence of "anomalous" softening; that is, softening is found only during the tensile part of the loading cycle. This behavior is rationalized on the basis of the material's flow properties. Crazing deformation is facilitated by tensile loading, but does not occur during compression. During tensile loading, craze-instigated crack growth occurs, the crack advancing on successive straining cycles. This is accompanied by a diminution in the tensile load-bearing capacity, but has no effect on the material's compressive flow strength. The result is the attainment of an invariant, stable compressive stress level, but a continually decreasing tensile stress level as shown in Fig. 12.40.

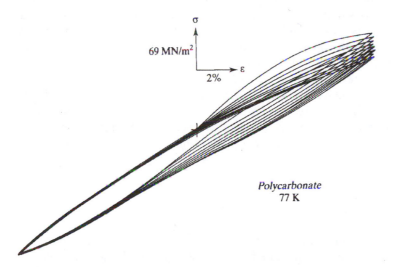

Figure 12.40

The cyclical stress-strain behavior of polycarbonate (PC) at 77 K. At this temperature, PC crazes, and this is manifested by a considerable "softening" during the tensile portion of the cycle. This behavior is consistent with crazing happening only when an algebraically positive mean pressure is imposed. (*From P. Beardmore and S. Rabinowitz, Treat. Matls. Sc. Tech., 6, 267, 1975.*)

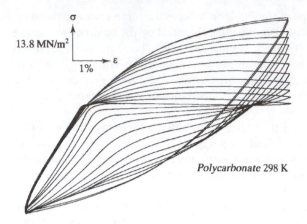

Figure 12.41
The cyclical stress-strain behavior of polycarbonate during the slow-crack-growth stage of fatigue. The crack advances during the tensile portion of the cycle, and the maximum tensile stress decreases with crack advance. The value of the maximum compressive stress is unaffected by this advance. (The test was conducted for a fixed strain amplitude.) (*From P. Beardmore and S. Rabinowitz, Treat. Matls. Sc. Tech., 6, 267, 1975.*)

For polymers, as for metals, fatigue-crack growth is caused by tensile, and not compressive, stresses. The cyclical stress-strain behavior of a polymer during the slow-crack-growth portion of the fatigue process is shown in Fig. 12.41.[15] The tensile softening is a manifestation of the decrease in material cross-sectional area as the crack grows. During compressive loading, the crack does not extend, and crack length has a relatively small effect on the material's compressive load-carrying capacity. This is clearly evident from the hysteretic stress-strain loop shapes. Of course, this kind of cyclical behavior is no reflection of the inherent cyclical flow response, but instead it results from the presence of fatigue cracks. The cyclical deformation response is represented by Fig. 12.39.

Our discussion of polymeric fatigue concludes in the next section where the interaction between fatigue and temperature, particularly with regard to the (sometimes substantial) rise in temperature caused by cyclical straining is considered. Since this is common to many practical situations, the topic has engineering relevance.

D. Temperature Effects

Polymers are subject to the same type of creep-fatigue interactions that metals are. However, these interactions are often more important in plastics because of their relatively low melting temperatures. In this section, however, we focus on a different

---

[15]Note that the conditions here differ from those apropos to Fig. 12.40. For the polycarbonate of Fig. 12.40, crazes, which are *not* cracks, are present. In Fig. 12.41, cracks, whether or not they are accompanied by crack-tip craze formation, are the cause of the behavior illustrated.

kind of fatigue-creep interaction, common to polymers but not to metals, which results from hysteretic heating resulting from the viscoelasticity exhibited by polymers over a relatively broad temperature range. The causes of this heat dissipation have been previously identified in Chap. 2. It can be shown (Prob. 12.21) that the heat dissipated per stress/strain cycle (δQ) is related to stress amplitude by

$$\delta Q = K\sigma_a^2 \tag{12.11}$$

where K is a constant (units of m^5/N). Multiplication of Eq. (12.11) by the cyclical frequency, ν, gives the power loss, $\dot{Q}$ (units of J/s)

$$\dot{Q} = K\nu\sigma_a^2 \tag{12.12}$$

For adiabatic heating (i.e., heating in which all of the heat generated remains within the material and none is lost to the surroundings) the time rate of change of the material's temperature (dT/dt) due to the hysteretic heating is

$$\frac{dT}{dt} = \frac{\dot{Q}}{C_p} = \frac{K\nu\sigma_a^2}{C_p} \tag{12.13}$$

where C_p (J/m^3) is the material's heat capacity.

Of course, no material is immune to heat loss to the surroundings, especially under "real-life" situations. Thus, Eq. (12.13) represents the maximum possible temperature-increase rate. By the same token, virtually all materials are subject to some degree of temperature rise due to these effects, and this causes thermal "softening" owing to the decrease in material strength with increasing temperature. Under certain circumstances, this can lead to instigation of massive flow in the material and/or enhancement of the fatigue process itself.

According to Eq. (12.13) this kind of thermal fatigue damage is frequency sensitive, with increasing frequencies leading to greater sample heating. This is observed experimentally (Fig. 12.42).[16] For the polymer whose behavior is illustrated in Fig. 12.42, its *inherent* fatigue response can only be determined by finding the stress amplitude–cycles to failure relationship at "zero" frequency; this is impossible, so the alternative is to cycle at a frequency so low that hysteretic heat is minimal and that which is generated is essentially immediately lost to the surroundings. We note that Fig. 12.42 shows that the kind of frequency-sensitive damage discussed here is manifested at ordinary, indeed relatively low, cyclic frequencies. This is in distinct contrast to fatigue behavior of metals, which is frequency sensitive only at very high frequencies.

As noted, Eq. (12.13) represents the temperature rise per unit time when all of the internally generated heat serves to increase the material's temperature. When some of this heat is lost to the surroundings, we expect fatigue behavior to be sensitive to the efficiency with which this is accomplished. The temperature increase

---

[16]Note that for the situation of Fig. 12.42, frequency adversely affects material fatigue behavior. This is in contrast to the effect it has on crack growth rates in PS (cf. Fig. 12.37). The difference arises because there is no macroscopic hysteretic heating taking place under the conditions appropriate to Fig. 12.37. Instead, the effect of strain rate on PS is to increase its craze yield stress, simultaneously enhancing its fatigue-crack growth resistance.

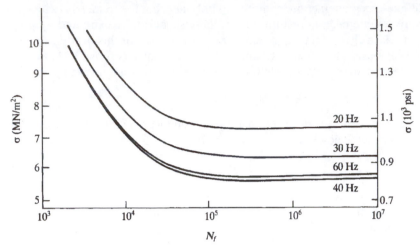

Figure 12.42
Stress amplitude–cycles to failure relationship in Halon TFE as affected by cyclical loading frequency. Increased frequency gives rise to greater hysteretic heating and thermal softening; this is manifested by a decrease in fatigue life and in the endurance limit. (*From M. N. Riddell, G. P. Koo, and J. L. O'Toole, Polymer Eng. Sc., 6, 363, 1966.*)

is expected to depend on sample size, being greater, for example, for thick samples and less for thin ones (external heat loss scales with sample surface area, and heat generated scales with sample volume). This effect, too, has been confirmed. For example, endurance limits are higher for thin samples compared to those for thick ones.

As is evident from this section, thermal effects can significantly impact fatigue behavior of plastics. We have seen that there are many important differences (and also certain similarities) between the fatigue behavior of metals and polymers. But the thermal-softening damage described here is of especial concern, for it can occur at ordinary temperatures and frequencies—those common to plastic use.

12.9
FATIGUE OF COMPOSITES

Not a word has yet been written about fatigue in composites. Indeed, their properties were not even displayed in the material property charts of Figs. 12.22 and 12.23. Why this omission? After all, composites are an important material class. There is a reason for their omission from our discussion of fatigue. It is provided below.

We use a common composite, an aligned fiber composite whose matrix is an epoxy resin, to illustrate. Cyclic damage accumulates in a material like this in the following manner. Fatigue cracks may form in the matrix. However, they propagate slowly since the matrix carries such a small fraction of the applied stress. However, the fibers in the wake of the crack experience fatigue damage. In many cases, the damage rate is accelerated by deleterious interactions with the environment (e.g., fiber oxidation or corrosion). As this crack wake damage increases, composite properties degrade. For example, the composite modulus decreases, and this is observed long before a crack has propagated sufficiently to lead to composite fracture. Thus, this progressive composite "weakening" constitutes engineering failure during cyclical loading. So from a practical perspective we need not be concerned with such matters as fatigue life and fatigue-crack growth rates in composites like this. Their "failure" under cyclic loading is defined by a different criterion than applies to the other material classes.

12.10
SUMMARY

Fatigue refers to mechanical failure (and the processes leading to it) when a material is subjected to a cyclical stress or strain. This kind of failure can take place in ceramics, metals, and polymers. However, it is not commonly thought important in ceramics; small cracks present during fatigue of them quickly result in catastrophic fracture owing to the low fracture toughnesses of most ceramics. Fatigue "failures" also occur in composites. However, from an engineering standpoint, composite properties are impaired by cyclical loading long before a fatigue crack propagates to failure in them. Because of these considerations, our discussion has centered on polymeric and metallic fatigue.

As with ductile fracture, fatigue fracture can be divided into three stages; crack nucleation, crack growth, and final fracture (the latter occurring by void coalescence for ductile fracture and by rapid crack advance for fatigue). However, there are important distinctions between the stages of ductile and fatigue fracture. Fatigue cracks (Stage I of the fatigue process) are formed initially at surfaces and/or defects associated with localized, inhomogeneous plastic flow. These types of fatigue-crack nucleation events are common to both metals and polymers. For metals, the localized surface plasticity leads to the development of a surface feature that, at some stage, can be characterized as a Stage I crack. In its initial growth, such a crack propagates in a direction determined by plastic-flow considerations, and this direction is not normal to the principal stress axis. After such a crack has progressed a certain distance, it alters its direction so that its long dimension becomes normal to the principal tensile axis. At this point, further crack advance depends on factors similar to those applying to tensile fracture.

Following this alteration of its course, continued advance of the fatigue crack takes place intermittently (the slow-crack-growth region of fatigue). The crack-advance rate, dc/dN (the increase in crack length, c, per cycle, N) is related to the "cyclical stress intensity," ΔK ($= \alpha \Delta \sigma c^{1/2}$, where α is a geometrical factor and $\Delta \sigma$ is the range of the cyclical stress). The higher the value of ΔK, the greater is dc/dN, and dc/dN of a given material is uniquely related to ΔK.[17] For most materials there exists a threshold cyclical stress intensity, ΔK_{th}, below which preexisting cracks will not propagate under cyclical loading. Using a material at ΔK values less than ΔK_{th}

---

[17]A caveat is that at high crack growth rates, dc/dN more fundamentally relates to the plastic strain range, $\Delta \varepsilon_{pl}$, than to $\Delta \sigma$.

constitutes "fail safe" fatigue-fracture design. In many cases, ΔK_{th} values are so low that this unduly restricts use of engineering materials in cyclical loading applications. Materials are then used in a non-fail-safe condition. Design in this instance involves use of a constitutive equation (e.g., Eq. (12.6)) to estimate fatigue life, employing the equation conservatively. Materials can fail by fatigue even when they contain no or inconsequentially small preexisting cracks. In this instance, "fail safe" design involves use of the material at a stress amplitude below the endurance or fatigue limit (or the equivalent stress amplitude at a large number of cycles for materials that do not display a true endurance limit). If the material is used in a non-fail-safe condition, constitutive equations relating fatigue life to stress and/or strain range (Eqs. (12.2)–(12.4)) are used to estimate fatigue life; these equations, too, are employed conservatively.

As ΔK is increased above ΔK_{th}, the Stage II fatigue-crack growth rate increases rapidly. At intermediate values of ΔK, crack growth rates relate to ΔK via the equation, $dc/dN = A(\Delta K)^m$, where A and m are constants determined empirically. (This is the equation that can be used in the design process described above.) At even greater values of ΔK, crack growth rate again increases rapidly, and growth rates are much in excess of those expected by extrapolation of the power-law equation to higher ΔK values.

The last stage of slow crack growth precedes final fast fracture (Stage III of the fatigue process). Fast fracture happens when one or the other of two criteria is met. One of these is a tensile fracture criterion. It states that fast fracture occurs when the crack has grown to an extent that the remaining cross-sectional area is unable to sustain the maximum tensile load. On this basis, Stage III fracture is a tensile overload fracture. The other criterion is that fast fracture ensues when the value of K_{max} ($= \alpha\sigma_{max}c^{1/2}$, where σ_{max} is the maximum applied tensile stress) exceeds the material's fracture toughness. In this case, Stage III fracture is dictated by fracture-mechanics considerations (cf. Chap. 9).

Fatigue-fracture surface appearances often reflect the above-noted aspects of Stage II crack growth. For intermediate ΔK values (those where the power-law relationship between dc/dN and ΔK holds), fatigue striations may be found on the fracture surfaces of metals and polymers when these are viewed at appropriate magnifications. The striation spacing corresponds to the crack advance per cycle in metals and, sometimes, polymers as well. Fatigue striae are not often found, or are at least less likely to be observed, at higher and lower values of ΔK. For example, Stage II crack advance in metals at high ΔK frequently involves ductile rupture reminiscent of tensile fracture, and this feature obscures and/or dominates fatigue striations that may be present. Likewise, Stage II cracks propagate discontinuously (discontinuous crack growth, DCG) at low values of ΔK in some polymers. During DCG the crack, rather than advancing a finite distance each cycle, remains dormant for several cycles, and then advances some distance during one cycle. This results in a characteristic fracture-surface-striation spacing much greater than the "average" rate of DCG advance; the latter is the observed spacing divided by the number of cycles between discontinuous advances. These fracture spacings in polymers visually resemble fatigue striae; clearly, however, they are not one and the same.

Stage II fatigue-crack propagation involves crack-tip plasticity and permanent deformation, irrespective of the nature of the material. However, the microscopic manifestation of this deformation differs between metals and polymers. For metals at intermediate ΔK levels, crack advance is concomitant with crack opening during the tensile portion of the cycle, subsequent blunting of the crack by crack-tip plasticity, and resharpening of it during stress reversal. The sequence results in the crack advancing a certain distance (the striation spacing) per cycle. Another example illustrating the intimate relation between Stage II crack growth and crack-tip deformation is provided in DCG of polymers. In most polymers displaying DCG, crazes develop at the crack tip owing to the stress concentration there. During the growth stasis, fatigue damage takes place at the crack tip. The damage results in craze lengthening and thickening, and the crack opening also increases. Finally, a critical crack-opening displacement is attained, and discontinuous crack advance is effected. In some polymers, shear-band formation accompanies the crazing common to DCG; the phenomenon is termed εDCG. In thermoplastics manifesting εDCG, crack advance can take place also solely by shear-band cracking. This kind of crack advance is found at relatively high applied stresses and temperatures.

Slow crack progression is more difficult when εDCG is the mode of crack growth rather than when DCG is associated only with crack-tip crazing. This is due to the work expended in forming shear bands, and also reflects that fracture toughness and polymer fatigue-fracture resistance are linked. Indeed, these parameters go hand in hand for engineering plastics, while this is not the case for metals and their alloys. In fact, Stage II crack growth rates are remarkably structure insensitive for metals. They can, however, be correlated with the structure-insensitive property, elastic modulus (the higher it is, generally the greater the material's slow-crack-growth resistance).

The macroscopic and microscopic features present on a fatigue-fracture surface reflect the above processes, particularly for metals. The slow-crack-growth surface in metals is characteristically smooth, unless obscured by features such as oxidation or abrasion. Clamshell, or beach, markings are typically seen on these surfaces, as well as occasionally the much finer fatigue striae. The spacing between beach markings is far greater than the distance the crack advances per cycle; the latter is measured by the striation spacing. The fast-fracture area of the fracture surface is dull and fibrous, bearing a distinct resemblance to a tensile ductile fracture surface.

A material's resistance to fatigue fracture can be ascertained by subjecting it to a cyclically varying stress (or strain)[18] and determining the number of cycles (N_f) required for it to fracture. The test results can be graphed with N_f as the abscissa and the cyclical stress (strain) amplitude as the ordinate. For a stress-controlled test, N_f depends on the value of the mean stress (σ_{mean}), decreasing as σ_{mean} increases. Engineering estimates, such as the Goodman diagram are useful (even though not precise) for correlating safe operating stress amplitudes for one mean stress with those for another. During high-cycle fatigue (where N_f is large, usually $> 10^3$), the material is macroscopically strained elastically, and the negative slope of the σ_a-N_f plot has a small absolute value. In fact, it is zero, or so close to it, for some materials,

---

[18]Equal-stress and -strain amplitude tests are equivalent for macroscopic linear elastic deformation for which the strain range, $\Delta\varepsilon$ ($= \Delta\varepsilon_{el}$) is related to stress range by $\Delta\varepsilon = \Delta\sigma/E$. When the strain is partly permanent (for metals) or partly nonlinear elastic (as for polymers) there are (sometimes substantial) differences between the results of an equal-stress amplitude test and an equal-strain amplitude test.

that there is a limiting stress amplitude (called the fatigue or endurance limit) below which the material will not fail by fatigue. As described previously, the endurance limit can be used in fatigue design for materials that can be considered initially flaw free. During high-cycle fatigue, N_f is related to either the elastic strain range ($\Delta\varepsilon_{el}$) *or* the stress range ($\Delta\sigma = E\Delta\varepsilon_{el}$), since the two are singularly related. For materials that fail at these low cyclical strains, it is found that nucleation of Stage I fatigue cracks (assuming the material is initially crack free) requires far more cycles than does the propagation of a Stage II crack to its critical length. This is true for high-cycle fatigue in both polymers and metals. Thus, resistance to crack nucleation is the dominant requirement for high-cycle fatigue in materials not containing preexisting flaws. As most fatigue cracks are then surface-originated as a result of localized plastic flow, surface hardening, which restricts permanent deformation, is an effective way of improving a material's high-cycle fatigue resistance. In metals this hardening may be accomplished in several ways, including cold-working the surface and, for steels, by thermal or chemical treatments that raise the surface yield strength. In this sense, high-cycle fatigue resistance is structure sensitive for a metal, even though Stage II crack-propagation rates are not.

As the cyclic stress or strain amplitude is increased, N_f decreases and, concomitantly, a component of cyclic plastic strain is developed, especially in metals. When N_f has a value of about 10^3, the elastic and plastic strain amplitudes are approximately equal, whereas $\Delta\varepsilon_{pl}$ exceeds $\Delta\varepsilon_{el}$ for lesser values of N_f. For this low-cycle fatigue, $\Delta\varepsilon_{pl}$ correlates better with N_f than does $\Delta\sigma$. Empirical correlations relating $\Delta\varepsilon_{pl}$ with N_f have been developed for low-cycle fatigue, and like equations correlate $\Delta\varepsilon_{el}$ with N_f during high-cycle fatigue. Use of these (alone or in combination when the elastic and plastic strain ranges are comparable) provides estimates of fatigue life in flaw-free materials.

Ductile metals are best suited to low-cycle fatigue applications. This relates to the fact that Stage II growth occupies most of the material's life during low-cycle fatigue. Since ductility facilitates crack-tip blunting and since crack advance often involves ductile rupture at the high values of ΔK common to low-cycle fatigue, it is clear that hard, brittle materials do not resist crack advance as well as less strong, but more ductile ones.

Knowledge of a material's fatigue behavior is facilitated by determining its response to an applied cyclic stress or strain. Applying a specified strain range involves monitoring the stress amplitude required to maintain the strain amplitude. The resulting varying stress amplitude is then measured as a function of the number of cycles (N). When this is done, it is found that some metals cyclically harden (the stress amplitude increases with N), whereas others soften (the stress amplitude decreases with N). The degree and the rate at which hardening/softening takes place depends on $\Delta\varepsilon_{pl}$ and the metal's mechanical properties. Typically, "hard" alloys of a given material soften and "soft" ones harden. Following a sufficient number of cycles (usually fewer than 100, and many times less than this), the stress amplitude attains a steady-state value. For metals that manifest wavy slip, the steady-state stress amplitude depends only on $\Delta\varepsilon_{pl}$, and *not* on the material's initial hardness or strength. This is not so for planar slip materials; initially hard materials that flow thusly are found to soften, and initially soft ones to harden, but the steady-state stress amplitude of the initially hard material is always greater than that of the initially soft one.

Polymers do not cyclically harden. They either soften or, less frequently, are cyclically stable. The degree to which softening occurs depends on polymer mechanical characteristics. Soft, pliable polymers soften more, and do so more rapidly, than relatively stiff ones.

The steady-state stress amplitude depends on $\Delta\varepsilon_{pl}$ for metals and on the total strain range for polymers. A plot of this amplitude against the corresponding strain amplitude defines the material's cyclical stress-strain curve. This differs (sometimes substantially) from the monotonic one, and a knowledge of the cyclical stress-strain curve is important in understanding fatigue. For metals, the cyclical stress-strain curve is often fitted empirically to an equation of the form $\Delta\sigma/2 = K'(\Delta\varepsilon_{pl}/2)^{n'}$, and the material constants K' and n' can be used in predicting fatigue lifetimes.

Temperature has an influence on fatigue response of both metals and polymers. This is important for creep-fatigue interactions in metals at high temperature. Guidelines for high-temperature fatigue (and for creep perturbed by a cyclical stress) remain empirical. Likewise, cumulative damage laws, similar to those used for creep and fatigue fracture, have been put forth. These are prone to considerable uncertainty, and must be used cautiously.

Temperature has a greater effect on polymer fatigue than on metal fatigue, particularly at ordinary temperatures. In part this reflects that ordinary (e.g., room-temperature) use of polymers takes place at relatively high homologous temperatures. Thus, creep is ubiquitous to polymers at ordinary temperatures. As such, even relatively low frequencies affect polymeric fatigue behavior. However, the effect of frequency can be complicated. In some cases increasing frequency increases the material's flow strength because of strain-rate sensitivity effects. This causes a concurrent increase in the polymer's fatigue-crack propagation resistance. In other polymers, hysteretic heating results in a temperature increase with an attendant diminution of the material's flow strength. This cyclical thermal softening can be extensive, even at ambient temperatures and modest cyclical frequencies. It often leads to premature "fatigue" failure even though it is clear that failure is instigated by temperature and not, per se, by the cyclical loading.

REFERENCES

Beardmore, P., and S. Rabinowitz: "Fatigue Deformation of Polymers," *Treat. Matls. Sc. Tech.*, ed. R. J. Arsenault, 6, 267, 1975.

Coffin, L. F., Jr.: "Fatigue," *Ann. Rev. Matls. Sc.*, 2, 313, 1972.

Fleck, N. A., K. J. Kang, and M. F. Ashby: "The Cyclic Properties of Engineering Materials," *Acta Metal. et Mat.*, 42, 365, 1994.

Laird, C.: "Fatigue," *Physical Metallurgy,* 4th ed., ed. R. W. Cahn and P. Haasen, North-Holland, Amsterdam, 2293, 1996.

Laird, C. M.: "Cyclic Deformation of Metals and Alloys," *Treat. Matls. Sc. Tech.*, ed. R. J. Arsenault, 6, 101, 1975.

Suresh, S.: *Fatigue of Materials,* Cambridge University Press, Cambridge, England, 1991.

Takemori, M. T.: "Polymer Fatigue," *Ann. Rev. Matls. Sc.*, 14, 171, 1984.

PROBLEMS

12.1 Show that the stress ratio, $R = \sigma_{min}/\sigma_{max}$, is uniquely defined if the mean stress and stress amplitude are specified.

12.2 The fatigue (endurance) limit of a 1045 (plain carbon) steel is about 300 MN/m^2 when the mean stress is zero. The tensile strength of this steel is 750 MN/m^2. Using the Goodman equation, estimate the safe stress amplitude for 1045 steel for the situation of a mean stress of 250 MN/m^2.

12.3 **a** Fatigue data extrapolation techniques other than the Goodman one exist. The Soderberg relation is a more conservative extrapolation. It is obtained by substituting the material yield strength, σ_y, for the tensile strength (T.S.) in the Goodman equation. The Gerber relation is more permissive than the Goodman one. It expresses the allowable stress amplitude (σ_a) at a mean stress, σ_{mean}, as

$$\sigma_a = \sigma_{fat}\left[1 - \left(\frac{\sigma_{mean}}{\text{T.S.}}\right)^2\right]$$

where σ_{fat} is the endurance (fatigue) limit at a mean stress of zero. An empirical equation that describes well the fatigue behavior of metals (see Fleck et al. reference) is

$$\sigma_a = \sigma_{fat}\left[1 - 0.4\left(\frac{\sigma_{mean}}{\sigma_y}\right)\right]$$

Make schematic plots of the permissible stress amplitudes as a function of σ_{mean}, using all four relationships mentioned above. Do this for two values (0.5 and 0.8) of the ratio, σ_y/T.S. For both situations list the extrapolation techniques in terms of increasing conservative nature.
b What would the ratio, σ_y/T.S., have to be for the Fleck et al. equation to be the same as the Goodman equation?

12.4 **a** Determine the parameter σ_f' (Eq. (12.2)) for the materials whose behavior is illustrated in Fig. 12.14. For all of the steels, $E = 210$ GPa.
b How do the values of σ_f' compare with the steel tensile strengths? (For the Maraging steel, T.S. $\cong 2000$ MPa. For the 4340 alloy steel, the tensile strength is 1800 MPa in the hardened condition and 1300 MPa in the annealed state.)

12.5 As mentioned, steel used for high-cycle fatigue applications frequently has its surface hardened by chemical or mechanical treatment. Very often the interior of the steel is soft and ductile; that is, the interior might be in an annealed rather than quenched and tempered condition. Explain why the soft and ductile interior is often preferred, even though a hard and relatively brittle surface is desired.

12.6 **a** Assume that Eq. (12.5) describes the variation of dc/dN with ΔK. Derive an expression for the number of cycles a material spends in Stage II of fatigue-crack growth. (Assume an initial crack size of c_0 and a final crack size, c_f.)
b The number of cycles spent in Stage II crack growth (N_{II}) is often used to estimate a material's fatigue life (N_f). Under what conditions (i.e., high- or low-cycle fatigue) would your expression in (a) reasonably approximate fatigue lives for "flaw-free" materials? Is it a reasonable approximation for high-cycle fatigue lifetimes in structures initially containing reasonably sized cracks? Give your reasoning.

12.7 A 30-cm-diameter 7079-T6 aluminum alloy, whose fatigue-crack growth behavior is shown in Fig. 12.20, fractured in a cyclic stress application. The applied stress was normal to and uniform across what was eventually the final fracture surface. The maximum value of this stress was specified as 40% of the material yield strength ($\sigma_y = 350$ MN/m^2). Post-failure analysis indicated that a fatigue crack grew from a small surface defect, and eventually produced a tensile-overload fracture. Striation spacing measurements of 5×10^{-5} and 13×10^{-5} cm were obtained at 0.75 and 3 cm, respectively, from the origin of the fatigue crack.

The shaft manufacturer claimed the user overstressed the part and was negligent by exceeding the maximum design stress. The user, claiming the part was operated within the design limits, sued the manufacturer for damages. You are retained to provide a technical judgment of liability. What is your verdict? Show your reasoning and state any assumptions.

12.8 In a plot similar to that of Fig. 12.16, plot c vs. N for a fixed R but for different values of the stress amplitude.

12.9 **a** Using a ruler, estimate the crack advance per cycle for the aluminum alloy whose fracture surface is shown in Fig. 12.2b.
b Assume this alloy is a 7075-T6 alloy whose fatigue-crack propagation behavior is shown in Fig. 12.18b. Estimate the ΔK range to which the material was subjected during the cyclic loading that gave rise to the striation spacing of Fig. 12.2b.

12.10 **a** Determine average values of A and m for Stage II fatigue-crack growth in the steels and the aluminum and titanium alloys whose behavior is displayed in Fig. 12.20a.
b Assume initial cracks of size 0.2 cm are present on the surfaces of these materials. Estimate their fatigue lifetimes for an R value of zero. Data: For the Al alloys, $K_{Ic} = 30$ MN/m$^{3/2}$, T. S. $= 520$ MN/m^2; for the Ti alloys, $K_{Ic} = 85$ MN/m$^{3/2}$, T.S. $= 1070$ MN/m^2; for the steels, $K_{Ic} = 55$ MN/m$^{3/2}$, T.S. $= 1720$ MN/m^2.

12.11 **a** Show that Eq. (12.8b) can be approximated by $N_f(\Delta\sigma)^m =$ constant.
b Refer to Fig. 12.8. Fit the fatigue behavior of the steel and the Al alloy to the relationship $N_f(\Delta\sigma)^m =$ constant, and determine the m value appropriate to each material. Do these m values agree with those obtained by analyzing Fig. 12.20a (see Prob. 12.10)?

12.12 Schematically sketch dc/dN vs. ΔK for different values of σ_{max} in the transition from Stage II to Stage III fatigue-crack growth. Explain the reasoning you used in constructing the curves.

12.13 Refer to Fig. 12.21a. Assume this figure is appropriate for an R value of -1. We now consider cyclic loading with $R = 0.7$. ΔK_{th} does not change much with variations in R for negative R values. However, ΔK_{th} decreases with increasing positive R values. Assume that ΔK_{th} decreases by 50% when R is increased from 0 to 0.7.
a Schematically sketch how Fig. 12.21a would change were it constructed for an R value of 0.7. How would the respective regions of intrinsic and extrinsic fatigue change?
b Now we quantitatively treat the problem. Assume we are dealing with a 7075-T6 aluminum alloy whose behavior is illustrated in Fig. 12.18. Further data for 7075-T6 are T.S. $= 570$ MPa, $\sigma_y = 500$ MPa, and $\sigma_{fat} = 160$ MPa. Construct diagrams like Fig. 12.21a for $R = -1$, $R = 0$, and $R = 0.7$. State any assumptions. (You may assume the Goodman equation describes the relationship between allowable stress amplitude and mean stress.) Comment on your results.

12.14 Discuss and contrast the development of a steady-state substructure during creep and during cyclic straining. What parameters are held constant in each type of test and which ones change? Can the steady-state substructure developed during cyclic straining be related to "recovery" processes, as it is in creep? Explain.

12.15 Suppose we have a history-dependent material and attempt to determine its cyclical stress-strain behavior by the "short cut" described in the text. That is, a small plastic strain range is applied and the steady-state stress amplitude is determined. Then the plastic strain range is progressively increased, with the procedure repeated at each strain range. Will the results of this procedure produce a valid cyclical stress-strain curve? If not, how will the values of K' and n' obtained from this procedure compare to their true values?

12.16 The Palmgren-Miner rule disregards "history" effects. That is, it assumes that it makes no difference whether a high stress amplitude is applied first and a lower one subsequently, or vice versa. Explain why this is not a good assumption.

12.17 **a** Refer to Fig. 12.28. Assume that the equilibrium dihedral angle is attained on each stress/strain reversal (i.e., that the situation of Fig. 12.28*d* is realized). How will the void growth per cycle depend on the relative grain-boundary displacement per cycle?
b Describe how the value of the cyclical frequency affects whether or not the equilibrium dihedral angle is attained on each load reversal.
c Estimate, in terms of relative grain-boundary displacement and appropriate diffusion coefficient(s), the critical cyclical frequency below which the equilibrium dihedral angle is attained on each load reversal. (Hint: Estimate the amount of mass that must be transferred on each such reversal (see Figs. 12.28*b–d*), and approximate the diffusion geometry.)

12.18 **a** Refer to Fig. 12.30, the fatigue-fracture map for copper alloy 175. Does the solid line (marked $N_f = 1/4$) or the dotted line (marked tensile instability) trace the *uniform* strain in a tension test?
b Estimate the dominant failure mechanism (i.e., tensile, creep, fatigue, etc.) for copper alloy 175 at the following plastic strain range–temperature combinations; $\Delta\varepsilon_{pl} = 5 \times 10^{-3}$, $T = 300$ K; $\Delta\varepsilon_{pl} = 5 \times 10^{-2}$, $T = 700$ K; $\Delta\varepsilon_{pl} = 10^{-2}$, $T = 500$ K. (The melting temperature of copper alloy 175 can be taken as 1350 K.)

12.19 **a** The polymer fatigue-crack map of Fig. 12.36 is drawn for uniaxial tensile loading. How will the respective regions of the map change if the loading is biaxial tension?
b How will crack-advance rates be changed for biaxial, as opposed to uniaxial, tensile loading?
c Compare the map of Fig. 12.36, which defines the conditions for craze/shear-band formation at fatigue crack tips, with the criterion determining the craze/shear-band transition in uniaxial tension (cf. Chap. 8).

12.20 Figure 12.40 shows the cyclic stress-strain behavior of a polycarbonate (PC) that crazes under tensile, but not compressive, loading. This behavior gives rise to tensile (but not compressive) fatigue "softening." Suppose the PC were to deform by shear banding during fatigue. Sketch the analog of Fig. 12.40 for this situation, and provide reasons for any differences between your sketch and Fig. 12.40.

12.21 Model polymer viscoelasticity on the basis of a Voigt model (cf. Chap. 2), and show that the energy loss per cycle is given by an expression of the form of Eq. (12.11). Identify the constant K of Eq. (12.11) in terms of the Voigt model parameters.

12.22 a Using the data of Fig. 12.42, estimate the endurance limit of Halon TFE at a cyclical stress frequency, $v = 0$. Is your estimate conservative? Explain.
b Can you think of a reason why the fatigue behavior of Halon TFE at $v = 60$ Hz is the same as that (or even slightly better than) at a cyclical frequency of 40 Hz?

CHAPTER 13

Embrittlement

13.1
INTRODUCTION

> "All engineering materials are reactive chemicals; the surprise is not that they fail, the surprise is that they work."

The above quotation is that of R. W. Staehle. It can be found in his paper referenced at the end of this chapter. What Staehle refers to is that given the "right" combination of environment and stress, just about any engineering material will fail. Environment here is construed broadly. By it we refer to the gaseous or fluid atmosphere in which a material operates. It is found that in many environments, material mechanical performance is degraded. This deleterious effect, in concert with an applied or residual stress, gives rise to what is termed "Environmentally Assisted Failure." While this is the preferred terminology, we shall employ the generic term "Embrittlement." This use is proper in the sense that many materials are "embrittled" as a result of the combined action of the environment and stress. Besides, embrittlement is easier to remember and it saves us a few words here and there.

The term "Embrittlement" is applied to a variety of phenomena causing mechanical performance degradation as a result of a stressed material's exposure to a "hostile" environment. Embrittlement may be manifested in a number of ways. Static fatigue or delayed fracture refers to material failure at a stress (or stress intensity) below that required to cause fracture in the absence of an embrittling agent. This kind of failure takes place by environmentally assisted slow crack growth to a stage where the crack is large enough to cause spontaneous fracture. This slow crack growth does not take place in the absence of the environment, and the material is immune to delayed failure. Static fatigue occurs in numerous materials, and in a variety of environments, and for a variety of reasons. For example, moisture causes slow crack growth of glass surface flaws; acetone does the same with stressed polystyrene; and zinc coated with mercury is also susceptible to static fatigue.

Embrittlement is also manifested by a reduction in tensile ductility, whether measured by uniform strain, strain to failure, or reduction in area. In a tensile test, the embrittled material's flow stress may be unchanged (as is often the case) or increased (which happens sometimes), but ductility is always reduced, sometimes markedly. Inherently ductile metals, for example, may have their reduction in area decreased substantially although still exhibiting ductile fracture if embrittlement is moderate. Alternatively, the metals may fracture in a macroscopically brittle way (%R.A. = 0) if embrittlement is severe.

Fracture work (toughness) is decreased, too, when embrittlement happens. For example, changes in impact energy often parallel tensile ductility loss and for those materials exhibiting a ductile-to-brittle transition temperature, this temperature is often raised by the embrittlement process.

As noted, embrittlement may result from a number of material-environment interactions. There are, however, some common features to embrittlement phenomena. For example, some metals and their alloys are prone to embrittlement when exposed to a fluid medium, whether it is a liquid metal (which gives rise to liquid-metal embrittlement—LME), an aqueous solution (which results in stress-corrosion cracking—SCC), or a gas which sometimes results in SCC and sometimes in impurity atom embrittlement (IAE). During these kinds of embrittlement, the interaction between the fluid and the tip of the crack that eventually causes failure is critical in determining the extent of embrittlement, and this is so even though the kinetics of LME and SCC often differ dramatically (LME slow-crack-growth rates are generally much greater than those of SCC). From an engineering standpoint a common approach to embrittlement is useful as it provides a conceptual basis for the phenomenon. Moreover, and as we shall see, the mechanisms responsible for the various kinds of embrittlement have not been clarified to the extent desirable, and this makes a catholic approach to embrittlement even more attractive. On the other hand, prevention of embrittlement devolves on a fundamental understanding of its mechanisms at the atomic or microstructural scale. This is the approach we basically follow here even though current theories on embrittlement often remain murky. When appropriate, however, commonalties as well as differences among the several forms of embrittlement are emphasized.

Many solids are embrittled by an external environment, be it gaseous, liquid, or solid. In these situations the embrittler first initiates a surface crack (if none is present to begin with) and then aids in the growth of such a crack until it reaches critical size. Examples of this kind of external embrittlement are many cases of metal embrittlement (ME) (LME being a special case of ME), SCC, and some forms of hydrogen embrittlement (HE).

Metal embrittlement refers to the degradation in mechanical properties of a higher-melting-point metal when it is in contact with a lower-melting metal. Generally, the contact takes place on the external surface of the embrittled metal, but ME can also happen when the embrittler is located within the bulk of the embrittled metal. For many years, it was thought that the lower-melting-point embrittler had to be in the liquid state to cause embrittlement. However, it was later shown that embrittlement can also take place if the lower-melting component is in solid form (solid-metal embrittlement—SME). Solid-metal embrittlement is often associated with external plating. An example is the occasionally found embrittlement of high-strength steel caused by cadmium plating, an especial irony in view of the fact that the plating is done to prevent corrosion of the steel in the first place. Solid-metal embrittlement is also sometimes caused by internal "plating." Lead is often added to high-strength steel to improve its machining characteristics. Under appropriate circumstances, however, the lead, located at grain boundaries, embrittles the steel to the same degree that it would were it plated on the steel surface.

Stress-corrosion cracking refers to the environmentally assisted fracture observed in many stressed materials when exposed to an aqueous (or sometimes gaseous), corrosive environment. The environment acts in concert with the stress to cause (in some cases) crack nucleation and subsequent slow crack growth. Clearly, there are some things common to LME and SCC, particularly with regard to environment–crack-tip interactions, and this is so even though crack-growth kinetics differ between LME and SCC. Some aqueous environments give rise to a special kind of SCC. These are those in which hydrogen is a byproduct of the crack-tip corrosion process, and in which this hydrogen enters into the material in front of the crack. This kind of environmentally aided failure is one form of hydrogen embrittlement (HE). In the situation just noted, hydrogen enters the metal via a surface reaction. However, hydrogen can also enter metals from a variety of other sources, including some associated with materials processing. There are several ways in which internal hydrogen can cause embrittlement. Sometimes, hydrogen in excess of its solid solubility enters the metal during processing, and subsequently precipitates as gas-filled voids (bubbles) that facilitate material fracture. When internal oxygen is also present, as it is in some steels, internal decarburization resulting in methane gas bubbles may take place. In certain other materials, the excess hydrogen precipitates in the form of a brittle hydride phase that reduces fracture toughness. Regardless of the specifics of the precipitate reaction, embrittlement to one degree or another is caused by it.

Impurity atoms, particularly "tramp" elements such as antimony, phosphorus, and sulfur, when present in small amounts frequently embrittle metals and ceramics. Current views relate this embrittlement to the segregation of the harmful element to grain or other internal boundaries where they reduce the cohesive energy and the plastic work accompanying crack propagation. This kind of embrittlement can cause a material prone to brittle fracture to fail in this way at temperatures at which it would ordinarily be ductile; that is, the ductile-to-brittle transition temperature is increased. Impurity-atom embrittlement is not found at high temperature owing to the inherent ductility of the metal or as a result of reduced grain-boundary segregation. Ductile metals can be embrittled by tramp elements, too. In some cases this is manifested by a modest decrease in ductility over the temperature range in which impurity segregation occurs; in others a fully embrittled state (e.g., %R.A. = 0 in a tensile test) can be attained.

Materials may also be embrittled by exposure to neutron radiation; we call this radiation embrittlement. Exposure to high-energy neutrons occurs in many conventional fission reactors, and this exposure is even greater in "fast" fission (e.g., breeder) reactors and is expected to be especially severe in materials proposed for use in fusion reactors (if these ever come into being). Neutron exposure induces "radiation damage" via collisions between neutrons and the atoms constituting the material. High-energy neutrons displace atoms over distances large in comparison to the interatomic spacing, and this results in the creation of vacancies and interstitials. These then react to form vacancy and interstitial loops that harden the material and make it more susceptible to brittle fracture. Moreover, if the temperature is sufficiently high, growth of vacancy loops takes place and this eventually leads to void formation. The effect causes material swelling that may be sufficient to impair structural integrity. Similarly, krypton and xenon—fission byproducts—are not soluble in the materials in which they form. They precipitate as internally pressurized gas bubbles that give rise to swelling and often also a reduction in fracture resistance. In proposed fusion reactors, nuclear transmutations will take place, and over the intended lifetime of a material its composition can be altered appreciably by this effect. While "embrittlement" is perhaps not the best term to describe these phenomena, the changes in material properties that result therefrom are similar to those observed for more conventional embrittlement mechanisms.

There are thus a number of features common to the embrittlement forms we consider. Conversely, there are important mechanistic and microstructural differences among them. Accordingly, here we separately discuss the different kinds of embrittlement, although at the same time points of commonality are emphasized if appropriate. The topics discussed deal primarily with embrittlement of metals and their alloys; these materials having been the most extensively studied. However, embrittlement of nonmetallics is becoming more common as their use increases. Accordingly, the environmental-induced degradation of nonmetallics is discussed following our treatment of the phenomena in metals. We start with ME. A reasonable consensus with respect to possible mechanisms causing it has emerged. Moreover, it appears to be a special case of brittle fracture, the principles governing which are on fairly firm footing (cf. Chaps. 9 and 10).

13.2
METAL EMBRITTLEMENT

A. Characteristics

Metal embrittlement (ME) may take place when a low-melting-point metal is placed in contact with a higher-melting-point one, the latter being embrittled as a consequence. For many years, it was thought the lower-melting material had to be in the liquid state if embrittlement was to be observed. However, this is not so. Embrittlement can, and does, occur in many cases when the embrittling agent is in solid form, although this solid-metal embrittlement (SME) is generally less pernicious and takes place at a lower rate than does liquid-metal embrittlement (LME). An example of metal embrittlement in which both SME and LME occurs is provided by the embrittlement of high-strength steels by lead,[1] as illustrated in Fig. 13.1. Here the ratio of the tensile reduction in area of leaded steels to that of unleaded steels is plotted as a function of the test temperature. Clearly, embrittlement, as marked by

---

[1] Here the lead is present as intergranular particles in the steel.

a decreased ductility in the leaded steel, happens at temperatures below lead's melting point (although the SME induced by lead is less severe than the LME). The temperature variation of most metal embrittlement is very much like that shown in Fig. 13.1. That is, there is a both a lower temperature below which, and a higher temperature above which, embrittlement is not observed. The lower temperature corresponds to one at which transport of the embrittling species to a crack tip is not rapid enough to allow environmentally assisted crack growth under the imposed stress or strain rate. Thus, increases in strain rate raise the lower embrittling temperature or, equivalently, reduce the tendency for SME. The high temperature at which ductility is restored also depends on strain rate, increasing with this rate. In many cases restoration of ductility relates to the increase in material toughness accompanying increases in temperature. This is consistent with viewing LME as a special case of brittle fracture, and is also in accord with the strain-rate variation of the upper "ductilizing temperature."

Other factors generally associated with an enhanced tendency for brittle fracture also tend to promote ME. For example, increases in grain size increase a material's susceptibility to ME, and so generally does an increase in material yield strength (e.g., by solid-solution strengthening). A possible exception to this generalization is the effect that cold work has on the susceptibility to ME; sometimes such working increases and sometimes it decreases this susceptibility.

The results of tension tests on two metals, Zn and Fe-Al alloys, prone to LME are shown in Fig. 13.2. The overall level of the flow stress is unaffected by the em-

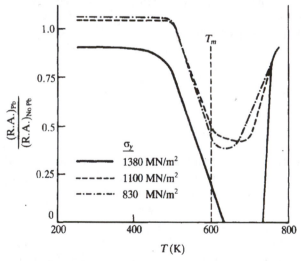

Figure 13.1
Temperature variation of tensile ductility of 4145 alloy steels (heat-treated to different strength levels) that are subject to embrittlement by lead. Relative ductility is measured by the ratio of the reduction in area of leaded steels to those not containing lead. Embrittlement is most severe above lead's melting point, but is still observed at temperatures less than this. Higher-strength steels are embrittled more severely than lower-strength ones. (*Adapted from S. Mostovy and N. N. Breyer, Trans. A.S.M., 61, 219, 1968.*)

brittling species, but fracture occurs prematurely in the embrittling environment. These are general features of LME; i.e., tensile ductilities and reductions in area are diminished by embrittlement. Since the macroscopic flow behavior of the material is unaffected, we conclude that the embrittler has no influence on material bulk properties. Instead, its effect is felt in small and localized regions where cracks nucleate (or are there to begin with) and then subsequently grow.

Figure 13.2*b* shows that embrittlement of Fe-Al alloys by Hg-In amalgams depends on the aluminum content of the alloy. Indeed, it is found that a critical aluminum content (about 4 at.% for Fe-Al alloys) must be exceeded for any diminution

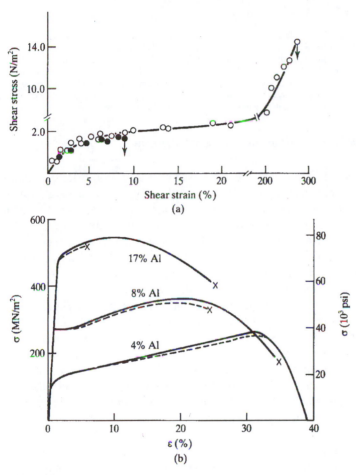

Figure 13.2
(a) Room-temperature shear stress–shear strain behavior of mercury coated (•) and uncoated (○) zinc single crystals. LME reduces the shear fracture strain from ca. 275% to about 9%. (b) Tensile stress-strain curves for polycrystalline Fe-Al alloys with (dashed curves) and without (solid curves) a liquid Hg-In coating. This coating causes embrittlement if the alloy aluminum content exceeds about 4 at.%. (*Part (a) from V. I. Likhtman and E. D. Shehckin, Sov. Phys.—U.S.P., 1, 91, 1958; (b) from N. S. Stoloff, R. G. Davies, and T. L. Johnston, Environment Sensitive Mechanical Behavior, Gordon and Breach, 1966, p. 613. Reproduced with permission of TMS.*)

Embrittler	Embrittled metal/alloy system						
	Al	Cu	Fe	Mg	Ni	Ti	Zr
Bi		L					
Cd			L,S			L,S	L,S
Ga	L	L	L				
Hg	L	L,S	L		L	L	
In	L,S	L	L,S				
Li		L	L		L		
Na	L,S	L		L	L		
Pb		L	L,S				
Sn	L	L	S				
Zn	L		L,S	L			

Notes: L = liquid-metal embrittlement; S = solid-metal embrittlement. The table includes only the most commercially important embrittled metal systems. Further, the classification for Fe includes Fe, low-alloy, and high-alloy steels. Embrittlement might be found in, for example, a high-alloy steel but not a low-alloy one. The absence of indicated embrittlement does not mean the metal combinations may not exhibit embrittlement under appropriate environmental conditions. Finally, it is reasonable that all combinations exhibiting solid-metal embrittlement will also manifest liquid-metal embrittlement, even though this might not be indicated in the table (e.g., Sn-Fe).

Table 13.1
Some metal-metal combinations prone to metal embrittlement

in the R.A. to be observed. Similar effects are found when the embrittler composition is varied. For example, mercury does not embrittle polycrystalline cadmium at room temperature, but the addition of relatively small amounts of indium (10 at.%) to mercury causes some reduction in tensile ductility, and the addition of 25 at.% indium results in a tensile R.A. equal to zero. It is not clear whether the effect is due to the overall liquid composition or, instead, to relatively "inert" mercury acting as a carrier for the embrittling agent; indium, in this case.

Results such as those mentioned above, show that ME is not ubiquitous, but occurs only between specific embrittler-metal combinations, and then only under certain temperature and stress or strain-rate conditions. Some of the known combinations are listed in Table 13.1. Many of these are of technological importance, cadmium-plated steel having been mentioned already. It must be cautioned that an absence of embrittlement as indicated by this table may only imply that conditions appropriate for it to be observed have not yet been identified. Indeed, over the last three decades the number of metal combinations prone to ME has grown many-fold, and combinations thought formerly to be innocuous have been found not to be so.

Metal embrittlement is often also manifested by static fatigue (Fig. 13.3). In such a test a material is subject to a stress level less than that required to cause plastic flow and the time required to cause fracture in the presence of an embrittling agent is measured. Often there is a stress (frequently referred to as the static fatigue limit) below which the material does not fracture regardless of the exposure time. Static fatigue curves of the kind shown in Fig. 13.3 often depend on the presence of initial flaws. For a material prone to brittle fracture, for example, such flaws reduce failure times and decrease the overall stress level in a diagram like Fig. 13.3. However, the static fatigue limit of a "flaw-free" material constitutes a design parameter for such materials in an embrittling environment.

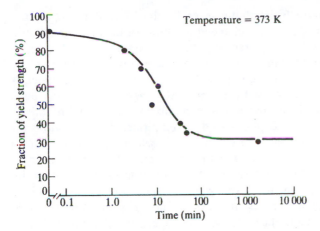

Figure 13.3
Static fatigue at 373 K of a 2024 aluminum alloy coated with mercury. In the test, a stress (some fraction of the tensile yield strength) is applied and the time required to fracture the material measured. The static-fatigue limit of this alloy is about 30% of its yield strength; applied stresses below this level do not result in delayed failure, regardless of the test duration. (*From W. Rostoker, J. M. McCaughey, and H. Markus, Embrittlement by Liquid Metals, Reinhold, New York, 1966.*)

Crack-growth rates can be measured during ME. When they are plotted versus the stress intensity (K), the crack-growth rate is found to vary with K as illustrated in Fig. 13.4. Slow crack growth takes place at stress intensities well below the material's fracture toughness (K_{Ic}). It appears there is a stress intensity below which crack growth will not occur (K_{IcME}), and this stress intensity relates to design with flawed materials in a metal embrittling environment similar to the way ΔK_{th} relates to fatigue design with similarly flawed materials. The plateau (Stage II of Fig. 13.4) in the crack-growth rate corresponds to steady-state environmentally assisted crack growth. The independence of dc/dt on K in Stage II indicates that although an applied stress may be necessary to drive crack growth, the rate at which it does is controlled by other processes (see Sect. 13.2B). Steady-state crack growth during ME, particularly LME, can be quite high (on the order of 0.1 m/s). As a crack continues to grow during Stage II, the stress intensity increases until the embrittled material fails at a K value (K_{Ic}) characteristic of the bulk material. The final fracture does not depend on the presence of the embrittling species; i.e., it is dictated by bulk material properties (e.g., K_{Ic}).

Clearly ME is a complex process involving material combinations, temperature, strain rate, stress state, and metallurgical and chemical factors. In spite of, or perhaps because of, the complexities of ME some empirical guidelines relating to it have been developed. First, a tensile stress component must be present for ME to happen. Second, the embrittler must "wet" the solid, irrespective of whether the embrittler is in the solid or liquid form. The wetting stipulation does not mean wetting must occur on a macroscopic level, but only at the atomically clean and sharp crack tip (its shape facilitates wetting via capillarity). In addition, metal combinations should be relatively insoluble in each other over the temperature range in which embrittlement is found. Enhanced solubility of the metal in a liquid embrittler, for example, would lead to crack-tip material dissolution and result in crack-tip blunting. It was also long thought that embrittlement would not take place if the two metals formed intermetallic compounds. Presumably such compounds would act as a barrier to crack-tip access by the embrittling agent. However, iron is embrittled by zinc and tin, and iron forms intermetallic compounds with both of these elements. Thus, there are exceptions to these empirical generalizations although why this is so is not yet clear. For example, protective compounds that do form may be fractured as a result of the crack-tip stress concentration, the process taking place periodically and allowing the embrittler periodic access to the crack tip. Such a scenario plays a role in some forms of stress-corrosion cracking (Sect. 13.3). Lastly, when preexisting surface cracks or flaws are not present, embrittlement is preceded by (usually localized) plastic flow. That is, crack initiation for ME is caused by such flow just as is ordinary low-temperature brittle fracture of metals. Having described the phenomenology of ME, we now consider the mechanisms by which it occurs.

B. Mechanisms

Fracture under metal-embrittling conditions takes place by crack initiation (provided critical flaws are not present initially) and subsequent slow crack growth to a

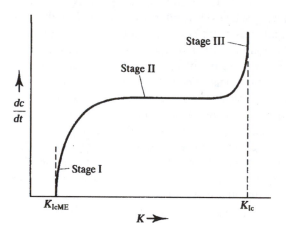

Figure 13.4
Schematic of ME crack-growth rate (dc/dt) variation with stress intensity. There is an apparent stress intensity below which crack growth does not occur (K_{IcME}). For stress intensities somewhat greater than this, crack-growth rate increases rapidly with stress intensity (Stage I). Steady-state crack growth (Stage II) takes place at a value of dc/dt specific to the metal-embrittler combination and test conditions (e.g., temperature). During Stage II, crack-growth rates are independent of K. As K_{Ic} is approached, crack-growth rates increase rapidly. Growth then is controlled by the stress and not the environment.

size sufficient to cause either tensile overload failure or, more often, to attain a stress intensity equal to the material fracture toughness. This last stage of fracture need not concern us. It is governed by criteria applying to fracture in the presence of cracks (cf. Chap. 9). Further, as has been noted, the embrittler plays no role in the final fracture.

Crack initiation is the least understood of the processes resulting in ME. As mentioned, plastic activity precedes crack nucleation, and this is reasonable when ME is viewed as a special case of brittle fracture. Moreover, crack nucleation is facilitated by localized plastic deformation, also in analogy with brittle fracture. What is not clear, however, is why certain ductile metals initiate cracks more easily in the presence of the embrittler than they do in its absence. Suggestions have been made that the embrittler diffuses into a thin layer of the metal, thereby facilitating crack nucleation. Presumably this locally increases the material's flow strength and enhances its susceptibility to fracture.

Crack growth during ME seems to be on firmer grounds, although this is not to say proposed mechanisms have been thoroughly verified but only that they are reasonable and conceptually appealing. One such model is the adsorption-induced reduction in cohesion model advanced by Stoloff and Johnson and Kamdar and Westwood (SJKW). The model is schematized in Fig. 13.5. It assumes that the embrittling species is adsorbed at the crack tip and that this alters (weakens) the atomic bonding there (Fig. 13.6a). The resulting stress-displacement relationship (Fig. 13.6b) is derived from the U-r curve of Fig. 13.6a (cf. Chap. 3). It is seen that the stress required to fracture atomic bonds is reduced by the adsorption process. The process reduces the fracture stress by reducing both the apparent surface energy (γ_s) and the local elastic modulus (Prob. 13.3). Since localized plastic flow often accompanies crack propagation in ME, just as it does in almost all forms of brittle fracture, the above-mentioned factors in and of themselves cannot account for the embrittlement phenomenon. However, the plastic work (γ_p) accompanying crack extension scales with γ_s and with the crack-tip radius of curvature (Chaps. 9 and 10). The adsorption serves to sharpen the crack tip, and when these additional factors are taken into account the SJKW model is plausible quantitatively as well as qualitatively appealing. For example, the toughness (i.e., the fracture work) of mercury-embrittled zinc is about half that of unembrittled zinc cleaved at low temperatures, and this kind of reduction in γ_p is expected on the basis of the SJKW model.

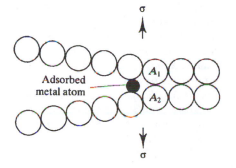

Figure 13.5
Atomic depiction of LME according to the SJKW reduction-in-cohesion model. The embrittling species is considered adsorbed at the crack tip, where it reduces the bond strength between the A metal atoms. This is manifested by a reduction in cohesive energy and fracture stress (see Fig. 13.6).

Some ME fracture surfaces exhibit ductile failure characteristics; i.e., crack advance apparently takes place by linkup of the crack tip via ductile failure of a ligament between it and a fractured or decohered particle in front of the tip. An alternative mechanism to account for ME crack growth in such situations has been advanced by Lynch, as illustrated in Fig. 13.7. The model assumes that plastic-flow resistance is *decreased* by the embrittling atoms. This gives rise to enhanced shear in the crack-tip vicinity and, in turn, easier crack growth with a reduced value of plastic work. In comparing the models, the adsorption decohesion model can be considered most pertinent in many ME fractures involving cleavage, whereas shear-

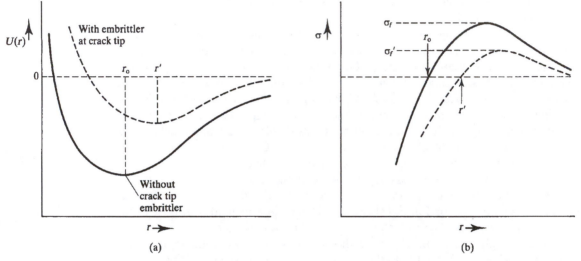

(a)   (b)

Figure 13.6
Schematic of (a) bonding energy and (b) bond strength as a function of the interatomic spacing for an unembrittled metal (solid line) and an embrittled one (dashed line). (a) The bond energy is decreased and the equilibrium interatomic spacing is increased when the embrittler is present. (b) This results in a decrease in the fracture stress (σ_f' vs. σ_f). The situation shown does not take into account plastic fracture work, but this also decreases in the presence of the embrittler.

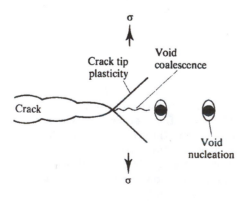

Figure 13.7
The adsorption-induced plasticity model for LME. Chemisorption facilitates crack-tip slip. This makes easier the linkup of the crack with a decohered particle in front of it; i.e., material toughness is reduced by the embrittler's presence. (*After S. P. Lynch, Mechanisms of Environment Sensitive Cracking of Materials, eds. P. R. Swann, F. P. Ford, and A. R. C. Westwood, The Metals Society, London, 1977, p. 201.*)

enhanced ME may be a better description when mixed-mode or ductile fracture features are associated with crack advance.

Both models view crack advance as taking place when embrittling atoms are present and adsorbed at the crack tip. This requires an adequate supply of the embrittling species in order to insure continued crack growth, since previously adsorbed atoms are presumably left behind on, and adhering to, the fracture surfaces. Additionally there must be a viable mechanism for transporting the embrittling species from its source to the crack tip. When the embrittler is in the liquid state, this material supply can be provided by fluid flow from the source to the crack tip. This would account for the rapid LME crack-growth rates (liquid metal viscosities are low and they flow easily) and for the relative temperature insensitivity of Stage II LME crack-growth rates (liquid-metal viscosities are not very temperature sensitive). During SME the embrittler must reach the crack tip by another mechanism. For example, it could be transported through the vapor phase. To provide an adequate supply of metal atoms by this mechanism requires that the embrittler have a reasonable vapor pressure at the embrittling temperature. This is not always the situation. For this (and other) reason(s) it is believed that transport takes place by surface diffusion along the fracture surface. Both the vapor-transport and surface-diffusion SME mechanisms are more sensitive to temperature than the fluid transport one of LME. The activation energy for vapor transport relates to the heat of sublimation, whereas the surface-diffusion mechanism has an activation energy apropos to this process. In either event, SME crack-growth rates will decrease with decreasing temperature and, as is observed experimentally, a temperature below which embrittlement vanishes is observed. Clearly, the critical lower temperature is that at which the embrittler is unable to be transported to the crack tip at a rate sufficient to aid crack growth. On this basis, increasing strain rate will, as is also observed, increase the critical lower temperature.

Even on the basis of our incomplete description of ME, it is evident that it is a complex phenomenon. Much light has been shed on it during the past several decades. It is expected that additional research will further clarify the mechanisms most responsible for ME. Liquid environments other than those containing metals can also embrittle materials. Stress-corrosion cracking is an example, and this topic is addressed next.

13.3
STRESS-CORROSION CRACKING

A. Characteristics

Stress-corrosion (SCC) cracking refers to the phenomenon in which failure of a stressed material exposed to an aqueous, corrosive fluid may take place.[2] There are similarities between SCC and ME. For example, static fatigue is often found in

---

[2]SCC can also happen in gaseous environments, although the occurrence of such is much less than that associated with liquid environments.

both embrittlement forms. Similarly, SCC slow-crack-growth rates as a function of stress intensity often display stages similar to those for ME slow crack growth.

There are also differences between ME and SCC. The former occurs only in metals exposed to a metal environment. However, SCC is observed in all material classes. The moisture-enhanced static fatigue of glass has been mentioned, and polymers below their glass transition temperature are often prone to static fatigue. Stress corrosion of such brittle materials occurs primarily by localized corrosion taking place at preexisting surface flaws.

The ductility of metals and their electrochemical activity make SCC much more complicated in them than in brittle solids. The corrosion processes taking place at a crack tip interact with the stress[3] in such a way that normally ductile metals can fracture in a macroscopically brittle manner. Because of the widespread use of metals and their alloys in hostile environments (e.g., the chemical and nuclear industries), SCC of metals is the focus of this section. In addition to chemical environment, temperature, stress/stress state, strain rate, etc., are also important in determining a material's susceptibility to SCC. Almost every metal or its alloys can be made prone to SCC if the proper environment and other conditions are present. However, there are some well-known and technologically important manifestations of SCC. Table 13.2 lists some of these.

Slow-crack-growth rates in SCC of metals are usually much less than analogous rates in LME; sometimes by orders of magnitude (Fig. 13.8). This indicates that the mechanisms governing SCC crack advance differ from those causing it in LME. For example, SCC involves transfer of atoms *away* from, as well as toward, the crack tip. Or there must be some barrier that prevents ready access of the embrittling agent to the crack tip in SCC. As described in the next section, models that take into account such considerations have been developed to explain the main features of SCC.

Metal/alloy	Aqueous Cl- solutions	Ammonia	Water vapor/steam	Nitrate solutions	Hot salts	Seawater	Caustic solutions
Al	X		X			X	
Cu		X	X	X			
Fe	X	X	X	X			X
Alloy Steels	X						
Stainless Steels	X		X		X	X	X
Mg	X		X				
Ni			X		X		X
Ti	X				X		
Zr	X						

Notes: X = indicates occurrence of stress-corrosion cracking. The terms Al, Cu, etc., encompass the pure metals and their alloys; however, SCC is far more common to the alloys of metals than to the pure elements that are their base. Fe in the above table denotes Fe and mild steels.

Table 13.2
Some technologically important environment-metal combinations leading to stress-corrosion cracking

---

[3]Many instances of SCC are caused by residual stresses. These stresses can be thermally induced, or they can result from deformation processing.

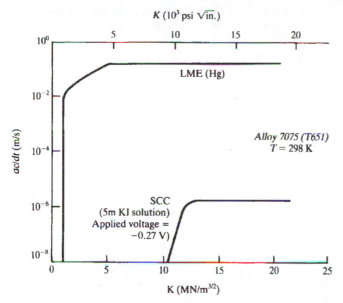

Figure 13.8
Stages I and II of room-temperature LME and SCC slow-crack growth in a 7075 Al alloy (T651 condition). K_{IcLME} (Hg is the embrittler in this case) is less than K_{IcSCC} (an aqueous iodide environment instigates SCC in this example) and steady-state crack-growth rates are also higher for LME. This kind of relative variation between LME and SCC of the Stage II crack-growth rates and the critical environmental stress intensity is common. (*Adapted from M. O. Speidel, The Theory of Stress Corrosion Cracking in Alloys, ed. J. C. Scully, NATO, Brussels, Belgium, 1971, p. 289.*)

EXAMPLE PROBLEM 13.1. The crack-growth rate data listed were obtained for an aluminum alloy subjected to a static tensile stress in an aqueous environment. If $K_{Ic} = 25$ MN/m$^{3/2}$ for this alloy, how long would it take to fracture a sample of it containing a 5-mm-deep crack if the sample is subjected to a stress of 60 MN/m^2?

Stress (MN/m^2)	c (mm)	dc/dt (m/s)
28	5.0	8×10^{-10}
28	10.0	26×10^{-9}
56	5.0	8×10^{-7}
56	7.5	8×10^{-7}

Solution. We first convert stress and crack length to a stress intensity through $K = \sigma(\pi c)^{1/2}$. (Recall Chap. 9. The factor π here is specific to a particular crack shape, and may not apply to the cracks under consideration. However, our answer is unaffected provided the crack shape is the same for all samples.)

Stress (MN/m^2)	c (mm)	K (MN/m$^{3/2}$)
28	5.0	3.50
28	10.0	4.96
56	5.0	7.00
56	7.5	8.57

A graph of dc/dt vs. K is shown below. Note that for the specified operating conditions, $K = 7.5$ MN/m$^{3/2}$. From the graph we see that these conditions correspond to steady-state crack growth with a crack velocity of 8×10^{-7} m/s.

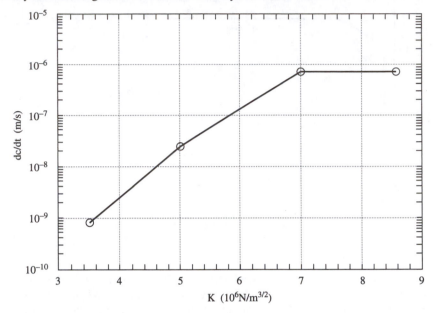

We now use fracture mechanics to determine the critical crack size (c_f) at which final fracture occurs. This is obtained using $c_f = K_{Ic}^2/\pi\sigma^2$. Substituting $\sigma = 60$ MN/m^2 and $K_{Ic} = 25$ MN/m$^{3/2}$, we find that $c_f = 0.055$ m. Hence, the anticipated material lifetime, t_f, is

$$t_f = 0.055 \text{ m}/(8 \times 10^{-7} \text{ m/s}) = 6.88 \times 10^4 \text{ s} = 19.1 \text{ hr}$$

B. Mechanisms

As with other fracture processes, if there are none present initially, cracks must nucleate for SCC to occur. In some cases corrosion may accomplish this by, for example, localized pitting or intergranular corrosion. When the latter occurs, the fracture path is intergranular although, in general, fracture paths may be intergranular, transgranular, or a mixture of the two for SCC just as they are for ME. In addition to this kind of crack initiation, localized strain may catalyze crack nucleation. Many metals form a protective surface barrier when exposed to environments causing SCC. The barrier may be quite thin (e.g., tens of nanometers if the film is a chemically adsorbed one) or relatively thick (say on the order of micrometers if the film is a true oxide). In either case, localized plastic flow of the metal may rupture this barrier, causing subsequent local corrosion that results in formation of the required initial flaw.

What causes a crack to grow after it is initiated? Stress is one factor; the term "stress corrosion cracking" explicitly recognizes this. Moreover, the stress must be tensile in nature; compressive stresses do not effect stress-corrosion crack advance just as they do not lead to fatigue or LME crack growth. But chemical factors are also important. For the crack to advance, metal at the crack tip must be removed at a more rapid rate than metal on the fracture surfaces in the crack wake. Were the tip of a crack to corrode at a lesser rate than the adjacent fracture surfaces, the crack would blunt and its advance would cease. Thus, we conclude that the fracture surfaces must be chemically less active than the crack tip for SCC to occur. While all that is required for the above to happen is that the chemical activity follow the stipulated rule, it is convenient to invoke the idea of a protective film on the crack surfaces to further elaborate SCC crack advance. This facilitates visualization of the process, although we caution that the "protective film" need not always be a film per se.

A schematic of the situation at the crack tip during SCC is given in Fig. 13.9. The situation is chemically, geometrically, and mechanically complex. In the schematic of Fig. 13.9, corrosion is envisaged to occur at the crack tip where bare metal is exposed to the corrosive fluid and the crack advances by metal dissolution there.[4] In contrast, the surfaces of the crack normal to the applied stress are coated by the protective film, and corrosion occurs at a much lower rate at these locations. Slow crack growth is presumed to take place as follows. Any protective film that forms at the crack tip is periodically ruptured by localized plastic flow (Fig. 13.10). Such flow is accentuated by the crack-tip stress concentration and can be further facilitated by prior interaction between the metal previously exposed at the crack tip and the chemical environment there. Comparison of Figs. 13.10a and 13.10b indicates that the area exposed to corrosion will depend on the film thickness developed between fracture events, thicker films being less detrimental for SCC. Likewise, the exposed metal area depends on its slip characteristics. Planar slip, for example, exposes more metal area than does dispersed (uniform) or wavy slip. This is one example of how structural features of the base metal (e.g., stacking-fault energy) influence crack advance rate in SCC. There is another variant of the processes illustrated in Figs. 13.10 that can cause crack advance. It assumes that when the (brittle) protective film is ruptured the crack advances slightly into the metal underlying the film. The transient access to the environment is thought to embrittle this metal so that when the film reforms and then subsequently fractures again, the process assumes a repetitive nature. We see that the primary difference between this mechanism and those shown in Figs. 13.10 has to do with whether the metal at the crack tip behaves in a ductile or brittle manner.

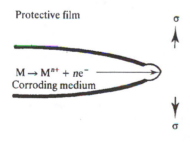

Protective film

$M \rightarrow M^{n+} + ne^-$
Corroding medium

σ

σ

Figure 13.9
Schematic of stress-corrosion cracking. Metal dissolves at the crack tip where plastic deformation causes protective-film rupture. The anode electrochemical reaction is $M \rightarrow M^{n+} + ne^-$, where n is the metal valence. The lateral sides of the crack are relatively passive, and do not corrode at the rate the exposed metal does. Processes taking place at the crack tip include reformation of the protective film, diffusion of the metal ions away from the crack tip (allowing the anode reaction to continue), and a cathode half reaction that depends on the fluid environment and temperature, but which often involves liberation of hydrogen.

[4]In the current discussion, we assume the mode of crack advance is corrosion (i.e., metal dissolution). There are alternative modes of crack advance. One involves material embrittlement in front of the crack by ion adsorption there. This description is similar to that of the SJKW model of ME. Other modes of crack advance have also been put forth. We return to them later.

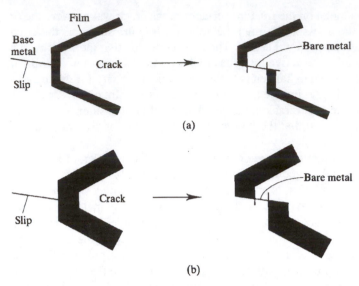

Figure 13.10
Film rupture, caused by crack-tip slip, permits a corrosive fluid to maintain access to bare metal. When the thickness of the film broken in this way is small (a), more base metal is exposed by a film rupture event than when the thickness is large (b).

The rate at which repassivation (i.e., reformation of the protective film) takes place at the fractured barrier determines both the crack-growth rate and also whether or not a material is prone to SCC. If repassivation is rapid, so that the time between film rupture events is long compared to the repassivation time, the crack advance rate is minimal. If repassivation is very slow, on the other hand, localized corrosion (e.g., pitting or intergranular corrosion) will control material failure, stress playing a secondary role in this situation. On this basis, strain rate is expected to influence a material's tendency for SCC, and this is found to be so (Fig. 13.11). We see that for this titanium alloy in a salt water environment, SCC is not manifested at low strain rates because the time between film rupture events is long in comparison to the repassivation time. At the higher strain rates, on the other hand, tensile ductility is the same in the corroding medium as it is in air, and this reflects that the corrosion reactions occur at a lesser rate than the mechanical factors causing fracture. Thus, in many cases, SCC is found only over a limited strain-rate range. This is in common with ME and other forms of environmentally assisted fracture discussed in this chapter.

This description can be made quantitative. The time variation of the crack-tip corrosion-current density (in units of A/m^2, and which is a measure of the metal dissolution rate) following a crack-tip film-rupture event is drawn schematically in Fig. 13.12. The maximum current density (I_{max}) is determined by solution and bare-metal chemistry and other factors (e.g., temperature). Following film rupture, the protective film begins to reform, and the corrosion current density decreases. The minimum current (I_{corr}) of Fig. 13.12 corresponds to the uniform corrosion-current that would take place if film rupture did not periodically occur; I_{corr} is a function of the material, the local crack-tip fluid composition, temperature, etc. The curve of Fig. 13.12 applies to the situation in which the passivation time is short compared

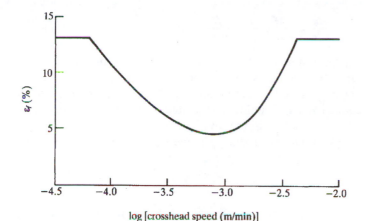

Figure 13.11
Tensile failure elongation of a Ti alloy in a 3% NaCl aqueous solution as it varies with strain rate. (The crosshead speed is the sample elongation rate; the strain rate is obtained by dividing the elongation rate by the sample length.) For reasons given in the text, SCC of this alloy in this environment occurs only over a limited strain-rate range. (*From J. C. Scully and D. T. Powell, Corr. Sc., 10, 719, 1970.*)

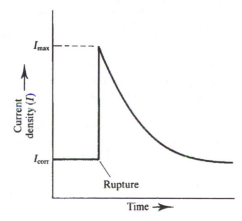

Figure 13.12
Time variation of the crack-tip corrosion-current density through a film rupture and repassivation sequence. Following rupture, the current density has a maximum value (I_{max}). It then decreases as repassivation takes place. After full repassivation, the current density has a minimum value (I_{corr}). This corresponds to a situation of uniform corrosion at the crack tip and on its lateral surfaces. Both I_{max} and I_{corr} are system- and environment-specific; i.e., they depend on the metal, its film, the corrosive fluid, and factors such as temperature and solution pH. For the situation depicted here, full repassivation occurs prior to a subsequent film-rupture event. (*After J. C. Scully, Corr. Sc., 15, 207, 1975.*)

to the time between film rupture events; that is, it pertains to the situations where crack advance takes place by corrosion per se. When SCC takes place at reasonable rates, the current density–time relationship for the condition of constant stress is illustrated in Fig. 13.13a. The crack-advance rate is (Prob. 13.7)

$$\frac{dc}{dt} = \frac{M}{n\rho F}\frac{Q_f}{t_f} = \frac{M}{n\rho F}\frac{Q_f}{\varepsilon_f}\dot{\varepsilon} \qquad (13.1)$$

In Eq. (13.1), M and ρ are the atomic weight and density of the corroding metal, F is Faraday's constant ($= 96,490$ C/mol), t_f is the time between film rupture events, n is the metal valence, and Q_f (units of C/m²) is the integral of the current density–time curve between these rupture events. The time t_f is related to the strain between rupture events (ε_f) and the local crack-tip strain rate ($\dot{\varepsilon}$) as $t_f = \varepsilon_f/\dot{\varepsilon}$. Equation (13.1) adequately expresses crack-growth rates during some stress corrosion processes

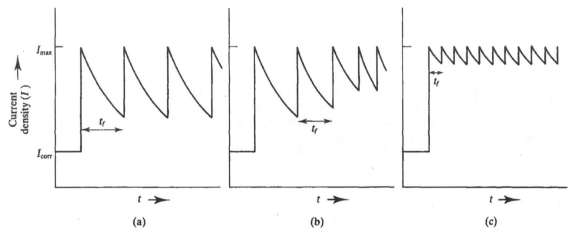

Figure 13.13
Current density–time behavior for three situations. (a) For constant stress, film rupture takes place periodically with a time between rupture events (t_f) less than the repassivation time. (b) For constant load, as the crack grows the interval between rupture events decreases and the average value of the current density increases. This corresponds to Stage I of the (dc/dt)-K curve. (c) When t_f becomes quite small, the average current density is close to I_{max}. This corresponds to Stage II of the crack-growth curve during which growth is controlled by crack-tip corrosion kinetics. (*After J. C. Scully, Corr. Sc., 15, 207, 1975.*)

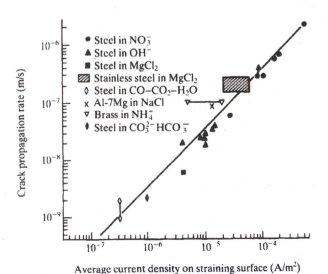

Average current density on straining surface (A/m²)

Figure 13.14
SCC crack-propagation (growth) rate vs. average current density ($= Q_f/t_f$) on the strained surfaces of several alloys in various aqueous environments. The approximate linear relationship between dc/dt and Q_f/t_f (on these logarithmic coordinates, the slope of the line drawn is close to one) is in accord with Eq. (13.1). (*From R. N. Parkins, Br. Corr. J., 14, 5, 1979.*)

(usually when the crack-tip stress intensity is relatively large). For example, dc/dt is found to vary linearly with Q_f (Fig. 13.14), and also to increase with strain rate, although more detailed considerations than we have employed indicate that the dependence of crack-growth rate on strain rate is less than linear (typically $dc/dt \sim (\dot{\varepsilon})^k$, where $k \cong 1/2$).

When a constant load (as would apply for a crack-growth rate measurement test), rather than constant stress, is applied the crack-tip stress increases with crack length. Under this kind of loading, the current density–time relationship is as shown in Fig. 13.13b. Both Q_f and t_f are decreased but the ratio Q_f/t_f increases (Prob. 13.6), and this leads to accelerated crack growth (cf. Eq. (13.1)). This causes a rapid increase in dc/dt (Prob. 13.6); the situation corresponds to Stage I of slow crack growth (e.g., Fig. 13.8). As the stress level continues to increase, the time between film rupture, events decreases, and eventually a situation is reached where an approximately steady-state and stress-independent crack-tip corrosion rate is attained (Fig. 13.13c). At this stage, crack-growth rates no longer relate to film rupture, but rather to the ability of the corroding medium to be replenished at the crack tip. These processes are discussed below in more detail. Here we only note that this type of crack-growth control corresponds to Stage II of slow crack growth (Fig. 13.15 and Fig. 13.8). Shown also in the schematic of Fig. 13.15 is the effect that cyclic loading has on SCC. The combined process is called corrosion fatigue, and we return to this phenomenon in Sect. 13.3C.

The above description of stress-corrosion crack growth is in accord with many observations. The model does not depend on the details of the corrosion reactions taking place at the crack tip. In a sense this is one of its strong points since these reactions can be complicated. However, since some idea of the reactions is required for a full description of SCC, we describe them further even though they are incompletely understood.

Two electrochemical reactions control crack-growth rates. In one—anodic dissolution—growth rates relate to the amount of metal removed by corrosion at the crack tip. The dissolution reaction is represented by the electrochemical half-reaction

$$M \rightarrow M^{n+} + ne^- \tag{13.2}$$

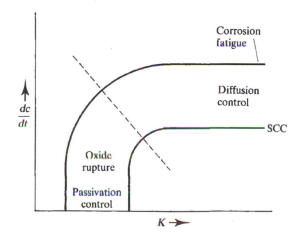

Figure 13.15
Schematic of dc/dt for SCC in a static and in a fatigue (corrosion fatigue) environment. Corrosion fatigue results in higher crack-growth rates. Stage I and Stage II crack growth remain controlled by film rupture-repassivation and crack-tip kinetics, respectively, but the transition between stages takes place at higher crack-growth rates and lower K values for corrosion fatigue.
(*After F. P. Ford, Embrittlement of Engineering Alloys, eds. C. L. Briant and S. K. Banerji, Treat. Matls. Sc. Tech., 25, 251, 1983.*)

where M is the metal atom and n its valence (the same n as in Eq. (13.1)). There is a clear link between the corrosion reaction of Eq. (13.2) and the crack-growth rate expression of Eq. (13.1), especially in those cases where anodic dissolution controls crack extension. However, we must also realize it is difficult to a priori predict crack-tip dissolution rates on the basis of bulk fluid chemistry which may differ substantially from crack-tip chemistry. As one example, a buildup of metal ions in the fluid in the crack-tip vicinity will reduce the forward reaction rate of Eq. (13.2) (the phenomenon is known as concentration polarization). Moreover, other chemical variations—such as varying pH levels at the crack tip—also prevent accurate prediction of crack-tip corrosion rates. And these are the types of factors that determine Stage II SCC crack-growth rates.

Crack-growth rates may also be controlled by the cathode reaction which takes place at the same rate as the anodic dissolution reaction. In many cases of SCC the cathode reaction involves hydrogen ($H^+ + e^- \rightarrow (1/2)H_2$), and the hydrogen so formed may control crack-growth rates through one or the other of the forms of hydrogen embrittlement (HE). In certain circumstances (ones that do not result in HE), cathodic hydrogen diffuses away from the corrodent-metal interface. However, some of the liberated hydrogen may enter into the metal lattice. If so, crack propagation may take place as a result of HE of the metal in front of the crack. Crack advance then relates to the hydrogen-diffusion distance within the metal, with crack propagation taking place periodically rather than quasi-continuously. Crack-advance rates for the HE form of SCC are generally greater than those expected on the basis of Eq. (13.1). Nevertheless, this equation is still useful when we realize that the hydrogen production rate is coupled to the metal dissolution rate. Moreover, film rupture will still be part and parcel of Stage I crack growth for HE. However, diffusion or surface reaction of hydrogen (instead of metal atoms within the liquid) will control the Stage II crack-growth rate for this form of embrittlement.

Whether anodic dissolution or HE is the cause of crack extension during SCC depends on the system and its chemical and thermal environment, and no clear-cut a priori predictions can be made in this regard. Indeed, small variations in crack-tip chemistry, including those that happen during crack extension, can change the controlling crack-growth mechanism. However, there is clear evidence that HE plays an important role in many observed cases of SCC of Al and Ti and their alloys.

Hydrogen embrittlement taking place during SCC is but one example of the deleterious effect that hydrogen can have on the mechanical properties of many metals. Hydrogen embrittlement of metals is becoming more commonplace as metals are used in increasingly hostile environments. Thus, we return to HE in Sect. 13.4. But before we do, we briefly consider corrosion fatigue. It, too, is a serious matter.

C. Corrosion Fatigue

Corrosion fatigue refers to material failure taking place in a corrosive environment under conditions of cyclic loading. Corrosion fatigue can be viewed as a form of SCC facilitated by a cyclic stress. Or it can be viewed as fatigue accelerated by a corrosive environment. We take the first viewpoint.

We first consider static fatigue of a material in an environment prone to cause SCC. Material behavior is manifested in a manner similar to that illustrated in

Fig. 13.3. That is, exposure to a hostile environment reduces the static stress level to which a material can be subjected absent failure. Further, there often is a stress level below which static fatigue does not occur. In fact, when the material is "flaw free" to begin with, the static fatigue limit can be used as a design parameter for SCC.

If the material is exposed to the same environment but a cyclic, rather than static, stress is imposed, material performance is degraded. This is manifested, in part, by a reduction in the fatigue limit.[5] In a figure like that of Fig. 13.3, for example, the permissible stress amplitude ($= \Delta\sigma/2$) would be plotted versus the time to failure, and this stress amplitude would be less than the corresponding static stress. If such a testing schedule is conducted on flaw-free materials, the corresponding "corrosion-fatigue" limit can be used in design as is discussed presently. The corrosion-fatigue limit depends on the material's chemical and thermal environment, as should be apparent.

As indicated in Fig. 13.15, the threshold stress intensity for SCC is also reduced when a cyclic stress is imposed.[6] As with the corrosion-fatigue limit, this material-specific critical K value (K_{cf}) depends on temperature and solution chemistry. (Because of the kinetic aspects of SCC, this critical stress is often found to be time sensitive. We neglect this factor in our discussion.) Thus, in a corrosion-fatigue environment K_{cf} represents a fail safe operating condition for SCC of materials containing preexisting flaws, just as does the corrosion-fatigue limit for materials not containing such flaws.

A picture may help in digesting this information. Figure 13.16 is a diagram very much like that of Fig. 12.21 which presented "fail safe" fatigue operating conditions. Three "fail safe" envelopes are shown in Fig. 13.16. One corresponds to ordinary tensile fracture in an environment in which embrittling effects are absent. Thus, K_{Ic} defines the failure criterion for materials containing preexisting flaws, and "Fracture Strength" is the fracture stress for flaw-free materials. A second envelope represents tensile fracture in an environment causing SCC. Here, the static fatigue limit substitutes for "Fracture Strength" in flaw-free materials, and K_{IcSCC} now defines fracture for flawed materials. Finally, the third envelope represents fail safe operation for corrosion-fatigue conditions. (The stress is now a stress amplitude, though.) The corrosion-fatigue limit, less than the static-fatigue limit, represents fail safe operation in flaw-free materials, and failure in flawed materials is now controlled by K_{cf}.

While Fig. 13.16 is conceptually useful, it does not have the direct bearing to engineering design that Fig. 12.21 does. In particular, the "design" parameters for corrosion fatigue—K_{cf} and the corrosion-fatigue limit—have not been identified to the same extent that the corresponding parameters for tensile and fatigue fracture have been. This is understandable given the wide variety of environment and material combinations that cause SCC. But it is still unfortunate.

---

[5]A true "corrosion-fatigue" limit is often not observed with cyclic loading. However, as with our discussion of fatigue endurance limits, an effective "corrosion-fatigue" limit can be defined. For example, it might be that stress amplitude that does not cause material failure during the intended lifetime of the structure.

[6]The abscissa of Fig. 13.15 (denoted as K) is taken proportional to $(\Delta\sigma)c^{1/2}$ for corrosion fatigue.

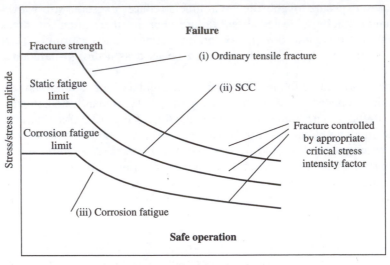

Figure 13.16
"Fail safe" stress (or stress amplitude)–crack length envelopes for three situations: (i) tensile loading in a benign environment, (ii) an environment causing SCC, and (iii) a corrosion-fatigue environment. For situation (i), failure at large crack sizes is controlled by the material fracture toughness; for "flaw-free" materials, the failure stress is the material fracture stress. For the SCC situation, fail-safe operation at large crack sizes is dictated by the threshold stress intensity, K_{IcSCC}, for stress corrosion. At small crack sizes (flaw-free materials,) the fail safe stress is the static fatigue limit. The fail safe envelope is further constricted for the corrosion-fatigue situation. In this instance, the ordinate is now the stress amplitude.

13.4
HYDROGEN EMBRITTLEMENT

A. Characteristics

Hydrogen embrittlement (HE) broadly describes the deleterious effects that hydrogen in several forms has on the mechanical properties of materials, particularly metals. Hydrogen can impair properties in several ways, and can enter into a material as a result of a number of causes. During melting of metals in humid environments, for example, hydrogen can enter into the liquid metal in amounts well in excess of its solid-state solubility. The subsequent solid-state precipitation of it, in one of several forms, can effect embrittlement. Hydrogen can also enter into a metal as a result of other processing operations (e.g., electroplating or pickling). Pickling involves immersion of the material into an acid for the purpose of cleaning its surface. The accompanying corrosion reactions can liberate hydrogen that is then absorbed into the metal in a manner analogous to the way it is in SCC. Hydrogen can also enter a part during service. This is often the situation for materials used in the petroleum and chemical industries.

Hydrogen embrittlement is often manifested in ways similar to those of other types of embrittlement. An example is provided by the delayed failure (static fa-

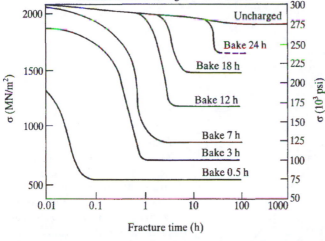

Figure 13.17
Static fatigue of notched bars of 4340 steel (unembrittled notched-bar strength = 2070 MN/m².). The steels were hydrogen-embrittled by cathodic charging. Some of the hydrogen was then removed by baking at 422 K for varying times; longer "bake" times result in lower residual hydrogen content. Increasing hydrogen content increases the material's susceptibility to static fatigue. (*From J. O. Morlett, H. Johnson, and A. Troiano, J. Iron. Steel Inst., **189**, 37, 1958.*)

tigue) of steels (Fig. 13.17). In the situation illustrated, hydrogen was introduced into the steel by cathodic charging, i.e., by evolving hydrogen at the steel surface through the hydrogen cathode reaction. The hydrogen introduced in this way can be removed by "baking" the steel in a dry environment. Thus, increases in "bakeout" time shown in Fig. 13.17 correspond to lower residual hydrogen contents. For the conditions applying here, the hydrogen effect can be effectively reversed if the steel is sufficiently dehydrogenated by such baking.

Hydrogen embrittlement also reduces tensile ductility in a manner similar to that found in ME; for example, HE and ME are both observed only over a limited temperature range. The severity of the ductility loss increases with increasing hydrogen content and with decreases in strain rate. The temperature and strain-rate effects associated with HE are rationalized in much the same way they are for other types of embrittlement. That is, at low temperature, diffusional processes resulting in embrittlement take place too slowly to permit its occurrence. At high temperatures, on the other hand, inherent material ductility precludes embrittlement.[7] As is often the case with other embrittlement forms, HE is most pronounced for high-strength materials; for example, hydrogen embrittles high-strength steels more than it does low-strength ones.

Slow-crack-growth rates relate to stress intensity for HE in much the same way they do for LME and SCC. That is, several stages of slow crack growth are often

---

[7]Other possibilities for high-temperature ductilization exist. For example, hydrogen may diffuse out of the metal at high temperature.

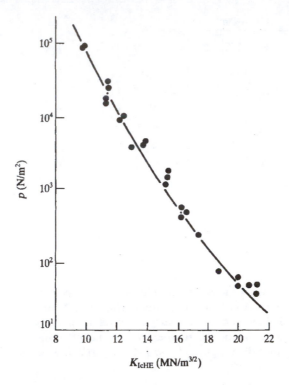

Figure 13.18
The hydrogen pressure depen-
dence of K_{IcHE} for a high-
strength martensitic steel
exposed to gaseous hydrogen.
Hydrogen pressure has a pro-
nounced effect on K_{IcHE}.
(*Adapted from R. A. Oriani and*
P. H. Josephic, "Equilibrium
Aspects of Hydrogen-Induced
Cracking of Steels," Acta
Metall., 22, 1065, Copyright
1974, with permission from
Elsevier Science.)

found for HE with two of them comprising the slow-growth regime. Stage II crack-growth rates [$(dc/dt)_{II}$] for HE are typically greater than for SCC, but less than for LME. The variation of $(dc/dt)_{II}$ among the several forms of embrittlement indicates that the rate of critical hydrogen transport (recall that embrittler diffusion is critical to Stage II crack growth) is less than the corresponding liquid-flow rate for LME but somewhat greater than, for example, the anion-diffusion rates that control some forms of SCC Stage II crack growth.

The threshold stress intensity, K_{IcHE}, is also a function of hydrogen content (Fig. 13.18). Since hydrogen transport controls Stage II crack growth, increasing hydrogen content also increases $(dc/dt)_{II}$ (a higher hydrogen content will provide the critical amount of mass transport in shorter times). Stage II crack-growth rates also depend on temperature. As indicated in Fig. 13.19, they reach a maximum at intermediate temperatures. The behavior is similar to that occurring in a tensile test of a hydrogen-embrittled steel, and takes place for the same reasons.

Material strength level is important in HE slow crack growth. A decrease in material strength reduces the stress-intensity range ($K_{Ic}-K_{IcHE}$) over which cracks propagate. Reduced strength also leads to lower crack-growth rates in general, and sometimes obscures the Stage I to Stage II crack-growth rate transition.

The examples cited to this point relate mainly to HE of ferritic steels. However, HE also occurs in many other common metals and their alloys (e.g., those of Al, Ti, Zr, Cu, and Ni) and in others not so widely used (e.g., those of Nb, Ta, V, and U). The mechanisms causing embrittlement, and its extent, are system- and environment-specific. Aluminum alloys, for example, are not susceptible to hydrogen embrittlement via exposure to an external gas. This is probably because the Al_2O_3 surface layer protects the underlying metal. However, when this film is ruptured, as

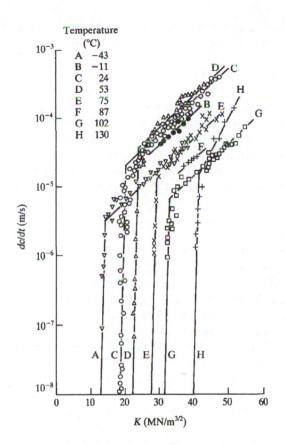

Temperature
(°C)

A	−43
B	−11
C	24
D	53
E	75
F	87
G	102
H	130

Figure 13.19
Stage II crack-growth rates in a 4130 martensitic steel (yield strength = 1330 MN/m²) vary with temperature when exposed to gaseous hydrogen ($p = 0.77$ MN/m² = 11.2 psi). For this situation $(dc/dt)_{II}$ increases with temperature in the range −43°C to 53°C, and then decreases for temperatures greater than 53°C. For this steel in this environment Stage II crack-growth rates also depend on the stress intensity. (*From H. G. Nelson and D. P. Williams, Stress Corrosion and Hydrogen Embrittlement of Iron Base Alloys, ed. R. W. Staehle, J. Hochmann, R. D. McCright, and J. E. Slater, NACE-5, Houston, TX, 1977, p. 390.*)

it is in SCC, then HE can take place. An extended discussion of all potential mechanisms of HE, particularly as they relate to specific alloy-environment combinations, is not possible. However, in the following section we review some proposed mechanisms of HE, and relate them to material behavior of the kinds just described.

B. Mechanisms

i. HYDROGEN PRESENT IN ATOMIC FORM. Embrittlement can occur when hydrogen is present as atomic hydrogen in the embrittled metal. This hydrogen can be located throughout the material volume (as, for example, it would be were it introduced during the original melting operation) or it can be preferentially located at the tip of a propagating crack (as it might be during some forms of SCC). Regardless of these details, HE due to atomic hydrogen primarily affects crack-propagation behavior.

Some basics first. Hydrogen is a small atom, and it diffuses rapidly over the temperature range in which HE is manifested. Further, hydrogen solubility in a metal depends on the applied stress level. In particular, a tensile stress increases atomic volume leading to an enhanced hydrogen solubility. Thus, because of the associated stress intensification at a crack tip, the region in front of the crack has a higher hydrogen solubility than regions removed from it. This causes hydrogen atoms to diffuse to the crack-tip vicinity. This hydrogen can come from an external source (as in SCC) or result from hydrogen distributed throughout the material volume. The effect that this hydrogen enrichment has on crack propagation does not depend much on the source of it.

How does this hydrogen promote crack growth? One of the first models that attempted to answer this question is similar to one applying to certain types of LME. That is, hydrogen is adsorbed at the crack tip, reducing there the material cohesive energy and the corresponding brittle fracture stress. Computational simulations—employing interatomic potentials—suggest this model has merit. However, features of it have yet to be confirmed experimentally. For example, ductile fracture features are often found in HE whereas the adsorption-induced cohesion reduction model would indicate that cleavage (or intergranular fracture if hydrogen preferentially segregates to grain boundaries) should be the dominant mode of crack propagation.

A second model, one emphasizing localized enhanced plasticity, appears to have more merit. Hydrogen, like other interstitial atoms, has a tendency to segregate to edge dislocation cores, and over the temperature range for which HE is manifested it does so quickly. In fact, the hydrogen concentration at the dislocation cores is predicted to saturate—i.e., every edge dislocation terminus has a hydrogen atom associated with it. In addition, the electronic screening provided by the hydrogen reduces the stress required for these dislocations to move. The hydrogen atoms move in concert with the dislocations during dislocation motion (provided the strain rate is not excessively high) and the reduced flow stress is maintained. The result is a diminished material yield strength and plastic-flow stress.

Because hydrogen segregates to the region in front of a crack tip, this hydrogen softening effect is manifested there in the form of localized plasticity. One might think, based on discussions in Chap. 10, that the reduced flow stress would be manifested by an increase in toughness. However, the localized plasticity of HE results in a relatively low energy fracture, somewhat analogous to that accompanying the last stages of ductile fracture by shear-band propagation. Enough experimental evidence has been obtained over the last 20 years or so to indicate that this kind of HE plays a major role in crack propagation in a wide variety of metals and their alloys; Al, Ni, Fe, and stainless steel are some of them. Before ending discussion of this type of HE, we note that the method of hydrogen introduction is immaterial to its manifestation. Hydrogen ingress can result from materials processing, from a fluid solution, or from a gaseous environment having a relatively high hydrogen partial pressure (recall Fig. 13.18).

On the basis of this discussion, HE taking place when hydrogen is present in solid solution (or when it has segregated to grain boundaries) can be considered a special case of impurity-atom embrittlement (IAE). There are many other atomic species that cause embrittlement and IAE is further discussed in Sect. 13.5.

ii. HYDROGEN PRESENT IN THE FORM OF A SECOND PHASE. Hydrogen embrittlement is often associated with hydrogen contents in excess of its solubility limit. The nonequilibrium amounts are introduced by thermal (e.g., melting) or other (e.g., electrochemical) processes, and the subsequent precipitation of the excess hydrogen is critical to the occurrence of some forms of HE. The precipitates formed can be either gaseous or solid, depending on system thermodynamics. We discuss these situations separately.

a. Embrittlement by gaseous precipitates. Excess hydrogen can precipitate in gaseous form. The chemical form of the hydrogen precipitated in this way depends on system thermodynamics. In hydrogen-containing copper exposed to oxygen, for example, water vapor bubbles—located typically at grain boundaries— form via the reaction between hydrogen and oxygen. These bubbles act as nucleated voids that accelerate creep fracture. The problem can be minimized by utilizing OFHC (oxygen-free high-conductivity) copper when conditions favor this kind of *hydrogen damage*. Another form of hydrogen damage takes place in steels employed in the chemical and petroleum industries. In this case, hydrogen locally decarburizes the steel, and methane bubbles are the reaction product. These bubbles, also located preferentially at grain or other internal boundaries, serve as crack nuclei. It should be noted that insofar as creep is concerned, bubbles of any kind not only enhance void nucleation, but also typically grow faster than "ordinary" voids because the bubbles are internally pressurized, and this facilitates cavity growth.

Gaseous hydrogen embrittlement at lower temperatures can be caused by precipitation of molecular hydrogen (H_2), also in the form of bubbles. These pressurized bubbles facilitate both crack nucleation and crack growth. For example, the dependence of K_{IcHE} of a martensitic steel on hydrogen pressure (Fig. 13.18) is in accord with this idea.

b. Hydride precipitation. Hydride precipitation can embrittle Ti, Zr, Nb, V, and some other metals and their alloys. Titanium and zirconium alloys are the most commercially important of this group, and we focus on their hydride embrittlement.

The body-centered cubic (β) phase is the high-temperature form of both titanium and zirconium. The solubility of hydrogen in this form of Ti and Zr is much greater than it is in their low-temperature hcp (α) form. On cooling a hydrogen-rich alloy from the β-phase field, the bcc material often transforms into a mixture of α phase and Ti or Zr hydride. Precipitation of the low-density hydride phase is accompanied by an increase in volume, and the internal stress field developed thereby is often sufficient to fracture the brittle hydride. If not, an applied stress easily accomplishes this. Fracture progression in such embrittled Ti and Zr alloys takes place by crack linkup between the fractured hydride particles. Hydrogen embrittlement of this kind, therefore, is a special case of ductile fracture wherein crack linkup takes place between fractured, brittle particles. On this basis it is expected that toughness and ductility will decrease with increasing amounts of the hydride phase. This is the case, as shown in Fig. 13.20. Given this description, HE due to hydride formation should differ phenomenologically from examples cited earlier. For example, embrittlement should be more pronounced at higher strain rates and should not "disappear" at low temperatures, and this is what is observed experimentally.

There is a special case of HE pertaining to hydride precipitation. It has to do with stress-generated hydride formation in front of a crack tip in a material containing atomic hydrogen. The crack-tip tensile stress facilitates hydride formation on thermodynamic grounds; i.e., because the volume of the hydride phase is greater than the matrix from which it forms, a tensile stress favors hydride formation. Following hydride formation, the crack-tip stress causes fracture of the brittle hydrides. We note that the situation is similar to that found in partially stabilized zirconias, materials that are *toughened* by a stress-induced martensite transformation. Indeed, similar "toughening" accompanies the hydride formation just discussed. However, the magnitude of this toughening is much less than the toughness reduction associated with the readier crack linkup when fractured hydrides are present in front of the crack tip. This type of HE has been observed in a number of metals (e.g., Ti, Zr, Nb, and V) prone to form hydrides, and for which the hydride density is less than the matrix from which it forms.

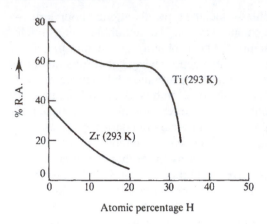

Figure 13.20
Room-temperature tensile reduction in area as a function of hydrogen content for αTi and αZr. The amount of the embrittling hydride phase increases with hydrogen content, accounting for the decreased ductility accompanying hydrogen content increases. (*Adapted from A. S. Tetelman and A. J. McEvily, Jr., Fracture of Structural Materials, Wiley, New York, 1967, p. 466. Original data from: (for Ti) R. I. Jaffe, J. Metals, 3, 247, 1955.*)

Certain alloying elements (e.g., molybdenum and vanadium) stabilize the body-centered cubic form of Ti and Zr and, on cooling Ti and Zr alloys containing these alloying elements from high temperature, a mixed α/β (hcp-bcc) structure results. Hydrogen embrittlement in these materials can take place absent hydride formation. It is believed that the embrittlement arises from the segregation of hydrogen to α/β interphase boundaries. The segregation reduces the boundary cohesive strength, promoting interphase-boundary fracture. That interphase-boundary fracture takes place during HE of α/β alloys supports this idea.

The idea of hydrogen segregation to internal boundaries plays a role in describing HE of other systems, particularly inherently ductile metals. Such segregation would be expected to reduce the stress required to decohere second-phase particles if segregation to particle-matrix interfaces took place. This leads to reduced ductility and fracture toughness by reducing the ductile fracture crack linkup distance, and also increases the number of fracture initiation sites. Fractographic measurement of dimple size on a number of "embrittled," inherently ductile fcc metals and alloys supports this viewpoint (Fig. 13.21.)

Hydrogen is but one element that can segregate to grain- or interphase boundaries. A number of other elements do likewise, and material embrittlement is often a consequence. Thus, in a sense, certain types of HE are a special case of impurity-atom embrittlement, the topic next addressed.

13.5
IMPURITY-ATOM EMBRITTLEMENT

A. Characteristics

We conclude our discussion of chemically induced metallic embrittlement by considering embrittlement caused by typical "tramp" elements, present ubiquitously to one degree or another in metals and their alloys. For lack of a better term, we call this impurity-atom embrittlement (IAE), although it is clear that other forms of embrittlement (e.g., SME and some manifestations of HE) can also be classified as such. Impurity-atom embrittlement in our context is caused by certain nonmetallic elements (e.g., S, P, O) that are present in a metal as a result of processing or through service exposure.

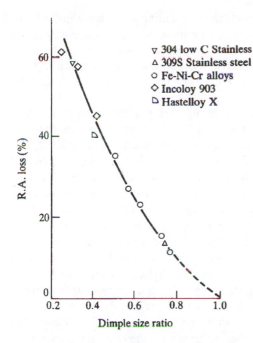

Figure 13.21
Loss of tensile reduction in area correlates with dimple size for a number of hydrogen-containing fcc alloys. The dimple size ratio is the dimple size on a fractured surface of a hydrogen-charged alloy compared to this size for an uncharged one (e.g., a ratio of 1.0 corresponds to the uncharged alloy with no reduction in ductility). Smaller dimple sizes correlate with a greater number of voids whose linkup results in fracture. (*From A. W. Thompson, Metall. Trans. A., 10A, 727, 1979.*)

Impurity-atom embrittlement is manifested by degradation of material mechanical properties, some of which are also found in other forms of embrittlement; e.g., tensile ductility and fracture toughness are reduced by IAE. When IAE takes place in metals prone to brittle fracture, such as ferritic steels, the DBTT is usually increased. Other similarities between IAE and other types of chemical embrittlement include the presence of a ductility minimum—that is, a temperature range over which embrittlement is observed—in inherently ductile metals. This is similar to what is found in SME and HE of some metals and alloys.

However, IAE differs phenomenologically in certain respects from other chemical embrittlement forms. For example, delayed failure and slow crack growth are not usually associated with IAE. Thus, while K_{Ic} is diminished by IAE, there is no IAE analog of a threshold stress intensity as is found, for example, for LME and SCC. This indicates that the atomic motion responsible for IAE takes place prior to, and not during, the material's exposure to stress.[8]

As mentioned, a number of metals and alloys are susceptible to IAE. Face-centered cubic nickel, for example, is commonly thought of as a very ductile metal, but prior to about 1870 it was considered brittle. It was discovered that sulfur,[9] when present in even minute amounts in Ni, was the culprit responsible for its embrittlement. Further, the effect of sulfur was most pronounced at intermediate

---

[8]An exception applies in some creep environments. In these situations, cavity/crack nucleation and/or growth is abetted by the impurity, and damage accumulation is accelerated and/or changes in fracture mode effected. Cavity/crack growth depends on impurity diffusion in these circumstances.

[9]Many elements embrittle Ni, but sulfur is one of the most potent. Other embrittlers are N, P, and O (the role of oxygen is discussed later). Additionally, certain metals (e.g., Se, Sn, Pb, Bi) when present in small amounts also embrittle nickel. However, these are special cases of metal embrittlement. It should be noted that not all "tramp" elements are deleterious to material performance. Boron, for example, enhances ductility in Ni and other nonferrous and ferrous metals.

temperatures. Nickel was subsequently ductilized by adding small amounts of manganese and/or magnesium to it. These elements combine with sulfur and prevent its segregation to grain boundaries where it instigates intergranular failure. This transition, from a transgranular to an intergranular fracture mode, is common to IAE, and reflects the importance of the interaction between the embrittling atom and the grain boundaries of the host metal.

Exposure of Ni and its alloys to oxygen at high temperatures can cause their oxygen embrittlement. An example is provided in Fig. 13.22 where we see that embrittlement is associated with intergranular oxygen penetration. And, as alluded to earlier and also demonstrated in Fig. 13.23, creep damage is often accelerated and fracture times decreased in oxygen environments. In the case illustrated in Fig. 13.23, oxygen reacts with carbon to form intergranular carbon monoxide gas bubbles, which are responsible for the increased creep damage. Oxygen can also react with hydrogen, as it does in hydrogen damage of copper, to form detrimental steam bubbles. However, oxygen in and of itself can also be harmful to a material's high-temperature performance. As is evident from Figs. 13.22 and 13.23, the grain boundary–embrittler atom interaction is important for oxygen embrittlement just as it is for sulfur embrittlement, and the interaction results in intergranular fracture in both cases. This is one example where IAE differs from previously discussed forms of embrittlement for which fracture can occur either transgranularly or intergranularly. Oxygen embrittlement of Ni and other materials can be eliminated by reducing or preventing oxygen ingress via grain-boundary penetration. Thus, single crystals can be used or protective coatings can be applied to the surfaces of polycrystals. Additionally, and for reasons not fully understood, small additions of boron reduce the tendency for this kind of embrittlement.

Impurity-atom embrittlement is also observed in body-centered cubic materials such as iron and steels, which are already prone to low-temperature brittle fracture.

 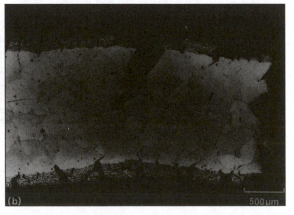

Figure 13.22
Profiles of fractured IN903A (an Fe-Ni-Cr alloy) samples after tensile testing at 1073 K. (a) The material was exposed to vacuum at 1273 K for 100 hours, and then tensile tested in air; this atmosphere accounts for the thin oxide surface film. (b) This alloy was exposed to air for 100 hours at 1273 K (the alloy has a thicker surface oxide as a result), and then tested in the same manner. The grain-boundary embrittlement due to oxygen (and not the thicker oxide scale) is the cause of this material's embrittlement. (*From R. H. Bricknell and D. A. Woodford, Metall. Trans., 12A, 1673, 1981.*)

The extent of embrittlement can be measured by changes in the DBTT or material toughness. In the form of Charpy impact energy the latter is often used to monitor the embrittlement of martensite tempered in the 575–625 K temperature range (Fig. 13.24). This form of IAE is reflected in a decrease in the impact energy of the tempered martensite, a decrease not consistent with that expected on the basis of changes wrought in the yield strength via tempering. Another form of temper embrittlement, this kind reversible unlike the 575–625 K kind, takes place in steels slowly cooled through, or tempered at, temperatures in the range of 650–850 K. This type of temper embrittlement increases the DBTT of the steel.

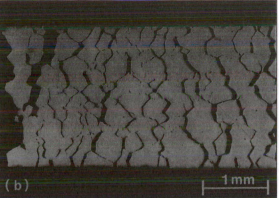

Figure 13.23
Gage sections of creep-tested Ni Alloy 270. (a) This low-carbon-content alloy was unloaded after 500 hours at a stress of 15.8 MN/m^2 and a temperature of 1073 K. Some slight creep cavitation has occurred. (b) This similar Ni 270 alloy contains carbon, which reacts with oxygen to form intergranular gas bubbles. The material failed at 23 hours under the same conditions that applied to the low-carbon alloy. (*From R. H. Bricknell and D. A. Woodford, Creep and Fracture of Engineering Materials, ed. B. Wilshire and D. R. J. Owens, Pineridge Press, Swansea, England, 1981, p. 249.*)

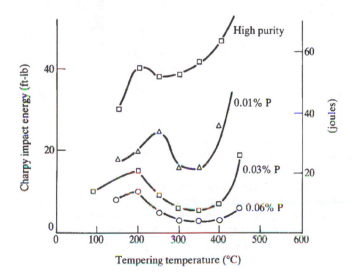

Figure 13.24
Tempered martensite embrittlement as monitored by impact energy. This energy is plotted vs. tempering temperature for a 3.5 Ni, 1.7 Cr, 0.3 C steel (compositions in wt.%.) The energy has a broad minimum for tempering in the 200–425°C range, but embrittlement is most pronounced for tempering between 300 and 350°C. Decreases in steel phosphorus content raise the overall impact-energy level and decrease the tendency for embrittlement. Thus, phosphorus is critical to this kind of embrittlement. (*From C. L. Briant and S. K. Banerji, Metall. Trans. A., 10A, 1729, 1979.*)

These examples of IAE are but a few of the many known of it. While the specifics of the embrittlement process differ among metal-embrittler combinations, the examples provided here serve as useful prototypes of IAE. We have seen that IAE has some features in common with other forms of embrittlement. The primary difference relates to the low mobility of the tramp elements within the embrittled metal in IAE, and this generally precludes it being manifested via static fatigue or slow crack growth. This differs from situations where the embrittler is fast moving, as it is during SCC, LME, and some forms of HE. There, slow crack growth and delayed failure are features of embrittlement.

B. Mechanisms

The characteristics of IAE described above can be considered "typical," and IAE mechanisms will be discussed on this premise. All cases of IAE involve an interaction between the embrittling species and the grain boundaries of the embrittled material. However, the manner in which embrittlement is manifested depends on the specific system and its thermal and chemical history. For example, sulfur in atomic form segregates to nickel grain boundaries at ordinary temperatures (Fig. 13.25). Presumably this lowers the intergranular fracture stress. This is similar, of course, to the mechanisms causing some forms of ME. It differs, though, in that *prior seg-*

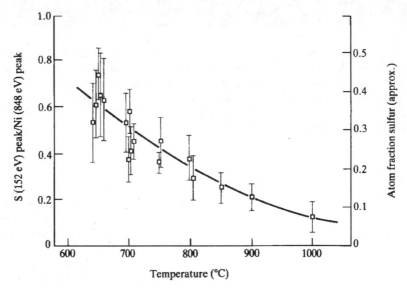

Figure 13.25
Equilibrium grain-boundary segregation of sulfur in a Ni–10% Cu alloy over the temperature range 600–1000°C. Substantial amounts of sulfur (see right-hand ordinate) segregate to grain boundaries even when the bulk concentration is only 0.007% as it is in this situation. The left-hand ordinate represents the "raw" data obtained from chemical analysis of grain boundaries; these data are converted to the approximate atomic fractions on the right ordinate. (*From R. A. Milford, Embrittlement of Engineering Alloys, ed. C. L. Briant and S. K. Banerji, Treat. Matls. Sc. Tech., 25, 1, 1983.*)

regation to grain boundaries is requisite for IAE, as atom-by-atom diffusion to a growing crack is precluded by the low tramp-element mobility. Embrittlement often disappears in ordinary nickel at high temperatures; however, when the embrittler is present in relatively high concentrations some loss in high-temperature ductility is still observed. In these cases, the cause is believed to be intergranular sulfide-containing precipitates, which make grain boundaries a likely fracture path. This situation often impairs the hot workability of nickel-base alloys.

The temperature variation of IAE in nickel and other ductile metals can be rationalized with kinetic/thermodynamic arguments (Probs. 13.16 and 13.17). On thermodynamic grounds, grain-boundary segregation is less at high temperatures, and the temperature at which ductility is restored corresponds to that at which such segregation is no longer sufficient to cause intergranular failure. At lower temperatures, kinetic factors preclude the degree of segregation required to embrittle grain boundaries. While this last explanation is conceptually appealing, and has been used previously to explain aspects of other kinds of embrittlement, there is some doubt as to its relevance to the low-temperature sulfur embrittlement of Ni. For example, some studies have thermally treated Ni-containing sulfur at temperatures at which segregation is known to take place, and then rapidly quenched the material to lower temperatures. One would expect, on the argument just used, that these materials would be more embrittled than those tested without the heat treatment. But this is not the case; the degree of embrittlement is the same for both situations. The results suggest that there may be a temperature dependence of the grain boundary–sulfur energy, although there does not seem to be a fundamental reason why this should be so. In closing our brief discussion of IAE in nickel and, by extension, other fcc metals, we note that factors generally promoting lesser ductility (e.g., an increased strain rate) also exacerbate IAE.

The forms of martensite embrittlement described are more complicated to explain. Intergranular fracture is found in martensite embrittlement, but it is held that the segregation responsible for it takes place during the austenitizing treatment. Thus, fracture takes place along prior austenite grain boundaries even though the martensite has many other internal boundaries that might be thought congenial for segregated species. Phosphorus is believed the element responsible for most manifestations of temper embrittlement, although metal atoms such as those of tin and antimony can produce an ME form of it. The fracture path observed as a result of 575–625 K embrittlement is also along prior austenite grain boundaries. In this case, sulfur segregation (as well as that of nitrogen and phosphorus) is held responsible for embrittlement, but the effect of the segregation is less direct than it is for the previous type of temper embrittlement we discussed. In particular, carbide precipitation along prior austenite grain boundaries causes the intergranular fracture observed in low-temperature martensite embrittlement. This precipitation does not take place in the absence of the embrittling species. Although several plausible explanations have been advanced for this complex precipitation–impurity atom interaction, none has yet been fully confirmed.

On the basis of our discussion, IAE is essentially another form of chemical embrittlement. We have attempted to note some similarities as well as differences between IAE and these other forms of embrittlement. We now direct our attention to radiation embrittlement. It differs in fundamental ways from chemical embrittlement.

13.6
RADIATION DAMAGE

Radiation damage is an encompassing term referring to the deleterious effects that radiation, particularly neutron radiation, can have on material properties. Embrittlement is only one manifestation of radiation damage. Others include the formation of internal voids or bubbles (causing material swelling), accelerated creep deformation, changes in material composition, unexpected chemical segregation, and the presence of phases not expected on the basis of chemical equilibrium. All of these are consequences of the interaction between neutrons and the atoms subject to their bombardment. To explain these phenomena requires some understanding of interactions between energetic neutrons and atoms. We thus initiate our discussion of radiation damage by considering this topic.

A. Neutron Interactions

High-energy neutrons interact with, and transfer energy to, atoms by means of elastic collisions. After such an event, the affected atom can be displaced from its lattice position provided sufficient energy is transferred to it, thereby producing a vacancy–self-interstitial pair (a Frenkel defect). If the energy transfer is sufficient to accomplish this, but not much in excess of it, the vacancy and interstitial recombine quickly and the defects are removed (the process is called athermal recovery). In order for recombination not to take place, the vacancy and interstitial must be separated initially by more than a few interatomic spacings, and this requires transfer of substantial amounts of energy from the neutron to the atom it collides with. The higher the initial neutron energy, the higher the transferred energy; thus, only higher-energy neutrons are responsible for this kind of radiation damage. How energetic such a neutron must be in order to be capable of radiation damage depends on several factors, but it is generally accepted that most damage is caused by neutrons having initial kinetic energies in excess of about one million electron volts (1 MeV = 1.6×10^{-13} J). Damage correlates with the total number of such high-energy neutrons that impact the material per unit area. This parameter, the *neutron fluence*, has units of neutrons per square meter (n/m^2). Fluence, in turn, is the time integral of the *neutron flux*, the latter being the number of incident high-energy neutrons per unit area per unit time. Again, only neutrons with energies greater than 1 MeV are considered in calculating the flux. This neutron-energy criterion is somewhat arbitrary. For example, the neutron spectrum (the distribution in neutron energies) is not considered in developing the criterion, nor is the irradiation temperature which, in some cases, is high enough to ameliorate damage by concurrent annealing.

High-energy neutrons are capable of transferring reasonable fractions of their energy (up to hundreds of kiloelectron volts) to the atoms they collide with. This results in two characteristics of neutron damage. First, even though the neutron energy loss per collision is large, the neutron still possesses a high kinetic energy following a collision. Thus, the neutron experiences a large number of primary "knock ons" before it eventually comes to rest in the lattice. Moreover, the distance over which the neutron dissipates its energy is large in comparison with atomic dimensions because the collision probability with an individual atom is rather small.

Further, the primary "knock on" atoms have energies high enough to displace other atoms from their lattice sites. The distance over which a "knock on" atom loses its energy is small in comparison to the analogous distance for high-energy neutrons. Nonetheless, these secondary collisions produce a cascade-like effect, and damage can be considered accrued along a cylindrical-like volume enclosing the paths of the knock on atoms (Fig. 13.26). The inner portion of the cylinder can be considered vacancy-rich while the region near the circumference is interstitial-rich (the cylinder radius is the distance traveled by the knock on atoms). Although many of the initially generated vacancy-interstitial pairs recombine by athermal recovery, many others do not. The resulting Frenkel defects harden the material. They also interact among themselves to form vacancy and interstitial loops—forms of dislocations— and these also increase material strength.

Depending on the neutron fluence and radiation temperature, vacancy loops can grow by addition of other vacancies to them, and eventually a void is formed. This produces a dilation causing material swelling, and the voids can also serve as nuclei for creep cavities. Bubbles (i.e., gas-filled voids) can also form during irradiation as a result of nuclear transmutation reactions. For example, xenon and krypton— insoluble in uranium and the oxide of it used as nuclear fuel—are common nuclear fission byproducts. Their precipitation as gas bubbles gives rise to fuel swelling, a factor that must be taken into account in nuclear reactor design. Additionally,

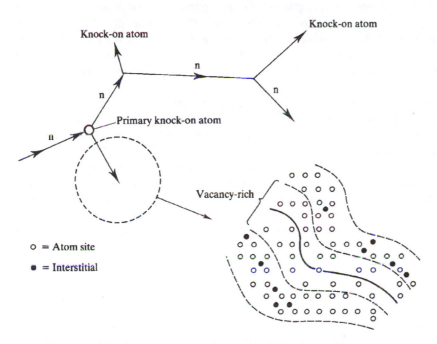

Figure 13.26
Schematic of lattice damage caused by high-energy neutrons. Each neutron has a number of collisions with the lattice atoms. Following almost every one, the affected atom has sufficient energy to be displaced from its lattice site, and to interact with other atoms causing further displacements of this kind. The displaced atoms produce a vacancy-rich volume immediately adjacent to the neutron path. This is surrounded by an interstitial-rich volume. The process illustrated is one of the fundamental means by which high-energy neutrons cause radiation damage.

Defect	Fission reactor*†	Fusion reactor*‡
Displacements per atom (dpa) per year	40	40
Helium production rate (atomic parts per million per year)	23	504
Hydrogen production rate (atomic parts per million per year)	290	1865

*Values are for a 316 stainless steel alloy.
†Appropriate for the Experimental Breeder Reactor (EBR-II).
‡Appropriate for a power density of 3.5 MW/m².
Source: Adapted from J. O. Steigler and L. K. Mansur, *Ann. Rev. Matls. Sc.,* **9,** 405, 1979.

Table 13.3
Defect production in fission and fusion reactor environments

hydrogen and helium are generated by nuclear reactions that occur between neutrons and the atoms of the base material. As with fission byproducts, these species can also precipitate as bubbles that can be harmful at relatively low (recall hydrogen embrittlement) and high temperatures (e.g., He bubbles can serve as cavities for creep fracture).

The extent of neutron damage is calculable, and the results of some such efforts are given in Table 13.3. The damage parameter, displacements per atom (dpa), is a measure of the number of times an atom is displaced from its lattice position by collisions with neutrons. Hydrogen and helium generation rates are expressed in atomic parts per million of these species produced over a specified time interval (e.g., one year); clearly, substantial amounts of these elements can be generated, especially when initial neutron energies are high as they would be, for example, in a fusion reactor. High-energy neutrons are truly a "hostile" environment for many materials!

As mentioned, creep rates can be increased (and creep-fracture times decreased) by exposure to neutron radiation. In some cases, creep deformation is beneficial (e.g., by relieving swelling-induced internal stresses), but often it is not. Creep deformation can be made easier under radiation as a result of the nonequilibrium defect density caused by it; the point is amplified later. We initiate discussion of radiation damage by describing the embrittling effect radiation can have on materials. While most of the data presented relate to neutron exposure of the kind found in ordinary fission reactors, similar (but more drastic) effects are observed through exposure to the higher fluences of breeder reactors and would also be found in proposed fusion reactors.

B. Radiation Embrittlement

Embrittlement resulting from neutron irradiation is associated with the increase in flow stress arising from neutron-generated defects. The strength increase is accompanied by a diminution in fracture toughness, and for ferrous body-centered cubic metals (used commonly for pressure vessels in nuclear-reactor systems) an increase in the DBTT. The change in fracture toughness can be assessed by direct measurement of K_{Ic} in the irradiated and nonirradiated conditions (Fig. 13.27), and the increase in the DBTT can be related to changes in impact energy (Fig. 13.28). Embrittlement is accompanied not only by an increase in the DBTT but also by a

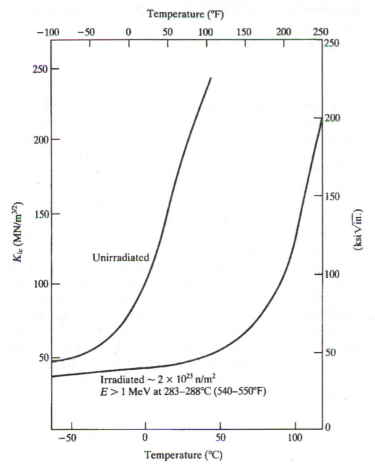

Figure 13.27
Temperature variation of fracture toughness for a pressure vessel steel, as affected by neutron radiation. This radiation increases the steel's ductile-to-brittle transition temperature. (*From S. H. Bush, J. Test. Eval., 2, 435, 1974.*)

decrease in the plateau (or shelf) value of the impact energy. Figure 13.28 additionally shows that the transition temperature and shelf energy correlate with neutron fluence.

The embrittling effect of neutron radiation can be viewed in a simple way. That is, as with most "strengthening" effects in metals, an increase in yield strength is associated with a decrease in fracture toughness and, for materials prone to a ductile-to-brittle transition, an increase in the DBTT as well. Irradiation temperature affects the degree of embrittlement. In particular, for a given neutron fluence, the changes in the DBTT and K_{Ic} are less the higher the irradiation temperature. This indicates that some recovery of neutron-generated defects takes place at higher irradiation temperatures.

Radiation embrittlement is also found in face-centered cubic metals such as austenitic stainless steels and nickel alloys. Its severest form is associated with accelerated creep fracture due to helium and/or hydrogen bubble precipitation. Other forms of damage—accelerated creep deformation and swelling—are also temperature sensitive, as described below.

C. Swelling

Perhaps swelling should not be classified an embrittlement phenomenon. But swelling does have an important effect on system behavior and reliability. Swelling takes place over a limited temperature range and its extent often relates to neutron fluence (Fig. 13.29). Swelling does not occur at low temperatures, because of the lack of mobility of the atoms or vacancies that constitute the voids or bubbles causing it; the lower temperature limit is about $0.3T_m$, where T_m is the material's absolute melting temperature. Likewise, swelling does not happen at high temperatures. There, defects generated via irradiation are removed rapidly by diffusion. The upper temperature limit depends on neutron flux and is material specific; however, this temperature is typically on the order of $0.55T_m$.

Damage causing void nucleation and growth can be modeled on the basis of the concentration of vacancies/vacancy loops, the concentration of interstitials/interstitial loops, the diffusivities of interstitials and vacancies, and the locations of sinks (grain boundaries, dislocations, the loops themselves) and traps (e.g., solute atoms) for these defects. This modeling is fundamentally sound, but the detailed ancillary data are woefully unavailable. The several diffusivities are often not known to the precision required to make accurate quantitative predictions. Moreover, diffusion coefficients may be increased (by virtue of the defect structure) or decreased (by reason of solute-atom trapping) in the radiation environment. Thus, effective diffu-

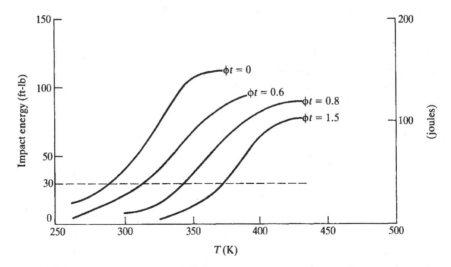

Figure 13.28

Impact energy vs. temperature for a steel irradiated at different neutron fluences (ϕt; expressed here in units of 10^{22} n/m^2 having energy > 1 MeV). Increasing fluence raises the DBTT (e.g., as measured by a 30 ft-lb energy criterion), and also decreases the upper shelf impact energy. (*After F. A. Brandt and A. J. Alexander, Radiation Effects on Metals and Neutron Dosimetry, ASTM STP 341, 162, 1963.*)

found useful for identifying the critical parameters controlling the extent and rate of swelling, and as the models become more accurate they will be put to even better use. Efforts like this are particularly important for a priori estimation of damage expected during potential fusion-reactor operation; experimental data here are sure to be slow coming and expensive to generate. Modeling has already proven fruitful in minimizing swelling taking place in fast breeder reactors. In particular, stainless steel is an important structural material in these reactors, and these steels are now processed so as to have a dislocation density propitious for minimizing swelling. This can be accomplished because the dislocations act as sinks for the point defects formed by radiation.

D. Radiation Creep

Creep is often enhanced by neutron radiation. As mentioned, cavities in the form of voids or bubbles nucleate more easily, with a concomitant decrease in fracture time, in a radiation environment. Beyond that, a material's creep rate can be significantly altered by such an environment.

Radiation creep differs from ordinary power-law creep, even though dislocation climb and glide are critical to both forms. For example, the steady-state substructure is dictated by neutron flux and temperature for radiation creep, and not by stress, with which it correlates for conventional dislocation creep. Because of the

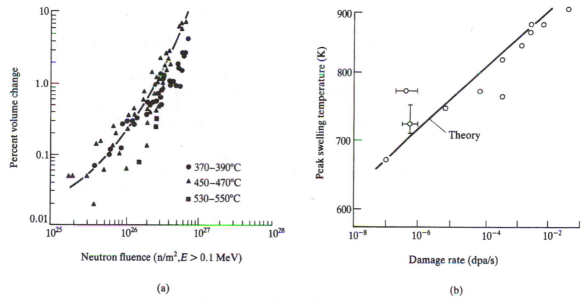

(a)   (b)

Figure 13.29
(a) Swelling relates to neutron fluence for irradiation over a temperature range of ca. 370–550°C. (Note, however, that there is some systematic variation in the amount of swelling at the different irradiation temperatures. Swelling is somewhat greater for the intermediate temperatures than it is at the higher and lower ones.) (b) The temperature for peak swelling in nickel depends on the fluence. (Here, fluence directly relates to the damage parameter—displacements per atom.) Higher damage rates increase the temperature of peak swelling. (*Part (a) from S. H. Bush, J. Test. Eval., 2, 435, 1974; (b) from N. H. Packen, K. Farrell, and J. W. Steigler, J. Nucl. Matls. 78, 143, 1978.*)

greater number of dislocation loops present in radiation environments, power-law creep rates are increased. Moreover, the stress dependence of the radiation creep rate differs from that for power-law creep. Since the structure is essentially stress-independent for radiation creep, strain rate depends less on stress (typically either linearly or quadratically) than it does for power-law creep. Creep mechanisms that depend quadratically on stress control creep rate at high stresses and those that depend linearly on it do so at lower stress levels.

The temperature dependence of radiation creep is also less than it is for power-law creep. In part this is due to the different substructures pertaining to each kind of creep. For example, the appropriate vacancy-diffusion activation energy for ordinary power-law creep involves both a formation and a motion energy. In contrast, for radiation creep the vacancy concentration is determined by fluence and temperature, and not only by temperature as it is for conventional creep. Thus, the vacancy motion activation energy is the only one governing, and a lesser temperature dependence of radiation creep is found.

In our treatment we have not covered all of the problems associated with material operation in high-energy neutron environments. For example, radiation-induced chemical inhomogeneity, phase solution and precipitation, and the presence of nonequilibrium phases have not been discussed. Not having done so does not diminish their importance, though. The area of radiation effects is scientifically and technologically important. It is hoped that further increases in our understanding of these phenomena will parallel those accomplished during the last half-century.

13.7
EMBRITTLEMENT OF INORGANIC GLASSES AND CERAMICS

The embrittlement of inorganic glasses and ceramics resembles embrittlement of metals. In some cases the mechanisms causing embrittlement are also similar for the different material classes. Because of the technological importance of glasses and ceramics, it is worthwhile to summarize briefly some specific features of their embrittlement.

Inorganic glass embrittlement can be manifested by static fatigue. The effect can be pronounced, as is shown in Fig. 13.30 which illustrates delayed failure of Pyrex glass as it is affected by humidity. Although static fatigue of Pyrex is observed in relatively dry environments, water is a clear culprit in instigating delayed failure in this and other inorganic glasses. In a number of ways, static fatigue of glasses is similar to stress corrosion of metals (SCC, Sect. 13.3). The embrittlement mechanisms are comparable in the sense that the environment promotes slow crack growth. This can be accomplished by "corrosive" effects that lead to advance of a crack having a constant crack-tip radius; or slow crack growth can be facilitated by environmental sharpening of the crack tip. And, as in SCC, if corrosion does not occur preferentially at the crack-tip (that is, if the corrosion rate is the same along the broad faces of the crack as at this tip) the environment does not promote premature material failure. An example is provided by the fracture behavior of glasses etched with hydrofluoric acid. Uniform "corrosion" of the glass surface occurs when it is immersed in such an acid. As a result, the lengths of preexisting surface cracks are reduced, and the uniform corrosion taking place within the crack blunts the crack tip. Both factors lead to a substantial increase (by as much as a factor of 100) in the fracture strengths of freshly etched glasses in comparison to those not so treated. Thus, in order for static fatigue to be observed, the crack-tip surface must dissolve more rapidly than the crack flat surfaces.

Such preferential dissolution can arise in two ways, both previously described in discussion of metal embrittlement. One possibility is that surface energy is decreased as a result of the adsorption of the "corroding" species at the crack tip. While this mechanism cannot be discounted, the currently accepted view is that static fatigue of glass is related to preferential crack-tip corrosion, i.e., it is similar to certain types of metallic SCC. The preferential-dissolution mechanism differs between metals and inorganic glasses, however. In contrast to metals, a passive layer does not form along the flat faces of a glass crack. Instead, preferential crack-tip dissolution is facilitated by the stress there. The enhanced corrosion apparently results from the stress-induced expansion of the glass network at the crack tip. For example, it is known that the corrosion rate of a glass depends on its specific volume; the higher it is (that is, the lower the glass density) the greater the corrosion rate. As with SCC of metals, slow crack growth in glass continues until the crack is sufficiently long so that the material's fracture toughness is attained; catastrophic fracture then occurs. Since glasses exhibit low fracture toughnesses, the extent of slow crack growth in them is less than in metals. In part this explains the low static-fatigue limits in glasses, which are often only ca. 20 percent of "virgin" glass strengths. The static-fatigue limit, of course, corresponds to a situation in which the crack-tip stress intensity does not increase with increased exposure time to the environment. For this condition the flat surfaces of the crack corrode at a rate such

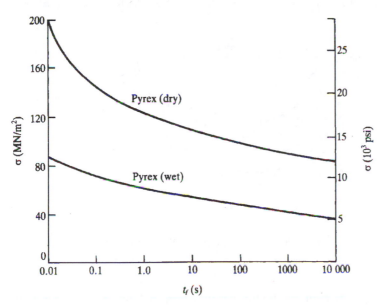

Figure 13.30
Static fatigue of Pyrex glass in wet and dry environments. The extent and rate of Pyrex embrittlement are much greater in the presence of water. (*Data from J. L. Glathart and F. W. Preston, J. Appl. Phy., 17, 189, 1946.*)

that the resulting crack-tip "blunting" compensates exactly for any increase in crack length resulting from stress-enhanced crack-tip corrosion.

The kinetics of glass static fatigue also are similar to those for many metal embrittlement phenomena. This is shown in Fig. 13.31 in which relative glass strength (i.e., the fracture strength divided by this strength in a liquid nitrogen environment ($T = 77$ K)) is plotted versus temperature for different exposure times in air. The embrittling agent here is water vapor. There is a (time-dependent) temperature below which static fatigue is not observed. This is due to the reduced atomic mobility at lower temperatures, i.e., to a lesser corrosion rate coming from a reduced rate of atomic detachment at the crack tip or by a lower rate of diffusion of H_2O molecules to the crack tip. As expected from this explanation, the temperature below which embrittlement is not observed increases with reduced exposure times. In the low-temperature regime, where such kinetic considerations control static fatigue, fracture times can be correlated with temperature via an Arrhenius-type relationship (Prob. 13.23). At higher temperatures, static fatigue is not as pronounced as at lower ones. The effect may be due to decreasing surface adsorption of atmospheric water or to increasing viscous deformation at the crack tip.

Static-fatigue crack-growth rates of glass correlate with stress intensity in much the same way these growth rates correlate with this factor in several types of metal embrittlement. Experimental results for soda-lime-silica glass are shown in Fig. 13.32a. As indicated there, crack growth can be divided into three stages. While not all stages are observed with all environment-glass combinations (a situation also holding for SCC of metals), the division into the three crack-growth stages is the one most generally observed. While it appears there is no critical environmental stress intensity below which crack growth does not take place for the glass whose behavior is illustrated in Fig. 13.32a, certain materials do display such a threshold

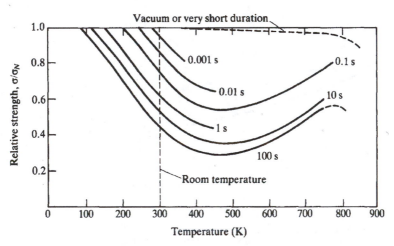

Figure 13.31
Relative glass strength vs. temperature for various exposure times in air. At low temperatures, static fatigue is not as pronounced because of the lack of thermal activation. At higher temperatures, it is also not as severe, owing to increased crack-tip viscous deformation and/or lesser adsorption of the embrittling species. (*Adapted from W. D. Kingery, H. K. Bowen, and D. R. Uhlmann, Introduction to Ceramics, 2d ed., Wiley, New York, 1976.*)

stress intensity. And, for all glasses, there is a stress intensity below which the material can be considered immune to static fatigue. This "critical" stress intensity decreases as the water content of the air increases. Higher humidity also increases the steady-state, or Stage II, crack-growth rate. During this stage, crack growth is controlled by the usual kinetic considerations. Unstable, final crack growth occurs at a K value that is nearly independent of the humidity level. Modeling of the crack-growth behavior in Stages I and II has been performed on the basis of reaction-rate kinetics. The attempts have met with considerable success, especially with regard to the dependence of crack-growth rates on stress, embrittling agent concentration, and factors such as the viscosity of the embrittling fluids.

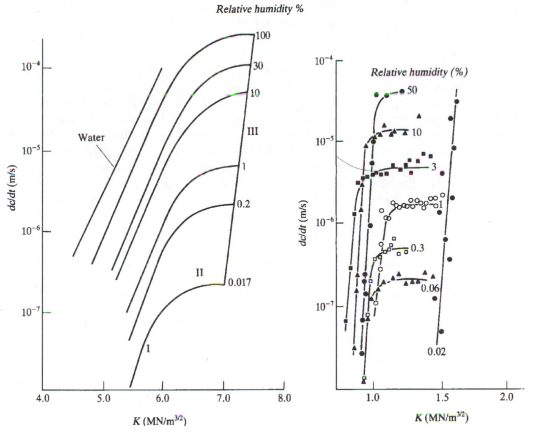

Figure 13.32

(a) Crack-growth rates in a soda-lime-silicate glass vs. stress intensity at various relative humidities. (Humidities are noted on the right-hand side of the figure.) The three crack-growth stages roughly parallel the similar stages in SCC and ME of metals. Steady-state (Stage II) crack-growth rates depend strongly on the humidity level, as does the stress intensity needed to cause a measurable crack-growth rate. Stage III (final) crack growth is independent of humidity, being only a function of stress intensity. (*Adapted with permission from S. M. Weiderhorn, J. Am. Ceram. Soc., 50, 407, 1967.*) (b) Crack-growth rates in an alumina single crystal as a function of stress intensity and relative humidity. The three stages of crack growth are similar to those found in glass. Stage II crack-growth rates increase with humidity content, but the correlation of these rates to Stage I behavior is not direct. (*After S. M. Weiderhorn, Int. J. Fract. Mech., 4, 171, 1968. Copyright 1968 Martinus Nijhoff Publishers. Reprinted by permission of Kluwer Academic Publishers.*)

Static fatigue of glass has been the most extensively investigated embrittlement phenomenon in inorganic solids other than metals. However, it and other manifestations of environmentally induced fracture, are also found in other oxides and ceramics. Most oxide ceramics contain a glassy silicate phase, so that their static fatigue behavior is similar to that of ordinary glasses. However, oxide ceramics not containing glassy phases are also prone to slow crack growth when exposed to embrittling agents. In many cases the cause is thought similar to impurity-atom embrittlement (IAE) of metals. That is, it is sensitive to impurity content, and is not controlled by the external environment. In these circumstances, crack-growth rates are dictated by the segregation tendencies of the harmful impurities to grain boundaries, and embrittlement kinetics are similar to those of IAE of metals. In other cases, though, embrittlement is environment sensitive. The behavior of aluminum-oxide single crystals (Fig. 13.32b) provides an example. The crack-growth behavior of this material in humid environments is similar to that found in glass in this same environment.

Static fatigue of inorganic glasses and ceramics can lead to substantial strength reductions; some data demonstrating this are provided in Table 13.4. The base measure of strength used in this table is the strength observed on testing in liquid nitrogen; this temperature is so low that diffusion is minimal and the environment can be considered inert. Testing at a higher temperature in gaseous nitrogen results in decreased strengths as a result of residual gas impurities; the strength decrease is system-specific. The most deleterious environment listed in the table is water vapor at 513 K. Strengths measured for this environment are much less than those observed in liquid nitrogen; in some cases (e.g., granite) the reductions in strength are on the order of 85%.

We conclude our description of environmentally assisted fracture of ceramics by first noting that the commonalties between it and similar fracture in metals are many. However, mechanistic differences can be substantial in some cases. The environmentally assisted fracture of ceramics is of special concern in view of their potential as high-temperature structural materials. One advantage of them in this re-

| Atmosphere | Liquid nitrogen | Dry nitrogen | Saturated water vapor | |
Temperature	77 K	513 K	298 K	513 K
Material		Strength (MN/m^2)		
Fused silica glass	453	445	384	253
Granite	258	136	162	41
Brazilian quartz	563	440	360	247
MgO crystal	210	184	98	55
Al$_2$O$_3$ crystal*	1050	804	759	471

*Transverse rupture strength.

Source: Adapted from W. D. Kingery, H. K. Bowen, and D. R. Uhlmann, *Introduction to Ceramics,* 2d ed., Copyright 1976. By permission of John Wiley & Sons, Inc.

Table 13.4

Compressive strengths of various glasses and ceramics in different environments: loading rate = 0.005 in/min

gard, noted previously, is their relative chemical inertness. While this is generally true, use of ceramics in environments that impair their integrity is an important technological problem. As such, it is currently being widely studied from both engineering and scientific perspectives.

EXAMPLE PROBLEM 13.2. Refer to Fig. 13.32. Using logarithmic coordinates, plot the steady-state crack-growth rates against the humidity content of the air. Fit the resulting curve to the relationship

$$\text{Crack-growth rate} \sim (\text{humidity content})^n$$

and determine the exponent n. Comment on your results.

Solution. The following Stage II crack-growth rates were obtained from Figs. 13.32a and b.

Glass		Alumina	
Humidity content (%)	dc/dt (m/s)	Humidity content (%)	dc/dt (m/s)
0.017	2.1×10^{-7}	0.06	2.1×10^{-7}
0.2	2.1×10^{-6}	0.3	4.8×10^{-7}
1.0	6.2×10^{-6}	1.0	1.7×10^{-6}
10.0	4.8×10^{-5}	3.0	2.8×10^{-6}
30.0	1.0×10^{-4}	10.0	1.4×10^{-5}
100	2.3×10^{-4}	50.0	3.9×10^{-5}

These data are plotted logarithmically in the accompanying figure. A reasonable straight line is found for both materials, indicating that the relationship stipulated is obeyed. To obtain the exponent n we rearrange this relationship;

$$n = \frac{\ln(r_2/r_1)}{\ln(h_2/h_1)}$$

where $r_{1,2}$ represents steady-state crack-growth rates at the humidity contents $h_{1,2}$. We take two datum from the respective lines of the graph constructed. For glass, we use crack-growth rates of 10^{-4} and 10^{-6} m/s for which $h_2 = 30$ and $h_1 = 0.10$; thus,

$$n_{\text{glass}} = \frac{\ln(100)}{\ln(30/0.10)} = \frac{4.605}{5.704} = 0.81$$

For the alumina, we use steady-state crack-growth rates of 4×10^{-5} and 4×10^{-7} m/s for which the pertinent humidity values are 50 and 0.18, respectively; thus,

$$n_{\text{alumina}} = \frac{\ln(100)}{\ln(50/0.18)} = \frac{4.605}{5.627} = 0.82$$

Note that the crack-growth rate exponents for both materials are essentially the same. Further, the growth rates vary *almost* linearly with humidity content. (The magnitude of these growth rates is less for the alumina, though.) A linear relationship between crack-growth rate and humidity content is what would be expected from first-order chemical kinetics which indicate reaction rates vary linearly with concentration.

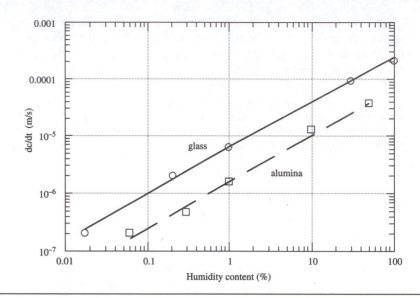

13.8
POLYMER EMBRITTLEMENT

Polymer embrittlement is manifested in a number of ways. Upon exposure to a particular environment a polymer may, for example, be prone to static fatigue, its strength level may be decreased, and/or its elongation to failure reduced.

A wide variety of environments can adversely affect polymeric mechanical performance. Oxygen, common to most environments, is the most prevalent embrittler of polymers. Water, in the form of liquid or as a constituent of humid air, is also a common embrittling agent, as is ultraviolet radiation coming from solar exposure. In addition, exposure to other kinds of fluids, such as acids or organic solvents, can result in diminished mechanical performance of polymers.

As with metals, the tendency for polymer embrittlement is environment- and system-specific. An example is provided by the relative resistance of polyether- and polyester-based polyurethanes to static fatigue via exposure to moisture. In Fig. 13.33, room-temperature strengths of the two kinds of polymers are plotted as a function of exposure time in water and humidified air at different temperatures. At 298 K, humid air substantially degrades the properties of the polyester-based polymer, but has a nil effect on the polyether one. And on exposure to water at a higher temperature, the strength of the polyester-based polyurethane is drastically reduced, but the strength of the polyether-based one is slightly increased.

The environments mentioned above can cause both polymer-chain scission and chain cross-linking. In addition, certain environments contain atoms or molecules that alter mechanical behavior even if scissioning or cross-linking does not occur. Scissioning and cross-linking both alter molecular structure, and change material mechanical characteristics as a result. Chain scission involves fragmentation of polymer chains via chemical reaction or by breaking of atomic bonds. The latter can be effected by the photonic energy associated with ultraviolet and other forms

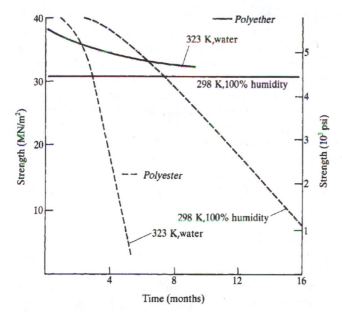

Figure 13.33

Environmental effects on polymeric behavior are system- and environment-specific, as shown here for the static fatigue of a polyether (solid curve) and a polyester-based polyurethane (dashed curve). The polyester is much more sensitive to water exposure than the polyether. Increasing the temperature and changing the environment from saturated air to liquid water enhances environmental degradation of the polyester. (*From R. J. Athey, Rubber Age, 96, 705, 1965.*)

of electromagnetic radiation, whereas chemically induced scission is somewhat analogous to a reverse polymerization reaction. A reduction in material strength accompanies the scission-induced reduction in polymer chain length. In contrast, cross-linking leads to polymer stiffening and a reduction in its ductility. An example is provided by the oxidation of rubber, which can take place in air at moderate temperatures. Oxidation promotes cross-linking and also expands the rubber volume. The combined effects lead to the development of a friable material following long-time service in oxidizing environments.

Chemical corrosion (e.g., by exposure of a polymer to some acids) often results in scissioning with a resulting material softening. However, a more dramatic effect on properties results from polymer swelling caused by exposure to a variety of gases and organic liquids. The phenomenon—called *solvent stress cracking*—is particularly acute with glassy polymers, and is associated with a tendency for enhanced craze formation. Solvent stress cracking is caused by absorption of the harmful species into the polymer. The absorption may have two effects. One is an increased resistance to shear flow, with a concurrent increase in the material's crazing tendency. (Recall that crazing and shear banding are competitive in glassy polymers.) Additionally, the absorbed material may ipso facto cause readier craze formation and more rapid craze growth. An interesting example is provided by the low-temperature nitrogen embrittlement of some glassy polymers. The tensile behavior at 77 K, at several different strain rates, of a polycarbonate immersed in helium gas (an inert environment) and in liquid nitrogen is shown in Fig. 13.34a. In the absence of environmental influence, the polycarbonate is moderately "ductile" at these temperatures, but its tensile strength is reduced appreciably when tested in liquid nitrogen; the effect is most pronounced at low strain rates. Similar behavior is observed in a brittle polymer (poly(methyl methacrylate), Fig. 13.34b). In both cases, the strength reduction is associated with an increased crazing tendency. Crazes are not only more pronounced in the presence of nitrogen, they are also longer and wider. The greater craze density and volume are most likely responsible for the still respectable ductilities exhibited by the environmentally degraded polymers. At somewhat higher temperatures, nitrogen exposure does not result in polymer strength reduction. This is attributed to the absence of absorption and is somewhat akin to the disappearance of IAE in metals and ceramics at high temperatures.

As mentioned, organic solvents also may initiate solvent stress cracking. As a result of the lower mobility of a fluid, as compared to a gas, this kind of stress-enhanced embrittlement is often observed at higher temperatures, including those in the vicinity of room temperature. Moreover, and in contrast to the like embrittlement due to gases, embrittlement kinetics are sufficiently slow that property deterioration is not observed during ordinary tensile testing. Instead, static fatigue is more often the manifestation of this kind of embrittlement. Craze growth at a stress level below the tensile strength is here limited by the diffusion of the ab-

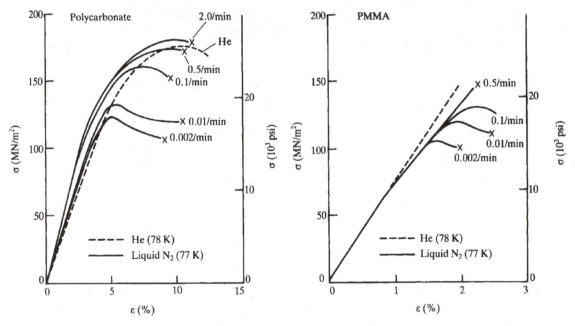

Figure 13.34
Liquid nitrogen embrittles both ductile [polycarbonate, (a)] and "brittle" [poly(methyl methacrylate), PMMA; (b)] polymers. The lesser ductilities and strengths are associated with increased craze tendencies in both polymers. (*Adapted from N. Brown and M. F. Parrish, Recent Advances in Science and Technology of Materials, Vol. 2A, ed. A. Bishay, Plenum Press, New York, 1974, p. 1.*)

sorbed species to the craze tip. That this diffusion model of craze advance is reasonable is confirmed by measurements of craze length as a function of time for poly(methyl methacrylate) immersed in methanol (Fig. 13.35). The square-root dependence of craze length on time indicates diffusion control of the process. Craze growth rate also varies with the fluid viscosity, viscous fluids resulting in lower growth rates. A low craze growth rate is desirable, on the one hand, since it reduces the tendency to premature fracture. On the other hand, it is undesirable since the embrittling tendency may remain undetected over relatively long service times, after which it manifests itself abruptly.

Finally, we recall that room and somewhat higher temperatures constitute creep environments for many polymers. Thus, a room-temperature static fatigue test displays aspects of creep deformation and the effect of the environment. The situation is somewhat analogous to the effect that environment has on high-temperature flow of metals.

In closing, we repeat that polymer embrittlement bears some phenomenological resemblance to metal and ceramic embrittlement. On the other hand, the complexity of polymer molecular architecture renders polymers susceptible to a number of diverse embrittlement manifestations. Consequently, extensive experimental and theoretical studies have been conducted in a quest to develop means to delay or prevent polymer degradation. Many of these approaches are based on polymer-chemistry principles that are outside the scope of this book. It is sufficient to note that various chemical treatments promote antioxidant and other environmental resistance in polymers. Given the remarkable improvements in polymer properties that have occurred in the last three decades, it is clear that protecting these materials from environmental degradation will continue to occupy the talents and efforts of materials scientists and engineers for some time to come.

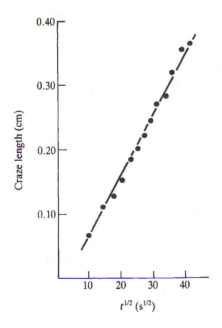

Figure 13.35
Craze length vs. square-root of time for crazes in poly(methyl methacrylate) immersed in methanol. The time dependence of craze length indicates craze growth is diffusion controlled. (*From E. J. Kramer, Developments in Polymer Fracture, Vol. 1, ed. E. H. Andrews, Applied Science Publishers, London, 1979, Chap. 3.*)

13.9

SUMMARY

In this chapter the phenomenology and mechanisms of several kinds of embrittlement were considered. With the exception of radiation damage, the other types of embrittlement we considered are some form of chemical embrittlement. While specific discussions have emphasized the differences as much as the similarities among the several chemical embrittlement forms, in this section we emphasize more strongly their commonalties.

All types of chemical embrittlement of metals cause a reduction in material ductility (e.g., as measured by reduction in area in a tensile test) and resistance to fracture (e.g., as assessed by a static fatigue test). Additionally, embrittlement is often, although not always, accompanied by a change in fracture mode from ductile transgranular fracture to brittle intergranular fracture or cleavage. Chemical embrittlement requires that the embrittler be present at the tip of the crack whose growth results in fracture. It is how these species get to this location, as well as their inherent embrittling capacity, that primarily defines the differences among the chemical embrittling forms.

Impurity-atom embrittlement (IAE) requires prior transport to internal boundaries of the elements causing it. When this occurs, intergranular fracture occurs along the segregated boundaries with a concurrent reduction in the fracture work. That prior segregation must take place for IAE indicates that the mobilities of the embrittling species are too low to allow them to migrate to a slowly growing crack tip for all reasonable crack-growth rates. Thus, static fatigue (delayed failure) and crack growth at a stress intensity less than the material fracture toughness (K_{Ic}) do not characterize IAE.

This is not so for other forms of chemical embrittlement of metals (ME, metal embrittlement; SCC, stress-corrosion cracking; and HE, hydrogen embrittlement). In these embrittlement forms, slow crack growth can occur at a stress intensity level less than K_{Ic}. In all of these situations there exists an apparent lower value of K (e.g., K_{IcSCC}, etc.) below which slow crack growth does not take place. As the stress intensity is increased above this critical value, crack-growth rate increases rapidly (Stage I of slow crack growth). During Stage I the stress and the environment interact to facilitate crack growth in a system- and environment-specific manner. At some (also specific) crack-growth rate there is a change in the controlling mechanism. In particular, a Stage II crack-growth region, in which dc/dt is approximately independent of K, is observed. This indicates that crack growth is limited by transport of the embrittling agent to the crack tip, and the Stage II crack-growth rate is limited by the rate of this transport. As an example, Stage II crack-growth rates of LME are often high because of the rapid flow of the liquid metal to the crack tip.

The mechanisms that cause crack-tip fracture depend on the embrittling species. That accepted for some forms of ME is like that held responsible for IAE, i.e., a lowering of the cohesive strength and a diminishment of the plastic fracture work. On the other hand, in one form of SCC (anodic dissolution) crack advance takes place by metal dissolution at the crack tip.

Since in a certain sense embrittlement can be considered to facilitate brittle fracture at the expense of ductile fracture, it is not surprising that many factors promoting brittle fracture also enhance the effect of chemical embrittlers and that those favoring ductile fracture minimize it. Thus, high temperatures eliminate, or diminish the effect of, chemical embrittlement. And for those materials—such as the bcc transition metals—susceptible to low-temperature brittle fracture, increases in strain rate raise the environmental DBTT. For ductile metals and alloys (e.g., most of the fcc ones), strain rate and temperature also influence embrittlement tendencies, but in ways different than for the low-temperature brittle metals. In ductile metals, ductility is often restored at low temperatures where the mobility of the embrittling atoms is reduced, and the temperature range over which brittleness is observed is displaced to higher temperatures by increases in strain rate. This is the situation, for example, for SME and HE.

Radiation damage caused by high-energy neutrons takes several forms. It can lead to an increased tendency to low-temperature embrittlement for bcc metals and alloys and a ductility reduction for some fcc ones. Over and above these effects, radiation damage can cause material swelling (due to void or bubble formation), increased creep rates, and decreased creep fracture times. The origin of these effects lies in the defects generated by neutron collisions or nuclear reactions between the neutrons and the atoms of the material. It is the subsequent (often complex) interactions among these defects (as well as their production rates) that determine the extent and rate of radiation damage. Physically plausible models can quantitatively account for the degree of damage. Until now, however, they have not been useful for predictive purposes because the ancillary data (e.g., diffusion coefficients, relevant diffusion distances, etc.) are not available. However, it is expected that their usefulness will increase as this information becomes available.

The above summary pertains to embrittlement of metals. However, many of the property alterations and the ways in which they depend on stress, temperature, and environment are similar in embrittlement of nonmetallics. As might be expected, microstructural and atomistic mechanisms of embrittlement differ among the material classes owing to fundamental differences in their atomic bonding and molecular architecture.

Our discussion of embrittlement has been incomplete because of the many forms it can take and the many ways in which it happens. Nonetheless, it is hoped that our treatment has shown that embrittlement is a common problem, and not to be taken lightly in view of the sometimes severe consequences attendant to it. For these reasons, the study of embrittlement remains intense within the materials engineering profession.

REFERENCES

Birnbaum, H. K.: "Mechanisms of Hydrogen-Related Fracture of Metals," *Proc. 1st Intl. Conf. on Environment-Induced Cracking of Metals,* ed. R. P. Gangloff and M. B. Ives, NACE-10, NACE, Houston, TX, 1988, p. 21.

Briant, C. L., and S. K. Banerji: "Intergranular Fracture in Ferrous Alloys in Nonaggressive "Environments," *Treat. Matls. Sc. Tech.,* ed. C. L. Briant and S. K. Banerji, 25, 21, 1983.

Ford, F. P.: "Stress Corrosion Cracking of Iron-Base Alloys in Aqueous Environments," *Treat. Matls. Sc. Tech.,* ed. C. L. Briant and S. K. Banerji, 25, 235, 1983.

Gesner, B. D.: *Polymer Stabilization,* ed. W. L. Hawkins, Wiley, New York, 1972, p. 353.

Jones, D. A.: *Principles and Prevention of Corrosion,* Macmillan, New York, 1992.

Lynch, S. P.: "Mechanisms of Fracture in Liquid-Metal Environments," *Embrittlement by Liquid and Solid Metals,* ed. M. H. Kamdar, TMS-AIME, Warrendale, Pa., 1984, p. 105.

Milford, R. A.: "Grain Boundary Embrittlement of Ni and Ni Alloys," *Treat. Matls. Sc. Tech.,* ed. C. L. Briant and S. K. Banerji, 25, 1, 1983.

Nelson, H. G.: "Hydrogen Embrittlement," *Treat. Matls. Sc. Tech.,* ed. C. L. Briant and S. K. Banerji, 25, 275, 1983.

Staehle, R. W.: "Understanding 'Situation-Dependent Strength': A Fundamental Objective in Assessing the History of Stress Corrosion Cracking," *Proc. 1st Intl. Conf. on Environment-Induced Cracking of Metals,* ed. R. P. Gangloff and M. B. Ives, NACE-10, NACE, Houston, TX, 1988, p. 561.

Steigler, J. O., and L. K. Mansur: "Radiation Effects in Structural Materials," *Ann. Rev. Matls. Sc.,* 9, 405, 1979.

Stoloff, N. S.: "Metal Induced Embrittlement—A Historical Perspective," *Embrittlement by Liquid and Solid Metals,* ed. M. H. Kamdar, TMS-AIME, Warrendale, Pa., 1984, p. 3.

Weiderhorn, S. M.: *Fracture Mechanics of Ceramics,* Vol. 2, ed. R. C. Bradt, D. P. H. Hasselman, and F. F. Lange, Plenum Press, New York, 1974, p. 613.

Woodford, D. A., and R. H. Bricknell, "Environmental Embrittlement of High Temperature Alloys by Oxygen," *Treat. Matls. Sc. Tech.,* ed. C. L. Briant and S. K. Banerji, 25, 157, 1983.

PROBLEMS

13.1 Metal embrittlement can take place between low-melting–high-melting metal combinations when the low-melting embrittler is either in the liquid or the solid form. However, for some metal combinations only LME is observed; SME is not. What is the cause of this behavior?

13.2 Schematically plot dc/dt during Stage II crack growth for LME and for SME. As K approaches K_{Ic}, do the relative LME and SME crack-growth rates change? Explain.

13.3 Predict the ratio of the fracture stress for metal embrittlement to the theoretical fracture strength for fully brittle fracture. (Your answer will contain the modulus, surface energy, and interatomic spacing for the embrittled and unembrittled conditions.) Do all of the critical factors defining the fracture strength change so as to facilitate fracture when an embrittler is present?

13.4 Metal embrittlement (ME) and stress-corrosion cracking (SCC) can be contrasted in a number of ways. Examples include the rate-controlling step leading to embrittlement. This step is either the rate of atomic detachment from an interface or a diffusion rate; the lesser of the rates controls the overall reaction rate. Complete the following statements.
a For SME involving surface diffusion, the rate-controlling step is a (diffusional/interface detachment) step involving transfer of (embrittler/embrittled metal) atoms from _____ to _____.
b For SME involving vapor transport, the rate-controlling step is a (diffusional/interface detachment) step involving transfer of (embrittler/embrittled metal) atoms from _____ to _____.
c For LME, the rate-controlling step is a (diffusional/interface detachment) step involving transfer of (embrittler/embrittled metal) atoms from _____ to _____.

d For SCC involving anodic dissolution (with no concentration polarization), the rate-controlling step is a (diffusional/interface detachment) step involving transfer of (embrittler/electrolyte) atoms from _____ to _____.

e For SCC accompanied by significant concentration polarization, the rate-controlling step is a (diffusional/interface detachment) step involving transfer of (embrittler/electrolyte) atoms from _____ to _____.

13.5 Construct a sketch similar to Fig. 13.10 for localized (i.e., strain constrained in a narrow band) and for uniform crack-tip plastic strain. Which kind of flow exposes a greater amount of metal at the crack tip for a given total plastic strain?

13.6 a Let the current-time relationship during repassivation be expressed by $(I - I_{corr})/(I_{max} - I_{corr}) = \exp(-t/\tau)$, where τ is a time constant associated with film reformation and t is the time following film rupture. If the time between rupture events is t_f, calculate the ratio Q_f/t_f. Discuss how this ratio changes as t_f varies from being a very short to a very long time.

b Using the results from (a), sketch dc/dt vs. t_f^{-1}. (Large values of t_f correspond to small values of stress-intensity factor and vice versa. Your plot corresponds to Stage I of crack growth.)

13.7 Derive Eq. (13.1). Hints: Consider a crack in a plate; the plate has an area A normal to the stress axis. The crack-tip material volume removal rate (dV/dt) is related to the crack advance rate by $dV/dt = A(dc/dt)$. Using the relation among density, mass, and volume, relate dc/dt to the mass removal rate. Then convert this to the charge removal rate using the metal valence.

13.8 Crack-growth rates for SCC of a brass were obtained for the following conditions:

Stress (MN/m²)	c (cm)	dc/dt (cm/yr)
350	0.05	0.025
350	0.10	0.05
700	0.05	0.10

For this brass, $\mathcal{G}_{Ic} = 70$ kN/m and $E = 84$ GN/m². It is proposed to use this material in this corrosive environment at an operating stress of 700 MN/m². Previous experience has shown that machining marks on the order of 10^{-3} cm are present on the metal surface. Estimate how long this brass will survive in this environment under the specified conditions. (Recall the relationship between $\mathcal{G}_{Ic}$, E, and K_{Ic}; see Chap. 9.)

13.9 Crack advance due to corrosion-related hydrogen embrittlement can take place discontinuously. What are the phenomenological similarities and differences between this kind of discontinuous crack growth and the εDCG of certain polymers during polymer fatigue?

13.10 a The text states that hydrogen embrittlement (HE) often disappears at low temperatures. What conditions allow this to happen?

b For many steels, the lower temperature at which HE is observed is higher than the ordinary DBTT. Thus, there are two temperature regions over which embrittlement is found. What strain rate and other conditions must apply so that HE would be manifested by an apparent increase in the ordinary DBTT? (That is, under what conditions is the lower temperature for HE higher than the ordinary DBTT?) What would be some interesting characteristics of this "double" embrittlement?

13.11 Use Fig. 13.19 to plot Stage II crack-growth rates (take these as dc/dt at $K = 1.5$ K_{IcHE}) as a function of temperature for the steel whose behavior is illustrated. Comment on your results.

13.12 Compare hydrogen embrittlement in α and $(\alpha + \beta)$ titanium alloys. Discuss the embrittlement mechanism(s) in the two alloy types, and how embrittlement is affected by strain rate and temperature. Would delayed failure take place as a result of HE in either of the alloy types?

13.13 Using physical reasoning, describe how hydrogen diffusion to a crack tip at low temperatures relates to hydrogen content, temperature, and other variables you consider pertinent.

13.14 a The text states that static fatigue is not found with impurity-atom embrittlement. Is this an "absolute" statement or can you conceive of conditions under which Stage I and Stage II crack growth would be observed for IAE? If so, what are these? And how would the values of any IAE Stage II crack-growth rates compare to those for SCC?
b Why is there no low-temperature ductilization associated with IAE?

13.15 Refer to Fig. 11.5a, the high-temperature fracture map for nickel. Schematically illustrate how a fracture map for nickel would differ between air and vacuum environments. If fracture times were placed on the maps, would the difference between air and vacuum environments be more pronounced for short or for long fracture times? Explain.

13.16 The equilibrium grain-boundary concentration (x_{be}) of a species that segregates to grain boundaries relates to the bulk concentration (x_0) and ΔH, the binding energy of the segregant to the boundary, through

$$\frac{x_{be}}{1 - x_{be}} = \frac{x_0}{(1 - x_0)} \exp\left[\frac{\Delta H}{RT}\right]$$

Use the curve drawn through the data of Fig. 13.25 to determine if the above relationship describes grain-boundary segregation of sulfur in this Ni-Cu alloy. If so, what is the value of ΔH?

13.17 Assume that a critical degree of grain-boundary segregation is required for IAE. Also assume that at high temperatures of IAE, diffusion of the embrittler to the grain boundary is rapid and the equilibrium grain-boundary concentration is quickly achieved. (This concentration is given by the expression in Prob. 13.16.)
a Estimate, as a function of ΔH, x_0, and the critical grain-boundary concentration, x_c, the temperature above which IAE is no longer observed. With our assumptions, would this upper temperature be a function of strain rate? Why or why not?
b Low-temperature IAE can be discussed on the basis of segregation kinetics. One expression for these kinetics is

$$\frac{x_b - x_{bo}}{x_{be} - x_{bo}} = 1 - \exp\left(-\frac{4Dt}{\alpha^2\delta^2}\right) \text{erfc}\left(\frac{2(Dt)^{1/2}}{\alpha\delta}\right)$$

where x_b is the grain-boundary concentration at time t, x_{be} is the equilibrium grain-boundary concentration, x_{bo} is the grain-boundary concentration at time $t = 0$, α is the ratio x_{be}/x_O, where x_O is the bulk concentration, D is the diffusion coefficient, δ is the grain-boundary thickness, and $\mathrm{erfc}(y)$ is a mathematical function whose values are tabulated in mathematical-table reference books. Estimate the lower temperature at which IAE is found. State and justify any assumptions. Will strain rate affect this temperature? If so, how?

13.18 Explain why neutron radiation embrittlement is not associated with low-temperature ductilization as is the situation, for example, with many forms of hydrogen embrittlement, stress-corrosion cracking, and metal embrittlement.

13.19 **a** Replot in greater detail the data of Fig. 13.29a so as to generate separate curves relating fluence and swelling at the three temperature ranges. (Do this by determining the average swelling at a given fluence for each of the temperature ranges.)
b Is there a temperature range for which swelling is generally greatest? Does your result agree with the statement in the text regarding the swelling-radiation temperature relationship?

13.20 Why is there a temperature below which swelling due to neutron irradiation does not happen, while there is no such lower temperature for neutron radiation embrittlement? (See also Prob. 13.18.)

13.21 **a** Consider Fig. 13.29b. Using physical reasoning, explain why the peak swelling temperature increases with the damage rate.
b For the nickel alloy whose behavior is illustrated, what is the range of peak swelling temperature when it is expressed as a fraction of nickel's absolute melting temperature? Is your answer in accord with statements in the text?

13.22 Schematically plot strain rate versus stress (logarithmic scale) for radiation creep when creep rate varies linearly, and when it varies quadratically, with stress. Show that at low stress levels the net creep rate varies linearly, and at higher stress levels quadratically, with stress.

13.23 Refer to Fig. 13.31. Relate the low-temperature kinetics of static fatigue to thermal activation. Do this by plotting the logarithm of the exposure time against the inverse of the absolute temperature below which static fatigue is not found. What is the apparent activation energy? (Hint: Recall that for a thermally activated process, the "reaction time," t, relates to the activation energy, Q, via $t \sim \exp(Q/RT)$ where T is the absolute temperature and R the gas constant.)

13.24 Use the data from Fig. 13.31 to construct static-fatigue curves for temperatures of 200, 300, and 400 K.

13.25 **a** Compare the enhanced craze formation of solvent stress cracking to the enhanced plastic activity model of metal embrittlement. Comment on similarities and differences.
b Would you expect LME crack-growth rates to be large or small compared with solvent stress cracking crack-growth rates? Explain your answer.
c Would apparent LME activation energies be large or small compared to activation energies for the solvent-induced crack growth? Explain.

CHAPTER 14

Cellular Solids

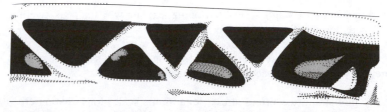

Frontispiece
A bird's wing—a section of one is illustrated here—is an example of a natural sandwich panel. The interior of the wing is porous, and the low density there combined with the solid wing skins provide a high ratio of wing stiffness and strength to weight. (*From D. W. Thompson, "On Growth and Form," abridged ed. (ed. J. T. Bonner), Cambridge University Press, Cambridge, 1961.*)

14.1
INTRODUCTION

If nature abhors a vacuum, she generally embraces empty spaces. The frontispiece of this chapter illustrates the construction of a bird's wing. It is primarily hollow; the only dense component of it is the skin. Much of what nature constructs is porous, not only bird's wings. A human skull bears more than a superficial resemblance to a bird wing; the differences relate primarily to the density of the porous core and the thickness of the bone (or skin) relative to this core thickness. Trees, a naturally occurring construction material, also contain lots of open spaces within their trunks and limbs.

The structures of the bird's wing and our other examples bear resemblance to those of sandwich panels which mankind employs as load-bearing structures. I-beams also bear such a resemblance. Solid beams, e.g., wooden beams used in residential construction, can support load. But an I-beam, which places the most material in locations subject to the highest stress, utilizes material far more efficiently

686

than a solid beam. Thus, human engineers mimic nature. Sandwich panels, like I-beams, are also material-efficient. Such a panel consists of a low-density (porous) core enclosed by skin panels (like our bird wing). Corrugated paper board is another example of a sandwich structure, and so are the floors of aircraft cabins. We shall see there are good reasons for constructing structures in the form of sandwich panels.

Sandwich panels are one application of cellular solids. Energy-absorbing devices are another one. Fragile components are protected during shipping by a surrounding (packaging) material, a cellular solid. Cellular solids are also used in more critical energy-absorbing applications; providing crash protection to automobile occupants, for example. In Sect. 14.4 we discuss how this protection is provided.

Cellular solids come in two basic forms; foams and honeycombs. Examples, manmade and natural, are provided in Figs. 14.1 and 14.2. Figures 14.1a and 14.1b are photographs of polyurethane foams. One of them is an open foam (Fig. 14.1a) and one of them (Fig. 14.1b) is a closed foam. An open foam can be viewed as a series of struts or beams connected in space. Sponge (Fig. 14.1c) is a naturally occurring open foam. A closed foam has a morphology comparable to grains in a polycrystal. Each cell has a shape similar to a grain. However, the "grain" interiors are hollow (in some foams, though, they contain a gas) while the grain "boundaries" constitute the solid portion of the foam.

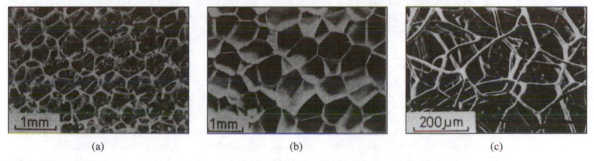

(a)   (b)   (c)

Figure 14.1
Photographs of (a) a manmade open-cell polyurethane foam, (b) a manmade closed-cell polyurethane foam, and (c) sponge, a natural open-cell foam. (*From M. F. Ashby, Metall. Trans. A, **14A,** 1755, 1983.*)

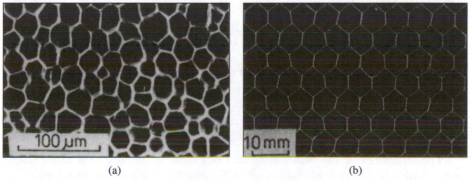

(a)   (b)

Figure 14.2
Photographs of (a) cork, a natural (sort of) honeycomb, and (b) a manmade aluminum honeycomb. (*From M. F. Ashby, Metall. Trans. A, **14A,** 1755, 1983.*)

Honeycombs are probably easier to visualize than foams. Honeycombs are, approximately, a two-dimensional version of a foam. However, honeycombs are typically more uniform. This holds for both natural honeycombs (e.g., cork, Fig. 14.2a)[1] and manmade honeycombs (e.g., the aluminum honeycomb of Fig. 14.2b). Honeycomb structures need not have the regular hexagonal arrangement displayed in Figs. 14.2. They also come in the form of triangles, squares, and irregular hexagons. In our discussion, however, we consider only the regular hexagonal variety. This makes the "arithmetic" easier. Discussion of other honeycomb arrangements is provided in the chapter references.

While cellular solids have other than structural applications (e.g., a polystyrene beverage cup used because of its thermal insulating characteristics), we focus on two important structural applications; their use as energy-absorbing devices and as cores in sandwich panels. We begin by first characterizing the density of a cellular solid. This property directly affects the cellular solid elastic modulus and its strength. As a consequence, the density also influences the cellular solid's energy-absorbing capacity and the properties of sandwich panels. Compressive behavior of cellular solids, energy-absorption mechanisms, and design protocols for energy absorption are then treated. We conclude with a discussion of the properties and failure mechanisms in sandwich panels.

Some mechanics is employed in our treatments. It is not at an advanced level, but the reader may wish to refresh herself here; references at the chapter end permit this. We do not bother ourselves with the details of the mechanics. Instead, we concentrate on how the results of the mechanics analyses impact the properties of cellular solids.

Before getting into the technical aspects, some remarks about the importance of cellular solids are in order. They are a big business. They command a market considerably greater than that for composite materials, for example. They are also of sufficient importance to have a technical journal (the *Journal of Cellular Solids*) devoted to them, and an excellent book dealing with them (*Cellular Solids: Structure and Properties,* by Lorna J. Gibson and Michael F. Ashby) is in its second edition. I have used this book extensively in developing this chapter. That will become apparent in the citations to many of the figures that follow. Nonetheless, an additional and more direct acknowledgment is also appropriate. It is provided here.

14.2
THE GEOMETRIES AND DENSITIES OF CELLULAR SOLIDS

Mimicking the geometry of a foam might appear difficult (e.g., Figs. 14.1a and b). Doing this for a honeycomb might appear easier (Figs. 14.2). There is also a geometrical distinction—dimensionality—between honeycombs and foams. Honeycombs are effectively two-dimensional structures. Their long dimension is much greater than those in the plane defining the honeycomb geometry. Foams, by contrast, are three-dimensional structures.

---

[1]We stretch things here. Cork is actually an anisotropic foam in which one dimension of the foam (the dimension into the plane of the photograph of Fig. 14.2a) is considerably greater than the other dimensions.

We model the respective geometries in a simple way (Figs. 14.3 and 14.4). Our modeling suffices, within a proportionality constant, to describe more complex geometrical as well as less-uniform arrangements within both foams and honeycombs. A regular honeycomb is viewed as a two-dimensional hexagonal arrangement of walls of side length, l, and thickness, t. Provided t is much less than l (i.e., provided, as is the usual case, the honeycomb density is low), the honeycomb density (ρ^*) relative to that of the solid material of the honeycomb (ρ_s) is (Prob. 14.1)[2]

$$\frac{\rho^*}{\rho_s} = \left(\frac{2}{\sqrt{3}}\right)\frac{t}{l} \qquad (14.1)$$

Other honeycomb configurations (e.g., triangular honeycombs, etc.) can be modeled similarly (Prob. 14.2). Because all honeycombs have the same dimensionality, all honeycomb relative densities vary linearly with t/l; only the associated proportionality constant is different for the different arrangements.

A similar exercise can be conducted for an open foam, which is taken as a cubic array of beams having thickness t and length l (Fig. 14.4). The relative density of this open foam is (Prob. 14.1).[3]

$$\frac{\rho^*}{\rho_s} = 3\frac{t^2}{l^2} \qquad (14.2)$$

Actual open foams have shapes different from the ideal version of Fig. 14.4. Nonetheless, the densities of these foams also vary with $(t/l)^2$. Moreover, densities of other idealized open foams (e.g., tetraikaidecahedra (a space-filling shape that

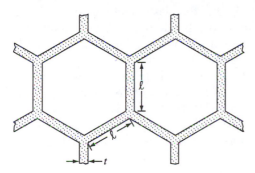

Figure 14.3
A regular honeycomb can be considered an array of hexagonal cells of edge length, l, and thickness, t. The density of the honeycomb is determined by the ratio, t/l (Eq. (14.1)). (*From M. F. Ashby, Metall. Trans. A, 14A, 1755, 1983.*)

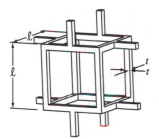

Figure 14.4
An open-cell foam can be mimicked with a cubic array of beams. The beams have length l (the cube edge length), and a cross-sectional area of t^2 where t is the beam thickness. The foam density relates to t/l via Eq. (14.2). (*From M. F. Ashby, Metall. Trans. A, 14A, 1755, 1983.*)

[2]Equation (14.1) overestimates the density at high relative densities. The error, though, is small; about 10% for a honeycomb having a relative density of 0.3 (a high relative density) and about 1% when the relative density is 0.1. We disregard these corrections. Carrying them through adds considerably to the complexity of the formulae we employ without adding much to the "physics" of honeycomb behavior.
[3]Equation (14.2) holds precisely only for low relative densities. As with honeycombs, we do not employ the (generally) minor correction that renders Eq. (14.2) precise.

mimics grains in a polycrystal), triangular prisms, and hexagonal prisms) also vary with $(t/l)^2$.

The density of a closed foam having a face thickness equal to that of its edges varies linearly with t/l. However, actual closed foams do not have faces this thick (e.g., Fig. 14.1); face thicknesses are typically much less than edge thicknesses. Because of this, the densities of actual closed foams usually also vary with $(t/l)^2$. Further, the mechanical response of closed-cell foams often is much like that of open foams; only minor alterations to properties such as modulus, strength, etc., are made by the faces. Thus, we do not differentiate between closed and open foams in our treatment.

Properties of cellular solids depend profoundly on their relative densities. Now that a protocol for describing these densities is developed, we describe how material properties relate to this variable. We first consider the compressive stress-strain behavior of cellular solids. This behavior is paramount in defining their energy absorbing capabilities.

14.3
COMPRESSIVE BEHAVIOR OF CELLULAR SOLIDS

A. Overview

A schematic of a compressive stress-strain curve for a cellular solid is given in Fig. 14.5. The deformation can be divided into three stages. In the first, linear elastic behavior is observed (we shall see, though, that the elastic modulus of a cellular solid lies well below that of a fully dense solid). In addition, and as will also become apparent, the linear elastic strain of which a cellular solid is capable generally well exceeds linear elastic strain limits of fully dense solids.

At a critical stress, the cellular solid "fails." The "failure" mode depends on the nature of the solid constituent. For example, failure can result from buckling of the beams of a foam or the walls of a honeycomb. This is common for elastomeric cellular solids. If the solid component is a metal (or certain thermoplastics), failure corresponds to plastic collapse of the beams or walls. And if the cellular solid is

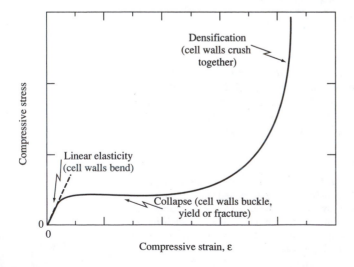

Figure 14.5
Schematic compressive stress-strain curve of a cellular solid. The deformation is divided into three regions. Initial linear elastic deformation is followed by a plateau which persists to fairly large strains. The plateau corresponds to cell-wall buckling (for elastomeric cellular solids), cell-wall yielding (for plastic cellular solids), or cell-wall fracture (if the cellular solid is constructed of a brittle material). At large strains, the stress increases rapidly with strain. This third deformation region corresponds to touching of the cell walls and material densification. (From M. F. Ashby, Metall. Trans. A, 14A, 1755, 1983.)

constructed from a ceramic or other brittle material, failure corresponds to fracture—"crushing"—of the beams or walls.

A plateau stress is observed in the second deformation stage that follows "failure." This stress persists to large compressive strains. During the plateau region,
an increasing fraction of the cellular solid buckles, plastically collapses, or is
crushed, as the case may be. The large strains accompanying the plateau stress are
one reason why cellular solids are good at absorbing energy. The energy absorbed
(per unit material volume) is the integral of the stress-strain curve. Thus, large extensions (even if accomplished at moderate stress levels) correlate with high energy
absorption.

At a certain strain, a third region—one in which the stress increases rapidly
with strain—is observed in the compressive stress-strain curve. This stage corresponds to material densification. That is, once a good fraction of the solid has buck-
led, collapsed, or fractured, further deformation forces the cell walls together. This
situation corresponds to "densification" of the cellular solid. All of the features of
the schematic of Fig. 14.5 are affected by the initial relative density and the properties of the material of the cell beams/walls. Thus, the cellular solid modulus, its
plateau stress, and the strain at which densification commences depend on the initial relative density. The modulus and plateau stress increase with increasing relative density, while the strain at which densification starts decreases with it. These
points are elaborated on in the following sections.

B. Compressive Elastic Behavior of Cellular Solids

We illustrate the compressive linear elastic behavior of cellular solids with a honeycomb (Fig. 14.6a). The arguments employed are dimensional. Further, because
elastic compression of a cellular solid results from beam/wall bending (Fig. 14.6b)
irrespective of whether a honeycomb or a foam is considered, our approach can be
extended to foams. Finally, in developing the equations describing elastic compression, we do not pay much attention to numerical coefficients. The appropriate
constants are inserted after we are finished with the mechanics.

Application of a force, F, (Fig. 14.6b) causes some of the walls of the honeycomb to bend. This results in a contraction (δ) in the direction of the applied load.
Beam theory tells us that δ is proportional to

$$\delta \sim \frac{Fl^3}{E_s I} \tag{14.3}$$

where E_s is the modulus of the honeycomb solid component and I is the moment of
inertia of the wall. For such a wall (Fig. 14.6), $I = bt^3$ where b is the depth of the
honeycomb into the plane of the drawing. Thus,

$$\delta \sim \frac{F}{E_s}\left(\frac{l^3}{t^3 b}\right) \tag{14.4}$$

The corresponding compressive strain, ε, is equal to δ/l so that

$$\varepsilon \sim \frac{F}{E_s}\left(\frac{l^2}{t^3 b}\right) \tag{14.5}$$

The nominal applied stress (σ) is proportional to F/lb so that the relation between the compressive stress and strain is expressed as

$$\varepsilon \sim \frac{\sigma}{E_s}\left(\frac{l}{t}\right)^3 \tag{14.6}$$

The honeycomb modulus is defined as $E^* = \sigma/\varepsilon$; thus

$$E^* \sim E_s\left(\frac{t}{l}\right)^3 \tag{14.7}$$

Substituting the relation Eq. (14.1) between relative density and t/l and also now introducing a previously neglected proportionality constant (C_h) we find

$$\left(\frac{E^*}{E_s}\right) = C_h\left(\frac{\rho^*}{\rho_s}\right)^3 \qquad \text{(honeycomb modulus)} \tag{14.8}$$

A more detailed treatment shows that C_h has a value very close to 1.5.

A similar analysis can be carried out for foams. The main alteration is that the moment of inertia for a beam in an open-cell foam is proportional to t^4. Using this in Eq. (14.3) and proceeding as above, we find that the modulus of an open foam relates to its density through (Prob. 14.3)

$$\left(\frac{E^*}{E_s}\right)_f = C_f\left(\frac{\rho^*}{\rho_s}\right)^2 \qquad \text{(open foam modulus)} \tag{14.9}$$

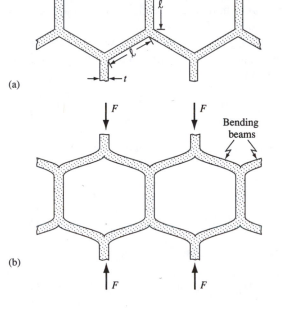

(a)

(b)

Figure 14.6
Schematic of linear elastic deformation of a compressed honeycomb. (a) The unstressed honeycomb. (b) Application of the force, F, bends the walls of the honeycomb, resulting in a compressive strain. Wall bending is the mechanism of linear elastic deformation of a honeycomb irrespective of the nature of the material constituting it. (*From M. F. Ashby, Metall. Trans. A, 14A, 1755, 1983.*)

with the constant C_f having a value close to one. Equation (14.9) has been found to adequately describe moduli of foamed solids over four orders of magnitude; i.e., for values of ρ^*/ρ_s ranging from 0.01 upward.

C. The Plateau Stress

Elastic deflection is supplanted by "failure" at a critical stress. As noted, the type of failure depends on the material of the cellular solid. For *elastomeric* cellular solids,[4] failure is due to elastic buckling of the cell walls/beams (Fig. 14.7a). If the solid component is ductile, the failure stress corresponds to plastic yielding of the beam/wall hinges (Fig. 14.7b). And when the solid component is brittle, "failure" corresponds to fracture of these hinges (Fig. 14.7c).

Failure comes when a critical moment (or load) is developed on the beams/walls of the cellular structure. The pertinent failure criteria are given in Table 14.1, which lists the respective plateau (failure) stresses for honeycombs and foams in terms of properties of the solid component (E_s, the modulus; σ_y, the plastic yield

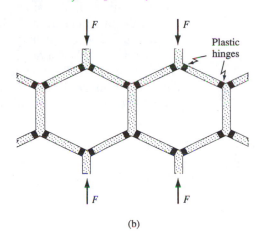

(b)

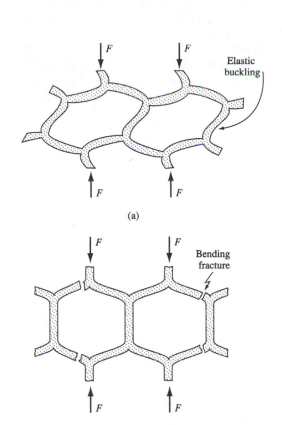

(a)

(c)

Figure 14.7
Failure modes of compressively stressed cellular solids depend on the nature of their solid constituent. (a) Elastomeric cellular solids fail by elastic buckling of their walls/beams. (b) If the material flows plastically, failure is by yielding of the "plastic" hinges in the walls. (c) If the solid is constructed from a brittle material, failure corresponds to fracture of the beam/wall hinges. The stress at which the various failures occurs is the plateau stress. (*From M. F. Ashby, Metall. Trans. A, 14A, 1755, 1983.*)

[4]An elastomeric cellular solid need not be constructed from a true elastomer. That is, as long as the solid's deformation is fully reversible, it is classified as an elastomeric cellular solid. For example, certain polyethylene (a thermoplastic) foams are elastomeric cellular solids on this definition.

Failure mode	Critical stress	Honeycomb	Foam
Elastic buckling	$\sigma_{el}*/E_s =$	$0.15\,(\rho*/\rho_s)^3$	$0.05\,(\rho*/\rho_s)^2$
Plastic collapse	$\sigma_{pl}*/\sigma_y =$	$0.5\,(\rho*/\rho_s)^2$	$0.3\,(\rho*/\rho_s)^{3/2}$
Crushing	$\sigma_f*/\sigma_f =$	$0.33\,(\rho*/\rho_s)^2$	$0.65\,(\rho*/\rho_s)^{3/2}$

Notes: E_s = modulus of solid constituent of cellular solid; σ_y = plastic yield strength of solid constituent; σ_f = fracture strength of solid constituent.

Table 14.1
Plateau stress for compressive loading of honeycombs and open foams

strength; and σ_f, the fracture stress for a brittle solid) and the cellular solid relative density. Because of their different dimensionalities, the plateau stress depends differently on this density for foams and honeycombs. We note that for elastomeric cellular solids the elastic strain at the plateau stress is independent of material properties. For example, $\sigma_{el}*$ (the elastic buckling stress for a foam) equals 0.05 $E_s(\rho*/\rho_s)^2$. However, since the modulus of such a foam (Eq. (14.9)) is $E_s(\rho*/\rho_s)^2$ the linear elastic limit of strain is 0.05 regardless of the foam material. Similarly (Prob. 14.4), the linear elastic strain limit for a honeycomb is 0.10. Both of these elastic strains are far greater than typical linear elastic strains for a fully dense solid.

During the plateau region, continued deformation occurs at a more or less constant stress level. However, depending on the material of the cellular solid, there are differences in the details of the plateau stress behavior (Figs. 14.8). The plateau stress in ceramic cellular solids displays many serrations (Fig. 14.8c). These corre-

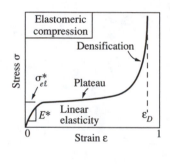

(a)

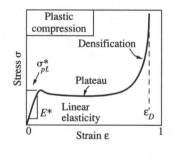

(b)

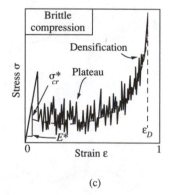

(c)

Figure 14.8
Schematics of the compressive behavior of a cellular solid as it varies with the nature of its solid material. Although the three deformation stages are exhibited with all material classes, there are differences in the shapes of the stress-strain curves, particularly at the failure-initiation point and in the plateau stress region. (a) Compressive stress-strain curves for elastomeric cellular solids are "smooth." The plateau persists until a strain at which densification begins. Then the curve gradually slopes upward until the densification strain is attained; after that the stress-strain curve is nearly vertical. (b) Following initiation of plastic collapse in a plastic cellular solid, the stress often decreases slightly. Although not shown in this schematic, the plateau often displays stress serrations cycling about the average plateau stress. The serrations correlate with collapse of individual cell walls/beams. (c) Such serrations are even more apparent during compression of a brittle cellular solid. They correspond to individual fractures of the cell walls/beams. In addition, the average plateau stress usually lies well below the fracture initiation stress (indicated by $\sigma_{cr}*$ in the figure and σ_f* in the text and Table 14.1). (*From "Cellular Solids—Structure and Properties," L. J. Gibson and M. F. Ashby, Pergamon Press, Oxford, 1988.*)

spond to individual beam/wall fractures. Further, failure by plastic collapse (Fig. 14.8b) also results in plateau stress fluctuations. These, too, are associated with collapse of individual "hinges." Figures 14.9 and 14.10 illustrate the progression of buckling in an elastomeric honeycomb and plastic collapse in a metal honeycomb during the plateau region.

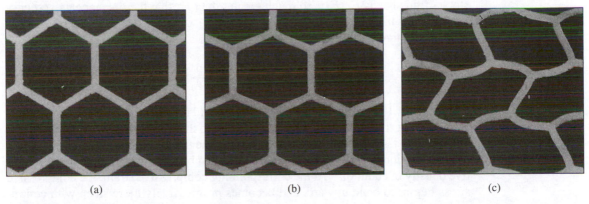

(a)   (b)   (c)

Figure 14.9
Photographs taken during compressive deformation of an elastomeric honeycomb (the stress axis is vertical). (a) Unstressed honeycomb. (b) Linear elastic deformation caused by bending of honeycomb walls. (c) Buckling at the plateau stress. (*From "Cellular Solids—Structure and Properties," L. J. Gibson and M. F. Ashby, Pergamon Press, Oxford, 1988.*)

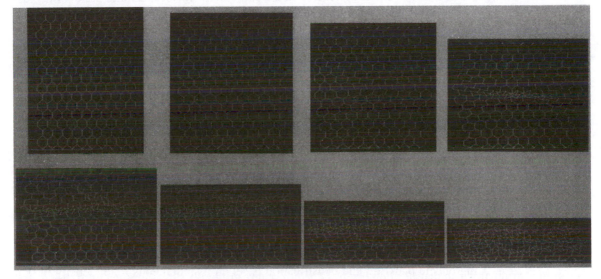

Figure 14.10
Photographs taken of sequential compression of an aluminum honeycomb. The strain increases from upper left (unstressed condition) to lower right. Upon application of stress the honeycomb first deforms elastically (upper row—second figure from left). With increasing deformation plastic collapse takes place. Heterogeneous deformation takes place at the plateau stress; the honeycomb consists of regions that have only deformed elastically and regions that have plastically collapsed. The fraction of material that has collapsed increases with strain (from upper right photographs through the three lower left photographs). The photograph on the far lower right-hand side is taken in the densification region. (*From S. D. Papka and S. Kyriakides, Acta Mater., 46, 2765, 1998.*)

D. Densification

With continued compression, the walls/beams of the cellular solid approach each other and eventually touch. This corresponds to the initiation of densification. At a certain critical strain (ε_D), the compressive stress increases rapidly. Indeed, its trace in the stress-strain curve appears almost vertical (cf. Fig. 14.5) although the actual slope of this line is E_s (E_s is so much greater than E^* that the line appears vertical). Experiments indicate that the critical strain approximately relates to the initial relative density through

$$\varepsilon_D \cong 1 - 1.4\left(\frac{\rho^*}{\rho_s}\right) \tag{14.10}$$

Equation (14.10) corresponds to the cellular solid having attained a relative density of ca. 0.7 via compressive deformation. We note that increasing initial relative densities lead to lesser plateau strains. Of course, the transition from the plateau region to densification is gradual and not abrupt. It can be considered to initiate at a relative density of about 0.5. At the corresponding compressive strain, the compressive stress begins to bend upward, and becomes progressively more steep with continued deformation.

E. Synopsis

All features of the compressive stress-strain behavior of a cellular solid—the elastic modulus, the plateau stress, and the densification strain—depend on its initial relative density. Figure 14.11 schematically illustrates the stress-strain behavior of open elastomeric foams having different initial relative densities.[5] The stress axis in Fig. 14.11 is normalized in terms of the modulus of the foam constituent. As a result, the "map" of Fig. 14.11 is generic; it applies to *all* elastomeric foams. Several features of this map are of interest. First, as noted previously, the linear elastic strain limit is 0.05 for elastomeric foams. Moreover, the plateau stress—when normalized by E_s—is solely a function of the initial relative density. So, too, is the densification strain. The stress asymptote occurs at a strain approximately equal to $(1 - 1.4(\rho^*/\rho_s))$ (cf. Eq. (14.10)).

Figure 14.11 is a form of a deformation-mechanism map for elastomer foams. The bold lines in the figure delineate the deformation modes (linear elasticity, elastic buckling, densification) that are observed at various compressive stress/strain combinations. The line separating linear elasticity and elastic buckling is nearly vertical; it is placed at a strain of 0.05. (The slight deviation from linearity at high stresses is observed only in foams with high relative densities.) The curve separating elastic buckling and densification is set an instantaneous foam density of 0.5. Such a density is achieved at a lower strain for foams with higher initial relative densities.

---

[5]Cellular solids having initial relative densities greater than 0.3 do not behave in the manner shown here. For example, they often do not display a true plateau stress because failure and densification overlap in them. However, most important cellular solids have initial relative densities less than (often substantially less than) 0.3.

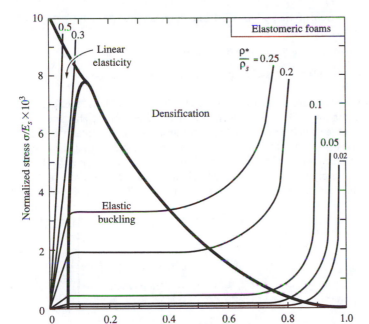

Figure 14.11

A deformation-mechanism map for an elastomeric foam in compression. Individual stress-strain curves for foams of different relative densities are shown. The solid bold lines separate the three deformation regions (linear elasticity, elastic buckling, and densification) depending upon the applied stress and the compressive strain. (*From M. F. Ashby, Metall. Trans. A, 14A, 1755, 1983.*)

Similar maps can be developed for plastic and brittle foams. However, they are not generic because the plateau stress (Table 14.1) and the linear elastic strain limit depend on the plastic yield strength (or fracture stress) as well as the modulus of the foam material. Problems at the end of this chapter illustrate these concepts. And, of course, compressive stress-strain curves (and, thus, deformation maps) can be developed for honeycomb compression. We show this in the following example problem.

EXAMPLE PROBLEM 14.1. Construct idealized compressive stress-strain curves for elastomeric honeycombs and open-cell foams having initial relative densities of 0.1. "Idealized" here means to assume that the plateau stress is constant until the densification strain, ε_D, is reached. Beyond this strain, the stress-strain curve is taken to be vertical.

Solution. **a.** *Elastic region*—The linear elastic strain limit is found by dividing the buckling stress by the cellular solid elastic modulus. We use Table 14.1 and Eqs. (14.8) and (14.9):

For the foam—$\sigma_{el}*/E_s = 0.05(\rho*/\rho_s)^2$; $E*/E_s = (\rho*/\rho_s)^2$. Thus, as anticipated, the linear elastic strain is 0.05.

For the honeycomb—$\sigma_{el}*/E_s = 0.15(\rho*/\rho_s)^3$; $E*/E_s = 1.5(\rho*/\rho_s)^3$. Thus, the honeycomb linear elastic strain is 0.10. As for elastomeric foams, this limit is the same for all elastomeric honeycombs.

b. *The plateau stress*— Substituting $\rho^*/\rho_s = 0.1$ into the relationships of Table 14.1, we obtain the plateau stresses:

$$\sigma_{el}^*/E_s \text{ (foam)} = 0.0005 \qquad \sigma_{el}^*/E_s \text{ (honeycomb)} = 0.00015$$

c. The *densification strain* for both the foam and the honeycomb is taken as $\varepsilon_D = 1 - 1.4(\rho^*/\rho_s)$ or $\varepsilon_D = 0.86$.

The compressive stress-strain curves are illustrated below. Note that the foam has the higher plateau stress for $\rho^*/\rho_s = 0.1$. However, this is almost invariably the situation. We can solve for the density at which a honeycomb would have a greater buckling stress by equating the σ_{el}^* values for the two constructions. We find that the honeycomb buckling stress exceeds that of the foam only if $0.15(\rho^*/\rho_s) \geq 0.05$ or $(\rho^*/\rho_s) \geq \frac{1}{3}$. Such a high density is rarely found in engineered cellular solids. This sample problem positions us to consider the next topic—energy absorption in cellular solids.

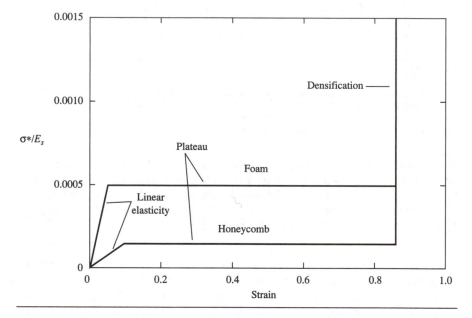

14.4
ENERGY ABSORPTION IN CELLULAR SOLIDS

Objects get tossed around in transit. Visit any airport baggage loading/unloading area and you will notice this. And sometimes the tossing around is not as deliberate as it appears to be with our luggage. Objects can be inadvertently dropped during shipping. If they have value some means must be found to protect them. Cellular solids are good at this. They can absorb significant energy per unit volume, and this characteristic is employed to protect objects during transit. Sometimes humans are the objects transported. Automobiles employ energy absorbing cellular solids to protect us in case of an accident.

In this section we discuss how cellular solids absorb energy, and we present design protocols when these solids are used for such purposes. While we focus on elastomeric foams (they are common packaging materials), our description can be readily extended to brittle and plastic packaging foam materials. Because their plateau stresses are typically higher than those of elastomeric foams, plastic metal foams and honeycombs are potential energy absorbing materials for automotive applications. That is, in certain situations plastic foams are capable of greater energy absorption than elastomeric foams. Problems provided at the end of this chapter illustrate this.

Why are cellular solids good at absorbing energy? Consider Fig. 14.12; it compares the compressive stress-strain curves of a fully dense solid and a cellular solid. The shaded areas correspond to the energy absorption per unit volume. For the same peak stress (σ_p), the energy absorbed is much greater in the cellular solid. The peak stress is an important parameter in protective packaging design. This stress is usually specified, and it corresponds to a "worst case" scenario during transport of the object to be protected. In addition, design requires a certain amount of energy absorption by the packaging material. Thus, we wish to have the maximum possible energy absorption consistent with the specified peak stress. On this criterion, the density of the cellular solid comes into play. Figure 14.13 schematically illustrates the compressive stress-strain behavior for three different density foams. If all of these are to absorb a *specified* energy (W, the area enclosed by the stress-strain curves up to the strains ε_1, ε_2, and ε_3, respectively) they must be subjected to

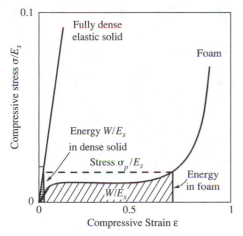

Figure 14.12
A schematic contrasting compressive stress-strain behavior of a cellular solid (a foam in this case) to that of a fully dense solid. The deformation energy per unit volume is the integral of the stress-strain curve. For a given peak stress, σ_p, a foam absorbs much more energy than its fully dense counterpart. (*From "Cellular Solids—Structure and Properties," L. J. Gibson and M. F. Ashby, Pergamon Press, Oxford, 1988.*)

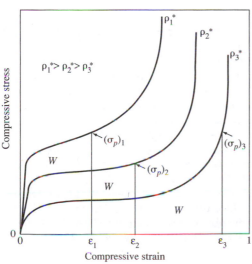

Figure 14.13
Schematic compressive stress-strain curves for cellular solids of different relative densities. The areas designated W correspond to equal energy absorbed per unit volume for the three materials. This occurs at different peak stresses and compressive strains for the different density materials. Thus, depending on the stipulated peak stress, there is an optimum foam density for absorbing energy. (*From "Cellular Solids—Structure and Properties," L. J. Gibson and M. F. Ashby, Pergamon Press, Oxford, 1988.*)

different peak stresses. The peak stress (σ_{p1}) in the highest-density foam of Fig. 14.13 exceeds that in the foam having the intermediate density. However, further decreases in foam density—to $\rho_3{}^*$—result in a foam that begins to densify before the stipulated energy absorption is attained. This results in a higher peak stress for the lowest-density foam. Thus, for a specified energy absorption, there is usually an optimum density foam for the also specified peak stress.

How are protective packaging materials selected? We illustrate the procedure in this section. We first describe how the energy absorbed is related to the maximum stress a cellular solid is subjected to (the peak stress) and to its relative density. Some simple examples illustrating design protocols are then provided. We illustrate with elastomeric foams. Problems at the chapter end extend the principles to other types of foams and also to honeycombs. We also employ the "idealized" foam whose behavior was described in Example Prob. 14.1.

As noted, a foam's compressive stress-strain curve displays three regions: a linear elastic region, a plateau region, and a densification region. Consider Fig. 14.14. We first integrate the area under the stress-strain curve to different maximum stresses $(\sigma_p,$ Fig. 14.14a). This area $(W;$ dimensions of energy per unit volume) is determined as a function of $\sigma_p.$ The exercise is repeated for different foam densities. We then plot W (after normalizing by E_s) vs. σ_p(also normalized by E_s). When

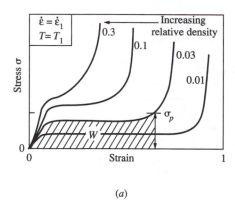

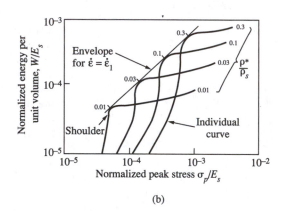

(a)   (b)

Figure 14.14

Procedure for constructing an energy-absorption diagram. (a) The compressive stress-strain curve is used as follows. For a specific relative density ($\rho^*/\rho_s = 0.03$ is used to illustrate), the area under this curve is determined as a function of the peak stress. After normalizing by the solid modulus (E_s), this area is plotted vs. a similarly normalized peak stress. (b) The resulting energy-absorption curve has three regions; they correspond to the three deformation regions. At low peak stresses, the energy absorbed–peak stress relation has a slope of ½ on the logarithmic coordinates employed. When the plateau stress is reached, the energy absorbed increases rapidly with strain (because the plateau stress is almost constant). This yields an almost vertical line in the energy-absorption curve. When densification starts, this curve is almost horizontal since little additional energy absorption is associated with the rapidly increasing densification stress. The characteristics of the energy-absorption curve depend on relative density (the plateau stress and densification strain depend on this parameter). A trace of the shoulders of the individual energy-absorption curves defines the maximum energy absorbed as it varies with the applied (peak) stress. (Note: As implied in these figures, the stress-strain curves, and thus the energy-absorption curve derived from it, are strain-rate sensitive. We have not considered this factor in our discussion.) (From "Cellular Solids—Structure and Properties," L. J. Gibson and M. F. Ashby, Pergamon Press, Oxford, 1988.)

we do this a family of curves for different density foams is generated (Fig. 14.14b). A diagram like that of Fig. 14.14b is called an energy-absorption diagram.

Actual energy-absorption diagrams for some polyurethane and polyethylene elastomeric foams are shown in Fig. 14.15. Such diagrams should be used in engineering design. However, we can approximate them using the idealized foam of Example Prob. 14.1. All of the essential physics are incorporated with such a procedure. Further, the "answers" obtained with the idealized diagram are good engineering approximations (Prob. 14.8). And, also important, the idealized diagram permits us to analytically express certain important parameters of an energy-absorption diagram.

The energy stored during linear elastic loading is equal to $\sigma_p^2/2E$ where σ_p is the applied stress. Substituting $E/E_s = (\rho^*/\rho_s)^2$ (appropriate to a foam) we obtain for the energy stored during linear elastic loading;

$$W = \left(\frac{\sigma_p^2}{2E_s}\right)\left(\frac{\rho_s}{\rho^*}\right)^2 \qquad (14.11)$$

Dividing by E_s;

$$\frac{W}{E_s} = 0.5\left(\frac{\sigma_p}{E_s}\right)^2\left(\frac{\rho_s}{\rho^*}\right)^2 \qquad \text{(linear elastic loading)} \qquad (14.12)$$

Equation (14.12) is manifested by the line having slope $= \frac{1}{2}$ in the energy-absorption diagrams of Figs. 14.14 and 14.16. The maximum stored linear elastic energy

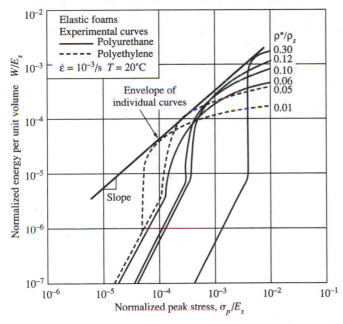

Figure 14.15

Energy-absorption curves for polyurethane and polyethylene elastomeric foams. The features of these experimentally determined curves closely parallel the schematic ones of Fig. 14.14. (*From "Cellular Solids—Structure and Properties," L. J. Gibson and M. F. Ashby, Pergamon Press, Oxford, 1988.*)

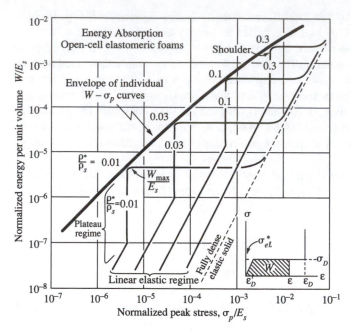

Figure 14.16
Idealized energy-absorption curves for open-cell elastomeric foams. The curves are calculated as described in the text (the corresponding compressive stress-strain curve is shown in the figure inset). The envelope through the shoulders of the individual curves corresponds to optimum design based on design-stipulated peak stresses and required energy-absorbing capability. (*From "Cellular Solids—Structure and Properties," L. J. Gibson and M. F. Ashby, Pergamon Press, Oxford, 1988.*)

is obtained by substituting $\sigma_{el}{}^* = 0.05 \, (\rho^*/\rho_s)^2$ into Eq. (14.11); thus, the maximum value of W/E_s for linear elastic loading is $0.00125(\rho^*/\rho_s)^2$.

The major energy absorption, however, takes place at the plateau stress. Here, this energy increases to a maximum value at a fixed stress (the elastic buckling stress). Thus, the plateau region is represented in an energy-absorption curve by a vertical line at the plateau stress. The energy absorbed during the plateau is the product of this stress and the strain (less the linear elastic limit of strain). Thus, for an elastomeric foam, the energy stored is

$$W = \sigma_{el}{}^*(\varepsilon - 0.05) \qquad \text{(plateau energy absorption)} \qquad (14.13)$$

(The 0.05 term in Eq. (14.13) represents the linear elastic strain limit.) The total energy absorbed at this deformation stage, though, also contains that due to linear elastic deformation. Using the expression just derived for this, the total energy absorbed during the plateau is more precisely expressed by

$$W = \sigma_{el}{}^* (\varepsilon - 0.025) \qquad \text{(plateau energy absorption)} \qquad (14.14)$$

Dividing Eq. (14.14) by E_s and substituting the expression for $\sigma_{el}{}^*$, we obtain

$$\frac{W}{E_s} = 0.05\left(\frac{\rho^*}{\rho_s}\right)^2 (\varepsilon - 0.025) \qquad \text{(plateau region)} \qquad (14.15)$$

The maximum energy stored during the plateau region is found at the densification strain (Eq. (14.10)). Thus,

$$\frac{W_{max}}{E_s} = 0.05\left(\frac{\rho^*}{\rho_s}\right)^2\left[0.975 - 1.4\left(\frac{\rho^*}{\rho_s}\right)\right] \qquad (14.16)$$

For strains exceeding the densification strain, the stress continues to rise but the energy absorbed does not; this accounts for the horizontal lines in Fig. 14.16 for stresses in excess of the plateau stress. This situation persists until the solid is fully dense. At this point the energy absorbed increases with the peak stress as it does for a fully dense solid (the dotted line of Fig. 14.16); i.e., $W/E_s = 0.5(\sigma_p/E_s)^2$.

As shown in Fig. 14.16, the shapes of the energy-absorption curves are similar for all foam densities. On the other hand, the maximum values of absorbed energy and the stress at which this occurs are highly density dependent. For a stipulated peak stress, the optimum energy-absorbing foam is that for which the peak stress equals the plateau stress. This corresponds to the "shoulder" in the energy-absorption curve for this foam. This can be illustrated with an example. Consider the energy-absorption curve of Fig. 14.16 for a foam with a relative density of 0.03. The plateau stress is about $4 \times 10^{-5} E_s$ (this is the stress corresponding to the vertical line for this foam). If the peak stress this foam experiences is the same as the plateau stress then the foam's energy absorption capacity is about $4 \times 10^{-5} E_s$ (this corresponds to the horizontal line for this foam). If we were to employ a foam having a density less than 0.03, it would have "bottomed" out at this stress and absorbed less energy. On the other hand, if we use a foam of greater density its plateau stress exceeds the peak stress. So the energy absorbed by it would be only that small amount associated with linear elastic deformation. Thus, the optimum energy-absorbing foam for a specific peak stress is that having a shoulder at this stress.

The curve (the bold line of Fig. 14.16) representing "optimum" energy absorption as a function of the peak stress is the envelope of the shoulders of the individual energy-absorption curves for foams of various densities. An equation can be developed for this envelope. It is a variant of Eq. (14.16) and is obtained by expressing the relative density in terms of the plateau stress (cf. Table 14.1). Doing this we obtain

$$\frac{W_{max}}{E_s} = \left(\frac{\sigma_p}{E_s}\right)\left[0.95 - 6.26\left(\frac{\sigma_p}{E_s}\right)^{1/2}\right] \qquad (14.17)$$

Equation (14.17) is employed in materials selection as follows. First, the peak stress is determined from engineering considerations. Then a foam having a plateau stress equal to this stress is chosen; this specifies the foam density. Equation (14.17) is subsequently employed to determine the energy-absorption capacity per unit foam volume; the energy absorption is also a specified design parameter. Knowing the necessary energy absorption per unit volume permits the dimensions of the foam to be determined. All this may seem a bit confusing. Perhaps some simple case studies might help to clarify.

CASE STUDY 1: DETERMINATION OF FOAM THICKNESS WITH A SPECIFIED FOAM MATERIAL. A computer is to be protected during shipping. The computer dimensions are $40 \times 40 \times 15$ cm, and its mass is 7 kg. If the case containing the computer were dropped during transit (e.g., during unloading it from a truck), the maximum drop height would be about 2 m. The maximum deceleration that the computer can experience without damage to it is 10 g (g = gravitational acceleration = 9.8 m/s^2). The packaging material is a polyurethane (PU) foam (E_s = 50 MN/m^2). Determine the foam thickness required to protect the computer during shipping. What is the optimum relative density of the PU foam?

Solution. First determine the kinetic energy ($U = mv^2/2$) of the computer dropped from a height of 2 m; this is the energy that the foam must absorb. The velocity, v, of a falling object is $(2xg)^{1/2}$ where x is the height the object has fallen from. Thus, v = $(2 \times 2 \times 9.8)^{1/2}$ = 6.3 m/s, and $U = (1/2)(7 \times (6.3)^2)$ = 139 J.

We now determine the peak stress. The peak force is $10mg$ = 690 N. The maximum stress occurs on the smallest lateral side of the protective foam. This area is 15 × 35 cm, or 5.25×10^{-2} m^2. Thus, the peak stress (σ_p) is 690/.0525 = 13 kN/m^2. Division by E_s gives σ_p/E_s = 2.6×10^{-4}. Substituting this into Eq. (14.17) we obtain W_{max}/E_s = 2.3×10^{-4} or W_{max} = 11,500 J/m^3. Dividing U by W_{max} gives the required foam volume (1.21×10^{-3} m^3). Further division by the smallest lateral area gives the foam thickness; t = 0.23 m. This is a fairly large thickness and this arises because of the computer's fragile nature. The foam density is found by equating the peak stress to the elastic buckling stress (Table 14.1). The relative foam density is 0.072.

CASE STUDY 2: DETERMINATION OF OPTIMUM FOAM MATERIAL. A 1-kg cubic object (edge length = 10 cm) is to be protected with a foam of specified thickness—3 cm. The object is unlikely to be dropped from a height greater than 1 m during shipping. The object can withstand a deceleration of $50g$ without damage. Select the optimum foam material (i.e., specify its modulus in solid form) and its relative density.

Solution. The velocity after falling a distance of 1 m is $v = (2 \times 1 \times 9.8)^{1/2}$ = 4.4 m/s. The corresponding kinetic energy is $(mv^2/2)$ = $(1 \times 19.4/2)$ = 9.7 J. Dividing this energy by the volume of one side of the protecting foam (= $(0.1 \times 0.1 \times 0.03)$ = 3×10^{-4} m^3), we get W_{max} = 3.2×10^4 J/m^3. The maximum force ($50mg$) is 490 N. Division by the foam area (10^{-2} m^2) gives σ_p = 4.9×10^4 N/m^2. Multiplying both sides of Eq. (14.17) by E_s we have

$$W_{max} = \sigma_p \left[0.975 - 6.26 \left(\frac{\sigma_p}{E_s} \right)^{1/2} \right]$$

or

$$3.2 \times 10^4 = 4.9 \times 10^4 \left[0.975 - 6.26 \left(\frac{\sigma_p}{E_s} \right)^{1/2} \right]$$

Solving this equation, we find σ_p/E_s = 2.6×10^{-3}, or E_s = 19 MN/m^2. The relative foam density (Table 14.1) is 0.23. (This is a fairly dense foam.)

We have used open foams as examples. But the same procedures can be used for closed foams and for honeycombs. Crash protection is a potentially important application of metal foams; Prob. 14.9 deals with such an application.

The discussion to this point has focused on the compressive behavior of foams. However, foams and (even more so) honeycombs are also used as core materials in sandwich panels. These panels are akin to an I-beam in that the "most" material is placed in locations subject to the greatest stresses. And, like I-beams, sandwich panels also are subject to tensile stresses. In the next section we attempt to simply describe sandwich panels and their properties.

14.5
SANDWICH PANELS

A sandwich panel consists of two solid skins separated by a porous (foam or honeycomb) core. Sandwich panels are employed in a variety of ways. Their use in aerospace structures is important. Here they find application as helicopter rotor blades and aircraft flooring panels to name just two uses of them. Sandwich structures are common, too, to sporting goods; modern skis, for example, are constructed from sandwich panels. Corrugated paper board is also a sandwich panel. Thus, uses of sandwich panels range from the mundane to the sophisticated.

There are good reasons to use sandwich panels. A sandwich provides good resistance to bending and buckling. Its strength and stiffness combined with a lightweight core provides excellent strength/stiffness per unit panel weight. Thus, as noted, a sandwich panel is structurally related to an I-beam.

A thorough discussion of sandwich panels is heavy going. Considerable mechanics, more mechanics than materials in fact, is involved. Here only a brief overview is provided. Many "details" are given short shrift or even neglected. Those interested in more—or even much more—can consult Gibson and Ashby.

A. Elastic Properties of Sandwich Panels

A sandwich panel loaded in three-point bending is illustrated in Fig. 14.17. The panel is composed of a core (a foam or honeycomb) enclosed by two solid skins, the faces. We use the following terminology. The core density is represented by ρ_c^*; the solid component of the core has the bulk density ρ_s. The subscript f designates the properties of the faces and the subscript c the core properties. The beam is of length l, and the thicknesses of the core and faces are c and t, respectively. Thus, the overall beam thickness is $d = 2t + c$. However, since $t \ll c$ usually, $d \cong c$. We consider only three-point bending. However, other types of loading (a uniformly distributed load, a cantilever end load, etc.) are treated similarly. The major differ-

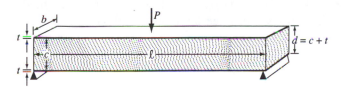

Figure 14.17
A sandwich panel loaded in three-point bending. The panel has a cellular solid core (a honeycomb or a foam) enclosed by two solid skins. The skin thickness (t) is typically much less than the core thickness (c). Sandwich panels can display high specific stiffnesses and strengths (i.e., a high stiffness or strength-to-weight ratio). (*From "Cellular Solids—Structure and Properties," L. J. Gibson and M. F. Ashby, Pergamon Press, Oxford, 1988.*)

ences in the results obtained lie in the numerical proportionality constants. Gibson and Ashby tabulate these constants for a number of common loading schemes.

We consider foamed cores, although in principle our treatment can be extended to honeycomb cores.[6] The Young's modulus of the foamed core is

$$E_c^* = E_s\left(\frac{\rho_c^*}{\rho_s}\right)^2 \tag{14.18}$$

Sandwich panels bend as well as deflect. Thus, we need to know the core shear modulus, G_c^*. This modulus approximately relates to the Young's modulus in the same way that tensile and shear moduli relate in fully dense solids. Thus,

$$G_c^* = 0.4E_s\left(\frac{\rho_c^*}{\rho_s}\right)^2 \tag{14.19}$$

The panel stiffness is calculated from its *equivalent flexural rigidity*, $(EI)_{eq}$, and its *equivalent shear rigidity*, $(AG)_{eq}$. $(EI)_{eq}$ represents resistance of the beam to bending; $(AG)_{eq}$ measures its resistance to shear. For our rectangular beam

$$(EI)_{eq} = \frac{1}{6}E_f bt^3 + \frac{1}{12}E_c bc^3 + \frac{1}{2}E_f btd^2 \tag{14.20}$$

The first two terms on the right-hand side of Eq. (14.20) represent the respective bending stiffnesses of the faces and core about their own centroids. In most sandwich panels these stiffnesses are much less than the third term on the right-hand side of Eq. (14.20), which represents the bending stiffness of the faces about the panel centroid. Thus, we consider only this last term. Further, since $d \cong c$, Eq. (14.20) simplifies to

$$(EI)_{eq} = \frac{1}{2}E_f btc^2 \tag{14.21}$$

The equivalent shear rigidity is

$$(AG)_{eq} = \frac{bd^2 G_c^*}{c} = bcG_c^* \tag{14.22}$$

where we have used $d \cong c$ in obtaining the final expression on the right side of Eq. (14.22).

When a load, P, is applied to the beam, it deflects. The deflection is the sum of bending (δ_b) and shear (δ_s) deflections. For three-point bend loading these are

$$\delta_b = \frac{Pl^3}{48(EI)_{eq}} \tag{14.23a}$$

and

---

[6]There is a further complication in considering honeycomb cores. In particular, the long axis of the honeycomb is often normal to the skin faces in a sandwich panel. Loading of the kind we consider stresses such a honeycomb arrangement along its long axes. Honeycomb elastic constants for this out-of-plane loading differ from those that apply to in-plane loading. The various failure stresses also differ between in- and out-of-plane loading; out-of-plane moduli and failure strengths are greater. Problems 14.11 and 14.12 treat some aspects of out-of-plane loading of a honeycomb.

$$\delta_s = \frac{Pl}{4(AG)_{eq}} \qquad (14.23b)$$

The beam compliance (inversely proportional to its stiffness) is the sum of Eqs. (14.23a) and (14.23b) divided by P. On substituting Eqs. (14.21) and (14.22) into Eqs. (14.23) we obtain

$$\frac{\delta}{P} = \frac{2l^3}{48E_f btc^2} + \frac{l}{4bcG_c^*} \qquad (14.24)$$

Beams are usually designed to have a specific compliance (or stiffness). Beyond that, a beam with a minimum weight for this compliance is desired. The beam weight (W) is the sum of the core and face weights; it is

$$W = 2\rho_f gblt + \rho_c^* gblc \qquad (14.25)$$

where g is the gravitational constant. If we wish to minimize W for a specified beam length, (l) and width (b) in a panel with specified skin and core materials, the variables are the skin (t) and core (c) thicknesses and the core density (ρ_c^*).

If the core density is also specified, the optimization procedure is straightforward. Equation (14.24) is rearranged to express t in terms of E_f and panel dimensions including c. This expression is then substituted into Eq. (14.25) which is differentiated with respect to c. Setting the result equal to zero, the optimum core thickness (c_{opt}) is obtained. Substituting c_{opt} into Eq. (14.24) permits the optimum thickness (t_{opt}) to also be determined. Problem 14.13 provides an opportunity to work with this procedure. However, optimization can also be performed graphically. We rearrange Eq. (14.25);

$$\left(\frac{t}{l}\right) = \frac{W}{2bl^2\rho_f g} - \left(\frac{\rho_c^*}{2\rho_f}\right)\left(\frac{c}{l}\right) \qquad (14.26)$$

A graph of t/l vs. c/l is constructed for different values of the weight, W (Fig. 14.18). This yields a series of straight lines. Solely on this criterion, beam weight is minimized with a beam having the minimum possible weight. But this cannot satisfy the stiffness requirement (Eq. (14.24)). Equation (14.24) can also be rearranged to express t/l in terms of c/l;

$$\frac{t}{l} = \left\{\frac{G_c^*}{6E_f(c/l)}\right\}\frac{1}{[(4\delta/P)(bG_c^*(c/l)] - 1} \qquad (14.27)$$

This relation is shown as a full line in Fig. 14.18. Both Eqs. (14.26) and (14.27) must be satisfied. This can be done for a number of beam weights. However, the *minimum* beam weight is that at which Eqs. (14.26) and (14.27) are satisfied at one point (in this figure this corresponds to the optimum design point and is designated by (t_{opt}, c_{opt})). For the beam geometry and material specifications noted in this figure's caption, this corresponds to a panel weight $W \cong 7.5$ N.

A more complete optimization is possible if the core density is also allowed to vary. In this case, the core shear modulus (G_c^*, Eq. (14.19)) is substituted into the compliance expression (Eq. (14.24)). The resulting equation is rearranged to express ρ_c^* in terms of c and t and the other (fixed) parameters. This equation is then

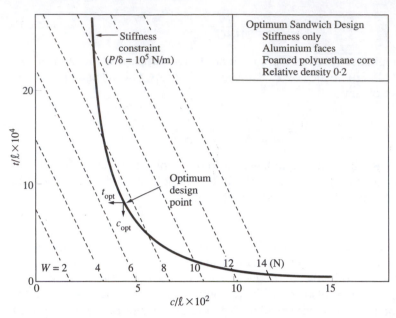

Figure 14.18
A graphical solution of Eqs. (14.26) and (14.27) for design of a panel having a specified stiffness with minimum weight. A series of straight lines (the dotted lines) represent Eq. (14.26) for various panel weights. The solid curve represents Eq. (14.27)
which contains the stiffness constraint. The two equations must be simultaneously
satisfied. Solutions are obtained at the intersection of the bold line with a series of
the dotted lines. However, only one solution (designated as c_{opt} and t_{opt}) provides
the minimum beam weight for the specified stiffness. (The graph is constructed
for a panel of aluminum skins ($\rho_f = 2700$ kg/m³, $E = 70$ GN/m²) and a foamed
polyurethane core ($\rho_c^*/\rho_s = 0.2$, $\rho_s = 1200$ kg/m³, $E = 1.6$ GN/m², $G_c^* = 45$ MN/m²).
The stiffness constraint is $P/\delta = 10^5$ N/m. The beam, of width $b = 50$ mm and length
$l = 1$ m, is loaded in three-point bending.) (*From "Cellular Solids—Structure and
Properties," L. J. Gibson and M. F. Ashby, Pergamon Press, Oxford, 1988.*)

substituted into Eq. (14.25) and the optimum values of c and t are found through solution to the equations

$$\frac{\partial W}{\partial c} = \frac{\partial W}{\partial t} = 0 \tag{14.28}$$

These optimum values are then resubstituted into the expression for the core density, yielding the optimum value of this density. The procedure is straightforward,
albeit cumbersome. Problem 14.14 provides an opportunity to see how adept you
are at it.

The optimum values of c, t, and ρ_c^* depend on the type of loading (e.g., three-
point bending vs. uniform loading) and the beam shape (e.g., circular vs. plate) as
well as the specified compliance and other parameters. However, there are some
general results. Regardless of the panel shape and its loading condition, the ratio of
the face to the core weight is ¼ for an optimum sandwich panel. And the ratio of the
bending to the shear deflection for such a panel is ½, also independent of loading
condition and panel shape.

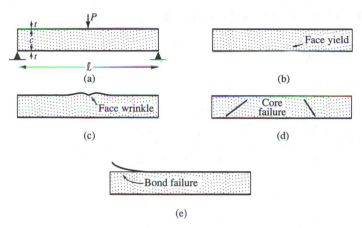

Figure 14.19
Possible sandwich panel failure modes. (a) Schematic of three-point bend loading.
(b) The panel may yield on the skin face in tension. (c) Or it may buckle on the com-
pressively loaded face (face-wrinkling failures). (d) The core can fail by shear.
(e) The skin may peel from the core to which it is joined (often by adhesives).
(From "Cellular Solids—Structure and Properties," L. J. Gibson and M. F. Ashby,
Pergamon Press, Oxford, 1988.)

This concludes our brief discussion of sandwich panel stiffness. But sandwich panels also must not "fail;" i.e., they must also be strong enough to sustain an applied stress without failing by one of several mechanisms. Treatment of sandwich panel strength is even more complicated in its details than is the comparable treatment of stiffness. Nevertheless, an abbreviated discussion of failure mechanisms in sandwich panels and the corresponding failure loads is provided in the following section.

B. Sandwich Panel Failure Modes

Figure 14.19 illustrates four possible types of sandwich panel failure. *Face yielding* corresponds to plastic-flow initiation in the face loaded in tension. *Face wrinkling* is a failure taking place on the compressively loaded panel skin; it is, in effect, a plastic buckling failure. *Core failure* is a plastic or brittle failure event in the core.[7] And *bond failure* consists of peeling of the skin from its core. (The skin and core are often joined with a resin adhesive.) While bond failures occur in sandwich panels, analysis of them is more difficult than it is for the other failure modes. Moreover, bond failure is typically caused by improper material joining. Were the bond to display the properties of which it is capable, bond failure would not occur. Thus, we restrict our discussion to the first three failure modes mentioned. We do not derive the equations that describe the failure criteria; only the results are presented. This permits us to focus on understanding the physics of failure, and allows us to spend more time in discussion of design for sandwich strength.

---

[7]Sandwich structures seldom employ elastomeric cores. This is because these panels require stiffness and strength, and polymeric foams cannot provide the required stiffness/strength.

The face yielding failure criterion can be expressed as;

$$\frac{Pl}{4btc} = \sigma_{fy} \qquad \text{(face yielding)} \qquad (14.29)$$

In Eq. (14.29) σ_{fy} is the yield strength of the skin material. The product Pl is proportional to the applied moment while bt is proportional to the tensile load bearing area of the skin. If we consider the moment as effectively amplified by the factor c, (the beam is bent about the panel centroid causing the skin strain to be "magnified" by this factor), Eq. (14.29) is essentially a tensile yield criterion. Note that face yielding depends only the properties of the face, and not the core. Only the core thickness impacts face yielding.

Face wrinkling depends on the properties of the core and the skin. The face-wrinkling failure criterion is

$$\frac{Pl}{4btc} = 0.57\, E_f^{1/3}\, E_s^{2/3} \left(\frac{\rho_c^*}{\rho_s}\right)^{4/3} \qquad \text{(face wrinkling)} \qquad (14.30)$$

This criterion also depends on a critical moment (this one causing a compressive failure) being reached.

Finally, the core shear failure criterion is expressed as

$$\frac{P}{2bc} = 0.15\sigma_{ys}\left(\frac{\rho_c^*}{\rho_s}\right)^{3/2} \qquad \text{(core shear)} \qquad (14.31)$$

where σ_{ys} is the plastic yield stress of the core. Equation (14.31) corresponds to a critical stress (P/bc) being reached in the core. Thus, core failure does not depend on the skin dimensions or on its properties. If the foam is constructed from a brittle material, foam fracture constitutes core failure. Consult Gibson and Ashby if you are interested in learning more about this failure mode. Equations (14.29)–(14.31) are enough for our purposes.

The three failure criteria depend differently on the panel dimensions (t and c), the properties of the skin and core materials, and the relative core density. These criteria can be represented graphically in a sandwich panel *failure map* (Fig. 14.20). Such a map has axes of t/l and ρ_c^*/ρ_s and is constructed for specific core and skin materials. The map of Fig. 14.20 applies to all panels having an aluminum skin and a rigid polyurethane core with specific properties. The different areas in the map define the controlling failure mode as it varies with ρ_c^*/ρ_s and t/l. The solid lines separating these areas are defined by equating two of the three failure criteria (Eqs. (14.29)–(14.31)). Along these lines failure by two modes occurs simultaneously. At the "triple" point—where the three lines intersect—failure by all three modes takes place at the same load.

Some discussion of the boundary lines and their significance is worthwhile. For small face thicknesses, failure of the face—by wrinkling in compression at low ρ_c^* and yielding in tension at high ρ_c^*—controls panel failure. The transition from one of these failure modes to the other depends only on the core density, as indicated by the horizontal line in Fig. 14.20 that defines the boundary between the two failure modes. For large face thicknesses, failure by wrinkling does not occur (the buckling stress is high at large t values). Instead, failure takes place in the core for low values of ρ_c^* or, when ρ_c^* is high, by face yielding. The transition from core to

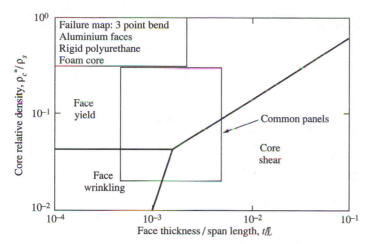

Figure 14.20

A sandwich panel failure map. The map delineates the expected failure mode as it depends on the relative core density (ρ_c^*/ρ_s) and face thickness to span-length ratio (t/l). Values of these parameters for common panels are indicated by the box in the figure. This diagram is constructed for a polyurethane foam–aluminum skin sandwich-beam in three-point bending. The boundary between the regions represent face thickness–core density combinations that lead to simultaneous failure by two failure modes. The point where the lines intersect represents a situation where all three failure modes are concurrent. (*From "Cellular Solids—Structure and Properties," L. J. Gibson and M. F. Ashby, Pergamon Press, Oxford, 1988.*)

face failure at large values of t depends on the values of ρ_c^* and t. The line drawn for this transition in Fig. 14.20 is obtained by equating Eqs. (14.29) and (14.31). Finally, for low core densities, failure occurs by face wrinkling (when t is small) or core shear (when t is large); the boundary line between these failure modes is obtained by equating Eqs. (14.30) and (14.31). Thus, the failure mode transitions appeal to common sense and have a physical basis.

Sandwich panel failure maps can be made more useful by plotting contours of constant failure load on them. This is done in Fig. 14.21 for the panel having the same characteristics as that of Fig. 14.20 but now with specific values of the panel dimensions b and c. The lines of constant failure load are horizontal in the core-shear region (Eq. (14.31)) and vertical in the face-yield region (Eq. (14.29)). The constant-failure-load lines in the face-wrinkling region are neither horizontal nor vertical. This is because a constant wrinkling load can be attained by increasing ρ_c^* while decreasing t or vice versa. Or, alternatively put, increasing the face thickness at a constant core density increases the buckling load. Figure 14.21 also contains some experimentally determined failure loads. These closely correspond to predicted failure loads, and verify the essential correctness of the approach described here.

However, Fig. 14.21 tells us nothing explicit about the optimum panel, one that maximizes failure load per unit panel weight. High failure loads are obviously found at large face thickness and core density values; e.g., the load contour of 5000 N in Fig. 14.21. But this load is associated with a relatively high panel weight.

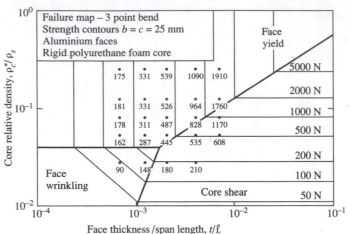

Figure 14.21
A sandwich panel failure map for the same material combinations as in Fig. 14.20. (Panel dimensions b and c are now specified as 0.25 mm.) Calculated failure loads (the solid lines) are shown. So, too, are some experimentally determined failure loads (the "numbers" and their associated points). There is good correspondence between the predicted and measured failure loads. Further, while not shown in the figure, there is also good correspondence between the predicted and observed failure modes. (*From "Cellular Solids—Structure and Properties," L. J. Gibson and M. F. Ashby, Pergamon Press, Oxford, 1988.*)

How do we maximize the failure load to panel weight ratio? Reconsider the face-wrinkling/face-yielding failure transition occurring at low values of t. We consider a value of t/l so low that the panel mass is almost all due to the core. For this situation, the panel weight scales directly with (ρ_c^*). The failure load (or its moment, Pl) does not depend on ρ_c^* for face yielding (Eq. (14.29)). However, Pl varies as ($\rho_c^{*4/3}$) for face wrinkling. Thus, if we plot the ratio of the failure moment to the beam weight we find that for low t/l values.

$$\left(\frac{Pl}{W}\right)_{wrinkling} \sim (\rho_c^*)^{1/3} \qquad \left(\frac{Pl}{W}\right)_{yielding} \sim (\rho_c^*)^{-1} \qquad (14.32)$$

Equations (14.32) are schematically drawn on logarithmic coordinates in Fig. 14.22. For low relative core densities, the face-wrinkling failure load is less than that for face yielding. The opposite holds for high ρ_c^* values. Since the failure load is determined by the lesser of the two loads, Fig. 14.22 shows that the maximum failure-load to panel-weight ratio occurs at the transition between face-wrinkling and core-yielding failure. The same situation applies to transitions from face-wrinkling to core-shear failure, and for core-shear to face-yielding failure (see Prob. 14.15). By extension, the sandwich structure that optimizes panel strength to density corresponds to the intersection of the three lines separating the several failure regions in Figs. 14.20 and 14.21.

Sandwich panels could be discussed at much greater length. However, if we did, the details (not to mention the algebra!) might become overwhelming. Need-

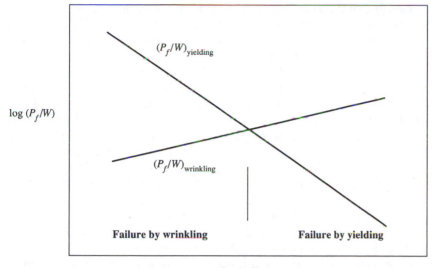

Figure 14.22
A schematic (for small *t/l*) on logarithmic coordinates of the failure-load to panel-weight ratio, P_f/W (Eqs. 14.32) as a function of the relative core density. For small *t/l*, the two possible failure modes are face wrinkling and face yielding. For the former, P_f/W varies as $(\rho_c^*)^{1/3}$; for the latter as $(\rho_c^*)^{-1}$. The failure mechanism is face wrinkling at low, and face yielding at high, ρ_c^*. The maximum P_f/W value is found at a core density corresponding to simultaneous failure by both mechanisms.

less to say, design for sandwich panels is intriguing. The number of variables—skin and core thicknesses, relative core density, moduli and strengths of the skin and core materials—is large. And the number of potential panel configurations and material combinations is extensive. Modern sandwich panels also often employ hierarchical panels; our earlier cited example of modern skis is a case in point.

14.6
SUMMARY

Cellular solids have two common forms—honeycombs and foams. The most important structural applications of cellular solids are as energy-absorbing devices and as the cores of sandwich panels. One important characteristic of a cellular solid is its relative density; i.e., its density compared to that of the fully dense solid that is its backbone. The mechanical properties of cellular solids closely relate to this density. For example, the elastic modulus varies with the square of the relative density (for a foam) or its cube (for in-plane loading of a honeycomb).

The energy-absorbing capabilities of cellular solids relate to their compressive stress-strain behavior. A cellular solid loaded in compression "fails" following a certain amount of linear elastic deformation. The failure mode and stress depend on the nature of the material of the solid. If it is an elastomer (or has characteristics common to glassy polymers above their glass transition temperature), the solid fails by elastic buckling of its walls/beams. The buckling stress, like the modulus, is a function of the relative density. If the material of the cellular solid is capable of plastic deformation, failure occurs by plastic bending of its beams/walls. And if the material is brittle, failure takes place by beam/wall fracture.

Following "failure," the material is compressed at a more-or-less constant stress (the plateau stress). This situation persists to relatively large strains, and accounts for the energy-absorbing capacity of cellular solids. At a certain critical strain, however, the material's walls/beams begin to touch, and for strains greater than this critical one, further compression is accompanied by material densification. During densification the compressive stress increases rapidly. The plateau stress and plateau strain limit depend on the cellular solid's relative density; the plateau stress increases with this density but the plateau strain decreases with it. Since the energy-absorbing capacity is proportional to the area under the stress-strain curve, and because the plateau stress and strain depend on the relative density, so does the energy absorption. Indeed, for a given applied stress there is an optimum cellular solid density that provides the greatest energy-absorbing capacity.

Design for optimum energy absorption can be discussed in terms of energy-absorption curves/diagrams. In such diagrams, the energy absorbed per unit volume (the area under the stress-strain curve) normalized by the elastic modulus of the solid material within the cellular solid is plotted vs. the maximum stress (also normalized by this modulus) the cellular solid experiences. This normalized energy increases very rapidly at the plateau stress. But, following the onset of densification, it barely increases. The trace of this energy-stress relationship depends on the material's relative density. Optimum design involves selecting a density corresponding to the plateau stress which is selected to be the maximum service stress. This also defines the maximum energy absorbed per unit volume, thereby permitting selection of the dimensions of protective packaging. While empirically determined energy-absorption diagrams should be employed in design, a good engineering approximation can be had with an idealized representation. The idealization divides the stress-strain curve into the three deformation regions and permits analytical expressions to be developed for the relationship between the energy stored and the maximum service stress.

Sandwich panels employ cellular solids as cores between fully dense skins (faces). The lower-density cellular interior permits the skins to be placed where the stresses on the panel are greatest. Because of this feature, sandwich panels can have high specific stiffnesses and strengths (i.e., high stiffness/strength to density ratios). Beam theory enables the panel stiffness (or its inverse, the compliance) to be expressed in terms of the skin and core thicknesses, the skin modulus, the modulus of the solid constituent of the core, and the core density. Optimization procedures permit design of panels having the highest specific stiffness. The procedure can be extended to panel strengths. In doing so, several different panel failure modes must be considered. These include face wrinkling (on the compressively loaded panel face), face yielding (on the panel face in tension), and core yielding. Analytical expressions for the several failure loads are available. These permit failure loads to be calculated in terms of panel geometry and properties of the panel constituents including the core density. Results are conveniently displayed in a failure map which delineates the dominant failure mode as it depends on the face thickness and the core density. Highest specific strengths are obtained at specific values of face thickness and core density that correspond to simultaneous failure by all three failure modes.

715

Problems

Cellular solids are important technologically. Our brief treatment of them has not done them justice. On the other hand, our approach can be considered another example of the interaction between mechanics and materials that exemplify descriptions of material mechanical behavior. Thus, the topic of this chapter is an appropriate way to (finally!) end this book.

REFERENCES

Allen, H. G.: *Analysis and Design of Structural Sandwich Panels,* Pergamon Press, Oxford, 1969.

Ashby, M. F.: "The Mechanical Properties of Cellular Solids," *Metall. Trans. A,* 14A, 1755, 1983.

Gibson, L. J., and M. F. Ashby: *Cellular Solids—Structure and Properties,* 2nd ed., Cambridge University Press, Cambridge, England, 1997.

Papka, S. D., and S. Kyriakides: "Experiments and Full-Scale Numerical Simulations of In-Plane Crushing of a Honeycomb," *Acta Mater.,* 46, 2765, 1998.

PROBLEMS

14.1 **a** Derive Eq. (14.1), the expression for the relative density of a regular honeycomb.
   b Derive Eq. (14.2), the expression for the relative density of an open foam. (Neglect terms in $(t/l)^2$ and higher.)

14.2 Consider honeycombs patterned on equilateral triangles and squares. Show that $(\rho^*/\rho_c) = 2(3)^{1/2}t/l$ for the triangular, and $(\rho^*/\rho_c) = 2t/l$ for the square, honeycomb. (Neglect terms in $(t/l)^2$ and higher.)

14.3 Use Eq. (14.3) along with the expression in the text for the moment of inertia of a square beam and the expression for the relative density of an open-cell foam (Eq. (14.2)) to derive Eq. (14.9)—the modulus of an open-cell foam.

14.4 Use Table 14.1 and Eq. (14.8) to show that the linear elastic strain limit for an elastomeric honeycomb is 0.10.

14.5 Construct figures comparable to Fig. 14.11 for plastic and brittle foams. Take the plastic foam to be constructed from aluminum having a yield strength of 70 MPa. Let the brittle material be a silicate glass with a fracture strength of 140 MPa. ($E_s = 70$ GPa for Al; = 80 GPa for the glass.)

14.6 **a** Schematically plot the maximum compressive energy stored per unit volume vs. the relative foam density for an elastomeric foam and for a plastic foam. (Simplify by considering only the energy absorbed in the plateau region.) Show that there is a relative density below which the plastic foam is a superior energy-absorbing material and above which the elastomeric foam is superior. Express the transition density in terms of the plastic material's yield strength and the elastomeric modulus. (Hint: Refer to Table 14.1.)

b Determine a numerical value for this transition density for a polyurethane (PU) and an aluminum (Al) foam. (Data: σ_y for Al is 70 MPa; $E_s = 45$ MPa for PU.) Comment on your answer.

c Sometimes it is desirable to have the maximum energy absorbed per unit mass, rather than per unit volume. Schematically plot the maximum compressive energy absorbed per unit mass vs. the relative foam density for an elastomeric foam and for a plastic foam.

d Repeat part (b) of this problem using the energy per unit mass criterion.

14.7 The solution to Prob. 14.6a shows that, regardless of the material relative density, metal foams typically absorb more energy per unit volume than do elastomeric foams.

a Why, then, are not metal foams employed in packaging materials to absorb energy?

b And why are not metal foams used in even more demanding energy-absorbing applications, e.g., as automobile bumpers?

14.8 Make several copies of Fig. 14.15. Draw on them several ideal energy-absorption curves for polyurethane (PU) and polyethylene (PE) foams. How do the ideal and actual energy-absorption curves compare? (Moduli: for PU, $E = 45$ MPa; for PE, $E = 250$ MPa.)

14.9 **a** Consider a polyurethane foam for use as an automobile bumper. The bumper is to absorb the kinetic energy of impact of a 3000-lb automobile traveling at 15 mph. The bumper area is 72 by 8 inches. In order to protect the vehicle's occupants, the maximum deceleration during impact cannot exceed $50g$. Use ideal energy-absorption diagrams to determine the required bumper thickness and its foam's relative density. Is your answer a practical one in terms of the required bumper thickness?

b Consider an automobile having a bumper with the characteristics corresponding to those you solved for in part (a). If the 3000-lb automobile were traveling at a velocity of 60 mph, how much would the bumper reduce the impact velocity?

c Aluminum foams are placed behind the bumper. They are intended to absorb impact energy in more serious crashes. Consider the residual velocity you found in part (b). The aluminum of the foam has a modulus and yield strength given in Probs. 14.5 and 14.6. Select an optimum Al foam that will absorb the residual kinetic energy of the higher-speed automobile. The maximum permissible deceleration remains at $50g$, and the Al foam has the same cross-sectional area as the bumper.

14.10 **a** Construct ideal energy-absorption curves for in-plane loading of a honeycomb. (Use Example Prob. 14.1 as a starting point.) Use specific densities (e.g., $\rho^*/\rho_s = 0.01, 0.03, 0.1,$ and 0.3) to do this.

b Plot—on the same diagram as in part (a)—the ideal energy-absorption curves for foams with these same relative densities. Comment on your results.

14.11 Consider an out-of-plane loaded elastomeric honeycomb as an energy-absorbing material. In out-of-plane loading the stress is applied along the long axis of the honeycomb. As a result, honeycomb properties are different from what they are for in-plane loading. In particular, the out-of-plane honeycomb modulus and elastic buckling stress are given by:

$$\frac{E^*}{E_s} = \left(\frac{\rho^*}{\rho_s}\right) \qquad \frac{\sigma^*_{el}}{E_s} = 3.4\left(\frac{\rho^*}{\rho_s}\right)^3$$

Note that the modulus varies linearly with the relative density (rather than with the cube of this density as it does for in-plane loading (Eq. (14.8)), and the elastic buckling stress is about a factor of 20 higher than for in-plane loading (Table 14.1).

a Construct ideal energy-absorption curves for out-of-plane loading of an elastomeric honeycomb of varying densities (consider the densities listed in Prob. 14.10).

b Compare these to comparable curves for in-plane loading, as determined in Prob. 14.10.

14.12 Consider an elastomeric honeycomb to be used as the automobile bumper which must function as described in Prob. 14.9. The honeycomb is constructed from polyurethane, but the honeycomb is subject to out-of-plane loading. Use the results from Prob. 14.11 to determine the required honeycomb thickness and relative density. Discuss which configuration—the foam or the out-of-plane loaded honeycomb—is better suited for use as an automobile bumper.

14.13 Use Eqs. (14.24) and (14.25) to obtain an analytical solution for the optimum core and face thicknesses graphically displayed in Fig. 14.18.

14.14 a Consider the core thickness, the face thickness, and the core relative density as variables in sandwich panel design. Follow the procedure described in the text (see the paragraph containing Eq. (14.28)), and determine expressions for c_{opt}, t_{opt} and ρ^*/ρ_s. Obtain numerical values for these parameters for sandwich constituents with the properties listed in the caption to Fig. 14.18 and for the beam width, length, specified stiffness, and loading condition given there.

b Construct a figure similar to Fig. 14.18, but one using the optimum relative density found in part (a). How are the optimum core and skin thickness of Fig. 14.18 changed when the optimum core density is employed?

14.15 Consider sandwich panels with high values of t/l so that the panel weight is almost entirely that of the faces. Use Eqs. (14.29) and (14.31) to develop equations for $P_f l$ (the failure load multiplied by the beam length) per unit panel weight for core-shear and face-yielding failure. Schematically sketch these expressions as a function of t/l (fixed (ρ_c^*/ρ_s)), and show that there is a transition from face-yield to core-shear failure with increasing t/l. Schematically sketch your expressions for $P_f l$ as a function of (ρ_c^*/ρ_s) (fixed t/l), and show that there should be a transition from core-shear to face-yield failure with increasing ρ_c^*/ρ_s. (Note: Your results are consistent with the transitions illustrated in Figs. 14.20 and 14.21.)

Name Index

Specific Substance Index

Subject Index